FIFTH EDITION

ELECTRICITY & ELECTRONICS: A SURVEY

DALE R. PATRICK

STEPHEN W. FARDO

Department of Technology
Eastern Kentucky University

Upper Saddle River, New Jersey
Columbus, Ohio

Library of Congress Cataloging in Publication Data

Patrick, Dale R.
Electricity and electronics : a survey / Dale R. Patrick, Stephen W. Fardo.--5th ed.
p. cm.
Includes index.
ISBN 0-13-019564-2
1. Electric engineering. 2. Electronics. I. Fardo, Stephen W.

TK146 .P34 2002
621.3—dc21

2001018547

Vice President and Editor in Chief: Stephen Helba
Assistant Vice President and Publisher: Charles E. Stewart, Jr.
Assistant Editor: Delia K. Uherec
Production Editor: Alexandrina Benedicto Wolf
Production Coordination: Carlisle Publishers Services
Design Coordinator: Robin G. Chukes
Cover Designer: Linda Fares
Cover Image: FPG International
Production Manager: Matthew Ottenweller

This book was set in Times Roman by Carlisle Communications, Ltd. It was printed and bound by Courier Kendallville, Inc. The cover was printed by Phoenix Color Corp.

10 9 8 7 6 5 4 3 2 1
ISBN 0-13-019564-2

To
“Our Three Sons”
STEVEN,
BRIAN,
and
DAVID

PREFACE

Electricity and Electronics: A Survey is an introductory text that explores many aspects of electricity and electronics in a basic and easy-to-understand manner. The key concepts are discussed using a "big picture" or systems approach that greatly enhances student learning. Many applications, testing procedures, and operational aspects of equipment and devices are discussed. Mathematics is used to support concepts, practical applications, and calculator use. Electrical safety is stressed throughout the text. Several additions have been made since the first edition, making this edition even more comprehensive than previous editions.

This fifth edition has been improved through the following features.

1. **Improved and updated text content:** Although the text organization remains the same, new topics have been added and content has been updated and revised to reflect changes in the field of electronics. The text has continually been updated through quality checks to ensure technical accuracy and clarity. Expanded content includes a brief history of electronics; new material on batteries (drastic changes to this technology have been brought about by computer applications and portable equipment); and communications advances such as cellular phones, lasers, and fiber optics.
2. **Improved illustrations and photos:** Numerous photos and illustrations have been updated to reflect changes in the field. Illustrations have been improved through the use of color. The second color has been added to show implied movement of electrical current and to add emphasis to figures.
3. **Two-color text:** The planned use of color adds life to the book and clarifies content. The second color emphasizes headings, provides distinct break points in text, makes figures with multiple parts easier to understand, and highlights review questions and problems at the end of each chapter for easy reference.
4. **Updated and improved ancillary package:** Supplements have been updated and improved to provide the students with a complete learning package. The accompanying Laboratory/Activities Manual now includes a CD-ROM with simulation labs. The Instructor's Resource Guide now includes a CD-ROM with PowerPoint illustrations. A Test Item File has been developed using Prentice-Hall Test Generator software.

We have continued to consider the needs of both students and instructors in preparing this comprehensive text. Chapter headings and the instructional order of chapters are the same as in previous editions to avoid confusion. Numbered sections provide a clear order of presentation. Technical reviewers have checked all example problems and end-of-chapter problems for accuracy.

ORGANIZATION OF THE TEXT

The book is divided into two main parts. Part I (Chapters 1 to 8) deals with the basics of ELECTRICITY, including electrical fundamentals, direct-current (dc) circuits, alternating-current (ac) circuits, and electrical applications. Part II (Chapters 9 to 18) deals with the basics of ELECTRONICS, including electronic devices, electronic circuits, and electronic systems applications. Figure I shows the organizational framework of the text.

The chapters are organized as follows to aid in student understanding:

1. Objectives
2. Introduction
3. Important Terms
4. Major Content
5. Review
6. Student Activities

Important terms are defined at the beginning of each chapter. The *review* section at the end of each chapter helps the students review and check their understanding of the major topics covered. The *student activities* stress the practical applications and problem solving used in the study of electricity and electronics, and are easy to understand. Equipment cost required for the laboratory activities is kept to a minimum. Most of the activities are low- or no-cost activities that can be done in a school laboratory or at home. The *analysis* sections are intended to provide thought provoking

Electricity

Chapter 1 Basics of Electricity
Chapter 2 Electrical Components and Measurements
Chapter 3 Electrical Circuits
Chapter 4 Magnetism and Electromagnetism
Chapter 5 Sources of Electrical Energy
Chapter 6 Alternating-Current Electricity
Chapter 7 Electrical Energy Conversion
Chapter 8 Electrical Instruments

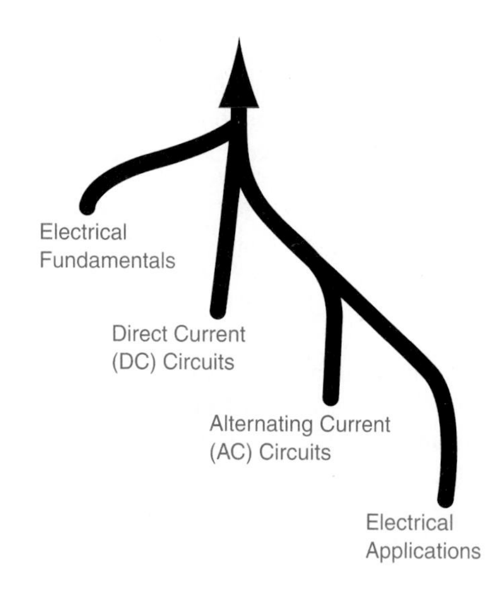

Electronics

Chapter 9 Electronic Basics
Chapter 10 Electronic Diodes
Chapter 11 Electronic Power Supplies
Chapter 12 Transistors
Chapter 13 Amplification
Chapter 14 Amplifying Systems
Chapter 15 Oscillators
Chapter 16 Communications Systems
Chapter 17 Digital Electronic Systems
Chapter 18 Electronic Power Control

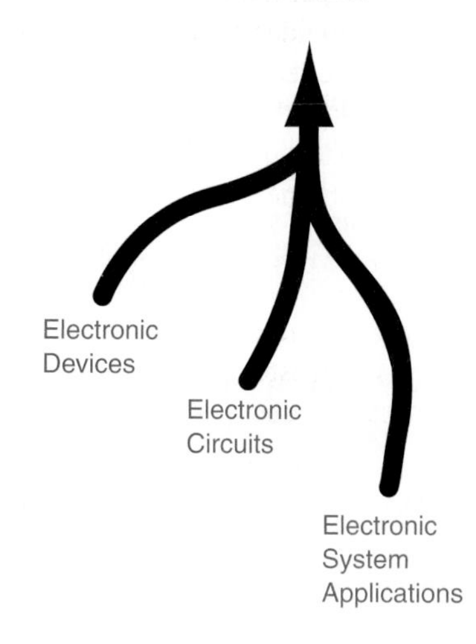

Figure I. Electricity and electronics framework.

problems and questions that are specifically focused on important topics in the chapter. *Answers* to all end-of-chapter questions and problems are provided in the *Instructor's Resource Guide.* Answers to odd-numbered questions are found at the back of the book.

The *appendices* are an important pedagogical feature, which make this book an excellent resource. Appendix A provides a comprehensive *glossary of electronic terms.* Appendix F stresses the importance of using a *calculator* to solve electrical circuit problems. Other appendices include a comprehensive list of electronic *symbols,* information on *soldering* techniques and electronic *tools,* and *acronyms and abbreviations.*

ANCILLARY PACKAGE

The comprehensive textbook and the following supplements provide a complete learning package for the study of electricity and electronics.

Laboratory/Activities Manual to accompany the text (0-13-091889-X): This accompanying manual stresses the practical applications of electricity and electronics, parallels the content of each chapter in the main text, reinforces the text material, and directs the learning process. To facilitate learning, activities are presented in a single-concept approach and require a short time to assemble and make the necessary measurements. The exercises involve some manipulative or hands-on activity that deals with circuit construction, testing operations, calculations, instrument use, and component identification. Equipment cost is kept at a minimum, and a few of the activities do not require any lab equipment.

The lab/activities manual contains over 150 activities, including approximately 50 new simulation labs using Electronics Workbench™ (EWB) software. The EWB activities provide the students with the option of using *circuit simulation* to understand electronic concepts. The lab/activities manual has a CD-ROM with the Electronics Workbench (EWB) simulation activities. Through this method, instructors can choose lab activities for simulation or hands-on activity for students. Labs with accompanying simulation are identified. Each experimental activity is organized as follows:

Introduction. Approximately one to two paragraphs: the first contains an overview of the activity and practical applications; the second gives the purpose of the activity and observations that the students should make.

Objective. Outlines what the students are expected to learn when the experiment is completed.

Equipment. Lists equipment and materials needed to perform the experiment.

Procedure. Provides logical step-by-step sequence to complete the learning activity. Charts and tables are provided to aid in data recording.

Analysis. Provides specific questions and problems that supplement the experimental activity.

Instructor's Resource Guide with CD-ROM (0-13-091365-0): This extensive resource guide contains (a) answers to the end-of-chapter review questions and problems, (b) suggested data for laboratory activities, (c) answers to test questions found in the laboratory/activities manual, and (d) answers to test questions found in the Test Item File. A CD-ROM containing PowerPoint™ slides of all figures in the text is also included.

Test Item File (ISBN 0-13-092225-0): A test item file, available to instructors using the text, contains comprehensive exam questions for each chapter and is developed using Prentice Hall Test Generator software.

ACKNOWLEDGMENTS

The authors would like to thank the many companies who have provided photographs and technical information for this edition. The authors would like to acknowledge the reviewers of this book: Sandra Herinckx, Texas State Technical College, TX; Florian Misoc, Pittsburg State University, KS; Alan Moltz, Naugatuck Valley Community Technical College, CT; and Murray Stocking, Ferris State University, MI. Finally, the authors express their sincere thanks to their wives, Kay and Helen, for their help and understanding during the manuscript preparation for the fifth edition of *Electricity and Electronics: A Survey.*

Dale R. Patrick
Stephen W. Fardo

BRIEF CONTENTS

PART I ELECTRICITY 1

1 BASICS OF ELECTRICITY 3
2 ELECTRICAL COMPONENTS AND MEASUREMENTS 37
3 ELECTRICAL CIRCUITS 75
4 MAGNETISM AND ELECTROMAGNETISM 129
5 SOURCES OF ELECTRICAL ENERGY 149
6 ALTERNATING-CURRENT ELECTRICITY 181
7 ELECTRICAL ENERGY CONVERSION 245
8 ELECTRICAL INSTRUMENTS 283

PART II ELECTRONICS 309

9 ELECTRONIC BASICS 311
10 ELECTRONIC DIODES 329
11 ELECTRONIC POWER SUPPLIES 357
12 TRANSISTORS 381
13 AMPLIFICATION 409
14 AMPLIFYING SYSTEMS 445
15 OSCILLATORS 475
16 COMMUNICATIONS SYSTEMS 503
17 DIGITAL ELECTRONIC SYSTEMS 551
18 ELECTRONIC POWER CONTROL 581

APPENDICES

A COMPREHENSIVE GLOSSARY OF TERMS 605
B SOLDERING 619
C ELECTRICAL TOOLS 623
D SYMBOLS FOR ELECTRICITY AND ELECTRONICS 631
E COMMON LOGARITHMS AND NATURAL LOGARITHMS 635
F CALCULATOR EXAMPLES 639
G UNITS OF MEASUREMENT 651
H PRINTED CIRCUIT BOARD (PCB) CONSTRUCTION 663
I ACRONYMS AND ABBREVIATIONS 665

ANSWERS TO SELECTED PROBLEMS 669

INDEX 675

CONTENTS

PART I ELECTRICITY 1

1 BASICS OF ELECTRICITY 3
Objectives, 4
Important Terms, 4
1.1 History of Electricity and Electronics, 5
Thales of Miletus, 5
The Compass, 6
William Gilbert, 6
Ben Franklin, 6
Volta's Battery, 6
Coulomb and Faraday, 6
Other Experiments, 7
1.2 Electrical Systems, 7
Examples of Electrical Systems, 9
1.3 Energy, Work, and Power, 12
1.4 Structure of Matter, 14
1.5 Electrical Charges, 19
1.6 Static Electricity, 19
1.7 Electrical Current, 20
Conductors, 21
Insulators, 21
Semiconductors, 22
Superconductors, 22
Current Flow, 23
1.8 Electrical Force (Voltage), 25
1.9 Resistance, 26
1.10 Voltage, Current, and Resistance, 27
1.11 Volts, Ohms, and Amperes, 29
1.12 Electrical Safety, 29
Lab or Shop Practices, 29
Electrical Hazards, 29
Fuses and Circuit Breakers, 30
1.13 Careers in Electronics, 30
Service Technician, 30
Field Engineering Technician, 30
Technologist, 31
Engineering Assistant, 31
Technical Writer, 31
Electronics Sales, 31
Computer Technician, 31
Review, 31
Student Activities, 32

2 ELECTRICAL COMPONENTS AND MEASUREMENTS 37
Objectives, 38
Important Terms, 38
2.1 Components, Symbols, and Diagrams, 39
2.2 Resistors, 39
Resistor Color Codes, 46
Power Rating of Resistors, 47
2.3 Electrical Units, 47
Small Units, 49
Large Units, 50
2.4 Scientific Notation, 50
2.5 Schematic Diagrams, 52
2.6 Block Diagrams, 52
2.7 Wiring Diagrams, 53
2.8 Measuring Resistance, 53
How to Measure Resistance, 55
2.9 Measuring Voltage, 57
How to Measure Dc Voltage, 57
2.10 Measuring Current, 58
How to Measure Direct Current, 59
2.11 Digital Meters, 60
Review, 62
Student Activities, 63

3 ELECTRICAL CIRCUITS 75
Objectives, 76
Important Terms, 76
3.1 Use of Calculators and Computers, 77
Recognizing Computer Parts, 78
Circuit Simulation Software, 78
Tutorial Software, 78
3.2 Ohm's Law, 79
3.3 Troubleshooting, 83
3.4 Series Electrical Circuits, 83
Summary of Series Circuits, 85
Series-Circuit Troubleshooting, 85
3.5 Parallel Electrical Circuits, 86
Parallel-Circuit Measurements, 87
Summary of Parallel Circuits, 88
Parallel-Circuit Troubleshooting, 88
3.6 Combination Electrical Circuits, 89
Combination-Circuit Measurements, 90

3.7 Kirchhoff's Laws, 90
3.8 More Examples of Series Circuits, 91
3.9 More Examples of Parallel Circuits, 92
3.10 More Examples of Combination Circuits, 94
3.11 Power in Dc Electrical Circuits, 95
3.12 Maximum Power Transfer in Circuits, 97
3.13 Voltage-Divider Circuits, 99
3.14 Voltage-Divider Design, 99
3.15 Voltage-Division Equation, 100
3.16 Negative Voltage Derived from a Voltage-Divider Circuit, 100
3.17 Voltage Division with a Potentiometer, 101
3.18 Problem-Solving Methods, 101
3.19 Kirchhoff's Voltage Method, 102
3.20 Superposition Method, 102
3.21 Equivalent Circuits, 106
3.22 Thevinin Equivalent Circuit Method, 106
Single-Source Problem, 106
Two-Source Problem, 107
3.23 Norton Equivalent Circuit Method, 107
3.24 Bridge-Circuit Simplification, 109
Review, 111
Student Activities, 112

4 MAGNETISM AND ELECTROMAGNETISM 129
Objectives, 130
Important Terms, 130
4.1 Permanent Magnets, 131
4.2 Magnetic Field around Conductors, 131
4.3 Magnetic Field around a Coil, 132
4.4 Electromagnets, 133
Magnetic Strength of Electromagnets, 134
4.5 Ohm's Law for Magnetic Circuits, 135
4.6 Domain Theory of Magnetism, 135
4.7 Electricity Produced by Magnetism, 136
4.8 Magnetic Devices, 137
Relays, 137
Solenoids, 139
Magnetic Motor Contactors, 139
Magnetic Circuit Breaker, 140
Electric Bell, 140
Reed Switches and Reed Relays, 140
Analog Meter Movement, 140
Magnetic Recording, 141
Electromagnetic Speakers, 141
4.9 Magnetic Terms, 142
4.10 Hall Effect, 143
4.11 Magnetic Levitation, 144
4.12 Rare Earth Magnets, 144
Review, 145
Student Activities, 145

5 SOURCES OF ELECTRICAL ENERGY 149
Objectives, 150
Important Terms, 150
5.1 Chemical Sources, 152
Primary Cells, 152
Secondary Cells, 152
5.2 Battery Connections, 154
Series Connection, 155
Parallel Connection, 155
Combination (Series-Parallel) Connection, 155
5.3 Light Sources, 155
5.4 Heat Sources, 155
5.5 Pressure Sources, 158
5.6 Electromagnetic Induction, 159
5.7 Generating a Voltage, 161
5.8 Electrical Generator Basics, 161
5.9 Single-Phase Ac Generators, 162
5.10 Three-Phase Ac Generators, 164
5.11 Direct-Current Generators, 166
Permanent Magnet Dc Generators, 168
Separately Excited Dc Generators, 169
Self-Excited Generators, 170
Generator Operating Characteristics, 174
Review, 175
Student Activities, 176

6 ALTERNATING-CURRENT ELECTRICITY 181
Objectives, 182
Important Terms, 182
6.1 Alternating-Current Voltage, 185
6.2 Single-Phase and Three-Phase Ac, 189
6.3 Measuring Ac Voltage, 191
6.4 Using an Oscilloscope, 191
6.5 Ohm's and Kirchhoff's Laws for Ac Circuits, 192
6.6 Inductance, 192
6.7 Capacitance, 193
6.8 Inductive Effects in Circuits, 196
Mutual Inductance, 196
Inductors in Series and Parallel, 196
Inductive Phase Relationships, 196
6.9 Capacitive Effects in Circuits, 197
Capacitors in Series, 197
Capacitors in Parallel, 197
6.10 Leading and Lagging Currents in Ac Circuits, 197
6.11 Capacitor Charging and Discharging, 198
6.12 Types of Capacitors, 199
6.13 Capacitor Testing, 200

6.14 Alternating-Current Circuits, 200
Resistive Circuits, 200
Inductive Circuits, 200
Resistive-Inductive (RL) Circuits, 201
Capacitive Circuits, 202
Resistive-Capacitive (RC) Circuits, 202
6.15 Vector Diagrams, 203
6.16 Mathematics for Ac Circuits, 204
Right Triangles and Trigonometry, 204
Rectangular Coordinates, 204
Quadrants, 205
Polar Coordinates, 205
Angular Velocity, 205
Complex Numbers, 207
Imaginary Numbers, 207
Rectangular Form of Complex Numbers, 207
Addition of Complex Numbers, 207
Polar and Trigonometric Forms, 208
6.17 Series Ac Circuits, 208
Series RL *Circuits, 208*
Series RC *Circuits, 208*
Series RLC *Circuits, 209*
General Procedure to Solve Series Ac Circuit Problems, 210
6.18 Parallel Ac Circuits, 212
General Procedure to Solve Parallel Ac Circuit Problems, 212
6.19 Power in Ac Circuits, 212
Power in Three-Phase Circuits, 217
6.20 Filters and Resonant Circuits, 219
Filter Circuits, 219
Resonant Circuits, 219
6.21 Transformers, 222
Transformer Operation, 222
Types of Transformers, 223
Transformer Efficiency, 227
Transformer Testing, 227
Review, 228
Student Activities, 229

7 ELECTRICAL ENERGY CONVERSION 245
Objectives, 246
Important Terms, 246
7.1 Lighting Systems, 247
Incandescent Lighting, 247
Fluorescent Lighting, 247
Vapor Lighting, 249
7.2 Heating Systems, 250
Resistance Heating, 251
Induction Heating, 251
Dielectric (Capacitive) Heating, 251
7.3 Mechanical Loads (Motors), 252
7.4 Direct-Current Motors, 253
Permanent Magnet Dc Motors, 254
Series-Wound Dc Motors, 254
Shunt-Wound Dc Motors, 254
Compound-Wound Dc Motors, 255
7.5 Three-Phase Ac Motors, 256
Three-Phase Ac Induction Motors, 256
Three-Phase Ac Synchronous Motors, 257
7.6 Single-Phase Ac Motors, 259
Universal Motors, 260
Induction Motors, 260
Single-Phase Synchronous Motors, 266
7.7 Synchro Systems and Servo Systems, 266
Dc Stepping Motors, 269
7.8 Motor Performance, 270
Effect of Load, 270
Effect of Voltage Variations, 271
Considerations for Mechanical (Motor) Loads, 271
7.9 Motor Control Basics, 272
Motor Starting Control, 272
Criteria for Selecting Motor Controllers, 274
Review, 276
Student Activities, 277

8 ELECTRICAL INSTRUMENTS 283
Objectives, 284
Important Terms, 284
8.1 Analog Instruments, 284
8.2 Measuring Direct Current, 285
8.3 Measuring Dc Voltage, 287
8.4 Measuring Resistance, 288
8.5 Measuring Electrical Power, 289
8.6 Measuring Electrical Energy, 290
8.7 Measuring Three-Phase Electrical Power, 291
8.8 Measuring Power Factor, 292
8.9 Measuring Power Demand, 293
8.10 Measuring Frequency, 294
8.11 Ground-Fault Indicators, 294
8.12 Measuring High Resistance, 295
8.13 Clamp-On Meters, 296
8.14 Wheatstone Bridge, 296
8.15 Cathode-Ray Tube Instruments, 296
General-Purpose Oscilloscopes, 298
Digital Storage Oscilloscopes, 298
8.16 Numerical Readout Instruments, 299
8.17 Chart Recording Instruments, 299
Review, 302
Student Activities, 303

PART II ELECTRONICS 309

9 ELECTRONIC BASICS 311
Objectives, 312
Important Terms, 312
9.1 Semiconductor Theory, 312
Atom Combinations, 312
Insulators, Semiconductors, and Conductors, 315
9.2 Semiconductor Materials, 319
N-*Type Material, 321*
P-*Type Material, 321*
9.3 Electron Emission, 323
Thermionic Emission, 324
Secondary Emission, 324
Photoemission, 325
Photoconduction, 326
Review, 327
Student Activities, 327

10 ELECTRONIC DIODES 329
Objectives, 330
Important Terms, 330
10.1 Junction Diode, 330
Depletion Zone, 331
Barrier Potential, 331
10.2 Junction Biasing, 332
Reverse Biasing, 332
Forward Biasing, 333
10.3 Diode Characteristics, 334
Forward Characteristic, 335
Reverse Characteristic, 335
Combined I-V *Characteristics, 336*
10.4 Diode Specifications, 337
Diode Temperature, 337
Junction Capacitance, 340
Switching Time, 340
10.5 Diode Packaging, 341
10.6 Diode Testing, 341
10.7 Semiconductor Diode Devices, 343
Zener Diodes, 343
Light-Emitting Devices, 345
Photovoltaic Cells, 345
Photodiodes, 346
Varactor Diodes, 347
10.8 Vacuum Tubes, 349
Diode Tubes, 351
Review, 354
Student Activities, 354

11 ELECTRONIC POWER SUPPLIES 357
Objectives, 358
Important Terms, 359
11.1 Power Supply Functions, 359
11.2 Transformer, 360
11.3 Rectification, 361
11.4 Filtering, 368
11.5 Voltage Regulation, 372
Review, 374
Student Activities, 375

12 TRANSISTORS 381
Objectives, 382
Important Terms, 382
12.1 Bipolar Transistors, 383
12.2 Transistor Biasing, 383
NPN *Transistor Biasing, 383*
Transistor Beta, 386
PNP *Transistor Biasing, 386*
12.3 Transistor Characteristics, 387
Operation Regions, 388
Characteristic Curves, 389
12.4 Transistor Packaging, 390
12.5 Bipolar Transistor Testing, 391
Transistor Junction Testing, 391
Lead Identification, 392
12.6 Unipolar Transistors, 394
12.7 Junction Field-Effect Transistors, 394
N-*Channel JFETs, 394*
P-*Channel JFETs, 395*
JFET Characteristic Curves, 395
12.8 MOS Field-Effect Transistors, 397
VMOS Field-Effect Transistors, 399
12.9 Unijunction Transistors, 400
12.10 Unipolar Transistor Testing, 401
Review, 404
Student Activities, 404

13 AMPLIFICATION 409
Objectives, 410
Important Terms, 410
13.1 Amplification Principles, 411
Reproduction and Amplification, 411
Voltage Amplification, 411
Current Amplification, 412
Power Amplification, 412
13.2 Bipolar Transistor Amplifiers, 412
13.3 Basic Amplifiers, 412
Signal Amplification, 414
Amplifier Bias, 415

Load-Line Analysis, 418
Linear and Nonlinear Operation, 420
13.4 Classes of Amplification, 421
13.5 Transistor Circuit Configurations, 423
Common-Emitter Amplifiers, 423
Common-Base Amplifiers, 424
Common-Collector Amplifiers, 426
13.6 Field-Effect Transistor Amplifiers, 427
Basic FET Amplifier Operation, 427
Dynamic JFET Amplifier Analysis, 427
MOSFET Circuit Operation, 429
FET Biasing Methods, 429
FET Circuit Configurations, 432
13.7 Integrated-Circuit Amplifiers, 434
Operational Amplifiers, 434
Review, 439
Student Activities, 440

14 AMPLIFYING SYSTEMS 445
Objectives, 446
Important Terms, 446
14.1 Amplifying System Functions, 447
14.2 Amplifier Gain, 448
14.3 Decibels, 449
14.4 Amplifier Coupling, 451
Capacitive Coupling, 451
Direct Coupling, 452
Transformer Coupling, 453
14.5 Power Amplifiers, 455
Single-Ended Power Amplifiers, 455
Push-Pull Amplifiers, 456
Crossover Distortion, 459
Class AB Push-Pull Amplifiers, 460
Complementary-Symmetry Amplifiers, 461
14.6 Integrated-Circuit Amplifying Systems, 462
14.7 Speakers, 464
14.8 Input Transducers, 465
Microphones, 465
Magnetic Tape Input, 465
Phonograph Pickup Cartridges, 466
Compact Discs, 468
Lasers, 468
Review, 470
Student Activities, 471

15 OSCILLATORS 475
Objectives, 476
Important Terms, 476
15.1 Oscillator Types, 477
15.2 Feedback Oscillators, 477
Oscillator Fundamentals, 477
LC *Circuit Operation, 478*
Armstrong Oscillator, 480
Hartley Oscillator, 483
Colpitts Oscillator, 484
Crystal Oscillator, 485
Pierce Oscillator, 485
15.3 Time Constant Circuits, 486
15.4 Relaxation Oscillators, 487
RC *Circuit, 488*
UJT Oscillator, 489
Astable Multivibrator, 491
Monostable Multivibrator, 492
Bistable Multivibrator, 493
IC Waveform Generators, 494
IC Astable Multivibrator, 495
Blocking Oscillators, 497
Review, 498
Student Activities, 498

16 COMMUNICATIONS SYSTEMS 503
Objectives, 504
Important Terms, 504
16.1 Systems, 505
16.2 Electromagnetic Waves, 506
16.3 Continuous-Wave Communication, 508
CW Transmitter, 509
CW Receiver, 511
16.4 Amplitude Modulation Communication, 517
Percentage of Modulation, 518
AM Communication System, 518
16.5 Frequency Modulation Communication, 526
FM Communication System, 526
16.6 Television Communication, 531
Picture Signal, 534
Television Transmitter, 536
Television Receiver, 537
Color Television, 539
16.7 Fiber Optics, 541
Review, 543
Student Activities, 544

17 DIGITAL ELECTRONIC SYSTEMS 551
Objectives, 552
Important Terms, 552
17.1 Digital Systems, 553
Decimal Numbering System, 553
Binary Numbering System, 554
17.2 System Operational States, 555
17.3 Binary Logic Functions, 556
AND Gates, 556
OR Gates, 556

NOT Gates, 557
Combination Logic Gates, 557
17.4 Gate Circuits, 558
17.5 Timing and Storage Elements, 558
17.6 Flip-Flops, 559
Counters, 559
17.7 Decade Counters, 561
17.8 Digital System Displays, 563
17.9 Decoding, 565
17.10 Digital Counting Systems, 565
17.11 Computers, 565
Hardware, 567
Data Information, 569
Memory, 569
Software, 572
Computer Operation, 572
Microcomputer Systems, 572
Review, 573
Student Activities, 574

18 ELECTRONIC POWER CONTROL 581
Objectives, 582
Important Terms, 582
18.1 Electrical Power Control System, 583
18.2 Silicon-Controlled Rectifiers, 584
SCR Construction, 585
SCR Operation, 586
SCR I-V Characteristics, 586
Dc Power Control with SCRs, 587
Ac Power Control with SCRs, 589
18.3 Triac Power Control, 591
Triac Construction, 591
Triac Operation, 591
Triac I-V Characteristics, 592
Triac Applications, 592
Start-Stop Triac Control, 593
Triac Variable Power Control, 594
Triac Phase Control, 595
18.4 Diac Power Control, 596
18.5 Electronic Control Considerations, 598
Review, 598
Student Activities, 599

APPENDIX A
Comprehensive Glossary of Terms 605

APPENDIX B
Soldering 619

APPENDIX C
Electrical Tools 623

APPENDIX D
Symbols for Electricity and Electronics 631

APPENDIX E
Common Logarithms and Natural Logarithms 635

APPENDIX F
Calculator Examples 639

APPENDIX G
Units of Measurement 651

APPENDIX H
Printed Circuit Board (PCB) Construction 663

APPENDIX I
Acronyms and Abbreviations 665

ANSWERS TO SELECTED PROBLEMS 669

INDEX 675

CHAPTER 1 Basics of Electricity

CHAPTER 2 Electrical Components and Measurements

CHAPTER 3 Electrical Circuits

CHAPTER 4 Magnetism and Electromagnetism

CHAPTER 5 Sources of Electrical Energy

CHAPTER 6 Alternating-Current Electricity

CHAPTER 7 Electrical Energy Conversion

CHAPTER 8 Electrical Instruments

CHAPTER

1

BASICS OF ELECTRICITY

OUTLINE

1.1 History of Electricity and Electronics
1.2 Electrical Systems
1.3 Energy, Work, and Power
1.4 Structure of Matter
1.5 Electrical Charges
1.6 Static Electricity
1.7 Electrical Current
1.8 Electrical Force (Voltage)
1.9 Resistance
1.10 Voltage, Current, and Resistance
1.11 Volts, Ohms, and Amperes
1.12 Electrical Safety
1.13 Careers in Electronics

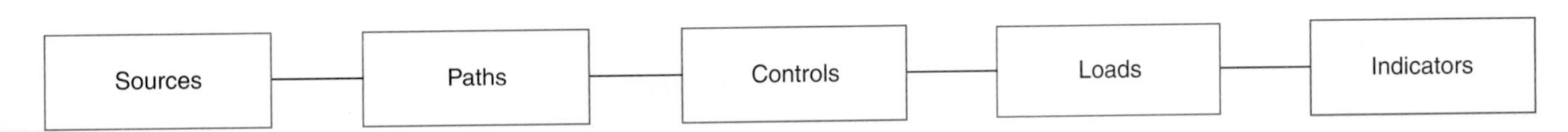

The Electrical Systems Model

The electrical systems model.

OBJECTIVES

Upon completion of this chapter, you will be able to:

1. Explain the parts of an electrical system.
2. Explain the composition of matter.
3. Explain the laws of electrical charges.
4. Define the terms *insulator, conductor,* and *semiconductor.*
5. Explain electrical current flow.
6. Define the terms *voltage, current,* and *resistance.*
7. Explain factors determining resistance.
8. Observe safe electrical practices in a lab or at home.

Electricity is a fascinating science that we use in many different ways. It would be difficult to think of the many ways that we use electricity each day. It is important to have an understanding of electricity and its applications.

This chapter deals with basic topics in the study of electricity. These include the discovery of electricity, basic electrical systems, energy and power, the structure of matter, electrical charges, static electricity, electrical current, voltage, resistance, and electrical safety. The beginning of this and other chapters has definitions of important terms. Preview these terms to gain a better understanding of what is discussed in the chapter. As you study the chapter, go back to the definitions when the need arises. There is also a review at the end of all chapters. These will aid in understanding the material in the chapter. Several student activities are also suggested at the end of all chapters. They may be completed at a lab or at home.

IMPORTANT TERMS

Before reading this chapter, review the following terms. These terms provide a basic understanding of some concepts that are discussed in the chapter. As you read other chapters, also review these terms as needed. In addition to these terms, a comprehensive glossary is included in App. A.

Ampere (A). The electrical charge movement, which is the basic unit of measurement for current flow in an electrical circuit.

Atom. The smallest particle to which an element can be reduced and still retain its characteristics.

Atomic number. The number of particles called protons in the nucleus (center) of an atom.

Closed circuit. A circuit that forms a complete path so that electrical current can flow through it.

Compound. The chemical combination of two or more elements to make an entirely different material.

Conductance (G). The reciprocal of resistance ($\frac{1}{R}$). The unit of measure is the siemens (S).

Conductor. A material that allows electrical current to flow through it easily.

Control. The part of an electrical system that affects what the system does; a switch to turn a light on and off is a type of control.

Conventional current flow. Current flow that is assumed to be in a direction from the positive (+) terminal to the negative (–) terminal of a source.

Coulomb (C). A unit of electrical charge that represents a large number of electrons.

Current. The movement of electrical charge; the flow of electrons through an electrical circuit.

Electromotive force (EMF). The pressure or force that causes electrical current to flow.

Electron. A negatively charged atomic particle; electrons cause the transfer of electrical energy from one place to another.

Electron current flow. Current flow that is assumed to be in the direction of electron movement from a negative (–) potential to a positive (+) potential.

Electrostatic field. The space or area around a charged body in which the influence of an electrical charge is experienced.

Element. The basic materials that make up all other materials; they exist by themselves (such as copper, hydrogen, carbon) or in combination with other elements (water is a combination of the elements hydrogen and oxygen).

Energy. Something that is capable of producing work, such as heat, light, chemical, and mechanical action.

Free electrons. Electrons located in the outer orbit of an atom which are easily removed and result in electrical current flow.

Indicator. The part of an electrical system that shows if it is on or off, or indicates a specific quantity.

Insulator. A material that offers high resistance to electrical current flow.

Kinetic energy. Energy due to motion.

Load. The part of an electrical system that converts electrical energy into another form of energy, such as an electric motor which converts electrical energy into mechanical energy.

Matter. Any material that makes up the world; anything that occupies space and has weight; can be a solid, a liquid, or a gas.

Metallic bonding. The method by which loosely held atoms are bound together in metals.

Molecule. The smallest particle to which a compound can be reduced before being broken down into its basic elements.

Neutron. A particle in the nucleus (center) of an atom which has no electrical charge or is neutral.

Nucleus. The core or center part of an atom, which contains protons having a positive charge and neutrons having no electrical charge.

Ohm (Ω). The unit of measurement of electrical resistance.

Open circuit. A circuit that has a broken path so that no electrical current can flow through it.

Orbit. The circular path along which electrons travel around the nucleus of an atom.

Path. The part of an electrical system through which electrons travel from a source to a load, such as the electrical wiring used in a building.

Potential energy. Energy due to position.

Power (P). The rate of doing work in electrical circuits, found by using the equation $P = I \times V$.

Proton. A particle in the center of an atom which has a positive (+) electrical charge.

Resistance (*R*). Opposition to the flow of current in an electrical circuit; its unit of measurement is the ohm (Ω).

Semiconductor. A material that has a value of electrical resistance between that of a conductor and an insulator.

Short circuit. A circuit that forms a direct path across a voltage source so that a very high and possibly unsafe electrical current flows.

Source. The part of an electrical system that supplies energy to other parts of the system, such as a battery that supplies energy for a flashlight.

Stable atom. An atom that will not release electrons under normal conditions.

Static charge. A charge on a material which is said to be either positive or negative.

Static electricity. Electricity with positive and negative electrical charges which is "at rest."

Valence electrons. Electrons in the outer shell or layer of an atom.

Volt (V). The unit of measurement of electrical potential.

Voltage. The electrical force or pressure that causes current to flow in a circuit.

Watt (W). The unit of measurement of electrical power; the amount of power converted when 1 A of current flows under a pressure of 1 V.

Work. The transforming or transferring of energy.

1.1 HISTORY OF ELECTRICITY AND ELECTRONICS

People are naturally curious. Most of us want to know what makes things work or why natural forces behave as they do. What makes lightning, hurricanes, or the rain? From simple curiosity about the origin of lightning, for example, we have developed the sciences of magnetism, electricity, and electronics.

Through magnetism, the invention of the compass made worldwide exploration possible. The ability to generate large quantities of electrical energy has provided us with a comfortable way of life. International communications, medical diagnoses, lights in our homes, heating, and cooling have all come from experiments that began with the study of the effects of magnetism and electricity.

Today, the science and technology of electronics makes computers work. These machines perform complex calculations in fractions of seconds that previously required hours of figuring. The step from magnetism to electricity and electronics was giant. This experimentation has enabled us to launch space vehicles that probe distant planets and transmit to us pictures of their moons. We can say, then, that the effects of magnetism eventually took us to the stars.

Early humans were astounded by certain "wonders" that we now associate with electricity and electronics, particularly lightning. Originally, many of these phenomena were explained by myths. For example, the Vikings believed that Thor, the god of thunder, owned a magic hammer. He would hurl it down to earth, creating lightning. Not until the eighteenth century was it realized that lightning was indeed electrical energy—millions of volts being discharged.

Electricity does not exist naturally in quantities that we can control and use for our benefit. We can generate electrical energy from several sources such as the sun, waterfalls, coal, and oil. Over the centuries, many people have experimented with electricity and magnetism. Through their work they have developed numerous theories.

Thales of Miletus

In an ancient city on the west coast of Asia Minor, lived a famous philosopher and mathematician named Thales

(640–546 B.C.). Thales was the first person to record interesting things about static electricity and magnetism. Certain people before Thales were probably aware of static electricity and magnetism as curiosities, but they left no writings to tell us about their experiments.

Thales found that amber would attract lightweight objects such as feathers, bits of dried grass, and straw. Thales noticed when he laid a stone, which had been rubbed with amber, near a straw that the straw would cling to the amber. Curious, he wanted to see what amber would do to other lightweight objects, and the results were the same. The lightweight objects were attracted to the amber. Thales concluded that it was his rubbing that made amber "magnetic."

Every good experimenter repeats his experiments several times. To be of value, the results must be the same each time. Thales noticed that although rubbed amber attracted lightweight objects, it did not attract metal. When he picked up a piece of lodestone (an iron ore with a silvery finish, also called magnetite), however, he found it could attract pieces of iron. He also noted that lodestone attracted iron without being rubbed.

The Compass

Eventually the use of lodestones as compasses became important; for example, one could be mounted inside a wagon. The lodestone-compass could be made to turn easily on a precision mounting to allow one end to point north. This type of iron material has permanent "poles." One end always points north and the other south. Later experiments showed that like poles repel each other and unlike poles attract. This concept is the basis of magnetism.

Compasses were used by military commanders during the Han dynasty, a ruling group that controlled China from 206 B.C. to A.D. 220. Lodestones were not used for ship navigation for 900 more years. Not until the thirteenth century A.D. was the compass used by Chinese navigators for the first time. By then, they had discovered that a needle could be magnetized and used in a compass by rubbing it with lodestone.

Arab sailors saw the advantages of the compass, adopted it, and brought it to Europe. This resulted in the great period of European exploration. For the first time, shipmasters could easily find their way across the sea without hugging the shore. Christopher Columbus undoubtedly used the compass when he left Spain trying to find an easier route to the Indies. This led to the discovery of America.

William Gilbert

The study of magnetic effects continued, but no person made specific progress until William Gilbert (1540–1603). He lived in England during the time of Shakespeare and was one of Queen Elizabeth's physicians. Like many doctors of his time, Gilbert was deeply interested in magnetism. He thought that since magnetism had an affect on objects, it could have healing powers for the human body.

He discovered that many substances besides amber could attract lightweight objects, but not all of them attracted objects equally. He carefully separated objects in the order of their ability to attract. He discovered other important facts during his studies. He found that it was not the heat given off by the rubbing that made amber attract other lightweight objects, but the friction.

Ben Franklin

In approximately 1746, a professor from Europe brought his lecture on electricity to the American colonies. At a lecture in Philadelphia sat a most-interested spectator who would be called the "first American scientist." Benjamin Franklin (1706–1790) was curious about everything. He is said to have started the first magazine in this country. He invented the stove that bears his name. He began the first circulating library and the first scientific society. He was also the American ambassador to France. As a result of his own experiments, Franklin concluded that there were two kinds of electricity: *positive* or plus (+) and *negative* or minus (−). He said that electricity was not created by rubbing a glass tube, but merely transferred. He went on to state that when an unelectrified object was rubbed, it did one of two things. It either gained electricity and reached a positive state, or it lost some of the "electric fluid," leaving the object in a negative state. This idea that electricity could be created and/or destroyed was very important.

Benjamin Franklin was a serious experimenter and made many worthwhile discoveries. He is best known for his kite experiment. In 1752, Franklin flew the famous kite on a day when a storm was about to break. At the top of the kite, he fastened a stiff wire that was pointing upward. At the other end of the string, he tied a metallic key. When it started to rain, the moistened string began to conduct electricity. It was fortunate for Franklin that there was no lightning!

Volta's Battery

Alessandro Volta (1745–1827), a physics professor at an Italian University, had a different idea as to the origin of electricity. Volta is known as the inventor of the "voltaic pile," more commonly known as the battery. On March 20, 1800, Volta sent a letter to the Royal Society of London describing his discovery. He created a stack of zinc and copper disks, with paper or leather disks in between. Volta soaked the middle disk in a salt solution or a mild acid such as vinegar or lemon juice. He built up a high stack—alternating the zinc, paper, and copper disks—and demonstrated the presence of electrical voltage.

Coulomb and Faraday

Charles de Coulomb (1736–1806) was the first person to measure the amount of electricity and magnetism generated in a circuit. Until then, only the flow of electricity, not the amount, could be detected. He invented several types of instruments for measuring electrical quantities.

Michael Faraday (1791–1867) was originally an instrument maker in London. When Faraday was 29 years old, he began a series of experiments on the relationship between electricity and magnetism. His pioneering work dealt with understanding how electric currents work. His experiments laid the groundwork for many practical inventions such as the motor, generator, transformer, telegraph, and telephone, but they would come 50 to 100 years later. Faraday created words such as *electrode, anode, cathode,* and *ion* to describe his work. We still use these terms today in electricity and electronics.

Other Experiments

As the early scientists continued to pursue electrical wonders of the day, more discoveries were uncovered. Heinrich Hertz (1857–1894), a German physicist, confirmed predictions of the possibility of radiating electromagnetic waves through space. As part of the apparatus used in demonstrating this remarkable phenomenon, he employed an induction coil and spark gap to create a transmitter that is the basis of radio communications today.

Guglielmo Marconi (1874–1937), an Italian inventor, while vacationing in the Italian Alps, read an announcement of the death of Hertz. The same article mentioned that Hertz demonstrated the possibility of transmitting electromagnetic waves through space. Marconi was struck with the potential such a discovery might hold for the communication of information across more than just the few feet of laboratory space which Hertz had done. He cut short his vacation and returned home to begin work on what was to become the wireless telegraph (today's radio). Marconi soon began experimenting with antenna placement and eventually invented the principle of the radio.

In 1819, Hans Christian Oersted, a Danish physicist, brought a small compass near a wire that was carrying an electric current. This compass consisted of a small, magnetized needle pivoted at the center so that it was free to rotate. As he brought the compass near the wire, Oersted noticed that the needle was deflected. This discovery started a chain of events that has helped shape our industrial civilization. Since a current flowing through a wire consists of electrons in motion, we may expect to find a magnetic field around the wire. It is this magnetic field, reacting with the magnetic field of the compass, that causes the needle to deflect. This magnet is called an electromagnet and is the basis of electrical power generation.

Thomas Edison (1847–1931) and Alexander Graham Bell (1847–1922) were American inventors who played significant roles in the history of electricity and electronics. Edison is credited with the invention of the telegraph, phonograph, incandescent lamp, and the first electrical-power generation system. Bell was the inventor of the telephone. Imagine what our society would be like without the inventions of these two individuals!

Another American, Lee De Forest (1873–1961), is called the "father of radio." He patented 300 inventions in wireless telegraphy, radio telephones, and even "talkies" with the sound track on film. He is best known for his improvements on the vacuum tube, which was used as an amplifier for many communications and industrial control applications. More improvements were added to the vacuum tube to achieve better results.

When there seemed to be no room for more advancements, another discovery made possible smaller electronic circuits that required less voltage but had longer life. Within a few years of this revolutionary discovery of the transistor, tubes were used only for special purposes.

In June 1948, John Bardeen, W. H. Brittain, and William Shockley developed the transistor. These men were American physicists who were at the time working for Bell Laboratories. This device revolutionized electronic communications and control. The beginning of the transistor age brought about many other discoveries.

Because of the transistor, equipment size began to shrink dramatically. Miniaturization has reached a point where thousands of transistors and resistors are packed tightly into one integrated circuit. A tiny chip may be no larger than a child's fingernail. We live in a high-tech society, a very advanced civilization. We now can spend more time improving technology for a safer and healthier environment and continued advances in the way we live and work. The history of electricity and electronics has had a major impact on the way we live in the modern world.

Today, we know that electricity is produced by the effect of tiny particles called *electrons.* Electrons are too small to be seen with the naked eye. They exist in all materials. The production of electricity has become an important part of our society. Electricity is one of the most important forms of energy in our world. Think of some of the things we could not have without electricity—lights, radios, TV sets, computers, telephones, home appliances—to name only a few. These things, which we take for granted, were not developed until rather recently. The use of large quantities of electricity began only in the twentieth century. Most electrical items that we use were invented less than 50 years ago. Now electricity is used practically everywhere.

1.2 ELECTRICAL SYSTEMS

A simple electrical *system* block diagram and pictorial diagram are shown in Fig. 1–1. Using a block diagram allows a better understanding of electrical equipment and provides a simple way to "fit pieces together." The system block diagram can be used to simplify many types of electrical circuits and equipment.

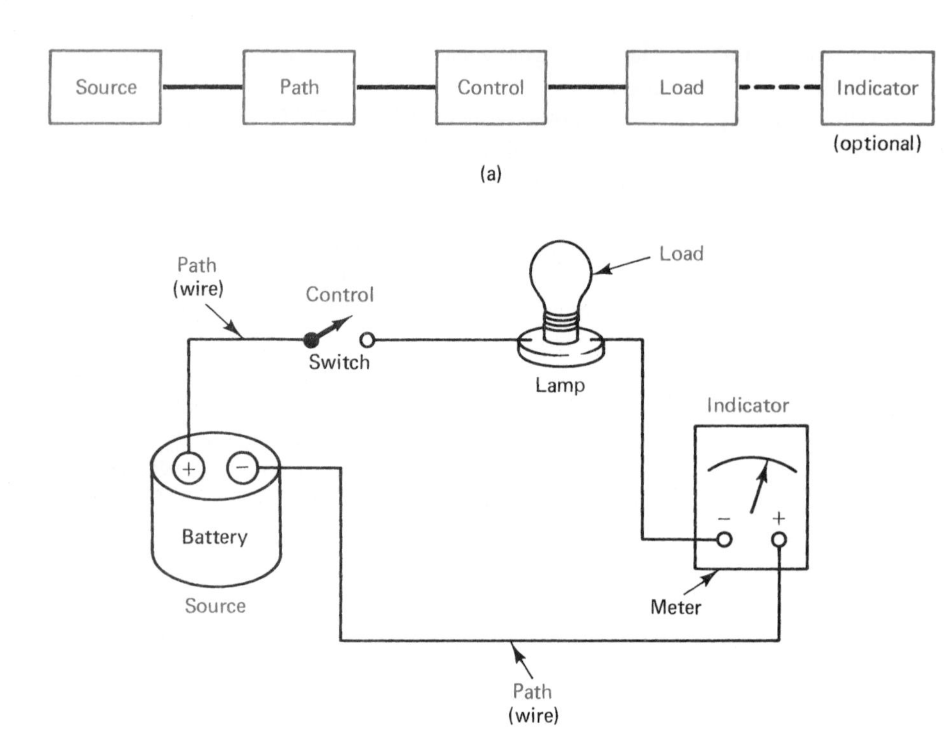

Figure 1–1. Electrical system: (a) block diagram; (b) pictorial diagram.

The parts of an electrical system are the source, path, control, load, and indicator. The concept of "electrical systems" allows discussion of some complex things in a simplified manner. This method is used to present much of the material in this book in order to make it easier to understand.

The *systems* concept serves as a "big picture" in the study of electricity and electronics. The role played by each part then becomes clearer. It is easy to understand the operation of a complete electrical system. When the function of the system is known, it is easier to understand the operation of each part. In this way, it is possible to see how the pieces of any electrical system fit together to make something that operates.

Each block of an electrical system has an important role to play in the operation of the system. Hundreds and even thousands of components are sometimes needed to form an electrical system. Regardless of the complexity of the system, each block must achieve its function when the system operates.

The *source* of an electrical system provides electrical energy for the system. Heat, light, chemical, and mechanical energy may be used as sources of electrical energy. Figure 1–2 shows some sources of electrical energy.

The *path* of an electrical system is simple compared with other system parts. This part of the system provides a path for the transfer of electrical energy. It starts with the energy source and continues through the load. In some cases this path is an electrical wire. In other systems a complex supply line is placed between the source and the load, and a return line from the load to the source is used. There are usually many paths within a complete electrical system. Figure 1–3 shows the distribution path of electrical power from its source to where it is used.

The *control* section of an electrical system is the most complex part of the system. In its simplest form, control is achieved when a system is turned on or off. Control of this type takes place anywhere between the source and the load. The term *full control* is used to describe this operation. A system may also use some type of partial control. *Partial control* causes some type of operational change in the system other than turning it on or off. A change in the amount of electrical current is a type of change that is achieved by partial control. Some common control devices are shown in Fig. 1–4 on page 11.

The *load* of an electrical system is the part or group of parts which do some type of work. Work occurs when energy goes through a transformation or change. Heat, light, and mechanical motion are forms of work produced by loads. Much of the energy produced by the source is changed to another type by the load. The load is usually the most obvious part of the system because of the work it does. An example is a light bulb which produces light. Some common loads are shown in Fig. 1–5 on page 12.

The *indicator* of an electrical system displays a particular operating condition. In some systems the indicator is an optional part that is not really needed; in other systems it is necessary for proper operation. In some cases adjustments are made by using indicators; in other cases an indicator is

(a)

(b)

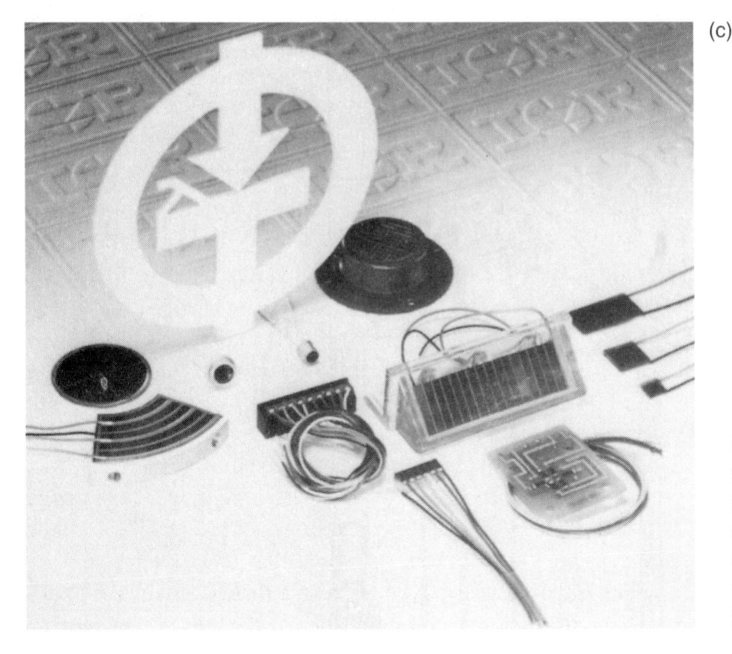

(c)

Figure 1–2. Sources of electrical energy: (a) batteries—convert chemical energy into electrical energy; (b) fossil-fueled steam plant in Paradise, Kentucky—converts mechanical energy and heat energy into electrical energy; (c) solar cells—convert light energy into electrical energy. [(a) Courtesy of Union Carbide Corp.; (b) courtesy of Tennessee Valley Authority; (c) courtesy of International Rectifier.]

attached temporarily to the system to make measurements. Test lights, panel meters, oscilloscopes, and chart recorders are common indicators used in electrical systems. Electrical indicators are shown in Fig. 1–6 on page 13.

Examples of Electrical Systems

Nearly everyone has used a flashlight. This device is designed to serve as a light source. Flashlights are a simple type of electrical system. Figure 1–7, shown on page 13, is a cutaway drawing of a flashlight with each part shown.

The battery of a flashlight serves as the energy *source* of the system. Chemical energy in the battery is changed into electrical energy to cause the system to operate. The energy source of a flashlight—the battery—may be thrown away. Batteries are replaced periodically when they lose their ability to produce energy.

The *path* of a flashlight is a metal case or a small metal strip. Copper, brass, or plated steel are used as paths.

The *control* of electrical energy in a flashlight is achieved by a slide switch or a pushbutton switch. This type of control closes or opens the path between the source and the load device. Flashlights have only a means of full control, which is operated manually by a person.

The *load* of a flashlight is a small lamp bulb. When electrical energy from the source passes through the lamp,

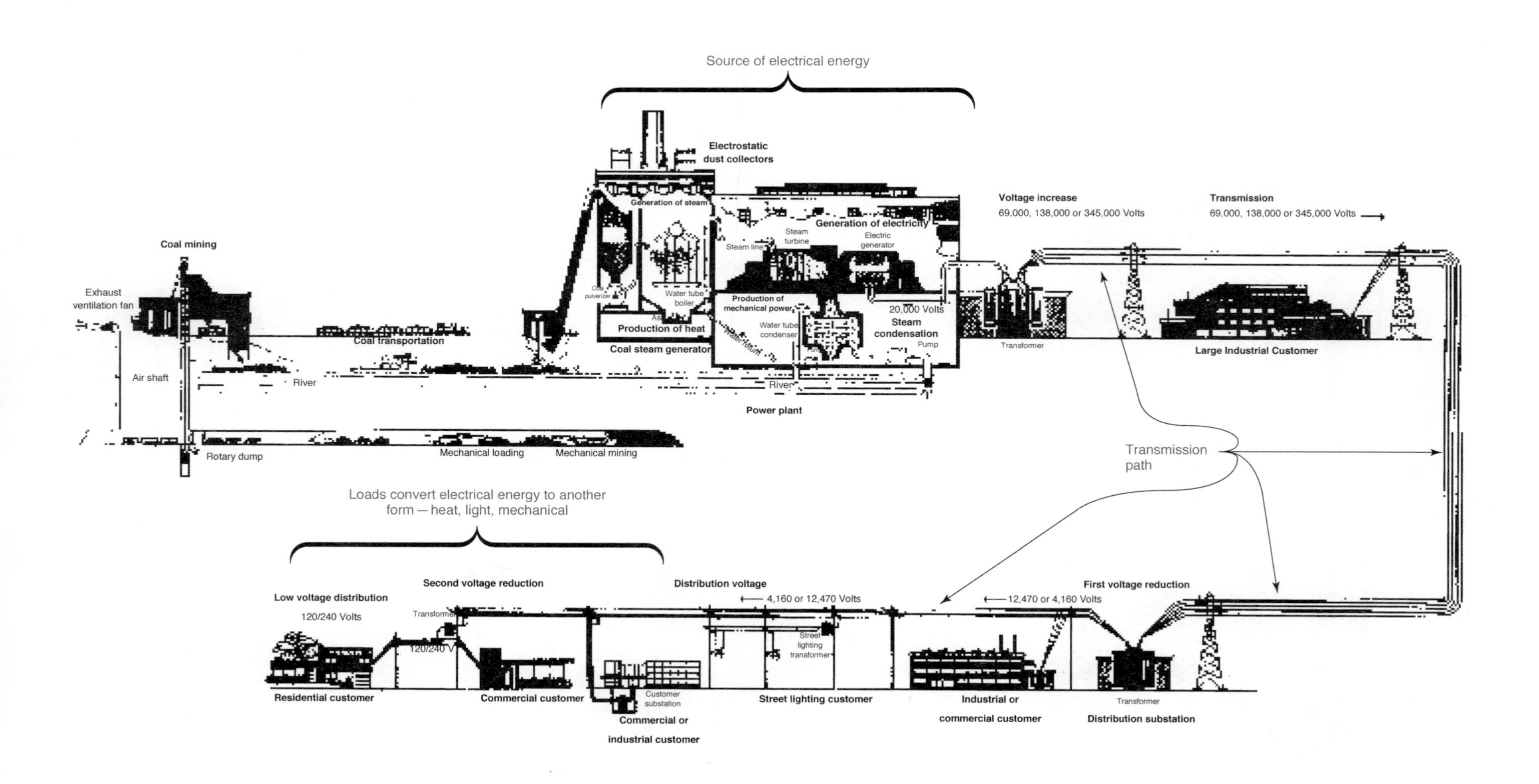

Figure 1–3. Distribution path for electrical power from its source to where it is used (Courtesy of Kentucky Utilities Co.).

(a)

(b)

Figure 1–4. Common control devices: (a) miscellaneous types of switches; (b) potentiometers—partial control. [(a) Courtesy of Eaton Corp., Cutler-Hammer Products; (b) courtesy of Allen-Bradley Co.]

the lamp produces a bright glow. Electrical energy is then changed into light energy. A certain amount of work is done by the lamp when this energy change takes place.

Flashlights do not use an *indicator* as part of the system. Operation is indicated, however, when the lamp produces light. The load of this system also acts as an indicator. In many electrical systems, the indicator is an optional part.

A more complex example of a system is the electrical power system that supplies energy to buildings such as our homes. Figure 1–8 on page 14 shows a sketch of a simple electrical power system.

The energy *source* of an electrical power system is much more complex than that of a flashlight. The source of energy may be coal, natural gas, atomic fuel, or moving water. This type of energy is needed to produce mechanical energy. The mechanical energy develops the motion needed to turn a turbine. Large generators are then rotated by the turbine to produce electrical energy. The energy conversion of this system is quite complex from start to finish. The functions of the parts of the system remain the same regardless of complexity.

In an electrical power system, the *path* consists of many electrical conductors. Copper wire and aluminum wire are ordinarily used as conductors. Metal, water, the earth, and the human body can all be made paths for electrical energy transfer.

(a)

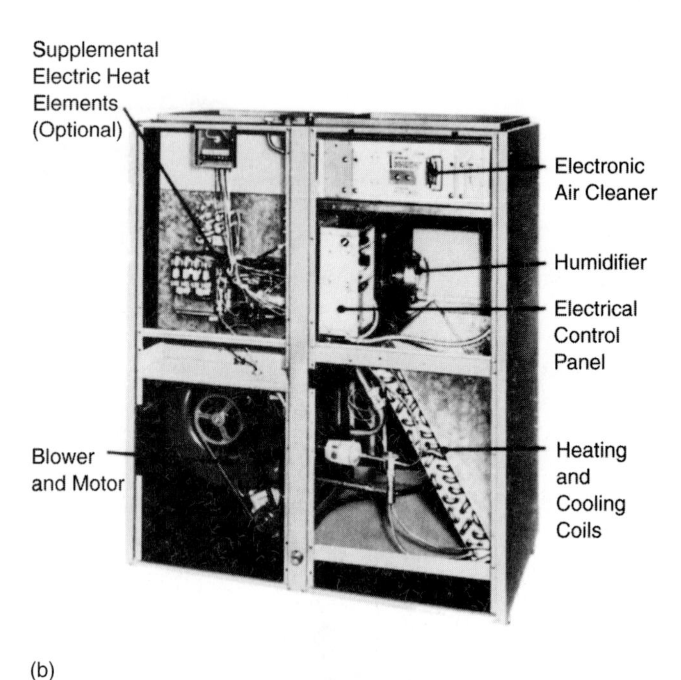

(b)

(c)

Figure 1–5. Common electrical loads: (a) light bulb—converts electric energy into light energy; (b) heat pump home heating system—converts electrical energy into heat energy; (c) electric motor—converts electrical energy into mechanical energy. [(a) Courtesy of Philips Lighting Co.; (b) courtesy of Williamson Co.; (c) courtesy of Delco Products Division—General Motors Corp.]

The *control* function of an electrical power system is performed in many different ways. *Full-control* devices include switches, circuit breakers, and fuses. *Partial control* of an electrical power system is achieved in many ways; for example, transformers are used throughout the system. These partial-control devices are designed to change the amount of voltage in the system.

The *load* of an electrical power system includes everything that uses electrical energy from the source. The total load of an electrical power system changes continually. The load is the part of the system that actually does work. Motors, lamps, electrical ovens, welders, and power tools are some common load devices. Loads are classified according to the type of work they produce. These include light, heat, and mechanical loads.

The *indicator* of an electrical power system is designed to show the presence of electrical energy. It may also be used to measure electrical quantities. Panel-mounted meters, oscilloscopes, and chart recording instruments are some of the indicators used in this type of system. Indicators of this type provide information about the operation of the system.

The systems concept is a method that may be used to study electricity and electronics. This method provides a common organizational plan that applies to most electrical systems. The energy source, path, control, load, and indicator are basic to all systems. An understanding of a basic electrical system helps to overcome some of the problems involved in understanding complex systems.

1.3 ENERGY, WORK, AND POWER

An understanding of the terms *energy, work,* and *power* is necessary in the study of electricity and electronics. The first term, *energy,* means the capacity to do work. For example, the capacity to light a light bulb, to heat a home, or to move something requires energy. Energy exists in many forms,

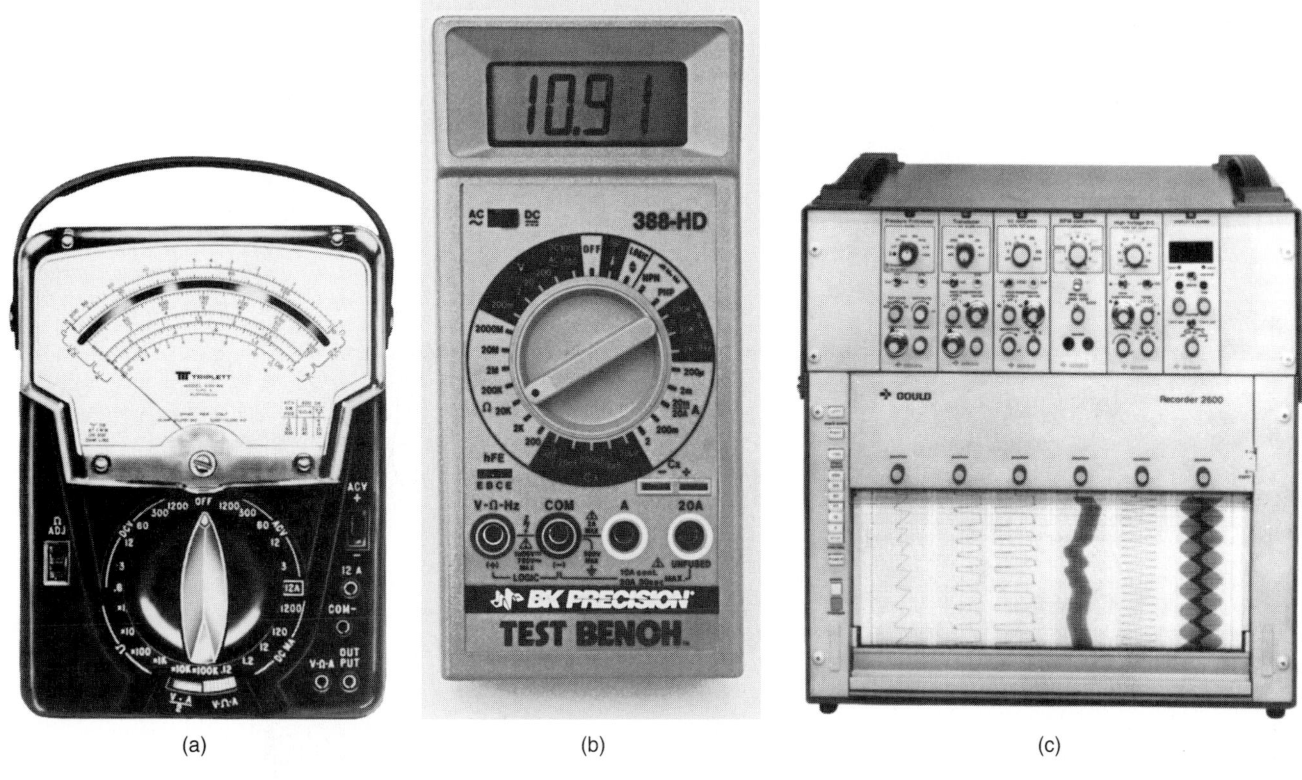

(a) (b) (c)

Figure 1–6. Electrical indicators: (a) analog meter; (b) digital multimeter; (c) chart recorder that makes a permanent record of some quantity. [(a) Courtesy of Triplett Corp.; (b) courtesy of B & K Precision Corp.; (c) courtesy of Gould Inc., Instruments Division, Cleveland, Ohio.]

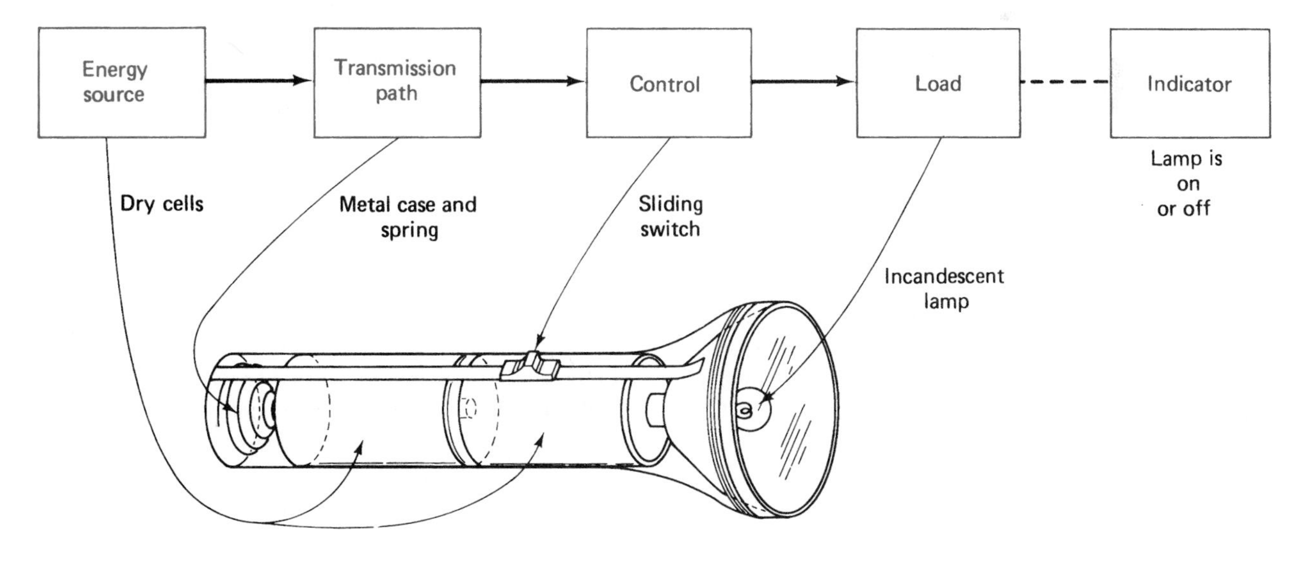

Figure 1–7. Cutaway drawing of a flashlight.

such as electrical, mechanical, chemical, and heat energy. If energy exists because of the movement of some item, such as a ball rolling down a hill, it is called *kinetic energy.* If it exists because of the position of something, such as a ball at the top of the hill but not yet rolling, it is called *potential energy.* Energy has become one of the most important factors in our society.

A second important term is *work.* Work is the transferring or transforming of energy. Work is done when a force is exerted to move something over a certain distance against

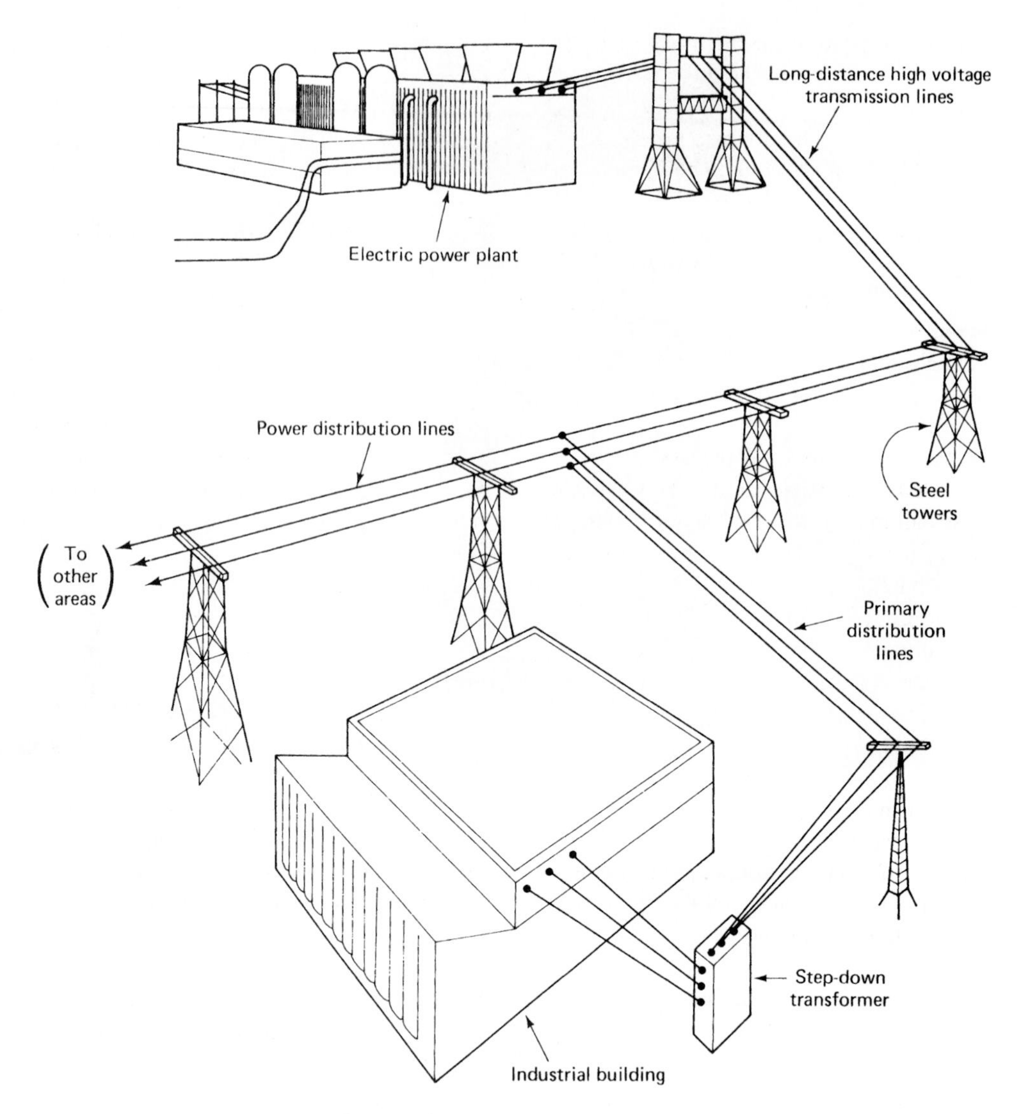

Figure 1–8. Electrical power system.

opposition. Work is done when a chair is moved from one side of a room to the other. An electrical motor used to drive a machine performs work. When force is applied to open a door, work is performed. Work is also done each time energy changes from one form into another.

A third important term is *power.* Power is the rate at which work is done. It considers not only the work that is performed, but the amount of time in which the work is done. For instance, electrical power is the rate at which work is done as electrical current flows through a wire. Mechanical power is the rate at which work is done as an object is moved against opposition over a certain distance. Power is either the rate of production or the rate of energy use. A *watt* is the unit of measurement of electrical power. Energy, work, and power will be discussed in greater detail in Chap. 3.

1.4 STRUCTURE OF MATTER

In the study of electricity and electronics, it is necessary to understand why electrical energy exists. To gain this understanding, let us look first at how certain natural materials are made. It is then easier to see why electrical energy exists.

We begin with some basic scientific terms. These terms are often used in the study of chemistry. They are also important in the study of electronics. First, we say that *matter* is anything that occupies space and has weight. Matter can exist in either a *solid,* a *liquid,* or a *gaseous* state. Solid matter includes such things as metal and wood; liquid matter is exemplified by water or gasoline; and gaseous matter includes such things as oxygen and hydrogen. Solids can be converted into liquids, and liquids can be made into gases.

For example, water can be a solid in the form of ice. Water can also be a gas in the form of steam. The difference is the particles of which they are made move when heated. As they move, they strike one another, causing them to move farther apart. Ice is converted into a liquid by adding heat. If heated to a high temperature, water becomes a gas. All forms of matter exist in their most familiar forms because of the amount of heat they contain. Some materials require more heat than others to become liquids or gases. However, all materials can be made to change from a solid to a liquid or from a liquid to a gas if enough heat is added. Also, these materials can change into liquids or solids if heat is taken from them.

The next important term in the study of the structure of matter is *element.* An element is considered to be the basic material that comprises all matter. Materials such as hydrogen, aluminum, copper, iron, and iodine are a few of the over 100 elements known to exist. A table of elements is shown in Fig. 1–9. Some elements exist in nature and some are manufactured. Everything around us is made of elements.

There are many more materials in our world than there are elements. Materials are made by combining elements. A combination of two or more elements is called a *compound.* For example, water is a compound made by the elements hydrogen and oxygen. Salt is made from sodium and chloride.

Another important term is *molecule.* A molecule is said to be the smallest particle to which a compound can be reduced before breaking down into its basic elements. For example, one molecule of water has two hydrogen atoms and one oxygen atom.

An even deeper look into the structure of matter shows particles called *atoms.* Within these atoms are the forces that cause electrical energy to exist. An atom is considered to be the smallest particle to which an element can be reduced and still have the properties of that element. If an atom were broken down any further, the element would no longer exist. The smallest particles found in all atoms are called *electrons, protons,* and *neutrons.* Elements differ from one another on the basis of the amounts of these particles found in their atoms. The relationship of matter, elements, compounds, molecules, atoms, electrons, protons, and neutrons is shown in Fig. 1–10.

The simplest atom, hydrogen, is shown in Fig. 1–11. The hydrogen atom has a center part called a *nucleus,* which has one proton. A proton is a particle which is said to have a positive (+) charge. The hydrogen atom has one electron, which orbits around the nucleus of the atom. The electron is said to have a negative (–) charge. Most atoms also have neutrons in the nucleus. A neutron has neither a positive nor a negative charge and is considered neutral. A carbon atom is shown in Fig. 1–12. A carbon atom has six protons (+), six neutrons (N), and six electrons (–). The protons and the neutrons are in the nucleus and the electrons *orbit* the nucleus. The carbon atom has two orbits or circular paths. In the first orbit, there are two electrons. The other four electrons are in the second orbit.

Review the table of elements in Fig. 1–9. Notice the different number of protons in the nucleus of each atom. This causes each element to be different. For example, hydrogen has 1 proton, carbon has 6, oxygen has 8, and lead has 82. The number of protons that each atom has is called its *atomic number.*

The nucleus of an atom contains protons (+) and neutrons (N). Because neutrons have no charge and protons have positive charge, the nucleus of an atom has net positive charge. Protons are believed to be about one-third the diameter of electrons. The *mass* or weight of a proton is thought to be over 1800 times more than that of an electron. Electrons move easily in their orbits around the nucleus of an atom. It is the movement of electrons that causes electrical energy to exist.

Early models of atoms showed electrons orbiting around the nucleus, in analogy with planets around the sun. This model is inconsistent with much modern experimental evidence. Atomic orbitals are very different from the orbits of satellites.

Atoms consist of a dense, positively charged nucleus surrounded by a cloud or series of clouds of electrons that occupy energy levels, which are commonly called *shells.* The occupied shell of highest energy is known as the *valence shell,* and the electrons in it are known as *valence electrons.*

Electrons behave as both particles and waves, so descriptions of them always refer to their *probability* of being in a certain region around the nucleus. Representations or orbitals are boundary surfaces enclosing the probable areas in which the electrons are found. All *s* orbitals are spherical, *p* orbitals are egg shaped, *d* orbitals are dumbbell shaped, and *f* orbitals are double dumbbell shaped.

Covalent bonding thus involves the *overlapping* of valence shell orbitals of different atoms. The electron charge then becomes concentrated in this region, thus attracting the two positively charged nuclei toward the negative charge between them. In ionic bonding, the ions are discrete units, and they group themselves in crystal structures, surrounding themselves with the ions of opposite charge.

An exact pattern is thought to be followed in the placement of electrons of an atom. The first orbit, or shell, contains up to 2 electrons. The next shell contains up to 8 electrons. The third contains up to 18 electrons, which is the largest quantity any shell can contain. New shells are started as soon as shells nearer the nucleus have been filled with the maximum number of electrons. Atoms with an incomplete outer shell are very *active.* When two unlike atoms with incomplete outer shells come together, they try to *share* their outer electrons. When their combined outer electrons are enough to comprise one complete shell, *stable* atoms are formed. For example, oxygen has 8 electrons: 2 in the first shell and 6 in its outer shell. There is room for 8 electrons in

PERIODIC CHART OF THE ELEMENTS

IA	IIA	IIIB	IVB	VB	VIB	VIIB	VIII	VIII	VIII	IB	IIB	IIIA	IVA	VA	VIA	VIIA	INERT GASES
1 **H** 1.00797 ±0.00001																1 **H** 1.00797 ±0.00001	2 **He** 4.0026 ±0.00005
3 **Li** 6.939 ±0.0005	4 **Be** 9.0122 ±0.00005											5 **B** 10.811 ±0.003	6 **C** 12.01115 ±0.00005	7 **N** 14.0067 ±0.00005	8 **O** 15.9994 ±0.0001	9 **F** 18.9984 ±0.00005	10 **Ne** 20.183 ±0.0005
11 **Na** 22.9898 ±0.00005	12 **Mg** 24.312 ±0.0005											13 **Al** 26.9815 ±0.00005	14 **Si** 28.086 ±0.001	15 **P** 30.9738 ±0.00005	16 **S** 32.064 ±0.003	17 **Cl** 35.453 ±0.001	18 **Ar** 39.948 ±0.0005
19 **K** 39.102 ±0.0005	20 **Ca** 40.08 ±0.005	21 **Sc** 44.956 ±0.0005	22 **Ti** 47.90 ±0.005	23 **V** 50.942 ±0.0005	24 **Cr** 51.996 ±0.001	25 **Mn** 54.9380 ±0.00005	26 **Fe** 55.847 ±0.003	27 **Co** 58.9332 ±0.00005	28 **Ni** 58.71 ±0.005	29 **Cu** 63.54 ±0.005	30 **Zn** 65.37 ±0.005	31 **Ga** 69.72 ±0.005	32 **Ge** 72.59 ±0.005	33 **As** 74.9216 ±0.00005	34 **Se** 78.96 ±0.005	35 **Br** 79.909 ±0.002	36 **Kr** 83.80 ±0.005
37 **Rb** 85.47 ±0.005	38 **Sr** 87.62 ±0.005	39 **Y** 88.905 ±0.0005	40 **Zr** 91.22 ±0.005	41 **Nb** 92.906 ±0.0005	42 **Mo** 95.94 ±0.005	43 **Tc** (99)	44 **Ru** 101.07 ±0.005	45 **Rh** 102.905 ±0.0005	46 **Pd** 106.4 ±0.05	47 **Ag** 107.870 ±0.003	48 **Cd** 112.40 ±0.005	49 **In** 114.82 ±0.005	50 **Sn** 118.69 ±0.005	51 **Sb** 121.75 ±0.005	52 **Te** 127.60 ±0.005	53 **I** 126.9044 ±0.00005	54 **Xe** 131.30 ±0.005
55 **Cs** 132.905 ±0.0005	56 **Ba** 137.34 ±0.005	57 ***La** 138.91 ±0.005	72 **Hf** 178.49 ±0.005	73 **Ta** 180.948 ±0.0005	74 **W** 183.85 ±0.005	75 **Re** 186.2 ±0.05	76 **Os** 190.2 ±0.05	77 **Ir** 192.2 ±0.05	78 **Pt** 195.09 ±0.005	79 **Au** 196.967 ±0.0005	80 **Hg** 200.59 ±0.005	81 **Tl** 204.37 ±0.005	82 **Pb** 207.19 ±0.005	83 **Bi** 208.980 ±0.0005	84 **Po** (210)	85 **At** (210)	86 **Rn** (222)
87 **Fr** (223)	88 **Ra** (226)	89 †**Ac** (227)															

*Lanthanum Series

58	59	60	61	62	63	64	65	66	67	68	69	70	71
Ce	**Pr**	**Nd**	**Pm**	**Sm**	**Eu**	**Gd**	**Tb**	**Dy**	**Ho**	**Er**	**Tm**	**Yb**	**Lu**
140.12 ±0.005	140.907 ±0.0005	144.24 ±0.005	(147)	150.35 ±0.005	151.96 ±0.005	157.25 ±0.005	158.924 ±0.0005	162.50 ±0.005	164.930 ±0.0005	167.26 ±0.005	168.934 ±0.0005	173.04 ±0.005	174.97 ±0.005

†Actinium Series

90	91	92	93	94	95	96	97	98	99	100	101	102	103
Th	**Pa**	**U**	**Np**	**Pu**	**Am**	**Cm**	**Bk**	**Cf**	**Es**	**Fm**	**Md**	**No**	**Lw**
232.038 ±0.0005	(231)	238.03 ±0.005	(237)	(242)	(243)	(247)	(247)	(249)	(254)	(253)	(256)	(253)	(257)

( ) Numbers in parentheses are mass numbers of most stable or most common isotope.

Atomic weights corrected to conform to the 1961 values of the Commission on Atomic Weights.

Figure 1–9. Periodic table.

(continued)

An Alphabetical List of the Elements

Element	Symbolic	Atomic No.	Atomic Weight	Element	Symbolic	Atomic No.	Atomic Weight
Actinium	Ac	89	277*	Molybdenum	Mo	42	95.95
Aluminum	Al	13	26.97	Neodymium	Nd	60	144.27
Americium	Am	95	243*	Neon	Ne	10	20.183
Antimony	Sb	51	121.76	Neptunium	Np	93	237*
Argon	Ar	18	39.944	Nickel	Ni	28	58.69
Arsenic	As	33	74.91	Niobium	Nb	41	92.91
Astatine	At	85	210*	Nitrogen	N	7	14.008
Barium	Ba	56	137.36	Nobelium	No	102	253
Berkelium	Bk	97	247*	Osmium	Os	76	190.2
Beryllium	Be	4	9.013	Oxygen	O	8	16.000
Bismuth	Bi	83	209.00	Palladium	Pd	46	106.7
Boron	B	5	10.82	Phosphorus	P	15	30.975
Bromine	Br	35	79.916	Platinum	Pt	78	195.23
Cadmium	Cd	48	112.41	Plutonium	Pu	94	244
Calcium	Ca	20	40.08	Polonium	Po	84	210
Californium	Cf	98	251*	Potassium	K	19	39.100
Carbon	C	6	12.01	Praseodymium	Pr	59	140.92
Cerium	Ce	58	140.13	Promethium	Pm	61	145*
Cesium	Cs	55	132.91	Protactinium	Pa	91	231*
Chlorine	Cl	17	35.457	Radium	Ra	88	226.05
Chromium	Cr	24	52.01	Radon	Rn	86	222
Cobalt	Co	27	58.94	Rhenium	Re	75	186.31
Copper	Cu	29	63.54	Rhodium	Rh	45	102.91
Curium	Cm	96	247	Rubidium	Rb	37	85.48
Dysprosium	Dy	66	162.46	Ruthenium	Ru	44	101.1
Einsteinium	E	99	254*	Rutherfordium	Rf	104	260*
Erbium	Er	68	167.2	or	or		
Europium	Eu	63	152.0	Kurchatonium	Ku		
Fermium	Fm	100	255*	Samarium	Sm	62	150.43
Fluorine	F	9	19.00	Scandium	Sc	21	44.96
Francium	Fr	87	233*	Selenium	Se	34	78.96
Gadolinium	Gd	64	156.9	Silicon	Si	14	28.09
Gallium	Ga	31	69.72	Silver	Ag	47	107.880
Germanium	Ge	32	72.60	Sodium	Na	11	22.997
Gold	Au	79	197.0	Strontium	Sr	38	87.63
Hafnium	Hf	72	178.6	Sulfur	S	16	32.066
Hahnium	Ha	105	262*	Tantalum	Ta	73	180.95
Helium	He	2	4.003	Technetium	Tc	43	97*
Holmium	Ho	67	164.94	Tellurium	Te	52	127.61
Hydrogen	H	1	1.0080	Terbium	Tb	65	158.93
Indium	In	49	114.76	Thallium	Tl	81	204.39
Iodine	I	53	126.91	Thorium	Th	90	232.12
Iridium	Ir	77	192.2	Thulium	Tm	69	168.94
Iron	Fe	26	55.85	Tin	Sn	50	118.70
Krypton	Kr	36	83.8	Titanium	Ti	22	47.90
Lanthanum	La	57	138.92	Tungsten	W	74	183.92
Lawrencium	Lr	103	257*	Uranium	U	92	238.07
Lead	Pb	82	207.21	Vanadium	V	23	50.95
Lithium	Li	3	6.940	Xenon	Xe	54	131.3
Lutetium	Lu	71	174.99	Ytterbium	Yb	70	173.04
Magnesium	Mg	12	24.32	Yttrium	Y	39	88.92
Manganese	Mn	25	54.94	Zinc	Zn	30	65.38
Mendelevium	Mv	101	256*	Zirconium	Zr	40	91.22
Mercury	Hg	80	200.61				

*Mass number of the longest-lived of the known available forms of the element.

Figure 1–9. (Continued) List of the elements (Courtesy of Sargent-Welch Scientific Co.).

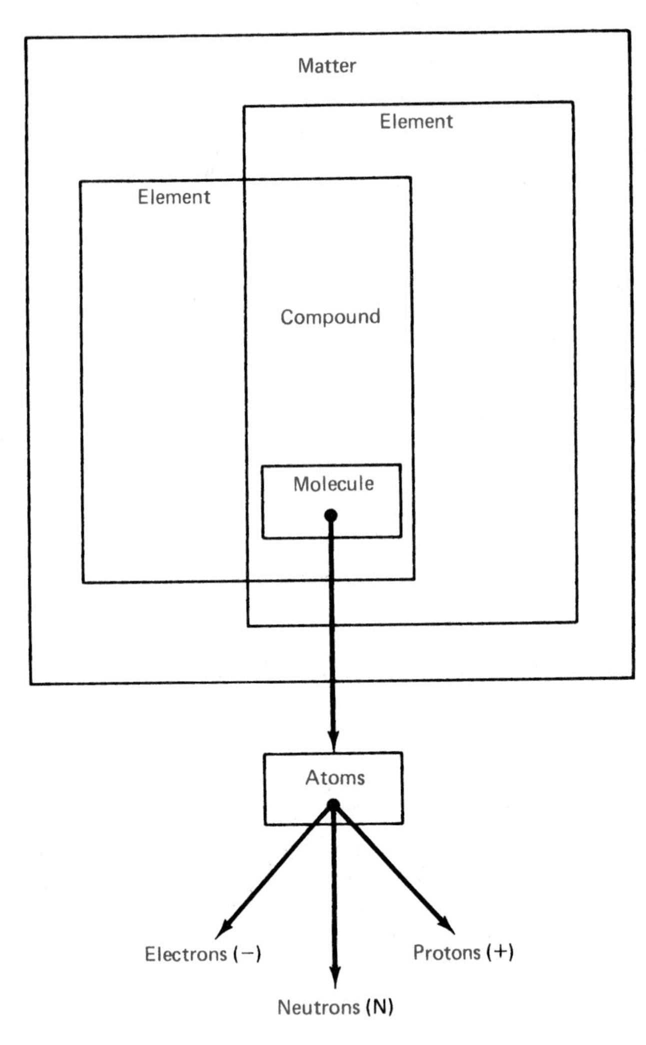

Figure 1–10. Structure of matter.

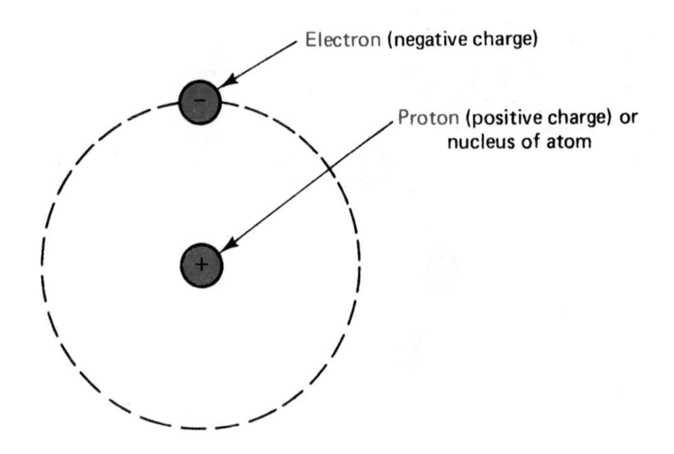

Figure 1–11. Hydrogen atom.

the outer shell. Hydrogen has 1 electron in its outer shell. When two hydrogen atoms come near, oxygen combines with the hydrogen atoms by sharing the electrons of the two hydrogen atoms. Water is formed, as shown in Fig. 1–13. All the electrons are then bound tightly together and a very stable water molecule is formed. The electrons in the incomplete outer shell of an atom are known as *valence electrons.* They are the only electrons that will combine with other atoms to form compounds. They are also the only electrons used to cause electric current to flow. For this reason it is necessary to understand the structure of matter.

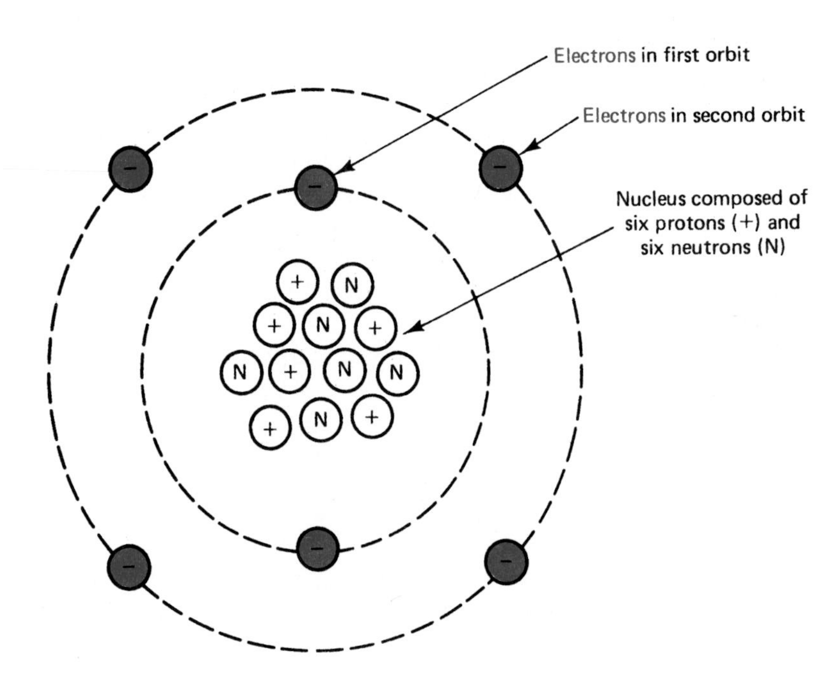

Figure 1–12. Carbon atom.

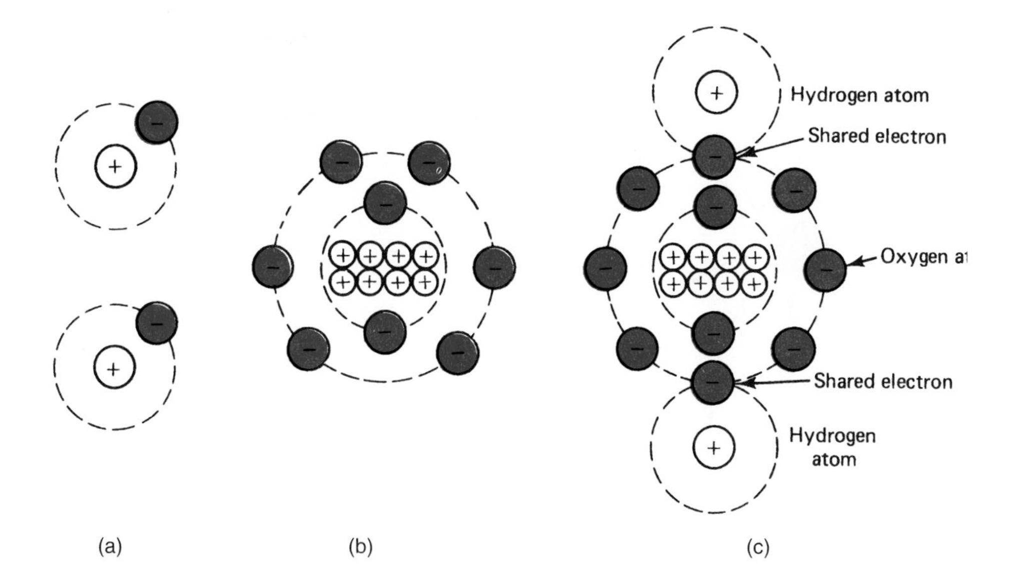

Figure 1–13. Water formed by combining hydrogen and oxygen: (a) hydrogen atoms; (b) oxygen atom; (c) water molecule.

1.5 ELECTRICAL CHARGES

Section 1.4 dealt with the positive and negative charges of particles called protons and electrons. These are parts of atoms that comprise all things in our world. The positive charge of a proton is similar to the negative charge of an electron. However, a positive charge is the *opposite* of a negative charge. These charges are called *electrostatic* charges. Figure 1–14 shows how electrostatic charges affect each other. Each charged particle is surrounded by an *electrostatic field.*

The effect that electrostatic charges have on each other is important. They either repel (move away) or attract (come together) each other. This action is as follows:

1. Positive charges repel each other [see Fig. 1–14(a)].
2. Negative charges repel each other [see Fig. 1–14(b)].
3. Positive and negative charges attract each other [see Fig. 1–14(c)].

Therefore, it is said that *like charges repel* and *unlike charges attract.*

The atoms of some materials can be made to gain or lose electrons. The material then becomes charged. One way to do this is to rub a glass rod with a piece of silk cloth. The glass rod loses electrons (–), so it now has a positive (+) charge. The silk cloth pulls away electrons (–) from the glass. Because the silk cloth gains new electrons, it now has a negative (–) charge. Another way to charge a material is to rub a rubber rod with fur.

It is also possible to charge other materials. If a charged rubber rod is touched against another material, it may become charged. Some materials are charged when they are brought close to another charged object. Remember that materials are charged due to the movement of electrons and protons. Also, remember that when an atom loses electrons (–), it becomes positive (+). These facts are important in the study of electricity and electronics.

Charged materials affect each other due to lines of force. Try to visualize these as shown in Fig. 1–14. These imaginary lines cannot be seen; however, they exert a force in all directions around a charged material. Their force is similar to the force of gravity around the earth. This force is called a *gravitational field.*

1.6 STATIC ELECTRICITY

Most people have observed the effect of *static electricity.* When objects become charged, it is due to static electricity. A common example of static electricity is lightning. Lightning is caused by a difference in charge (+ and –) between the earth's surface and the clouds during a storm. The arc produced by lightning is the movement of charges between the earth and the clouds. Another common effect of static electricity is being "shocked" by a doorknob after walking across a carpeted floor. Static electricity also causes clothes taken from a dryer to cling together and hair to stick to a comb.

There are two types of electricity: (1) static electricity and (2) current electricity. The use of electricity today depends mostly on current electricity. Current electricity is the controlled movement of electrical charges, and is discussed in detail in this book. Static electricity has some practical uses, such as the Van de Graaff generator shown in Fig. 1–15 used in nuclear research. A Van de Graaff generator is a piece of equipment that produces electrical charges. A small electrostatic generator is shown in Fig. 1–16. Electrical charges are used to filter dust and soot in devices called electrostatic filters.

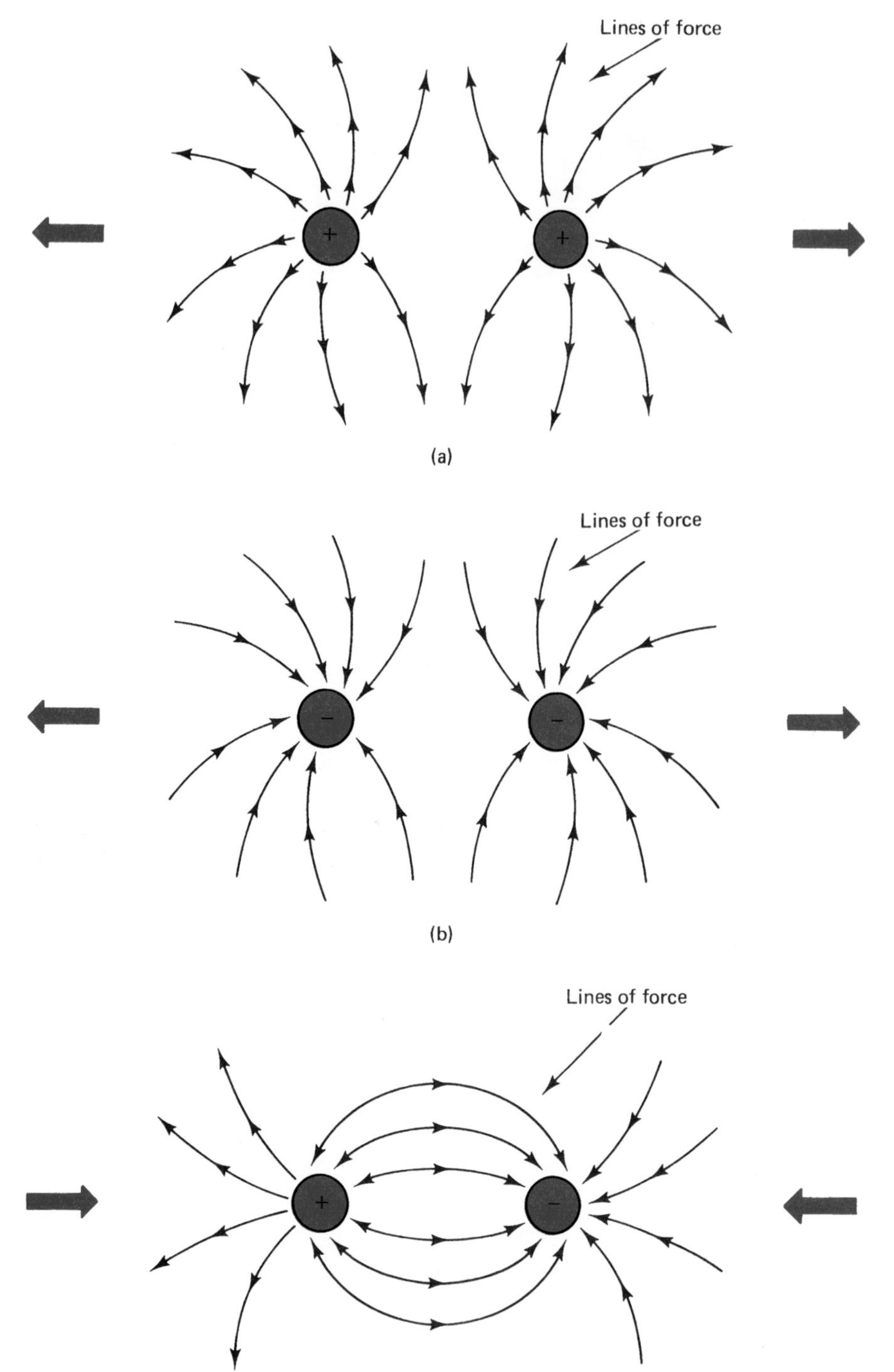

Figure 1–14. Electrostatic charges: (a) positive charges repel; (b) negative charges repel; (c) positive and negative charges attract.

Electrostatic precipitators are used in power plants to filter the exhaust gas that goes into the air. Static electricity is also used in the manufacture of sandpaper and in the spray painting of automobiles. A device called an *electroscope,* shown in Fig. 1–17, is used to detect a negative or positive charge.

1.7 ELECTRICAL CURRENT

Static electricity is caused by stationary charges; however, electrical current is the motion of electrical charges from one point to another. Electrical current is produced when elec-

Figure 1–15. Van de Graaff generator (Courtesy of Westinghouse Electric Corp.).

Figure 1–16. Electrostatic generator—Van de Graaff type (Courtesy of Sargent-Welch Scientific Co.).

trons (–) are removed from their atoms. Some electrons in the outer orbits of the atoms or certain elements are easy to remove. A force or pressure applied to a material causes electrons to be removed. The movement of electrons from one atom to another is called *electrical current flow.*

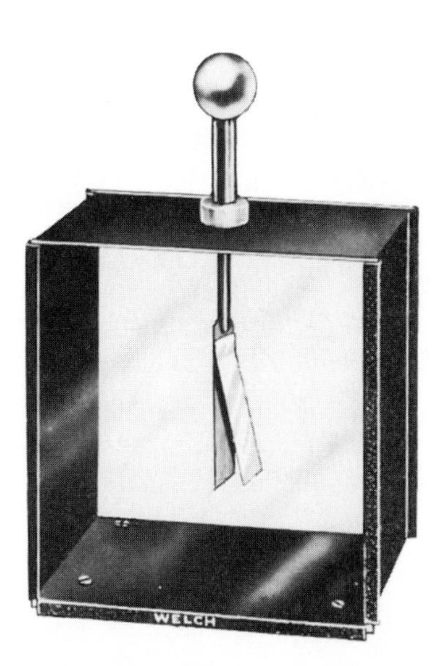

Figure 1–17. Electroscope (Courtesy of Sargent-Welch Scientific Co.).

Conductors

A material through which current flows is called a *conductor.* A conductor passes electrical current very easily. Copper and aluminum wire are commonly used as conductors. Conductors are said to have low *resistance* to electrical current flow. Conductors usually have three or fewer electrons in the outer orbit of their atoms. Remember that the electrons of an atom orbit around the nucleus. Many metals are electrical conductors. Each metal has a different ability to conduct electrical current. For example, silver is a better conductor than copper. Silver is too expensive to use in large amounts. Aluminum does not conduct electrical current as well as copper. The use of aluminum is common, as it is cheaper and lighter than other conductors. Copper is used more than any other conductor. Materials with only one outer orbit or valence electron (gold, silver, copper) are the best conductors. Notice that these elements are located in column IB of the periodic table (Fig. 1–9).

Insulators

Some materials do not allow electrical current to flow easily. The electrons of materials that are insulators are difficult to release. Some insulators have their outer orbits filled with eight electrons. Others have outer orbits that are over half filled with electrons. The atoms of materials that are insulators are said to be stable. Insulators have high resistance to the movement of electrical current. Some examples of insulators are plastic and rubber. Figure 1–18 shows some types of insulators. A test facility for large electrical power line insulators is shown in Fig. 1–19.

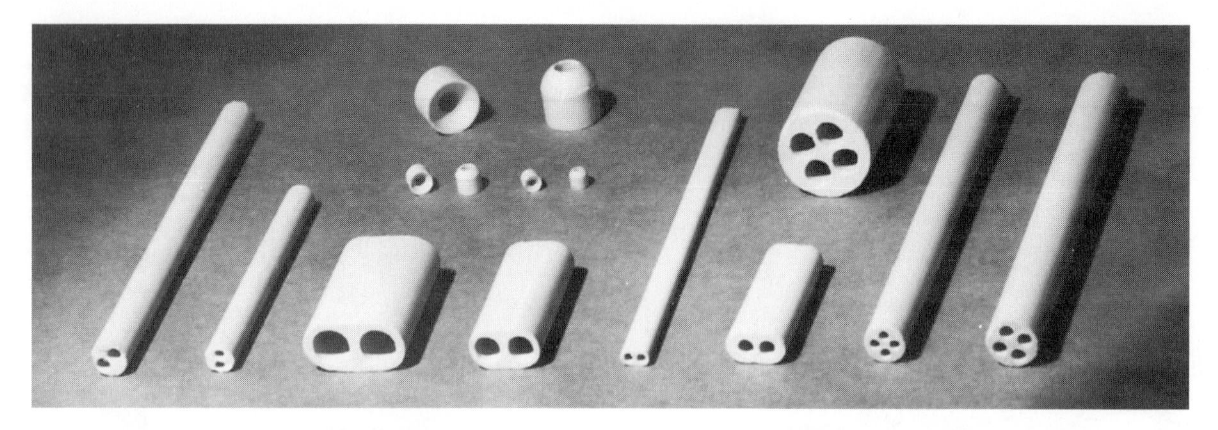

Figure 1–18. Types of insulators (Courtesy of Foxboro Co.).

Figure 1–19. Testing electrical power line insulators (Courtesy of Lapp Insulator, Leroy, NY).

Semiconductors

Materials called *semiconductors* have become extremely important in electronics. Semiconductor materials will not conduct electrical current easily and are not good insulators. Their classification also depends on the number of electrons their atoms have in the outer orbit. Semiconductors have four electrons in their outer orbits. Remember that conductors have outer orbits less than half full and insulators ordinarily have outer orbits more than half full. Figure 1–20 compares conductors, insulators, and semiconductors. Some common types of semiconductor materials are silicon, germanium, and selenium. Notice that these elements are located in column IVA of the periodic table in Fig. 1–9. Semiconductors are discussed in detail in Chap. 9.

Superconductors

Much effort is being put forth in superconductor research. Research scientists would like to create a room temperature superconductor. A *superconductor* is a conductor that has no resistance to the flow of electrical current. This phenomenon was first observed in the early 1900s by a Dutch scientist, Kamerlingh Onnes. Onnes discovered that if a mercury crystal was cooled to a temperature just above absolute zero (−452°F), it lost all resistance to the flow of current. Only recently has this process begun to be understood.

Superconduction has been observed in certain metals when they were cooled to temperatures near absolute zero and in some ceramic compounds. The idea of room temperature superconductors now seems to be a real possibility. The basic idea of superconductors is simple. A normal conductor has resistance to electrical current. This resistance converts energy to heat. If current continues to flow, a source of energy must be present to replace energy lost as heat. A superconductor has no resistance and therefore does not generate heat. If an electrical current flows in a superconductor, it will flow without ceasing.

Superconductive power lines could save enormous amounts of energy in the transmission of electrical power. Typically, 15% to 20% of the energy produced by electric power plants is lost in transmission. Superconducting transmission lines would lower the cost of producing electricity, conserve natural resources, and reduce pollution. Other applications have also been proposed for superconductors, such as high-speed trains levitated by magnetic fields created around superconductive rails.

In a normal conductor, electrons are continually colliding with atoms which make up the conductor. Each collision

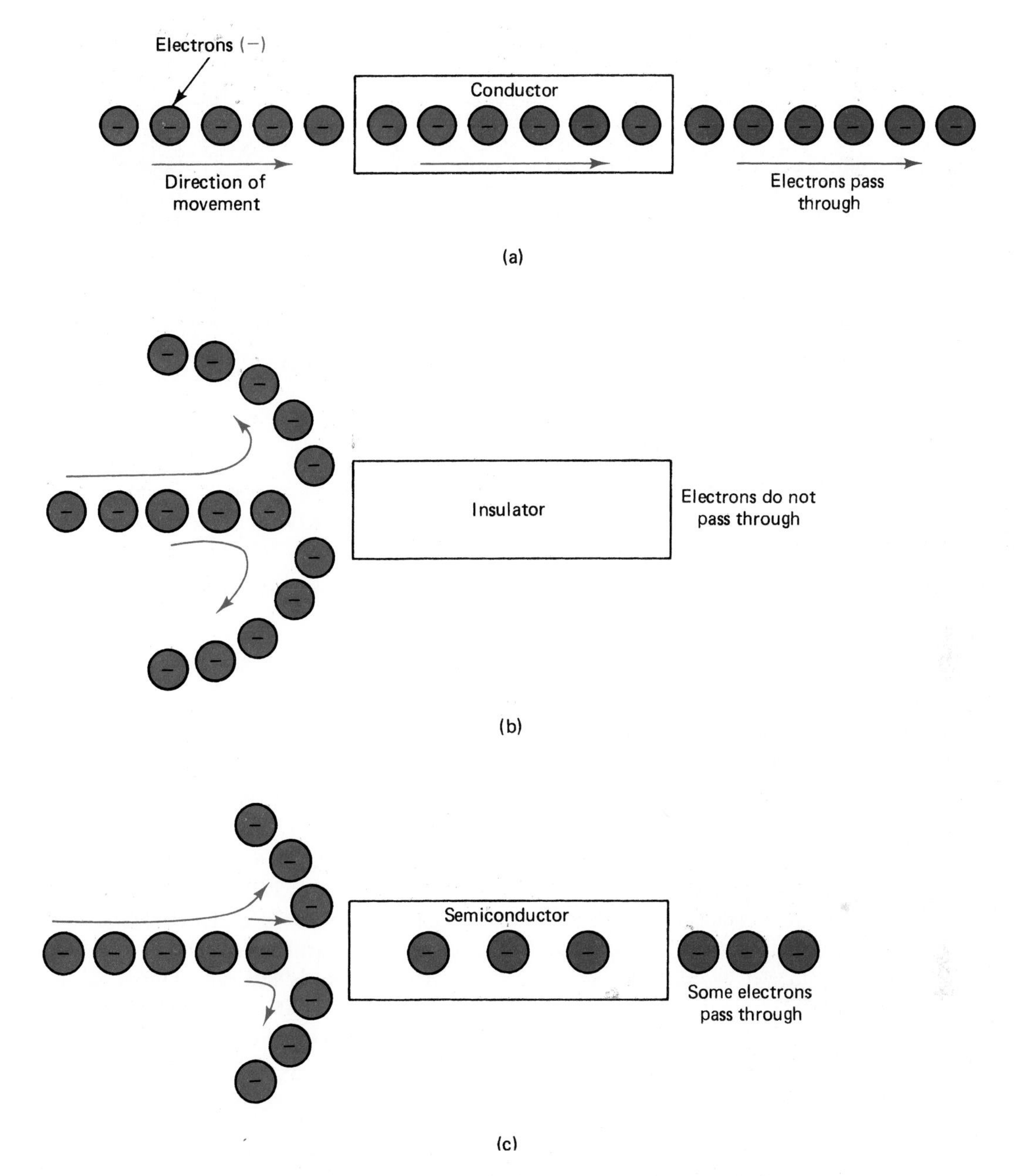

Figure 1–20. Comparison of (a) conductors; (b) insulators; and (c) semiconductors.

causes energy to be lost and heat to be generated. In a superconductor, each passing electron causes a small vibration in the conductor. This action clears a path for more electrons. As the temperature rises, the motion of the conductor's atoms increases. Eventually, the bond between electrons is broken and superconduction stops. For this reason, superconduction requires an extremely low operating temperature.

Current Flow

The usefulness of electricity is due to *electrical current flow.* Current flow is the movement of electrical charges along a conductor. Static electricity or electricity at rest has some practical uses due to electrical charges. Electrical current flow allows us to use electrical energy to do many types of work.

The movement of outer orbit electrons of conductors produces electrical current. The electrons on the outer orbit of the atoms of a conductor are called *free electrons.* Energy released by these electrons as they move allows work to be done. As more electrons move along a conductor, more energy is released. This is called an increased electrical current flow. The movement of electrons through a conductor is shown in Fig. 1–21.

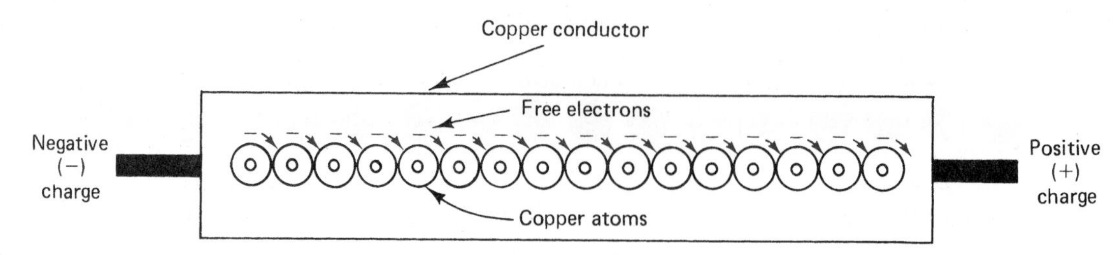

Figure 1–21. Current flow through a conductor.

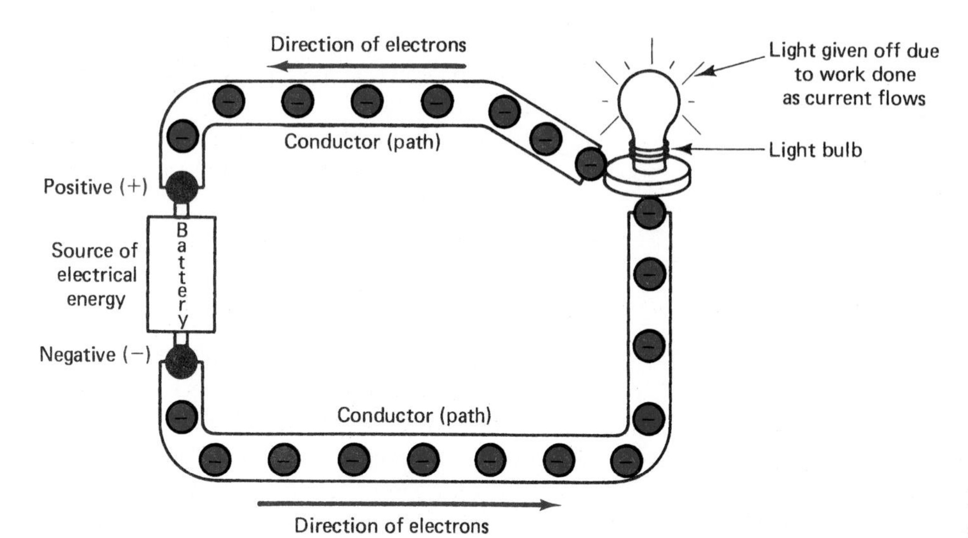

Figure 1–22. Current flow in a closed circuit.

To understand how current flow takes place, it is necessary to know about the atoms of conductors. Conductors such as copper have atoms which are loosely held together. Copper is said to have atoms connected by *metallic bonding.* A copper atom has one outer orbit electron which is loosely held to the atom. These atoms are so close together that their outer orbits overlap each other. Electrons can move easily from one atom to another. In any conductor, the outer orbit electrons continually move in a random manner from atom to atom.

The random movement of electrons does *not* result in current flow. Electrons must move in the same direction to cause current flow. If electrical charges are placed on each end of a conductor, the free electrons move in one direction. Figure 1–21 shows current flow through a conductor caused by negative (–) and positive (+) electrical charges. Current flow takes place because of the *difference* in the charges at each end of the conductor. Remember that like charges *repel* and unlike charges *attract.*

When an electrical charge is placed on each end of the conductor, the free electrons move. Free electrons have a negative (–) charge, so they are repelled by the negative (–) charge on the left of Fig. 1–21. The free electrons are attracted to the positive (+) charge on the right. The free electrons move to the right from one atom to another. If the charges on each end of the conductor are increased, more free electrons will move. This increased movement will cause more electrical current flow.

Current flow is the result of electrical energy caused as electrons change orbits. This impulse moves from one electron to another. When one electron (–) moves out of its orbit, it enters another atom's orbit. An electron (–) is then *repelled* from that atom. This action goes on in all parts of a conductor. Remember that electrical current flow is a *transfer* of energy.

Electrical Circuits. Current flow takes place in electrical circuits. A *circuit* is a path or conductor for electrical flow. Electrical current flows only when it has a complete or *closed-circuit* path. There must be a source of electrical energy to cause current to flow along a closed path. Figure 1–22 shows a battery used as an energy source to cause current to flow through a light bulb. Notice that the path or circuit is *complete.* Light is given off by the light bulb due to the work done as electrical current flows through a closed circuit. Electrical energy produced by the battery is changed to light energy in this circuit.

Electrical current cannot flow if a circuit is open. An *open circuit* does not provide a complete path for current flow. If the circuit of Fig. 1–22 became open, no current would flow. The light bulb would not glow. Free electrons of the conductor would no longer move from one atom to another. An example of an open circuit is when a light bulb "burns out." Actually, the filament (part that produces light) has become open. The open filament of a light bulb stops current flow from the source of electrical energy. This causes the bulb to stop burning or producing light.

Another common circuit term is *short circuit.* In electrical work, a short circuit can be very harmful. It occurs when a conductor connects directly across the terminals of an electrical energy source. If a wire is placed across a battery, a short circuit occurs. For safety purposes, a short circuit should never happen. Short circuits cause too much current to flow from the source. The battery would probably be destroyed and the wire could get hot or possibly melt due to a short circuit.

Direction of Current Flow. Electrical current flow is the movement of electrons along a conductor. Electrons are negative charges. Negative charges are attracted to positive charges and repelled by other negative charges. Electrons move from the negative terminal of a battery to the positive terminal. This is called *electron current flow.* Electron current flow is in a direction of electron movement from negative to positive through a circuit.

Another way to look at electrical current flow is in terms of charges. Electrical charge movement is from an area of high charge to an area of low charge. A high charge can be considered positive and a low charge, negative. Using this method, an electrical charge is considered to move from a high charge to a low charge. This is called *conventional current flow.* Conventional current flow is the movement of charges from positive to negative.

Electron current flow and conventional current flow should not be confusing. They are two different ways of looking at current flow. One deals with electron movement and the other deals with charge movement. For most applications, the assumed direction does not matter. In this book, *electron current flow* is used.

Amount of Current Flow (the Ampere). The amount of electrical current that flows through a circuit depends on the number of electrons that pass a point in a certain time. The *coulomb* is a unit of measurement of electrical charge. In electricity, many units of measurement are used. A coulomb is a large quantity of electrons. It is estimated that one coulomb is 6,250,000,000,000,000,000 electrons (6.25×10^{18} in scientific notation form). Because electrons are minute, it takes many to comprise one unit of measurement. When one coulomb passes a point on a conductor in one second, one *ampere* of current flows in the circuit. The unit is named for A. M. Ampere, an eighteenth-century scientist who studied electricity. Current is commonly measured in units called *milliamperes* and *microamperes.* These are smaller units of current. A milliampere is one thousandth (1/1000) of an ampere and a microampere is one millionth (1/1,000,000) of an ampere. Electrical units of measurement such as these are discussed in detail in Chap. 2.

Current Flow Compared with Water Flow. An electrical circuit is a path in which an electrical current flows. Current flow is similar to the flow of water through a pipe. Electrical current and water flow can be compared in some ways. Water flow is used to show how current flows in an electrical circuit. When water flows through a pipe, something causes it to move. The pipe offers opposition or resistance to the flow of water. If the pipe is small, it is more difficult for the water to flow.

In an electrical circuit, current flows through wires (conductors). The wires of an electrical circuit are similar to the pipes through which water flows. If the wires are made of a material that has high resistance, it is difficult for current to flow. The result is the same as water flow through a pipe which has a rough surface. If the wires are large, it is easier for current to flow in an electrical circuit. In the same way, it is easier for water to flow through a large pipe. Electrical current and water flow are compared in Fig. 1–23. Current flows from one place to another in an electrical circuit. Similarly, water that leaves a pump moves from one place to another. The rate of water flow through a pipe is measured in gallons per minute. In an electrical circuit, the current is measured in *amperes.* The flow of electrical current is measured by the number of *coulombs* that pass a point on a conductor each second. A gallon of water is a certain number of molecules. A coulomb is a certain number of electrons. A current flow of one coulomb per second is defined as one ampere of current flow.

1.8 ELECTRICAL FORCE (VOLTAGE)

Water pressure is needed to force water along a pipe. Similarly, electrical pressure is needed to force current along a conductor. Water pressure is usually measured in pounds per square inch (psi). Electrical pressure is measured in *volts.* If a motor is rated at 120 volts, it requires 120 volts of electrical pressure applied to the motor to force the proper amount of current through it. More pressure would increase the current flow and less pressure would not force enough current to flow. The motor would not operate properly with too much or too little voltage. Water pressure produced by a pump causes water to flow through pipes. Pumps produce pressure that causes water to flow. The same is true of an electrical energy source. A source such as a battery or generator causes current to flow through a circuit. As voltage is increased, the amount of current in a circuit is also increased. Voltage is also called *electromotive force (EMF).* This term is largely

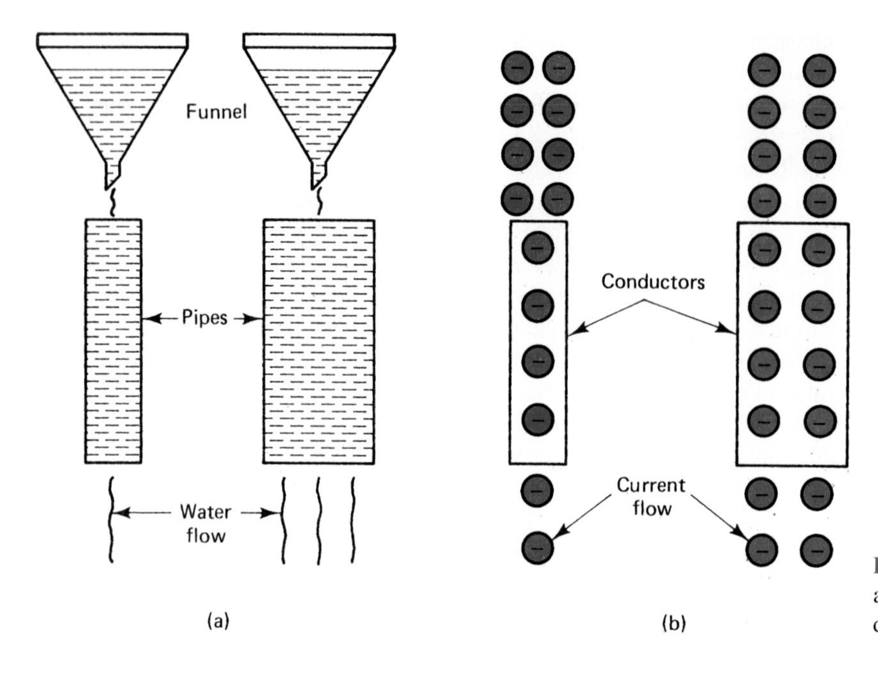

Figure 1–23. Comparison of electrical current and water flow: (a) water pipes; (b) electrical conductors.

responsible for the usage of *E* as a symbol for voltage. With the development of solid-state and computer electronics, *E* has other meanings. To avoid confusion, the letter *V* is now commonly used as the symbol for voltage.

1.9 RESISTANCE

The opposition to current flow in electrical circuits is called *resistance.* Resistance is not the same for all materials. The number of free electrons in a material determines the amount of opposition to current flow. Atoms of some materials give up their free electrons easily. These materials offer low opposition to current flow. Other materials hold their outer electrons and offer high opposition to current flow.

Electrical current is the movement of free electrons in a material. Electrical current needs a source of electrical pressure to move the free electrons through a material. An electrical current will not flow if the source of electrical pressure is removed. A material will not release electrons until enough force is applied. With a constant amount of electrical force (voltage) and more opposition (resistance) to current flow, the number of electrons flowing (current) through the material is smaller. With constant voltage, current flow is increased by decreasing resistance. Decreased current results from more resistance. By increasing or decreasing the amount of resistance in a circuit, the amount of current flow can be changed.

Materials that are good conductors have many free electrons. Insulating materials do not easily give up the electrons in the outer orbits of their atoms. Metals are the best conductors, with copper, aluminum, and iron wire being the most common. Carbon and water are two nonmetals which are also conductors. Materials such as glass, paper, rubber, ceramics, and plastics are common types of insulators.

Even very good conductors have some resistance that limits the flow of electrical current through them. The resistance of any material depends on four factors:

1. The material of which it is made
2. The length of the material
3. The cross-sectional area of the material
4. The temperature of the material

The material of which an object is made affects its resistance. The ease with which different materials give up their outer electrons is very important in determining resistance. Silver is an excellent conductor of electricity. Copper, aluminum, and iron have more resistance, but are more commonly used as they are less expensive. All materials conduct an electrical current to some extent, even though some (insulators) have extremely high resistance.

Length also affects the resistance of a conductor. The longer a conductor, the *greater* the resistance. The shorter a conductor, the *lower* its resistance. A material resists the flow of electrons because of the way in which each atom holds onto its outer electrons. The more material in the path of an electrical current, the less current flow the circuit will have. If the length of a conductor is doubled, there would be *twice* as much resistance in the circuit. Another factor that affects resistance is the cross-sectional area of a material. The greater the cross-sectional area of a material, the *lower* the resistance. The smaller this area, the *higher* the resistance of

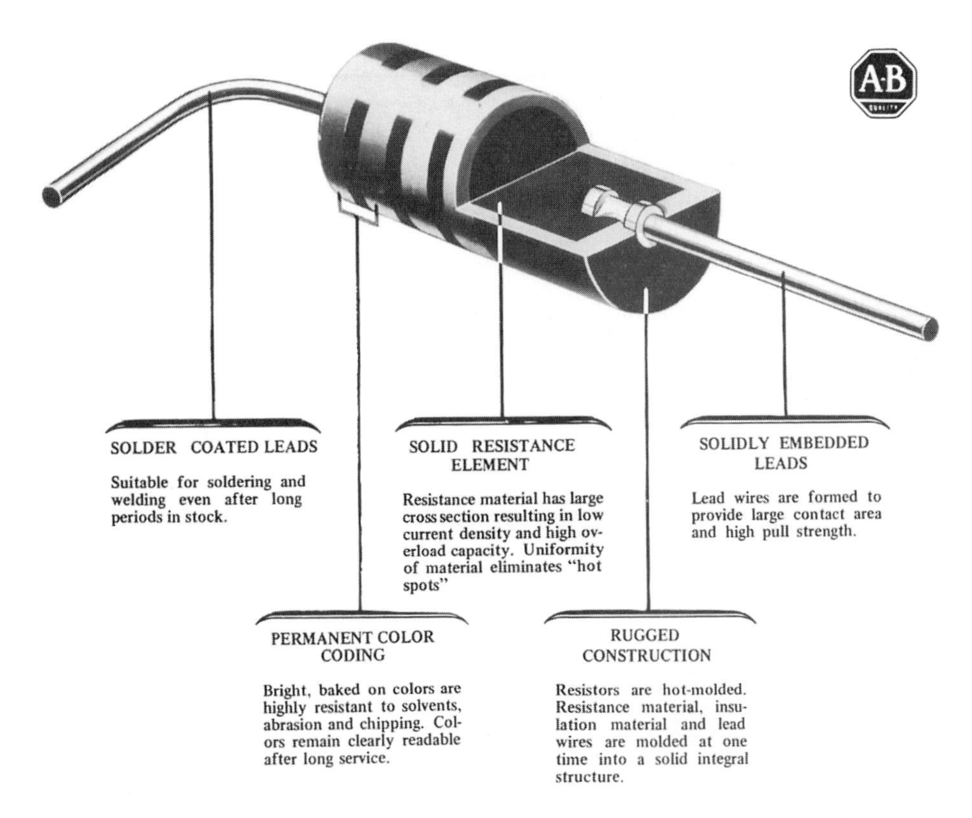

Figure 1–24. Resistor used in electrical circuits (Courtesy of Allen-Bradley Co.).

the material. If two conductors have the same length but one has twice the cross-sectional area, the current flow would be twice as much through the wire with the larger cross-sectional area. This happens because there is a wider path through which the electrical current can flow. Twice as many free electrons are available to allow current flow.

Temperature also affects resistance. For most materials, the higher the temperature, the more resistance the material offers to the flow of electrical current. The colder the temperature, the less resistance a material offers to the flow of electrical current. This effect is produced because a change in the temperature of the material changes the ease with which the material releases its outer electrons. A few materials, such as carbon, have lower resistance as the temperature increases. The effect of temperature on resistance is the least important of the factors that affect resistance. A device called a resistor, which is used in electrical circuits, is shown in Fig. 1–24.

1.10 VOLTAGE, CURRENT, AND RESISTANCE

We depend on electricity to do many things which are sometimes taken for granted. It is important to learn some of the basic electrical terms that are commonly used in the study of electricity and electronics. Three of these basic electrical terms are *voltage, current,* and *resistance.* The term *voltage* is best understood by looking at a flashlight battery. The battery is a source of voltage. It is capable of supplying electrical energy to a light connected to it. The voltage the battery supplies should be thought of as electrical "pressure." The battery has positive (+) and negative (–) terminals.

A battery is commonly used to supply electrical pressure to circuits. An electrical circuit in its simplest form has a source, a conductor, and a load. An electrical circuit is shown in Fig. 1–22. The battery is a source of electrical pressure or voltage. The conductor is a path to allow the electrical current to pass the load. The lamp is called a *load,* because it changes electrical energy into light energy. When the source, conductor, and load are connected, a complete circuit is made. When the battery or voltage source is connected to the light bulb by a conductor, a current will flow. Current flows because of the electrical pressure produced by the battery. The battery is similar to a water pump, which also supplies pressure. Water in pipes is somewhat similar to the flow of current through a conductor. When the conductor is connected to the lamp, a current flows. The current flow causes the lamp to light. Electrical current is the flow of electrons through the conductor. Electrons move because of the pressure produced by the battery. Remember that electrons have a negative charge (–).

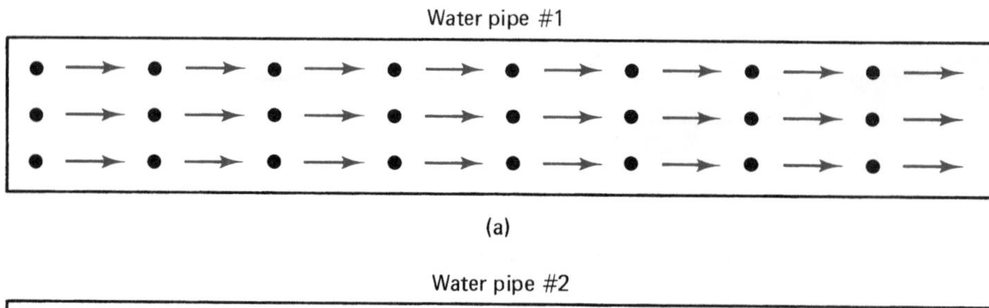

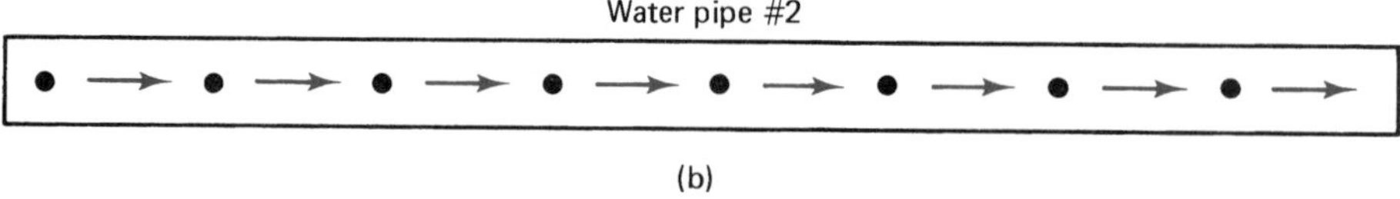

Figure 1–25. Water pipes showing the effect of resistance: (a) water pipe 1—many drops of water flow through; (b) water pipe 2—only a few drops of water flow through.

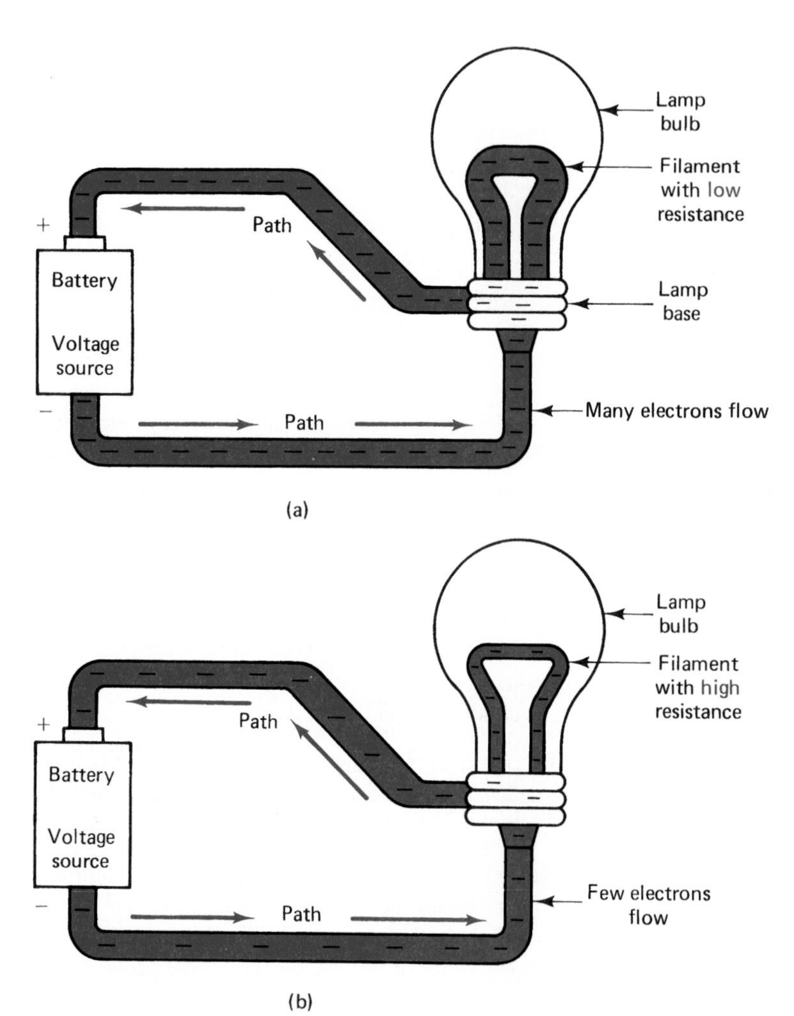

Figure 1–26. How lamp filament size affects current flow: (a) low resistance; (b) high resistance.

The movement of electrons through a conductor takes place at a rate based on the resistance of a circuit. A lamp offers resistance to the flow of electrical current. Resistance is opposition to the flow of electrical current. More resistance in a circuit causes less current to flow. Resistance can be explained by using the example of two water pipes shown in Fig. 1–25. If a water pump is connected to a large pipe, such as pipe 1, water flows easily. The pipe offers a small amount of resistance to the flow of water. However, if the same water pump is connected to a small pipe, such as pipe 2, there is more opposition to the flow of water. The water flow through pipe 2 is less.

Inside a lamp bulb, the part that glows is called a filament. The filament is a wire that offers resistance to the flow of electrical current. If the filament wire of a lamp is made of large wire, much current flows as shown in Fig. 1–26(a). The filament offers a small amount of resistance to the flow of current. However, the circuit in Fig. 1–26(b) shows a lamp with a filament of small wire. The small wire has more resistance or opposition to current flow. Therefore, less current flows in circuit (b) because it has higher resistance.

The terms *voltage, current,* and *resistance* are important to understand. Voltage is electrical pressure that causes current to flow in a circuit. Current is the movement of particles called electrons through a conductor in a circuit. Resistance is opposition to the flow of current in a circuit.

1.11 VOLTS, OHMS, AND AMPERES

Many similarities exist between water and electrical systems. These similarities help in understanding basic electrical quantities. The volt (unit of electrical pressure) is compared with the pressure that causes water to flow in pipes. Because the volt is a unit of electrical pressure, it is always measured *across* two points. An electrical pressure of 120 volts exists across the terminals of electrical outlets in the home. This value is measured with an electrical instrument called a *voltmeter.*

The ampere, or amp, is a measure of the rate of flow of electrical current or electron movement. Electrical current is similar to the rate of water flow in a pipe. Water flow is measured in gallons per minute. An ampere is the number of coulombs per unit of time flowing in an electrical circuit. An ampere is a measure of the *rate* of flow. An *ammeter* is used to measure the number of electrons that flow in a circuit.

When pressure is applied to a water pipe, water flows. The rate of flow is limited by friction in the pipe. When an electrical pressure (voltage) is applied to an electrical current, the number of electrons (current) that flows is limited by the resistance of the path. Resistance is measured with a meter called an *ohmmeter.* The basic unit of resistance is the *ohm.*

1.12 ELECTRICAL SAFETY

Electrical safety is extremely important. You should be aware that many potential dangers are not easy to see. Safety should be based on the understanding of basic electrical principles and common sense. The physical arrangement of equipment in the electrical lab or work area should be done in a safe manner. Well-designed electrical equipment should always be used. For economic reasons, electrical equipment is sometimes improvised. It is important that all equipment be made as safe as possible, especially equipment and circuits that are designed and built in the lab or shop.

Work surfaces in the shop or lab should be covered with a material that is nonconducting, and the floor of the lab or shop should also be nonconducting. Concrete floors should be covered with rubber tile or linoleum. A fire extinguisher that has a nonconducting agent should be placed in a convenient location. Extinguishers should be used with caution. Their use should be explained by the instructor or lab supervisor.

Electrical circuits and equipment in the lab or shop should be plainly marked. Voltages at outlets require special plugs for each voltage. Several voltage values are ordinarily used with electrical lab work. Storage facilities for electrical supplies and equipment should be kept neat. Neatness encourages safety and helps keep equipment in good condition. Tools and small equipment should be maintained in good condition and stored in a tool panel or marked storage area. Tools should have insulated handles. Tools and equipment plugged into convenience outlets should be wired with three-wire cords and plugs. The purpose of the third wire is to prevent electrical shocks by grounding all metal parts connected to the outlet.

Soldering irons are often used in the electrical shop or lab. They can be a fire hazard. They should have a metal storage rack. Irons should be unplugged when not in use. Soldering irons can also cause burns if not used properly. *Rosin-core* solder should always be used in the electrical lab or shop.

Adequate laboratory space is needed to reduce the possibility of accidents. Proper ventilation, heat, and light also provide a safe working environment. Wiring in the electrical lab or shop should conform to the specifications of the National Electrical Code® (NEC). The NEC governs all electrical wiring in buildings.

Lab or Shop Practices

All student activities should be performed with low voltages when possible. Instructions should be written with clear directions for performing lab activities. All lab or shop work should emphasize safety. Students should always have the teacher check experimental circuits before they are plugged into the power source. Electrical lab projects should be constructed to provide maximum safety when used. Horseplay of any type in the lab or shop should not be allowed.

Electrical equipment should be disconnected from the source of power before working on it. When testing electronic equipment, such as TV sets or other 120-volt devices, use an *isolation transformer* to isolate the chassis ground from the ground of the equipment. This reduces the shock hazard when working with 120-volt equipment.

Electrical Hazards

A good first-aid kit should be located in every electrical shop or lab. Any accident should be reported immediately to the proper school officials. Teachers should be proficient in the treatment of minor cuts and bruises. They should also be able to apply artificial respiration. In case of electric shock, when breathing stops, artificial respiration must be immediately started. Extreme care should be used in moving a shock victim from the circuit that caused the shock. An insulated material should be used so that someone else does not come in contact with the same voltage. It is not likely that a high-voltage shock will occur; however, students should know what to do in case of emergency.

Normally, the human body is not a good conductor of electrical current. When wet skin comes in contact with an electrical conductor, the body is a better conductor. A slight shock from an electrical circuit should be a sign that something is wrong. Equipment that causes a shock should be checked immediately and repaired or replaced. Proper grounding is important in preventing shock.

Safety devices called ground-fault circuit interrupters (GFCIs) are now used for bathroom and outdoor power receptacles. They have the potential of saving many lives by preventing shock. GFCIs immediately cut off power if a shock occurs. The NEC specifies where GFCIs should be used.

Electricity causes many fires each year. Electrical wiring with too many appliances connected to a circuit overheats wires. Overheating may set fire to nearby combustible materials. Defective and worn equipment can allow electrical conductors to touch one another to cause a *short circuit.* A short circuit causes a blown fuse, or a spark or arc, which might ignite insulation or other combustible materials or burn electrical wires.

Fuses and Circuit Breakers

Fuses and circuit breakers are important safety devices. When a fuse "blows," it means that something is wrong with the circuit. Causes of blown fuses include:

1. A short circuit caused by two wires touching
2. Too much equipment on the same circuit
3. Worn insulation allowing bare wires to touch grounded metal objects such as heat radiators or water pipes

After correcting the problem, a new fuse of *proper size* should be installed. Power should be turned off to replace a fuse. Never use a makeshift device in place of a new fuse of the correct size. This destroys the purpose of the fuse. Fuses are used to cut off the power and prevent overheated wires.

Circuit breakers are now quite common. Circuit breakers operate on spring tension. They can be turned on or off like wall switches. If a circuit breaker "opens," something is wrong in the circuit. Locate and correct the cause and *then* reset the breaker.

Always remember to use common sense when working with electrical equipment or circuits. Follow safe practices in the electrical lab and in the home. Detailed safety information is available from the National Safety Council and other organizations. *It is always wise to be safe.*

1.13 CAREERS IN ELECTRONICS

Preparation for employment in the field of electronics begins with a study of the theory of electricity and electronics. This includes the theory and operation of electronic circuit devices and test equipment. Testing and troubleshooting in the field of electronics are important. A knowledge of equipment operation is basic for employment in the electronics industry. In electronics, the completion of training is the beginning of a lifetime of learning. The basic principles of electronics remain the same. Students must develop skills in reasoning and analysis. Electronics technicians must be capable of growth. As the technology rapidly changes, skills quickly become obsolete.

Programs should provide a thorough study of the basics. Technicians enter the job market with entry-level skills, and the employer provides training on specific items of equipment and procedures. There are many job classifications in electronics, each with technicians using specific skills. Some careers are described in the following sections.

Service Technician

A service technician is involved in servicing and repairing consumer products such as televisions, video cassette recorders (VCRs), and audio systems. A new technician usually trains on the job with an experienced technician. Instruction on complex equipment is usually accomplished in schools and at seminars sponsored by manufacturers.

Items to be repaired are usually brought to the technician. Some stores make housecalls during which minor repairs are made. In these situations, the technician must deal directly with the customer. Technical skills and good interpersonal skills must be developed. Some technicians begin their careers working in a store and then become self-employed.

Field Engineering Technician

A field engineering technician works for an equipment manufacturer and provides service and repair at the equipment user's location. The equipment may include automatic bank teller machines, environmental control systems, large computers, copying machines, and radar systems. Training on the equipment is usually provided by the manufacturer. The technician assigned to service a system may assist in its installation and testing. Field service technicians may work out of the corporate headquarters or at service centers. Technicians from each center service equipment within a region, usually making short trips.

Technologist

Technologists work closely with scientists and engineers who are involved in research and development. As new circuit devices and equipment are developed, they must be tested, modified, and refined before they can be sold. Technologists assist in fabricating devices and circuits as well as in testing them in electronic circuits. The test results are reported to scientists, usually in written form. More than merely technical skills are required. Scientists and technologists must communicate well to succeed in developing test procedures. Reporting requires skills in written communication, thus a technologist should take courses in oral and written communication.

Engineering Assistant

Engineers are typically assigned tasks in the areas of system development and design. Engineers select technicians to assist them. Working from drawings supplied by the engineer, the technician selects parts, builds and tests the circuits, and reports the results. The technician is usually involved with the project through the early stages of the manufacturing process.

Technical Writer

Technical writers assemble technical data on electronic systems and write manuals which cover theory, operation, and servicing. In addition to electronics skills, technical writers possess the ability to develop and communicate the material needed by users and technicians. This material is a critical part of the development process. A well-documented electronic system is easier to sell than one in which the instructions are difficult to understand.

Electronics Sales

Electronics sales can be an interesting career. Technicians who enter sales careers must possess technical knowledge and the ability to communicate with customers. A successful salesperson understands psychology and has ambition and a cheerful personality. Courses in communication and sales are needed.

Computer Technician

The most rapidly growing field of employment related to electronics is that of computer technician. Computers are now used at home, in offices, and in industry. Each of these settings requires skilled technicians to install, service, and maintain the computer systems. A strong, technical foundation in electricity and electronics is a must to becoming a skilled computer technician. This type of technician is also concerned with local area networks (LANs), the Internet (World Wide Web), and other communications links that keep the systems operating. A knowledge of both hardware and software (computer programming) is needed to be able to design, repair, install, or maintain computer systems. This area should be a rewarding career for anyone interested in electronics and computer systems. The knowledge gained and skills developed by using this book provide a foundation for many electrical and electronic occupations.

REVIEW

1. How was electricity discovered?
2. What are the five parts of any electrical system?
3. Discuss each of the parts of an electrical system.
4. What are some types of energy?
5. What is *work?*
6. What is *power?*
7. What are the three types of matter?
8. What are elements, compounds, and molecules?
9. Discuss the three types of particles that make up atoms.
10. Discuss the terms *orbit* and *valence electrons.*
11. Discuss electrostatic charges.
12. What are some applications of static electricity?
13. What are conductors, insulators, semiconductors, and superconductors?
14. What is (a) an open circuit, (b) a closed circuit, and (c) a short circuit?

15. Discuss conventional current flow and electron current flow. Which type is used in this book?
16. How does electrical current flow compare with water flow in a pipe?
17. What is *voltage?*
18. What is *resistance?*
19. What factors determine the resistance of a material?
20. List at least five important things to remember about electrical safety.
21. List at least five careers in which a knowledge of electricity and electronics is needed.

STUDENT ACTIVITIES

Activities are included at the end of each chapter in this book. A separate *Laboratory and Activities Manual* is also available. This manual contains lab activities, self-study questions, and chapter examinations. The "ANALYSIS" items focus on specific topics. They are thought-provoking questions and problems. In addition, simulation labs are available in the manual. They utilize Electronics Workbench® software.

• *Electrical Systems (see Sec. 1.2)*

1. Obtain one or more flashlights.
2. Examine the construction of the flashlight.
3. On a sheet of paper, discuss the following parts of the flashlight:
   - **a.** Source of energy
   - **b.** Path of current flow
   - **c.** Control method
   - **d.** Load

• *Structure of Matter (see Sec. 1.4)*

1. Obtain some golf balls and rigid wire.
2. Make a model that shows the arrangement of electrons, protons, and neutrons in a simple atom, such as hydrogen.
3. Paint the golf balls different colors to represent electrons, protons, and neutrons.
4. Drill holes in the golf balls or attach the wire in some other way to make the model sturdy.
5. In writing, discuss the arrangement of electrons, protons, and neutrons in an atom.

• *Static Electricity (see Sec. 1.6)*

1. Obtain a small piece of plastic and some fur or wool cloth.
2. Tear off six to eight small pieces of paper.
3. Rub the plastic with the fur or cloth.
4. Hold the plastic near the pieces of paper.
5. In writing, explain what happens and why it happens.

• *Soldering (Study App. B before proceeding or complete Activity 1–1, page 1 in the manual.)*

1. Obtain a soldering iron and some rosin-core solder.

2. Complete any of the following soldering activities assigned by your teacher.
   a. Component mounting
      (1) Obtain a component-mounting strip and a resistor or some other component from your teacher.
      (2) Allow the soldering iron to heat to its proper temperature.
      (3) Apply a small amount of solder to the tip of the iron to check its temperature.
      (4) Solder the resistor onto the component-mounting strip.
   b. Printed circuit board (PCB) soldering
      (1) Obtain a used PCB and a 3-in. length of insulation wire (No. 22 AWG size will work) to practice soldering.
      (2) Place the wire near the holes in the PCB where you intend to solder it. Do this to find the length of wire required.
      (3) Strip $\frac{3}{8}$ to $\frac{1}{2}$ in. of insulation from one end of the wire with a wire stripper.
      (4) Cut off any excess wire and strip the insulation off the other end of the wire. The wire should lie flat on the board when soldered.
      (5) Solder both ends of the wire to the PCB.
   c. Printed circuit board (PCB) desoldering
      (1) Obtain a desoldering tool or use a soldering iron.
      (2) Using the proper desoldering technique, remove a component from a PCB without damaging the PCB.
   d. Installing a terminal connector
      (1) Obtain a "solderless" terminal connector, crimping tool, and a piece of insulated wire from your instructor.
      (2) Strip enough insulation from the wire so that the terminal connector fits properly.
      (3) Use the crimping tool to fasten the terminal connector to the wire.

• *Electricity in the Home*

1. Complete this activity at home: List several electrical devices and equipment in your home. See if the completed list is longer than those of others in the class.
2. Turn in your list to the teacher when finished.

• *Electrical Conductors*

1. Obtain an American Wire Gauge (see Fig. 1–27) and several conductors from your teacher.
2. Measure the diameter of each conductor with the gauge.
3. Record the values on a sheet of paper.
4. Record whether the conductor is copper or aluminum.

• *Electrical Tool Identification*

1. Study the illustrations in App. C.
2. Learn the names of the electrical tools shown.
3. Discuss the uses of at least five tools assigned by your teacher.
4. Your teacher can now test you to see how many electrical tools you can identify properly.

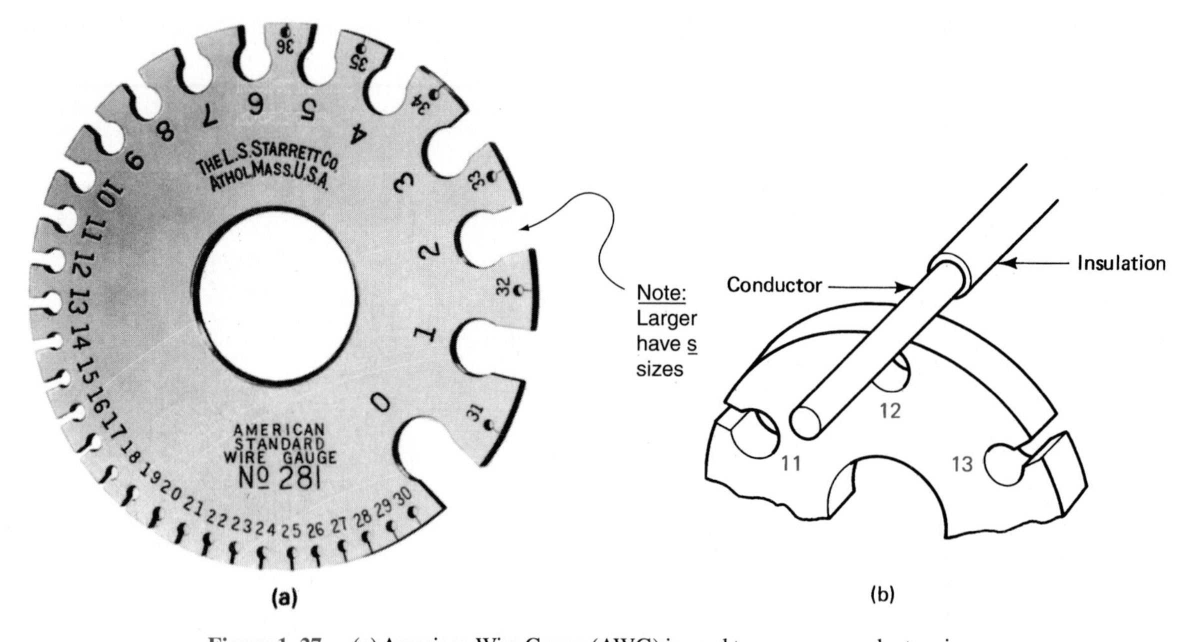

Figure 1–27. (a) American Wire Gauge (AWG) is used to measure conductor size; (b) using the AWG. [(a) Courtesy of L.S. Starrett Co.]

• *Electrical Safety (See Sec. 1.12 or complete Activity 1–2, page 3 in the manual.)*

1. Use the inspection checklist shown in Fig. 1–28.
2. Examine your classroom or shop area in teams of two or three to locate any safety hazards.
3. Make notes on the checklist on each part of the inspection.

• • ANALYSIS

Discuss safety as it relates to lab or shop practices, electrical hazards, and electrical wiring.

• *Self-Examination*

1. Complete the Self-Examination for Chap. 1, page 5 in the manual.
2. Check your answers on page 408 in the manual.
3. Study the questions that were answered incorrectly.

• *Chapter 1 Examination*

Complete the Chap. 1 Examination on page 7 in the manual.

Use this checklist for a laboratory or shop area. A safe and healthy environment is always necessary. Inspections that are similar to this should be done periodically. This is not a detailed inspection.

For the items on the checklist, place the number to the left of the item that describes its condition as follows:

0 = Does not apply to this area
1 = Very good (needs no attention)
2 = Good (needs little attention)
3 = Acceptable (needs some attention)
4 = Bad (needs attention soon)
5 = Very bad (needs immediate attention)

Place comments to the right of each item that needs attention.

A. General room environment
___ 1. Temperature
___ 2. Ventilation
___ 3. Lighting
___ 4. Condition of equipment
___ 5. Condition of floors
___ 6. Condition of walls and windows
___ 7. Availability of fire extinguishers
___ 8. Condition of fire extinguishers
___ 9. Fire escape plan
___ 10. Hanging objects on walls or ceilings
___ 11. Electrical power outlets
___ 12. General appearance (neatness)
___ 13. Condition of benches
___ 14. Condition of supply rooms
___ 15. Storage of materials (especially hazardous ones)
___ 16. Other

B. Equipment
___ 1. High-voltage and danger areas identified
___ 2. Machines and equipment well guarded and protected
___ 3. All moving gears, belts, etc., protected properly
___ 4. Tools in good condition
___ 5. Machines in good condition
___ 6. Equipment controls easy to reach
___ 7. Main switch or circuit breaker to turn off all equipment easy to reach
___ 8. Other

C. Electrical distribution
___ 1. No temporary wiring used
___ 2. Controls are enclosed properly
___ 3. Power outlet voltages (other than 120 V) are identified
___ 4. Equipment is wired properly
___ 5. Overload protection is used on equipment
___ 6. Master switch available to turn off equipment
___ 7. Other

D. Miscellaneous
___ 1. First-aid kit available
___ 2. Accident reports available
___ 3. Evidence of safety awareness
___ 4. Safety glasses and protective equipment available where needed
___ 5. Other

E. General safety rating of the area checked:
Comments:

Figure 1–28. Electrical safety inspection checklist.

CHAPTER 2

ELECTRICAL COMPONENTS AND MEASUREMENTS

OUTLINE

2.1 Components, Symbols, and Diagrams
2.2 Resistors
2.3 Electrical Units
2.4 Scientific Notation
2.5 Schematic Diagrams
2.6 Block Diagrams
2.7 Wiring Diagrams
2.8 Measuring Resistance
2.9 Measuring Voltage
2.10 Measuring Current
2.11 Digital Meters

Digital multimeters are often used for electrical measurements.

OBJECTIVES

Upon completion of this chapter, you will be able to:

1. Diagram a simple electrical circuit.
2. Identify schematic electrical symbols.
3. Convert electrical quantities from metric units to English units and English units to metric units.
4. Use scientific notation to express numbers.
5. Identify different types of resistors.
6. Identify resistor value by color code and size.
7. Explain the operation of potentiometers (variable resistors).
8. Construct basic electrical circuits.
9. Connect an ammeter in a circuit and measure current.
10. Demonstrate how the voltmeter, ammeter, and ohmmeter are connected to a circuit.
11. Measure current, voltage, and resistance of basic electrical circuits.
12. Demonstrate safety while making electrical measurements.
13. Demonstrate proper, safe use of an ohmmeter to measure resistance.

Most electrical equipment is made of several parts or components that work together. It would be almost impossible to explain how electrical equipment operates without using symbols and diagrams. Electrical diagrams show how the component parts of equipment fit together. Common electrical components are easy to identify. It is also easy to learn the symbols used to represent electrical components. The components of electrical equipment work together to form an electrical system.

Another important activity in the study of electricity and electronics is *measurement.* Measurements are made in many types of electrical circuits. Learning the proper ways of measuring resistance, voltage, and current are important as these are the three most commonly measured quantities.

IMPORTANT TERMS

Before reading this chapter, review the following terms. These terms provide a basic understanding of electrical components and measurements.

Ammeter. A meter used to measure current flow.

Battery. An electrical energy source consisting of two or more cells connected together.

Block diagram. A diagram used to show how the parts of a system fit together.

Cell. An electrical energy source that converts chemical energy into electrical energy.

Component. An electrical device used in a circuit.

Conductor. A material that allows electrical current to flow through it easily.

Continuity check. A test to see if a circuit is an open or closed path.

Lamp. An electrical load device that converts electrical energy to light energy.

Multifunction meter. A meter that measures two or more electrical quantities, such as a volt-ohm-milliammeter (VOM), which measures voltage, resistance, and current. Such meters are also commonly called *multimeters.*

Multirange meter. A meter that has two or more ranges to measure an electrical quantity.

Ohmmeter. A meter used to measure resistance.

Polarity. The direction of an electrical potential (– or +) or a magnetic charge (north or south).

Potentiometer. A variable-resistance component used as a control device in electrical circuits.

Precision resistor. A resistor used when a high degree of accuracy is needed.

Resistor. A component used to control either the amount of current flow or the voltage distribution in a circuit.

Schematic diagram. A diagram used to show how the components of electrical circuits are wired together.

Scientific notation. The use of "powers of 10" to simplify large and small numbers.

Switch. A control device used to turn a circuit on or off.

Symbol. Used as a simple way to represent a component on a diagram or an electrical quantity in a formula.

Voltage drop. The electrical potential (voltage) that exists across two points of an electrical circuit.

Voltmeter. A meter used to measure voltage.

Volt-ohm-milliammeter (VOM). A multifunction, multirange meter which is usually designed to measure voltage, current, and resistance. (Also called a multimeter.)

Wiring diagram. A diagram that shows how wires are connected by showing the point-to-point wiring and the path followed by each wire.

2.1 COMPONENTS, SYMBOLS, AND DIAGRAMS

Anyone who studies electricity and electronics should be able to identify the components used in simple electrical circuits. *Components* are represented by symbols. *Symbols* are used to make diagrams. A *diagram* shows how the components are connected in a circuit. For example, it is easier to show symbols for a battery connected to a lamp than to draw a pictorial diagram of the battery and the lamp connected together. There are several symbols that are important to recognize. These symbols are used in many electrical diagrams. Diagrams are used for installing, troubleshooting, and repairing electrical equipment. Using symbols makes it easy to draw diagrams and to understand the purpose of each circuit. Common electrical symbols are listed in App. D.

Most electrical equipment uses wires (conductors) to connect its components or parts together. The symbol for a conductor is a narrow line. If two conductors cross one another on a diagram, they may be shown by using symbols. Figure 2–1(a) shows two conductors crossing one another. If two conductors are connected together, they may also be identified by symbols, as shown in Fig. 2–1(b). Figure 2–1(c) shows some common types of conductors and connectors which are used to secure them to a circuit.

The symbols for two lamps connected across a battery are shown in Fig. 2–2 using symbols for the battery and lamps. Notice the part of the diagram where the conductors are connected together.

A common electrical component is a switch such as the toggle switch shown in Fig. 2–3. The simplest switch is a single-pole single-throw (SPST) switch. This switch turns a circuit on or off. Figure 2–4(a) shows the symbol for a switch in the OFF or open position. There is no path for current to flow from the battery to the lamp. The lamp will be off when the switch is open. Figure 2–4(b) shows a switch in the ON or closed position. This switch position completes the circuit and allows current to flow.

Many electrical circuits use a component called a resistor. Resistors are usually small, cylinder-shaped components such as those shown in Fig. 2–5. They are used to control the flow of electrical current. A typical color-coded resistor and its symbol are shown in Fig. 2–6 on page 42. The most common type of resistor uses color coding to mark its value. Resistor value is always in *ohms*. For instance, a resistor might have a value of 100 ohms. The symbol for ohms is the Greek capital letter omega (Ω). Each color on the resistor represents a specific number. Resistor color-code values are easy to learn.

Another type of resistor is called a potentiometer or "pot." A pot is a variable resistor whose value can be changed by adjusting a rotary shaft. For example, a 1000-Ω pot can be adjusted to any value from zero to 1000 Ω by rotating the shaft. The pictorial and symbol of this component are shown in Fig. 2–7(a) and (b) on page 42. In the example shown in Fig. 2–7(c), potentiometer 1 is adjusted so that the resistance between points A and B is zero. The resistance between points B and C is 1000 Ω. By turning the shaft as far in the opposite direction as it will go, the resistance between points B and C becomes zero (see potentiometer 2). Between points A and B, the resistance is now 1000 Ω. By rotating the shaft to the center of its movement, as shown by potentiometer 3, the resistance is split in half. Now the resistance from point A to point B is about 500 Ω and the resistance from point B to point C is about 500 Ω.

The symbol for a battery is shown in Fig. 2–2. The symbol for any battery over 1.5 volts (V) is indicated by two sets of lines. A 1.5-V battery or cell is shown with one set of lines. The voltage of a battery is marked near its symbol. The long line in the symbol is always the positive (+) and the short line is the negative (–) side of the battery. Some common types of batteries are shown in Fig. 2–8 on page 43.

A simple circuit diagram using symbols is shown in Fig. 2–9 on page 43. This diagram shows a 1.5-V battery connected to an SPST switch, a 100-Ω resistor, and a 1000-Ω potentiometer. Because symbols are used, no words have to be written beside them. Anyone using this diagram should recognize the components represented and how they fit together to form a circuit.

2.2 RESISTORS

There is some resistance in all electrical circuits. Resistance is added to a circuit to control current flow. Devices that are used to cause proper resistance in a circuit are called *resistors.* A wide variety of resistors are used; some have a fixed value, and others are variable. Resistors are made of resistance wire, special carbon material, or metal film. Wire-wound resistors are ordinarily used to control large currents, and carbon resistors control currents which are smaller. Some types of resistors are shown in Fig. 2–10 on pages 43–45.

Wire-wound resistors are constructed by winding resistance wire on an insulating material. The wire ends are attached to metal terminals. An enamel coating is used to protect the wire and to conduct heat away from it. Wire-wound resistors may have fixed taps which can be used to change the resistance value in steps. They may also have sliders which can be adjusted to change the resistance to any fraction of their total resistance. *Precision-wound resistors* are used where the resistance value must be accurate, such as in measuring instruments.

Carbon resistors are constructed of a small cylinder of compressed material. Wires are attached to each end of the cylinder. The cylinder is then covered with an insulating coating.

Variable resistors are used to change resistance while equipment is in operation. They are called potentiometers or rheostats. Both carbon and wire-wound *variable* resistors are

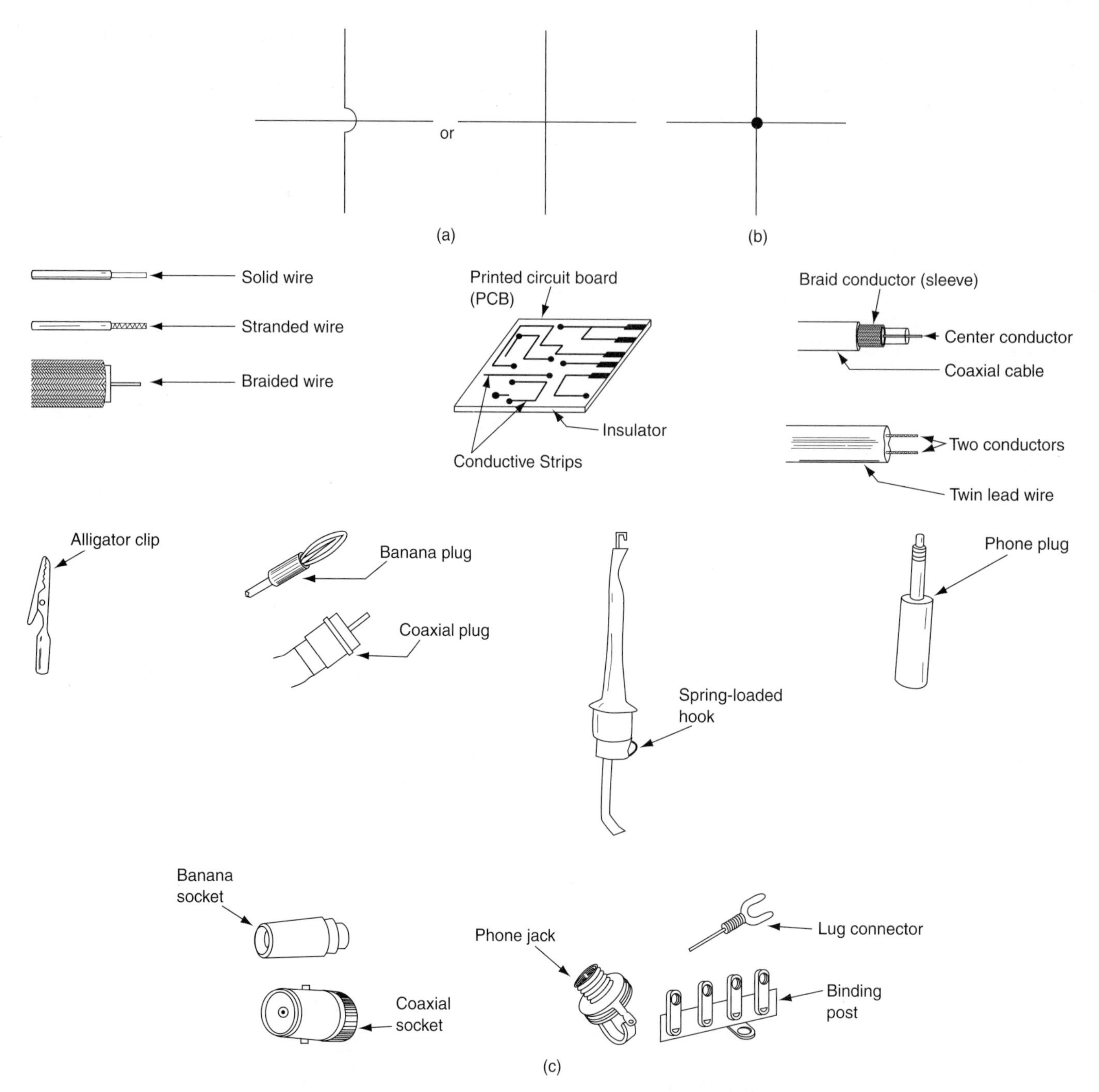

Figure 2–1. Symbols for electrical conductors: (a) conductors crossing; (b) conductors connected; (c) common types of conductors and connectors.

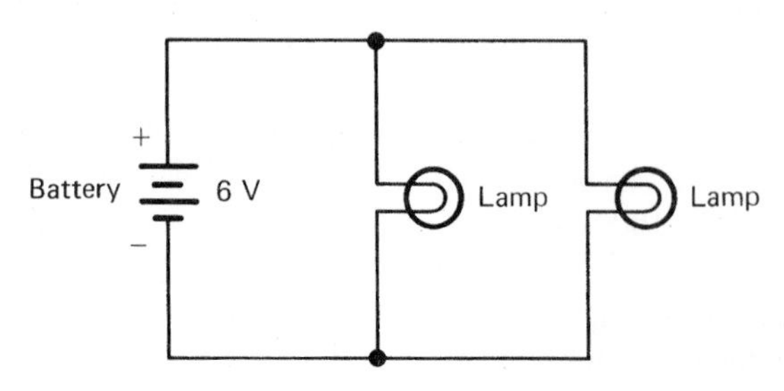

Figure 2–2. Symbols for a battery connected across two lamps.

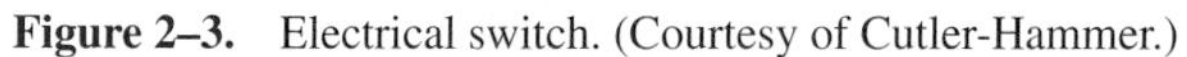

Figure 2–3. Electrical switch. (Courtesy of Cutler-Hammer.)

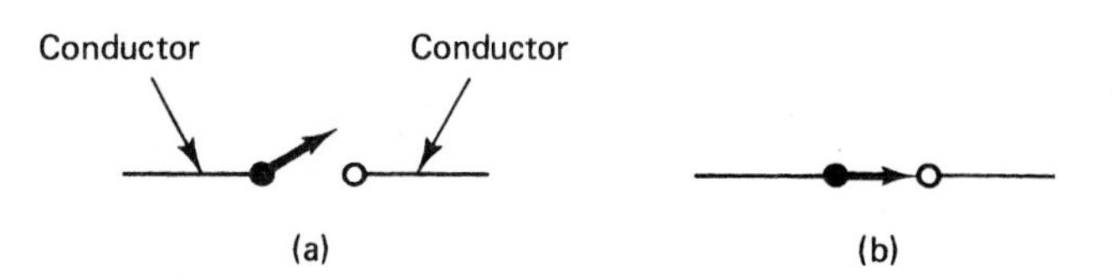

Figure 2–4. Symbol for a single-pole single-throw (SPST) switch: (a) OFF or open conditions; (b) ON or closed condition.

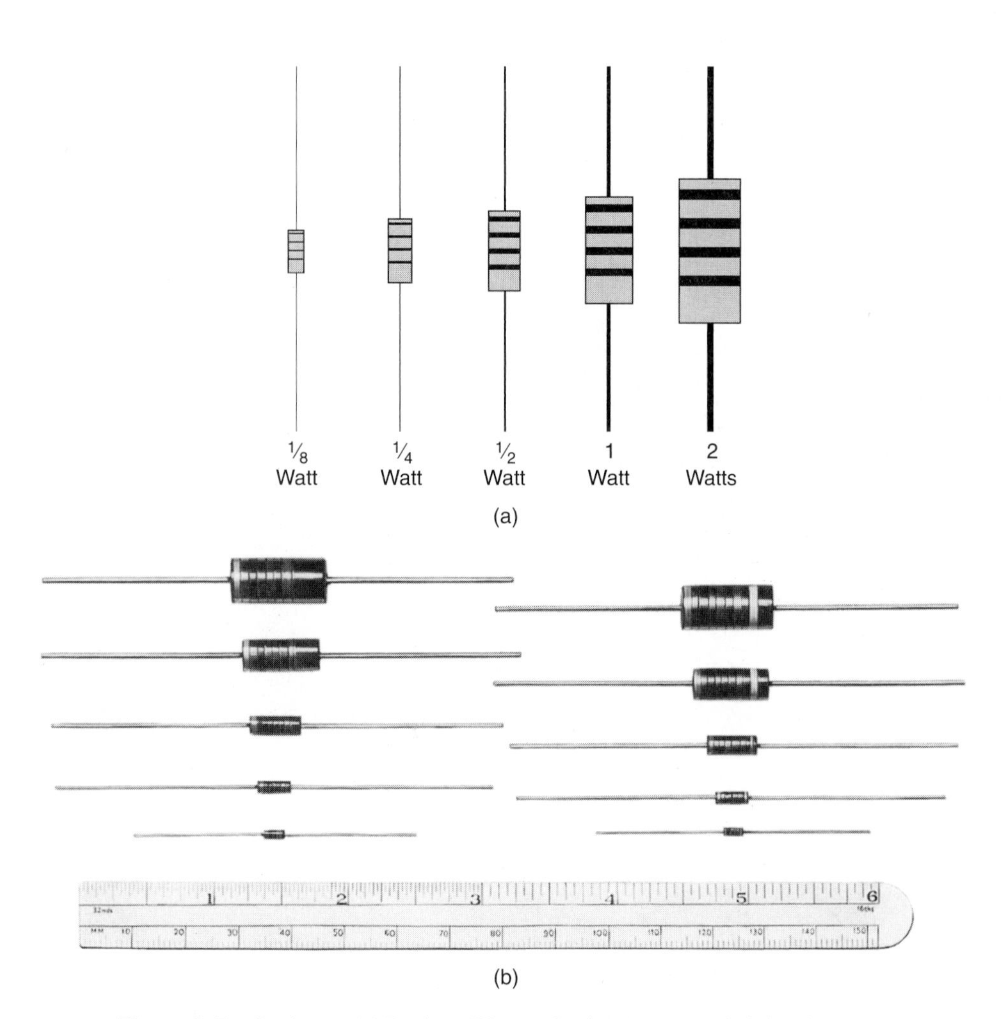

Figure 2–5. Resistors: (a) Resistor Wattage Rating (near actual size)—larger resistors are used for higher wattages. (b) Various sizes of resistors. (Courtesy of Allen-Bradley Co.)

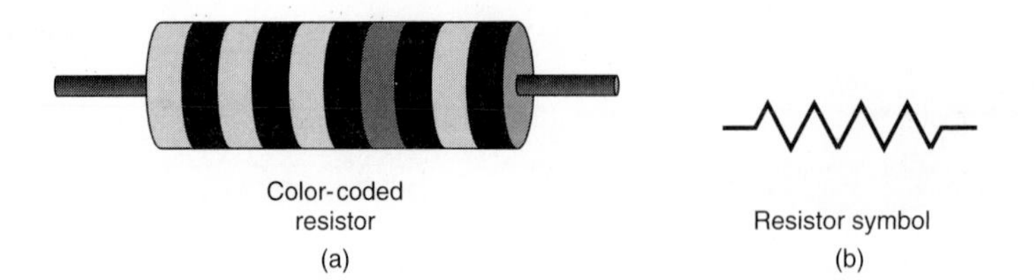

Figure 2–6. Color-coded resistor.

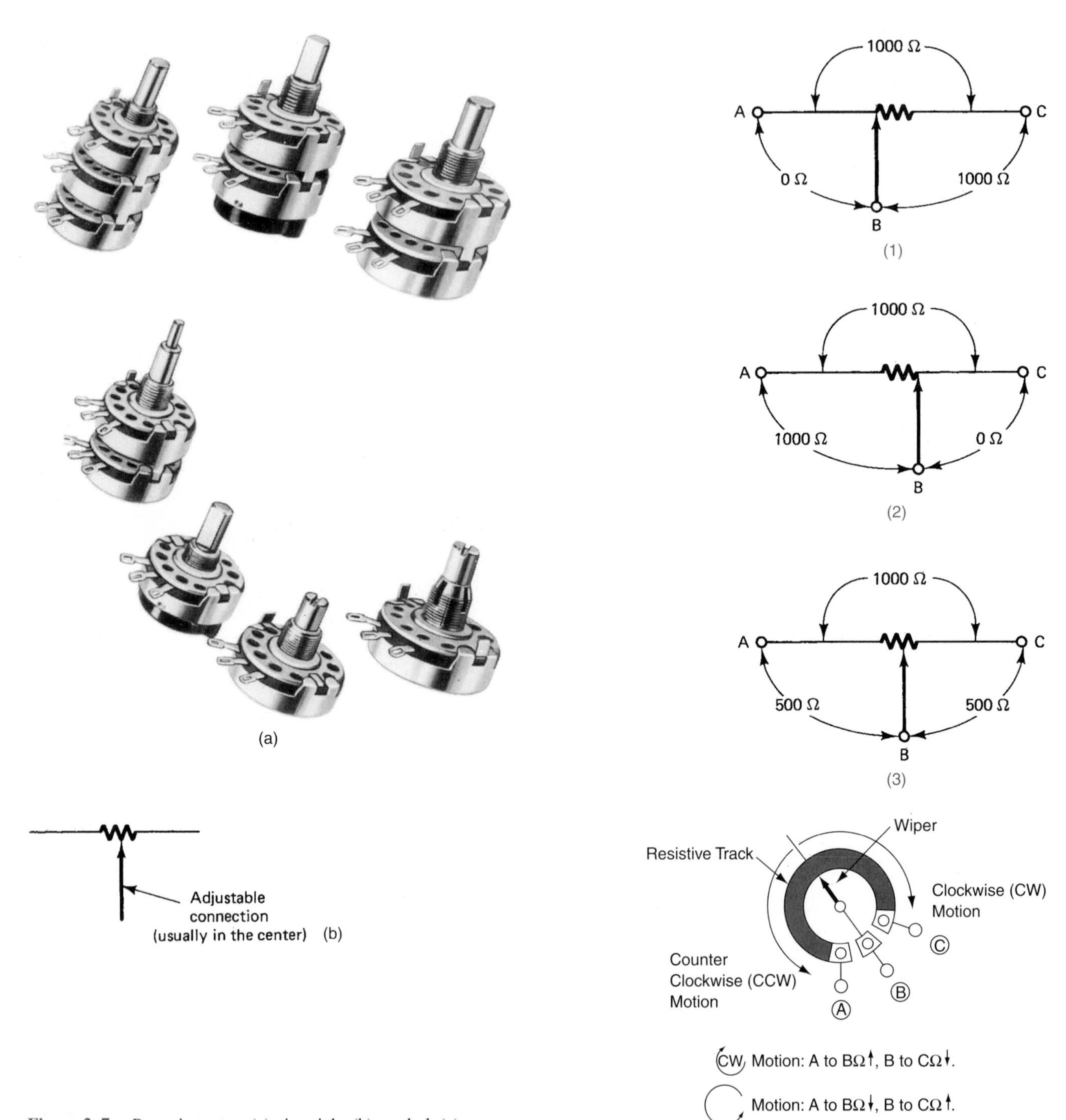

Figure 2–7. Potentiometers: (a) pictorials; (b) symbol; (c) examples. [(a) Courtesy of Allen-Bradley Co.]

Figure 2–8. Common types of batteries. (Courtesy of Union Carbide Corp.)

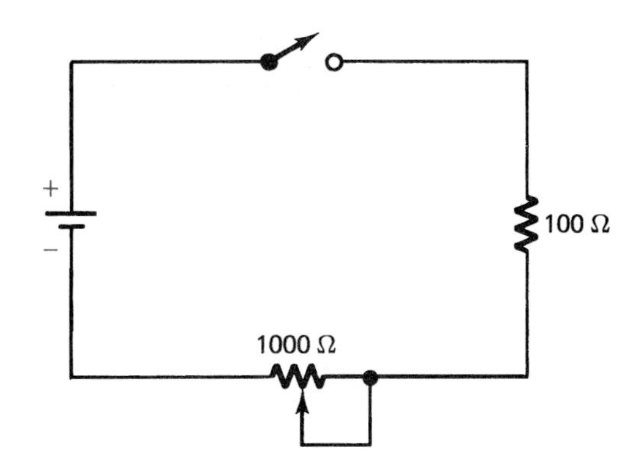

Figure 2–9. Simple circuit diagram.

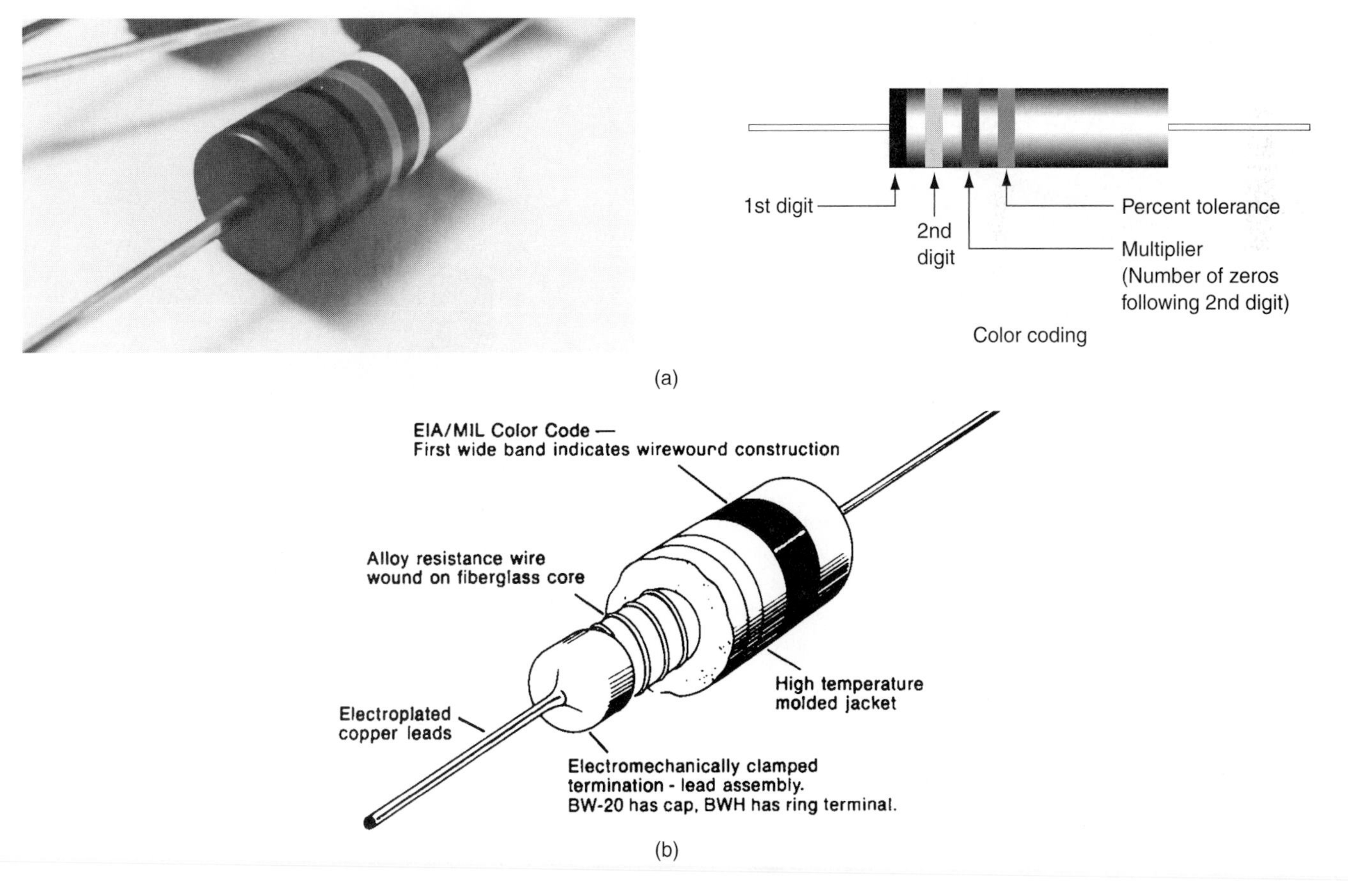

Figure 2–10. Types of resistors: (a) carbon composition resistor; (b) molded wire-wound resistor. (All courtesy of TRW/UTC Resistors.)

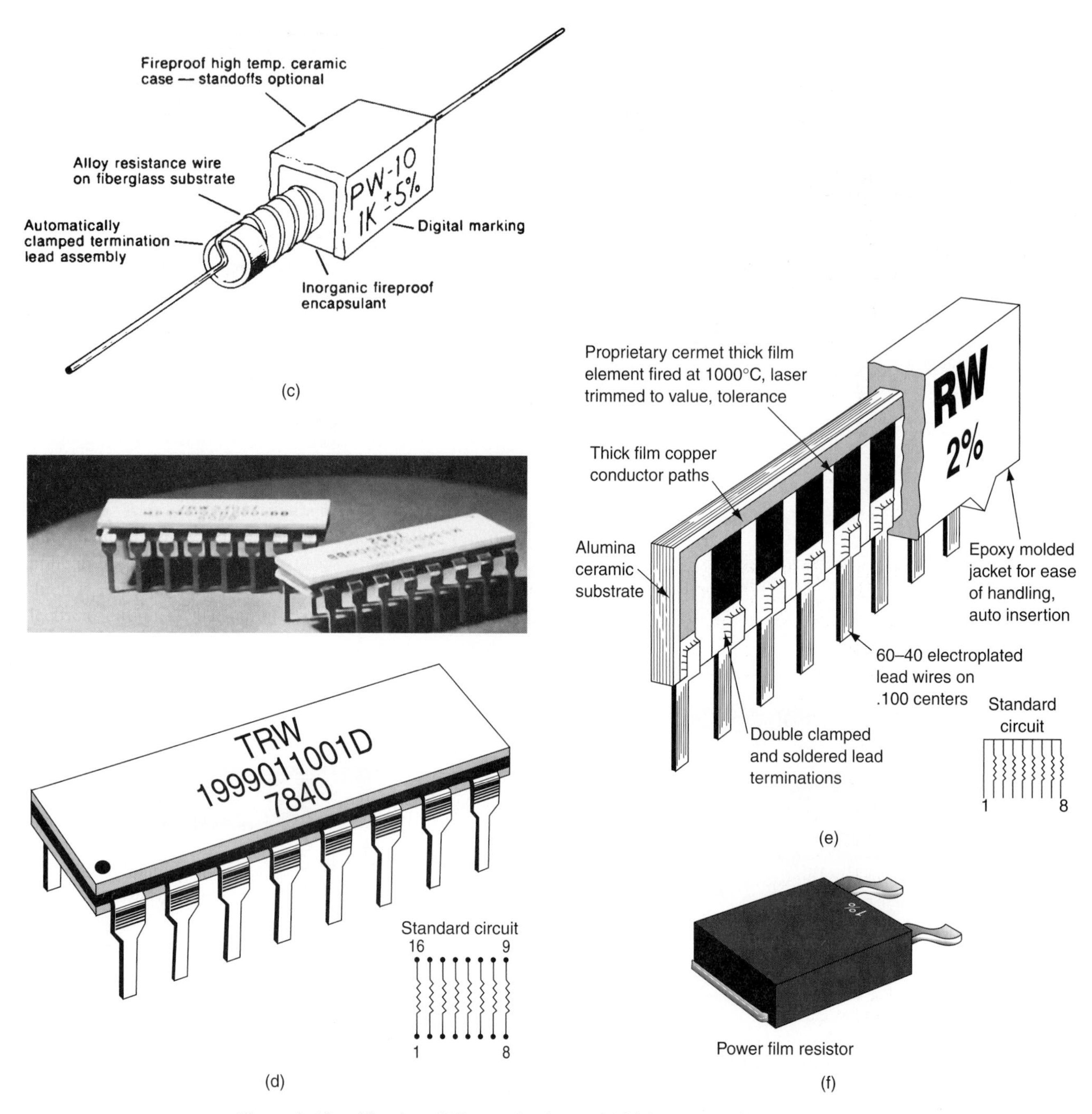

Figure 2–10. (Continued) Types of resistors: (c) high-wattage wire-wound resistor; (d) 16-pin dual-in-line package (DIP) resistor network used in printed circuit boards (PCBs); (e) 8-pin single-in-line package (SIP) resistor network used in PCBs; (f) power metal-film resistor. (All courtesy of TRW/UTC Resistors.)

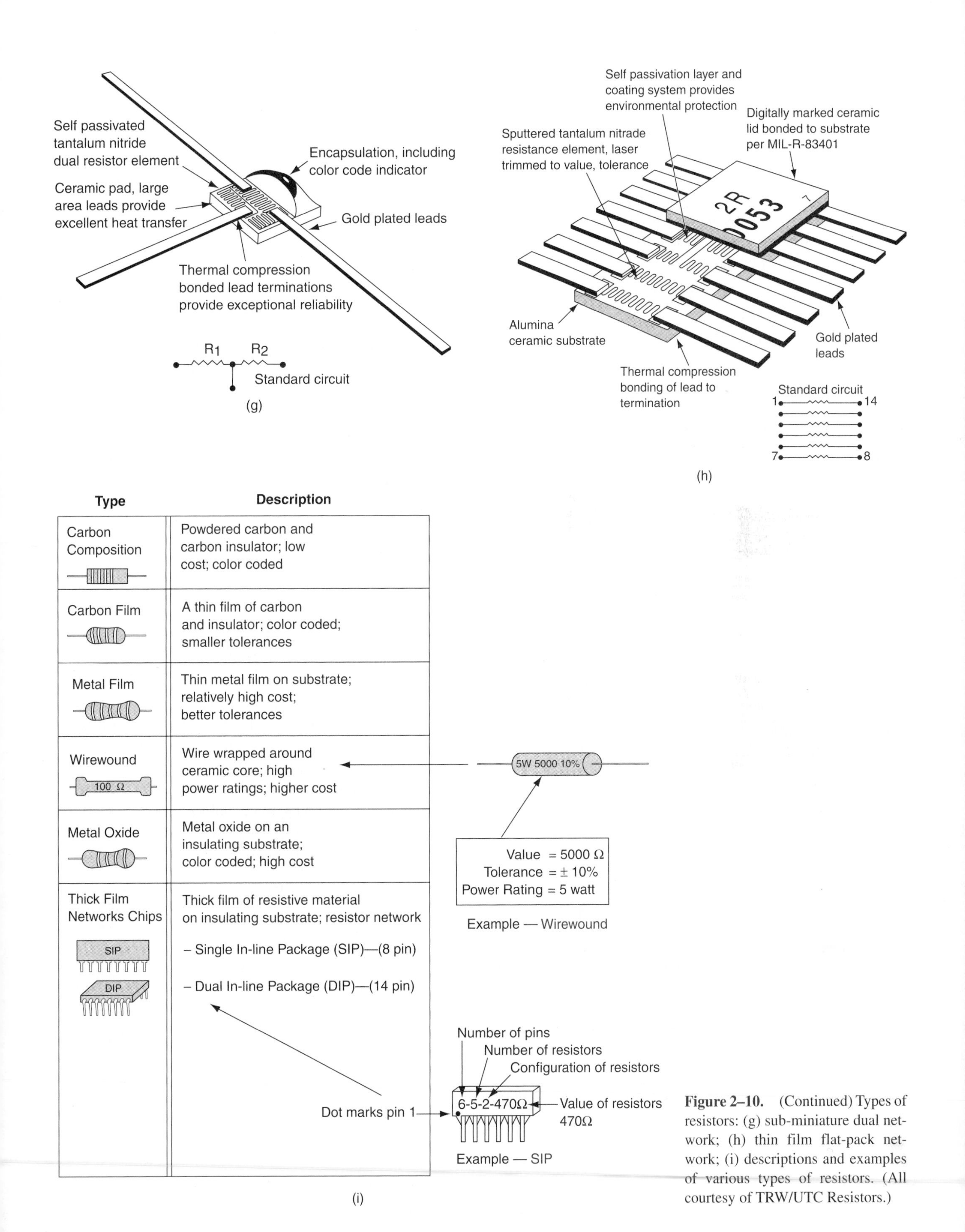

Type	Description
Carbon Composition	Powdered carbon and carbon insulator; low cost; color coded
Carbon Film	A thin film of carbon and insulator; color coded; smaller tolerances
Metal Film	Thin metal film on substrate; relatively high cost; better tolerances
Wirewound	Wire wrapped around ceramic core; high power ratings; higher cost
Metal Oxide	Metal oxide on an insulating substrate; color coded; high cost
Thick Film Networks Chips	Thick film of resistive material on insulating substrate; resistor network – Single In-line Package (SIP)—(8 pin) – Dual In-line Package (DIP)—(14 pin)

Figure 2–10. (Continued) Types of resistors: (g) sub-miniature dual network; (h) thin film flat-pack network; (i) descriptions and examples of various types of resistors. (All courtesy of TRW/UTC Resistors.)

made by winding on a circular form [see Fig. 2–11(a)]. A contact arm is attached to the form to make contact with the wire. The contact arm can be adjusted to any position on the circular form by a rotating shaft. A wire connected to the movable contact is used to vary the resistance from the contact arm to either of the two outer wires of the variable resistor. For controlling smaller currents, carbon variable resistors are made by using a carbon compound mounted on a fiber disk [see Fig. 2–11(b)]. A contact on a movable arm varies the resistance as the arm is turned by rotating a metal shaft.

Resistor Color Codes

It is usually easy to find the value of a resistor by its color code or marked value. Most wire-wound resistors have resistance values (in ohms) printed on the resistor. If they are not marked in this way, an ohmmeter must be used to measure the value. Most carbon resistors use color bands to identify their value. Resistors are of two common types: four-color band and five-color band, as shown in Fig. 2–12.

Resistors are color coded with an end-to-center color band system of marking. In the color-coding system, colors are used to indicate the *resistance value* in ohms. A color band also is used to indicate the *tolerance* of the resistor. The colors are read in the correct order from the end of a resistor. Numbers from the resistor color code, as shown in Fig. 2–13, are substituted for the colors. Through practice using the resistor color code, the value of a resistor may be determined at a glance.

It is difficult to manufacture a resistor to the exact value required. For many uses, the actual resistance value may be as much as 20% higher or lower than the value marked on the resistor without causing any problem. In most uses, the actual resistance does not need to be any closer than 10% higher or lower than the marked value. This percentage of variation between the marked color code value and the actual value is called *tolerance*. A resistor with a 5% tolerance should be no more than 5% higher or lower than the marked value. Resistors with tolerances of lower than 5% are called *precision* resistors.

Resistors are marked with color bands starting at one end of the resistor. For example, a carbon resistor that has three color bands (yellow, violet, and brown) at one end has the color bands read from the end toward the center, as shown in Fig. 2–14. The resistance value is 470 Ω. Remember that black as the third color means that *no* zeros are to be added to the digits. A resistor with a green band, a red band, and black band has a value of 52 Ω.

An example of a resistor with the five-band color-code system is shown in Fig. 2–15. Read the colors from left to right from the end of the resistor where the bands begin. Use the resistor color chart of Fig. 2–13 to determine the value of the resistor in ohms, and its tolerance. Notice that a six-band resistor is included. These are not used as often.

The first color of the resistor in Fig. 2–15 is orange. The first digit in the value of the resistor is 3. The second color is black, so the second digit in the value of the resistor is 0. The next color is yellow, so the third digit is 4. The first three digits of the resistor value are 304. The fourth color is silver, which indicates the number by which the first three digits are to be multiplied. The color silver is a multiplier of 0.01. Multiply 304×0.01 to obtain the value of the resistor ($304 \times 0.01 = 3.04$).

The fifth color is brown. The tolerance of the resistor is given with this band. The tolerance shows how close the actual value of the resistor should be to the color-code value. Tolerance is a percentage of the actual value. In this example, brown shows a tolerance of $\pm 1\%$, which means that the resistor should be within 1% of 3.04 Ω in either direction.

Sometimes there are only three colors on the body of the resistor. Then the tolerance is 20%. Using the color code, it is easy to list the ohms value and tolerance of resistors.

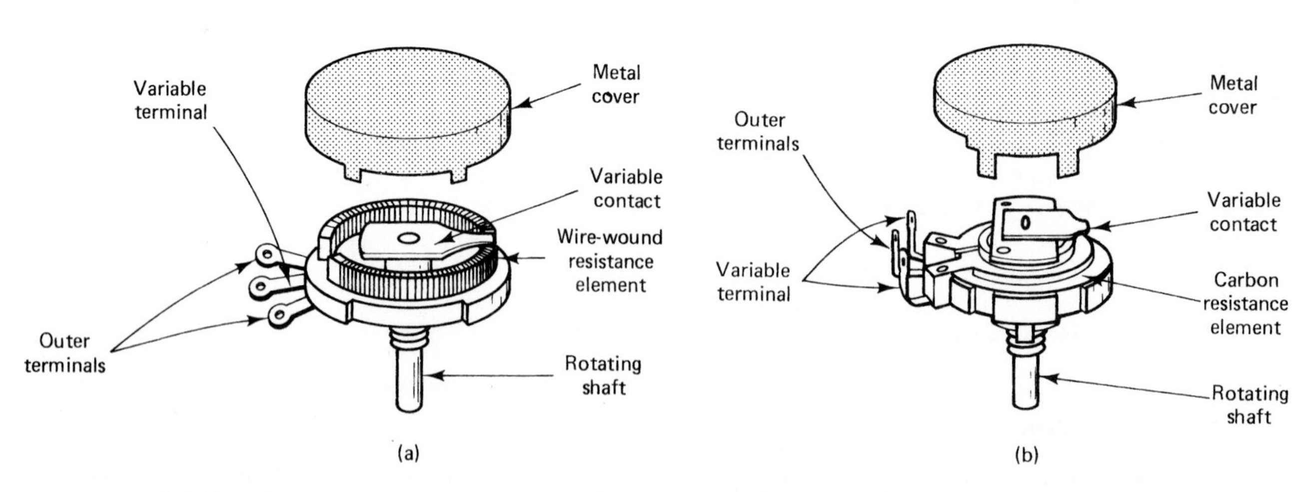

Figure 2–11. Variable-resistor construction: (a) wire-wound variable resistor; (b) carbon variable resistor.

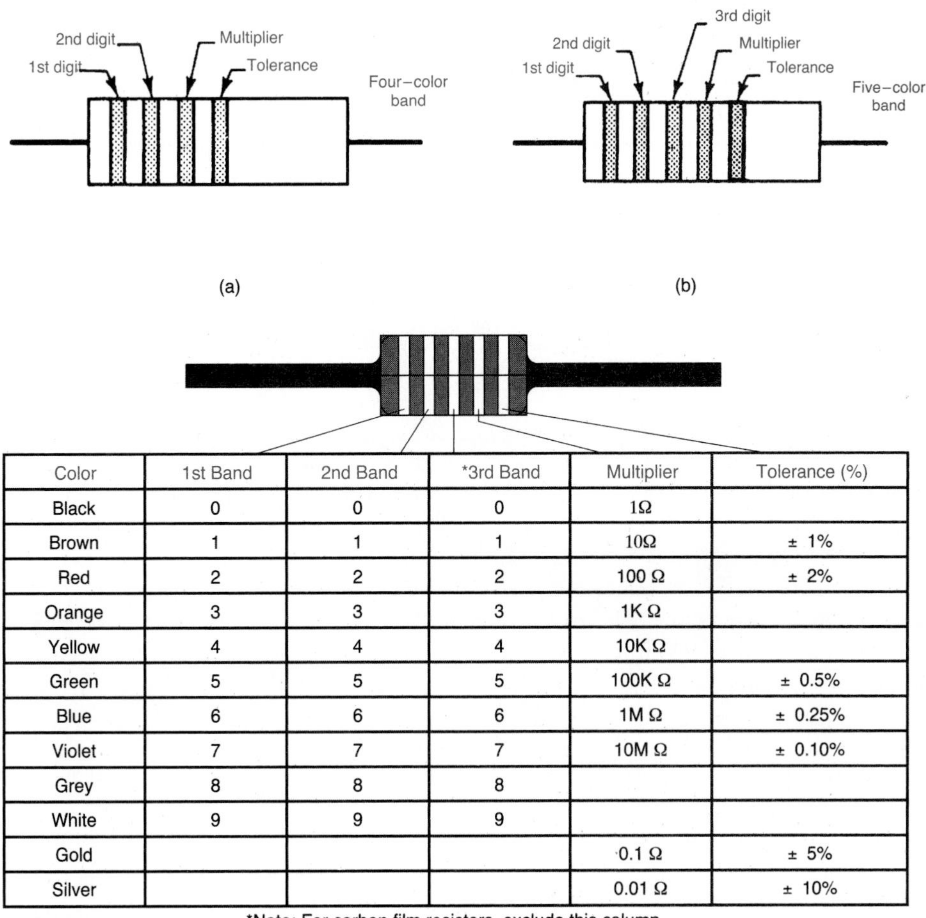

Color	1st Band	2nd Band	*3rd Band	Multiplier	Tolerance (%)
Black	0	0	0	1Ω	
Brown	1	1	1	10Ω	± 1%
Red	2	2	2	100 Ω	± 2%
Orange	3	3	3	1K Ω	
Yellow	4	4	4	10K Ω	
Green	5	5	5	100K Ω	± 0.5%
Blue	6	6	6	1M Ω	± 0.25%
Violet	7	7	7	10M Ω	± 0.10%
Grey	8	8	8		
White	9	9	9		
Gold				0.1 Ω	± 5%
Silver				0.01 Ω	± 10%

*Note: For carbon film resistors, exclude this column.

(c)

Figure 2–12. Carbon resistors: (a) four-color band; (b) five-color band; (c) resistor color-code bands.

Standard values of 5% and 10% tolerance color-coded resistors are listed in Table 2–1.

Another way to determine the value of color-coded resistors is to remember the following mnemonic statement:

*B*ig *B*rown *R*abbits *O*ften *Y*ield *G*reat *B*ig *V*ocal *G*roans *W*hen *G*ingerly *S*lapped.

The first letter of each word in the mnemonic statement is the same as the first letter in each of the colors used in the color code. The words of the device are counted (beginning with *zero*) to find the word corresponding to each digit or the multiplier (refer to Fig. 2–16). Use a method such as this mnemonic statement to remember the color code.

Power Rating of Resistors

The size of a resistor helps to determine its power rating. Larger resistors are used in circuits that have high power ratings. Small resistors will become damaged if they are put in high-power circuits. The power rating of a resistor indicates its ability to give off or dissipate heat. Common power (wattage) ratings of color-coded resistors are ⅛, ¼, ½, 1, and 2 watts (W). Notice the size differences of the resistors in Fig. 2–5. Resistors that are larger in physical size will give off more heat and have higher power ratings.

2.3 ELECTRICAL UNITS

All quantities can be measured. The distance between two points may be measured in meters, kilometers, inches, feet, or miles. The weight of an object may be measured in ounces, pounds, grams, or kilograms. Electrical quantities may also be measured. The more common electrical units of measurement are discussed in this section.

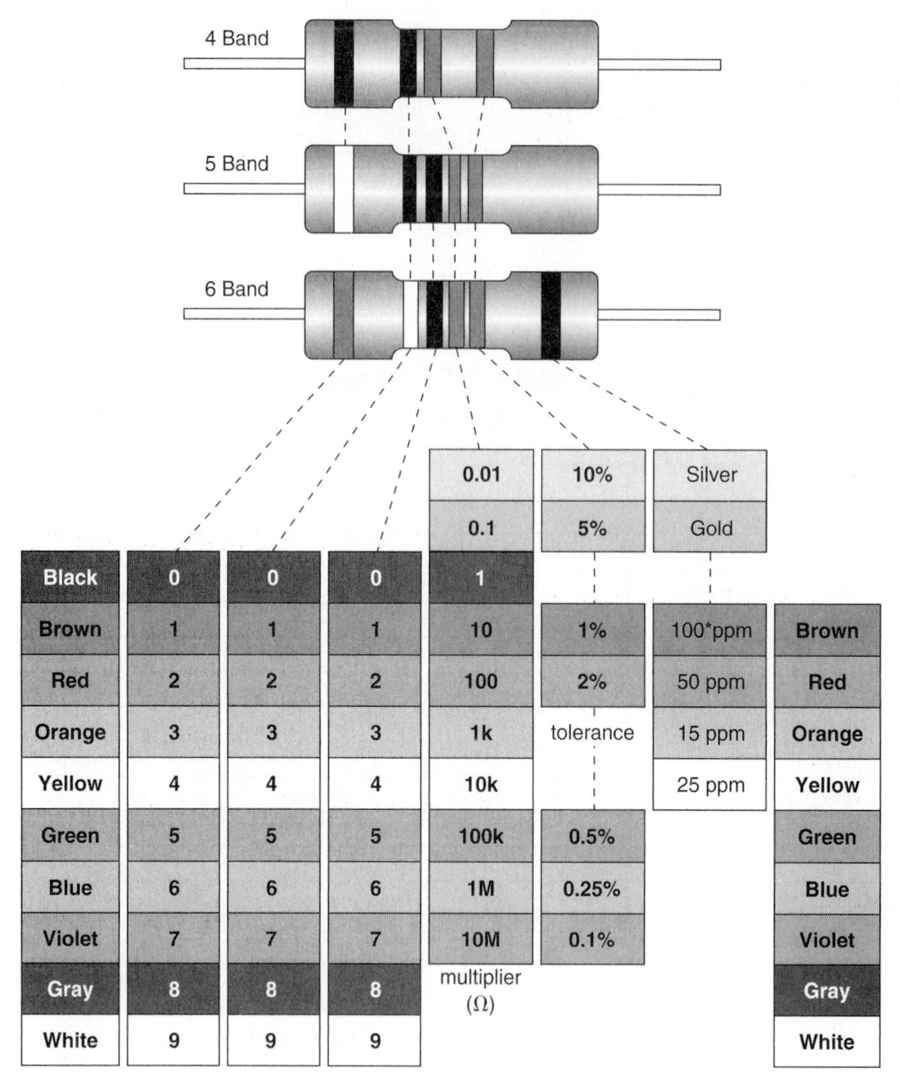

Figure 2–13. Resistor color code (four band, five band, and six band).

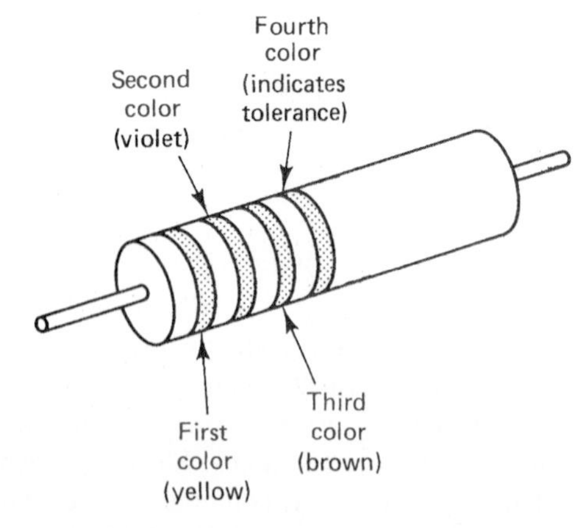

Figure 2–14. End-to-center system for carbon resistors—resistor value = 470 Ω.

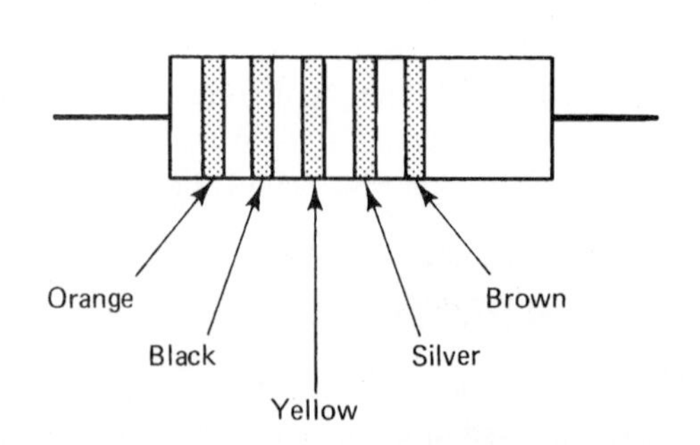

Figure 2–15. Resistor color-code example (five band).

TABLE 2–1 Standard Values of Color-Coded Resistors (Ω)

10% Tolerance											
0.27	1.2	5.6	27	120	560	2,700	12,000	56,000	0.27M	1.2M	5.6M
0.33	1.5	6.8	33	150	680	3,300	15,000	68,000	0.33M	1.5M	6.8M
0.39	1.8	8.2	39	180	820	3,900	18,000	82,000	0.39M	1.8M	8.2M
0.47	2.2	10	47	220	1,000	4,700	22,000	0.1M	0.47M	2.2M	10M
0.56	2.7	12	56	270	1,200	5,600	27,000	0.12M	0.56M	2.7M	12M
0.68	3.3	15	68	330	1,500	6,800	33,000	0.15M	0.68M	3.3M	15M
0.82	3.9	18	82	390	1,800	8,200	39,000	0.18M	0.82M	3.9M	18M
1.0	4.7	22	100	470	2,200	10,000	47,000	0.22M	1.0M	4.7M	22M
5% Tolerance											
0.24	1.1	5.1	24	110	510	2,400	11,000	51,000	0.24M	1.1M	5.1M
0.27	1.2	5.6	27	120	560	2,700	12,000	56,000	0.27M	1.2M	5.6M
0.30	1.3	6.2	30	130	620	3,000	13,000	62,000	0.30M	1.3M	6.2M
0.33	1.5	6.8	33	150	680	3,300	15,000	68,000	0.33M	1.5M	6.8M
0.36	1.6	7.5	36	160	750	3,600	16,000	75,000	0.36M	1.6M	7.5M
0.39	1.8	8.2	39	180	820	3,900	18,000	82,000	0.39M	1.8M	8.2M
0.43	2.0	9.1	43	200	910	4,300	20,000	91,000	0.43M	2.0M	9.1M
0.47	2.2	10	47	220	1,000	4,700	22,000	0.1M	0.47M	2.2M	10M
0.51	2.4	11	51	240	1,100	5,100	24,000	0.11M	0.51M	2.4M	11M
0.56	2.7	12	56	270	1,200	5,600	27,000	0.12M	0.56M	2.7M	12M
0.62	3.0	13	62	300	1,300	6,200	30,000	0.13M	0.62M	3.0M	13M
0.68	3.3	15	68	330	1,500	6,800	33,000	0.15M	0.68M	3.3M	15M
0.75	3.6	16	75	360	1,600	7,500	36,000	0.16M	0.75M	3.6M	16M
0.82	3.9	18	82	390	1,800	8,200	39,000	0.18M	0.82M	3.9M	18M
0.91	4.3	20	91	430	2,000	9,100	43,000	0.20M	0.91M	4.3M	20M
1.0	4.7	22	100	470	2,200	10,000	47,000	0.22M	1.0M	4.7M	22M

Big	Brown	Rabbits	Often	Yield	Great	Big	Vocal	Groans	When	Gingerly	Slapped
Black	Brown	Red	Orange	Yellow	Green	Blue	Violet	Gray	White	Gold	Silver
0	1	2	3	4	5	6	7	8	9	5%	10%

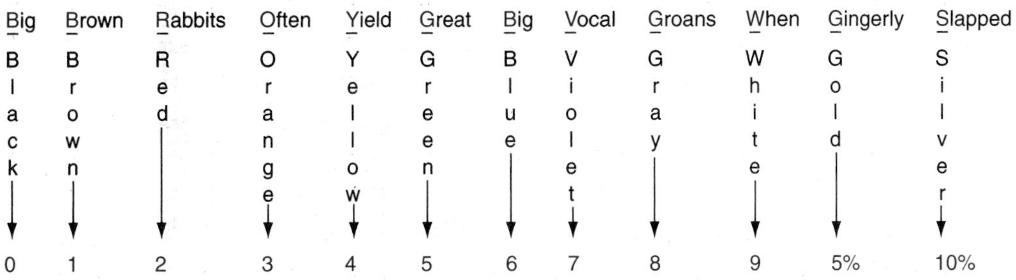

Figure 2–16. Mnemonic device used to memorize resistor color code.

There are four common units of electrical measurement. *Voltage* is used to indicate the force that causes electron movement. *Current* is a measure of the amount of electron movement. *Resistance* is the opposition to electron movement. The amount of work done or energy used in the movement of electrons in a given period is called *power.* Table 2–2 shows the four basic units of electrical measurement. Appendix G includes a discussion of basic units of measurement for future reference.

Small Units

The electrical unit used to measure a certain value is often less than a whole unit (less than 1). Examples of this are 0.6 V, 0.025 A, 0.0550 W. When this occurs, *prefixes* are used. Some prefixes are shown in Table 2–3.

For example, a millivolt is 1/1000 of a volt and a microampere is 1/1,000,000 of an ampere. The prefixes in Table 2–3 may be used with any electrical unit of measurement. The unit is divided by the fractional part of the unit. For example, if 0.6 V is to be changed to millivolts, 0.6 V is divided by the fractional part of the unit. So 0.6 V equals 600 millivolts (mV) or 0.6 divided by 0.001 = 600 mV. If 0.0005 A is changed to microamperes, 0.0005 A is equal to 500 microamperes (μA) or 0.0005 divided by 0.000001 = 500 μA. When changing a basic electrical unit to a prefix unit, move the decimal point of the unit to the right by the same number

TABLE 2–2 Basic Units of Electrical Measurement

Electrical Quantity	*Unit of Measurement*	*Symbol*	*Description*
Voltage	Volt (V)	*V*	Electrical pressure that causes current flow
Current	Ampere (A)	*I*	Amount of electron movement through a circuit
Resistance	Ohm (Ω)	*R*	Opposition to current flow
Power	Watt (W)	*P*	Rate of use of energy as current flows through a circuit

TABLE 2–3 Prefixes of Units Smaller than 1

Prefix	*Abbreviation*	*Fractional Part of a Whole Unit*
milli	m	1/1,000 or 0.001 (3 decimal places)
micro	μ	1/1,000,000 or 0.000001 (6 decimal places)
nano	n	1/1,000,000,000 or 0.000000001 (9 decimal places)
pico	p	1/1,000,000,000,000 or 0.000000000001 (12 decimal places)

of places in the fractional prefix. To change 0.8 V to millivolts, the decimal point of 0.8 V is moved three places to the right (800) since the prefix *milli* has three decimal places. So 0.8 V equals 800 mV. The same method is used for converting any electrical unit to a unit with a smaller prefix.

Often an electrical unit with a prefix is converted back to the basic unit. For example, milliamperes may be converted back to amperes. Microvolts sometimes are converted back to volts. When a unit with a prefix is converted back to a basic unit, the prefix must be multiplied by the fractional part of the whole unit of the prefix. For example, 68 mV converted to volts is equal to 0.068 V. When 68 mV is multiplied by the fractional part of the whole unit (0.001 for the prefix *milli*), this equals 0.068 V (68 mV $\times$ 0.001 = 0.068 V).

When changing a fractional prefix unit into a basic electrical unit, move the decimal in the prefix unit to the left the same number of places of the prefix. To change 225 millivolts to volts, move the decimal point in 225 three places to the left (0.225) since the prefix *milli* has three decimal places. So 225 mV equals 0.225 V. This same method is used when changing any fractional prefix unit back to the original electrical unit.

TABLE 2–4 Prefixes of Large Units

Prefix	*Abbreviation*	*Number of Times Larger than 1*
kilo	k	1,000
mega	M	1,000,000
giga	G	1,000,000,000

Large Units

Sometimes electrical units of measurement are quite large, such as 20,000,000 W, 50,000 Ω, or 38,000 V. When this occurs, prefixes are needed to make these large numbers easier to use. Some prefixes used for large electrical values are shown in Table 2–4. To change a large value to a smaller unit, divide the large value by the number of the prefix. For example, 48,000,000 Ω is changed to 48 megohms (MΩ) by dividing (48,000,000 divided by 1,000,000 = 48 MΩ). To convert 7000 V to 7 kilovolts (kV), divide 7000 by 1000 (7000 divided by 1000 = 7 kV). To change a large value to a smaller value, move the decimal point in the large value to the *left* by the number of zeros of the prefix; thus, 3600 V equals 3.6 kV (3600). To convert a prefix unit back to a large number, move the decimal point to the right by the same number of places in the unit. Also, the number may be multiplied by the number of the prefix. If 90 MΩ is converted to ohms, move the decimal point six places to the right (90,000,000). The 90-MΩ value may also be multiplied by the number of the prefix, which is 1,000,000. Thus, 90 MΩ $\times$ 1,000,000 = 90,000,000 Ω.

The simple conversion scale shown in Fig. 2–17 is useful when converting large and small units to units of measurement with prefixes. This scale uses either powers of 10 or decimals to express the units. Refer to the calculator procedures for conversion of electrical units in App. F (Fig. F–10).

2.4 SCIENTIFIC NOTATION

Using powers of 10 or scientific notation greatly simplifies math operations. A number that has many zeros to the right or to the left of the decimal point is made simpler by putting it in the form of scientific notation (powers of 10). For example, 0.0000035 $\times$ 0.000025 is difficult to multiply. It can be put in the form $(3.5 \times 10^{-6}) \times (2.5 \times 10^{-5})$. Notice the number of places that the decimal point is moved in each number.

Table 2–5 lists some of the powers of 10. In a whole number, the power to which the number is raised is *positive*. It equals the number of zeros following the 1. In decimals, the power is *negative* and equals the number of places the decimal point is moved to the left of the 1. Easy powers of 10 to remember are $10^2 = 100$ (10×10) and $10^3 = 1000$ $(10 \times 10 \times 10)$.

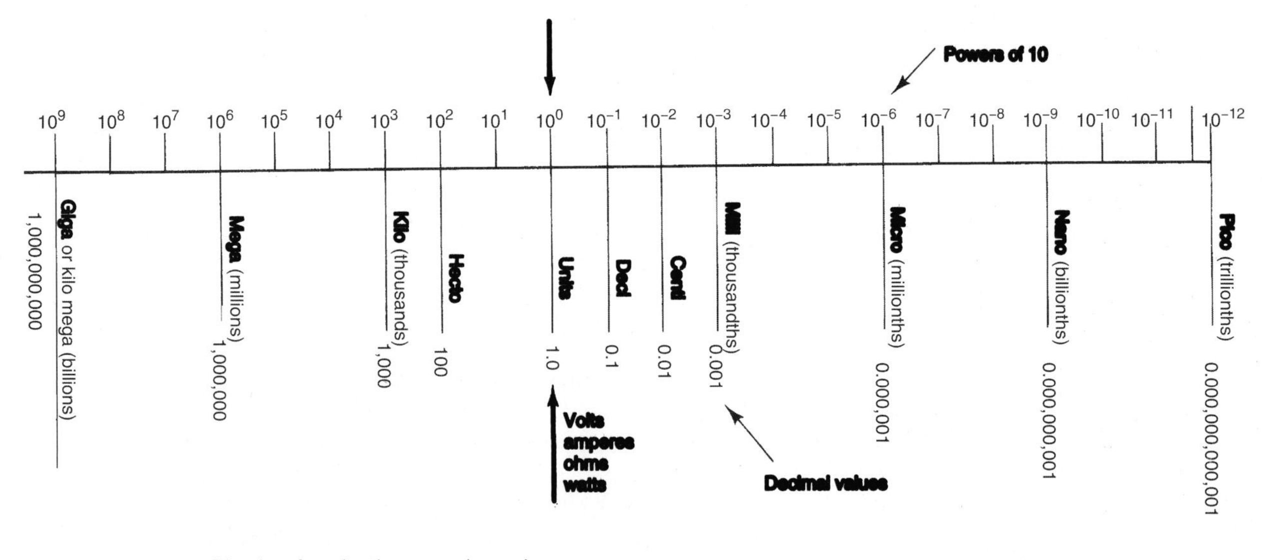

Figure 2–17. Simple conversion scale for large and small numbers.

TABLE 2–5 Powers of 10

	Number	Power of 10
Whole numbers	1,000,000	10^6
	100,000	10^5
	10,000	10^4
	1,000	10^3
	100	10^2
	10	10^1
	1.0	10^0
Decimals	0.1	10^{-1}
	0.01	10^{-2}
	0.001	10^{-3}
	0.0001	10^{-4}
	0.00001	10^{-5}
	0.000001	10^{-6}

Any number written as a multiple of a power of 10 and a number between 1 and 10 is said to be expressed in *scientific notation.* For example,

$$81{,}000{,}000 = 8.1 \times 10{,}000{,}000 \text{ or } 8.1 \times 10^7$$
$$500{,}000{,}000 = 5 \times 100{,}000{,}000 \text{ or } 5 \times 10^8$$
$$0.0000000004 = 4 \times 0.0000000001 \text{ or } 4 \times 10^{-10}$$

Scientific notation simplifies multiplying and dividing large numbers of small decimals. For example,

$$\begin{aligned} &4800 \times 0.000045 \times 800 \times 0.0058 \\ &= (4.8 \times 10^3) \times (4.5 \times 10^{-5}) \times (8 \times 10^2) \\ &\quad \times (5.8 \times 10^{-3}) \\ &= (4.8 \times 4.5 \times 8 \times 5.8) \times (10^{3-5+2-3}) \\ &= 1002.24 \times 10^{-3} \\ &= 1.00224 \end{aligned}$$

$$\begin{aligned} &95{,}000 \div 0.0008 \\ &= \frac{9.5 \times 10^4}{8 \times 10^{-4}} \\ &= \frac{9.5 \times 10^{4-(-4)}}{8} \\ &= \frac{9.5 \times 10^8}{8} = 1.1875 \times 10^8 \\ &= 118{,}750{,}000 \end{aligned}$$

With some practice the use of scientific notation becomes easy. Refer also to the calculator procedures for scientific notation in App. F (Fig. F–11).

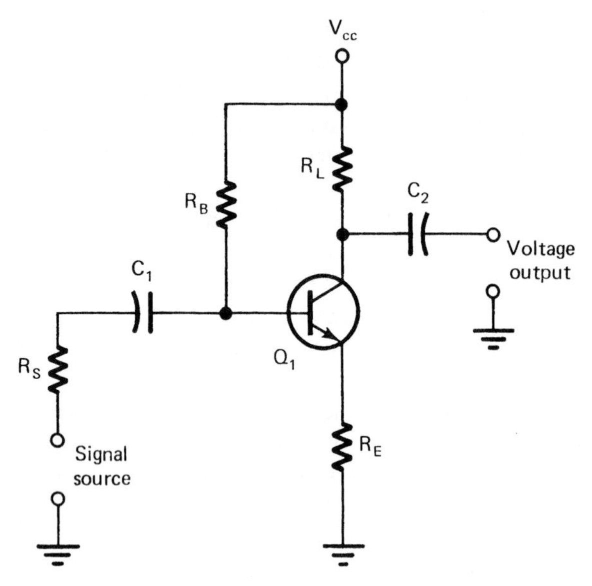

Figure 2–18. Schematic diagram of a transistor amplifier.

2.5 SCHEMATIC DIAGRAMS

Schematic diagrams are used to represent the parts of electrical equipment or circuits. They show how the components or parts of each circuit fit together. Schematic diagrams are used to show the *details* of the electrical connections of any type of circuit or system. Schematics are used by manufacturers of electrical equipment showing operation and as an aid in servicing the equipment. A typical schematic diagram is shown in Fig. 2–18. Notice the symbols that are used. Symbols are used to represent electrical components in schematic diagrams. Standard electrical symbols are used by all equipment manufacturers. Some common basic electrical symbols are shown in App. D. These symbols should be memorized.

2.6 BLOCK DIAGRAMS

Another way to show how electrical equipment operates is to use block diagrams. Block diagrams show the functions of the subparts of any electrical system. A block diagram of an electrical system was shown in Fig. 1–1(a). The same type of diagram is used to show the parts of a radio in Fig. 2–19(a).

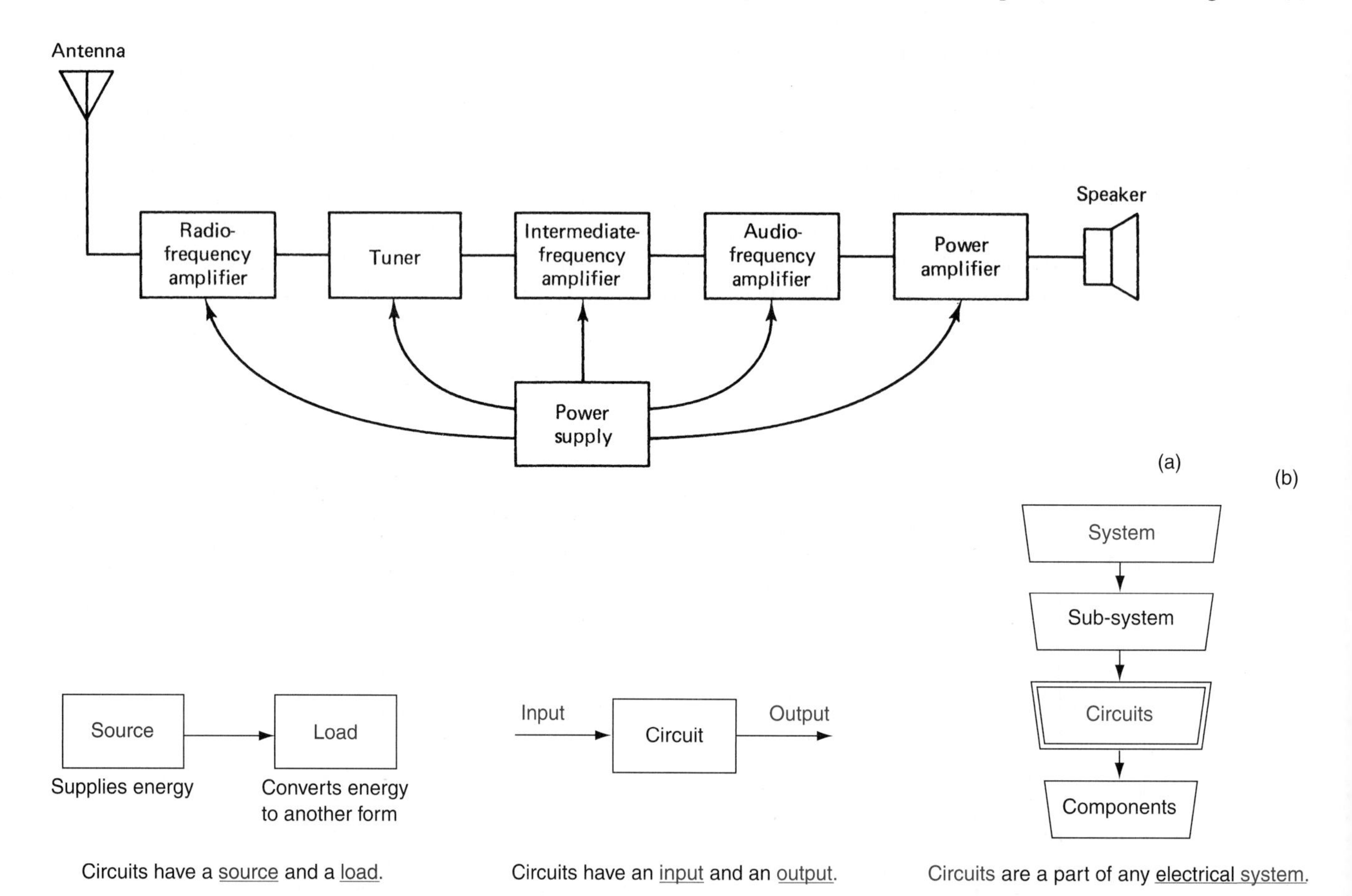

Figure 2–19. (a) Block diagram of the parts of a radio. (b) Electrical circuits and systems: Circuits have a *source* and a *load;* Circuits have an *input* and an *output;* Circuits are a part of any *electrical system.*

Inside the blocks, symbols or words are used to describe the function of the block. Block diagrams usually show the operation of the whole system. They provide an idea of how a system operates; however, they do not show detail like a schematic diagram. It is easy to see the major subparts of a system by looking at a block diagram. Fig. 2–19(b) shows block diagrams that represent electrical circuits and systems.

2.7 WIRING DIAGRAMS

Another type of electrical diagram is called a wiring diagram (sometimes called a cabling diagram). Wiring diagrams show the actual location of parts and wires on equipment. The details of each *connection* are shown on a wiring diagram. Schematic and block diagrams show only how parts fit together *electrically.* Wiring diagrams show the details of actual connections. A simple wiring diagram is shown in Fig. 2–20.

2.8 MEASURING RESISTANCE

Many important electrical tests may be made by measuring resistance. *Resistance* is opposition to the flow of current in an electrical circuit. The current that flows in a circuit depends on the amount of resistance in that circuit. Measure resistance in an electrical circuit by using a meter, such as an ohmmeter or multimeter. Multimeters such as those shown in Fig. 1–6 are the most used meters for doing electrical work. A multimeter is used to measure resistance, voltage, or current. The type of measurement is changed by adjusting the "function select switch" to the desired measurement.

Two popular types of meters used to measure electrical quantities are an analog meter and a digital meter. Figure 2–21(a) compares the two types of meters. Notice that the analog meter has a scale which is used to make a measurement. The digital meter provides a direct reading of the quantity measured, making it easier to interpret. Both types of meters are used by electrical technicians. Analog meters will be discussed in this chapter to provide the basic rules of interpreting analog scales. The same rules of meter use apply for analog and digital meters. Figure 2–21(b) shows the controls of a common type of VOM or multimeter which will also be used to learn analog scales in this chapter. Even if digital meters are used, analog scales should be studied.

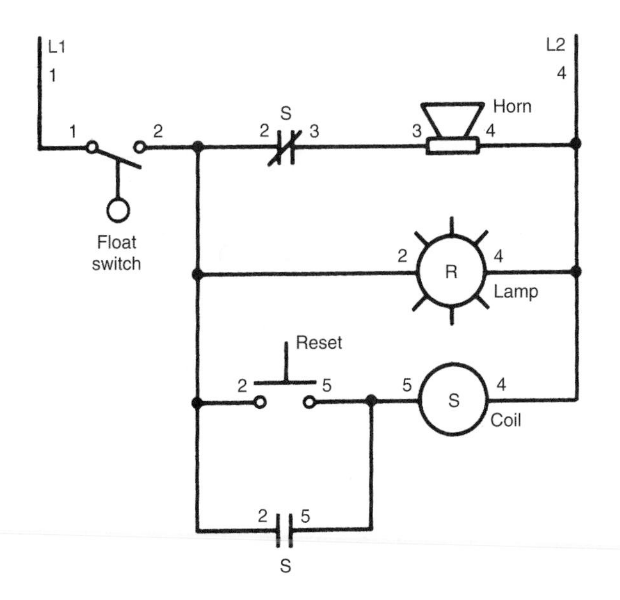

Figure 2–20. Simple wiring diagram.

Notice that the function select switch of Fig. 2–21(b) is in the center of the meter. Also notice that the lower right section contains ranges for measuring ohms or resistance. This type of meter is called a multirange, multifunction meter. The ohms measurement ranges are divided into four portions: ×1, ×10, ×1000, and ×100,000. Most multimeters or VOMs are similar to the example shown. The meter may be adjusted to any of the four positions for measuring resistance. The test leads used with the VOM are ordinarily black and red. These colors are used to help identify which lead is the positive or negative side of the meter. This is important when measuring direct-current (dc) values. Red indicates positive (+) polarity and black indicates negative (–) polarity.

Refer again to the diagram of the meter controls shown in Fig. 2–21(b). The red test lead is put into the hold or "jack" marked with "V-Ω-A" or volts-ohms-amps. The black test lead is put into the hold or jack labeled "–COM" or negative common. The function selector switch should be placed on one of the resistance ranges. When the test leads are touched together or "shorted," the meter needle moves from the left side of the meter to the right side. This test shows that the meter is operational.

The meter's *scale* is also important. Figure 2–22 shows the scale of a common type of VOM. Notice that the top scale, from zero to infinity (∞), is labeled "Ohms." This scale is used for measuring ohms *only.* On most VOMs, the top scale is the resistance or ohms scale. To measure any resistance, first select the proper meter range. On the meter range shown in Fig. 2–21(b), there are four ranges: ×1, ×10, ×1000, and ×100,000. These values are called *multipliers.* The ohmmeter must be properly "zeroed" before attempting to measure resistance accurately. To "zero" the ohmmeter properly, touch the two test leads together. This should cause the needle to move from infinity (∞) on the left to zero (0) on the right. Infinity represents a very high resistance. Zero represents a very low resistance. If the needle does not reach zero or goes past zero when the test leads are touched or shorted, the control marked "Ohms Adjust" is used. The needle is adjusted to zero when the test leads are touched together. The ohms adjust control is indicated by "ΩADJ" on the diagram. The ohmmeter should always be zeroed prior to each resistance measurement and after changing ranges. If the meter is not zeroed, measurements will be incorrect.

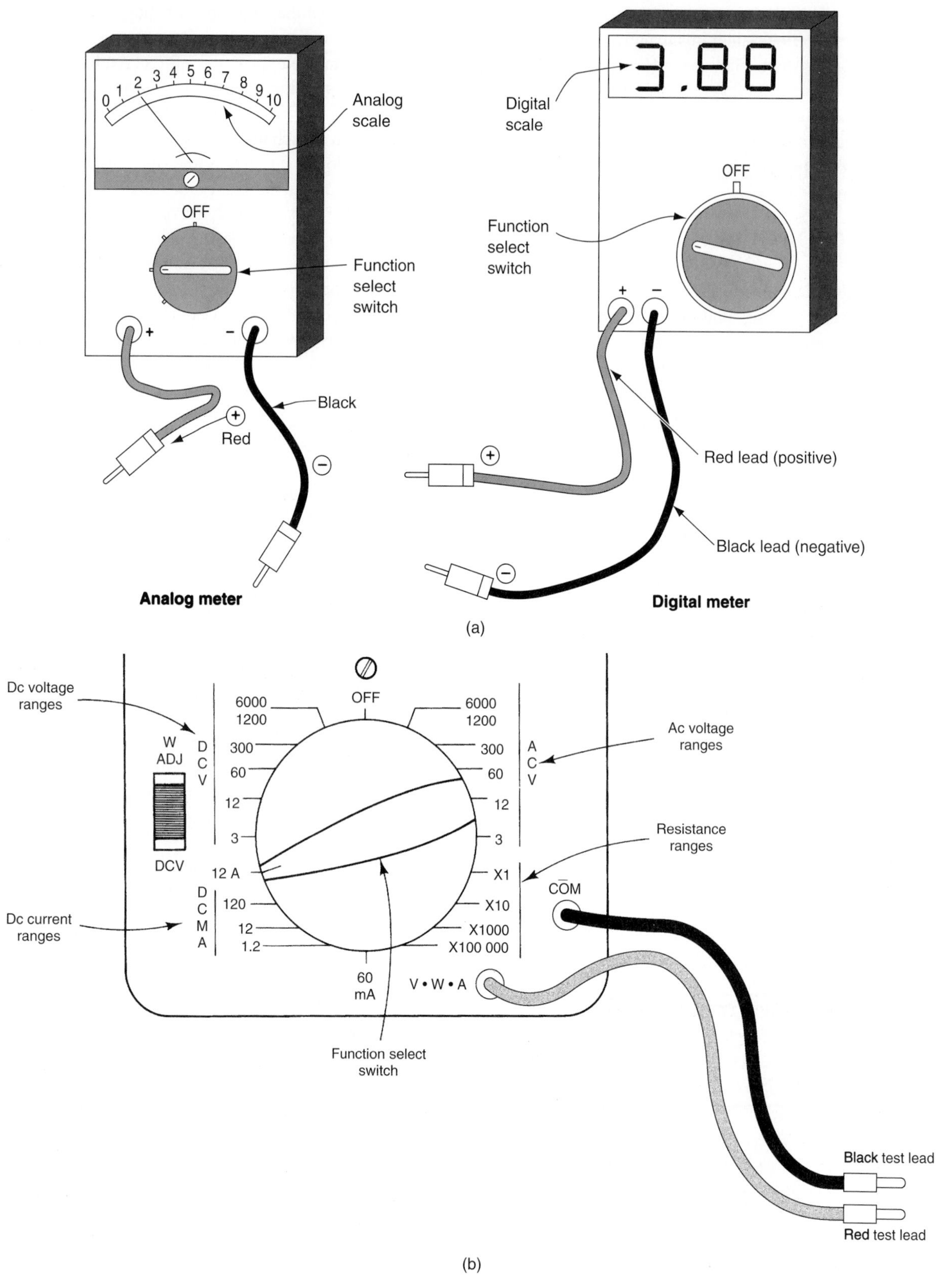

Figure 2–21. Multimeters. (a) Multimeters used to measure resistance, voltage, or current; (b) controls of a typical VOM (multimeter).

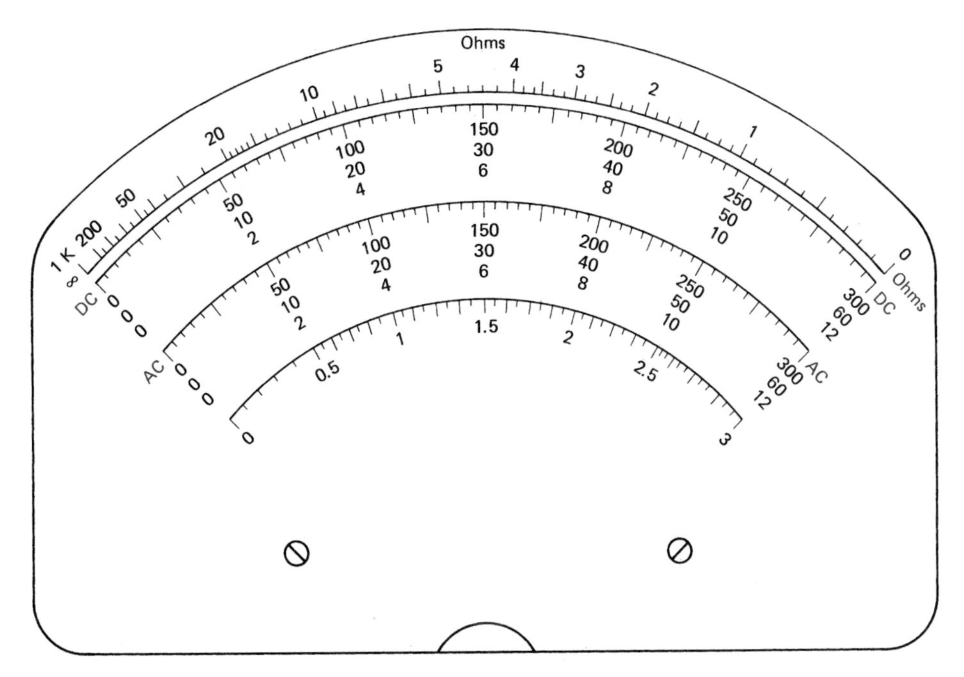

Figure 2–22. VOM (multimeter) scale.

A more accurate measurement of resistance occurs when the meter's needle stops somewhere between the middle of the ohms scale and zero. Choosing the proper range adjustments controls how far the needle moves. If the same range selected is $\times 1$, this means that the number to which the needle points must be multiplied by 1. If the function select switch is adjusted to the $\times 100{,}000$ range, this means that the number the needle points to is multiplied by 100,000. The meter needle should always move to near the center of the scale. Always zero the meter when changing ranges, and always multiply the number indicated on the scale by the multiplier of the range. Never measure the resistance of a component until it has been disconnected or the reading may be wrong. Voltage should *never* be applied to a component when measuring resistance.

A VOM may also be used to measure the resistance of a potentiometer, as shown in Fig. 2–23. If the shaft of the pot is adjusted while the ohmmeter is connected to points A and C, no resistance change will take place. The resistance of the potentiometer is measured in this way. Connecting to points B and C or to points B and A allows changes in resistance as the shaft is turned. The potentiometer shaft may be adjusted both clockwise and counterclockwise. This adjustment affects the measured resistance across points B and C or B and A. The resistance varies from zero to maximum and from maximum back to zero as the shaft is adjusted.

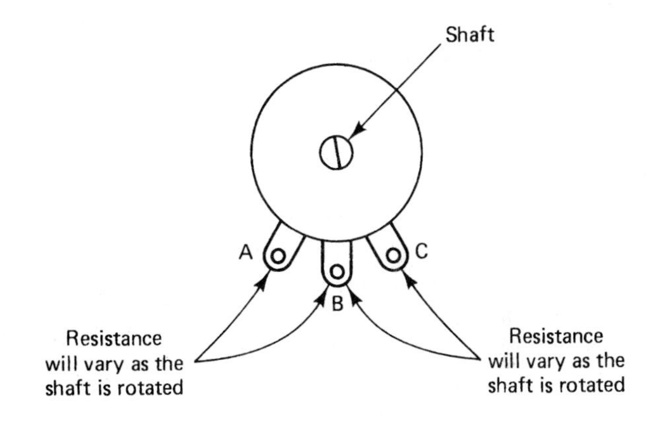

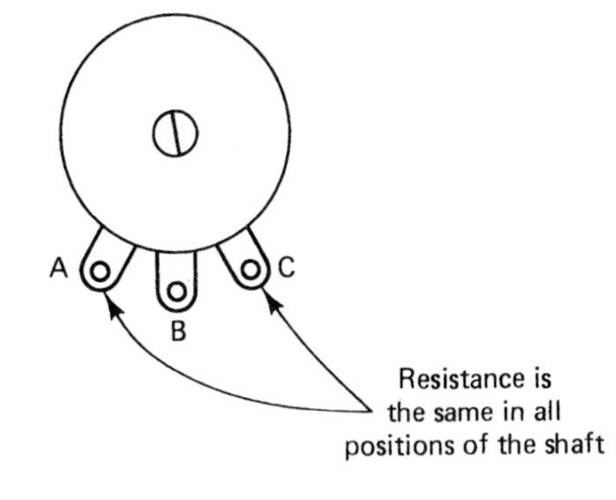

Figure 2–23. Measuring the resistance of a potentiometer.

How to Measure Resistance

Remember that resistance is the opposition to flow of electrical current. For example, a lamp connected to a battery has resistance. Its resistance value is determined by the size of the filament wire. The filament wire opposes the flow of electrical current from the battery. The battery causes current to flow through the lamp's filament. The amount of current through the lamp depends on the filament resistance. If it offers little opposition to current flow from the battery, a *large* current flows in the circuit. If the lamp filament has high resistance, it offers much opposition to current flow from the battery. Then a *small* current flows in the circuit.

Resistance tests are sometimes called *continuity checks.* A continuity check is made to see if a circuit is open or closed (a "continuous" path). An ohmmeter is also used to measure exact values of resistance. Resistance must always be measured with no voltage applied to the component being measured. The ohmmeter ranges of a VOM are used to measure resistance. This type of meter is often used by electrical technicians since it measures resistance, voltage, and current. By adjusting the rotary function select switch, the meter can be set to measure either resistance, voltage, or current. The meter switch shown in Fig. 2–21(b) has settings for:

1. Direct-current (dc) voltage
2. Direct-current (dc) amps and milliamps
3. Alternating-current (ac) voltage
4. Resistance (ohms)

Notice that the lower right part of the function select switch is for measuring resistance or ohms. The ohms measurement settings are marked as ×1, ×10, ×1000, and ×100,000. When measuring resistance with an ohmmeter, first put the test leads into the meter. The test leads are usually black and red wires that plug into the meter. The red wire is plugged into the hole marked "volts-ohms-amps" (V-Ω-A). The black wire is plugged into the hole marked "negative common" (–COM).

The scale of the meter is used to indicate the value of resistance in ohms. Notice that the right side of the scale is marked with a zero and the left side is marked with an infinity (∞) sign.

When the test leads are touched together, the needle on the scale of the meter should move to the right side of the scale. This indicates zero resistance. The needle of the meter is adjusted so that it is exactly over the zero mark. This is called zeroing the meter. This must be done to measure any resistance accurately. The "ohms adjust" (Ω ADJ) control is used to zero the needle of the meter. The meter should be zeroed before each resistance measurement is made.

It is important to be able to read the scale of the meter. Notice that the top scale shown in Fig. 2–22 is labeled with a zero on the right side and an infinity (∞) sign on the left side. This scale is used for measuring resistance only. Remember that the most accurate readings are made when the meter's needle moves to somewhere between the center of the scale and zero.

To measure any resistance accurately, first select the proper range. The ranges for measuring resistance are on the lower right part of the function select switch. This switch has resistance ranges marked as ×1, ×10, ×1000, and ×100,000. If the meter's range setting is on the ×1, this means that the reading on the meter's scale must be multiplied by 1.

Some examples with the meter range set on different multipliers are done as follows. The meter must be zeroed when a range is changed. The test leads are then placed across a resistance. Assume that the needle of the meter moves to point A on the scale of Fig. 2–24. The resistance equals $7.5 \times 1 = 7.5\ \Omega$. Now change the meter range to ×1000. The reading at point B equals $5.5 \times 1000 = 5500\ \Omega$. At point C, the reading is $0.3 \times 1000 = 300\ \Omega$. The same procedure is used for the ×100,000 range. If the needle moves to 2.2 (point D) on the scale, the reading would be equal to $2.2 \times 100{,}000$, or 220,000 Ω. If the meter range is set on ×100,000 and the needle moves to 3.9 (point E) on the scale, the reading is $3.9 \times 100{,}000$, or 390,000 Ω.

Remember to zero the meter by touching the test leads together and using the ΩADJ control before making a resistance measurement. Each time the meter range is changed, the meter needle must be zeroed on the scale. If this procedure is not followed, the meter reading will not be accurate.

To learn to measure resistance, it is easy to use color-coded resistors. These resistors are small and easy to handle. Practice in the use of the meter to measure several values of resistors makes reading the meter much easier.

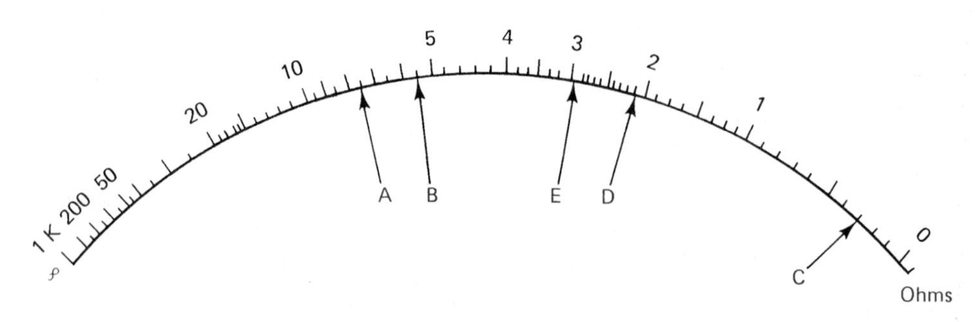

Figure 2–24. Ohm's scale of a VOM (multimeter).

2.9 MEASURING VOLTAGE

Voltage is applied to electrical equipment to cause it to operate. It is important to be able to measure voltage to check the operation of equipment. Many electrical problems develop due to either too much or too little voltage being applied to the equipment. A voltmeter is used to measure voltage in an electrical circuit. A VOM is also used to measure voltage. Refer to the controls of the VOM shown in Fig. 2–21(b). The voltage ranges shown are 3, 12, 60, 300, 1200, and 6000 V. Other VOMs have different ranges and scales. When the function select switch is adjusted to 3 V on the *dc volts range,* the meter measures *up to* 3 V. The same is true for the other ranges of dc voltage. The voltage value of each range is the maximum value of voltage that may be measured with the VOM set on that range.

When making voltage measurements, adjust the function select switch to the highest range of dc voltage. Connect the red and black test leads to the meter by putting them into the proper jacks. The red test lead should be put into the jack labeled "V-Ω-A." The black test lead should be put into the jack labeled "–COM."

It is easy to become familiar with the part of the meter's scale that is used to measure dc voltage. Refer to the VOM scale of Fig. 2–22. Notice that the part of the scale below the ohms scale is the dc voltage scale. This scale is usually black. Notice that there are three dc voltage scales: 0 to 12 V, 0 to 60 V, and 0 to 300 V. All dc voltages are measured using one of these scales. Notice that each of the dc voltage ranges on the function select switch corresponds to a number on the right side of the meter scale *or* a number that can be easily multiplied or divided to equal the number on the function switch.

Notice that when the 12-, 60-, or 300-V range is used, the scale is read directly. On these ranges, the number to which the needle points is the actual value of the voltage being measured. When the 3-, 1200-, or 6000-V range is used, the number to which the needle points must be multiplied or divided. If the meter's needle points to the number 50 while the meter is adjusted to the 60-V range, the measured voltage is 50 V. If the meter's needle points to the number 250 while the meter is adjusted to the 3-V range, the measured voltage is 2.5 V ($250 \div 100 = 2.5$). When the 1200-V range is used, the numbers on the 0- to 12-V scale are read and then multiplied by 100. Most VOMs have several scales. Some of these scales are read directly, whereas others require multiplication or division.

Before making any measurements, the proper dc voltage range is chosen. The value of the range being used is the maximum value of voltage that can be measured on that range. For example, when the range selected is 12 V, the maximum voltage the meter can measure is 12 V. Any voltage above 12 V could damage the meter. To measure a voltage that is unknown (no indication of its value), start by using the highest range on the meter. Then slowly adjust the range downward until a voltage reading is indicated on the right side of the meter scale.

Matching the meter polarity to the voltage polarity is important when measuring dc voltage. The meter needle moves backward, possibly damaging the meter if the polarities are not connected properly. Meter polarity is simple to determine. The positive (+) red test lead is connected to the positive side of the dc voltage being measured. The negative (–) black test lead is connected to the negative side of the dc voltage being measured. The meter is always connected across (in parallel with) the dc voltage being measured.

How to Measure Dc Voltage

Voltage is the electrical pressure that causes current to flow in a circuit. A common voltage source is a battery. Batteries come in many sizes and voltage values. The voltage applied to a component determines how much current will flow through it.

A dc voltmeter or the dc voltage ranges of a VOM are used to measure dc voltage. The upper left part of the VOM function select switch of Fig. 2–21(b) is used for measuring dc voltage. The dc voltage ranges are 3, 12, 60, 300, 1200, and 6000 V.

When measuring voltage with a VOM, first put the test leads into the meter. The red test lead is plugged into the hole marked "volts-ohms-amps" (V-Ω-A). The black test lead is plugged into the hole marked "negative common" (–COM). The scale of the meter is used to indicate voltage (in volts). Notice that the left side of the dc voltage ranges is marked zero and the right side is marked 300, 60, and 12. The meter needle rests on the side until some voltage is measured. These three scales are used to measure dc voltages on the sample meter scale.

To measure a dc voltage, first select the proper range. The ranges for measuring dc voltage are on the upper left part of the meter's function select switch. If the meter's range setting is on the 3-V range, then the voltage being measured cannot be larger than 3 V. If the voltage is greater than 3 V, the meter would probably be damaged. Be careful to use a meter range that is larger than the voltage being measured. Notice that each of the dc voltage ranges on the function select switch corresponds to a number on the right side of the meter scale or a number that can be easily multiplied or divided to equal the number on the function select switch. When the 12-, 60-, or 300-V range is used, the number that the needle points to is the actual value of the voltage being measured. When the 3-, 1200-, or 6000-V range is used, the number that the needle points to is multiplied or divided by 100. If the meter's needle points to the number 850 while the 3-V range is being used, the measured voltage is 8.5, because 850 divided by 100 equals 8.5.

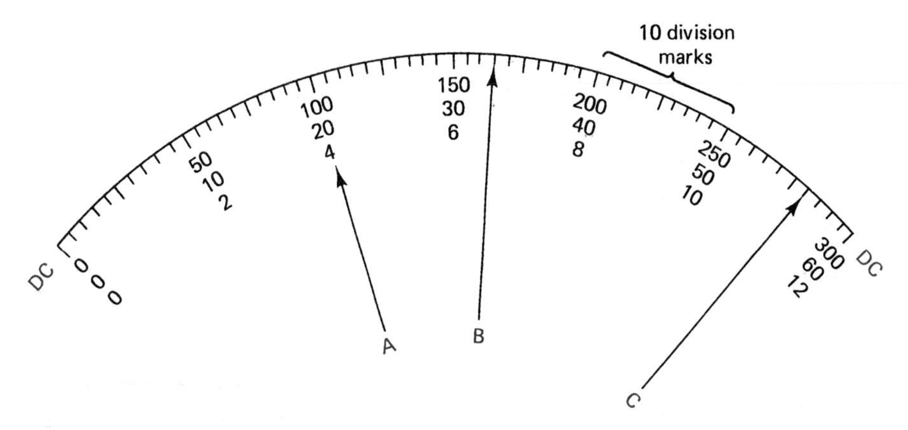

Figure 2–25. Dc voltage scale of a VOM (multimeter).

Some examples of dc voltage measurements with the meter set on the 3-V range are given below. If the test leads of the meter are placed across a voltage source and the meter's needle moves to point A on the scale of Fig. 2–25, the dc voltage is equal to 100 divided by 100, or 1 V. The reading at point B is 165 divided by 100, or 1.65 V. At point C, the reading is 280 divided by 100, or 2.8 V. There is some difficulty in reading the voltage divisions on the scales. Look at the division marks from 200 to 250. The difference between 200 and 250 is 50 units (250 minus 200 equals 50). There are 10 division marks between 200 and 250. The voltage per division mark is 50 divided by 10, or 5 V per division. So each division mark between 200 and 250 V is equal to 5 V. This procedure is like reading a ruler or other types of scales.

If the range switch is changed to the 12-V position, the voltage is read directly from the meter scale. For example, if the range is set on 12 V and the meter needle moves to point A on Fig. 2–25, the voltage equals 4 V. The reading at point B equals 6.6 V. At point C, the reading is 11.2 V. The same procedure is used for all other ranges.

When measuring voltage, always be sure to select the proper range. The range used is the *maximum* value of voltage that can be measured on that range. For example, when the range selected is 12 V, the maximum voltage that the meter can measure is 12 V. Any voltage above 12 V would damage the meter. When measuring an unknown voltage, start with the highest range setting on the meter. Then slowly adjust the range setting to lower values until the meter needle moves to somewhere between the center and right side of the meter scale.

When measuring direct-current voltage, polarity is very important. The proper matching of meter polarity and voltage source polarity must be ensured. The negative (black) test lead of the meter is connected to the negative polarity of the voltage being measured. The positive (red) test lead is connected to the positive polarity of the voltage. If the polarities are reversed, the meter needle will move backward and possibly damage the meter.

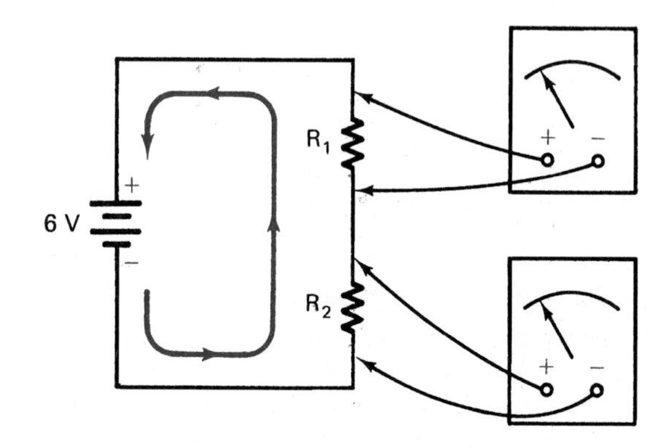

Figure 2–26. Measuring voltage drop in a dc circuit.

A certain amount of voltage is needed to cause electrical current to flow through a resistance in a circuit. This voltage is called *voltage drop.* Voltage drop is measured across any component through which current flows. The polarity of a voltage drop depends on the direction of current flow. Current flows from the negative polarity of a battery to the positive polarity. In Fig. 2–26, the bottom of each resistor is negative. The negative test lead of the meter is connected to the bottom of the resistor. The positive test lead is connected to the top. The meters are connected as shown to measure each of the voltage drops in the circuit. If the meter polarity is reversed, the meter needle would move in the wrong direction.

Alternating-current (ac) voltage is measured in the same way as dc voltage. The ac voltage scales and ranges on the VOM are used. Measuring ac voltage with a VOM is discussed in Chap. 6.

2.10 MEASURING CURRENT

Current flows through a complete electrical circuit when voltage is applied. Many important tests are made by measuring current flow in electrical circuits. The current values in an electrical circuit depend on the amount of resistance in the circuit. Learning to use an ammeter to measure current in an electrical circuit is important.

VOMs will also measure dc current. Refer to the controls of the VOM shown in Fig. 2–21(b). The function select switch may be adjusted to any of five ranges of direct current: 12 A, 120 mA, 12 mA, 1.2 mA, and 60 μA. For example, when the function select switch is placed in the 120-mA range, the meter is capable of measuring up to 120 mA of current. The value of the current set on the range is the maximum value that can be measured on that range. The function select switch should first be adjusted to the highest range of direct current. Current is measured by connecting the meter into a circuit as shown in Fig. 2–27, which is referred to as connecting the meter in *series* with the circuit. Series circuits are discussed in detail in Chap. 3.

Current flows from a voltage source when some device that has resistance is connected to the source. When a lamp is connected to a battery, current flows from the battery through the lamp. In the circuit of Fig. 2–27, electrons flow from the negative battery terminal, through the lamp, and back to the positive battery terminal. Electrons are so small that the human eye cannot see them, but their movement can be measured with an ammeter.

As the voltage applied to a circuit increases, the current also increases. So, if 12 V is applied to the lamp in Fig. 2–27, a larger current will flow through the lamp. If 24 V is applied to the same lamp, an even larger current will flow. As resistance gets smaller, current increases. Resistance is the opposition to current flow. When a circuit has more resistance, it has less current flow.

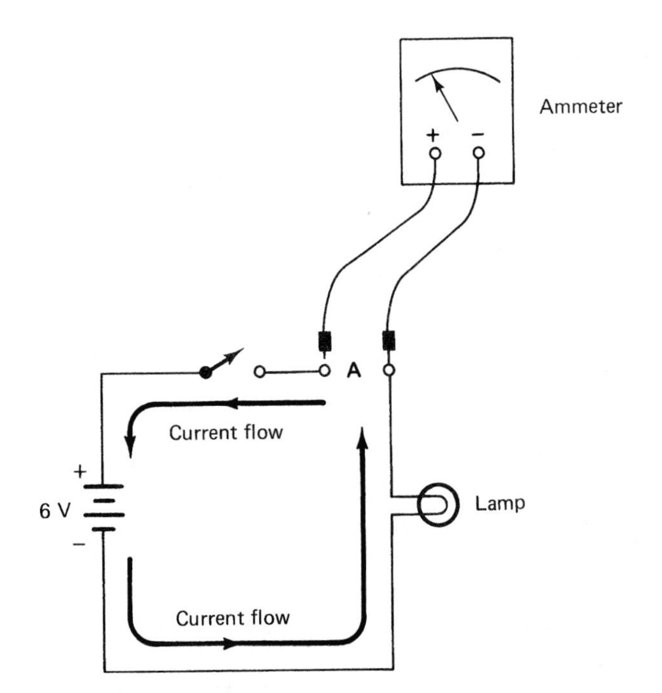

Figure 2–27. Meter connection for measuring direct current.

How to Measure Direct Current

Refer to the direct-current ranges of the VOM shown in Fig. 2–21(b). Notice that the ranges begin with 12 A. The next ranges are for measuring 120 mA, 12 mA, 1.2 mA, and 60 μA. There is a total of five current ranges. The function select switch is adjusted to any of these five ranges for measuring direct current. When measuring current, always start with the meter set on its highest range. By practicing this procedure, it becomes a habit. This habit helps in using a meter properly. Always start on the highest range. Then move the range setting to a lower value if the meter needle only moves a small amount. The most accurate reading is when the meter needle is between the center of the scale and the right side. The same scales on the VOM are often used for measuring direct-current or dc voltage.

If the meter range is set on the 12-A range, the scale is read directly. The bottom dc scale, which has the number "12" on the right side, is used. Some examples are shown in Fig. 2–28 with the meter set on the 12-A range. At point A on the scale, the reading is 4.6 A. The reading at point B is 8.8 A.

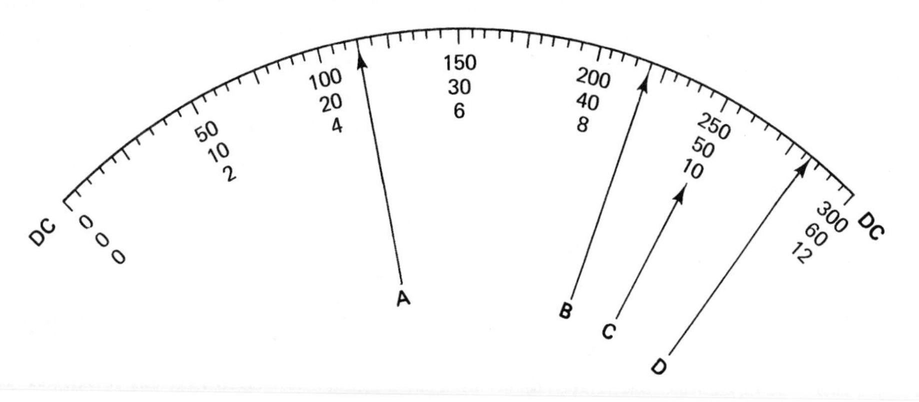

Figure 2–28. Direct-current scale of a VOM (multimeter).

The 60-μA range on the meter is for measuring very small currents. This range is also read directly on the meter scale. Notice that the number "60" is the middle number on the right of the dc scale.

When the meter is set on the 120-mA range, the meter will measure up to 120 mA of direct current. The readings on the scale are multiplied by 10 on this range setting. The readings at the points shown in Fig. 2–28 for the 120-mA range are:

Point A = 46 (4.6 times 10) mA
Point B = 88 (8.8 times 10) mA
Point C = 100 (10 times 10) mA
Point D = 113 (11.3 times 10) mA

Notice that the reading at point D is between two of the scale divisions. Point D is halfway between the 11.2 and 11.4 divisions on the scale. So the reading is 11.3 times 10, or 113 mA.

The test lead polarity of the VOM is also important when measuring direct current. The VOM is connected to allow current to flow through the meter in the right direction. The negative test lead is connected nearest to the negative side of the voltage source. The meter is then connected *into* the circuit. To measure current, a wire is removed from the circuit to place the meter into the circuit. No voltage should be applied to the circuit when connecting the meter. The meter is placed in series with the circuit. Series circuits have one path for current flow.

The proper procedure for measuring current through point A in the circuit of Fig. 2–27 is:

1. Turn off the circuit's voltage source by opening the switch.
2. Set the meter to the highest current range (12 A).
3. Remove the wire at point A.
4. Connect the negative test lead of the meter to the terminal nearest to the negative side of the voltage source.
5. Connect the positive lead to the end of the wire which was removed from point A.
6. Turn on the switch to apply voltage to the circuit.
7. Look at the meter needle to see how far it has moved up the scale.
8. Adjust the meter range until the needle moves to between the center of the scale and the right side.

Always remember the following safety tips when measuring current:

1. Turn off the voltage before connecting the meter, so as not to get an electrical shock. This is an important habit to develop. Always remember to turn off the voltage before connecting the meter.
2. Set the meter to its highest current range. This ensures that the meter needle will not move too far to the right of the scale and possibly damage the meter.
3. A wire is disconnected from the circuit and the meter is put in series with the circuit. Always remember to disconnect a wire and reconnect the wire to one of the meter test leads. If a wire is not removed to put the meter into the circuit, the meter will not be connected properly.
4. Use the proper meter polarity. The negative test lead is connected so that it is nearest the negative side of the voltage source. Similarly, the positive test lead is connected so that it is nearest the positive side of the voltage source.

2.11 DIGITAL METERS

Many digital meters are now in use. They employ numerical readouts to simplify the measurement process and to make more accurate measurements. Instruments such as digital counters, digital multimeters, and digital voltmeters are commonly used. Digital meters such as the ones shown in Fig. 2–29 rely on the operation of digital circuitry to produce a numerical readout of the measured quantity.

The readout of a digital meter is designed to transform electrical signals into numerical data. Both letter and number readouts are available, as are seven-segment, discrete-number, and bar-matrix displays. Each method has a device designed to change electrical energy into light energy on the display.

Digital multimeters (DMMs) have numerical readouts which are called seven-segment displays. Figure 2–30(a) shows the seven-segment display and Fig. 2–30(b) indicates the parts of the display which are illuminated when each number is displayed. The *resolution* of a digital multimeter is the smallest change of a quantity that the meter can measure. The smaller the change, the better the meter's resolution. The resolution of a meter is determined by the number of digits in the display. Many meters have 3½-digits in the display. A 3½-digit multimeter has three digit positions that can indicate from 0 through 9 and one digit position that can indicate only a value of 1. This latter digit is called the half-digit and is always the digit on the left of the display. For example, if the reading of 0.999 V, as shown in Fig. 2–31(a), increased by 0.001 volt to 1 volt, the display shows 1.000 volt. A change of 0.001 volt is the resolution of the 3½-digit multimeter shown in Fig. 2–31(b).

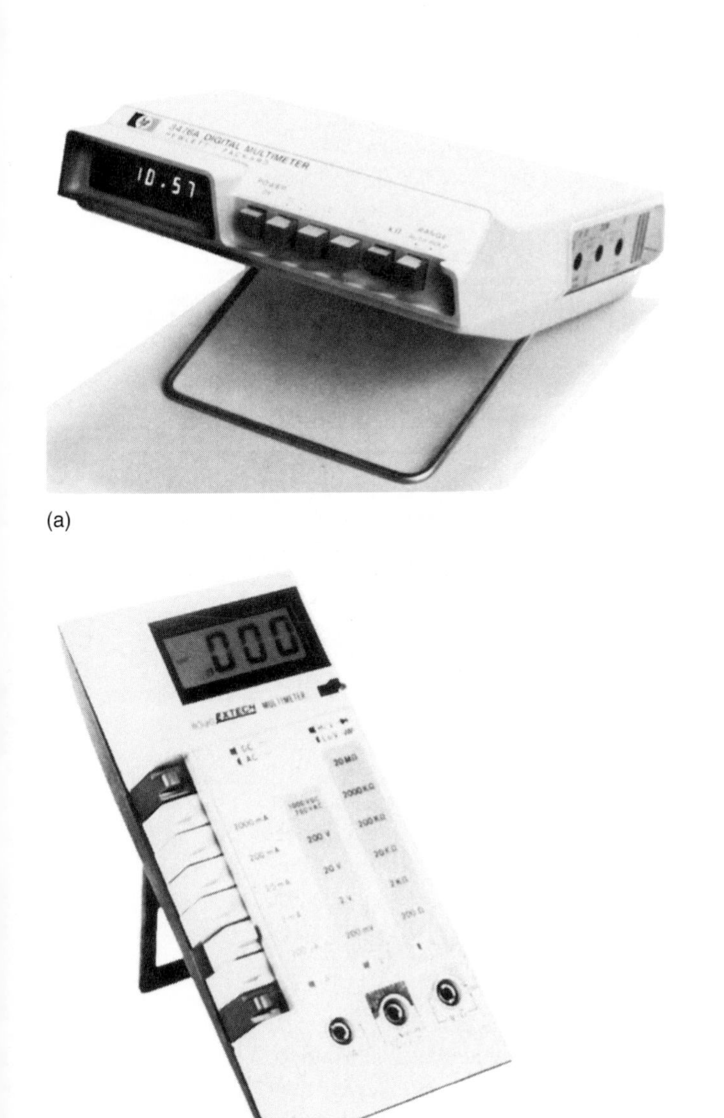

Figure 2–29. Digital multimeters (DMMs). [(a) Courtesy of Hewlett-Packard Corp.; (b) courtesy of Extech International Corp.; (c) courtesy of Triplett Corporation.]

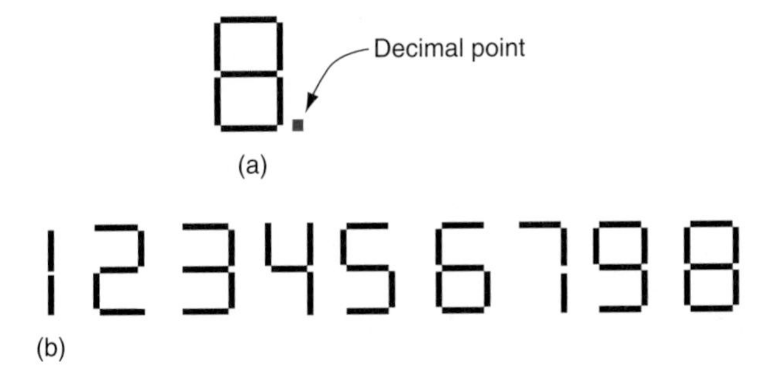

Figure 2–30. Seven-segment display used with digital multimeters (DMMs).

(a)

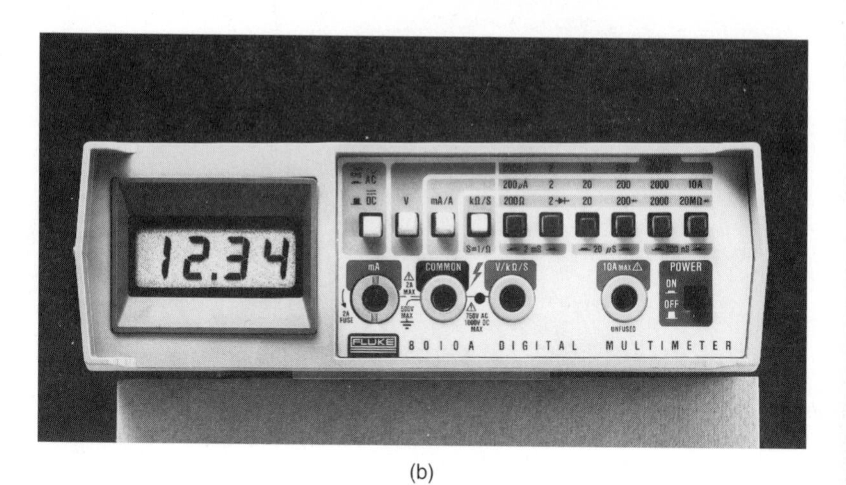

(b)

Figure 2–31. 3½-digit digital multimeter: (a) sample reading; (b) digital multimeter used for electrical measurements—resistance, voltage, and current. [(b) Courtesy of John Fluke Mfg. Co., Inc.]

REVIEW

1. Why are symbols used to represent electrical components?
2. Draw the symbols for the components listed here.
   a. Resistor
   b. Ground connection
   c. SPST switch
   d. Potentiometer
   e. Lamp
   f. Fuse
3. Draw the symbols for (a) conductors crossing and (b) conductors connected.
4. Draw the symbol for a battery placing the positive (+) sign and the negative (–) sign at the proper sides of the symbol.
5. Identify each of the meter symbols illustrated here.

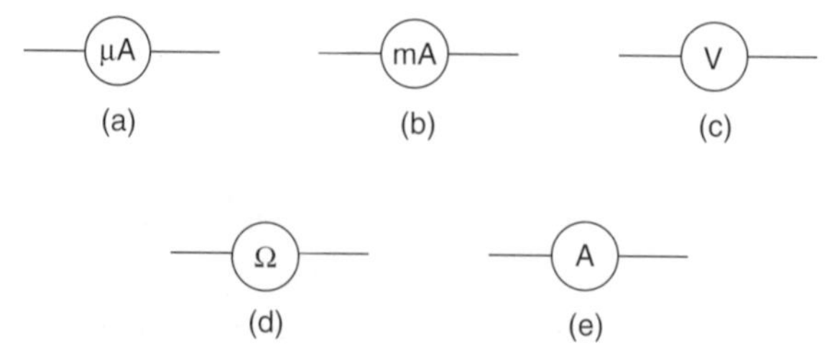

6. Draw a diagram of a circuit that has one battery, two resistors, a single-throw switch, and a current meter connected to form one path.

7. Why is scientific notation used?
8. Why is it necessary to convert large or small electrical units by using prefixes?
9. On which part of the ohms scale of a meter are the most accurate resistance measurements made?
10. What is the $\times 1000$ range of the ohmmeter?
11. What is meant by *zeroing* the meter?
12. If the range of the ohmmeter is adjusted to $\times 100{,}000$ and the needle points to 10.6 on the ohms scale, what is the value of the resistance being measured?
13. What values are shown by the first and second color bands of a color-coded resistor?
14. What value is shown by the third color band of a color-coded resistor?
15. What value is shown by the fourth color band of a color-coded resistor?
16. What unit of electrical measurement is used for resistance?
17. If a fourth color band is not indicated on a resistor, what is its tolerance value?
18. When measuring an unknown voltage, what meter range should be used?
19. What determines how the meter is connected to measure the voltage drop across a resistor?
20. When is it necessary to multiply or divide a scale reading on a VOM?
21. How is the proper voltage range of a VOM selected?
22. Could 4 V be measured on the 3-V range of a VOM? Why?
23. How is the VOM properly connected into a circuit to measure current flow?
24. Why is proper polarity important when measuring direct current?
25. What is the proper procedure to be followed when attempting to measure an unknown current?

STUDENT ACTIVITIES

• *Lab Meter Use (see Figs. 2–21 and 2–22)*

1. Obtain a meter from the lab.
2. Study the meter and try to answer correctly each of the following questions.
   a. What company manufactured the meter?
   b. What is the model number of the meter?
   c. The meter will measure up to _____ amperes of direct current.
   d. Alternating current (ac) is read on the _____ colored scales.
   e. The "Ohms Adjust" control must be checked each time the resistance range is changed. (True or False)
   f. To measure a direct current greater than 1 A, the range switch is placed in the _____ position.
   g. For measuring current in a circuit, the meter should be connected in series or parallel?
   h. For measuring voltage, the meter should be connected in series or parallel?
   i. To measure 18 mA of current, the range switch should be placed in the _____ position.
   j. To measure 10 μA of current, the range switch should be placed in the _____ range.
   k. To measure resistance, the red test lead must be placed in the jack marked _____ and the black test lead in the jack marked _____.
   l. The most accurate resistance reading is on the left side of the meter scale. (True or False) (Omit for digital meter.)

m. Polarity is important when measuring dc voltage. (True or False)

n. Polarity is *not* important when measuring resistance. (True or False)

o. Up to _____ volts can be measured with the meter.

p. For measuring a resistor valued at 10 Ω, the _____ range should be used.

q. Polarities must be observed when measuring direct current. (True or False)

r. It is correct to measure the resistance of a circuit with voltage applied. (True or False)

s. When measuring an unknown value of voltage, one should start at the highest scale and work down to the correct scale. (True or False)

t. Meters should be handled with care. (True or False)

• Electrical Symbols

1. Memorize the electrical symbols in App. D.
2. The teacher can give a test on these after they have been memorized or they can be learned as they are used in circuits later.

• Component Identification

Identify the components shown in Fig. 2–32. List the name of the component, its symbol, and its use.

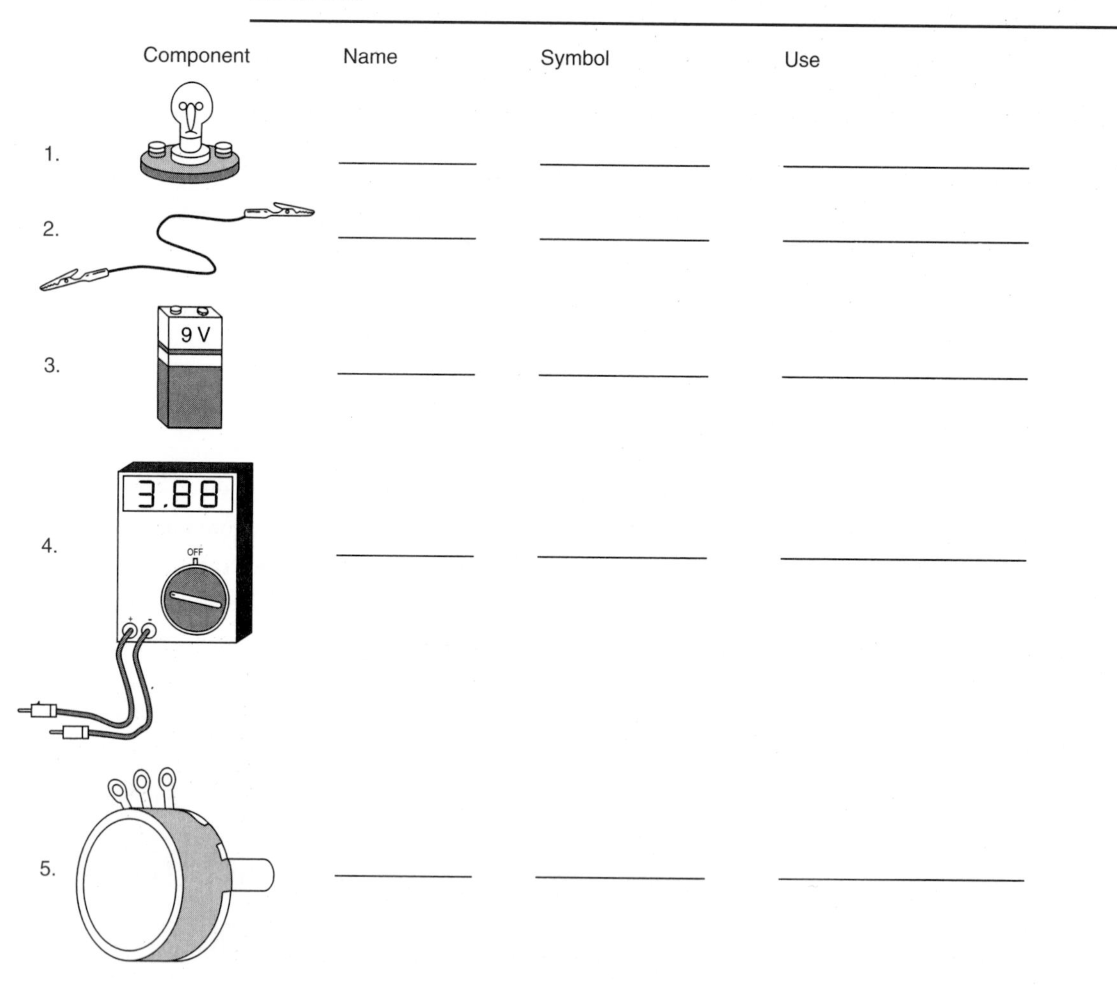

Figure 2–32. Component identification.

• Electrical Units

Complete Table 2–6 by entering the abbreviations, units of measure, and symbols for each of the electrical units.

TABLE 2–6 Electrical Units of Measurement

Unit	*Abbreviation*	*Unit of Measurement*	*Symbol*
Voltage			
Current			
Resistance			
Power			

• Components, Equipment, and Symbols

Complete Activity 2–1, page 9 in the manual.

ANALYSIS

1. Draw the symbols for these components:
   Fixed resistor; Potentiometer; Open SPST switch
2. Draw the symbols for these conductors:
   Conductors crossing; Conductors connected
3. Draw the symbol for a battery, placing the positive (+) sign and the negative (–) sign at the proper sides of the symbol.
4. Identify each of the meter symbols illustrated in Fig. 2–33.

• Resistor Color Code

Complete Activity 2–2, page 13 in the manual.

ANALYSIS

1. What are two systems of resistor color coding?
2. What is represented by the first and second colors of a color-banded resistor?
3. What is indicated by the third color of a color-banded resistor?
4. What unit of electrical measurement is used to indicate resistance?
5. What is indicated by the fourth and fifth colors of a five-color banded resistor?
6. If no fourth color is indicated on the body of a resistor, what is its tolerance value?

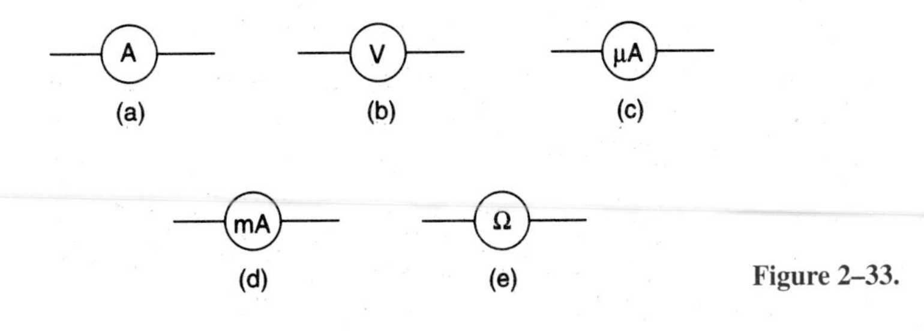

Figure 2–33.

Resistance Measurement Problems

Refer to the figure below and fill in the blanks with the values that correspond to the pointer locations.

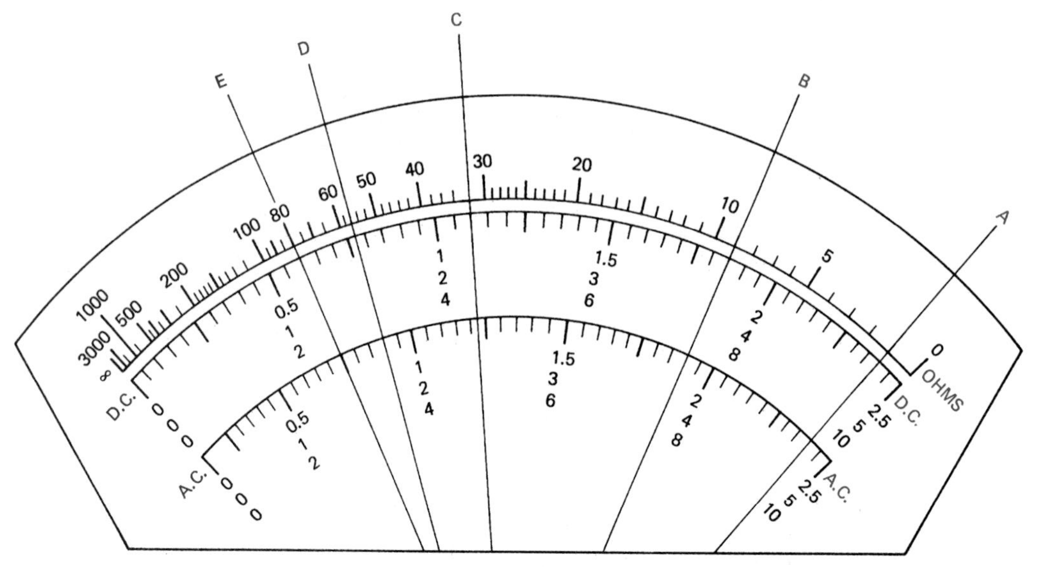

VOM (multimeter) scale.

Ohms Range	A	B	C	D	E
$R \times 1$	(1) ____	(6) ____	(11) ____	(16) ____	(21) ____
$R \times 10$	(2) ____	(7) ____	(12) ____	(17) ____	(22) ____
$R \times 100$	(3) ____	(8) ____	(13) ____	(18) ____	(23) ____
$R \times 1{,}000$	(4) ____	(9) ____	(14) ____	(19) ____	(24) ____
$R \times 10{,}000$	(5) ____	(10) ____	(15) ____	(20) ____	(25) ____

Measuring Resistance

Complete Activity 2–3, page 17 in the manual.

ANALYSIS

1. Why is the ohms scale of an analog meter considered to be nonlinear?
2. Where on the ohms scale of an analog meter are the most accurate measurements found?
3. What is meant by the $\times 1000$ range on the ohmmeter?
4. What is meant by *zeroing* the ohmmeter?
5. Why is it necessary to zero the ohmmeter?
6. If the range of an analog ohmmeter is set to $\times 100{,}000$ and the needle points to 0.6 on the ohms scale, what is the value of the resistance being measured?

Voltage Measurement Problems

Determine the voltage values using the following figure.

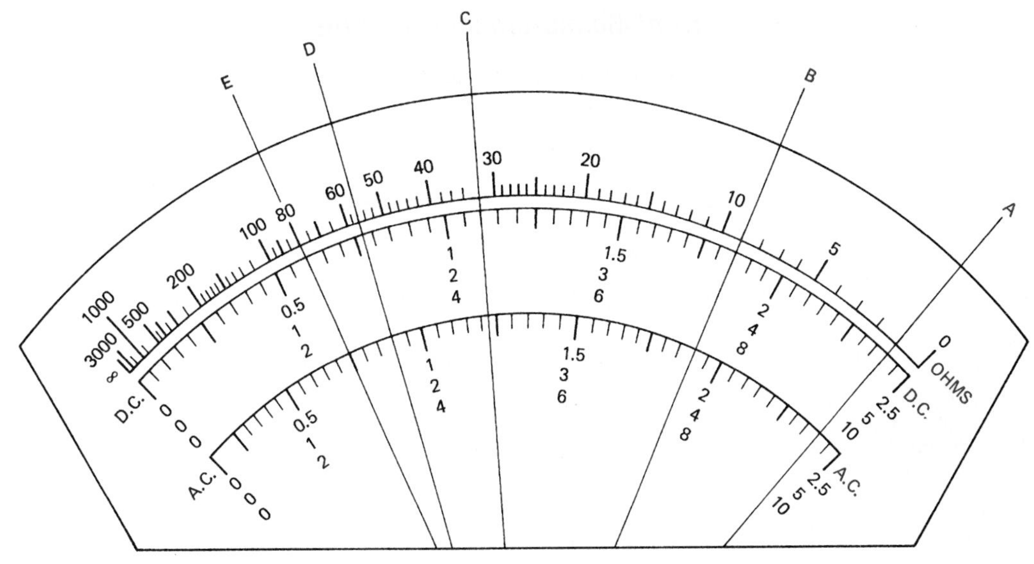

VOM scale.

Dc Volts Range	A	B	C	D	E
2.5 V dc	(26) ____	(32) ____	(38) ____	(44) ____	(50) ____
10 V dc	(27) ____	(33) ____	(39) ____	(45) ____	(51) ____
50 V dc	(28) ____	(34) ____	(40) ____	(46) ____	(52) ____
250 V dc	(29) ____	(35) ____	(41) ____	(47) ____	(53) ____
500 V dc	(30) ____	(36) ____	(42) ____	(48) ____	(54) ____
1000 V dc	(31) ____	(37) ____	(43) ____	(49) ____	(55) ____

Measuring Voltage

Complete Activity 2–4, page 21 in the manual.

ANALYSIS

1. What is *voltage?*
2. If you were measuring an unknown voltage, what meter range would you choose? Why?
3. What determines the polarity of a voltage drop developed by a resistor through which electrons are moving?
4. What causes the voltage drop developed across any component through which electrons are moving?
5. When is it necessary to use a multiplier or divisor with a scale on an analog meter?
6. How is the proper range of a multimeter selected?
7. Could 4 V be measured on the 3-V range of a multimeter? Why?
8. Indicate the proper range of the meter shown in Fig. 2–31(b) for measuring the following dc voltages:

   10 V = _____ range   6 V = _____ range   65 V = _____ range

   2.8 V = _____ range   100 V = _____ range   40 V = _____ range

6. 45,000 ohms = __________ megohms

7. 0.85 megohm = __________ ohms

8. 6500 watts = __________ kilowatts

9. 68,000 volts = __________ kilovolts

10. 9200 watts = __________ megawatts

Scientific Notation (See Table 2–5 or complete Activity 2–7, page 35 in the manual.)

Write the following numbers as powers of 10 on a sheet of paper.

1. 0.00001

2. 0.00000001

3. 10,000,000

4. 1000

5. 10

6. 0.01

7. 10,000

8. 0.0001

9. 1.0

10. 1,000,000

Write the following numbers in scientific notation (as a number between 1 and 10 times a power of 10).

11. 0.00128

12. 1520

13. 0.000632

14. 0.0030

15. 28.2

16. 7,300,000,000

17. 52.30

18. 8,800,000

19. 0.051

20. 0.000006

ANALYSIS

1. Discuss the procedure used to convert a quantity to scientific notation.

2. Discuss the procedure used to convert electrical units to a smaller or larger unit.

Metric Conversions (See App. G or complete Activity 2–8, page 37 in the manual.)

Answer each of the following metric conversion problems on a sheet of paper.

1. 1 meter = __________ centimeters

2. 1 centimeter = __________ millimeters

3. 5000 grams = __________ kilograms

4. 2 liters = __________ milliliters

5. 1 meter = __________ inches

6. 1 centimeter = __________ inches

7. 1 mile = __________ kilometers

8. 1 gram = __________ ounces

9. 30 ounces = __________ grams

10. 1 kilogram = __________ pounds

11. 3 liters = __________ quarts

12. 1 gallon = __________ liters

13. 10 cups = __________ liters

14. 50 miles = __________ kilometers

15. 2 cubic centimeters (cm^3) = __________ cubic inches ($in.^3$)

16. 4 cubic feet (ft^3) = __________ cubic meters (m^3)

ANALYSIS

Discuss the use of conversion tables to convert English units to metric units and metric units to English units.

Resistor Color Code

1. Using Fig. 2–13 for four-color band resistors, write the resistance value and tolerance of each example here on a sheet of paper, as _____ ohms; _____ tolerance.

a. (1) Violet
(2) Green
(3) Orange
(4) Gold

b. (1) Yellow
(2) Violet
(3) Green

c. (1) Green
(2) Blue
(3) Red

d. (1) Red
(2) Red
(3) Blue

e. (1) Black
(2) Brown
(3) Brown
(4) Gold

f. (1) Gray
(2) Red
(3) Black
(4) Silver

g. (1) White
(2) Brown
(3) Orange
(4) Gold

h. (1) Yellow
(2) Violet
(3) Orange

i. (1) Brown
(2) Green
(3) Orange

j. (1) Blue
(2) Gray
(3) Black
(4) Gold

k. (1) Green
(2) Blue
(3) Black
(4) Gold

l. (1) Brown
(2) Black
(3) Black

m. (1) Orange
(2) White
(3) Black
(4) Gold

n. (1) Red
(2) Violet
(3) Gold
(4) Gold

o. (1) Brown
(2) Brown
(3) Black
(4) Gold

• *Circuit Construction*

Construct the simple electrical circuit shown in Fig. 2–35 using the materials listed here.

1 1.5-V lamp and socket*
1 sheet metal strip ½ in. wide × 5 in. long
1 sheet metal strip ½ in. wide × 2 in. long
1 battery holder (can be made with sheet metal)
2 small nails
1 piece of wood ½ in. × 3 in. × 5 in.
3 pieces of small wire
1 1.5-V battery

• *Printed Circuit Construction*

Complete Activity 2–9, page 39 in the manual.

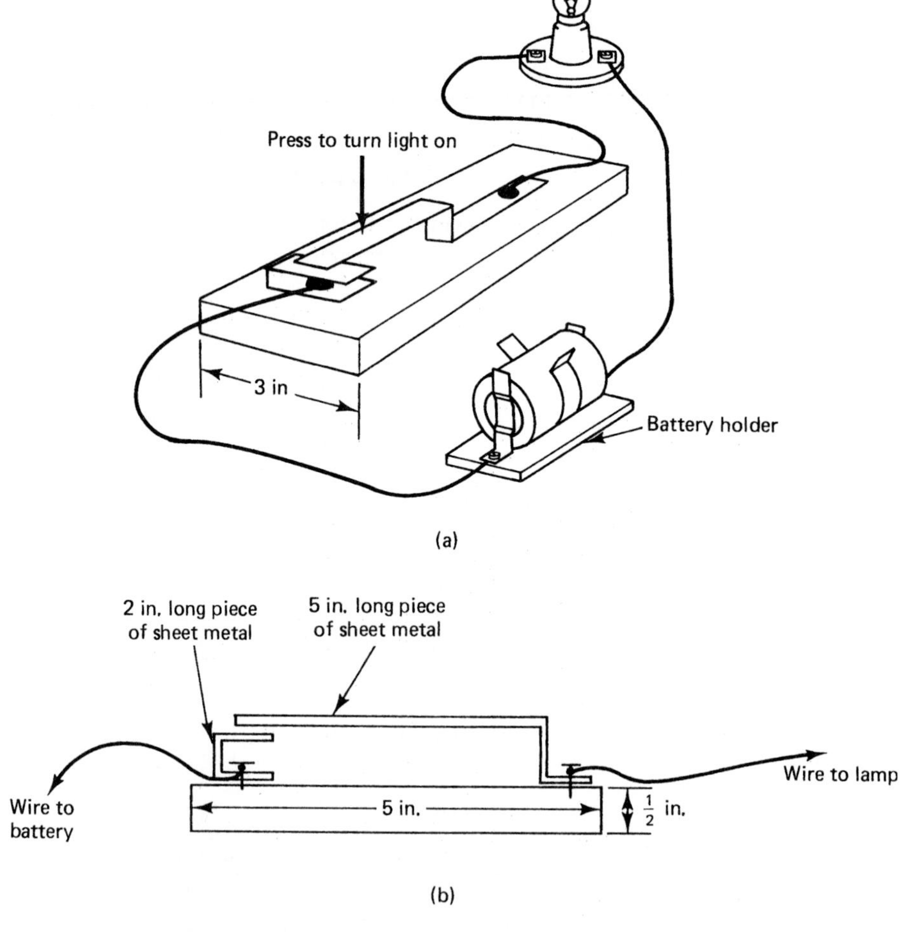

Figure 2–35. Electrical circuit to build: (a) pictorial view; (b) side view.

*Wires can be carefully soldered to the bulb if a socket is not available.

ANALYSIS

1. What are some advantages of printed circuits?
2. Which method of printed circuit board (PCB) construction would be most desirable for mass production? Why?
3. Discuss briefly each of the four printed circuit construction methods that were outlined in this activity.

Self-Examination

1. Complete the Self-Examination for chap. 2, page 41 in the manual.
2. Check your answers on page 408 in the manual.
3. Study any questions that were answered incorrectly.

Meter Exercises

1. Complete the Meter Exercises, page 43 in the manual.
2. Check your answers on page 409 in the manual.

Chapter 2 Examination

Complete the Chap. 2 Examination on page 45 in the manual.

CHAPTER

3

ELECTRICAL CIRCUITS

OUTLINE

3.1 Use of Calculators and Computers
3.2 Ohm's Law
3.3 Troubleshooting
3.4 Series Electrical Circuits
3.5 Parallel Electrical Circuits
3.6 Combination Electrical Circuits
3.7 Kirchhoff's Laws
3.8 More Examples of Series Circuits
3.9 More Examples of Parallel Circuits
3.10 More Examples of Combination Circuits
3.11 Power in Dc Electrical Circuits
3.12 Maximum Power Transfer in Circuits
3.13 Voltage-Divider Circuits
3.14 Voltage-Divider Design
3.15 Voltage-Division Equation
3.16 Negative Voltage Derived from a Voltage-Divider Circuit
3.17 Voltage Division with a Potentiometer
3.18 Problem-Solving Methods
3.19 Kirchhoff's Voltage Method
3.20 Superposition Method
3.21 Equivalent Circuits
3.22 Thevinin Equivalent Circuit Method
3.23 Norton Equivalent Circuit Method
3.24 Bridge-Circuit Simplification

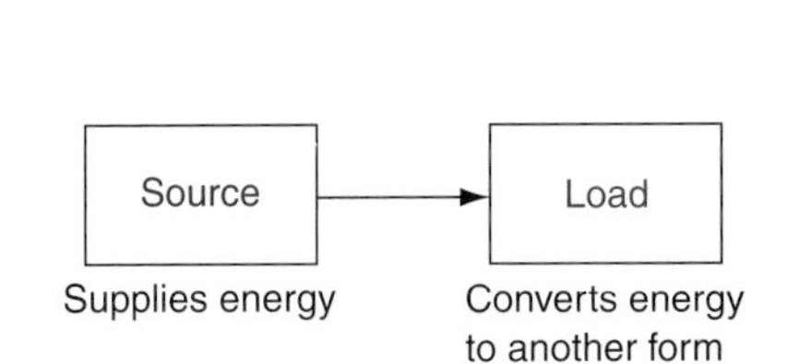

Circuits have a source and a load.

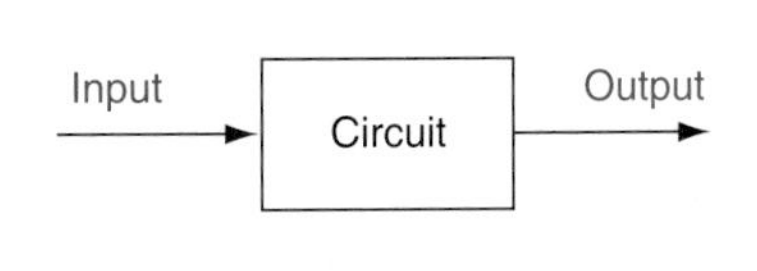

Circuits have an input and an output.

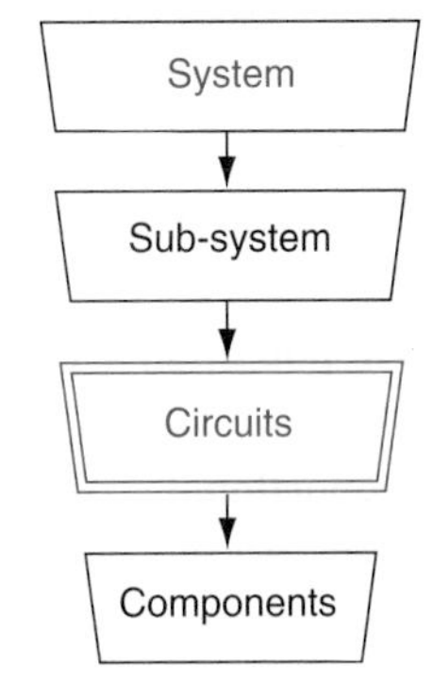

Circuits are a part of any electrical system.

Electrical circuits are basic to all electrical systems.

- **Binary number system:** A system using only two digits (0 and 1) to represent all numbers. Used in computers because an electrical circuit has two states: on and off.
- **Bit:** The smallest unit of information in a computer; an abbreviation for *b*inary dig*it*.
- **Byte:** A string of bits, usually eight, that represent a number or character.
- **CD-ROM:** Short for compact disc read-only memory.
- **Central processing unit:** The part of a computer that executes instructions.
- **Chip:** An electronic circuit on a tiny piece of semiconductor material, usually silicon.
- **Clone:** A computer that can run the same software as namebrand home computers.
- **Floppy disk:** A flexible disk covered in cardboard or plastic; it is used to store data or instructions. Most programs are sold on floppy disks.
- **Gigabyte:** One billion bytes.
- **Hard disk:** A sealed disk used to store data or instructions. It costs more than a floppy disk but stores more information and works faster.
- **Hardware:** The physical parts of a computer system, such as the keyboard.
- **Keyboard:** Typewriter-like device for putting data into a computer or giving it commands.
- **Kilobyte:** 1024 bytes.
- **Laser printer:** A printer that uses a laser to print.
- **Magneto-optical (MO) disk:** A disk, usually removable, that is read and written to by both laser light and magnetic technology. Though slower than magnetic disks, they are considered more durable.
- **Megabyte:** One million bytes.
- **Memory:** The part of the computer where data or instructions are stored in binary form.
- **Microprocessor:** A single chip that has all the functions of a computer's central processing unit.
- **Modem:** A device that moves information between computers, usually over telephone lines.
- **Monitor:** A television-like screen that displays data.
- **Mouse:** A handheld device that moves the cursor on a computer screen.
- **Operating system:** A program that controls all other programs run on a computer system.
- **Peripherals:** External devices used with a computer, such as monitors and modems.
- **Personal computer:** A desktop or portable computer intended for use by an individual.
- **Printer:** A device that prints numbers, letters, or graphic images from a computer.
- **Program:** Instructions that tell a computer how to do something. For instance, a word-processing program tells a computer how to work with text.
- **ROM:** Read-only memory. Permanent data or instructions that users cannot alter.
- **RAM:** Random access memory. Temporary memory that users can retrieve and alter.
- **Software:** Computer programs.

Figure 3–2. Computer terms and phrases.

letter *V*, current with the letter *I*, and resistance with the letter *R*. The mathematical relationship of the three electrical quantities is shown in the following formulas. These should be memorized. The Ohm's law circle in Fig. 3–4 is helpful to remember the formulas.

$$V = I \times R$$

$$I = \frac{V}{R}$$

$$R = \frac{V}{I}$$

Voltage (*V*) is measured in *volts*. Current (*I*) is measured in *amperes*. Resistance (*R*) is measured in *ohms*. If two electrical values are known, the third value can be calculated by using one of the formulas. Look at Fig. 3–5. Using the Ohm's law current formula, $I = V/R$, the calculated value of I in the circuit is

$$I = \frac{V}{R} = \frac{10\ \text{V}}{10\ \Omega} = 1\ \text{A}$$

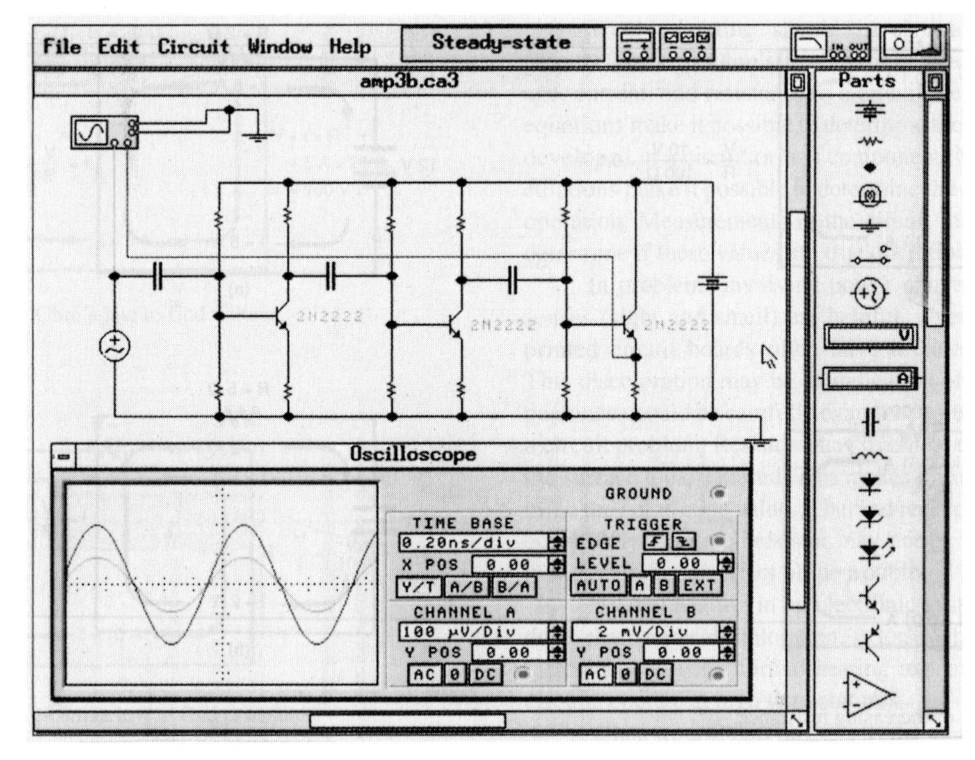

Figure 3–3. Computer screen for a circuit simulation software.

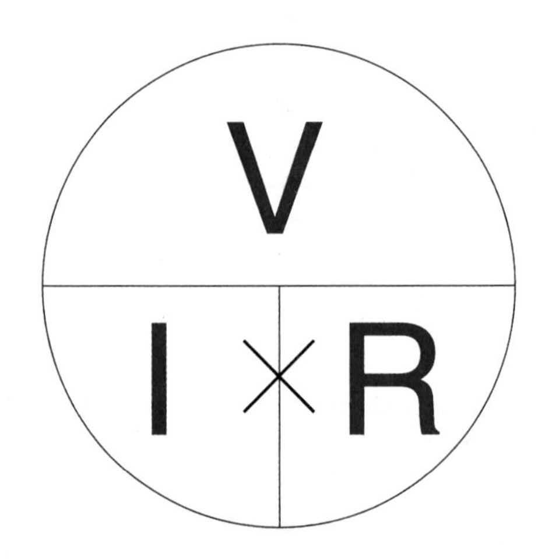

Figure 3–4. Ohm's law circle: *V*—voltage; *I*—current; *R*—resistance. To use the circle, cover the value you want to find and read the other values as they appear in the formula: $V = I \times R$; $I = V/R$; $R = V/I$.

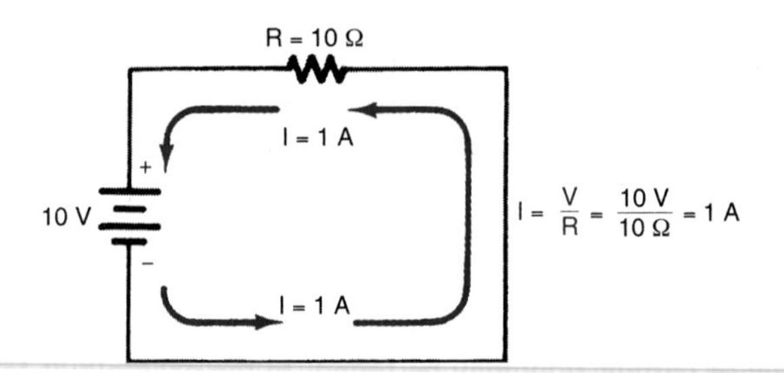

Figure 3–5. Ohm's law example.

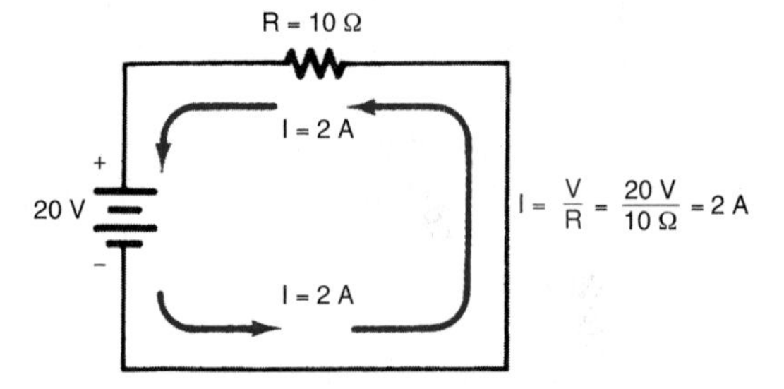

Figure 3–6. Effect of doubling the voltage.

If the voltage in the circuit is doubled as shown in Fig. 3–6, the calculated current using the Ohm's law current formula ($I = V/R$) is

$$I = \frac{V}{R} = \frac{20\text{ V}}{10\ \Omega} = 2\text{ A}$$

From this example, notice that as voltage is increased, current also increases if resistance remains the same. Also, if voltage is doubled, current is doubled. If voltage is 10 times larger, the current becomes 10 times larger.

Now look at Fig. 3–7(a). The calculated current flow in the circuit is

$$I = \frac{V}{R} = \frac{10\text{ V}}{100\ \Omega} = 0.1\text{ A}$$

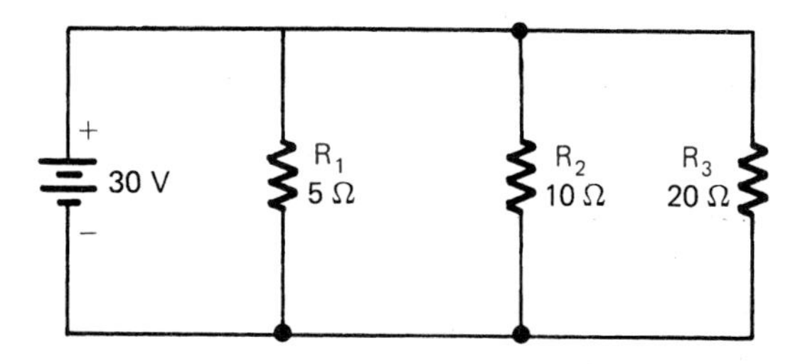

Figure 3–37. Finding power values in a parallel circuit.

3. Power converted by resistor R_3:

$$P_3 = \frac{V^2}{R_3} = \frac{30^2}{20} = \frac{900}{20} = 45 \text{ W}$$

4. Total power converted by the circuit:

$$\begin{aligned} P_T &= P_1 + P_2 + P_3 \\ &= 180 \text{ W} + 90 \text{ W} + 45 \text{ W} = 315 \text{ W} \end{aligned}$$

Refer also to the calculator procedures for electrical power in App. G (Fig. G–16).

The *watt* is the basic unit of electrical power. To determine an actual quantity of electrical energy, a factor that indicates *how long* a power value continued must be used. Such a unit of electrical energy is called a *watt-second.* It is the product of watts (W) and time (in seconds). The watt-second is a very small quantity of energy. It is more common to measure electrical energy in *kilowatt-hours* (kWh). It is the kWh quantity of electrical energy that is used to determine the amount of electric utility bills. A kilowatt-hour is 1000 W in 1 h of time or 3,600,000 W per second.

As an example, if an electrical heater operates on 120 V and has a resistance of 20 Ω, what is the cost to use the heater for 200 h at a cost of 5 cents per kWh?

1. $$P = \frac{V^2}{R} = \frac{120^2}{20\ \Omega} = \frac{14{,}400}{20\ \Omega}$$

$$= 720 \text{ W} = 0.72 \text{ kW}$$

2. There are 1000 W in a kilowatt (1000 W = 1 kW).
3. Multiply the kW that the heater has used by the hours of use:

$$\begin{aligned} \text{kW} \times 200 \text{ h} &= \text{kilowatt-hours (kWh)} \\ 0.72 \times 200 \text{ h} &= 144 \text{ kWh} \end{aligned}$$

4. Multiply the kWh by the cost:

$$\text{kWh} \times \text{cost} = 144 \text{ kWh} \times 0.05 = \$7.20$$

Energy, work, and power were discussed in Chap. 1. Electrical circuit operation depends on the relationship of these quantities. A basic law of physics states that *energy* (the capacity to do work) can be neither created nor destroyed. It can, however, be converted from one form to another. The process of converting energy from one form to another is called *work.* The rate at which work is accomplished, or at which energy is converted, is called *power.* Power may be expressed in an energy unit called the *watt* (W) or in the power unit called *horsepower* (hp). For electrical circuits, the watt is used exclusively. However, many electric motors are rated by horsepower and must be converted from horsepower to watts, which is accomplished by the following:

$$1 \text{ hp} = 746 \text{ W}$$

The energy-power-work relationship, given in units of watts for power and in joules for work, is as follows:

$$P = \frac{W}{t} \text{ or power} = \frac{\text{energy}}{\text{time}}$$

where: P = power (W)
W = work (energy converted) (J)
t = time (s)

As shown by this equation, 1 W of power is equal to 1 J of energy converted in 1 s of time. If the same amount of work is accomplished in less time, then more power is produced. If time is increased for a given amount of work, then the result is less power produced.

In electrical circuits, power is developed when current flows through a resistance. The source of energy in a circuit provides the energy to do the work of setting the electrons in motion and producing current flow. When electrons pass through a resistance, the moving electrons collide with the atoms in the resistor, creating friction. In the process, energy is transferred from the electrons to the resistor. The resistor must give off, or *dissipate,* a corresponding amount of energy. Some energy is dissipated in the form of heat. The rate at which dissipation occurs determines the power developed for the circuit.

Voltage is equal to work energy (in joules) divided by the charge in coulombs. This equation is expressed as

$$V = \frac{W}{Q} \text{ or voltage} = \frac{\text{work (energy)}}{\text{charge}}$$

Current is the charge that flows, divided by time. Current is expressed as

$$I = \frac{Q}{t} \text{ or current} = \frac{\text{charge}}{\text{time}}$$

Rearranging each of these equations shows the following:

$$W = VQ \text{ or work} = \text{voltage} \times \text{charge}$$

and

$$t = \frac{Q}{I} \text{ or time } = \frac{\text{charge}}{\text{current}}$$

Power is calculated as

$$P = \frac{W}{t}$$

Substituting from the rearranged equations shows the following:

$$P = (VQ)\left(\frac{I}{Q}\right) \text{ or power } = (\text{work} \times \text{charge}) \times \left(\frac{\text{current}}{\text{charge}}\right)$$

Canceling terms gives the following power formula:

$$P = VI \text{ or power } = \text{voltage} \times \text{current}$$

In some cases, it is desirable to determine the amount of power developed by a resistance when only its value and the amount of current flow are known. In this case, the quantity V must be eliminated from the equation and an equivalent value substituted. To determine power when the resistance and current are known, use the power formula

$$P = VI$$

Because $V = IR$, we may substitute:

$$P = (I \times R)I$$

Simplifying yields

$$P = (I \times I)R$$

Because the value $I \times I$ is stated as I^2, this equation is written as

$$P = I^2R \text{ or power } = (\text{current})^2\ (\text{resistance})$$

SAMPLE PROBLEM: WORK

Work is done when a force (F) is moved a distance (d), or:

$W = F \times d$, where
W = work in joules
F = force in newtons
d = distance the force moves in meters

Given: An object with a mass of 22 kg is moved 55 meters.

Find: The amount of work done when the object is moved.

Solution: The force of gravity acting on the object is equal to 9.8 (a constant that applies to objects on earth) multiplied by the mass of the object, or:

$$F = 9.8 \times 22 \text{ kg} = 215.6 \text{ newtons}$$
$$W = F \times d$$
$$= 215.6 \times 55$$
$$W = 11{,}858 \text{ joules}$$

SAMPLE PROBLEM: POWER

Power is the time rate of doing work, which is expressed as:

$$P = \frac{W}{t}$$

where

P = power in watts
W = work done in joules
t = time taken to do the work in seconds

Given: An electric motor is used to move an object along a conveyor line. The object has a mass of 150 kg and is moved 28 meters in 8 seconds.

Find: The power developed by the motor in watts and horsepower units.

Solution:

$$\text{Force } (F) = 9.8 \times \text{mass}$$
$$= 9.8 \times 150 \text{ kg}$$
$$F = 1470 \text{ newtons}$$

$$\text{Work } (W) = F \times d$$
$$= 1470 \times 28 \text{ m}$$
$$W = 41{,}160 \text{ joules}$$

$$\text{Power } (P) = \frac{W}{t}$$
$$= \frac{41{,}160}{8}$$

$$P = 5145 \text{ watts}$$

$$\text{Horsepower} = \frac{P}{746}, \text{ because}$$

$$1 \text{ horsepower} = 746 \text{ W}$$

$$\text{hp} = 5145/746 = 6.9 \text{ hp}$$

Some simple electrical circuit examples have been discussed in this chapter. They become easy to understand after practice with each type of circuit. It is important to understand the characteristics of series, parallel, and combination circuits.

3.12 MAXIMUM POWER TRANSFER IN CIRCUITS

An important consideration in electrical circuits is called *maximum power transfer.* Maximum power is transferred

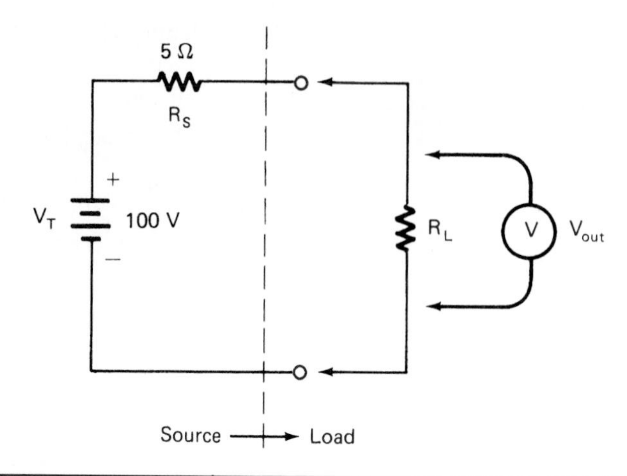

R_L	I_L	V_{out}	Power Output (W)
0	$\frac{100\text{ V}}{5\ \Omega} = 20\text{ A}$	20 A × 0 Ω = 0 V	20 A × 0 V = 0 W
2.5 Ω	$\frac{100\text{ V}}{7.5\ \Omega} = 13.3\text{ A}$	13.3 A × 2.5 Ω = 33.3 V	13.3 A × 33.3 V = 444 W
5 Ω	$\frac{100\text{ V}}{10\ \Omega} = 10\text{ A}$	10 A × 5 Ω = 50 V	10 A × 50 V = 500 W
7.5 Ω	$\frac{100\text{ V}}{12.5\ \Omega} = 8\text{ A}$	8 A × 12.5 Ω = 60 V	8 A × 60 V = 480 W
10 Ω	$\frac{100\text{ V}}{15\ \Omega} = 6.7\text{ A}$	6.7 A × 10 Ω = 67 V	6.7 A × 67 V = 444 V

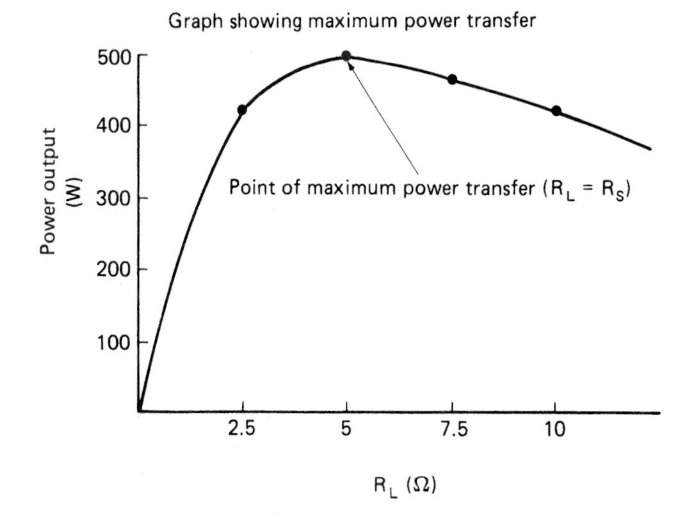

Figure 3–38. Problem that shows maximum power transfer.

from a voltage source to a load when the load resistance (R_L) is equal to the internal resistance of the source (R_S). The source resistance limits the amount of power that can be applied to a load. For example, as a flashlight battery gets older, its internal resistance increases. This increase in the internal resistance causes the battery to supply less power to the lamp load. Thus, the light output of the flashlight is reduced.

Figure 3–38 shows an example that illustrates maximum power transfer. The source is a 100-V battery with an internal resistance of 5 Ω. The values of I_L, V_{out}, and power output (P_{out}) are calculated as follows:

$$I_L = \frac{V_T}{R_S + R_L} \qquad V_{out} = I_L \times R_L \qquad P_{out} = I_L \times V_{out}$$

Notice the graph in Fig. 3–38 showing that maximum power is transferred from the source to the load when $R_L = R_S$. This is an important circuit design consideration for power sources, amplifier circuits, microphones, or practically any type of electronic circuit.

3.13 VOLTAGE-DIVIDER CIRCUITS

The simple series circuit of Fig. 3–39 is a voltage divider. Voltage division takes place due to voltage drops across the three resistors. Because each of the three resistors has the same value (1 kΩ), the voltage drop across each is 3 V. Thus, a single voltage source is used to derive three separate voltages.

Another method used to accomplish voltage division is the *tapped resistor.* This method relies on the use of a resistor which is wire wound and has a tap onto which a wire is attached. The wire is attached so that a certain amount of the total resistance of the device appears from the tap to the outer terminals. For example, if the tap is in the center of a 100-Ω wire-wound resistor, the resistance from the tap to either outer terminal is 50 Ω. Tapped resistors often have two or more taps to obtain several combinations of fixed-value resistance. Figure 3–39(b) shows a tapped resistor used as a voltage divider. In the example, the voltage outputs are each 3 V, derived from a 6-V source.

A common method of voltage division is shown in Fig. 3–39(c). Potentiometers are used as voltage dividers in volume control circuits of radios and televisions. They may be used to vary voltage from zero to the value of the source voltage. In the example, the voltage output may be varied from 0 to 1.5 V. It is also possible to use a voltage-divider network and a potentiometer to obtain many variable voltage combinations, as discussed in the following section.

3.14 VOLTAGE-DIVIDER DESIGN

The design of a voltage-divider circuit is a good application of basic electrical theory. Refer to the circuit of Fig. 3–40. Resistors R_1, R_2, and R_3 form a voltage divider to provide the proper voltage to three known loads. The loads could be transistors of a 9-V portable radio, for example. The operating voltages and currents of the load are constant. The values of R_1, R_2, and R_3 are calculated to supply proper voltage to each of the loads. The value of current through R_1 is selected as 10 mA. This value is ordinarily 10% to 20% of the total current flow to the loads (10 mA + 30 mA + 60 mA = 100 mA and 10% of 100 mA = 10 mA).

To calculate the values of R_1, R_2, and R_3, the voltage across each resistor and the current through each resistor must be known. Start with R_1 at the bottom of the circuit. The current through R_1 is given (10 mA). The voltage across R_1 is 2 V since the ground is a zero-voltage reference and 2 V must be supplied to load 1. The value of R_1, as shown in the procedure of Fig. 3–40, must be 200 Ω (2 V ÷ 10 mA).

Resistor R_2 has a voltage of 3 V across it. Point A has a potential of +2 V and point B has a potential of +5 V for load 2. The *difference in potential* or voltage drop is therefore 5 V − 2 V = 3 V. The current through R_2 is 20 mA. A current of 10 mA flows up through R_1 and 10 mA flows to point A from load 1. These two currents (10 mA + 10 mA = 20 mA) combine and flow through R_2. The value of R_2 must be 150 Ω (3 V ÷ 20 mA).

Resistor R_3 has a voltage of 4 V across it (9 V − 5 V = 4 V). The current through R_3 is 50 mA, because 20 mA flows upward through R_2 and 30 mA flows from load 2 to point B (20 mA + 30 mA = 50 mA). The value of R_3 must be 80 Ω (4 V ÷ 50 mA).

With the calculated values of R_1, R_2, and R_3 used as a voltage-divider network, the proper values of 0.02 W, 0.06 W,

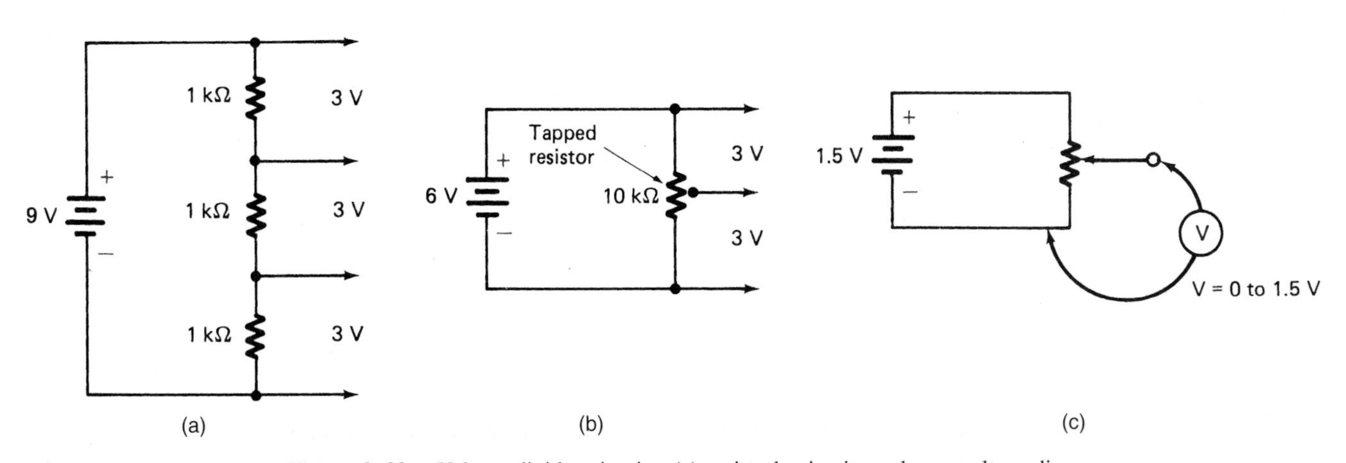

Figure 3–39. Voltage-divider circuits: (a) series dc circuit used as a voltage divider; (b) tapped resistor used as a voltage divider; (c) potentiometer used as a voltage divider.

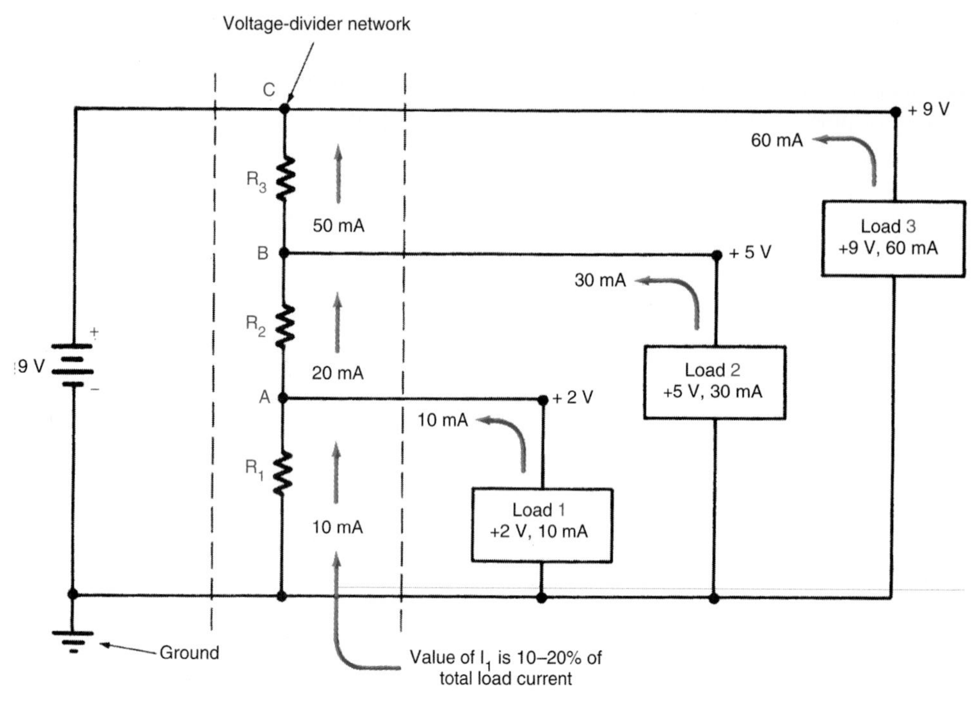

Procedure

$$R_1 = \frac{V_1}{I_1} = \frac{2\text{ V}}{10\text{ mA}} = 200\ \Omega$$

$$R_2 = \frac{V_2}{I_2} = \frac{3\text{ V}}{20\text{ mA}} = 150\ \Omega$$

$$R_3 = \frac{V_3}{I_3} = \frac{4\text{ V}}{50\text{ mA}} = 80\ \Omega$$

$$P_1 = V_1 \times I_1 = 2\text{ V} \times 10\text{ mA} = 20\text{ mW} = 0.02\text{ W}$$

$$P_2 = V_2 \times I_2 = 3\text{ V} \times 20\text{ mA} = 60\text{ mW} = 0.06\text{ W}$$

$$P_3 = V_3 \times I_3 = 4\text{ V} \times 50\text{ mA} = 200\text{ mW} = 0.2\text{ W}$$

Figure 3–40. Voltage-divider design.

and 0.2 W are calculated in Fig. 3–40. Often, a *safety factor* is used to ensure that power values are large enough. A safety factor is a multiplier used with the minimum power values. For example, if a safety factor of 2 is used, the minimum power values for the circuit would become $P_1 = 0.02\text{ W} \times 2 = 0.04\text{ W}$, $P_2 = 0.06\text{ W} \times 2 = 0.12\text{ W}$, and $P_3 = 0.2\text{ W} \times 2 = 0.4\text{ W}$.

3.15 VOLTAGE-DIVISION EQUATION

The voltage-division equation, often called the *voltage-divider rule,* is convenient to use with voltage-divider circuits when the current delivered from the voltage divider is negligible. The voltage-divider equation and a sample problem are shown in Fig. 3–41. This equation applies to series circuits. The voltage (V_x) across any resistor in a series circuit is equal to the ratio of that resistance (R_x) to total resistance (R_T) multiplied by the source voltage (V_T).

3.16 NEGATIVE VOLTAGE DERIVED FROM A VOLTAGE-DIVIDER CIRCUIT

Voltage-divider circuits are often used as power sources for other types of electronic circuits. In circuit design and analysis, reference is often made to *negative* voltage. The concept of a negative voltage is made clear in Fig. 3–42. Voltage is ordinarily measured with respect to a ground reference point. The circuit ground is shown at point A. Point E in Fig. 3–42 is connected to the negative side of the power source. Point A, where the ground reference is connected, has a higher potential than point E. Therefore, the voltage across points A and E is −50 V.

$$V_x = \frac{R_x}{R_T} \times V_T$$

V_x is voltage across a resistance
V_T is total voltage applied
R_x is the resistance where V_x is measured
R_T is the total resistance of the voltage-divider network

(a)

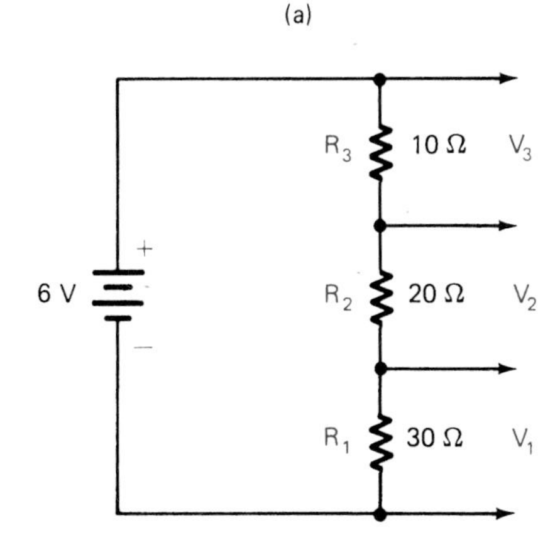

Procedure

$$V_1 = \frac{R_1}{R_T} \times V_T = \frac{30\ \Omega}{60\ \Omega} \times 6\ \text{V} = \frac{1}{2} \times 6\ \text{V} = 3\ \text{V}$$

$$V_2 = \frac{R_2}{R_T} \times V_T = \frac{20\ \Omega}{60\ \Omega} \times 6\ \text{V} = \frac{1}{3} \times 6\ \text{V} = 2\ \text{V}$$

$$V_3 = \frac{R_3}{R_T} \times V_T = \frac{10\ \Omega}{60\ \Omega} \times 6\ \text{V} = \frac{1}{6} \times 6\ \text{V} = 1\ \text{V}$$

(b)

Figure 3–41. Voltage division: (a) equation for problem solving; (b) sample problem.

3.17 VOLTAGE DIVISION WITH A POTENTIOMETER

A typical circuit design problem using a potentiometer is shown in Fig. 3–43. A given value of 10 kΩ is used as the potentiometer. The desired variable voltage from the potentiometer center terminal to ground is 5 to 10 V. The values of R_1 and R_3 are calculated to derive the desired variable voltage from the potentiometer.

The current flow in a voltage-divider network is established by the value of R_2 (10 kΩ) and the range of voltage variation (5 V to 10 V = 5 V variation). The current flow calculation in the circuit of Fig. 3–43 is shown in the procedure. Since $I = V/R$, the current through R_2 and the other parts of this series circuit is 0.5 mA. Once the current is found, values of R_1 and R_3 may be found as shown in the procedure. Figure 3–43 shows an easy method of determining voltage drops. A network of resistances in series can be thought of as a scale. In the example, the voltage at point A is +5 V and

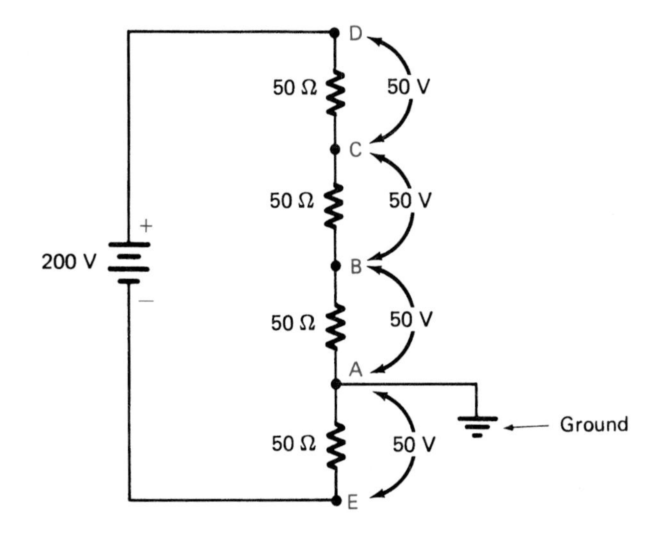

Voltage combinations (taken from ground reference)
V_{AB} = 50 V
V_{AC} = 100 V
V_{AD} = 150 V
V_{AE} = −50 V

Figure 3–42. Negative voltage derived from a voltage divider.

the voltage at point B is +10 V. The difference in potential is 5 V (10 V − 5 V = 5 V). This is similar to reading a scale. Refer also to the calculator procedures for voltage-divider circuits in App. G (Fig. G–17).

3.18 PROBLEM-SOLVING METHODS

Several different techniques may be applied to solve electrical circuit problems. Some of these methods include the following:

1. Kirchhoff's voltage law (KVL)—an algebraic procedure which may be used to find current flow in electrical circuits for single voltage sources and for circuits which have two or more voltage sources;
2. Superposition—a nonalgebraic procedure which may be used to find current flow in electrical circuits for single voltage sources and for circuits which have two or more voltage sources;
3. Equivalent circuits—simplified circuits, which include Thevinin and Norton equivalent circuit applications, may be used to solve circuit problems; and
4. Bridge circuit simplification—a procedure which may be used to make problem solving with bridge circuits easier to accomplish.

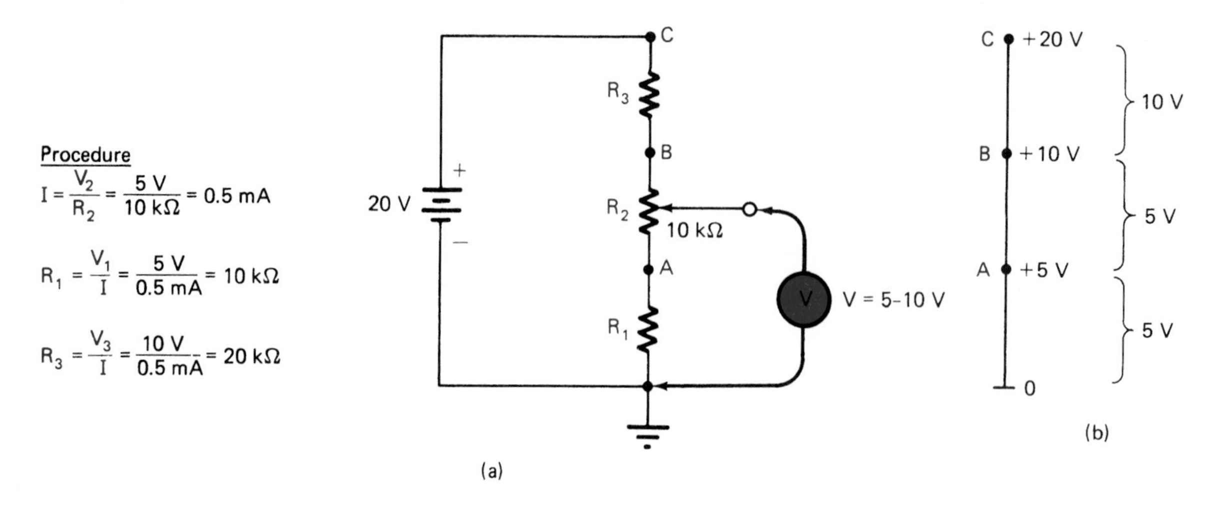

Figure 3–43. Voltage-divider design problem: (a) circuit; (b) voltage values.

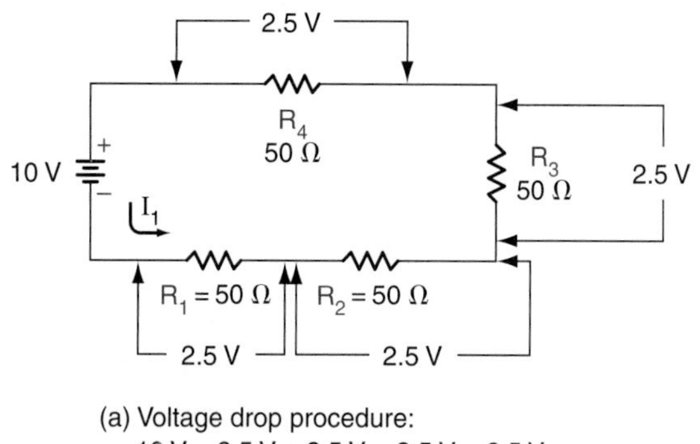

(a) Voltage drop procedure:

$$10\text{ V} = 2.5\text{ V} + 2.5\text{ V} + 2.5\text{ V} + 2.5\text{ V}$$

(b) Algebraic procedure:

$$10\text{ V} - 50\,I_1 - 50\,I_1 - 50\,I_1 - 50\,I_1 = 0$$

$$10\text{ V} - 200\,I_1 = 0$$

$$I_1 = \frac{10\text{ V}}{200} = 0.05\text{ A}$$

Figure 3–44. Kirchhoff's voltage law: (a) voltage drop procedure; (b) algebraic procedure.

3.19 KIRCHHOFF'S VOLTAGE METHOD

Kirchhoff's voltage law is illustrated in two different ways in Fig. 3–44. The voltage law may be stated in two ways: (1) The sum of voltage drops in a closed-looped circuit is equal to the source voltage; and (2) the algebraic sum of the voltage sources and voltage drops in a closed-loop path is equal to zero. The first method deals with the voltage drops in a closed-loop, or series, path. The sum of the voltage drops across the components is equal to the source voltage. In the example of Fig. 3–44, voltage drops are written as $R \times I$, such as $50\ \Omega \times I_1 = 50I_1$. The current in a loop is given an algebraic value (I_1). Remember that any voltage drop is equal to $I \times R$. The algebraic procedure of Kirchhoff's voltage law involves setting up a simple equation for a circuit loop. Values of current flow in circuits may be found by using this procedure. In a circuit that has only one voltage source, it is easier to use Ohm's law for series circuits to find current flow.

The advantage of the algebraic method for problem solving is that currents in multiple-source circuits may be easily calculated. Ohm's law cannot be used to find the current flow through each of the paths shown in the circuit of Fig. 3–45(a). The algebraic procedure derived from Kirchhoff's voltage law allows the calculation of current in a circuit with more than one voltage source [see Fig. 3–45(b)].

The method used in Fig. 3–45 may be used for multiple-voltage-source problems that have two current loops. The first step in this method is to assign directions of current flow (from − to +) in the circuit. When there are several sources, start with the largest voltage source. To avoid confusion, the current paths should be marked so that they appear different. The example uses a solid line for path 1 (I_1) and a dashed line for path 2 (I_2). When both paths pass through a resistance, the current is called $I_1 + I_2$.

An equation is developed for each of the circuit loops, based on Kirchhoff's voltage law. Each current is followed from the largest voltage source in the direction of the current arrow. Voltage sources must be given the proper sign when setting up an equation. When the current direction is to + through the source, a negative (−) sign is used in the equation. A positive (+) sign is used when the direction of the current arrow is from + to − through the source. The equation for each loop is developed using simple algebraic procedures, as shown. Practice in using this method makes it a convenient way to calculate current flow in a circuit with two current loops and two or more voltage sources. Another example is shown in Fig. 3–46.

3.20 SUPERPOSITION METHOD

An alternate method for finding current flow in circuits with two or more voltage sources is called the *superposition*

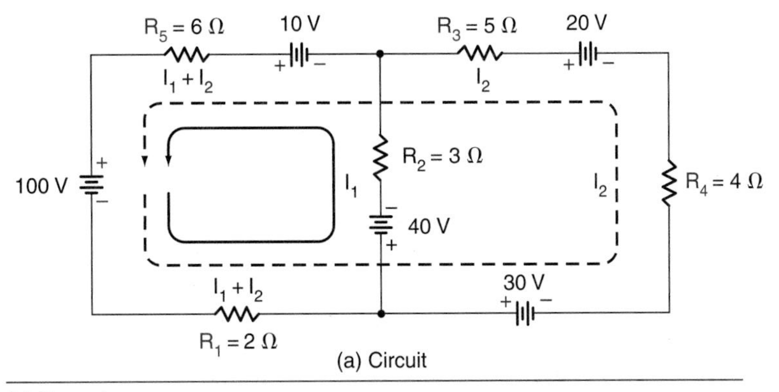

(b) Procedure

1. Assign a direction of current flow (– to +) through each path. Start with the largest voltage source.

2. Develop an equation for loop 1 (solid line):

$$100\text{ V} - 2(I_1 + I_2) + 40\text{ V} - 3\,I_1 - 10\text{ V} - 6(I_1 + I_2) = 0$$
$$100\text{ V} - 2\,I_1 + 2\,I_2 + 40\text{ V} - 3\,I_1 - 10\text{ V} - 6\,I_1 + 6\,I_2 = 0 \quad \text{(Multiply } I_1 + I_2 \text{ terms)}$$

Equation for loop 1 ⟶ $\boxed{130\text{ V} - 11\,I_1 - 8\,I_2 = 0}$ (Combine like terms)

3. Develop an equation for loop 2 (dashed line):

$$100\text{ V} - 2(I_1 + I_2) + 30\text{ V} - 4\,I_2 - 20\text{ V} - 5\,I_2 - 10\text{ V} - 6(I_1 + I_2) = 0$$
$$100\text{ V} - 2\,I_1 - 2\,I_2 + 30\text{ V} - 4\,I_2 - 20\text{ V} - 5\,I_2 - 10\text{ V} - 6\,I_1 - 6\,I_2 = 0$$

(Multiply $I_1 + I_2$ terms)

Equation for loop 2 ⟶ $\boxed{100\text{ V} - 8\,I_1 - 17\,I_2 = 0}$ (Combine like terms)

4. Place the two equations together and eliminate either I_1 or I_2 to solve for one unknown term:

$$(\times 8)\ 130\text{ V} - 11\,I_1 - 8\,I_2 = 0$$
$$(\times 11)\ 100\text{ V} - 8\,I_1 - 17\,I_2 = 0$$

Multiply top equation by 8 and bottom equation by 11 to eliminate I_1

5. Rewrite the equations after they have been multiplied:

$$1040\text{ V} - 88\,I_1 - 64\,I_2 = 0$$
$$(-)\ 1100\text{ V} \overset{(+)}{-} 88\,I_1 \overset{(+)}{-} 187\,I_2 = 0$$
$$-\ 60\text{ V} \qquad +123\,I_2 = 0$$

(These cancel)

6. Subtract bottom equation from top equation

7. Solve for I_2:

$$-60\text{ V} + 123\,I_2 = 0$$
$$+\,123\,I_2 = 60\text{ V}$$
$$I_2 = \frac{60\text{ V}}{123} = 0.488\text{ A}$$

8. Substitute value of I_2 in one of the original equations to solve for I_1:

$$100\text{ V} - 8\,I_1 - 17\,I_2 = 0$$
$$100\text{ V} - 8\,I_1 - 17(0.488) = 0$$
$$100\text{ V} - 8\,I_1 - 8.3\text{ V} = 0$$
$$91.7\text{ V} - 8\,I_1 = 0$$
$$-\,8\,I_1 = -\,91.7\text{ V}$$
$$I_1 = \frac{-\,91.7\text{ V}}{8}$$
$$I_1 = 11.46\text{ A}$$

9. Add I_1 and I_2 to get I_T:

$$I_1 + I_2 = I_T$$
$$11.46\text{ A} + 0.488\text{ A} = 11.948\text{ A}$$

Figure 3–45. Voltage law example.

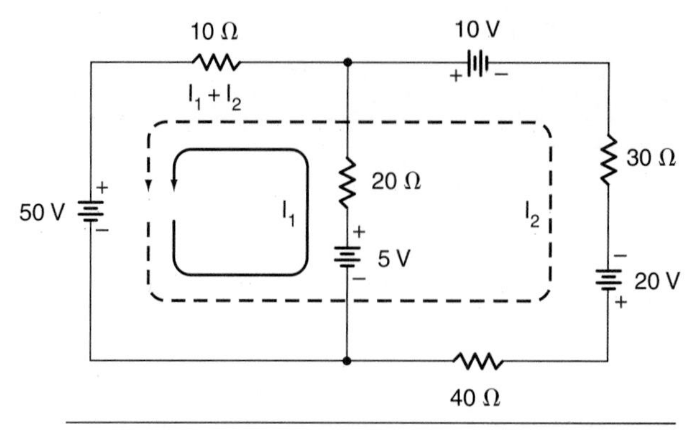

1. Equation for loop 1:

$$50\text{ V} - 5\text{ V} - 20\,I_1 - 10(I_1 + I_2) = 0$$
$$45\text{ V} - 20\,I_1 - 10\,I_1 - 10\,I_2 = 0$$
$$\boxed{45\text{ V} - 30\,I_1 - 10\,I_2 = 0}$$

2. Equation for loop 2:

$$50\text{ V} - 40\,I_2 + 20\text{ V} - 30\,I_2 - 10\text{ V} - 10(I_1 + I_2) = 0$$
$$60\text{ V} - 40\,I_2 - 30\,I_2 - 10\,I_1 - 10\,I_2 = 0$$
$$\boxed{60\text{ V} - 80\,I_2 - 10\,I_1 = 0}$$

3. Combine equations and cancel one unknown term:

$$45\text{ V} - 30\,I_1 - 10\,I_2 = 0$$
$$(\times 3)\ 60\text{ V} - 10\,I_1 - 80\,I_2 = 0 \longleftarrow \text{Multiply this equation by 3}$$

$$45\text{ V} - 30\,I_1 - 10\,I_2 = 0$$
$$(-)\ 180\text{ V} \overset{(+)}{-} 30\,I_1 \overset{(+)}{-} 240\,I_2 = 0 \longleftarrow \text{Subtract bottom equation from top equation}$$
$$-135\text{ V} \qquad +230\,I_2 = 0$$
$$230\,I_2 = 135\text{ V}$$
$$I_2 = \frac{135\text{ V}}{230} = 0.587\text{ A}$$

4. Substitute value of I_2 in one equation to solve for I_1:

$$60\text{ V} - 10\,I_1 - 80\,I_2 = 0$$
$$60\text{ V} - 10\,I_1 - 80(0.587) = 0$$
$$60\text{ V} - 10\,I_1 - 47\text{ V} = 0$$
$$13\text{ V} - 10\,I_1 = 0$$
$$-10\,I_1 = -13\text{ V}$$
$$I_1 = \frac{-13\text{ V}}{10} = 1.3\text{ A}$$

5. Add $I_1 + I_2$ to get I_T:

$$I_1 + I_2 = I_T$$
$$1.3\text{ A} + 0.587\text{ A} = 1.887\text{ A}$$

Figure 3–46. Kirchhoff's voltage law example problem.

method. This nonalgebraic method involves some rather lengthy, but simple, calculations. Multiple-voltage-source circuits may be broken down into as many individual circuits as there are voltage sources. For instance, a circuit with two voltage sources is reduced to two individual circuits. Each voltage source is considered separately, with other voltage sources *short-circuited* for making current calculations. In this way, the contribution of each voltage source to the current flow in the circuit may be determined. For a two-source circuit, one of the individual circuits is *superimposed* onto the other using the following procedure.

Figure 3–47 shows a circuit with two voltage sources. The procedure for finding current flow through each component in the circuit is as follows:

1. Short-circuit one power source and use basic Ohm's law procedure to find current flow through each component.
2. Record the amount of current and the direction of flow through each component on this circuit.
3. Short-circuit the other power source and use basic Ohm's law procedures to find current flow through each component.

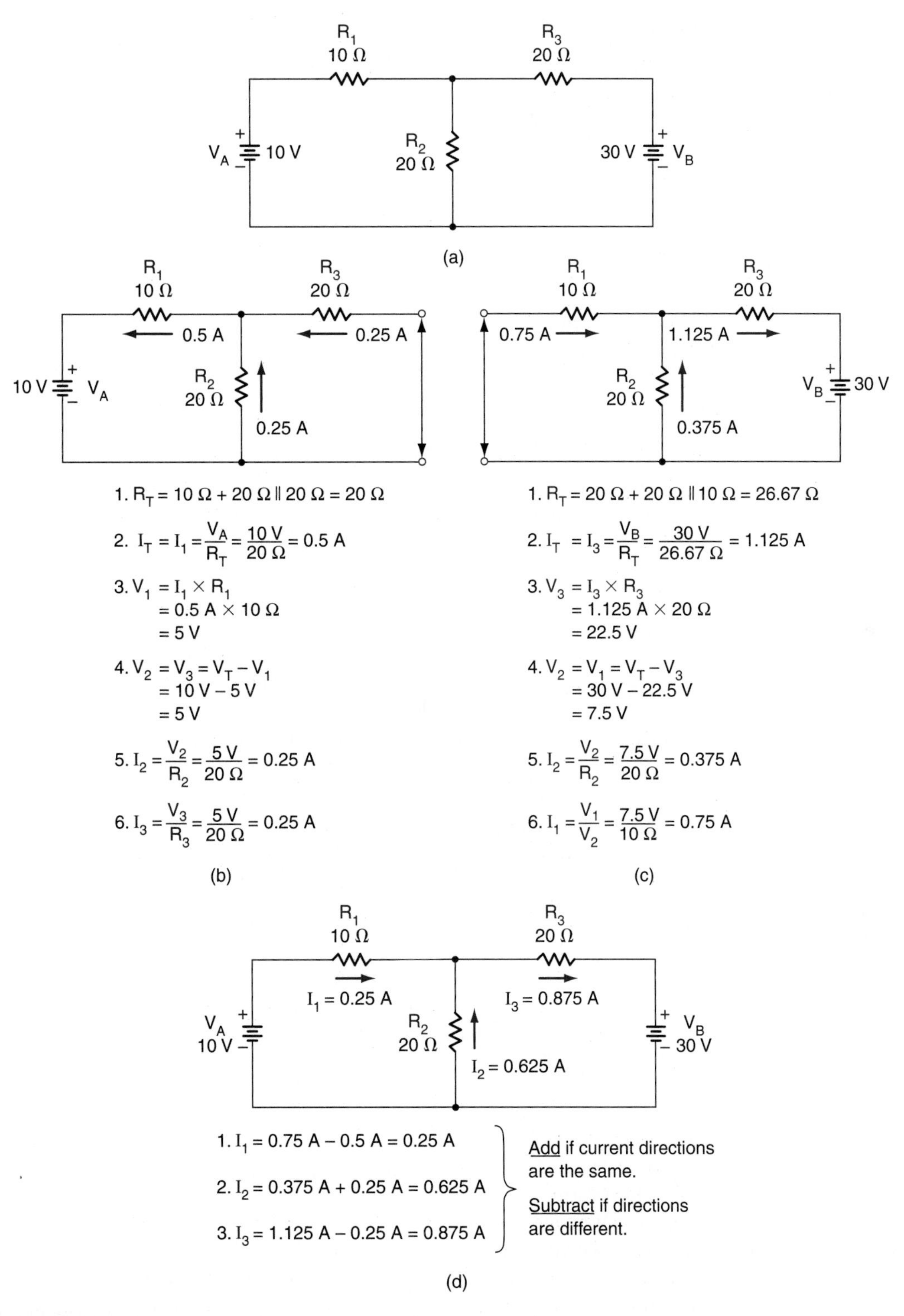

Figure 3–47. The superposition method: (a) original circuit; (b) circuit with 30-V source shorted; (c) circuit with 10-V source shorted; (d) original circuit with currents recorded.

4. Record the amount of current and the direction of flow through each component on this circuit.
5. Find the current flow through each component by looking at the direction of flow through each individual circuit. If the directions through components of both circuits are the same, *add* the values. If the directions of current flow are opposite, *subtract* the values.
6. Record the amount and direction of current flow on the original circuit. The current flows in the direction of the *largest* flow in an individual circuit.

The superposition method can also be used for circuits with more than two sources. A four-source circuit, for example, would require four individual circuits superimposed to find resultant current values. Current flows in the same directions would be added and those in opposite directions would be subtracted. Direction of current flow is in the direction of the largest sum of currents through the path.

3.21 EQUIVALENT CIRCUITS

The previous sections have dealt with relatively simple circuit applications. Simplification of more complex circuits may be accomplished by applying equivalent circuit methods. Several equivalent circuit methods, sometimes called *complex circuit theorems* or *network theorems,* may be utilized to simplify complex circuits. This unit deals primarily with the Thevinin and Norton equivalent circuit applications for solving complex electrical circuit problems.

3.22 THEVININ EQUIVALENT CIRCUIT METHOD

The Thevinin equivalent circuit method is used to simplify electrical circuits. A French engineer, M. L. Thevinin, developed this method, which allows a complex circuit to be reduced to one equivalent voltage source and series resistance for purposes of calculation or lab experimentation. It is a practical method that is used to calculate load currents and load voltages for any value of load resistance. Working with varying values of load resistance is greatly simplified by using the Thevinin equivalent circuit method.

The Thevinin equivalent circuit is shown in Fig. 3–48. It is called an equivalent circuit because it is equivalent to a more complex circuit (as seen by a load connected to the circuit). Remember that circuits have a *source* and a *load.* A complex circuit is reduced to one with a single voltage source (V_{TH}) and a series resistance (R_{TH}). These values are called the equivalent voltage and equivalent resistance. The

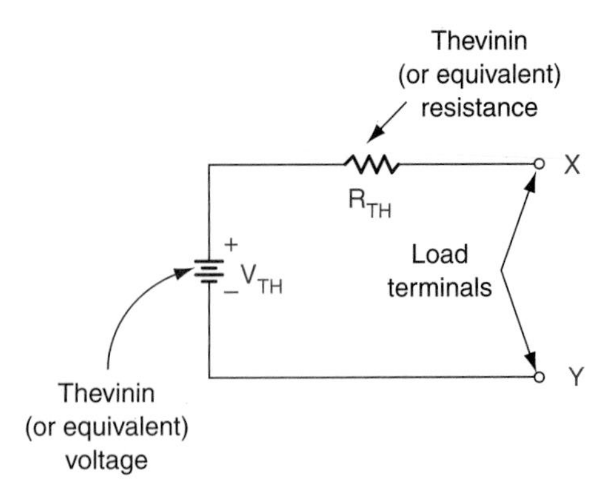

Figure 3–48. Thevinin equivalent circuit.

load is connected to the load terminals of the circuit, which are labeled as points X and Y.

The procedure for simplifying circuits using the Thevinin method is explained next. Some examples are shown in Figs. 3–49 through 3–52 on pages 107–109.

Single-Source Problem

The Thevinin equivalent circuit method may be used for simplifying circuits which have one voltage source. Figure 3–49 shows a circuit with one voltage source and the calculations used to obtain an equivalent circuit. The procedure for finding V_{TH} and R_{TH} is as follows:

1. Find V_{TH}.
   a. Remove the load from the circuit, leaving terminals X and Y open.
   b. Use basic Ohm's law procedures to find the voltage across the load (X and Y) terminals. The voltage across the load is the equivalent voltage (V_{TH}). When resistances are in series with the load they are disregarded when finding V_{TH}. V_{TH} is an open-circuit voltage; therefore, maximum voltage in the current loop containing the load appears across the load terminals (X and Y).
2. Find R_{TH}.
   a. Replace the source with a short circuit.
   b. Remove the load from the circuit, leaving terminals X and Y open.
   c. "Look into" the circuit *from the load terminals* to determine the circuit configuration as seen by a load connected to the load terminals. Some examples of determining circuit configurations as seen from the load are shown in Fig. 3–50.

After the equivalent circuit has been developed, it is simple to calculate values of load current (I_L) and voltage

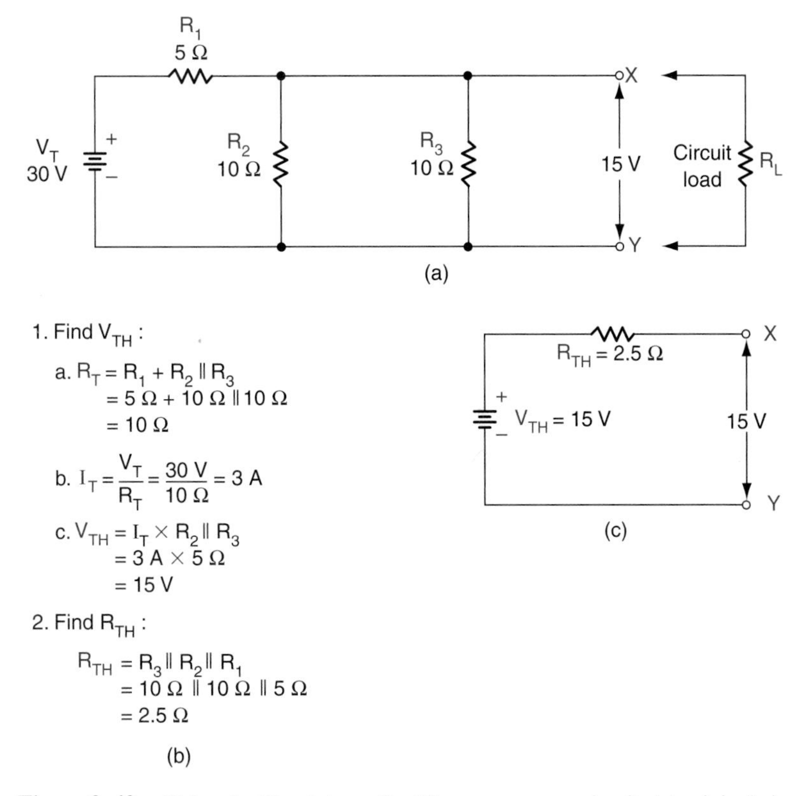

Figure 3–49. Using the Thevinin method for a one-source circuit: (a) original circuit; (b) problem-solving procedure; (c) Thevinin equivalent circuit.

output across to load resistance (V_{out}). Figure 3–51 shows the calculations of several values of I_L and V_{out} with a Thevinin equivalent circuit.

Two-Source Problem

The Thevinin equivalent circuit is easy to apply to a circuit that has two voltage sources. Consider the circuit of Fig. 3–52, which has 10-V and 2-V sources. Notice that the load terminals (X and Y) are in the center of the diagram. To find the R_{TH}, look from the load terminals into the circuit. R_1 and R_2 are in parallel, as seen from the load terminals.

The Thevinin equivalent voltage (V_{TH}) is found by looking at the difference in potential across the circuit resistances. The same procedure as single-source circuits is used; however, the potential at point X must be found. The potential at point X in the example is the V_{TH}. The difference in potential across points A and B is 8 V (10 V − 2 V). The 8-V value is then used to find the current that would flow through R_1 and R_2. Once the current is calculated, the voltage across either resistance may be found. The voltage across R_1 may be *subtracted* from the potential at point A to find V_{TH}. Also, the voltage across R_2 may be *added* to the potential at point B to determine V_{TH}.

Notice that if the polarity of V_2 is reversed, the difference in potential across R_1 and R_2 would become 12 V [10 V − (−2 V)]. This would cause the value of V_{TH} to change also.

Remember that Thevinin equivalent circuits are used to reduce more complex circuits into a circuit that has one equivalent voltage source and one series resistance. They are very helpful in simplifying the procedure for calculating load current and voltage output of circuits that have several values of load resistance.

3.23 NORTON EQUIVALENT CIRCUIT METHOD

Another method of simplifying circuits is the Norton equivalent circuit method, which is shown in Fig. 3–53. The Norton current (I_N) is the maximum current that will flow from the source. It is calculated when $R_L = 0\ \Omega$. The Norton resistance (R_N) is calculated in the same way as Thevinin resistance (R_{TH}). The Norton method allows the reduction of a circuit to a constant current source (I_N) and an equivalent resistance (R_N) in parallel. Figure 3–54(a) shows a sample procedure for applying the Norton method. The procedure for developing a Norton equivalent circuit is as follows:

1. Find I_N.
   a. Short-circuit the load (X-Y) terminals.

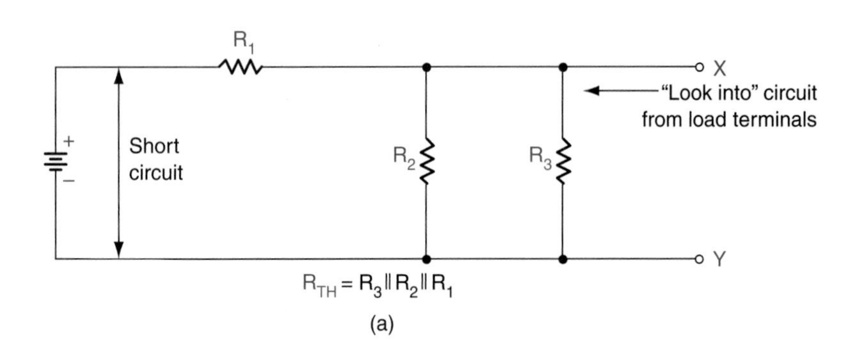

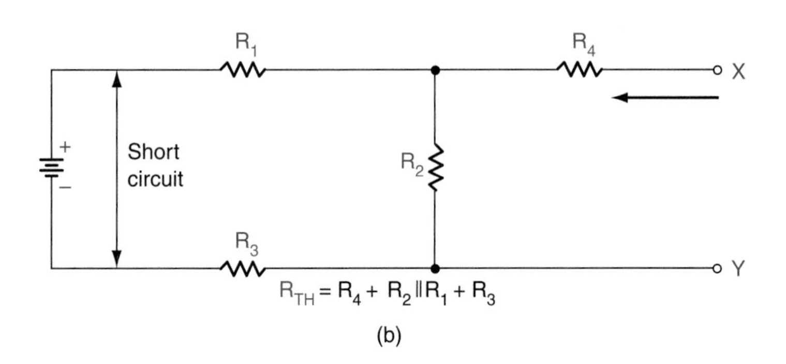

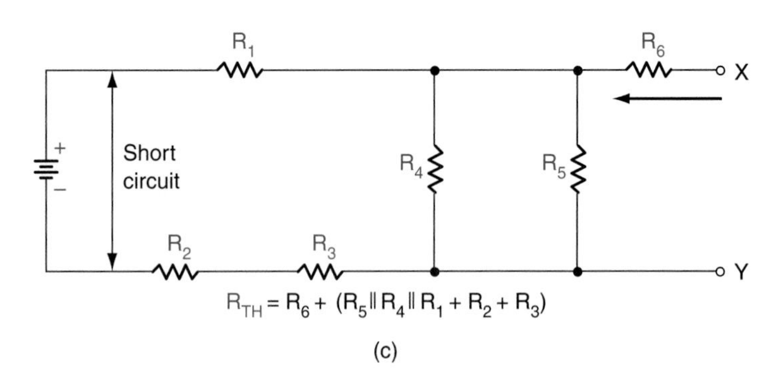

Figure 3–50. Determining circuit configuration for finding R_{TH}.

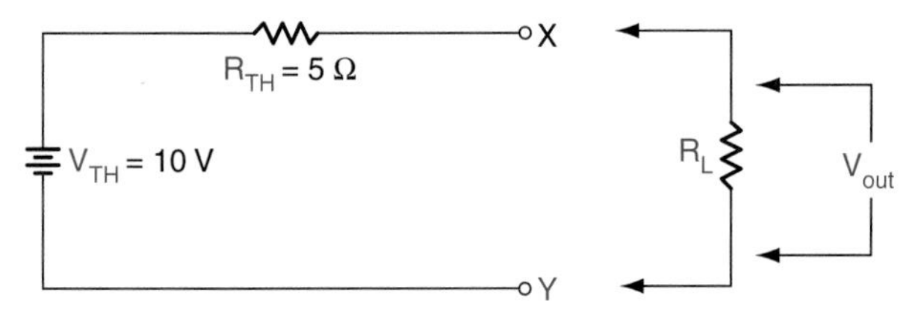

R_L	$I_L = \frac{V_{TH}}{R_{TH} + R_L}$	$V_{out} = I_L \times R_L$
5 Ω	$\frac{10\ V}{10\ \Omega} = 1\ A$	1 A × 5 Ω = 5 V
10 Ω	$\frac{10\ V}{15\ \Omega} = 0.667\ A$	0.667 A × 10 Ω = 6.67 V
15 Ω	$\frac{10\ V}{20\ \Omega} = 0.5\ A$	0.5 A × 15 Ω = 7.5 V
20 Ω	$\frac{10\ V}{25\ \Omega} = 0.4\ A$	0.4 A × 20 Ω = 8 V

Figure 3–51. Calculating load current and voltage output.

b. Calculate the current that will flow from the source when the load resistance is equal to 0 Ω (short circuit). This is the Norton current (I_N).

c. Label the direction of current flow from the source with an arrow on the equivalent circuit diagram.

2. Find R_N.

a. Use the same procedure as outlined for finding R_{TH}.

b. Label the value of R_N on the equivalent circuit diagram.

The load current (I_L) that will flow from a circuit may easily be calculated by applying the Norton equivalent circuit. The formula used to calculate load current values from the equivalent values of R_N and I_N is

$$I_L = \frac{I_N \times R_N}{R_N + R_L}$$

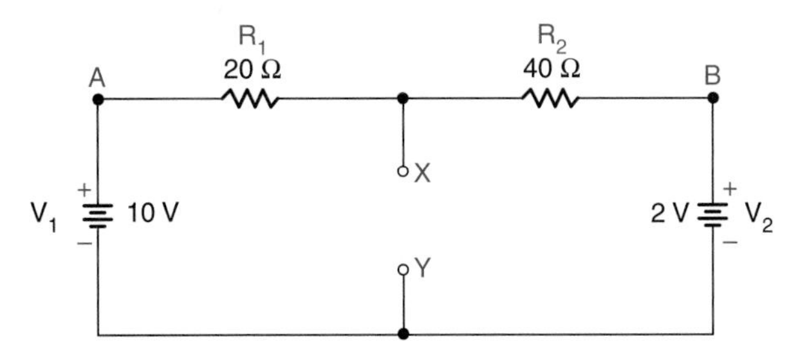

1. Find R_{TH}:

$$R_{TH} = R_1 \parallel R_2 = 20\ \Omega \parallel 40\ \Omega = 13.33\ \Omega$$

2. Find V_{TH}:

Potential at point A = +10 V
Potential at point B = +2 V
Difference in potential = 10 V − 2 V = 8 V

$$\text{Current flow} = \frac{8\ V}{60\ \Omega} = 0.133\ A$$

$$V_1 = I \times R_1 = 0.133\ A \times 20\ \Omega = 2.67\ V$$
$$V_{TH} = 10\ V - V_1 = 10V - 2.67\ V = 7.33\ V$$

or

$$V_2 = I \times R_2 = 0.133\ A \times 40\ \Omega = 5.33\ V$$
$$V_{TH} = 2\ V + V_2 = 2\ V + 5.33\ V = 7.33\ V$$

(a)

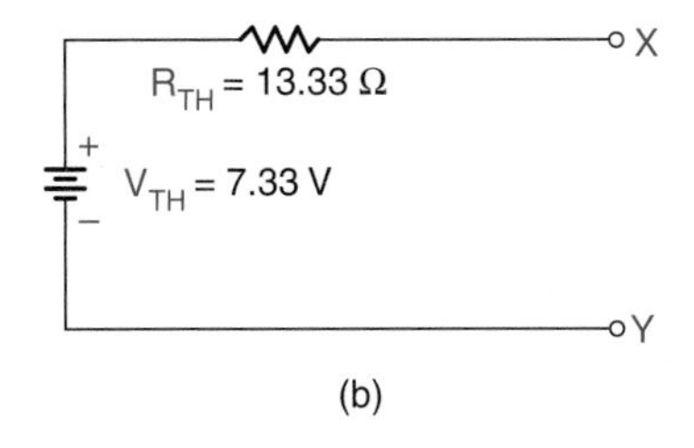

(b)

Figure 3–52. Two-source Thevinin equivalent circuit: (a) problem-solving procedures; (b) Thevinin equivalent circuit.

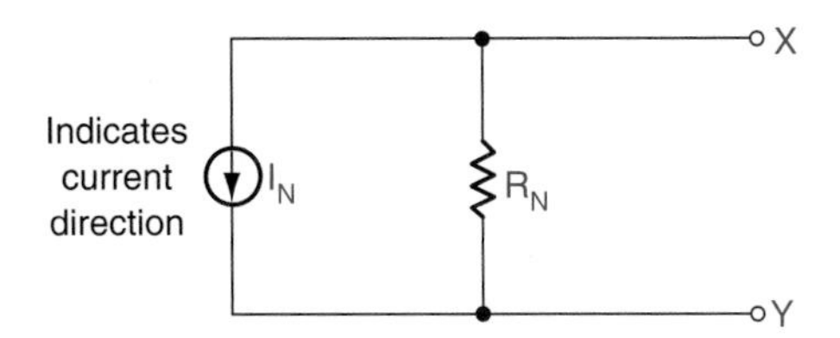

Figure 3–53. Norton equivalent circuit.

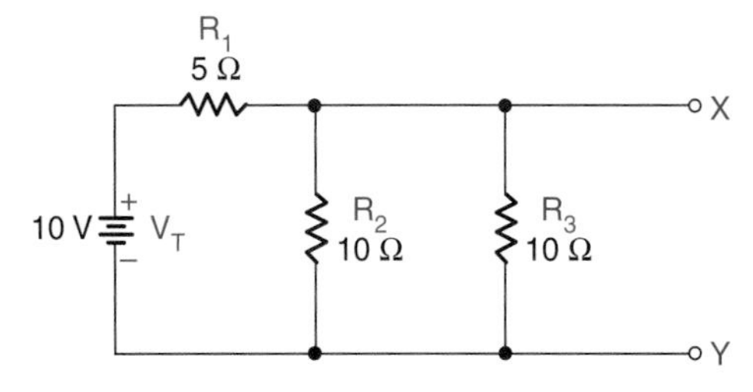

1. Find I_N:

$R_L = 0\ \Omega$, so R_2 and R_3 are eliminated.

$$I_N = \frac{V_T}{R} = \frac{10\ V}{5\ \Omega} = 2\ A$$

2. Find R_N:

$$R_N = R_3 \parallel R_2 \parallel R_1 = 10\ \Omega \parallel 10\ \Omega \parallel 5\ \Omega = 2.5\ \Omega$$

(a)

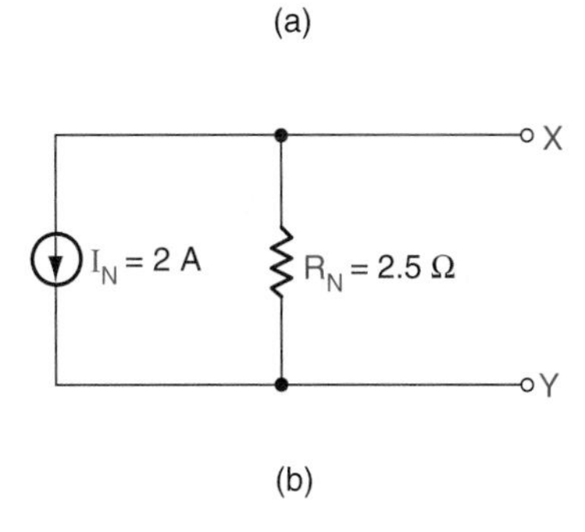

(b)

Figure 3–54. Norton equivalent circuit procedure: (a) problem-solving procedure; (b) Norton equivalent circuit.

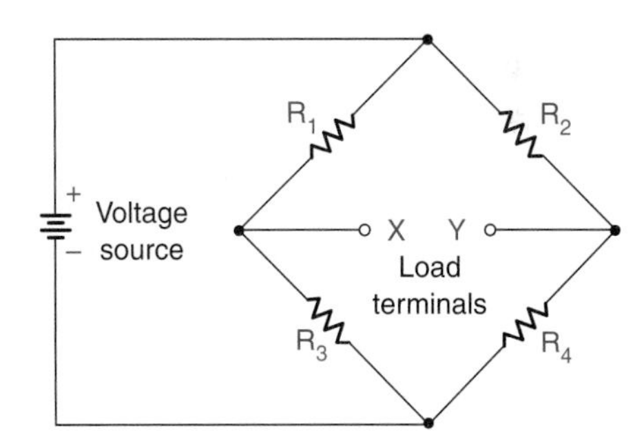

Figure 3–55. Bridge circuit.

3.24 BRIDGE-CIRCUIT SIMPLIFICATION

A bridge circuit is shown in Fig. 3–55. Bridge circuits are used for several applications, particularly in electrical measurement. A bridge circuit may be designed to measure electrical component values by comparing an *unknown* value with a known or standard value. Other applications of bridge circuits include rectification circuits, which convert alternating current into direct current.

Bridge circuits are difficult to analyze using Ohm's law techniques. The easiest method to use in developing a Thevinin equivalent circuit is to simplify the analysis of this type of circuit. Use the following procedure to calculate the value of Thevinin equivalent voltage (V_{TH}) and equivalent resistance (R_{TH}) for the bridge circuit shown in Fig. 3–56.

1. Find R_{TH}.
   a. Remove the load resistance from the circuit.
   b. Look into the circuit from the load (X and Y) terminals to determine R_{TH}. The power supply should be replaced by a short circuit.
   c. The circuit arrangement of the four resistors is as shown in Fig. 3–56(b).
   d. Calculate the R_{TH} of this arrangement and label its value on the equivalent circuit diagram.
2. Find V_{TH}.
   a. Disregard R_2 and R_4 and calculate the current that would flow through R_1 and R_3 if R_2 and R_4 were disconnected from the circuit.
   b. Disregard R_1 and R_3 and calculate the current that would flow through R_2 and R_4 if R_1 and R_3 were disconnected from the circuit.
   c. Calculate the voltage drop across R_3 ($V_3 = I_3 \times R_3$) with the current determined in step (a).
   d. Calculate the voltage drop across R_4 due to the current determined in step (b) ($V_4 = I_4 = R_4$).
   e. Subtract V_3 from V_4. This is the difference in potential (voltage drop) across points X and Y. This value is the V_{TH} of the circuit, which should be labeled on the equivalent circuit diagram.

The use of a Thevinin equivalent circuit greatly simplifies the calculation of load current and voltage output of bridge circuits. Notice in Fig. 3–57 that addition of a load resistance to a bridge circuit produces a complex circuit configuration. Simplification of a bridge circuit using the Thevinin method provides an easy way to analyze bridge circuits.

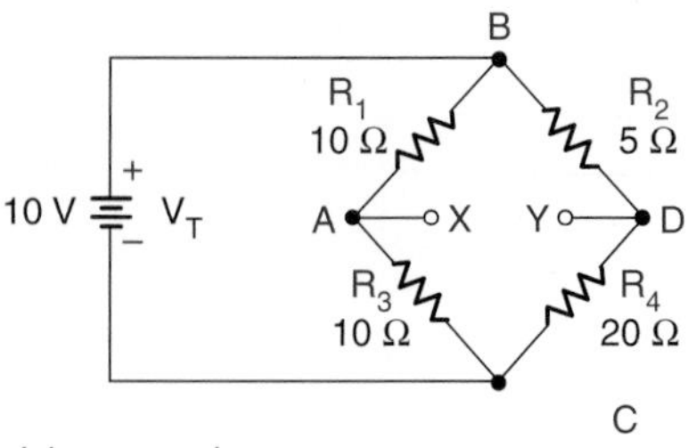

Problem-solving procedure

1. Find R_{TH}:

a. Resistor configuration looking from load terminals with voltage source circuited is:

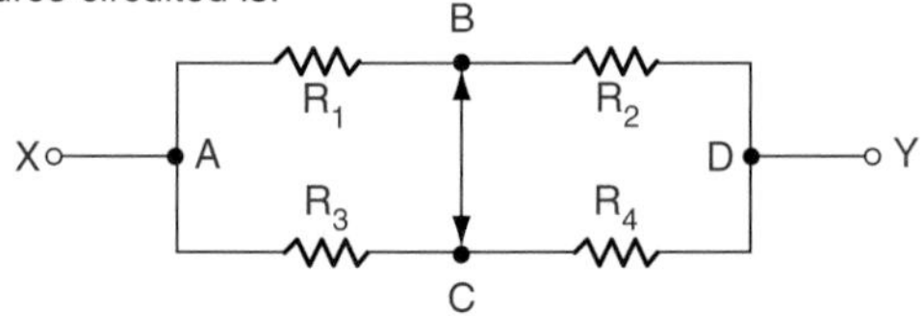

b. This configuration is the same as:

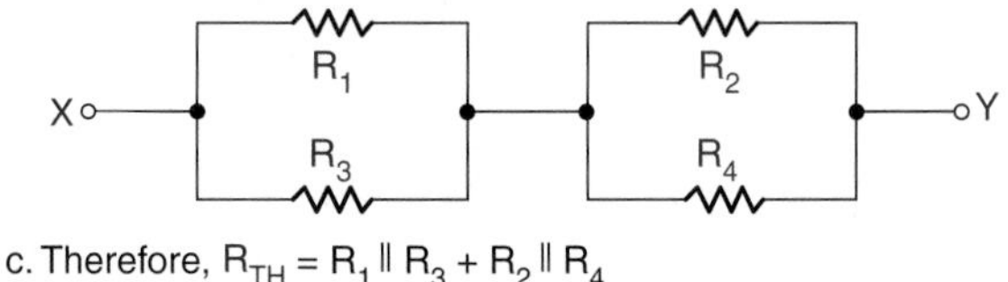

c. Therefore, $R_{TH} = R_1 \parallel R_3 + R_2 \parallel R_4$
$= 10\ \Omega \parallel 10\ \Omega + 5\ \Omega \parallel 20\ \Omega$
$= 5\ \Omega + 4\ \Omega = 9\ \Omega$

2. Find V_{TH}:

a. Disregard $R_2 - R_4$ path to find $I_1 = I_3$:

$$I_1 = I_3 = \frac{V_T}{R_1 + R_3} = \frac{10\ V}{20\ \Omega} = 0.5\ A$$

b. Disregard $R_1 - R_3$ path to find $I_2 = I_4$:

$$I_2 = I_4 = \frac{V_T}{R_2 + R_4} = \frac{10\ V}{25\ \Omega} = 0.4\ A$$

c. Calculate V_3:
$V_3 = I_3 \times R_3$
$= 0.5\ A \times 10\ \Omega$
$= 5\ V$

d. Calculate V_4:
$V_4 = I_4 \times R_4$
$= 0.4\ A \times 20\ \Omega$
$= 8\ V$

e. Subtract V_3 from V_4 to find V_{TH}:
$V_{TH} = V_4 - V_3$
$= 8\ V - 5\ V$
$= 3\ V$

3. Complete the Thevinin equivalent circuit diagram:

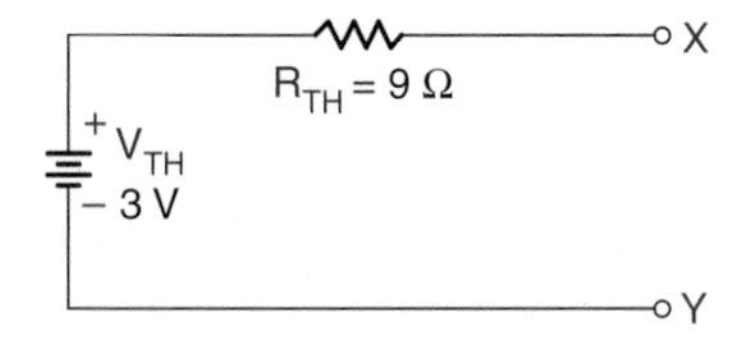

Figure 3–56. Simplification of a bridge circuit.

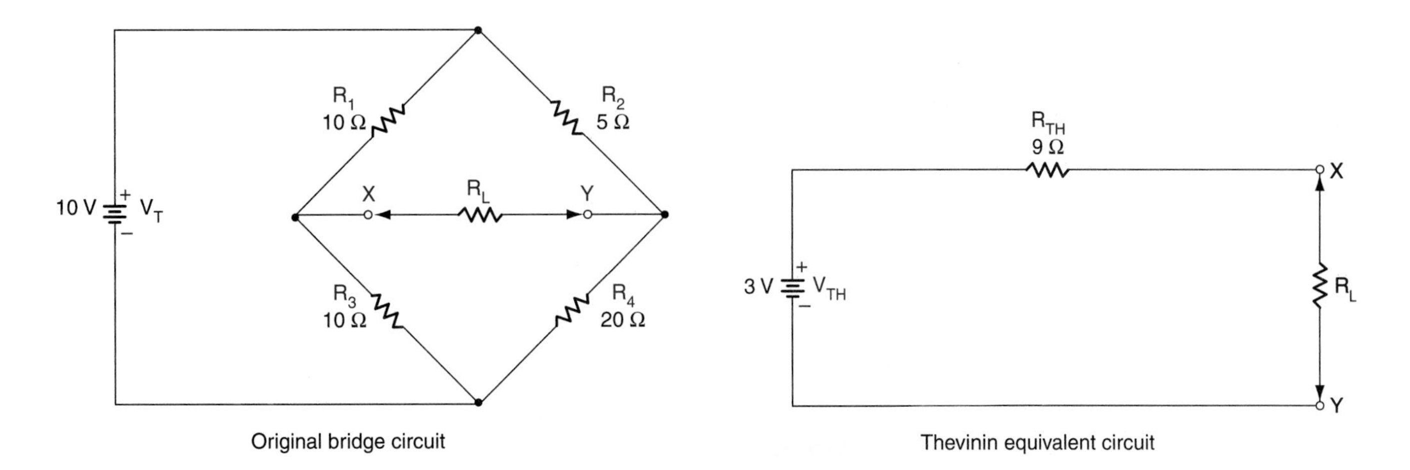

Load current and output voltage calculations:

R_L	$I_L = \frac{V_{TH}}{R_{TH} + R_L}$	$V_{out} = I_L \times R_L$
3 Ω	$\frac{3\text{ V}}{12\ \Omega} = 0.25\text{ A}$	0.25 A × 3 Ω = 0.75 V
6 Ω	$\frac{3\text{ V}}{15\ \Omega} = 0.2\text{ A}$	0.2 A × 6 Ω = 1.2 V
9 Ω	$\frac{3\text{ V}}{18\ \Omega} = 0.16\text{ A}$	0.16 A × 9 Ω = 1.44 V
12 Ω	$\frac{3\text{ V}}{21\ \Omega} = 0.143\text{ A}$	0.143 A × 12 Ω = 1.71 V

Figure 3–57. Calculating load current and voltage output of a bridge circuit.

REVIEW

1. What is Ohm's law?
2. What are the symbols used for (a) voltage, (b) current, and (c) resistance?
3. What is the relationship of voltage and current in a circuit?
4. What is the relationship of resistance and current in a circuit?
5. What is a series circuit?
6. What is a parallel circuit?
7. What is a combination circuit?
8. What are the voltage, current, and resistance characteristics of (a) series circuits and (b) parallel circuits?
9. How is total resistance of a series circuit measured?
10. How is total resistance of a parallel circuit measured?
11. How is total current of a series circuit measured?
12. How is total current of a parallel circuit measured?
13. How is voltage drop measured for a series circuit?
14. How is voltage drop measured for a parallel circuit?
15. What are Kirchhoff's laws? Explain them.

16. Explain the three ways used to find total resistance of parallel circuits.
17. What are three ways to find the electrical *power* of a circuit?
18. What is meant by kilowatt-hour?
19. What is a voltage-divider circuit?
20. What is meant by a negative voltage?
21. Discuss each of the following problem-solving methods: (a) Kirchhoff's voltage law; (b) Superposition; (c) Thevinin's theorem; (d) Norton's theorem; and (e) Bridge-circuit simplification.

STUDENT ACTIVITIES

• *Applying Basic Electrical Theory (see Fig. 3–4)*

Use Ohm's law to solve the following problems. Record your answers on a sheet of paper.

1. A doorbell requires 0.5 A of current to ring. The voltage applied to the bell is 12 V. What is the resistance? $R =$ _____ Ω.
2. A relay used to control a motor has a 50-Ω resistance. It draws a current of 0.5 A. What voltage is required to operate the relay? $V =$ _____ V.
3. An automobile battery supplies a current of 10 A to the starter. It has a resistance of 1.25 Ω. What is the voltage delivered by the battery? $V =$ _____ V.
4. What voltage is needed to light a lamp if the current required is 3 A and the resistance of the lamp is 80 Ω? $V =$ _____ V.
5. If the resistance of a radio receiver circuit is 200 Ω and it draws 0.6 A, what voltage is needed? $V =$ _____ V.
6. A television draws 0.25 A. The operating voltage is 120 V. What is the resistance of the TV circuit? $R =$ _____ Ω.
7. The resistance of the motor of a vacuum cleaner is 30 Ω. For a voltage of 120 V, find the current. $I =$ _____ A.
8. The magnet of a speaker carries 0.1 A when connected to a 50-V supply. Find its resistance. $R =$ _____ Ω.
9. How much current is drawn from a 12-V battery when operating an automobile horn of 16-Ω resistance? $I =$ _____ A.
10. How much current is drawn by a 2600-Ω clock that plugs into a 120-V outlet? $I =$ _____ A.

• *Using Ohm's Law (see Fig. 3–4)*

Use Ohm's law to solve these problems.

1. $I = 2$ A, $R = 500$ Ω, $V =$ _____ V.
2. $R = 20$ Ω, $I = 3$ A, $V =$ _____ V.
3. $I = 1.2$ A, $R = 1000$ Ω, $V =$ _____ V.
4. $R = 1.5$ kΩ, $I = 130$ mA, $V =$ _____ V.
5. $R = 120$ kΩ, $I = 20$ mA, $V =$ _____ V.
6. $V = 2$ V, $R = 1.5$ kΩ, $I =$ _____ A.
7. $R = 1.8$ kΩ, $V = 10$ V, $I =$ _____ A.
8. $R = 6.8$ kΩ, $V = 15$ V, $I =$ _____ A.
9. $R = 5.6$ kΩ, $V = 12$ V, $I =$ _____ A.

10. $V = 240$ V, $R = 1$ MΩ, $I =$ _____ A.
11. $V = 20$ V, $I = 5$ A, $R =$ _____ Ω.
12. $V = 35$ V, $I = 5$ mA, $R =$ _____ Ω.
13. $I = 10$ μA, $V = 120$ V, $R =$ _____ Ω.
14. $I = 260$ mA, $V = 120$ V, $R =$ _____ Ω.
15. $V = 5$ V, $I = 20$ μA, $R =$ _____ Ω.

• *Application of Ohm's Law*

Complete Activity 3–1, page 49 in the manual.

• • ANALYSIS

1. How is current affected by a doubled voltage when resistance remains the same?
2. How is current affected by a doubled resistance when voltage remains the same?
3. What voltage would cause 10 mA of current through a resistance of 10 Ω?
4. What resistance would limit the current to 1 mA when 10 V is applied?
5. What current would result when 1 V is applied to 1 Ω of resistance?

• *Working with Series Circuits (see Fig. 3–15)*

Solve each of the following series-circuit problems.

- Find each of the following values for Fig. 3–58:
  1. Current (I) = _____ A.
  2. Voltage across R (V_R) = _____ V.

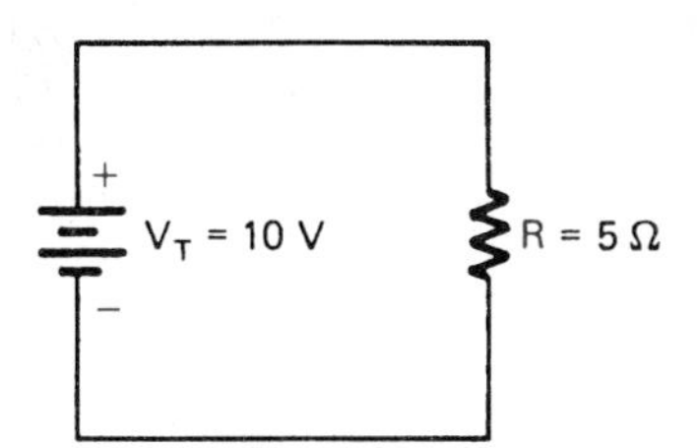

Figure 3–58.

- Find each of the following values for Fig. 3–59:
  3. Total resistance (R_T) = _____ Ω.
  4. Current through R_1 (I_1) = _____ A.
  5. Current through R_2 (I_2) = _____ A.
  6. Voltage across R_2 (V_2) = _____ V.

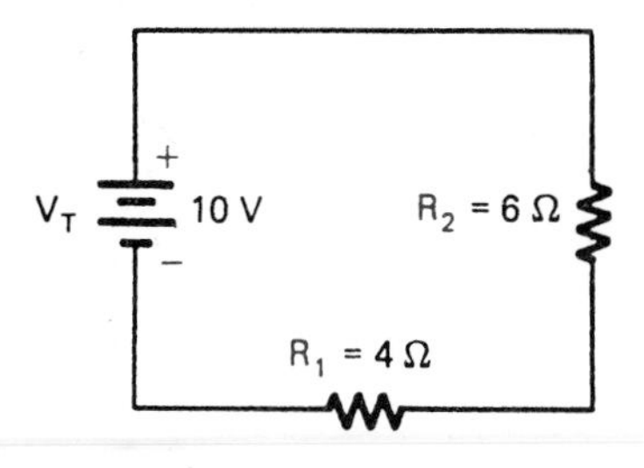

Figure 3–59.

- Find each of the following for Fig. 3–60:
  7. Voltage across R_1 (V_1) = _____ V.
  8. Resistance of R_2 = _____ Ω.
  9. Voltage across R_2 (V_2) = _____ V.
  10. Resistance of R_3 = _____ Ω.

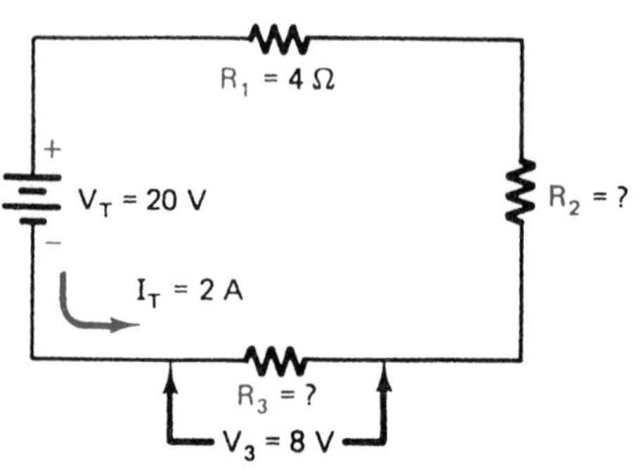

Figure 3–60.

- Find each of the following for Fig. 3–61:
  11. V_1 = _____ V.
  12. R_2 = _____ Ω.
  13. V_3 = _____ V.
  14. R_3 = _____ Ω.
  15. I_3 = _____ A.

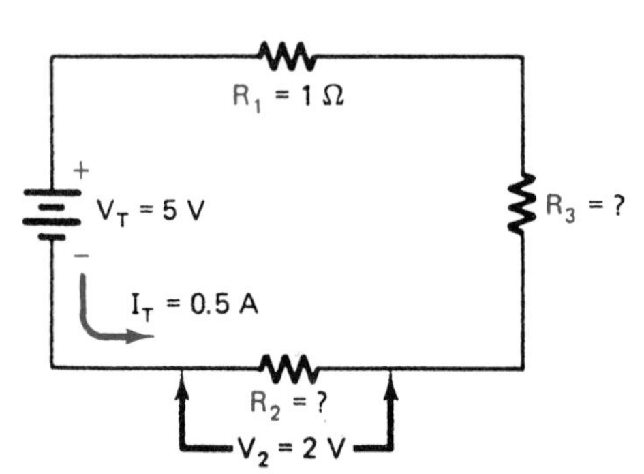

Figure 3–61.

- Find each of the following for Fig. 3–62:
  16. I_T = _____ A.
  17. V_1 = _____ V.
  18. V_2 = _____ V.
  19. V_3 = _____ V.
  20. Power converted by R_1 (P_1) = _____ W.
  21. Power converted by R_2 (P_2) = _____ W.
  22. Power converted by R_3 (P_3) = _____ W.
  23. Total power converted by circuit (P_T) = _____ W.

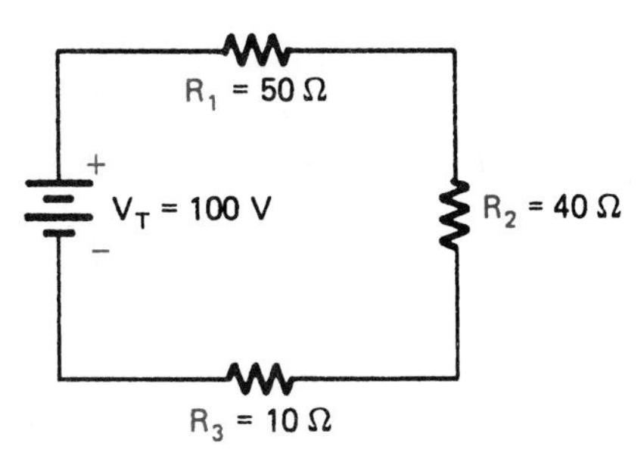

Figure 3–62.

- Find each of the following for Fig. 3–63:
  24. V_T = _____ V.
  25. R_T = _____ Ω.
  26. V_1 = _____ V.
  27. I_T = _____ A.
  28. Total power (P_T) = _____ W.

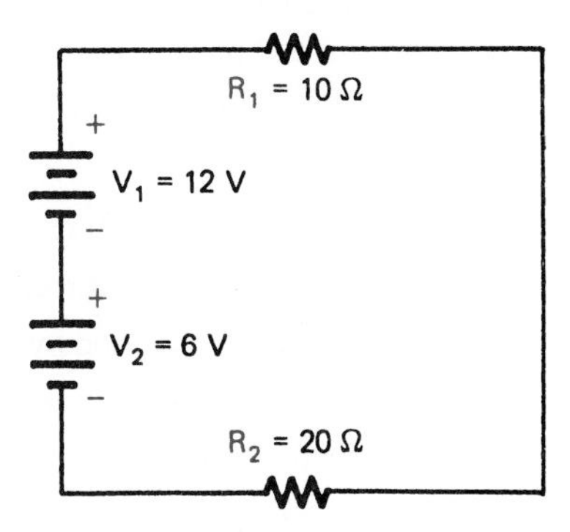

Figure 3–63.

Working with Parallel Circuits (see Fig. 3–22)

Solve each of the following parallel-circuit problems.

- Find each of the following for Fig. 3–64:
  1. Total resistance (R_T) = _____ Ω.
  2. Total current (I_T) = _____ A.
  3. Current through resistor R_1 (I_1) = _____ A.
  4. Current through resistor R_2 (I_2) _____ A.

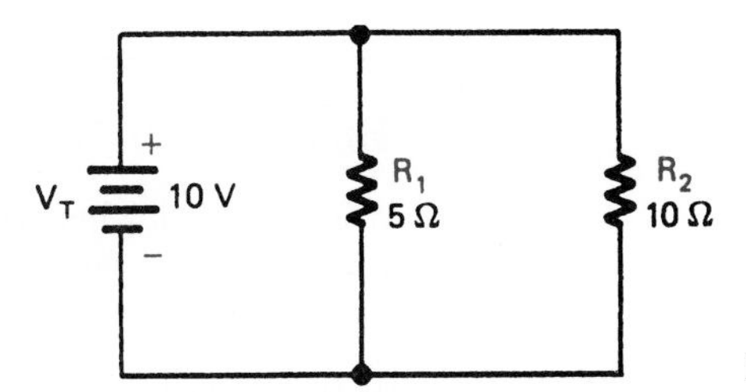

Figure 3–64.

- Find each of the following for Fig. 3–65:
  5. Total resistance (R_T) = _____ Ω.
  6. Total current (I_T) = _____ A.

7. Current through resistor R_1 (I_1) = _____ A.
8. Current through resistor R_2 (I_2) = _____ A.
9. Current through resistor R_3 (I_3) = _____ A.
10. Current through resistor R_4 (I_4) = _____ A.
11. Total power (P_T) = _____ W.

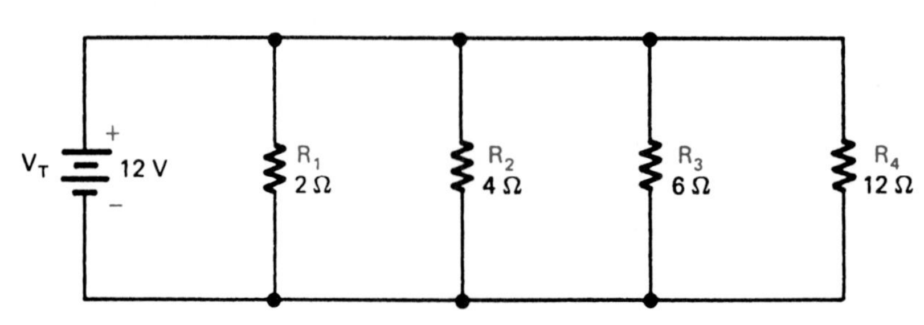

Figure 3–65.

- Find each of the following for Fig. 3–66:

12. Total resistance (R_T) = _____ Ω.
13. Total current (I_T) = _____ A.
14. Current through resistor R_1 (I_1) = _____ A.
15. Current through resistor R_2 (I_2) = _____ A.
16. Current through resistor R_3 (I_3) = _____ A.
17. Power converted by resistor R_1 (P_1) = _____ W.
18. Power converted by resistor R_2 (P_2) = _____ W.
19. Power converted by resistor R_3 (P_3) = _____ W.
20. Total power (P_T) = _____ W.

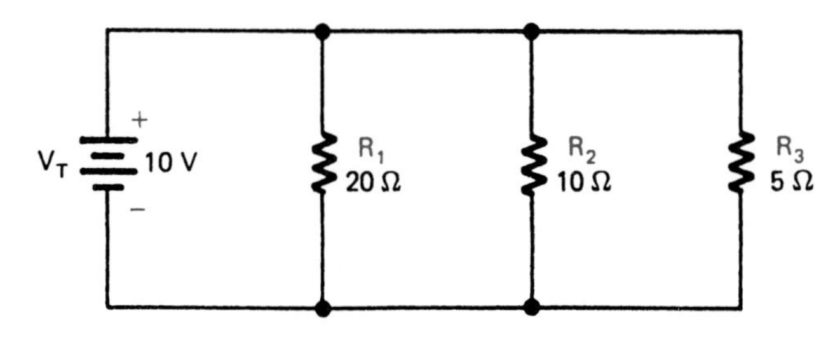

Figure 3–66.

• *Working with Combination Circuits (see Fig. 3–27)*

Solve each of the following combination-circuit problems.

- Find each of the following for Fig. 3–67:

1. Total resistance (R_T) = _____ Ω.
2. Total current (I_T) = _____ A.
3. Voltage across R_1 (V_1) = _____ V.
4. Total power (P_T) = _____ W.
5. Current through R_2 (I_2) = _____ A.
6. Voltage across R_2 (V_2) = _____ V.

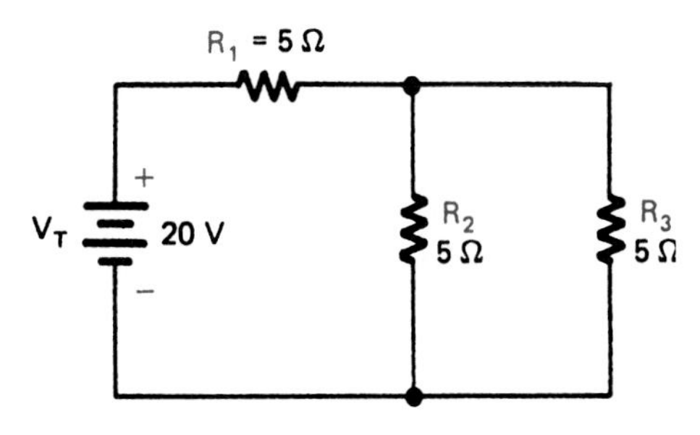

Figure 3–67.

- Find each of the following for Fig. 3–68:

7. Total resistance (R_T) = _____ Ω.

8. Total current (I_T) = _____ A.

9. Total power (P_T) = _____ W.

10. Voltage across R_1 (V_1) = _____ V.

11. Current through R_2 (I_2) = _____ A.

12. Current through R_4 (I_4) = _____ A.

13. Voltage across R_5 (V_5) = _____ V.

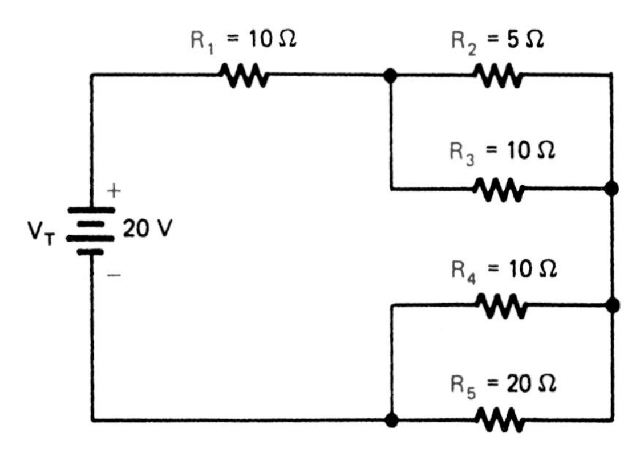

Figure 3–68.

- Find each of the following for Fig. 3–69:

14. Total resistance (R_T) = _____ Ω.

15. Total current (I_T) = _____ A.

16. Voltage across resistor R_1 (V_1) = _____ V.

17. Voltage across resistor R_4 (V_4) = _____ V.

18. Current through resistor R_2 (I_2) = _____ A.

19. Current through resistor R_3 (I_3) = _____ A.

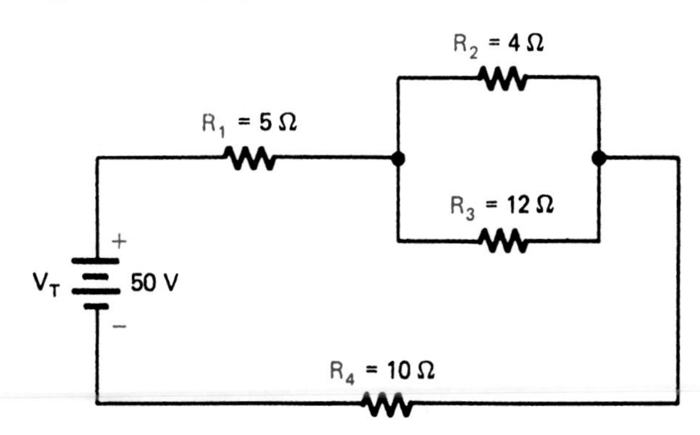

Figure 3–69.

- Find each of the following for Fig. 3–70:

20. Total resistance (R_T) = _____ Ω.

21. Total current (I_T) = _____ A.

22. Voltage across resistor R_1 (V_1) = _____ V.

23. Voltage across resistor R_2 (V_2) = _____ V.

24. Voltage across resistor R_3 (V_3) = _____ V.

25. Current through resistor R_3 (I_3) = _____ A.

26. Current through resistor R_2 (I_2) = _____ A.

27. Total power (P_T) = _____ W.

28. Power converted by R_3 (P_3) = _____ W.

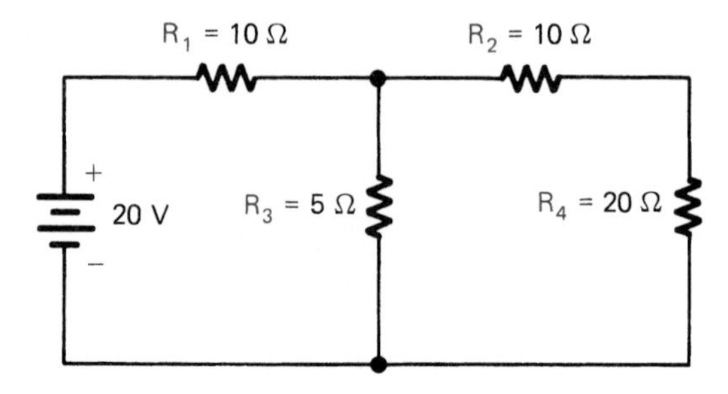

Figure 3–70.

Series-Circuit Measurement

1. Construct a series circuit similar to Fig. 3–16. Use a 6-V lantern battery or a 1.5-V battery with a holder and any values of resistors R_1, R_2, and R_3 (from 100 Ω to 10 kΩ).
2. Make the following measurements with a multimeter.
   a. *Before* connecting the circuit, measure:
      (1) R_1 = _____ Ω.
      (2) R_2 = _____ Ω.
      (3) R_3 = _____ Ω.
   b. *After* connecting the battery, measure:
      (1) V_1 = _____ V.
      (2) V_2 = _____ V.
      (3) V_3 = _____ V.
      (4) I_1 = _____ mA.
      (5) I_2 = _____ mA.
      (6) I_3 = _____ mA.
      (7) I_T = _____ mA.
3. Calculate the following power values and record them:
   a. $P_1 = V_1 \times I_1 =$ _____ W.
   b. $P_2 = V_2 \times I_2 =$ _____ W.
   c. $P_3 = V_3 \times I_3 =$ _____ W.
   d. $P_T = V_T \times I_T = P_1 + P_2 + P_3 =$ _____ W.

Series Dc Circuits

Complete Activity 3–2, page 51 in the manual.

ANALYSIS

1. List the characteristics of a series circuit.
2. What is the voltage drop across R_2 in the circuit of Fig. 3–71?

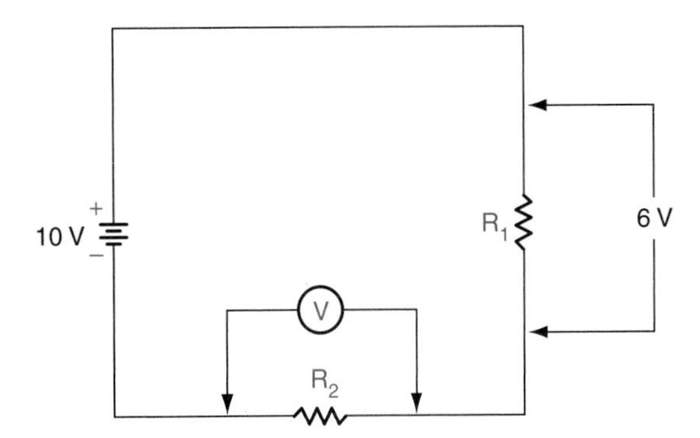

Figure 3–71.

3. If three resistors valued at 6 Ω, 8 Ω, and 10 Ω, respectively, were connected in series, what would be their total resistance?
4. In Fig. 3–71, if 100 mA of current is flowing through R_1, how much current is flowing through R_2, R_3, and R_4?
5. How many paths for current are illustrated in the circuit of Fig. 3–71?
6. What is the total voltage drop in the circuit illustrated in Fig. 3–72?

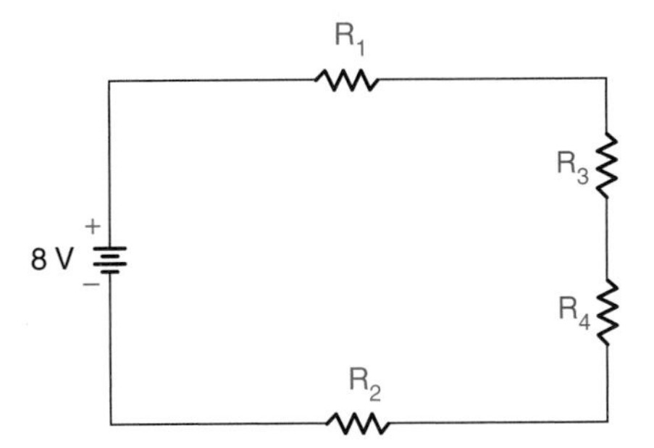

Figure 3–72.

7. If three resistors valued at 10 Ω each were connected in series to a voltage source, what portion of the source voltage would appear across each resistor?

Parallel-Circuit Measurement

1. Construct a parallel circuit similar to Fig. 3–17. Use a 6-V lantern battery or a 1.5-V flashlight battery with a holder and any values of resistors R_1, R_2, and R_3 (from 100 Ω to 10 kΩ).
2. Make the following measurements with a multimeter.
   a. *Before* connecting the circuit, measure:
      (1) R_1 = _____ Ω.
      (2) R_2 = _____ Ω.
      (3) R_3 = _____ Ω.
   b. Connect R_1, R_2, and R_3 as shown in Fig. 3–17. *Do not* connect the battery. Measure: R_T = _____ Ω (point A to point B in Fig. 3–17).

c. *After* connecting the battery, measure:

(1) $V_1 =$ _____ V.

(2) $V_2 =$ _____ V.

(3) $V_3 =$ _____ V.

(4) $I_1 =$ _____ A.

(5) $I_2 =$ _____ A.

(6) $I_3 =$ _____ A.

(7) $I_4 =$ _____ A.

3. Calculate the following power values and record them:

a. $P_1 = V_1 \times I_1 =$ _____ W.

b. $P_2 = V_2 \times I_2 =$ _____ W.

c. $P_3 = V_3 \times I_3 =$ _____ W.

d. $P_T = V_T \times I_T =$ _____ W.

• Parallel Dc Circuits

Complete Activity 3–3, page 53 in the manual.

• • ANALYSIS

1. List the characteristics of a parallel circuit.

2. What is the current through R_3 in the circuit illustrated in Fig. 3–73?

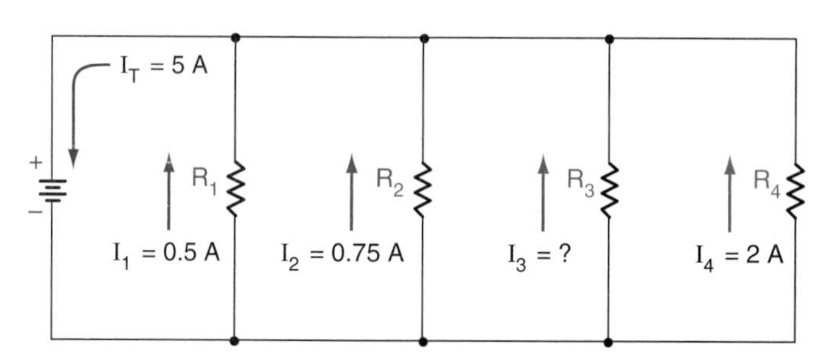

Figure 3–73.

3. What is the total current in the circuit of Fig. 3–74?

4. What is the voltage across R_2 in the circuit of Fig. 3–74?

5. What is the total resistance of the circuit shown in Fig. 3–74?

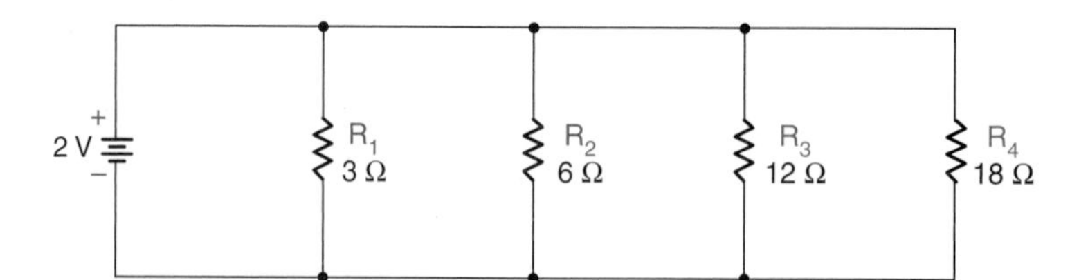

Figure 3–74.

6. If the resistance of each path of a parallel circuit is equal, how will the current in one path compare with the current in other paths?

7. In the circuit of Fig. 3–75, explain where the VOM should be placed to measure total current.

8. In the circuit of Fig. 3–75, explain where the VOM should be placed to measure current through R_2.

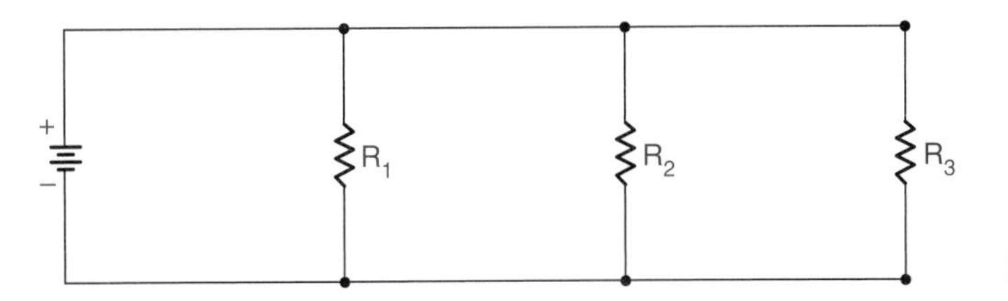

Figure 3–75.

Combination-Circuit Measurement

1. Construct a combination circuit similar to Fig. 3–24. Use a 6-V lantern battery or a 1.5-V flashlight battery with holder and any values of resistors R_1, R_2, and R_3 (from 100 Ω to 10 kΩ).
2. Make the following measurements with a VOM and record them on a sheet of paper.
   a. *Before* connecting the circuit, measure:
   (1) R_1 = _____ Ω.
   (2) R_2 = _____ Ω.
   (3) R_3 = _____ Ω.
   b. Connect R_1, R_2, and R_3 as shown in Fig. 3–24. Measure:
   (1) R_1 = _____ Ω.
   (2) R_2 = _____ Ω.
   (3) R_3 = _____ Ω.
   c. *After* connecting the battery to the circuit, measure:
   (1) V_1 = _____ V.
   (2) V_2 = _____ V.
   (3) V_3 = _____ V.
   (4) I_1 = _____ mA.
   (5) I_2 = _____ mA.
   (6) I_3 = _____ mA.
   (7) I_T = _____ mA.

Combination Dc Circuits

Complete Activity 3–4, page 57 in the manual.

ANALYSIS

1. How does the total current of a combination circuit compare with the current through its series components?
2. How does the sum of the currents in the parallel paths of a combination circuit compare with the total current of the circuit?
3. How does the sum of the currents in the parallel paths of a combination circuit compare with the current through the series components?
4. How does the sum of the voltage across the series components and the parallel paths of a combination circuit compare with the source voltage?

5. If you subtracted the voltage across the parallel paths of a combination circuit from the source voltage, how would the results compare with the voltage across the series components of the circuit?
6. If you subtracted the sum of the voltages across the series components of a combination circuit from the source voltage, how would the results compare with voltage across the parallel paths of the circuit?
7. If you added the sum of the resistance of the series components of a combination circuit to the total parallel resistance of the circuit, how would the results compare with the total resistance of the circuit?

Power in Dc Circuits

Complete Activity 3–5, page 59 in the manual.

ANALYSIS

1. How much current flows through a 100-W light bulb when 100 V is connected to it?
2. How much power is converted by a 12-V lamp with a filament resistance of 2 Ω?
3. What is the maximum current that could safely pass through a 1000-Ω, 1-W resistor?
4. What power is converted by a 5-kΩ resistor that has 1 mA of current through it?
5. What power is converted by a light bulb that drops 6 V because of its 10-Ω filament?
6. What is the maximum voltage drop that could be developed by a 1.2-kΩ, ½-W resistor?

Voltage-Divider Circuits

Complete Activity 3–6, page 63 in the manual.

ANALYSIS

1. Two transistors are to be supplied the following dc voltages and currents: 5 V at 10 mA and 3 V at 5 mA. Use the circuit in Fig. 3–76 to determine the values of R_1, R_2, and R_3 required for this voltage-divider circuit.

   R_1 = _____ Ω.

   R_2 = _____ Ω.

   R_3 = _____ Ω.

2. Determine the minimum power rating of each resistor in the voltage divider of Fig. 3–76: P_{R1} = _____ W; P_{R2} = _____ W; P_{R3} = _____ W.

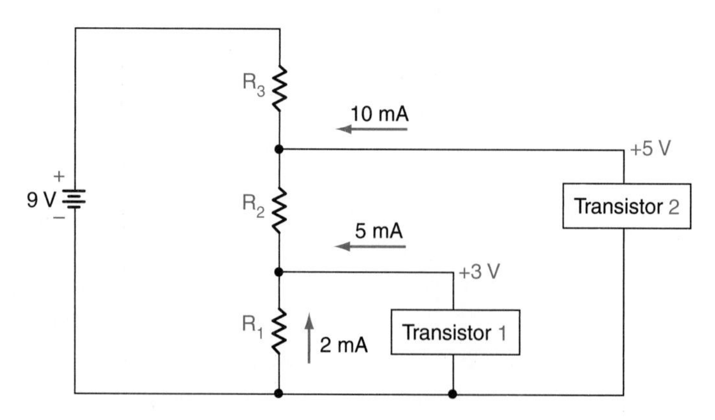

Figure 3–76.

Kirchhoff's Voltage Law (KVL) Problems

Solve each of the following problems by applying Kirchhoff's voltage law.

1. Use the algebraic method to find values of I_1, I_2, and $I_1 + I_2$ for the circuit of Fig. 3–77.

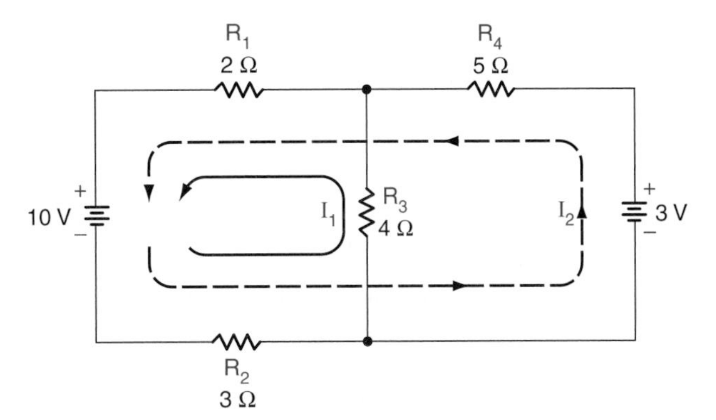

Figure 3–77.

2. Solve for values of I_1, I_2, and $I_1 + I_2$ for the circuit of Fig. 3–78.

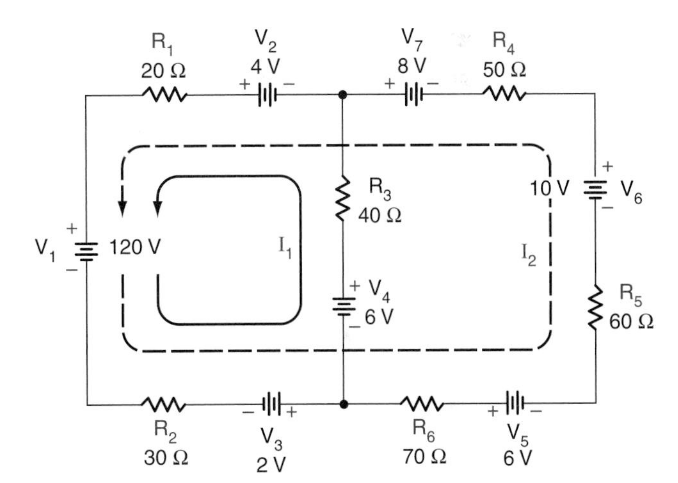

Figure 3–78.

3. Refer to Fig. 3–79. Set up equations for loops 1, 2, and 3 of the circuit.

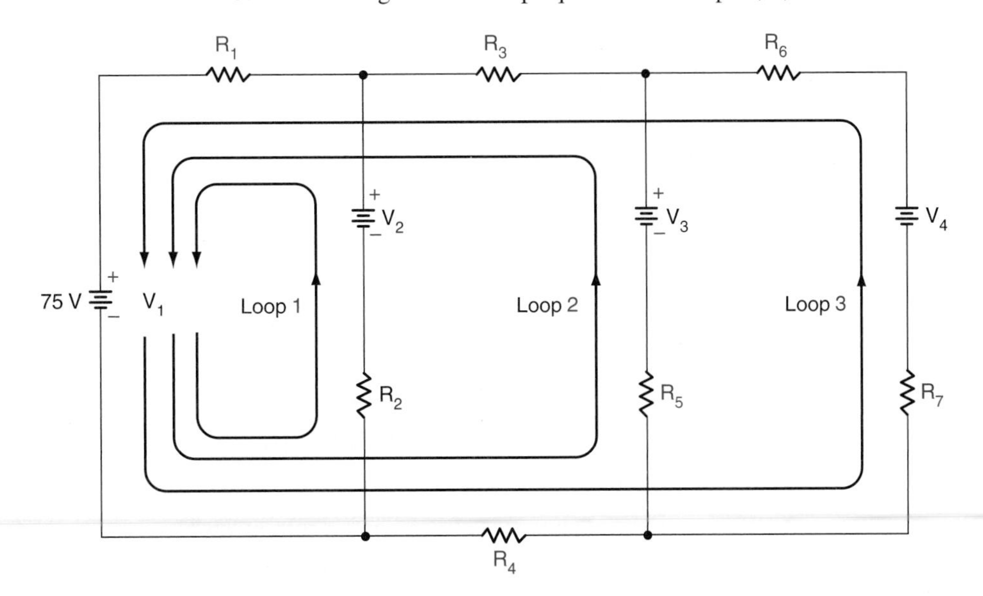

Figure 3–79.

Kirchhoff's Voltage Law

Complete Activity 3–7, page 67 in the manual.

ANALYSIS

1. State Kirchhoff's voltage law.
2. Why is it important to be able to apply Kirchhoff's voltage law?

Kirchhoff's Current Law (KCL) Problems

Solve each of the following problems by applying Kirchhoff's current law to the circuit shown in Fig. 3–80.

1. I_1 in (a) = _____ A.
2. I_2 in (b) = _____ A.
3. I_3 in (c) = _____ A.
4. I_4 in (d) = _____ A.

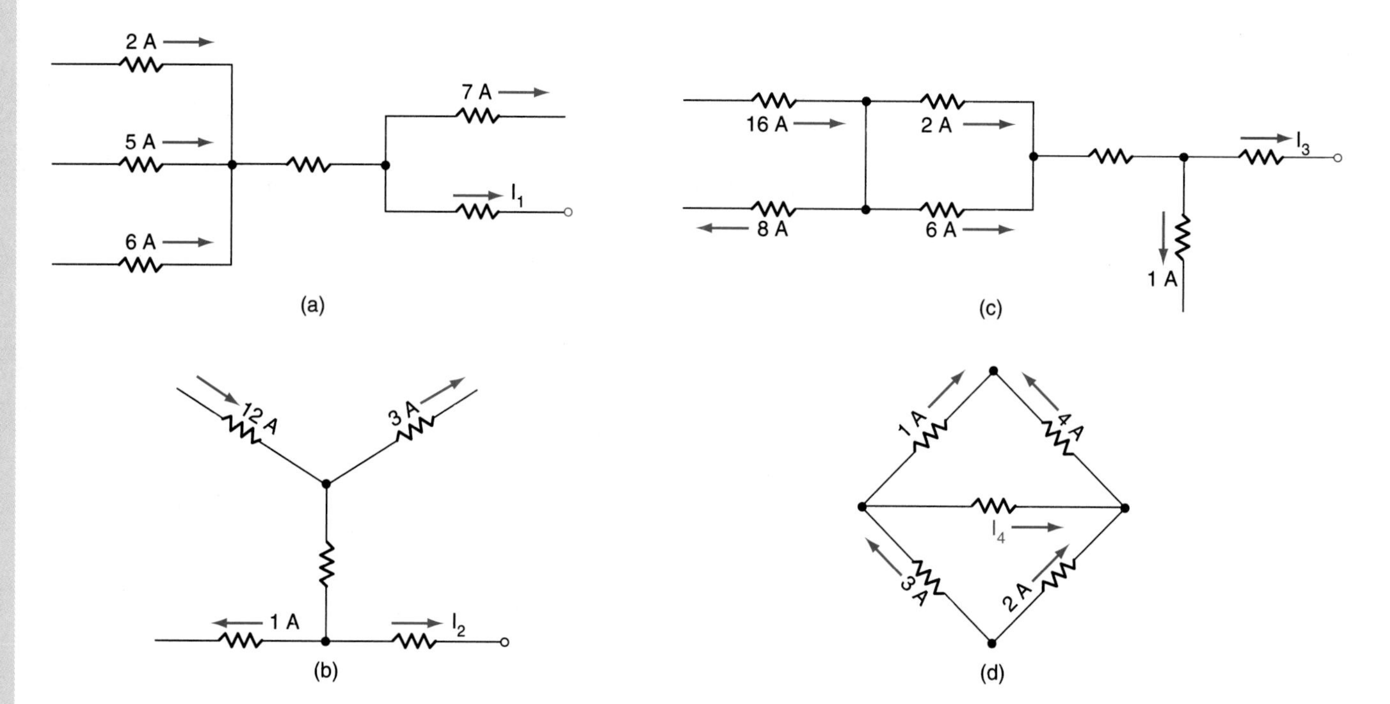

Figure 3–80.

Kirchhoff's Current Law

Complete Activity 3–8, page 69 in the manual.

ANALYSIS

1. State Kirchhoff's current law.
2. What are some applications of Kirchhoff's current law?
3. Solve the problems of Fig. 3–81 by applying Kirchhoff's current law.

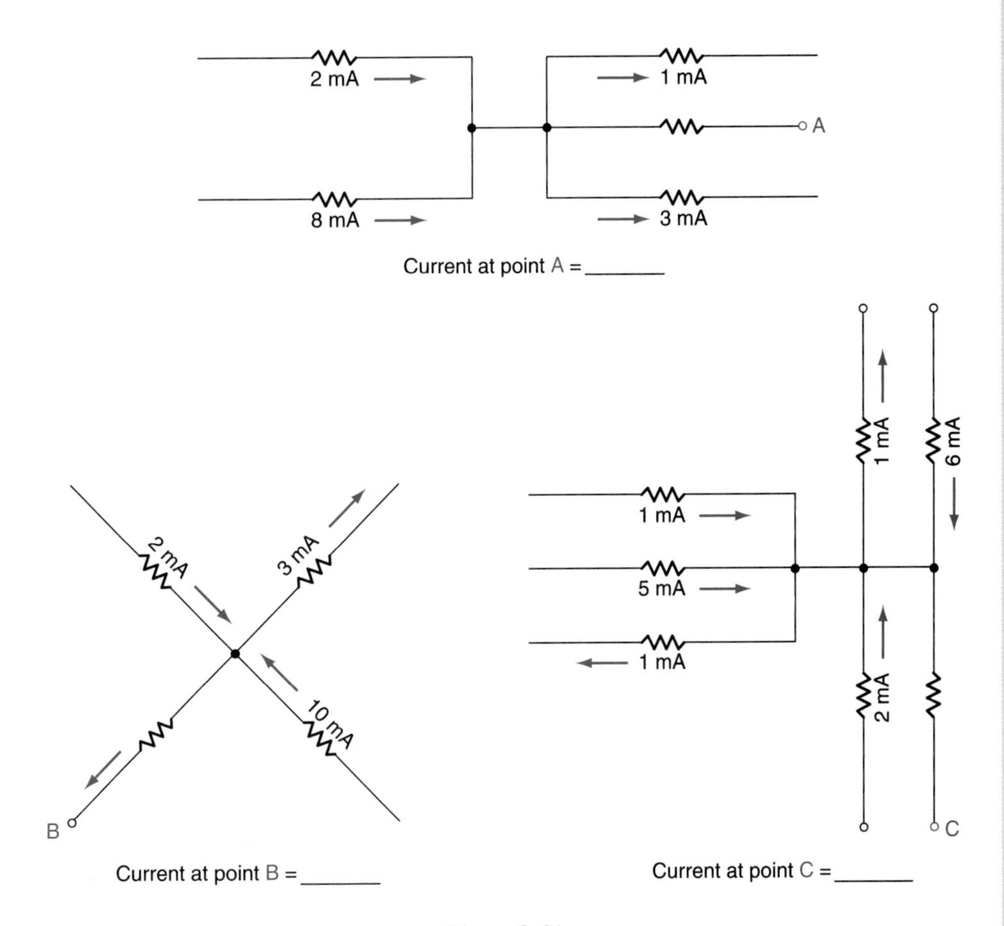

Figure 3–81.

Maximum Power Transfer Problems

Solve the following maximum power transfer problems.

1. Find the values of load current (I_L), voltage output (V_{out}), and power output (P_{out}) for the circuit of Fig. 3–82, using load resistance values of 0, 1, 2, 3, 4, 5, 6, 7, and 8 Ω.
2. Draw a power transfer curve using the values obtained for Fig. 3–82. Plot power output (watts) on the vertical axis and load resistance (R_L) on the horizontal axis.
3. Find the values of V_{out} and P_{out} for the circuit of Fig. 3–82 if the value of input voltage is changed to 20 V, and R_S is changed to 3 Ω. Use the same R_L values as the preceding problem.

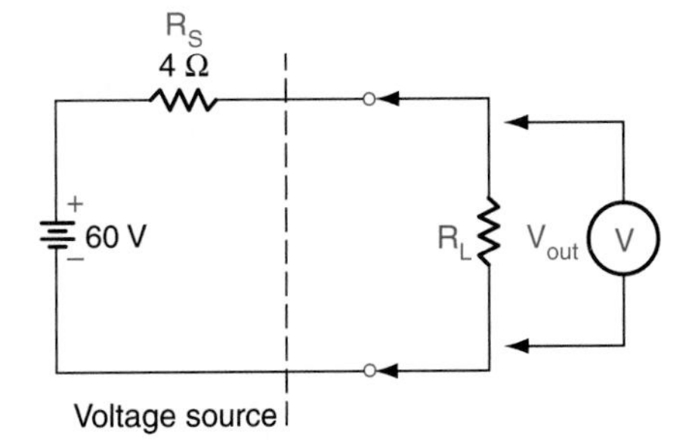

Figure 3–82.

Maximum Power Transfer

Complete Activity 3–9, page 71 in the manual.

ANALYSIS

1. On a sheet of graph paper, plot the relationship of R_L (horizontal axis) and power (vertical axis) for the values given here:

R_L	P_{out}
0 Ω	0 W
250 Ω	4500 W
500 Ω	5000 W
750 Ω	4800 W
1000 Ω	4400 W

2. State the maximum power transfer theorem.

Superposition Problems

Solve each of the following problems by applying the superposition method.

1. Find the current flow through R_1, R_2, and R_3 in the circuit of Fig. 3–83.

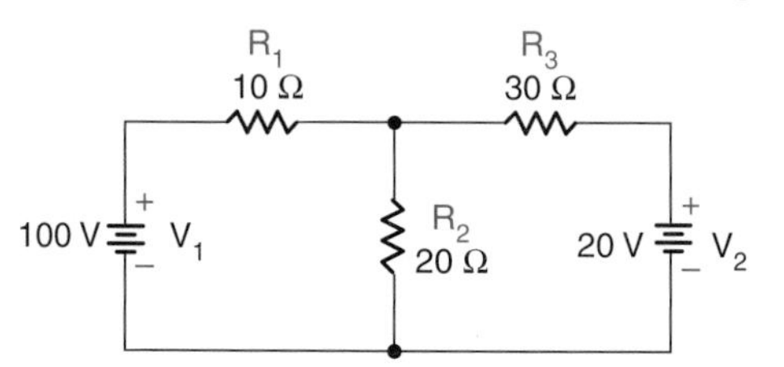

Figure 3–83.

2. Find the current flow through R_1, R_2, R_3, R_4, and R_5 in the circuit of Fig. 3–84.

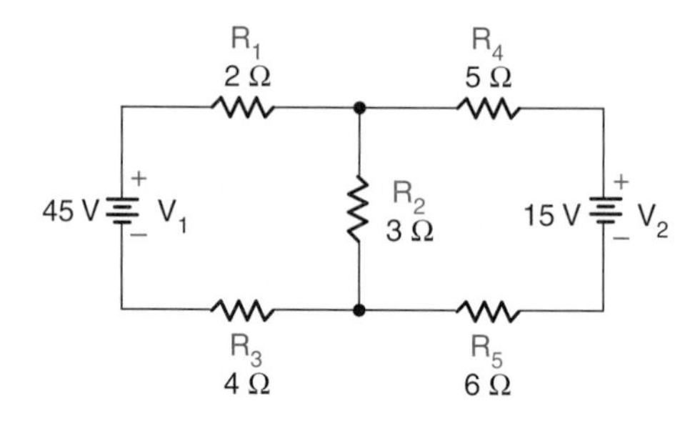

Figure 3–84.

Thevinin Equivalent Circuit Problems

Solve each of the following problems by applying Thevinin's equivalent circuit method.

1. Find the Thevinin voltage (V_{TH}) and Thevinin resistance (R_{TH}) for the circuit of Fig. 3–85. Sketch the Thevinin equivalent circuit.
2. Refer to the values obtained for Fig. 3–85. Calculate the values of load current (I_L) and voltage output (V_{out}) for load resistance values of: (a) 2 Ω, (b) 3 Ω, (c) 4 Ω, and (d) 5 Ω.

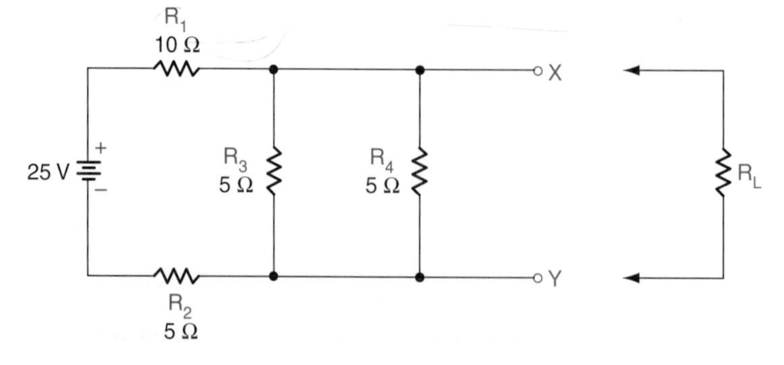

Figure 3–85.

3. Find the Thevinin voltage (V_{TH}) and Thevinin resistance (R_{TH}) for the circuit of Fig. 3–86. Sketch the Thevinin equivalent circuit.
4. Refer to the values obtained for Fig. 3–86. Calculate the values of load current (I_L) and output voltage (V_{out}) for load resistance values of: (a) 20 Ω, (b) 30 Ω, and (c) 50 Ω.

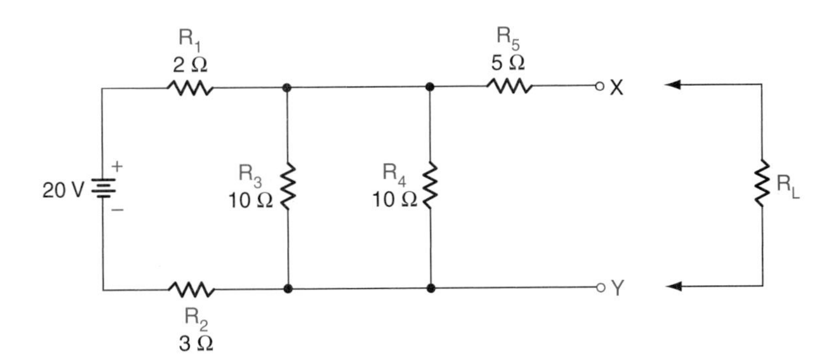

Figure 3–86.

5. Find the values of V_{TH} and R_{TH} for the two-source circuit of Fig. 3–87.
6. Calculate I_L and V_{out} for R_L values of: (a) 10 Ω, (b) 20 Ω, (c) 30 Ω (for Fig. 3–87).

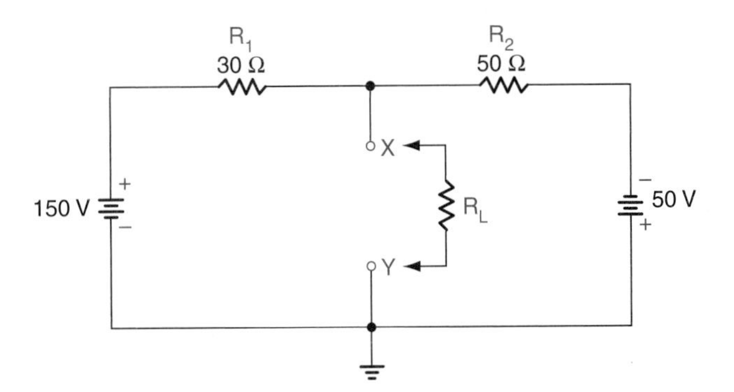

Figure 3–87.

Norton Equivalent Circuits

Solve the following problem by applying Norton's equivalent circuit method.
Find the Norton current (I_N) and Norton resistance (R_N) for the circuit of Fig. 3–88. Sketch the Norton equivalent circuit.

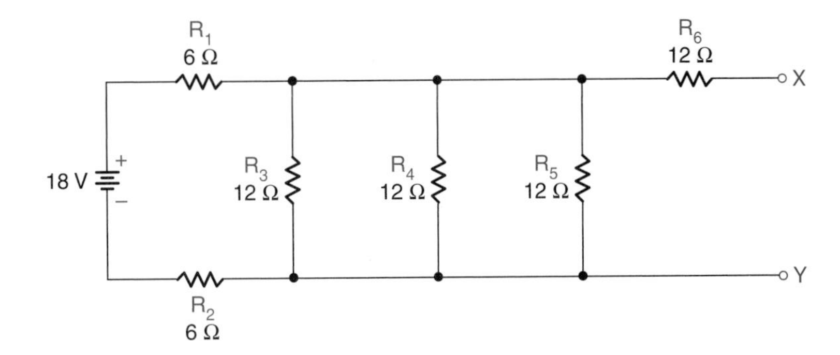

Figure 3–88.

Bridge-Circuit Simplification

Solve the following problems by applying bridge-circuit simplification.
1. Find the values of V_{TH} and R_{TH} for the bridge circuit of Fig. 3–88.

2. Refer to the values obtained for Fig. 3–89. Calculate the values of load current (I_L) and output voltage for R_L values of: (a) 3 Ω, (b) 5 Ω, and (c) 8 Ω.

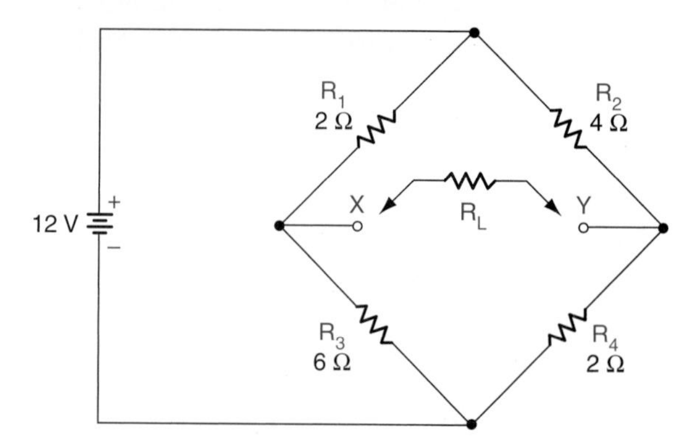

Figure 3–89.

Self-Examination

1. Complete the Self-Examination for Chap. 3, page 73 in the manual.

2. Check your answers on pages 410–411 in the manual.

3. Study the questions which were answered incorrectly.

Chapter 3 Examination

Complete the Chap. 3 Examination on page 81 in the manual.

CHAPTER

4

MAGNETISM AND ELECTROMAGNETISM

OUTLINE

4.1 Permanent Magnets
4.2 Magnetic Field around Conductors
4.3 Magnetic Field around a Coil
4.4 Electromagnets
4.5 Ohm's Law for Magnetic Circuits
4.6 Domain Theory of Magnetism
4.7 Electricity Produced by Magnetism
4.8 Magnetic Devices
4.9 Magnetic Terms
4.10 Hall Effect
4.11 Magnetic Levitation
4.12 Rare Earth Magnets

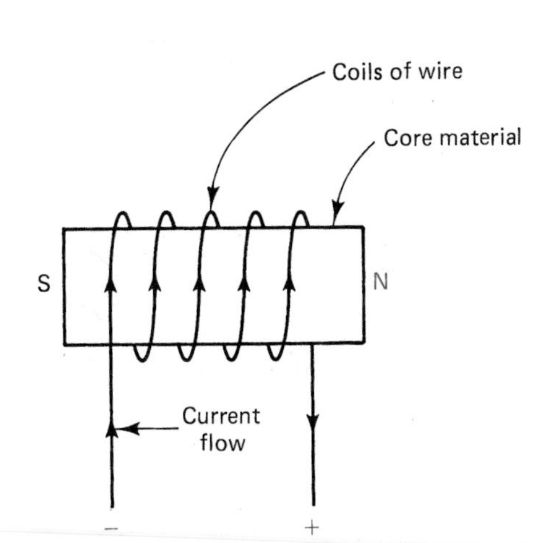

Magnetism and electromagnetism have many applications.

OBJECTIVES

Upon completion of this chapter, you will be able to:

1. Define the various terms relative to magnetism.
2. Explain the operation of various magnetic devices.
3. State Faraday's law for electromagnetic induction.
4. List three factors that affect the strength of electromagnets.
5. Apply the left-hand rule for polarity.
6. Describe the construction of a relay and solenoid.
7. Define the terms *residual magnetism, permeability, retentivity, magnetic saturation,* and *magnetizing force.*
8. Describe the domain theory of magnetism.

Magnetism has been a topic of study for many years. Some metals in their natural state attract small pieces of iron. This property is called *magnetism.* Materials that have this ability are called natural magnets. The first magnets used were called lodestones. Now, artificial magnets are made in many different strengths, sizes, and shapes. Magnetism is important because it is used in electric motors, generators, transformers, relays, and many other electrical devices. The earth itself has a magnetic field like a large magnet.

Electromagnetism is magnetism which is brought about due to electrical current flow. There are many electrical machines which operate because of electromagnetism. This chapter deals with magnetism, electromagnetism, and some important applications.

IMPORTANT TERMS

This chapter will discuss magnetism and electromagnetism, and their many important applications. The terms that follow help to define some of the specifics involved in the study of magnetism and electromagnetism.

Alnico. An alloy of aluminum, nickel, iron, and cobalt used to make permanent magnets.

Ampere-turn. The unit of measurement of magnetomotive force (MMF); amperes of current times the number of turns of wire.

Armature. The movable part of a relay.

Coefficient of coupling (k). A decimal value that indicates the amount of magnetic coupling between coils.

Core. Iron or steel materials of internal sections of electromagnets around which coils are wound.

Core saturation. When the atoms of a metal core material are aligned in the same pattern so that no more magnetic lines of force can be developed.

Coupling. The amount of mutual inductance between coils.

Domain theory. A theory of magnetism that assumes groups of atoms produced by movement of electrons align themselves in groups called "domains" in magnetic materials.

Electromagnet. A coil of wire wound on an iron core so that as current flows through the coil it becomes magnetized.

Flux (ϕ). Invisible lines of force that extend around a magnetic material.

Flux density. The number of lines of force per unit area of a magnetic material or circuit.

Gauss. A unit of measurement of magnetic flux density.

Gilbert. A unit of measurement of magnetomotive force (MMF).

Hysteresis. The property of a magnetic material that causes actual magnetizing action to lag behind the force that produces it.

Laws of magnetism. (1) Like magnetic poles repel; (2) unlike magnetic poles attract.

Lines of force. Same as magnetic flux. *See* Flux.

Lodestone. The name used in early times for natural magnets.

Magnet. A metallic material, usually iron, nickel, or cobalt, which has magnetic properties.

Magnetic circuit. A complete path for magnetic lines of force from a north to a south pole.

Magnetic field. Magnetic lines of force that extend from a north pole and enter a south pole to form a closed loop around the outside of a magnetic material.

Magnetic flux. *See* Flux.

Magnetic materials. Metallic materials such as iron, nickel, and cobalt which exhibit magnetic properties.

Magnetic poles. Areas of concentrated lines of force on a magnet which produce north and south polarities.

Magnetic saturation. A condition that exists in a magnetic material when an increase in magnetizing force does not produce an increase in magnetic flux density in the material.

Magnetomotive force (MMF). A force that produces magnetic flux around a magnetic device.

Magnetostriction. The effect that produces a change in shape of certain materials when they are placed in a magnetic field.

Natural magnet. Metallic materials that have magnetic properties in their natural state.

Permanent magnet. Bars or other shapes of materials that retain their magnetic properties.

Permeability (μ)**.** The ability of a material to conduct magnetic lines of force as compared with air.

Polarities. *See* Magnetic poles.

Relay. An electromagnetically operated switch.

Reluctance ($\Re$)**.** The opposition of a material to the flow of magnetic flux.

Residual magnetism. The magnetism that remains around a material after the magnetizing force has been removed.

Retentivity. The ability of a material to retain magnetism after a magnetizing force has been removed.

Solenoid. An electromagnetic coil with a metal core that moves when current passes through the coil.

4.1 PERMANENT MAGNETS

Magnets are made of iron, cobalt, or nickel materials, usually in an alloy combination. An alloy is a mixture of these materials. Each end of the magnet is called a *pole.* If a magnet were broken, each part would become a magnet. Each magnet would have two poles. Magnetic poles are always in pairs. When a magnet is suspended in air so that it can turn freely, one pole will point to the North Pole of the earth. The earth is like a large permanent magnet. This is why compasses can be used to determine direction. The north pole of a magnet will *attract* the south pole of another magnet. A north pole *repels* another north pole and a south pole *repels* another south pole. The two laws of magnetism are: (1) like poles repel, and (2) unlike poles attract.

The magnetic field patterns when two permanent magnets are placed end to end are shown in Fig. 4–1. When the magnets are farther apart, a smaller force of attraction or repulsion exists. This type of permanent magnet is called a *bar magnet.*

Some materials retain magnetism longer than others. Hard steel holds its magnetism much longer than soft steel. A magnetic field is set up around any magnetic material. The field is made up of lines of force or magnetic flux. These magnetic flux lines are invisible. They never cross one another, but they always form individual closed loops around a magnetic material. They have a definite direction from the north to the south pole along the outside of a magnet. When magnetic flux lines are close together, the magnetic field is strong. When magnetic flux lines are farther apart, the field is weaker. The magnetic field is strongest near the poles. Lines of force pass through all materials. It is easy for lines of force to pass through iron and steel. Magnetic flux passes through pieces of iron as shown in Fig. 4–2.

When magnetic flux passes through a piece of iron, the iron acts like a magnet. Magnetic poles are formed due to the influence of the flux lines. These are called *induced poles.* The induced poles and the magnet's poles attract and repel each other. Magnets attract pieces of soft iron in this way. It is possible to temporarily magnetize pieces of metal by using a bar magnet. If a magnet is passed over the top of a piece of iron several times in the same direction, the soft iron becomes magnetized. It will stay magnetized for a short time.

When a compass is brought near the north pole of a magnet, the north-seeking pole of the compass is attracted to it. The polarities of the magnet may be determined by observing a compass brought near each pole. Compasses detect the presence of magnetic fields.

Horseshoe magnets are similar to bar magnets. They are bent in the shape of a horseshoe, as shown in Fig. 4–3. This shape gives more magnetic field strength than a similar bar magnet, because the magnetic poles are closer. The magnetic field strength is more concentrated into one area. Many electrical devices use horseshoe magnets.

A magnetic material can lose some of its magnetism if it is jarred or heated. People must be careful when handling equipment that contains permanent magnets. A magnet also becomes weakened by loss of magnetic flux. Magnets should always be stored with a *keeper,* which is a soft-iron piece used to join magnetic poles. The keeper provides the magnetic flux with an easy path between poles. The magnet will retain its greatest strength for a longer period of time if keepers are used. Bar magnets should always be stored in pairs with a north pole and a south pole placed together. A complete path for magnetic flux is made in this way.

4.2 MAGNETIC FIELD AROUND CONDUCTORS

Current-carrying conductors produce a magnetic field. It is possible to show the presence of a magnetic field around a current-carrying conductor. A compass is used to show that the magnetic flux lines are circular in shape. The conductor is in the center of the circular shape. The direction of the current flow and the magnetic flux lines can be shown by using the *left-hand rule* of magnetic flux. A conductor is held in the left hand as shown in Fig. 4–4(a). The thumb points in the direction of current flow from negative to positive. The fingers will then encircle the conductor in the direction of the magnetic flux lines.

A circular magnetic field is produced around a conductor. The field is stronger near the conductor and becomes weaker farther away from the conductor. A cross-sectional end view of a conductor with current flowing toward the observer is shown in Fig. 4–4(b). Current flow toward the observer is shown by a circle with a dot in the center. Notice

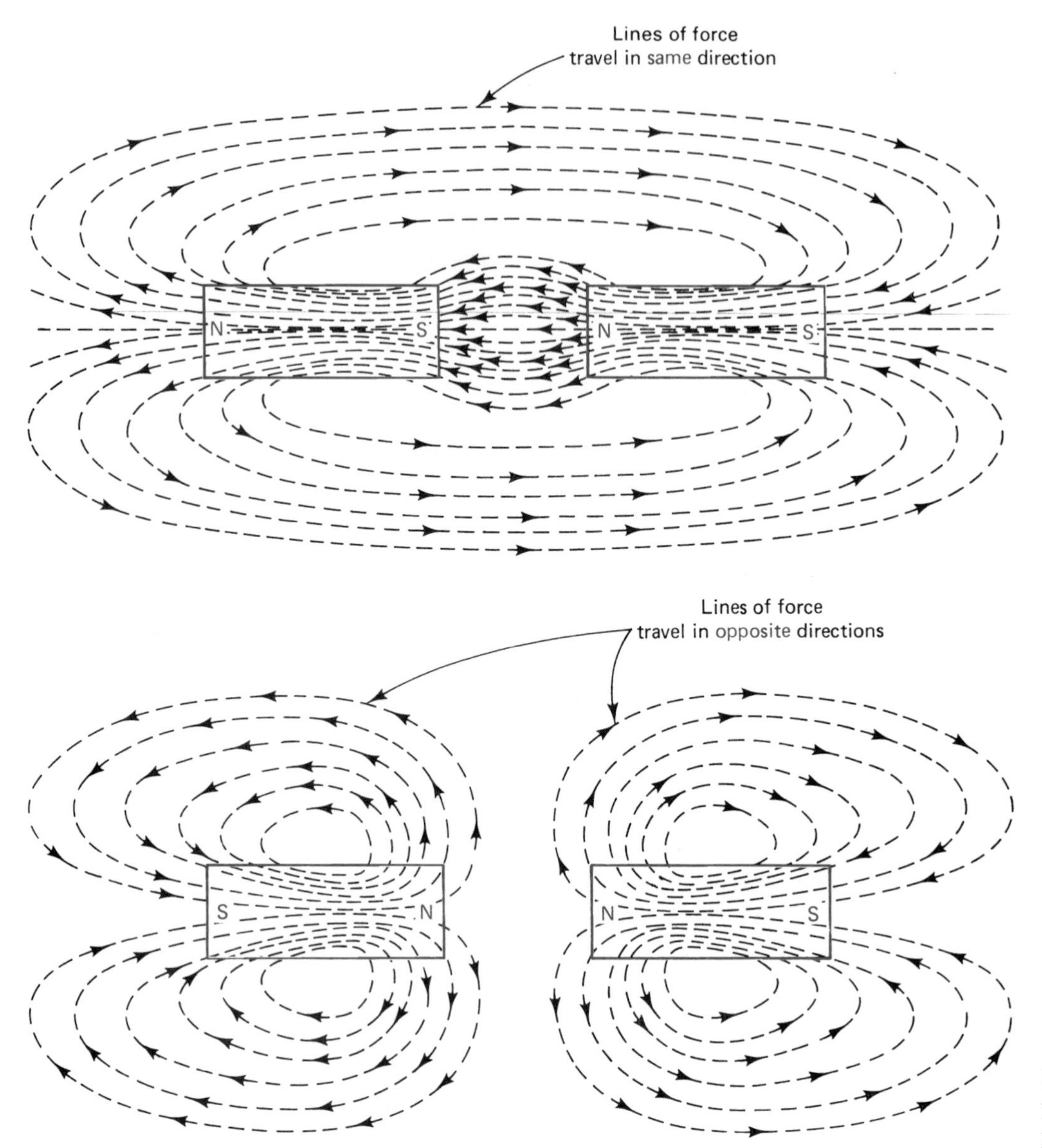

Figure 4–1. Magnetic field patterns when magnets are placed end to end.

that the direction of the magnetic flux lines is clockwise. This can be verified by using the left-hand rule.

When the direction of current flow through a conductor is reversed, the direction of the magnetic lines of force is also reversed. The cross-sectional end view of a conductor in Fig. 4–4(c) shows a current flow in a direction away from the observer. Notice that the direction of the magnetic lines of force is now counterclockwise.

The presence of magnetic lines of force around a current-carrying conductor can be observed by using a compass. When a compass is moved around the outside of a conductor, the needle will align itself tangent to the lines of force as shown in Fig. 4–4(d). The needle will not point toward the conductor. When current flows in the opposite direction, the compass polarities reverse. The compass needle will align itself tangent to the conductor.

4.3 MAGNETIC FIELD AROUND A COIL

The magnetic field around one loop of wire is shown in Fig. 4–5. Magnetic flux lines extend around the conductor as shown. Inside the loop, the magnetic flux is in one direction. When many loops are joined together to form a coil, the magnetic flux lines surround the coil as shown in Fig. 4–6. The field around a coil is much stronger than the field of one loop of wire. The field around the coil is the same shape as the field around a bar magnet. A coil that has an iron or steel core inside it is called an *electromagnet.* A core increases the magnetic flux density of a coil.

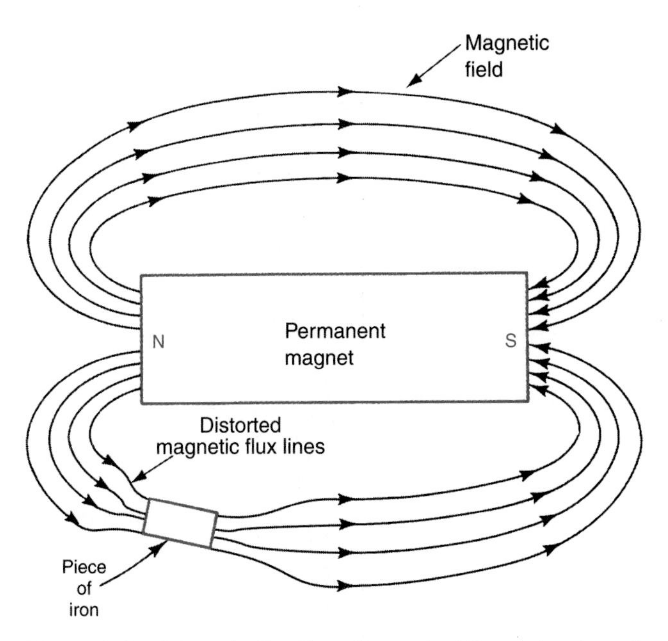

Figure 4–2. Magnetic flux lines distorted while passing through a piece of iron.

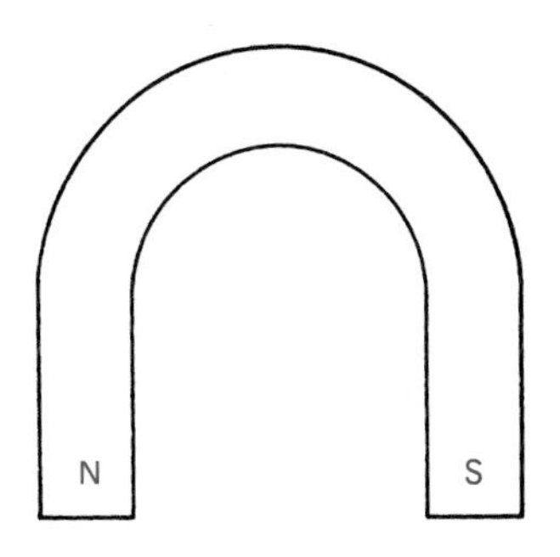

Figure 4–3. Horseshoe magnet.

4.4 ELECTROMAGNETS

Electromagnets are produced when current flows through a coil of wire as shown in Fig. 4–7 on page 135. The north pole of a coil of wire is the end where the lines of force exit. The south pole is the end where the lines of force enter the coil. This is like the field of a bar magnet. To find the north pole of a coil, use the left-hand rule for polarity, as shown in Fig. 4–8 on page 135. Grasp the coil with the left hand. Point the fingers in the direction of current flow through the coil. The thumb points to the north polarity of the coil.

When the polarity of the voltage source is reversed, the magnetic poles of the coil will also reverse. The poles of an electromagnet can be checked with a compass. The compass is placed near a pole of the electromagnet. If the north-seeking pole of the compass points to the coil, that side is the north side.

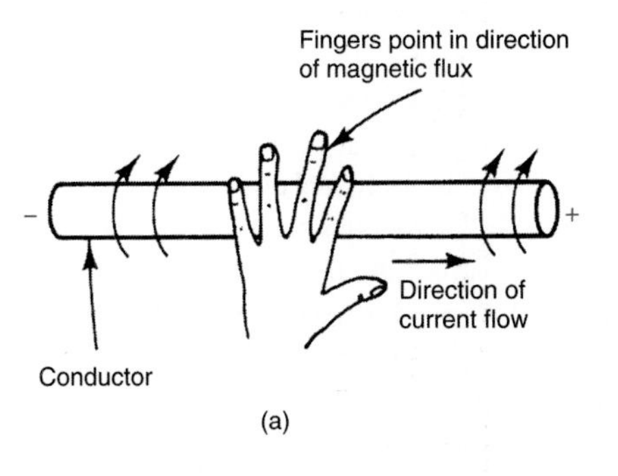

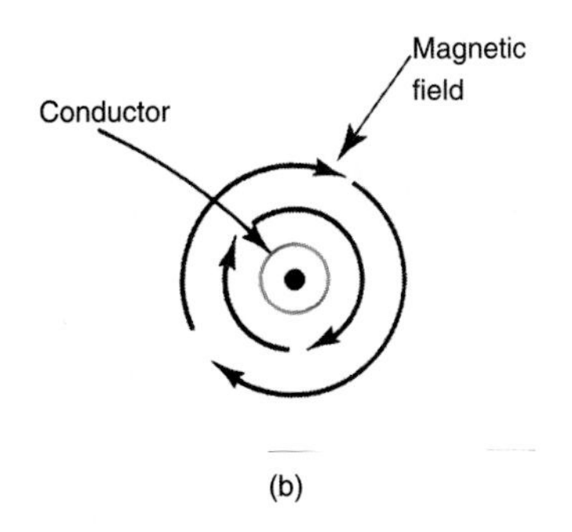

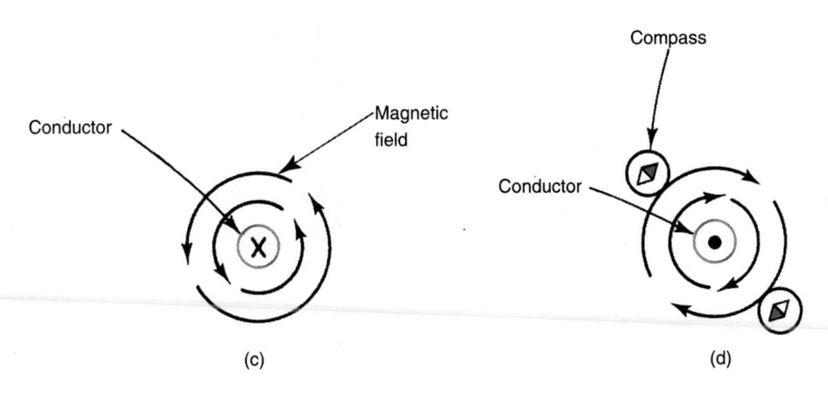

Figure 4–4. Magnetic fields: (a) left-hand rule of magnetic flux; (b) cross section of a conductor with current flow *toward* the observer; (c) cross section of a conductor with current flow *away from* the observer; (d) compass aligns *tangent* to the circular magnetic lines of force.

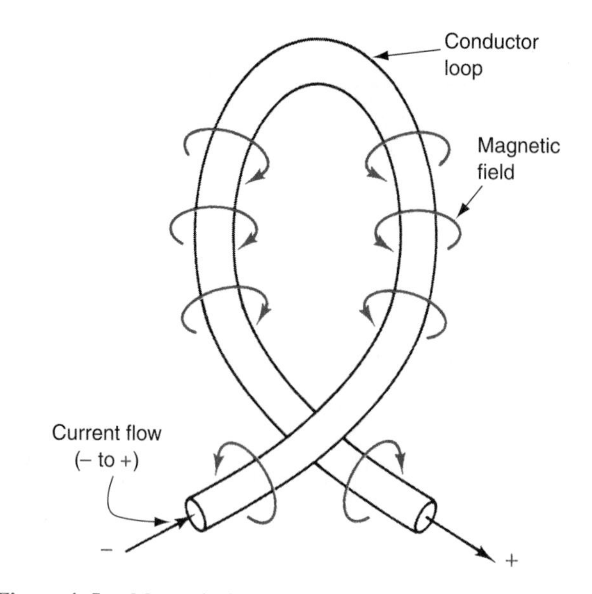

Figure 4–5. Magnetic field around a loop of wire.

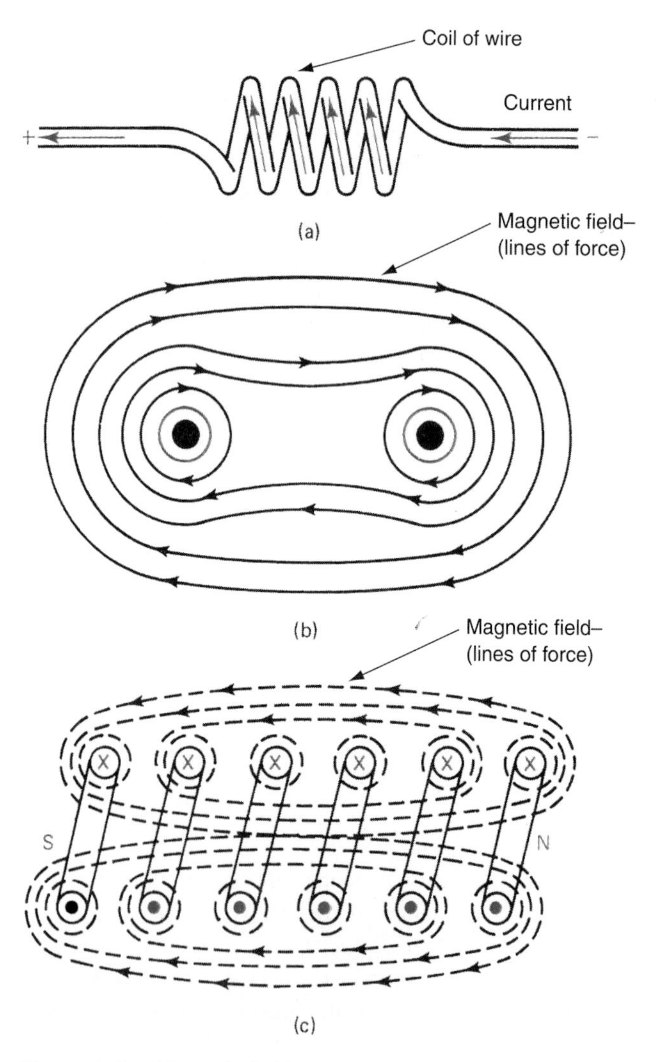

Figure 4–6. Magnetic field around a coil: (a) coil of wire showing current flow; (b) lines of force around two loops that are parallel; (c) cross section of a coil showing lines of force.

Electromagnets have several turns of wire wound around a soft-iron core. An electrical power source is then connected to the ends of the turns of wire. When current flows through the wire, magnetic polarities are produced at the ends of the soft-iron core. The three basic parts of an electromagnet are (1) an iron core, (2) wire windings, and (3) an electrical power source. Electromagnetism is made possible by electrical current flow which produces a magnetic field. When electrical current flows through the coil, the properties of magnetic materials are developed.

Magnetic Strength of Electromagnets

The magnetic strength of an electromagnet depends on three factors: (1) the amount of current passing through the coil, (2) the number of turns of wire, and (3) the type of core material. The number of magnetic lines of force is increased by increasing the current, by increasing the number of turns of wire, or by using a more desirable type of core material. The magnetic strength of electromagnets is determined by the *ampere-turns* of each coil. The number of ampere-turns is equal to the current in amperes multiplied by the number of turns of wire ($I \times N$). For example, 200 ampere-turns is produced by 2 A of current through a 100-turn coil. One ampere of current through a 200-turn coil would produce the same magnetic field strength. Figure 4–9 shows how the magnetic field strength of an electromagnet changes with the number of ampere-turns.

The magnetic field strength of an electromagnet also depends on the type of core material. Cores are usually made of soft iron or steel. These materials will transfer a magnetic field better than air or other nonmagnetic materials. Iron cores increase the *flux density* of an electromagnet. Figure 4–10 shows that an iron core causes the magnetic flux to be more dense.

An electromagnet loses its field strength when the current stops flowing. However, an electromagnet's core retains a small amount of magnetic strength after current stops flowing. This is called *residual magnetism* or leftover magnetism. It can be reduced by using soft-iron cores or increased by using hard-steel core material. Residual magnetism is very important in the operation of some types of electrical generators.

In many ways, electromagnetism is similar to magnetism produced by natural magnets such as bar magnets; however, the main advantage of electromagnetism is that it is easily controlled. It is easy to increase the strength of an electromagnet by increasing the current flow through the coil, which is done by increasing the voltage applied to the coil. The second way to increase the strength of an electromagnet is to have more turns of wire around the core. A greater number of turns produces more magnetic lines of force around the electromagnet. The strength of an electromagnet is also affected by the type of core material used. Different alloys of iron are used to make the cores of electromagnets. Some ma-

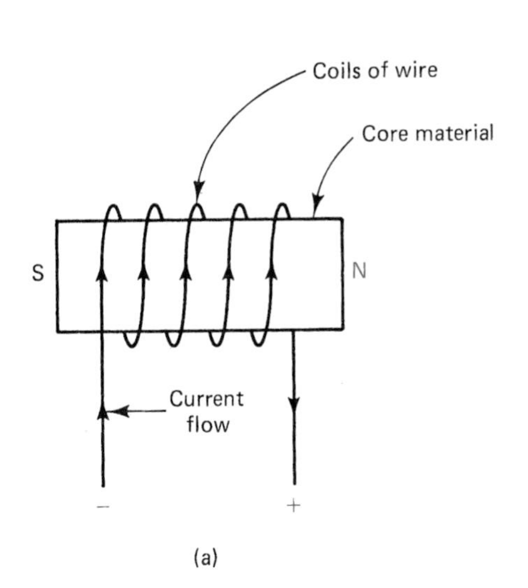

Figure 4–7. Electromagnets: (a) pictorial; (b) an electromagnet in operation. [(b) Courtesy of O.S. Walker Co.]

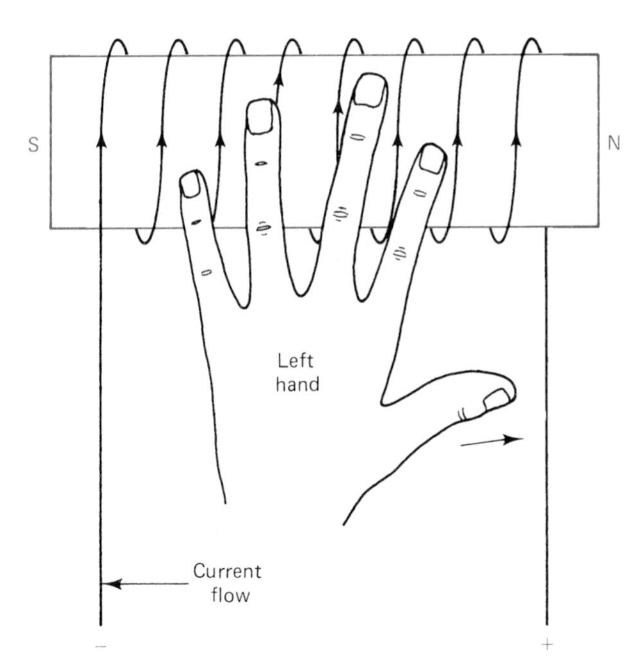

Figure 4–8. Left-hand rule for finding the polarities of an electromagnet.

terials aid in the development of magnetic lines of force to a greater extent. Other types of core materials offer greater resistance to the development of magnetic flux around an electromagnet.

4.5 OHM'S LAW FOR MAGNETIC CIRCUITS

Ohm's law for electrical circuits was studied in Chap. 3. A similar relationship exists in magnetic circuits. Magnetic circuits have *magnetomotive force* (MMF), *magnetic flux* (ϕ), and *reluctance* (R). MMF is the force that causes a magnetic flux to be developed. Magnetic flux is the lines of force around a magnetic material. Reluctance is the opposition to the flow of a magnetic flux. These terms may be compared with voltage, current, and resistance in electrical circuits, as shown in Fig. 4–11. When MMF increases, magnetic flux increases. Remember that in an electrical circuit, when voltage increases, current increases. When resistance in an electrical circuit increases, current decreases. When reluctance of a magnetic circuit increases, magnetic flux decreases. The relationship of magnetic and electrical terms in Fig. 4–11 is important to learn.

4.6 DOMAIN THEORY OF MAGNETISM

A theory of magnetism was presented in the nineteenth century by a German scientist named Wilhelm Weber. Weber's theory of magnetism was called the molecular theory. It dealt with the alignment of molecules in magnetic materials. Weber believed that molecules were aligned in an *orderly* arrangement in magnetic materials. In nonmagnetic materials, he believed that molecules were arranged in a *random* pattern.

Weber's theory has now been modified somewhat to become the domain theory of magnetism. This theory deals with the alignment of "domains" in materials rather than molecules. A *domain* is a group of atoms (about 10^{15} atoms). Each domain acts like a tiny magnet. The rotation of electrons around the nucleus of these atoms is important. Electrons have a negative charge. As they orbit around the nucleus of atoms, their electrical charge moves. This moving electrical field produces a magnetic field. The polarity of the magnetic field is determined by the direction of electron rotation.

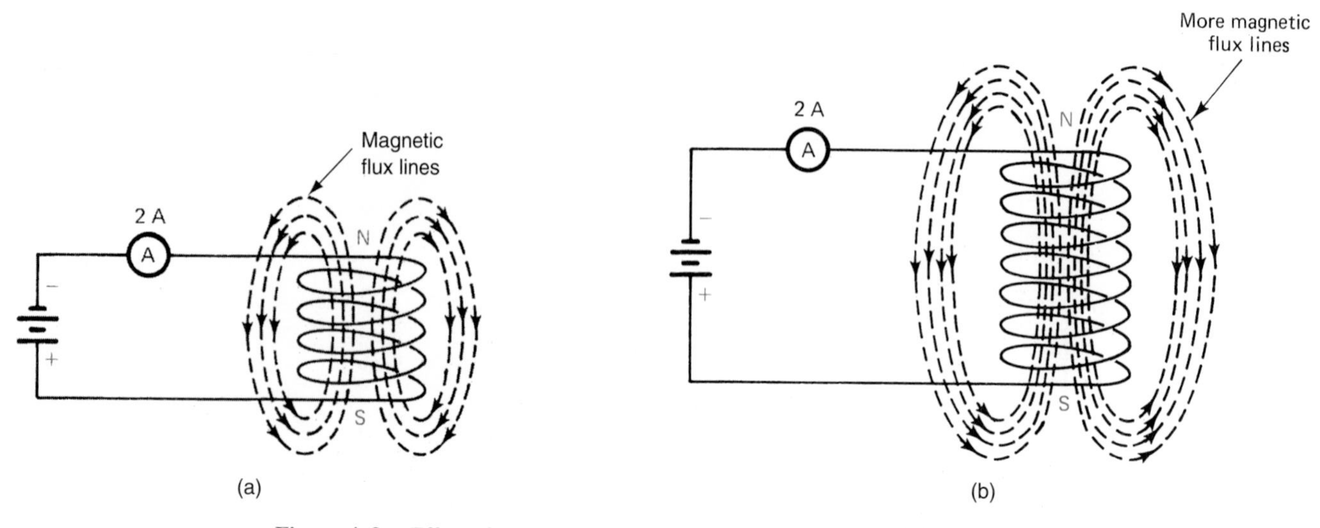

Figure 4–9. Effect of ampere-turns on magnetic field strength: (a) five turns and two amperes = 10 ampere-turns; (b) eight turns and two amperes = 16 ampere-turns.

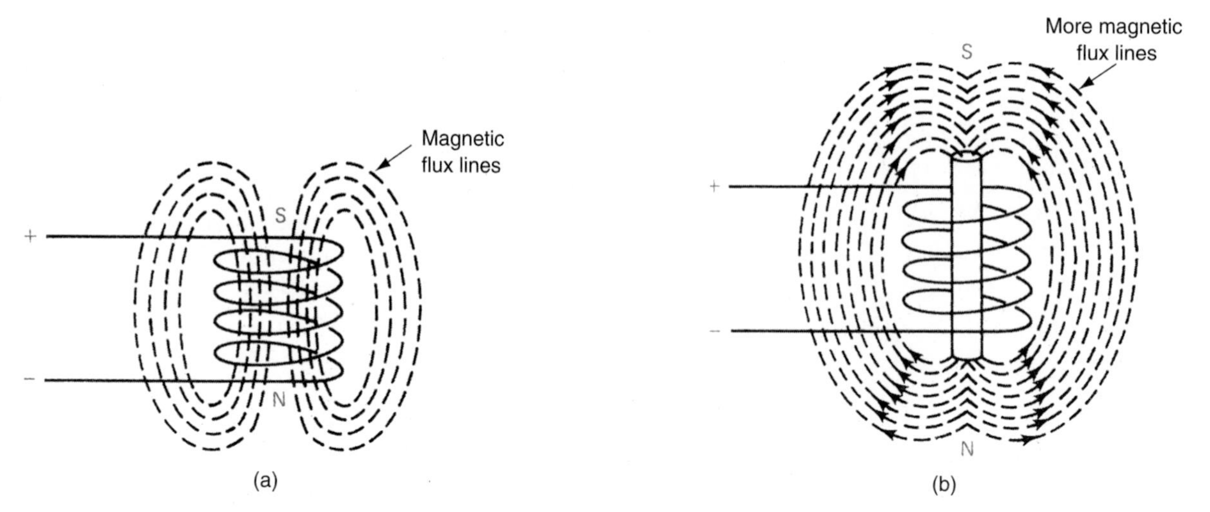

Figure 4–10. Effect of an iron core on magnetic strength: (a) coil without a core; (b) coil with a core.

The domains of magnetic materials are atoms grouped together. Their electrons are believed to spin in the same directions. This produces a magnetic field due to electrical charge movement. Figure 4–12 shows the arrangement of domains in magnetic, nonmagnetic, and partially magnetized materials. In nonmagnetic materials, half of the electrons spin in one direction and half in the other direction. Their charges cancel each other out. There is no magnetic field produced because the charges cancel. Electron rotation in magnetic materials is in the same direction. This causes the domains to act like tiny magnets that align to produce a magnetic field.

4.7 ELECTRICITY PRODUCED BY MAGNETISM

A scientist named Michael Faraday discovered in the early 1830s that electricity is produced from magnetism. He found that if a magnet is placed inside a coil of wire, electrical current is produced when the magnet is *moved.* Faraday found that electrical current is caused by magnetism and motion.

Faraday's law is stated as follows: When a coil of wire moves across the lines of force of a magnetic field, electrons flow through the wire in one direction. When the coil of wire moves across the magnetic lines of force in the opposite

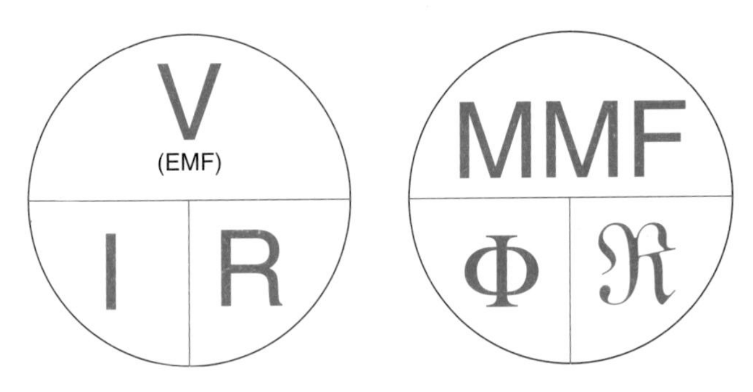

Figure 4–11. Relationship of magnetic and electrical terms.

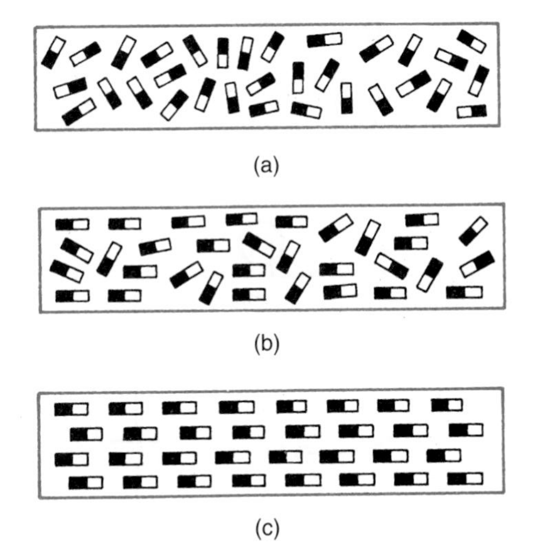

Figure 4–12. Domain theory of magnetism: (a) unmagnetized; (b) slightly magnetized; (c) fully magnetized—saturation.

direction, electrons flow through the wire in the opposite direction.

This law is the principle of electrical power generation produced by magnetism. Figure 4–13 shows Faraday's law as it relates to electrical power generation.

Current flows in a conductor placed inside a magnetic field only when there is motion between the conductor and the magnetic field. If a conductor is stopped while moving across the magnetic lines of force, current stops flowing. The operation of electrical generators depends on conductors moving across a magnetic field. This principle, called electromagnetic induction, is discussed in more detail in Chap. 5.

4.8 MAGNETIC DEVICES

Many types of electrical devices operate due to the effects of magnetism or electromagnetism. Among these devices are relays, solenoids, and magnetic motor contactors.

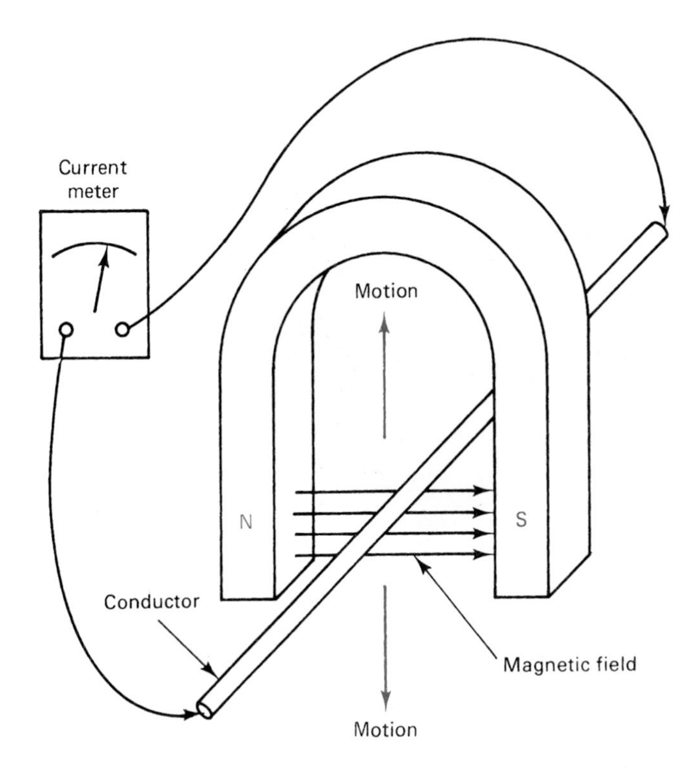

Figure 4–13. Faraday's law—electrical current is produced when there is relative motion between a conductor and a magnetic field.

Relays

Relays are electrical devices that rely on magnetism to operate. They control other equipment such as motors, lights, or heating elements. Relays are important devices. They are used to start the operation of other equipment. They use a small amount of electrical current to control a larger current, such as the current through a motor.

The basic construction and symbols of a relay are shown in Fig. 4–14. A relay has an electromagnetic coil with electrical power applied to its two external leads. When the power is turned on, the electromagnet is energized. The electromagnet part of the relay controls a set of contacts. The contacts are called *normally open* (NO) or *normally closed* (NC), depending on their conditions when the electromagnet is not energized. There is also *common* contact.

If a lamp and its power source are connected in series with the common and normally open contact as shown in Fig. 4–14(b), the lamp will be off when the relay is not energized. Notice that the lamp or any load connected to the relay contacts requires a separate power source. If the relay is energized by applying power to it, the common contact is attracted to the normally open contact by magnetic energy. The common contact is built onto a movable armature that moves when the electromagnet is energized. When the relay is energized, the light connected to the normally open contact will turn on.

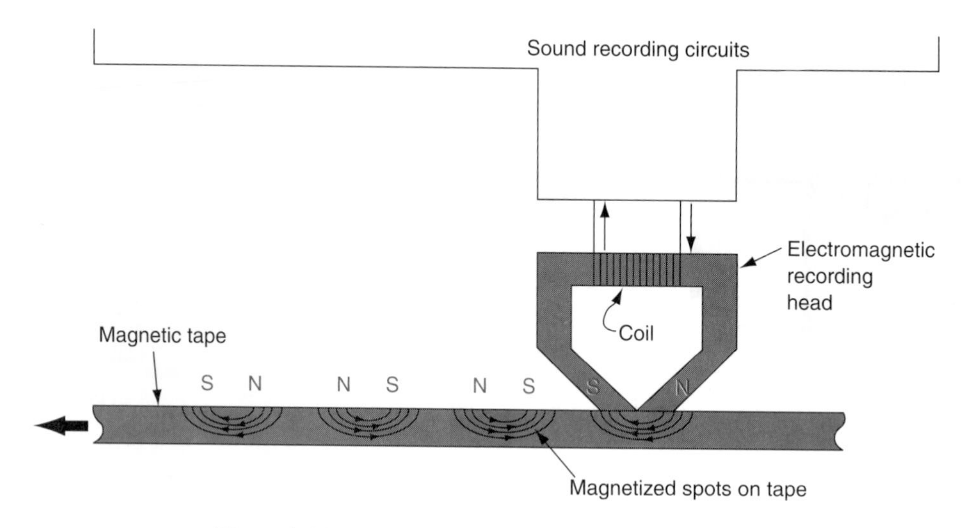

Figure 4–22. Electromagnetic tape recording principle.

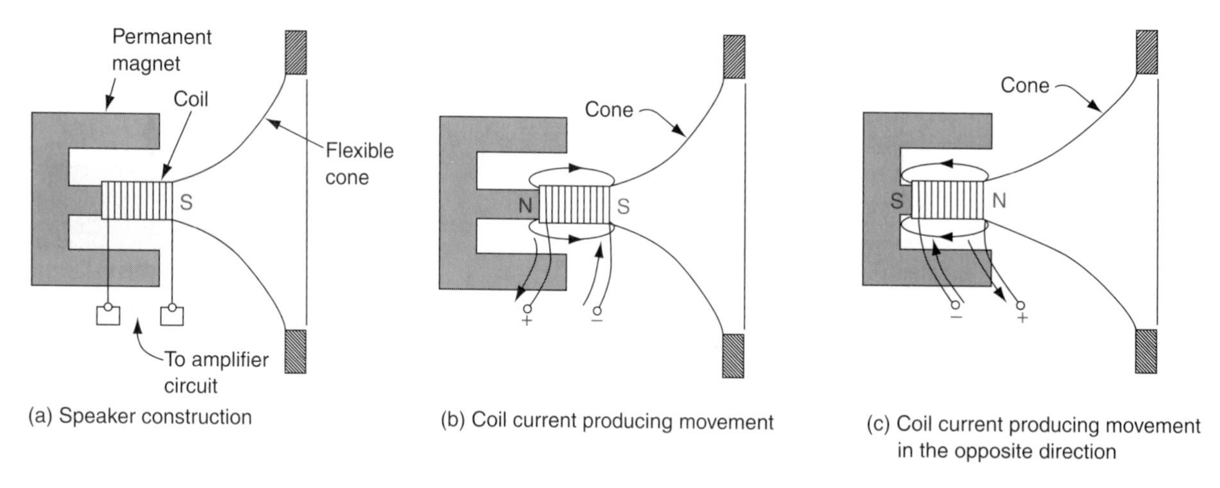

Figure 4–23. Electromagnetic speaker.

to a hollow cylinder with a coil wound around it. A permanent magnet is positioned within the electromagnetic coil. When current flows through the coil in one direction, the coil will move to the right. Current flow in the opposite direction causes the coil to move in the other direction. Movement of the coil causes the flexible diaphragm to move in a direction based on current flow direction. More coil current causes a stronger electromagnetic field, which will cause the diaphragm to move a greater distance. The diaphragm vibrates in and out as the intensity of the sound input changes. The air vibrations which occur due to this action cause sound waves to be produced.

4.9 MAGNETIC TERMS

Several basic terms are important for understanding electromagnetic principles. One magnetic line of force is called a *maxwell.* The amount of magnetic flux (ϕ) is measured in a unit called the *weber.* A weber is equal to 10^8 (100,000,000) lines of force. *Flux density* is equal to the number of lines of force per square meter and is measured in the unit *tesla.*

$$\text{magnetic flux } (\phi) = \text{number of lines of force in webers}$$

and

$$\text{flux density } (B) = \frac{\text{magnetic flux } (\phi)}{\text{area } (A)}$$

Magnetomotive force (MMF) is the magnetic effect that causes a magnetic field to be produced. MMF or ampere-turns is equal to the current through a coil multiplied by the number of turns of wire in the coil.

$$\text{MMF} = I \times N \text{ (ampere-turns)}$$

where MMF = magnetomotive force, in ampere-turns
I = current, in amperes
N = number of turns in the coil

The length of a coil is also a factor that affects the field strength. The term *magnetizing force* (H) is used to express the magnetic field strength and is calculated as follows:

$$H = \frac{\text{MMF}}{l}$$

where H = magnetizing force in ampere-turns/meter (m)
MMF = magnetomotive force
l = length of coil, in meters (m)

Reluctance ($\Re$) is the opposition to the development of a magnetic field in an electromagnet.

$$\text{reluctance } (\Re) = \frac{\text{MMF (magnetomotive force)}}{\phi \text{ (magnetic flux)}}$$

(measured in ampere-turns/weber)

The relationship of MMF, ϕ, and $\Re$ in magnetic circuits to V, I, and R in electrical circuits shown in Fig. 4–11 should be reviewed.

Residual magnetism is an important effect in the operation of some types of electrical equipment. Residual magnetism is the ability of an electromagnet to hold a small magnetic field after electrical current is turned off. A small magnetic field remains around an electromagnet after it is demagnetized. This magnetic field is very weak.

Permeability (μ) is the ability of a magnetic material to transfer magnetic flux. It is the ability of a material to magnetize and demagnetize. Soft-iron material has a *high* permeability—it transfers magnetic flux easily. Soft iron magnetizes and demagnetizes rapidly. This makes soft iron a good material to use in the construction of generators, motors, transformers, and other electromagnetic devices. Permeability is similar to electrical conductance, which is a measure of how well a material allows current flow.

Magnetic

$$\text{permeability } (\mu) = \frac{1}{\Re \text{ (reluctance)}}$$

Electric

$$\text{conductance } (G) = \frac{1}{\text{R (resistance)}}$$

A related term is *relative permeability* (μ_r), which is a comparison of the permeability of a material to the permeability of air (1.0). Suppose that a material has a relative permeability of 1000. This means that the material will have 1000 times more magnetic flux than an equal amount of air. The relative permeability of materials is shown in Table 4–1.

Another magnetic term is *retentivity.* The retentivity of a material is its ability to retain a magnetic field after a magnetizing force is removed. Some materials will retain a magnetic flux for a long time. Other materials lose their magnetic flux almost immediately after the magnetizing force is removed.

TABLE 4–1 Permeabilities (μ) and relative permeabilities (μ_r) of various materials

Material	*Relative Permeability* (μ_r)	*Permeability* (μ)
Air	1	1.26×10^{-6}
Nickel	50	6.28×10^{-5}
Cobalt	60	7.56×10^{-5}
Cast iron	90	1.1×10^{-4}
Machine steel	450	5.65×10^{-4}
Transformer iron	5,500	6.9×10^{-3}
Silicon iron	7,000	8.8×10^{-3}

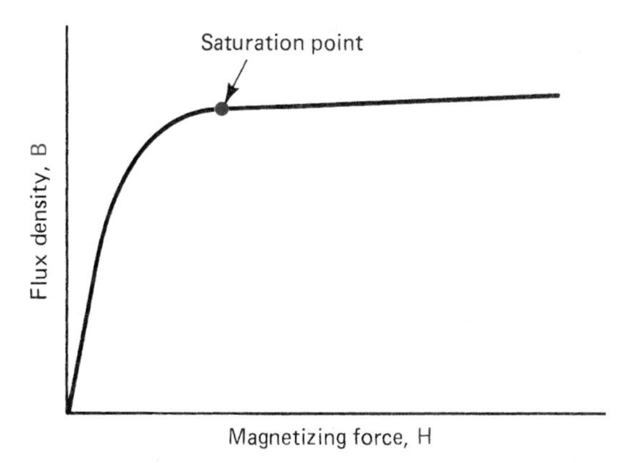

Figure 4–24. Magnetization or B-H curve.

Magnetic saturation is important in the operation of electrical equipment that has electromagnets, especially generators. Saturation is best explained by the curve shown in Fig. 4–24, which is called a magnetization or B-H curve. The curve shows the relationship between a magnetizing force (H) and flux density (B). Notice that as a magnetizing force increases, flux density also increases. *Flux density* is the amount of lines of flux per unit area of a material. An increase in flux density occurs until magnetic saturation is reached. At the saturation point, the maximum alignment of domains within the material has taken place. Beyond saturation an electromagnet is not capable of more magnetic field strength. The B-H curve is a straight line beyond the saturation point. Notice the shape of the B-H curve. The magnetization or setting up of the magnetic field (B) lags the magnetizing force (H) due to friction of the molecules of a material. This time lag between magnetizing force (H) and the development of magnetic flux (B) is called *hysteresis.*

4.10 HALL EFFECT

Magnetism can be used other than in generators to produce voltage. When a magnetic field is placed at right angles to a

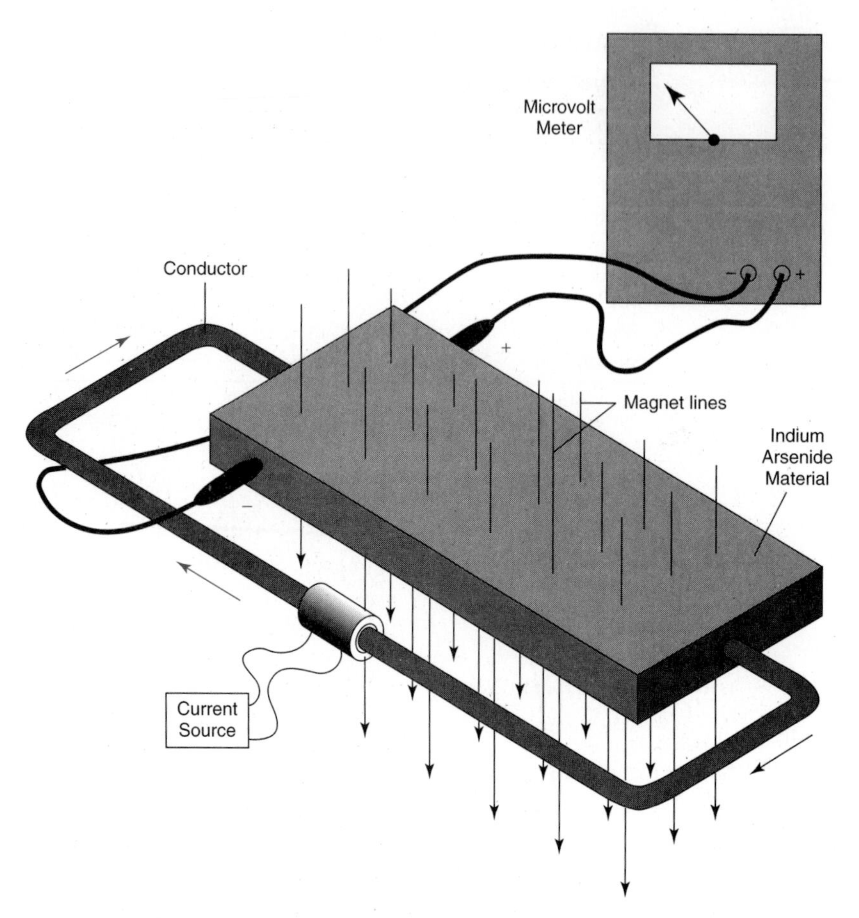

Figure 4–25. Illustration showing the Hall effect.

current-carrying conductor, a small voltage is produced across the conductor known as the *Hall effect*. This effect has many applications in switching and measurement circuits. Voltages produced by this process are small (in the microvolt range) when conductors are used. When a semiconductor material such as indium arsenide is used, the voltage produced can be as much as 100 mV.

The Hall effect is illustrated in Fig. 4–25. A block of indium arsenide has a small current flowing through it. If a magnetic field is placed perpendicular to the direction of current flow, a voltage is produced across the width of the semiconductor. The amount of voltage produced is directly proportional to the flux density of the magnetic field.

An application of the Hall effect is as a sensor in instruments designed to measure the strength of magnetic fields. One such device is known as a *gauss meter.* Within its measuring probe is a Hall effect device. When placed within a magnetic field, it produces a voltage directly proportional to the flux density of the field. This voltage is sensed and displayed on the meter, which is calibrated in gauss. There are many other uses for Hall effect devices, including position sensors for machines, switches for computer keyboards, and electronic ignition systems of automobiles.

4.11 MAGNETIC LEVITATION

Magnetic levitation is an interesting phenomenon. "Maglev" has been used for developing trains that do not ride on rails, but "levitate" above the rails on a magnetic cushion. Experimental models of this type of train have reached speeds of 300 miles per hour with a smooth ride. The forward motion of the train is produced by the attraction of fields with opposite magnetic polarities and the repulsion of fields with like magnetic polarities. For this type of application, the magnetic fields must be quite strong.

4.12 RARE EARTH MAGNETS

New rare earth permanent magnets are used for many applications today. They may be used to produce a uniform, variable magnetic field with about one tenth the size of iron-core electromagnets. They also do not require special power supplies. Magnetic fields play an important role in numerous industrial processes. For many industrial applications, rare earth permanent magnets offer advantages of smaller size, lower cost, and ease of operation. A permanent or "hard" magnet is one that remains constant in terms of its magnetization and direction. Rare earth permanent magnets, unlike steel magnets, can be formed into any shape without demagnetizing.

Rare earth magnets are typically made of a material called neodymium; however, other materials have been tested to ensure proper magnetic characteristics. Applications include brushless direct-current (dc) motors and actuators, cordless tools, and computer drive systems.

REVIEW

1. What were the first magnets called?
2. What is electromagnetism?
3. What three materials are used in the construction of permanent magnets?
4. What are the two laws of magnetism?
5. How can a piece of iron be temporarily magnetized?
6. Why should magnets be stored in a "keeper"?
7. What is the left-hand rule of magnetic flux around a conductor?
8. What is the left-hand rule for determining the polarity of an electromagnet?
9. What are the three basic parts of an electromagnet?
10. What are three ways to increase the strength of an electromagnet?
11. What is residual magnetism?
12. Discuss the relationship of magnetomotive force, magnetic flux, and reluctance in a magnetic circuit.
13. Discuss the domain theory of magnetism.
14. What is Faraday's law?
15. How is Faraday's law important in the generation of electrical power?
16. What is a relay?
17. What are NO and NC contacts of relays?
18. What are pickup current and dropout current ratings of relays?
19. What is a solenoid?
20. What is a magnetic contactor?
21. What is permeability? retentivity? saturation? magnetizing force?

STUDENT ACTIVITIES

• *Working with Permanent Magnets (see Fig. 4–1)*

1. Obtain two permanent magnets, a sheet of paper, and some iron filings.
2. Place one magnet on a flat surface with a sheet of paper placed over it.
3. Carefully pour iron filings onto the paper above the magnet.
4. Carefully lift the paper and place the second magnet under the paper with the two north poles about 1 in. apart. Make a sketch of the magnetic field pattern on a sheet of paper.
5. Place the north and south poles about 1 in. apart. Make a sketch of the magnetic field patterns.
6. Discuss the laws of magnetism.

• The Nature of Magnetism

Complete Activity 4–1, page 83 in the manual.

• • ANALYSIS

1. Describe the reaction of like poles and unlike poles of magnetic fields.
2. Why can a compass be used to detect the presence of a magnetic field?
3. Define the following terms:
   a. Flux
   b. Magnetic field
   c. Magnetic lines of force
4. In what direction does magnetic flux travel externally around a magnet?
5. Sketch the magnetic field of a permanent magnet.

• Magnetizing an Iron Bar

1. Obtain a magnet, an iron bar (about 6 in. long and ½ to 1 in. in diameter), and a compass.
2. Place the iron bar on a flat surface.
3. Gently stroke the iron bar with the magnet, moving in the same direction each time.
4. Place the compass near one end of the iron bar. Describe what happens.

• Making an Electromagnet (see Fig. 4–7)

1. Obtain the following materials:
   ⅜- to ¾-in.-diameter bolt, about 4 in. long
   10 to 20 ft. of solid copper wire (No. 18 to No. 24 diameter)
   6-V lantern battery
   Compass
2. Use the bolt as an iron core. Start at one end and wind the wire around the core in the same direction. Leave 6 to 10 in. of wire on each end, for connecting to the battery. Wrap tape around the wire to secure it in place. The finished coil should have between 100 and 200 turns of wire.
3. Momentarily connect the ends of the coil to a 6-V lantern battery. It may be necessary to remove about ½ in. of insulation from the ends of the wire. This will allow good electrical contact to be made. Do not keep the battery connected for over a few seconds at a time.
4. Test the electromagnet with a compass to see if it has a north polarity at one end and a south polarity at the other end.
5. Try picking up metal objects with the electromagnet.
6. Let the teacher check the electromagnet.
7. On a sheet of paper, discuss the construction and operation of an electromagnet. Be sure to discuss ways to make the electromagnet stronger.

• Electromagnetic Relays

Complete Activity 4–2, page 85 in the manual.

ANALYSIS

1. What are normally open contacts?
2. What are normally closed contacts?
3. What is meant by the term *pickup current?*
4. What is meant by the term *dropout current?*

Principles of Electromagnetism

Complete Activity 4–3, page 87 in the manual.

ANALYSIS

1. What are the magnetic polarities of the coil of wire shown in Fig. 4–26?
2. What are the three factors that affect the strength of an electromagnet?

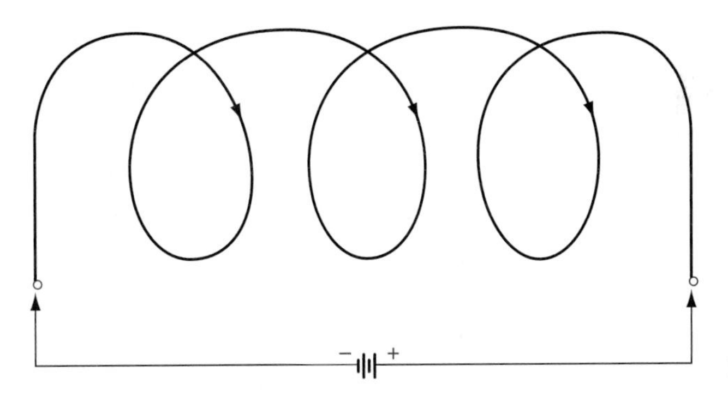

Figure 4–26.

Solenoids

Complete Activity 4–4, page 91 in the manual.

ANALYSIS

1. What is a solenoid?
2. Where are solenoids used?
3. Why are solenoids sometimes called "pusher" coils?
4. Why is the plunger of solenoids spring-loaded?

Self-Examination

1. Complete the Self-Examination for Chap. 4, page 93, in the manual.
2. Check your answers on page 411 in the manual.
3. Study the questions which were answered incorrectly.

Chapter 4 Examination

Complete the Chap. 4 Examination on page 95 in the manual.

CHAPTER

5

SOURCES OF ELECTRICAL ENERGY

OUTLINE

5.1 Chemical Sources
5.2 Battery Connections
5.3 Light Sources
5.4 Heat Sources
5.5 Pressure Sources
5.6 Electromagnetic Induction
5.7 Generating a Voltage
5.8 Electrical Generator Basics
5.9 Single-Phase Ac Generators
5.10 Three-Phase Ac Generators
5.11 Direct-Current Generators

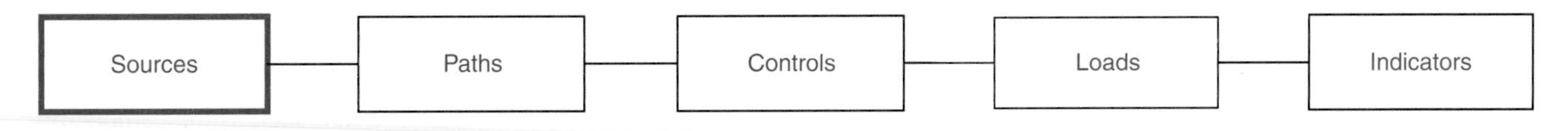

Sources supply electrical energy to other parts of an electrical system.

OBJECTIVES

Upon completion of this chapter, you will be able to:

1. Describe basic types of batteries.
2. Properly connect batteries in series, parallel, and combination configurations.
3. Explain the purposes of different configurations of battery connections.
4. Describe light, heat, pressure, and mechanical sources of electrical energy.
5. State Faraday's law for electromagnetic induction.
6. Explain the factors that affect the generation of voltage.
7. Describe single-phase ac, three-phase ac, and direct-current generators.

Electrical energy sources convert some other form of energy into electrical energy. Batteries and electrical generators are two major sources of electrical energy. Batteries convert chemical energy into electrical energy. The types of electrical generators include single-phase ac generators, three-phase ac generators, and dc generators. Electrical generators rely on the principle of electromagnetic induction to convert mechanical energy into electrical energy. Batteries, generators, and other sources of electrical energy are discussed in this chapter. Some sources produce direct-current (dc) energy whereas others produce alternating-current (ac) energy.

IMPORTANT TERMS

The following terms are used to aid in the understanding of electrical energy sources.

Ac. An abbreviation for alternating current.

Alternating current. The current produced when electrons move first in one direction and then in the opposite direction.

Alternator. A rotating machine that generates ac voltage.

Ampere-hour. A capacity rating of storage batteries which refers to the maximum current that can flow from the battery during a specified time.

Armature. The part of a generator into which current is induced.

Armature reaction. The effect in direct-current generators and motors in which current flow through the armature conductors causes a magnetic field which reacts with the main magnetic field and increases sparking between the brushes and commutator.

Battery. An electrical energy source consisting of two or more cells connected together.

Brush. A sliding contact made of carbon and graphite, connected between the commutator and the power source of a dc motor or load of a dc generator.

Commutation. The process of changing ac induced into the rotor of a dc generator into dc applied to the load circuit.

Commutator. An assembly of copper segments that provide a method of connecting rotating coils to the brushes of a dc generator.

Copper losses. The power losses that occur when heat is produced as electrical current flows through the copper windings of an electrical machine; also called I^2R losses.

Core. Laminated iron or steel used in the internal construction of the magnetic circuit of electrical machines.

Cycle. A sequence of events that causes one complete pattern of alternating current from a zero reference, in a positive direction, back to zero, then in a negative direction and back to zero.

Dc. An abbreviation for direct current.

Dc generator. A rotating machine that produces a form of direct-current (dc) voltage.

Delta connection. A method of connecting three-phase alternator stator windings in which the beginning of one phase winding is connected to the end of the adjacent phase winding; power lines extend from the three beginning-end connections.

Direct current. The flow of electrons in one direction from negative (−) to positive (+).

Dry cell. A nonliquid cell that produces dc voltage by chemical action.

Eddy current. Induced current in the metal parts of electrical machines which causes heat losses.

Efficiency. The ratio of output power to input power.

$$\text{efficiency} = \frac{\text{power output (watts)}}{\text{power input (watts)}}$$

Electrode. A specific part of a unit such as the cathode of a semiconductor cell.

Electrolyte. The solution used in a cell which produces ions.

Field coils. Electromagnetic coils that develop the magnetic fields of electrical machines.

Field pole. Laminated metal that serves as the core material for field coils.

Frequency. The number of ac cycles per second, measured in hertz (Hz).

Generator. A rotating electrical machine that converts mechanical energy into electrical energy.

Hydrometer. An instrument used to measure the specific gravity or "charge" of the electrolyte of a storage battery.

Induced current. The current that flows through a conductor due to magnetic transfer of energy.

Induced voltage. The potential that causes induced current to flow through a conductor which passes through a magnetic field.

Ion. An atom that has lost or gained electrons, making it a negative or positive charge.

Laminations. Thin sheets of soft iron or steel used in the construction of electrical machines to reduce heat losses due to eddy currents.

Lead-acid cell. A secondary cell that has positive and negative plates made of lead peroxide and lead and has a liquid electrolyte of sulfuric acid mixed with water.

Left-hand rule. (1) To determine the *direction of the magnetic field* around a single conductor, point the thumb of the left hand in the direction of current flow (− to +) and the fingers will extend around the conductor in the direction of the magnetic field; (2) to determine the *polarity of an electromagnetic coil,* extend the fingers of the left hand around the coil in the direction of current and the thumb will point to the north polarity; (3) to determine the *direction of induced current flow* in a generator conductor, hold the thumb, forefinger, and middle finger of the left hand at right angles to one another, point the thumb in the direction of motion of the conductor, the forefinger in the direction of the magnetic field (N to S), and the middle finger will point in the direction of induced current.

Lenz's law. The induced voltage in any circuit is always in such a direction that it will oppose the force that produces it.

Neutral plane. The theoretical switching position of the commutator and brushes of a dc generator or motor which occurs when no current flows through the armature conductors and the main magnetic field has least distortion.

Photovoltaic cell. A cell that produces dc voltage when light shines onto its surface.

Piezoelectric effect. The property of certain crystal materials to produce a voltage when pressure is applied to them.

Primary cell. A cell that cannot be recharged.

Prime mover. A system that supplies the mechanical energy to rotate an electrical generator.

Pulsating dc. A voltage or current value that rises and falls with current flow always in the same direction.

Regulation. A measure of the amount of voltage change that occurs in the output of a generator due to changes in load.

Rotating-armature method. The method used when a generator has dc voltage applied to produce a field to the stationary part (stator) of the machine and voltage is induced into the rotating part (rotor).

Rotating-field method. The method used when a generator has dc voltage applied to produce a field to the rotor of the machine and voltage is induced into the stator coils.

Rotor. The rotating part of an electrical generator or motor.

Running neutral plane. The actual switching position of the commutator and brushes of a dc generator or motor which shifts the theoretical neutral plane due to armature reaction.

Secondary cell. A cell that can be recharged by applying dc voltage from a battery charger.

Sine wave. The waveform of ac voltage.

Single-phase ac generator. A generator that produces single-phase ac voltage in the form of a sine wave.

Slip rings. Copper rings mounted on the end of a rotor shaft and connected to the brushes and rotor windings.

Specific gravity. The weight of a liquid as compared with the weight of water, which has a value of 1.0.

Split-ring commutator. *See* Commutator.

Stator. The stationary part of an electrical generator or motor.

Storage battery. *See* Secondary cell.

Thermocouple. A device that has two pieces of metal joined together so that when its junction is heated, a voltage is produced.

Three-phase ac generator. A generator that produces three ac sine-wave voltages that are separated in phase by 120°.

Voltaic cell. A cell that produces voltage due to two metal electrodes that are suspended in an electrolyte.

Wye connection. A method of connecting three-phase alternator stator windings in which the beginnings *or* ends of each phase winding are connected to form a common or neutral point; power lines extend from the remaining beginnings or ends.

5.1 CHEMICAL SOURCES

Conversion of chemical energy into electrical energy occurs through chemical cells. When two or more cells are connected in series or parallel (or a combination of both), they form a battery. A cell is made of two different metals immersed in a liquid or paste called an *electrolyte.* Chemical cells are either primary or secondary cells. Primary cells are usable only for a certain time. Secondary cells are renewed after being used to produce electrical energy once again, which is known as *charging.* Both primary and secondary cells have many uses.

Primary Cells

The operation of a primary cell involves the placing of two unlike materials called *electrodes* into the solution, or electrolyte. When the materials of the cell are brought together, their molecular structures change. During this chemical change, atoms may either gain additional electrons or leave behind some electrons. These atoms then have either a positive or a negative electrical charge. They are called *ions.* Ionization of atoms allows a chemical solution of a cell to produce an electrical current.

A load device such as a lamp may be connected to a cell. Electrons flow from one of the cell's electrodes to the other through the electrolyte material. This creates an electrical current flow through the load. Current leaves the cell through its negative electrode. It passes through the load device and then goes back to the cell through its positive electrode. A complete circuit exists between the cell (source) and the lamp (load).

The voltage output of a primary cell depends on the electrode materials used and the type of electrolyte. The familiar carbon-zinc cell shown in Fig. 5–1 produces approximately 1.5 V. The negative electrode of this cell is the zinc container. The positive electrode is a carbon rod. A paste material acts as the electrolyte. It is placed between the two electrodes. This type of cell is called a *dry cell.*

Many types of primary cells are used today. The carbon-zinc cell is the most used type. This cell is low in cost and available in many sizes. Applications are mainly for portable equipment and instruments. For uses that require higher voltage or current than one cell can deliver, several cells are combined in series, parallel, or series-parallel connections. Carbon-zinc batteries are available in many voltage ratings.

An alkaline (zinc–manganese dioxide) cell is shown in Fig. 5–2. Alkaline cells have a voltage per cell of 1.5 V. They supply higher-current electrical loads. They have much longer lives than carbon-zinc cells of the same types.

Another type of primary cell is the lithium type. Lithium batteries have an extremely long life and provide a leakproof, high-energy source for a wide range of applications. Lithium batteries typically operate at approximately 1.9-V output. New types of lithium batteries are being developed to make them compatible with 1.5-V applications.

Table 5–1 summarizes primary cell types.

Secondary Cells

Chemical cells that may be reactivated by charging are called secondary cells or storage cells. Common types are the lead-acid, nickel-cadmium, and nickel–metal-hydride (NiMH) cells.

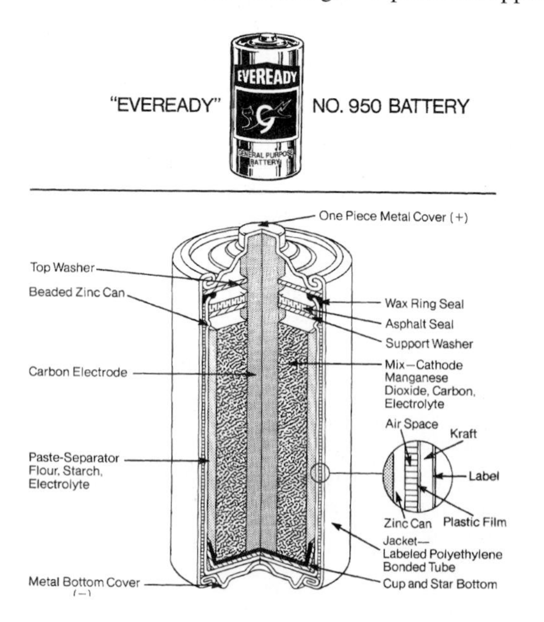

Figure 5–1. Carbon-zinc cell. (Courtesy of Union Carbide Corp.)

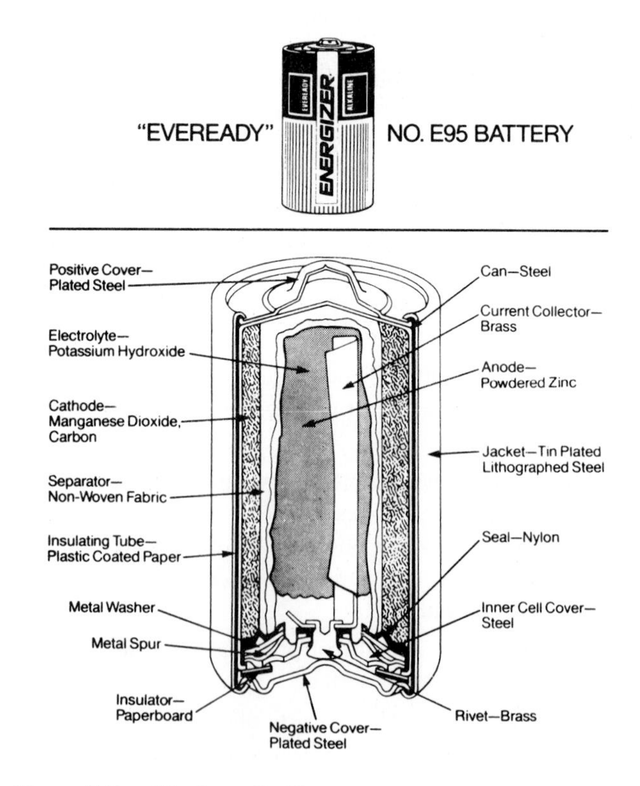

Figure 5–2. Alkaline cell. (Courtesy of Union Carbide Corp.)

TABLE 5–1 Primary Cell Summary

Battery Type	*Cell Voltage*	*Uses*
Carbon-Zinc	1.5	Portable radios, tape players, televisions, toys (most popular, low cost)
Alkaline	1.4	Portable radios, tape players, televisions, toys (has long life)
Mercuric Oxide	1.35 and 1.4	Watches, hearing aids, pacemakers, cameras, test equipment
Silver Oxide	1.5	Watches, hearing aids, pacemakers, cameras (expensive)
Lithium	1.9	Liquid crystal watches, computer memories, calculators, sensors

LEAD-ACID CELLS. The lead-acid cell of Fig. 5–3 is a secondary cell. The electrodes of lead-acid cells are made of lead and lead peroxide. The positive plate is lead peroxide (PbO_2). The negative plate is lead (Pb). The electrolyte is sulfuric acid (H_2SO_4). When the lead-acid cell supplies current to a load, the chemical process is written as

$$Pb + PbO_2 + 2H_2SO_4 \rightarrow 2PbSO_4 + 2H_2O$$

The sulfuric acid ionizes to produce four positive hydrogen ions (H^+) and two negative sulfate ($SO^-{}_4$) ions. A negative charge is developed on the lead plate when an $SO^-{}_4$ ion combines with the lead plate to form lead sulfate ($PbSO_4$). The positive hydrogen ions (H^+) combine with electrons of the lead peroxide plate. They become neutral hydrogen atoms. The H^+ ions also combine with the oxygen (O) of the lead peroxide plate to become water (H_2O). The lead peroxide plate then has a positive charge. A lead-acid cell has a voltage between electrodes of about 2.1 V when fully charged.

Cells discharge when supplying current for a long time. They are no longer able to develop an output voltage when discharged. Cells may be charged by causing direct current to flow through the cell in the opposite direction. The chemical process of charging is written as

$$2PbSO_4 + 2H_2O \rightarrow Pb + PbO_2 + 2H_2SO_4$$

or

2 parts lead sulfate

+ 2 parts water yields lead + lead peroxide

+ 2 parts sulfuric acid

The original condition of the chemicals is reached by charging. The chemical reaction is reversible.

The amount of charge of a lead-acid cell is measured by a *specific gravity* test. A *hydrometer* is used to test the electrolyte solution. The specific gravity of a liquid is an index of how heavy a liquid is compared with water. Pure sulfuric acid has a specific gravity of 1.840. The dilute sulfuric acid of a fully charged lead-acid cell varies from 1.275 to 1.300. During the discharge of the cell, water is formed, which reduces the specific gravity of the electrolyte. A specific gravity of between 1.120 and 1.150 indicates a fully charged cell. A hydrometer used to test lead-acid cells is shown in Fig. 5–4.

The capacity of a battery made of lead-acid cells is given by an ampere-hour rating. A 50-ampere-hour battery is rated to deliver 50 A for 1 h, 25 A for 2 h, or 12.5 A for 4 h. The ampere-hour rating is an approximate value. It depends on the rate of discharge and the operating temperature of the battery.

NICKEL-CADMIUM CELLS. Another type of secondary cell is a nickel-cadmium cell. These cells are available in many sizes. Figure 5–5, on page 156, shows nickel-cadmium cells. They are often used in portable equipment. The positive plate of this cell is nickel hydroxide. The negative plate is cadmium hydroxide. The electrolyte is made of potassium hydroxide. These cells have a long life. A fully charged nickel-cadmium cell has a voltage of approximately 1.25 V.

NICKEL–METAL-HYDRIDE CELLS. Nickel–metal-hydride (NiMH) batteries are another type of rechargeable cell. Compared with nickel-cadmium types, NiMH batteries charge faster. In a nickel–metal-hydride battery, the positive electrode is made of nickel and the negative electrode is made of hydrogen-storing metal alloys. This type of battery will typically permit 500 charge-discharge cycles and recharge in 1.5 h. They operate at 1.2 V, which is approximately the same as nickel-cadmium batteries.

Secondary cells have many uses. Storage batteries are used in some buildings to provide emergency power when a power failure occurs. Standby systems are needed, especially for lighting when power is off. Automobiles use storage batteries for their everyday operation. Many types of instruments and portable equipment use batteries for power. Some instruments use rechargeable secondary cells and others use primary cells. The use of batteries for portable equipment is becoming increasingly important due to the increased use of electronic equipment such as personal computers (PCs), facsimiles (FAXs), headphone stereos, portable compact disc (CD) players, calculators, security systems, cordless and cellular telephones, power tools, and many other applications. Table 5–2 on page 157 summarizes secondary cells.

BATTERIES OF THE FUTURE. The demand for smaller and lighter cellular phones, computers, and cordless

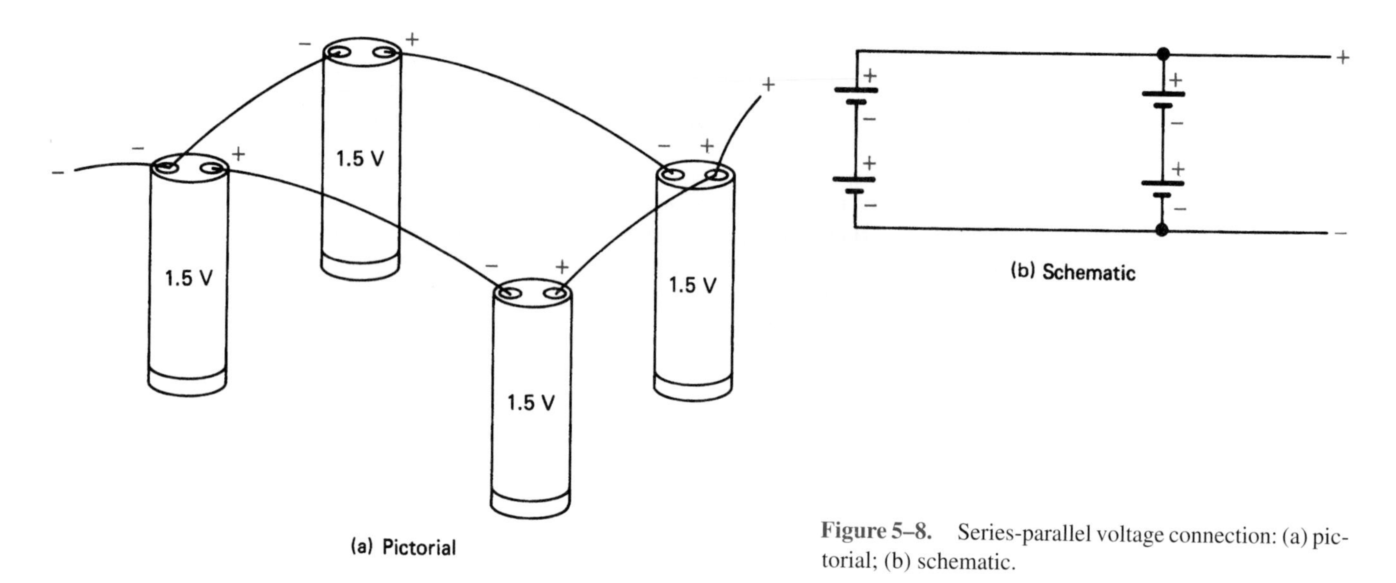

Figure 5–8. Series-parallel voltage connection: (a) pictorial; (b) schematic.

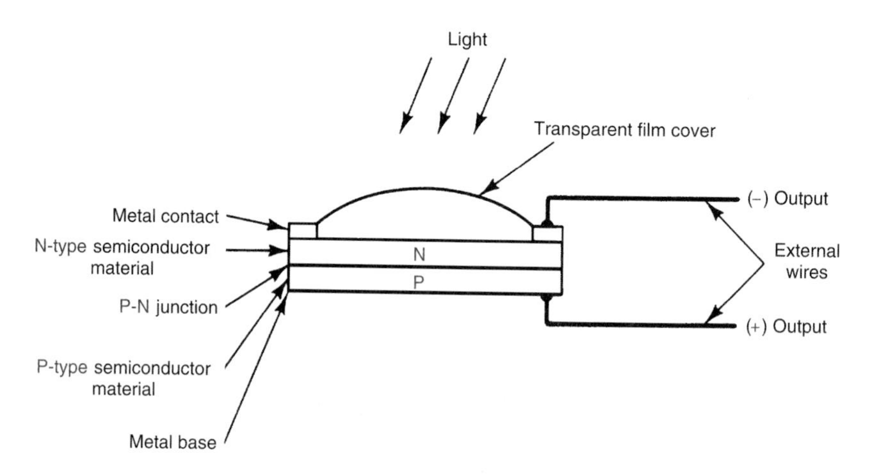

Figure 5–9. Solar cell.

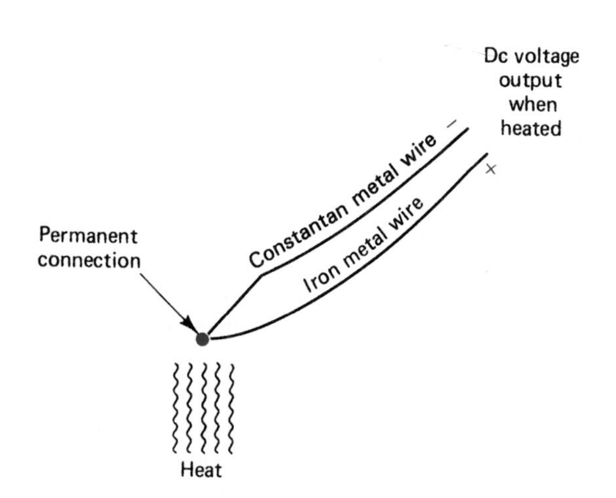

Figure 5–10. Operating principle of a thermocouple.

5.5 PRESSURE SOURCES

Electrical energy is produced by mechanical energy in electrical generators. Mechanical energy is used to rotate prime movers that drive electrical generators. Mechanical energy in the form of pressure is also used as a source of electrical energy.

The change of mechanical pressure into electrical energy is called the *piezoelectric effect.* Certain crystal materials may be compressed as pressure is applied to the surfaces. A voltage is created between their top and bottom surfaces. The amount of voltage is determined by the amount of pressure—the greater the pressure, the greater the voltage; the less the pressure, the less the voltage will be for any piezoelectric crystal. These pressure-sensitive crystals have been used as phonograph cartridges to change the pressure applied by the grooves of the record into a voltage. They are also

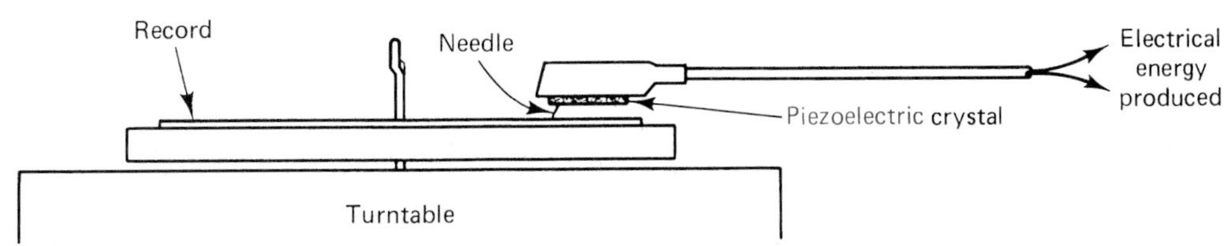

Figure 5–11. Piezoelectric principle of a phonograph cartridge used for vinyl recording sound systems.

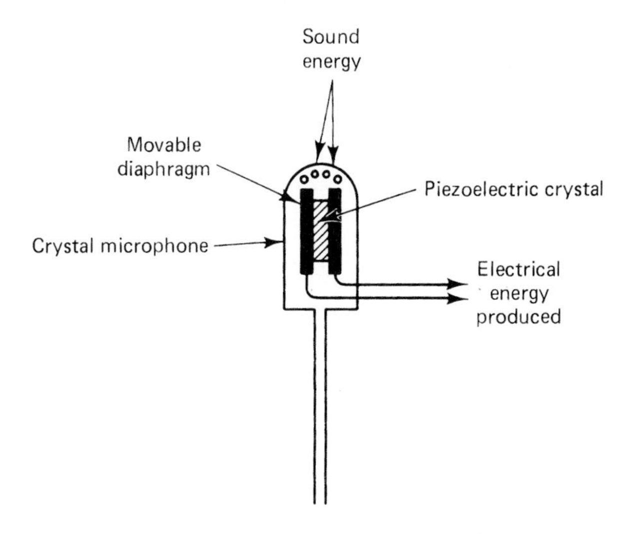

Figure 5–12. Piezoelectric principle of a microphone.

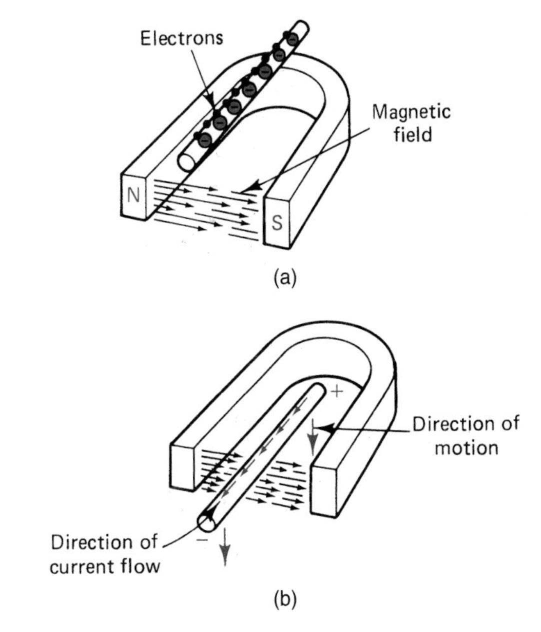

Figure 5–13. Principle of electromagnetic induction: (a) conductor placed inside a magnetic field; (b) conductor moved downward through the magnetic field.

used as pressure sensors to sense and measure pressure in security and industrial systems.

When these crystalline materials are subjected to a mechanical pressure, electrical energy is developed across the material. Crystals such as quartz and Rochelle salt have this characteristic. An application of the piezoelectric principle is the cartridge and needle assembly used for vinyl recording sound systems (see Fig. 5–11). The cartridge contains a crystalline material. The crystal vibrates according to the size of the grooves of a phonograph record. The needle is attached to the cartridge to connect the record grooves to the crystal. The crystalline material produces a voltage due to the mechanical vibrations or pressure changes. These small voltage changes are then amplified by a sound system.

It is also possible to convert pressure in the form of sound into electrical energy with piezoelectric crystals. This is done with crystal microphones, as shown in Fig. 5–12. Sound waves are used to cause vibration of a piezoelectric crystal. A voltage is developed across the crystal and amplified by the sound system. Higher sound pressure causes more voltage output to be produced.

5.6 ELECTROMAGNETIC INDUCTION

Electrical energy can be produced by placing a conductor inside a magnetic field. An experiment by Faraday was quite important in showing the following principle: When a conductor moves across the lines of force of a magnetic field, electrons in the conductor tend to flow in a certain direction. When the conductor moves across the lines of force in the opposite direction, electrons in the conductor tend to flow in the opposite direction. This is the principle of electrical power generation. Most of the electrical energy used today is produced by magnetic induction.

Figure 5–13 shows the principle of electromagnetic induction. Electrical current is produced only when there is motion. When the conductor is brought to a stop while crossing lines of force, electrical current stops.

If a conductor or a group of conductors is moved through a strong magnetic field, induced current will flow and a voltage will be produced. Figure 5–14 shows a loop of wire rotated through a magnetic field. The position of the loop inside the magnetic field determines the amount of induced current and voltage. The opposite sides of the loop move across the magnetic lines of force in opposite directions. This movement causes an equal amount of electrical current to flow in opposite directions through the two sides of the loop. Notice each position of the loop and the resulting output voltage in

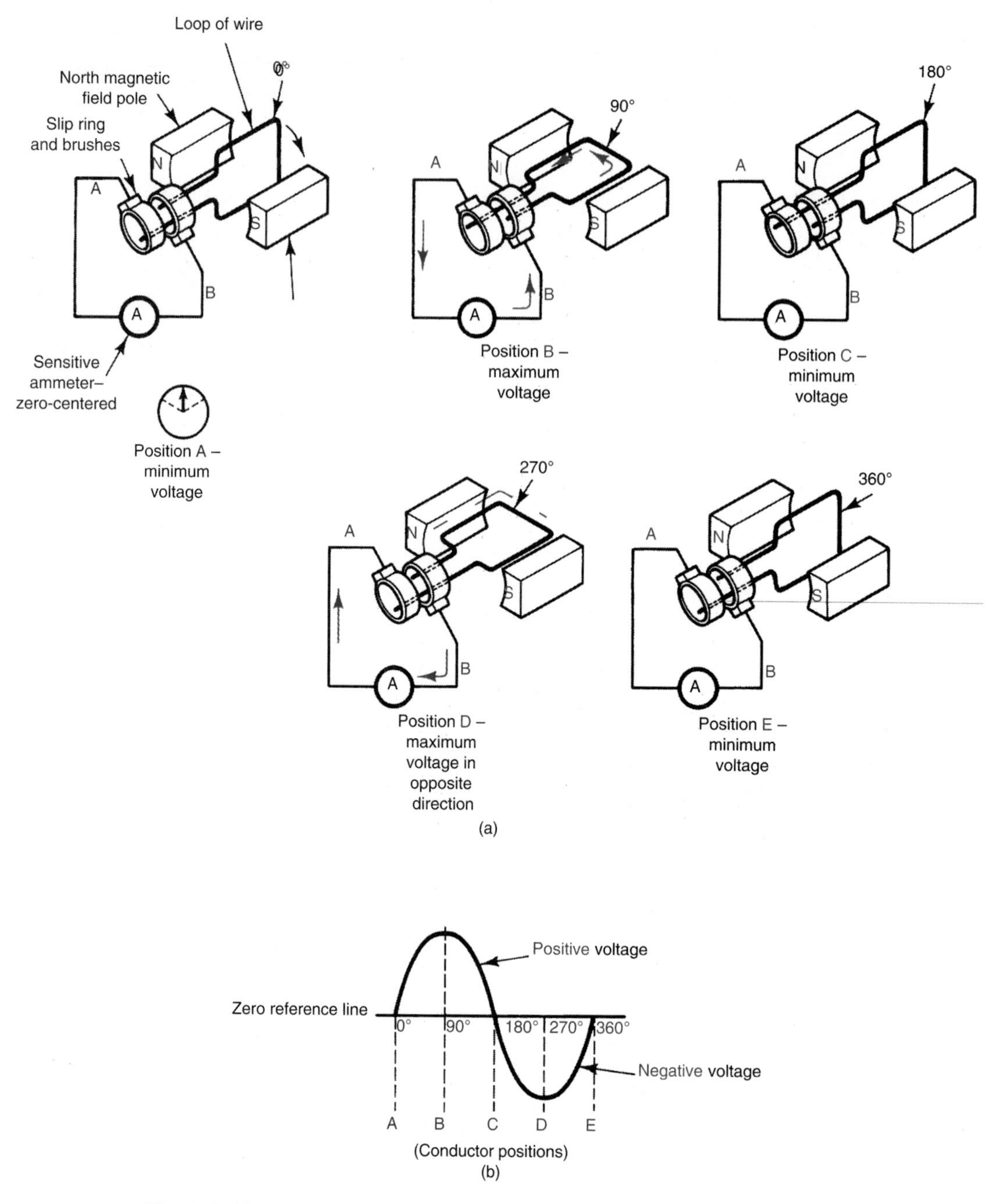

Figure 5–14. Voltage generation: (a) loop of wire rotated through a magnetic field; (b) voltage produced by the movement of the loop of wire.

Fig. 5–14. The electrical current flows in one direction and then in the opposite direction with every complete revolution of the conductor. This method produces alternating current (ac). One complete rotation is called a *cycle.* The number of cycles per second is known as the *frequency.* Most ac generators produce 60 cycles per second.

The ends of the conductor that move across the magnetic field of the generator shown in Fig. 5–14 are connected to slip rings and brushes. The slip rings are mounted on the same shaft as the conductor. Carbon brushes are used to make contact with the slip rings. The electrical current induced into the conductor flows through the slip rings to the brushes. When the conductor turns half a revolution, electrical current flows in one direction through the slip rings and the meter. During the next half-revolution of the coil, the positions of the two sides of the conductor are opposite. The direction of the induced current is reversed. Current now flows through the meter in the opposite direction.

The conductors that make up the rotor of a generator have many turns. The generated voltage is determined by the number of turns of wire used, the strength of the magnetic field, and the speed of the prime mover used to rotate the machine.

Direct current can also be produced by electromagnetic induction. A simple dc generator has a split-ring commutator instead of two slip rings. The split rings resemble one full ring, except that they are separated by small openings. Induced electrical current still flows in opposite directions to each half of the split ring. However, current flows in the same direction in the load circuit due to the action of the split rings. Dc generator operation is discussed later in this chapter.

5.7 GENERATING A VOLTAGE

Electromagnetic induction occurs when a conductor passes through a magnetic field and cuts across lines of force. As a conductor passes through a magnetic field, it cuts across the magnetic flux lines. As the conductor cuts across the flux lines, the magnetic field develops a force on the electrons of the conductor. The direction of the electron movement determines the polarity of the induced voltage. The left-hand rule is used to determine the direction of electron flow. This rule for generators is stated as follows: Hold the thumb, forefinger, and middle finger of the left hand perpendicular to each point. Point the forefinger in the direction of the magnetic field from north to south. Point the thumb in the direction of the motion of the conductor. The middle finger will then point in the direction of electron current flow.

The amount of voltage induced into a conductor cutting across a magnetic field depends on the number of lines of force cut in a given time. This value is determined by the following three factors.

1. The *speed* of the relative motion between the magnetic field and the conductor
2. The *strength* of the magnetic field
3. The *length* of the conductor passed through the magnetic field

If the speed of the conductor cutting the magnetic lines of force is increased, the generated voltage increases. If the strength of the magnetic field is increased, the induced voltage also increases. A longer conductor allows the magnetic field to induce more voltage into the conductor. The induced voltage increases when each of the three quantities listed is increased.

SAMPLE PROBLEM: VOLTAGE INDUCED INTO A CONDUCTOR

In electrical generators, the coils move with respect to a magnetic field or flux. Electromagnetic induction occurs in accordance with Faraday's law. This law states that (1) if a magnetic flux that links a conductor loop has relative motion, a voltage is induced; and (2) the value of the induced voltage is proportional to the rate of change of flux.

The voltage induced in a conductor of a generator is defined by Faraday's law as follows:

$$V_i = B \times L \times v$$

where

V_i = induced voltage in volts
B = magnetic flux in teslas
L = length of conductor within the magnetic flux in meters
v = relative speed of the conductor in meters per second

Given: the conductors of the stator of a generator have a length of 0.5 M. The conductors move through a magnetic field of 0.8 tesla at a rate of 60 m/s.

Find: the amount of induced voltage in each conductor.

Solution:

$$V_i = B \times L \times v$$
$$= 0.8 \times 0.5 \times 60$$
$$V_i = 24 \text{ volts}$$

5.8 ELECTRICAL GENERATOR BASICS

Electrical generators are used to produce electrical energy. They require some form of mechanical energy input. The mechanical energy is used to move electrical conductors (turns of wire) through a magnetic field inside the generator. All generators operate due to electromagnetic induction. A generator has a stationary part and a rotating part housed inside a machine assembly. The stationary part is called the *stator* and the rotating part is called the *rotor.* The generator has magnetic field poles of north and south polarities. Generators must have a method of producing rotary motion (mechanical energy). This system is called a *prime mover* and is connected to the generator shaft. There must also be a method of electrically connecting the rotating conductors to an external circuit. This is done by a slip ring or commutator and brush assembly. Slip rings are used with ac generators and commutators with dc generators. Slip rings are made of solid sections of copper and the commutator is made of several copper sections which are separated from each other. These assemblies are permanently mounted on the shaft of a generator. They connect to the ends of the conductors of the rotor. When a load is connected to a generator, a complete circuit is made. With all the generator parts working together, electrical power is produced.

5.9 SINGLE-PHASE AC GENERATORS

Single-phase electrical power is often used, particularly in homes. However, little electrical power is produced by single-phase generators. Alternating-current generators are usually referred to as "alternators." Single-phase electrical power used in homes is usually produced by three-phase generators at power plants. It is then converted to single-phase electrical energy before it is distributed to homes. Single-phase generators have several uses.

The current produced by single-phase generators is in the form of a sine wave, so called due to its mathematical origin. It is based on the trigonometric sine function used in mathematics. Two cycles of single-phase ac voltage are shown in Fig. 5–15. This voltage is known as a sine-wave voltage.

The voltage induced into the conductors of the armature varies as the sine of the angle of rotation between the conductors and the magnetic field (see Fig. 5–16). The voltage induced at a specific time is called *instantaneous voltage* (V_i). Voltage induced into an armature conductor at a specific time is found by using the formula

$$V_i = V_{\max} \times \sin \theta$$

$V_{\max}$ is the maximum voltage induced into the conductor. The symbol theta (θ) is the angle of conductor rotation.

For example, at the 60° position, assume that the maximum voltage output is 100 V. The instantaneous voltage induced at 60° is $V_i = 100 \times \sin \theta$.

The frequency of the voltage produced by alternators is usually 60 hertz (Hz). A cycle of ac is generated when the rotor moves one complete revolution (360°). *Cycles per second* or *hertz* refers to the number of revolutions per second. For example, a speed of 60 revolutions per second (3600 rpm) produces a frequency of 60 Hz. The frequency of an alternator is found by using the formula

$$f(\text{Hz}) = \frac{\text{number of magnetic poles} \times \text{speed of rotation (rpm)}}{120}$$

The frequency is measured in hertz. If the number of poles (field coils) is increased, the speed of rotation can be reduced and still produce a 60-Hz frequency.

For a generator to convert mechanical energy into electrical energy, three conditions must exist:

1. There must be a *magnetic field.*
2. *Conductors* must be placed adjacent to the magnetic field.
3. There must be *relative motion* between the magnetic field and conductors.

The two methods used to accomplish these conditions are the *rotating-armature* method and the *rotating-field* method, both shown in Fig. 5–17.

In the rotating-armature method, shown in Fig. 5–17(a), ac voltage is induced into the rotor conductors. The magnetic field is developed by a set of stationary field poles. Relative motion between the conductors and the magnetic field comes from a prime mover connected to the generator shaft. Prime movers can be gasoline engines, diesel engines, steam turbines, or electric motors. Remember that generators convert mechanical energy into electrical energy. The rotating-armature method can only be used to produce small amounts of electrical power. The major disadvantage is that the ac voltage passes through the slip-ring/brush assembly. High voltages would cause sparking or arc-over between the brushes and slip rings. The maintenance involved in replacing brushes and repairing slip-ring assemblies would be expensive. This method is used for alternators with low power outputs.

The rotating-field method shown in Fig. 5–17(b) is used for alternators with larger power outputs. The dc excitation voltage is used to develop the magnetic field. Dc voltage is applied to the rotor of the generator. The ac voltage output is induced into the stationary conductors of the machine. Because the dc excitation voltage is much smaller than the ac voltage output, maintenance problems are reduced. The conductors of the stationary part of the machine can be made larger. They will carry more current because they do not rotate.

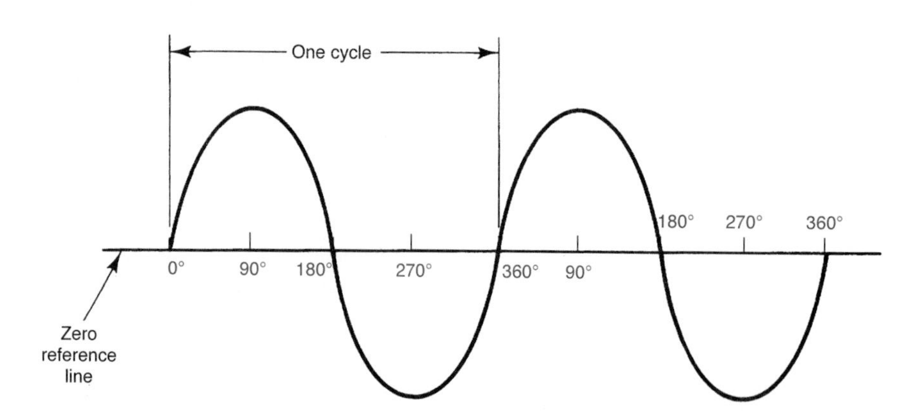

Figure 5–15. Two cycles of an alternating-current (ac) sine wave.

In the wye
or ends of each w
windings are the
The voltage acros
Line voltage is h
Line voltage (V_L
multiplied by the

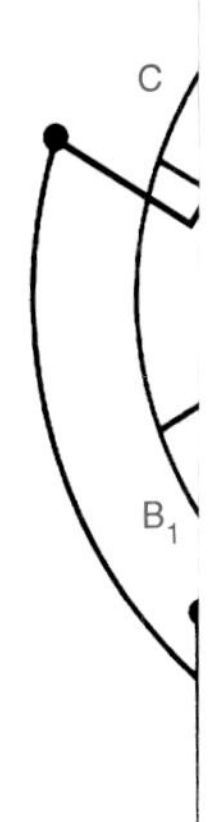

Degrees	Sine
0/360	0.000
30	0.500
60	0.866
90	1.000
120	0.866
150	0.500
180	0.000
210	−0.500
240	−0.866
270	−1.000
300	−0.866
330	−0.500

(a)

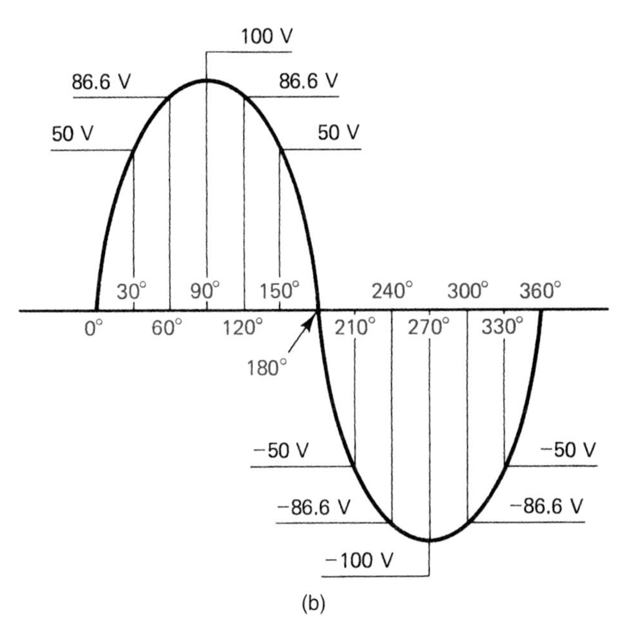

(b)

Figure 5–16. Generation of an ac sine wave: (a) sine values of angles from 0° to 360°; (b) sine wave produced.

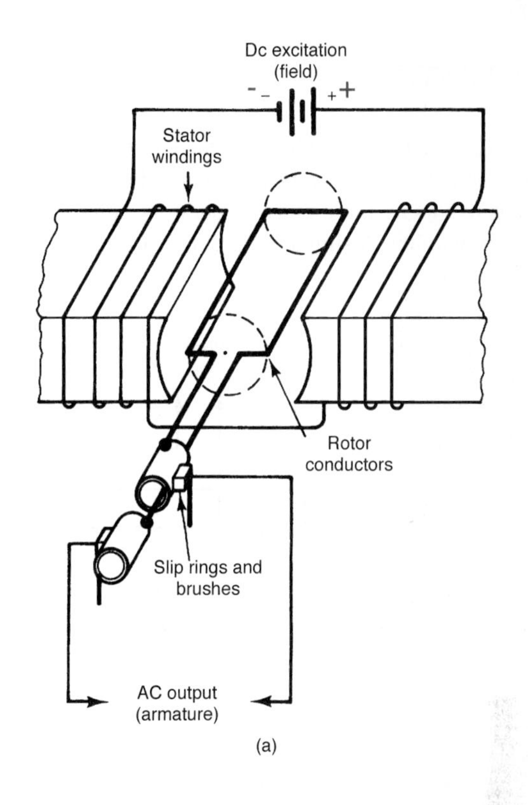

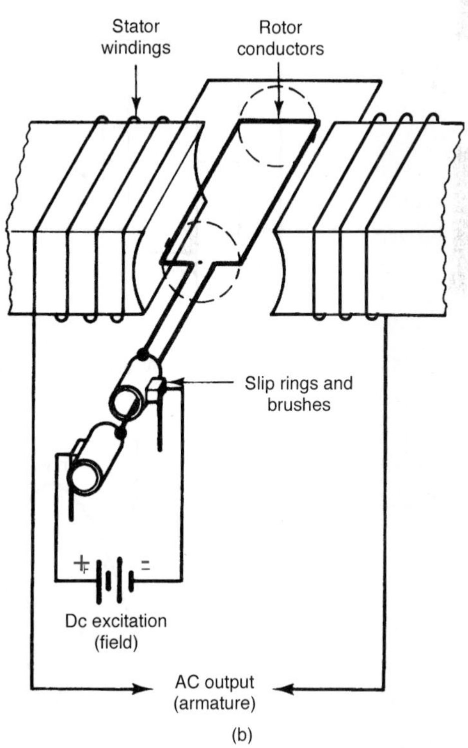

Figure 5–17. Generating voltage: (a) rotating-armature method; (b) rotating-field method.

SAMPLE PROBLEM:

Given: a six-pole three-phase alternator rotates at a speed of 3600 rpm.

Find: the frequency of the alternator.

Solution:

$$f = \frac{N/3 \times \text{rpm}}{120}$$

$$= \frac{6/3 \times 3600}{120}$$

$$= 60 \text{ Hz}$$

Note that if the number of *poles* is increased, the *speed of rotation* may be reduced while still maintaining a 60-Hz frequency.

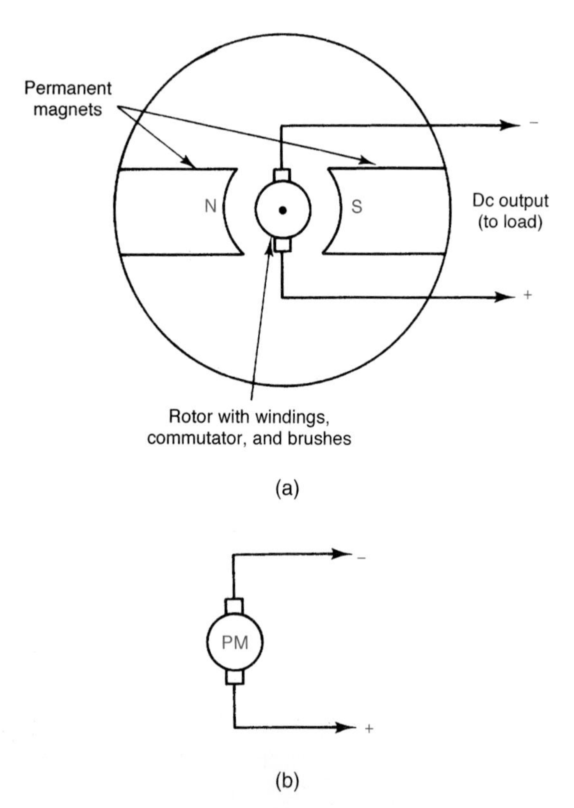

Figure 5–24. Permanent magnet dc generator: (a) pictorial; (b) schematic.

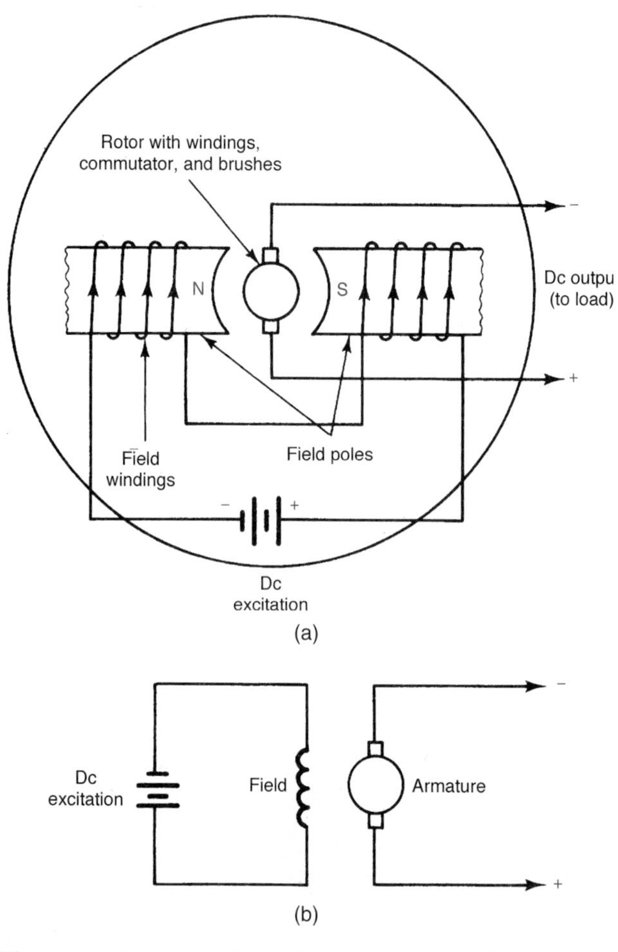

Figure 5–25. Separately excited dc generator: (a) pictorial; (b) schematic.

current, but they do not change the strength of the magnetic field. The output voltage of a separately excited dc generator is varied by adjusting the current flow through the field coils. A large rheostat in series with the field coils can be used to control current flow and adjust output voltage.

Separately excited dc generators are used when precise voltage control is needed. Certain industrial processes require this precision. The cost of separately excited dc generators is usually high. Another disadvantage is that a separate direct-current electrical energy source is needed.

Self-Excited Generators

SELF-EXCITED, SERIES-WOUND DC GENERATORS. Dc generators produce direct current, so it is possible to take part of a generator's output to use as exciting current for the field coils. Generators that use part of their own output to supply dc exciting current are called *self-excited* dc generators. The method of connecting the armature windings and field windings together determines the type of generator. The armature and field windings may be connected in either series, parallel (shunt), or series parallel (compound). These are the three types of self-excited dc generators which could be designed.

Series-wound dc generators have armature windings connected in series with the field windings and the load as shown in Fig. 5–26. In the series-wound dc generator, the total current flows through the load and through the field coils and armature. The field coils are wound with low-resistance wire which has a few turns of large-diameter wire. An electromagnetic field is produced by the current flow through the coils. Remember that current flow is the same in all parts of a series circuit.

If the load is disconnected, no current would flow through the generator. The field coils retain a small amount of magnetism after they are deenergized, which is called *residual magnetism*. Due to residual magnetism, current begins to flow as soon as the generator operates again. As the current increases, the magnetic flux of the field also increases. The output voltage rises as current flow increases. An output graph of a series-wound dc generator is shown in Fig. 5–27. The peak of the curve shows magnetic saturation of the field coils. At this point, the "domains" of the coils have maximum alignment. Beyond this point, an increase in load current causes a decrease in output voltage due to energy losses which occur. The output of a series-wound dc generator varies with changes in load current. Self-excited, series-wound generators have only a few applications.

SELF-EXCITED, SHUNT-WOUND DC GENERATORS. When the field coils, armature circuit, and load are con-

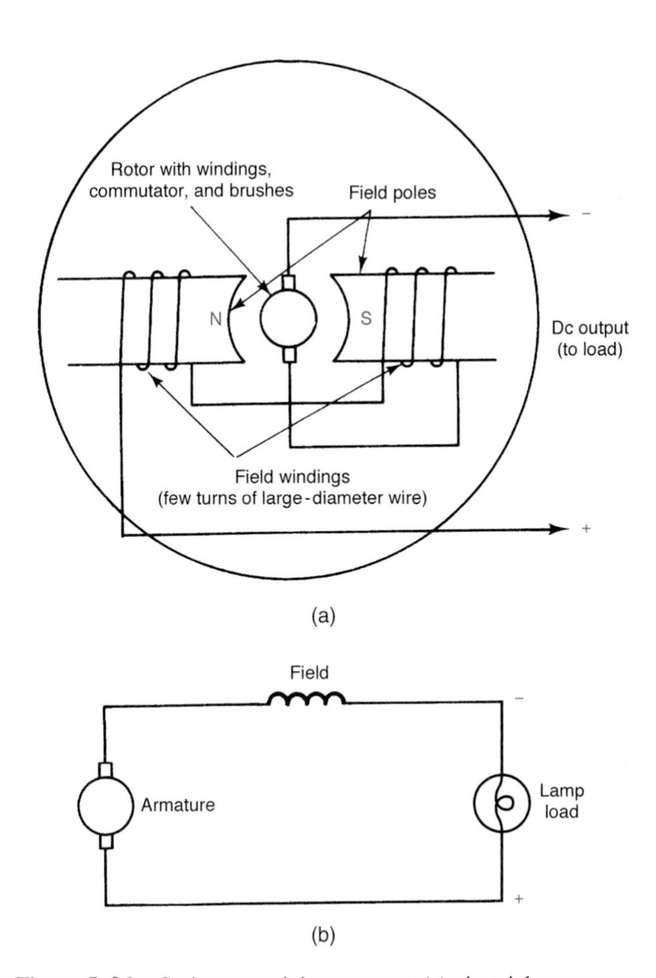

Figure 5–26. Series-wound dc generator: (a) pictorial; (b) schematic.

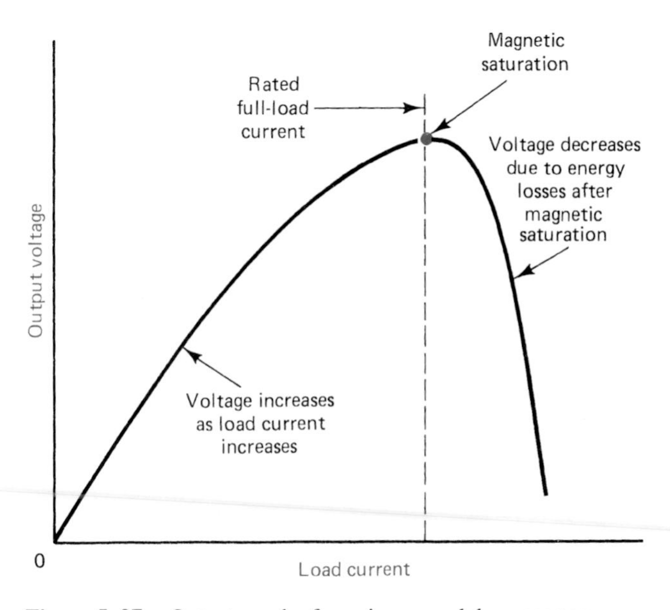

Figure 5–27. Output graph of a series-wound dc generator.

nected in parallel, a shunt-wound dc generator is formed. Figure 5–28 shows a shunt-wound dc generator. The armature current developed by the generator (I_A) has two paths. One path is through the load (I_L) and the other is through the field coils (I_F). The shunt-wound dc generator is designed so that the field current is not more than 10% of the total armature current (I_A). This is so that most of the generated armature current (I_A) will flow to the load.

A strong electromagnetic field must be produced. Also, the field current must be low. The field coils are wound with many turns of wire. They rely very little on the amount of field current to produce a strong magnetic field. The small-diameter wires limit the field current to a low value due to their high resistance.

When no load is connected to a shunt-wound dc generator, a voltage is still generated. The voltage supplies energy to the field coils. Residual magnetism in the field coils is important for shunt-wound dc generators also. When a shunt generator is turned on, current flows in the armature and field circuit due to residual magnetism. As current increases, the output voltage increases until magnetic saturation occurs.

When a load is connected to a dc shunt generator, the armature current (I_A) increases. The current increases the voltage ($I \times R$) drop of the armature. This causes a slightly smaller output voltage. Increases in load current cause slight decreases in output voltage as shown in the graph of Fig. 5–29. With load currents less than the rated value, the voltage is nearly constant. Large load currents cause the output voltage to drop sharply due to energy losses. Self-excited, shunt-wound dc generators are used when a fairly constant output voltage is needed.

SELF-EXCITED, COMPOUND-WOUND DC GENERATORS. Compound-wound dc generators have two sets of field windings. One set is made of low-resistance coils connected in series with the armature circuit. The other set is made of high-resistance coils connected in parallel with the armature circuit. A compound-wound dc generator is shown in Fig. 5–30.

The output voltage of a series-wound dc generator increases with increases in load current. The output voltage of a shunt-wound dc generator decreases slightly with increases in load current. A compound-wound dc generator has both series and shunt windings. Its output voltage is almost constant regardless of load current. The series field windings set up a magnetic field to counteract the voltage reduction caused by the voltage ($I \times R$) drop of the armature circuit. This produces a constant voltage.

A constant output voltage is produced by a *flat-compounded* dc generator. The no-load voltage is equal to the rated full-load voltage of a flat-compounded generator. No-load voltage is the output when there is no load connected to the generator. Full-load voltage is the output when the rated value of load is connected to the circuit. A compound-wound dc generator with full-load voltage

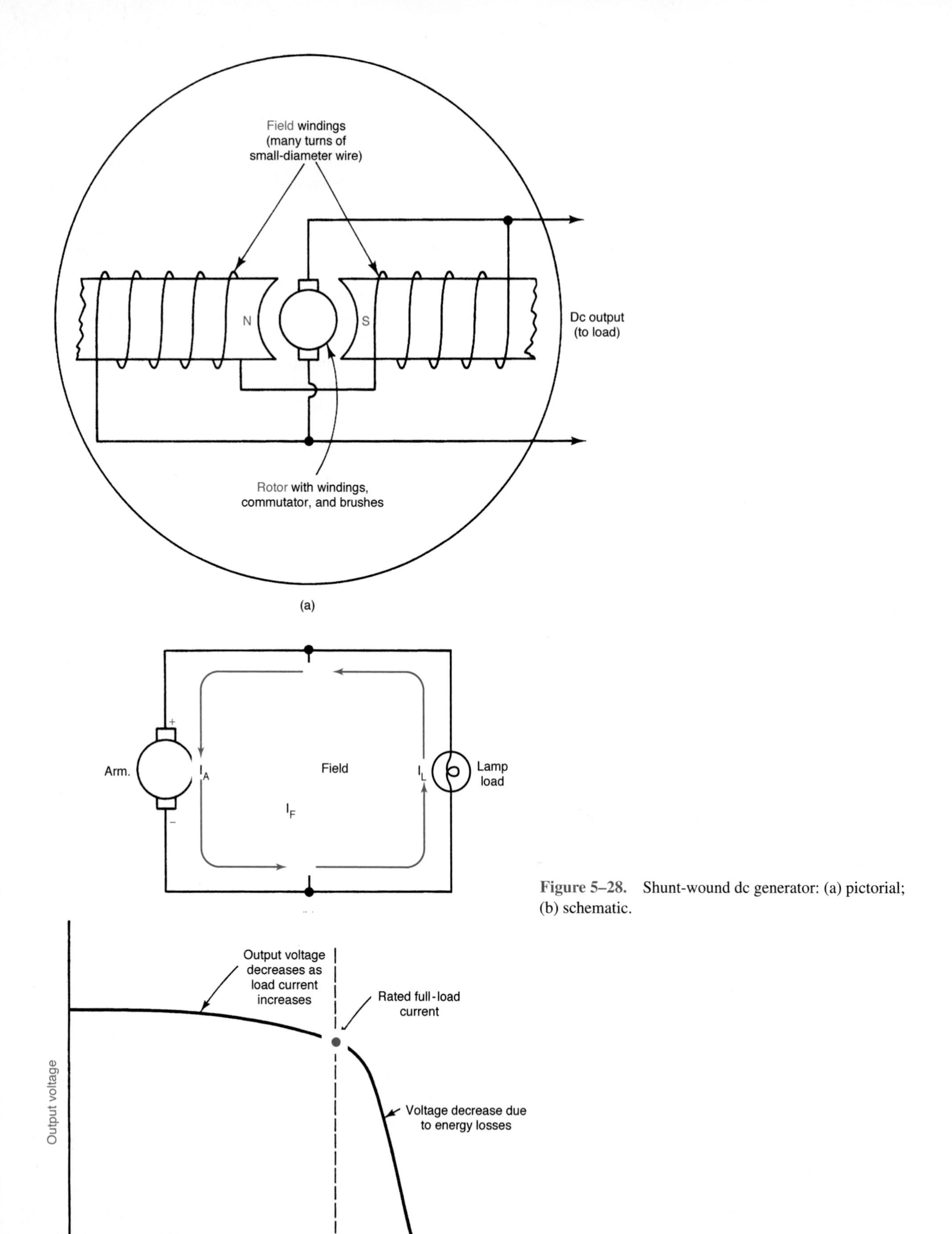

Figure 5–28. Shunt-wound dc generator: (a) pictorial; (b) schematic.

Figure 5–29. Output graph of a shunt-wound dc generator.

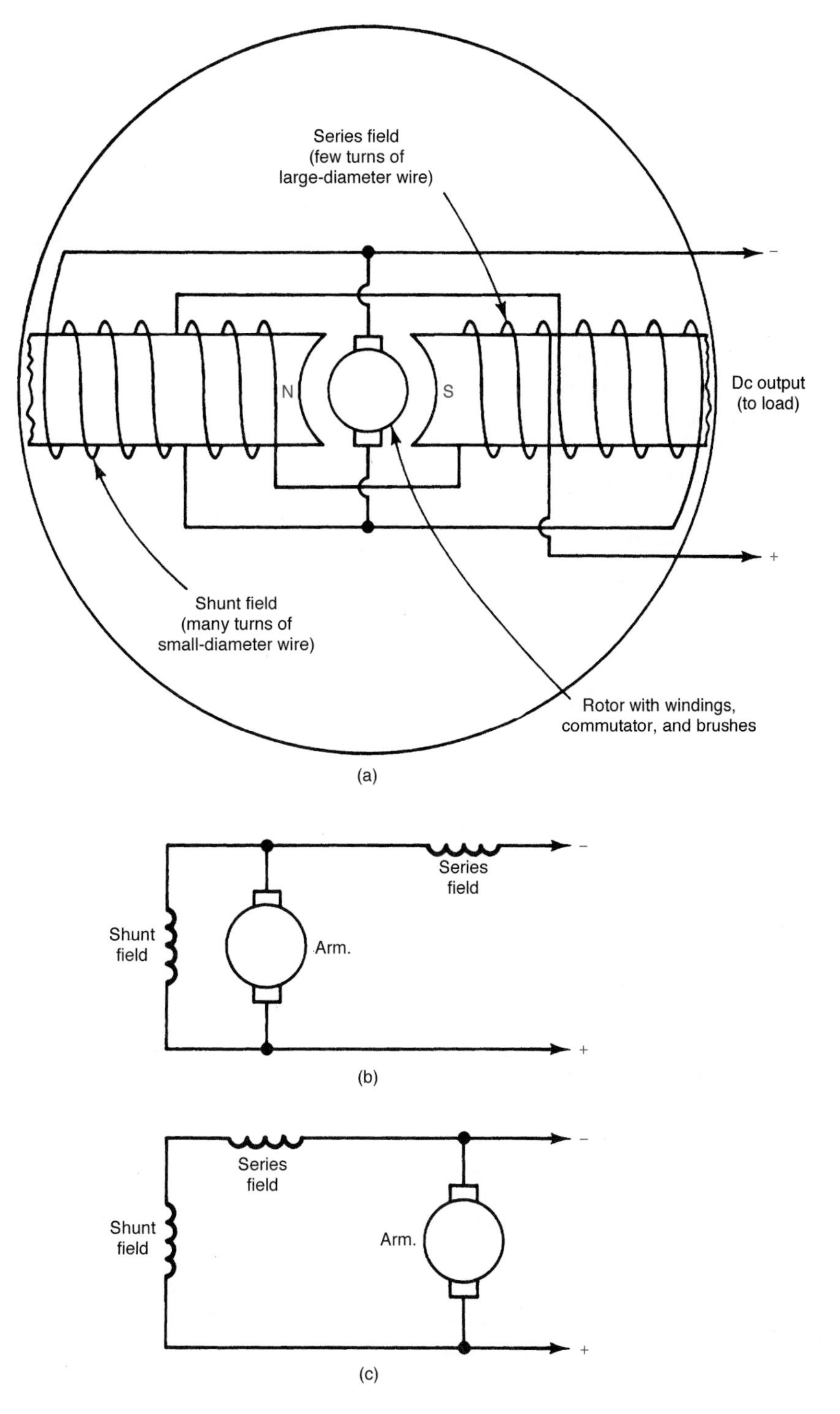

Figure 5–30. Compound-wound dc generator: (a) pictorial; (b) short shunt generator—shunt field is connected across the armature only; (c) long shunt generator—shunt field is connected across the armature and series field.

greater than no-load voltage is called an *overcompounded* generator. A generator with full-load voltage less than no-load voltage is called an *undercompounded* generator. Output graphs for the three types of compound generators are shown in Fig. 5–31.

Compound-wound dc generators can be made so that the series and shunt fields either aid or oppose each other. If the polarities of the coils on one side are the same, the magnetic fields aid each other. This type is called a *cumulative* compound dc generator. A *differential* compound dc motor has opposite polarities on both sides. The cumulative generator is most used. Compound-wound dc generators are used for applications that require constant voltage output.

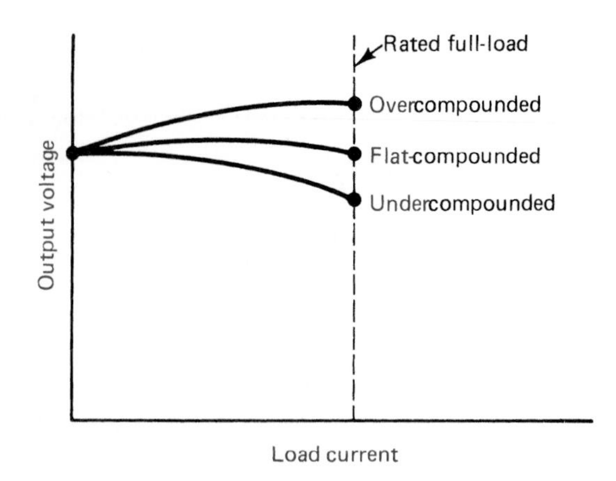

Figure 5–31. Output graph for three types of compound-wound dc generators.

Generator Operating Characteristics

Dc generators have a characteristic known as *armature reaction.* Current flow through the armature windings produces a circular magnetic field. These fields react with the main field as shown in Fig. 5–32. A magnetic field is produced which tends to distort the main magnetic field. As load current increases, armature current also increases. The increase in armature current causes more armature reaction to occur. Armature reaction causes sparking between the brushes and commutator. The theoretical switching point of the current to the load occurs at the generator's *neutral plane.* This is a plane or position where no voltage is induced into the armature conductor connected to the commutator. Also, no distortion of the main magnetic field takes place. The switching position is perpendicular to the main magnetic field. Armature reaction causes the main magnetic field to become distorted. The new switching position occurs when a small voltage is induced into an armature coil. This new switching position is called the *running neutral plane.*

Armature reaction is reduced by using windings called *interpoles* between the main field coils. These coils are connected in series with the armature circuit. An increase in armature current causes a stronger magnetic field around the interpoles. Its field counteracts the distortion of the main field caused by armature reaction.

Generator power output is usually rated in kilowatts. This rating is the electrical power generating capacity of a generator. Ratings are specified by the manufacturer on the nameplate of a generator. Other ratings include output voltage, speed, and temperature limits. Generators are made in many sizes.

As the load of a generator is increased, the voltage drop due to increased current flowing through the armature resistance increases. The output voltage then decreases. The amount of voltage change depends on the type of generator. The amount of change in output voltage from no-load value to rated full-load value is called *voltage regulation.* Voltage regulation is found by using the formula:

$$\%\ \mathrm{VR} = \frac{V_{NL} - V_{FL}}{V_{FL}} \times 100$$

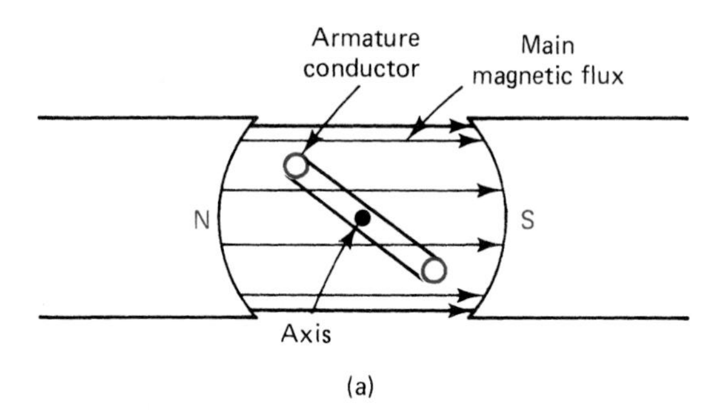

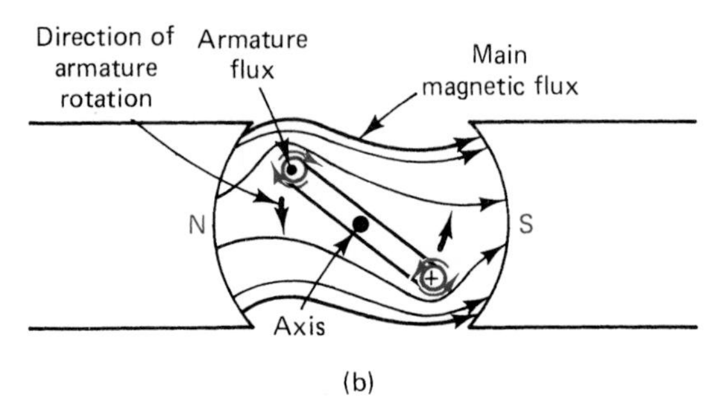

Figure 5–32. Effect of armature reaction in dc generators: (a) main magnetic field with no current flow through the armature windings; (b) distortion of the main magnetic field with current flow through the armature windings.

where

% VR is the voltage regulation,
V_{NL} is the voltage with no load connected, and
V_{FL} is the rated full-load voltage of the generator.

The efficiency of a generator is the ratio of its power output in watts and its power input in horsepower. The efficiency of a generator is found by using this formula:

$$\%\ \text{efficiency} = \frac{P_{out}}{P_{in}} \times 100$$

where P_{in} is the power input in horsepower and P_{out} is the power output in watts. Horsepower must be converted to watts. Since 1 hp = 746 W, multiply the horsepower by 746.

SAMPLE PROBLEM:

Given: a single-phase alternator has a no-load output voltage of 122.5 volts and a rated full-load voltage of 120.0 volts.

Find: the voltage regulation of the alternator.

Solution:

$$\%VR = \frac{V_{NL} - V_{FL}}{V_{FL}} \times 100$$

$$= \frac{122.5 - 120}{120} \times 100$$

$$= 0.02 = 2\%$$

SAMPLE PROBLEM:

Given: a three-phase alternator has a power output of 22 MW and a power input of 35,000 horsepower.

Find: the efficiency of the alternator.

Solution:

$$\%\ Eff = \frac{P_{out}}{P_{in}} \times 100$$

$$= \frac{22{,}000{,}000\ W}{35{,}000\ hp \times 746} \times 100$$

$$\%\ Eff = 84\%$$

To convert horsepower to watts, remember that 1 hp = 746 watts. The efficiency of a generator usually ranges from 70% to 85%.

REVIEW

1. List eight sources of electrical energy.
2. What are the two types of electrical current produced by energy sources called?
3. What are two classifications of chemical cells, and how are they different?
4. Discuss the operation of a carbon-zinc cell.
5. Discuss the operation of a lead-acid cell.
6. How do nickel-cadmium cells differ from carbon-zinc cells?
7. How is light energy used to produce electrical energy?
8. How is heat energy used to produce electrical energy?
9. How is pressure used to produce electrical energy?
10. Discuss electromagnetic induction.
11. What three factors determine the amount of voltage induced into a conductor inside a magnetic field?
12. What are the differences between slip rings and a commutator used for electrical generators?
13. What is a sine wave?
14. If the maximum voltage of a generator is 50 V and the angle of rotation is 240°, what is the instantaneous voltage?
15. If the speed of rotation of a four-pole generator is 1800 rpm, what is its output frequency?
16. What are the two methods used to produce voltage output in generators, and how do they differ?
17. Discuss the two types of three-phase connections used for stator windings in three-phase generators.
18. Discuss the construction of a dc generator.
19. What are the types of dc generators?
20. Discuss armature reaction.
21. If a generator has a no-load voltage output of 100 V and a full-load voltage output of 90 V, what is its voltage regulation percentage?
22. If a generator has a power output of 2000 W and a power input of 3 hp, what is its efficiency?

STUDENT ACTIVITIES

• Chemical Cells (see Fig. 5–33)

1. Obtain a lemon, two pieces of metal (copper and zinc, if available), and a meter that measures low values of direct current.
2. Connect the meter as shown.
3. Squeeze the lemon until the opening where the metal strips are placed is juicy.
4. Exchange the positions of the two pieces of metal and watch the meter.
5. On a sheet of paper, discuss what occurs.

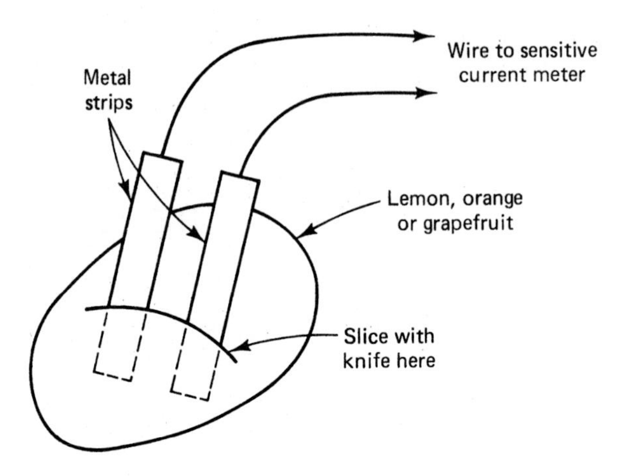

Figure 5–33. Making a chemical cell.

• Light Sources of Electrical Energy (see Fig. 5–9)

1. Obtain a solar cell.
2. Connect a VOM, set on a 1-V dc range, across the leads of the solar cell. Be sure to observe polarity.
3. Shine a flashlight onto the solar cell.
4. What is the maximum dc voltage output? Maximum voltage = _______ V dc.
5. On a sheet of paper, record your findings of this activity. Explain the relationship of light and voltage output and the method used to convert light energy into electrical energy.

• Sources of Electrical Energy—Light

Complete Activity 5–1, page 97 in the manual.

• • ANALYSIS

1. List five sources of electrical energy.
2. Where might photovoltaic cells be used?
3. What are some advantages of converting light to electrical energy?

• Sources of Electrical Energy—Heat

Complete Activity 5–2, page 99 in the manual.

ANALYSIS

1. What is a thermocouple?
2. Explain in detail the Seebeck effect.
3. How does the amount of heat applied to the junction of a thermocouple affect its millivolt output?
4. For what would a thermocouple be used?
5. If a thermocouple exhibited an output of 1 millivolt for each 50° temperature difference between its hot and cold ends, what would be the heated junction temperature when its output is 5 millivolts?

• *Electromagnetic Induction (see Fig. 5–34)*

1. Obtain the following materials:

   Electromagnet coil constructed in Student Activity of Chap. 4.
   Galvanometer (zero centered)
   Permanent magnet
   Piece of string (about 2 ft. long)
   Wood to construct a platform
2. Connect the electromagnet coil to the terminals of the galvanometer. Place them on a flat surface (see Fig. 5–34).
3. Attach the permanent magnet to a string. Connect the other end of the string to the platform so that the magnet is about 0.25 in. above the coil.
4. Push the magnet and allow it to move back and forth across the coil. Observe the galvanometer movement.

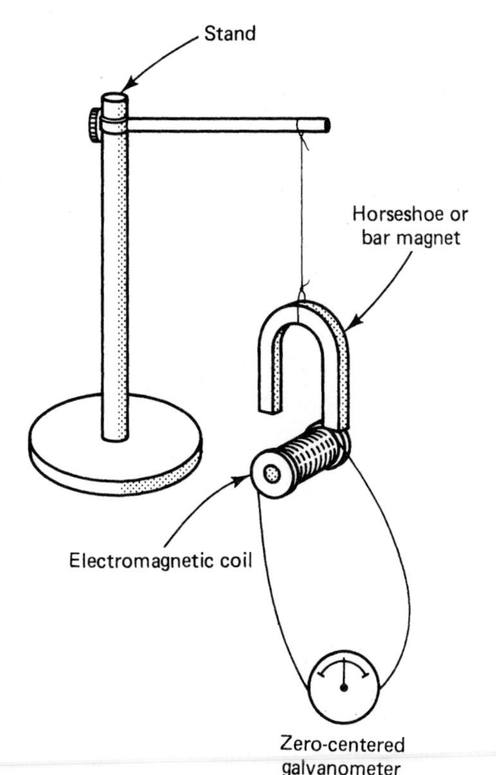

Figure 5–34. Setup for studying electromagnetic induction.

5. Let the teacher check your setup.
6. On a sheet of paper, discuss the electromagnetic induction. Be sure to discuss how speed and magnetic field strength affect current.
7. Turn in your paper to the teacher.

• *Electromagnetic Induction*

Complete Activity 5–3, page 103 in the manual.

• • ANALYSIS

1. How could the strength of the induced EMF and current in a coil increase?
2. Why would the needle of the meter in Fig. 5–34 deflect in the opposite direction when the magnet is positioned on the opposite side of the coil?
3. What factors control the strength of any induced EMF and current?
4. What are some types of electrical equipment which rely on electromagnetic induction to operate?

• *Generator Problems*

On a sheet of paper, work the following problems dealing with electrical generators.

- If the peak voltage of a single-phase ac generator is 320 V, what are the instantaneous induced voltages after an armature conductor has rotated the following number of degrees?
  1. 25°
  2. 105°
  3. 195°
  4. 290°
  5. 335°
- A three-phase ac (wye-connected) generator has a phase voltage of 120 V and phase current of 10 A. What is its:
  6. Line voltage
  7. Line current
  8. Total power
- A three-phase ac (delta-connected) generator has a line voltage of 480 V and a line current of 7.5 A. Find:
  9. Phase voltage
  10. Phase current
  11. Total power
- Find the operating frequencies of the following ac generators:
  12. Single-phase, two-pole, 3600 rpm
  13. Single-phase, four-pole, 1800 rpm
- Find the voltage regulation of ac generators with the following no-load and full-load voltage values:
  14. $V_{NL} = 120$ V, $V_{FL} = 115$ V
  15. $V_{NL} = 220$ V, $V_{FL} = 210$ V
  16. $V_{NL} = 480$ V, $V_{FL} = 478$ V

- Find the efficiency of the following ac generators:
  17. Power input = 20 hp, power output = 12 kW
  18. Power input = 2.5 hp, power output = 1200 W
  19. Power input = 4.25 hp, power output = 3 kW
- A series-wound dc generator has an armature resistance of 1.5 Ω and a field resistance of 4.5 Ω. The voltage across a 12-Ω load is 100 V. What is the value of the following?
  20. Armature current
  21. Power output
- A shunt-wound dc generator has an armature resistance of 2.5 Ω and a field resistance of 40 Ω. Its terminal voltage is 120 V. Calculate:
  22. Armature current
  23. Field current
  24. Current delivered to the load
  25. Power output
- A 120-V dc generator is rated at 20 kW. What is its maximum load current?

• *Automobile Alternator (see Fig. 5–20)*

1. Obtain an automobile alternator.
2. Disassemble the alternator to observe the parts.
3. Locate the stator, rotor, brushes, and slip rings.
4. On a sheet of paper, discuss the construction and operation of an automobile alternator.

• *Sources of Electrical Energy—Mechanical*

Complete Activity 5–4, page 105 in the manual.

• • ANALYSIS

1. What factors determined the output of any mechanical generator?
2. What type of voltage output is normally associated with a mechanical generator using collector rings?
3. What type of voltage output is normally associated with a mechanical generator using a commutator?
4. How does the armature rotation speed affect the generator output?
5. What determines the output frequency of any mechanical generator?
6. What is the relation between pressure and voltage output of a piezoelectric crystal?

• *Self-Examination*

1. Complete the self-examination for Chap. 5, page 109 in the manual.
2. Check your answers on p. 411 in the manual.
3. Study the questions that were answered incorrectly.

• *Chapter 5 Examination*

Complete the Chap. 5 Examination on page 111 in the manual.

CHAPTER

6

ALTERNATING-CURRENT ELECTRICITY

OUTLINE

6.1 Alternating-Current (Ac) Voltage
6.2 Single-Phase and Three-Phase Ac
6.3 Measuring Ac Voltage
6.4 Using an Oscilloscope
6.5 Ohm's and Kirchhoff's Laws for Ac Circuits
6.6 Inductance
6.7 Capacitance
6.8 Inductive Effects in Circuits
6.9 Capacitive Effects in Circuits
6.10 Leading and Lagging Currents in Ac Circuits
6.11 Capacitor Charging and Discharging
6.12 Types of Capacitors
6.13 Capacitor Testing
6.14 Alternating-Current Circuits
6.15 Vector Diagrams
6.16 Mathematics for Ac Circuits
6.17 Series Ac Circuits
6.18 Parallel Ac Circuits
6.19 Power in Ac Circuits
6.20 Filters and Resonant Circuits
6.21 Transformers

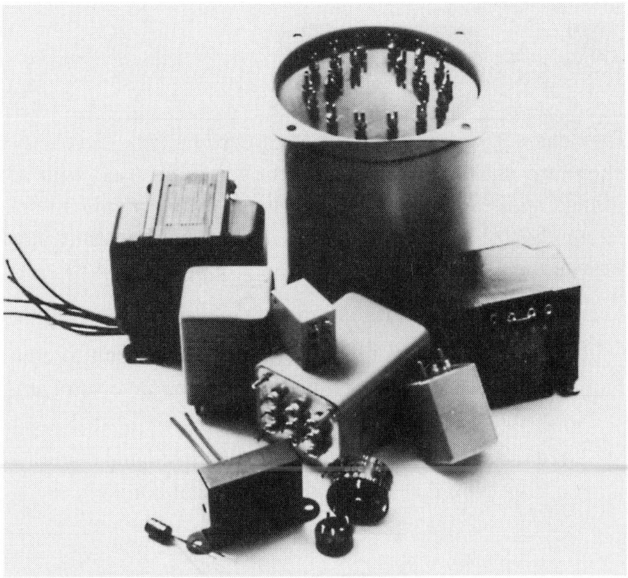

Transformers play an important role in ac system operation.

OBJECTIVES

Upon completion of this chapter, you will be able to:

1. Explain the difference between ac and dc.
2. Define the process of electromagnetic induction.
3. Describe factors affecting induced voltage.
4. Draw a simple ac generator and explain ac voltage generation.
5. Convert peak, peak-peak, average, and rms/effective values from one to the other.
6. Explain the relationship between period and frequency of an ac waveform.
7. Recognize the different types of ac waveforms.
8. Measure alternating current with an ac ammeter.
9. Use a multimeter to measure ac voltage.
10. Explain the basic operation of an oscilloscope.
11. Demonstrate how an oscilloscope can be used to measure the amplitude, period, frequency, and phase relationships of ac waveforms.
12. Demonstrate how an ac voltmeter is used to measure ac voltages.
13. Perform ac circuit measurements and calculations.
14. Define and calculate impedance.
15. Draw diagrams illustrating the phase relationship between current and voltage in a capacitive circuit.
16. Define capacitive reactance.
17. Solve problems using the capacitive reactance formula.
18. Define impedance.
19. Calculate impedance of series and parallel resistive-capacitive circuits.
20. Determine current in *RC* circuits.
21. Explain the relationship between ac voltages and current in a series-resistive circuit.
22. Understand the effect of capacitors in series and parallel.
23. Explain the characteristics of a series *RC* circuit.
24. Solve Ohm's law problems for ac circuits.
25. Understand and solve problems involving true power, apparent power, power factor, and reactive power.
26. List several purposes of transformers.
27. Describe the construction of a transformer.
28. Explain transformer action.
29. Calculate turns ratio, voltage ratio, current ratio, power, and efficiency of transformers.
30. Explain the purpose of isolation transformers and autotransformers.
31. Explain factors that cause losses in transformer efficiency.
32. Identify types of filter circuits.
33. Draw response curves for basic filter circuits.
34. Investigate the characteristics of series resonant and parallel circuits.
35. Investigate the frequency response of high-pass, low-pass, and bandpass filters.
36. Define resonance.
37. Calculate resonant frequency.
38. List the characteristics of series and parallel resonant circuits.
39. Define and calculate Q and bandwidth.

Much of the electrical energy used today is called alternating current (ac). Most of the electrical equipment and appliances used in homes operate from the alternating-current energy delivered by power lines. Alternating-current electricity has many applications in homes, industries, and commercial buildings. Electrical power plants in our country produce alternating current or ac electricity. Most power plants have huge steam turbines that rotate ac generators. These generators produce three-phase ac which is distributed by long-distance power transmission lines to the places where the electrical power is used. Industries and large commercial buildings use three-phase ac. Homes use single-phase ac power. Alternating current is the most common form of electrical energy used in the United States.

IMPORTANT TERMS

This chapter discusses alternating-current (ac) electricity. There are many devices and circuits that are used with ac voltage applied to them. Review these terms to gain an understanding of some of the topics of this chapter. Many new terms are introduced here.

Ac. An abbreviation for alternating current.

Admittance (Y). The total ability of an ac circuit to conduct current, measured in siemens; the inverse of impedance (Z): $Y = 1/Z$.

Air-core inductor. A coil wound on an insulated core or a coil of wire that does not have a metal core.

Amplitude. The vertical height of an ac waveform.

Angle of lead or lag. The angle between applied voltage and current flow in an ac circuit, in degrees; in an inductive (L) circuit, voltage (V) leads current (I); in a capacitive (C) circuit, current (I) leads voltage (V).

Apparent power (volt-amperes). The applied voltage times current delivered to an ac circuit.

Attenuation. A reduction in value.

Average voltage (V_{avg}). The value of an ac sine-wave voltage which is found by the formula $V_{avg} = V_{peak} \times 0.637$.

Bandpass filter. A frequency-sensitive ac circuit that allows incoming frequencies within a certain band to pass through but attenuates frequencies that are below or above this band.

Bandwidth. The band (range) of frequencies that will pass easily through a bandpass filter or resonant circuit.

Capacitance (C). The property of a device to oppose changes in voltage due to energy stored in its electrostatic field.

Capacitive reactance (X_C). The opposition to the flow of ac current caused by a capacitive device (measured in ohms).

$$X_C = \frac{1}{2\pi fC}$$

Capacitor. A device that has capacitance and is usually made of two metal plate materials separated by a dielectric material (insulator).

Center tap. An electrical connection point at the center of a wire coil or transformer winding.

Choke coil. An inductor coil used to block the flow of ac current and pass dc current.

Condenser. A term occasionally used to mean *capacitor.*

Conductance (G). The ability of a resistance of a circuit to conduct current, measured in siemens; the inverse of resistance: $G = 1/R$.

Cycle. A sequence of events that causes one complete pattern of alternating current from a zero reference, in a positive direction, back to zero, then in a negative direction, and back to zero.

Decay. A term used for a gradual reduction in value of a voltage or current.

Decay time. The time required for a capacitor to discharge to a certain percentage of its original charge or the time required for current through an inductor to reduce to a percentage of its maximum value.

Decibel (dB). A unit used to express an increase or decrease in power, voltage, or current in a circuit; one tenth of a bel.

Dielectric. An insulating material placed between the metal plates of a capacitor.

Dielectric constant. A number that represents the ability of a dielectric to develop an electrostatic field, as compared with air, which has a value of 1.0.

Effective voltage (V_{eff}). The value of an ac sine-wave voltage which has the same heating effect as an equal value of dc voltage.

$$V_{eff} = V_{peak} \times 0.707$$

Electrolytic capacitor. A capacitor that has a positive plate made of aluminum or tantalum and a dry paste or liquid used to form the negative plate.

Electrostatic field. The space or area around a charged body in which the influence of an electrical charge is experienced.

Farad (F). The unit of measurement of capacitance that is required to contain a charge of 1 C when a potential of 1 V is applied.

Filter. A circuit used to pass certain frequencies and attenuate all other frequencies.

Frequency. The number of ac cycles per second, measured in hertz (Hz).

Frequency response. A circuit's ability to operate over a range of frequencies.

Henry (H). The unit of measurement of inductance that is produced when a voltage of 1 V is induced when the current through a coil is changing at a rate of 1 A per second.

Hertz (Hz). The international unit of measurement of frequency equal to one cycle per second.

Impedance (Z). The total opposition to current flow in an ac circuit which is a combination of resistance (R) and reactance (X) in a circuit; measured in ohms.

$$Z = \sqrt{R^2 + X^2}$$

Inductance (L). The property of a circuit to oppose changes in current due to energy stored in a magnetic field.

Inductive circuit. A circuit that has one or more inductors or has the property of inductance, such as an electric motor circuit.

Inductive reactance (X_L). The opposition to current flow in an ac circuit caused by an inductance (L), measured in ohms.

$$X_L = 2\pi fL$$

Inductor. A coil of wire that has the property of inductance and is used in a circuit for that purpose.

In phase. Two waveforms of the same frequency which pass through their minimum and maximum values at the same time and polarity.

Instantaneous voltage (V_i). A value of ac voltage at any instant (time) along a waveform.

Isolation transformer. A transformer with a 1:1 turns ratio used to isolate an ac power line from equipment with a chassis ground.

Lagging phase angle. The angle by which current *lags* voltage (or voltage *leads* current) in an inductive circuit.

Leading phase angle. The angle by which current *leads* voltage (or voltage *lags* current) in a capacitive circuit.

Maximum power transfer. A condition that exists when the resistance of a load (R_L) equals that of the source which supplies it (R_S).

Mho. *Ohm* spelled backward; the old unit of measurement for conductance, susceptance, and admittance; now replaced by siemens.

Mica capacitor. A capacitor made of metal foil plates separated by a mica dielectric.

Mutual inductance (M). When two coils are located close together so that the magnetic flux of the coils affect one another in terms of their inductance properties.

Parallel resonant circuit. A circuit that has an inductor and capacitor connected in parallel to cause response to frequencies applied to the circuit.

Peak-to-peak voltage ($V_{p\text{-}p}$). The value of ac sine-wave voltage from positive peak to negative peak.

Peak voltage (V_{peak}). The maximum positive or negative value of ac sine-wave voltage.

$$V_{\text{peak}} = V_{\text{eff}} \times 1.41$$

Period (time). The time required to complete one ac cycle; time = 1/frequency.

Phase angle (θ). The angular displacement between applied voltage and current flow in an ac circuit.

Power (P). The rate of doing work in electrical circuits, found by using the equation $P = I \times V$.

Power factor (PF). The ratio of true power in an ac circuit and apparent power:

$$PF = \frac{\text{true power } (W)}{\text{apparent power } (VA)}$$

Primary winding. The coil of a transformer to which ac source voltage is applied.

Quality factor (Q). The "figure of merit" or ratio of inductive reactance and resistance in a frequency-sensitive circuit.

Reactance (X). The opposition to ac current flow due to inductance (X_L) or capacitance (X_C).

Reactive circuit. An ac circuit that has the property of inductance or capacitance.

Reactive power (VAR). The "unused" power of an ac circuit has inductance or capacitance, which is absorbed by the magnetic or electrostatic field of a reactive circuit.

Resistance (R). Opposition to the flow of current in an electrical circuit; its unit of measurement is the ohm (Ω).

Resistive circuit. A circuit whose only opposition to current flow is resistance; a nonreactive circuit.

Resonant circuit. *See* Parallel resonant circuit *and* Series resonant circuit.

Resonant frequency (f_r). The frequency that passes most easily through a frequency-sensitive circuit when $X_L = X_C$ in the circuit:

$$f_r = \frac{1}{2\pi \sqrt{L \times C}}$$

Root mean square (rms) voltage. *See* Effective voltage.

Sawtooth waveform. An ac waveform shaped like the teeth of a saw.

Secondary winding. The coil of a transformer into which voltage is induced; energy is delivered to the load circuit by the secondary winding.

Selectivity. The ability of a resonant circuit to select a specific frequency and reject all other frequencies.

Series resonant circuit. A circuit that has an inductor and capacitor connected in series to cause response to frequencies applied to the circuit.

Siemens (S). *See* Mho.

Signal. An electrical waveform of varying value which is applied to a circuit.

Sine wave. The waveform of ac voltage.

Step-down transformer. A transformer that has a secondary voltage lower than its primary voltage.

Step-up transformer. A transformer that has a secondary voltage higher than its primary voltage.

Susceptance (B). The ability of an inductance (B_L) or a capacitance (B_C) to pass ac current; measured in siemens:

$$B_L = \frac{1}{X_L} \quad \text{and} \quad B_C = \frac{1}{X_C}$$

Tank circuit. A parallel resonant LC circuit.

Theta (θ). The Greek letter used to represent the phase angle of an ac circuit.

Transformer. An ac power control device that transfers energy from its primary winding to its secondary winding by mutual inductance and is ordinarily used to increase or decrease voltage.

True power (W). The power actually converted by an ac circuit, as measured with a wattmeter.

Turns ratio. The ratio of the number of turns of the primary winding (N_P) of a transformer to the number of turns of the secondary winding (N_S).

Vector. A straight line whose length indicates magnitude and position indicates direction.

Volt-ampere (VA). The unit of measurement of apparent power.

Volt-amperes reactive (VAR). The unit of measurement of reactive power.

Waveform. The pattern of an ac frequency derived by looking at instantaneous voltage values that occur over time; on a graph, a waveform is plotted with instantaneous voltages on the vertical axis and time on the horizontal axis.

Wavelength. The distance between two corresponding points that represents one complete wave.

Working voltage. A rating of capacitors which is the maximum voltage that can be placed across the plates of a capacitor without damage occurring.

6.1 ALTERNATING-CURRENT (AC) VOLTAGE

When an ac source is connected to some type of load, current direction changes several times in a given unit of time. Remember that direct current (dc) flows in one direction only. A diagram of one cycle of alternating current is compared with a dc waveform in Fig. 6–1(a). This waveform is called an ac sine wave. When the ac generator shaft rotates one complete revolution, or 360°, one ac sine wave is produced. Ac and dc voltage generation is illustrated in Fig. 6–1. Note that the ac sine wave has a positive peak at 90°, then decreases to zero at 180°. It then increases to a peak negative voltage at 270°, then decreases to zero at 360°. The cycle then repeats itself. Current flows in one direction during the positive part and in the opposite direction during the negative half-cycle.

Dc voltage [Fig. 6–1(b)] is a straight line or unidirectional voltage. The direction of electron current flow is from negative to positive through a dc circuit. A commonly used source of dc voltage is a variable dc power supply [see Fig. 6–1(c)]. This type of dc power supply can be used to adjust dc voltage to the value desired. They are used primarily in laboratory or test facilities. Batteries are also a source of dc voltage.

Alternating current (ac) is produced by generators at electrical power plants. This ac voltage is in the form of sine waves [Fig. 6–1(a)]. In a laboratory, variable ac is often provided by a *function generator* [see Fig. 6–1(d)]. The function generator produces variable voltage output at a wide range of frequencies.

Figure 6–2 shows five cycles of alternating current. If the time required for an ac generator to produce five cycles were 1 s, the frequency of the ac would be 5 cycles per second. Ac generators at power plants in the United States operate at a frequency of 60 cycles per second, or 60 Hz. Hertz is the international unit for frequency measurement. If 60 ac sine waves are produced every second, a speed of 60 revolutions per second is needed. This produces a frequency of 60 cycles per second.

Ac voltage is measured with either an analog or a digital voltmeter (multimeter). The polarity of the meter leads is not important, because ac changes direction. Remember that polarity is important when measuring dc, because direct current flows only in one direction. Many analog multimeters do not measure ac current. They have ranges for ac voltage only.

Shown in Fig. 6–3 are several voltage values associated with alternating current. Among these are peak positive, peak negative, and peak-to-peak ac values. Peak positive is the maximum positive voltage reached during a cycle of ac. Peak negative is the maximum negative voltage reached. Peak-to-peak is the voltage value from peak positive to peak negative. These values are important to know when working with radio and TV amplifier circuits. For example, the most important ac value is called *effective* or measured value. This value is less than the peak positive value. A common ac voltage is 120 V, which is used in homes and is an effective value voltage. Its peak value is about 170 V. The effective value of ac is defined as the ac voltage that will do the same amount of work as a dc voltage of the same value. For instance, in the circuit of Fig. 6–4, if the switch is placed in position 2, 10-V ac effective value is applied to the lamp. The lamp should produce the same amount of brightness with 10-V ac effective value as with 10-V dc applied. When ac voltage is measured with a meter, the reading indicated is effective value.

In some cases, it is important to convert one ac value to another. For instance, the voltage rating of electronic devices must be greater than the peak ac voltage applied to them. If 120 V ac is the measured voltage applied to a device, the peak voltage is about 170 V. So the device must be rated over 170 V rather than 120 V.

To determine peak ac, when the measured or effective value is known, the formula

$$\text{peak} = 1.41 \times \text{effective value}$$

is used. When 120 V is multiplied by the 1.41 conversion factor, the peak voltage is found to be about 170 V.

Two other terms that should be mentioned are *rms value* and *average value*. Rms stands for "root mean square" and is equal to 0.707 × peak value. Rms refers to the mathematical method used to determine effective voltage. Rms voltage and effective voltage are the same. Average voltage

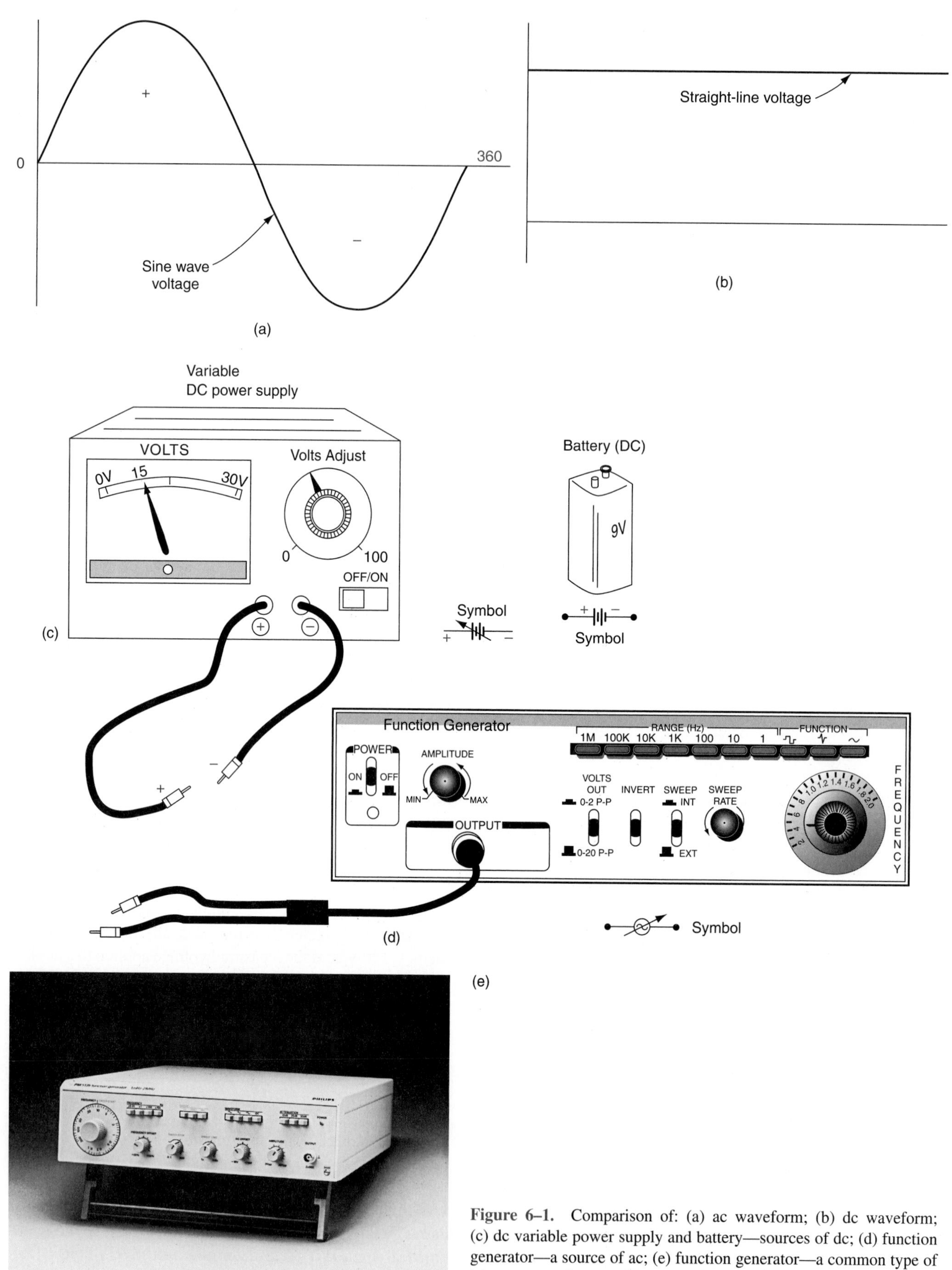

Figure 6–1. Comparison of: (a) ac waveform; (b) dc waveform; (c) dc variable power supply and battery—sources of dc; (d) function generator—a source of ac; (e) function generator—a common type of laboratory equipment. (Courtesy Fluke Corp.)

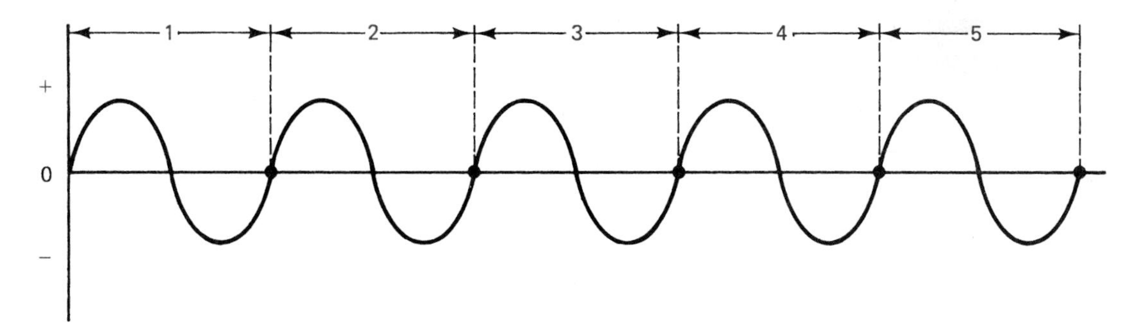

Figure 6–2. Five cycles of alternating current.

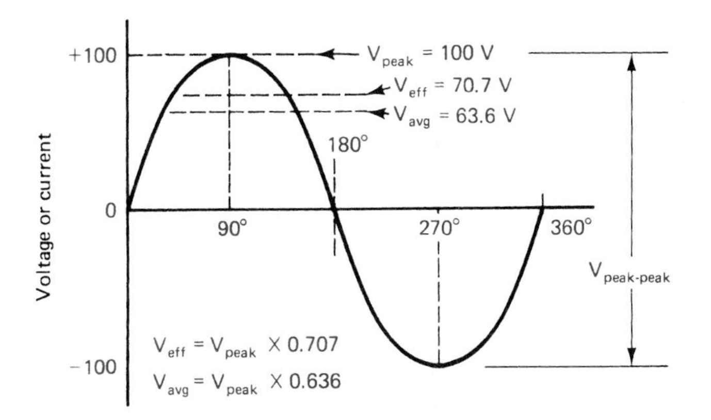

Figure 6–3. Voltage values of an ac waveform.

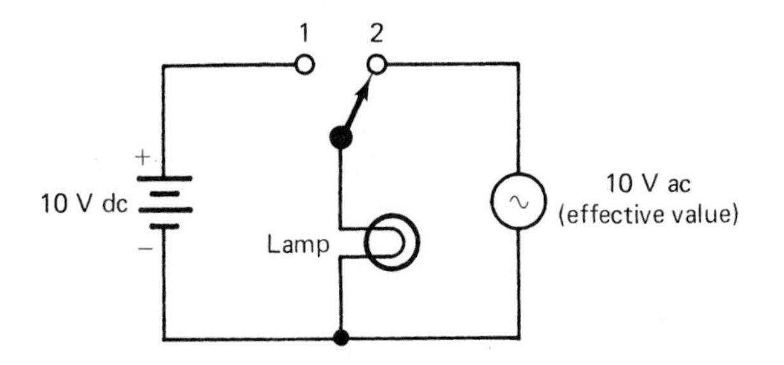

Figure 6–4. Comparison of effective ac voltage and dc voltage.

is the mathematical average of all instantaneous voltages that occur at each period of time throughout an alternation. The average value is equal to 0.636 times the peak value for one-half cycle (alternation) ac voltage.

Up to now, the only ac waveform that has been discussed is the sine wave. Except for dc, the sine wave is the simplest of all waveforms. In the study of electronics, you will encounter many waveforms that do not have the simple structure of the sine wave. These are known as *nonsinusoidal* or *complex* waves. A nonsinusoidal wave does not follow the sine curve in amplitude variations. Its form is not necessarily symmetrical, and it may be composed of more than one frequency.

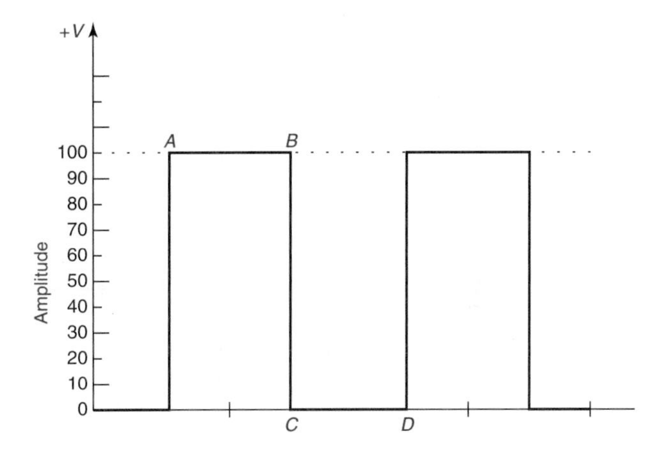

Figure 6–5. A square-wave voltage.

One common nonsinusoidal wave is the *square wave.* Figure 6–5 shows the graph of a square wave. At point A of Fig. 6–5, the voltage has risen from zero to a positive 100-V level. The voltage remains at the positive level through the period from A to B and then drops to zero at point C. The voltage remains zero for a time identical to the period that it was positive. At point D, the wave period is complete and the cycle begins again. The period of the square wave is the time from point A to point D. Its frequency, known as its *fundamental frequency,* is the reciprocal of the period:

$$f = \frac{1}{t}$$

or frequency equals 1 divided by time. Like the sine wave, a square wave has a peak and a peak-to-peak value.

A square wave contains a fundamental frequency and odd harmonics. A *harmonic* is a sine wave whose frequency is a (whole number) multiple of the fundamental frequency. For example, for a frequency of 5 kHz, the first harmonic is 5 × 1 kHz, which is the fundamental frequency. The second harmonic is 2 × 5 kHz, or 10 kHz. The third harmonic is 15 kHz. The fourth harmonic is 20 kHz, and so on. The even multiples are known as *even harmonics,* and the odd multiples are known as *odd harmonics.* Thus, the harmonic

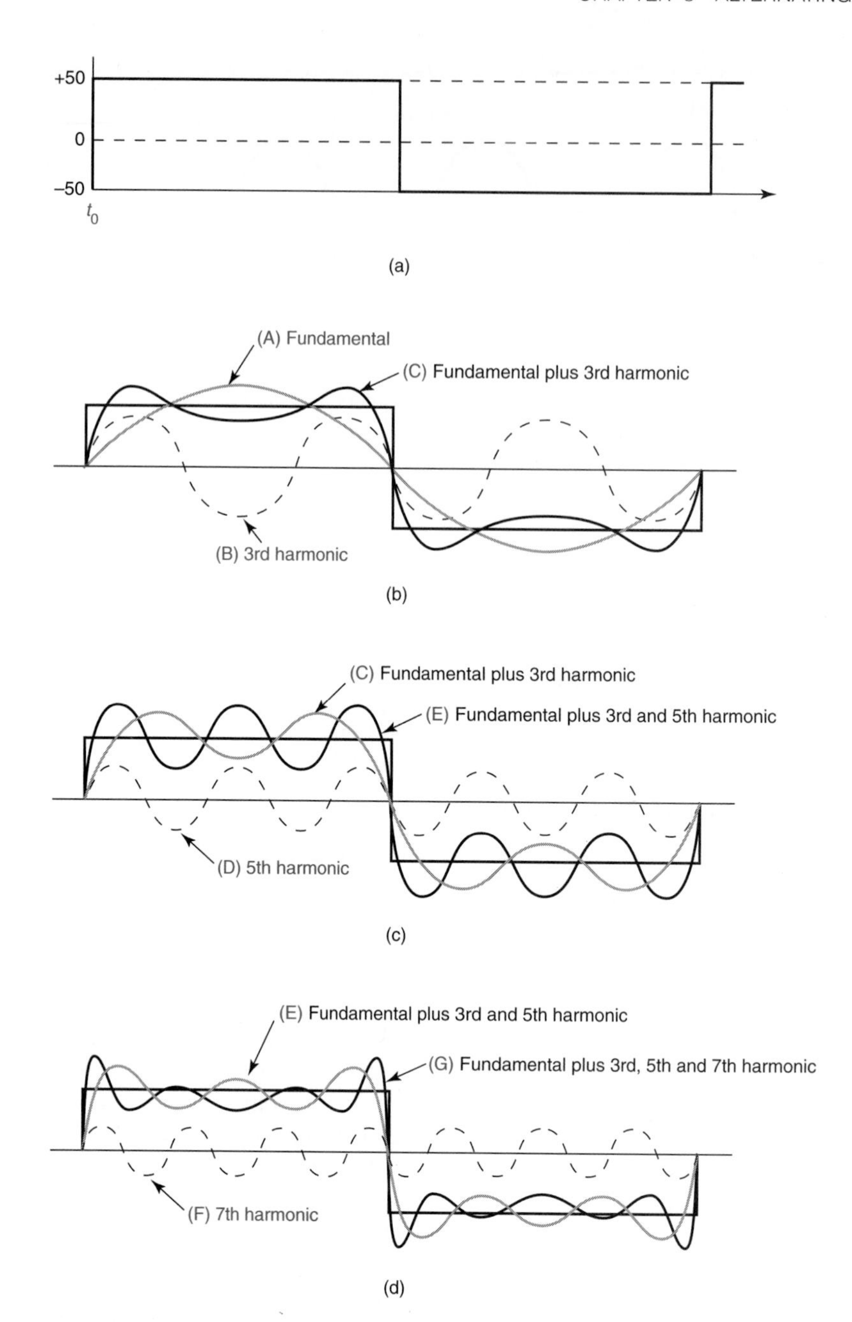

Figure 6–6. Harmonics: (a) square wave to be produced; (b) fundamental plus third harmonic; (c) fundamental plus third and fifth harmonic; (d) fundamental plus third, fifth, and seventh harmonic.

content of a square wave is the first, third, fifth, seventh, ninth, and so on.

A square wave can be formed by adding the odd harmonics to a fundamental frequency. The harmonics will not have the same amplitude as the fundamental. The third harmonic's amplitude is one-third of the fundamental, the fifth harmonic's amplitude one-fifth the fundamental, and so on. The square wave to be produced is shown in Fig. 6–6(a). In Fig. 6–6(b), a sine wave with the same frequency as the fundamental frequency of the square wave is labeled (A). Its third harmonic is labeled (B). The algebraic sum of the fundamental and the third harmonic is labeled (C). Notice that the sides begin to show a sharper rise and the top is beginning to flatten. In Fig. 6–6(c), the fifth harmonic is added algebraically to waveform (C) to form waveform (D), which is the fundamental plus the third and fifth harmonics. In Fig. 6–6(d), the seventh harmonic is added algebraically to waveform (D) to form waveform (E), which is the fundamental plus the third, fifth, and seventh harmonics. Notice that the sides of waveform (D) are becoming vertical and the positive and negative peaks are flattening out. As more harmonics are added, this process will continue until a nearly perfect square wave is produced.

The harmonic content of a square wave, or any nonsinusoidal wave, may be viewed on an instrument known as a *spectrum analyzer.* The trace of a spectrum analyzer represents frequency. Its presentation is in the frequency domain. The graph of a sine wave is in the time domain as would be displayed on an oscilloscope.

A *pulse* is a voltage or current that momentarily makes a sharp change in amplitude. It remains at this value for a time and then returns to its original value. As shown in Fig. 6–7, a pulse may be of a short, medium, or relatively long duration. The difference between the lower and upper voltage levels of the pulse is known as its *amplitude.* The waveforms in this figure are shown in the time domain, with time increasing from left to right. Thus, the left rise of the pulse is known as the *leading edge,* while the right rise is known as the *trailing edge.* The *pulse width* is the time between the leading and trailing edges. The pulse may repeat itself over a definite period. The *pulse repetition time* (PRT) is the time from the leading edge of one pulse to the leading edge of the next. The rate at which the pulses occur is known as the *pulse repetition frequency* (PRF) and is equal to the reciprocal of the PRT. The PRF is also referred to as the pulse repetition rate (PRR). PRF and PRT are related as follows:

$$\text{PRF} = \frac{1}{\text{PRT}}$$

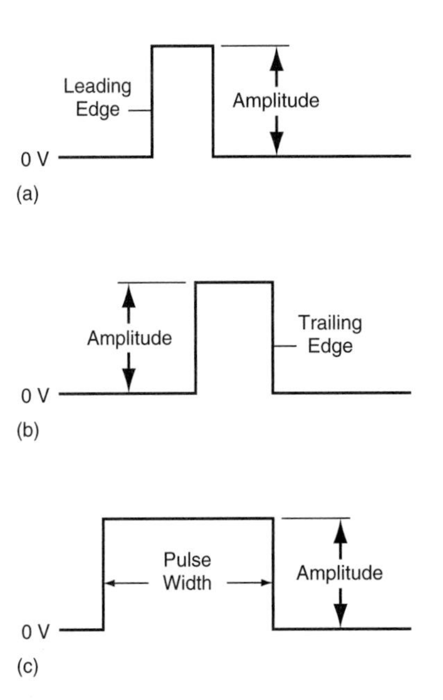

Figure 6–7. Pulses: (a) short duration; (b) medium duration; (c) long duration.

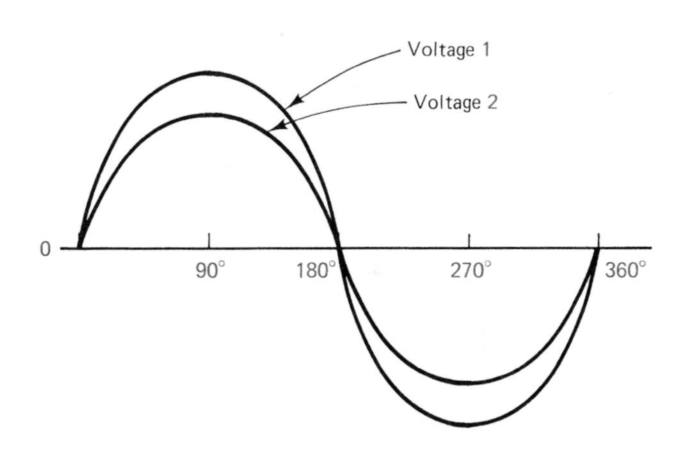

Figure 6–8. Two ac voltages that are in phase.

6.2 SINGLE-PHASE AND THREE-PHASE AC

Single-phase ac voltage is produced by single-phase ac generators or it can be obtained across two power lines of a three-phase system. A single-phase ac source has a hot wire and a neutral wire to carry electrical current. The neutral is grounded to help prevent electrical shocks. Single-phase power is the type of power distributed to our homes. A three-phase ac source has three power lines which carry electrical current. Three-phase voltage is produced by three-phase generators at power plants. Three-phase voltage is a combination of three single-phase voltages which are electrically connected. This voltage is similar to three single-phase ac sine waves separated in phase by 120°. Three-phase ac is used to power large equipment in industry and commercial buildings. It is not distributed to homes. There are three current-carrying power lines on a three-phase system. A three-phase wye system has a neutral connection and a three-phase delta system does not. Chapter 5 discussed the voltage and current relationships of three-phase wye and delta systems.

The term *phase* refers to time or the difference between one point and another. If two sine-wave voltages reach their zero and maximum values simultaneously, they are "in phase." Figure 6–8 shows two alternating-current voltages that are in phase. If two voltages reach their zero and maximum values at different times, they are "out of phase." Figure 6–9 shows two alternating-current voltages that are out of phase. Phase difference is given in degrees. The voltages shown are out of phase by an angle of 90°.

Remember that single-phase ac voltage is in the form of a sine wave. Single-phase ac voltage is used for low-power applications, primarily in the home. Almost all electrical power is generated and transmitted over long distances as three-phase ac. Three coils are placed 120° apart in a generator to produce three-phase ac voltage. Most ac motors

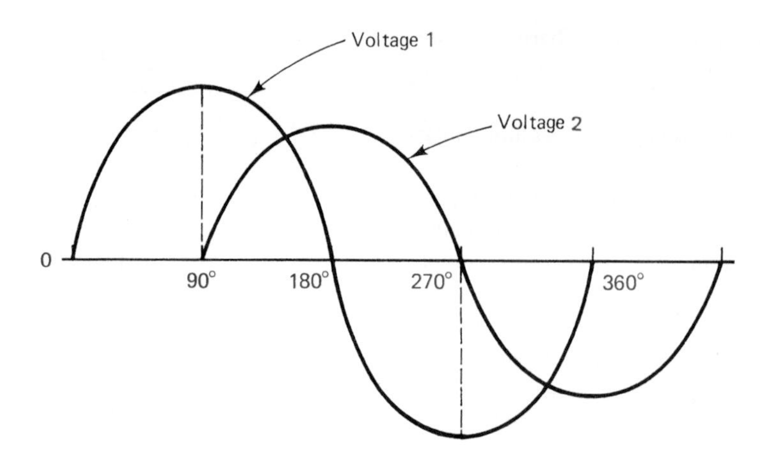

Figure 6–9. Two ac voltages that are out of phase by an angle of 90°.

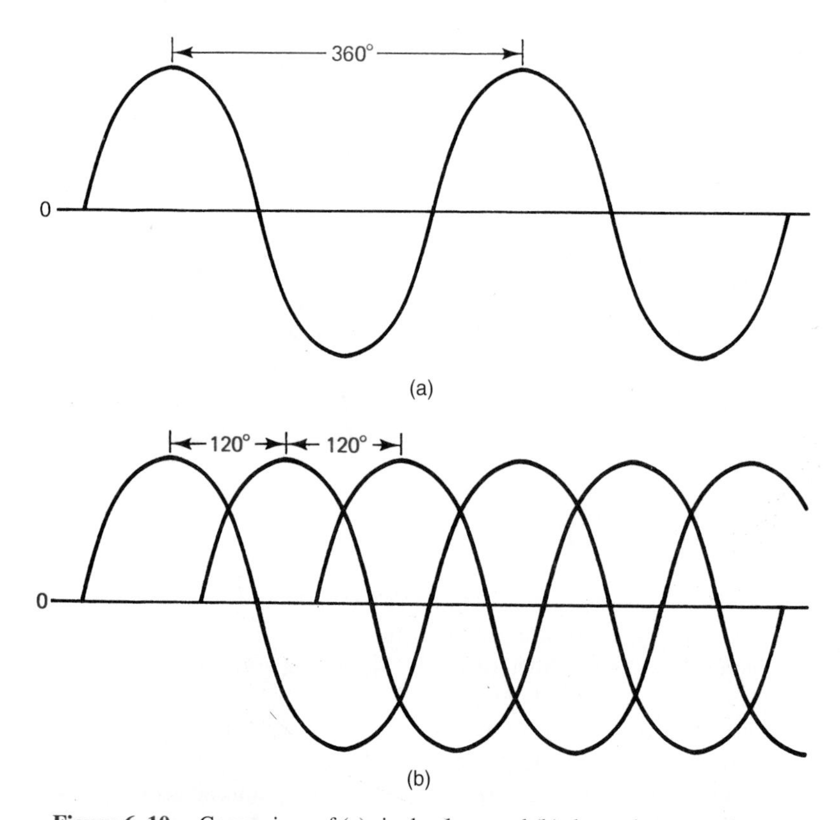

Figure 6–10. Comparison of (a) single-phase and (b) three-phase ac voltages.

over 1 hp in size operate with three-phase ac power applied. Most industries and commercial buildings have three-phase equipment.

Three-phase ac systems have several advantages over single-phase systems. In a single-phase system, the power is said to be "pulsating." The peak values along a single-phase ac sine wave are separated by 360°, as shown in Fig. 6–10(a). This is similar to a one-cylinder gas engine. A three-phase system is somewhat like a multicylinder gas engine. The power is more steady. One cylinder is compressing when the others are not, which is similar to the voltages in three-phase ac systems. The power of one separate phase is pulsating, but the total power is more constant. The peak values of three-phase ac are separated by 120°, as shown in Fig. 6–10(b). This makes three-phase ac power more desirable to use.

The power ratings of motors and generators are greater when three-phase ac power is used. For a certain frame size, the rating of a three-phase ac motor is almost 50% larger than a similar single-phase ac motor.

6.3 MEASURING AC VOLTAGE

A multimeter (VOM) may be used to measure ac voltage. Ac voltage is measured in the same way as dc voltage with two exceptions:

1. When measuring ac voltage, proper polarity does not have to be observed.
2. When measuring ac voltage, the ac voltage ranges and scales of the meter must be used.

Figure 6–11 shows the scale of an analog (multimeter) with the section used to measure ac voltage marked. Ac voltages may also be measured with a digital multimeter (DMM).

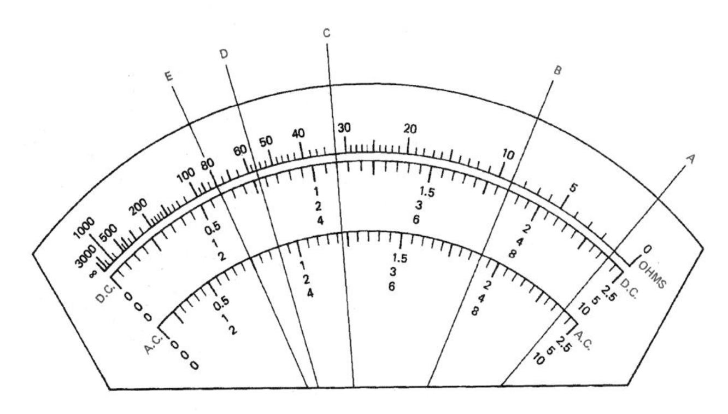

Figure 6–11. Scale of an analog multimeter with the ac voltage section shown.

6.4 USING AN OSCILLOSCOPE

Another way to measure ac voltage is with an oscilloscope, such as the one shown in Fig. 6–12. Oscilloscopes provide a visual display of waveforms on their screen. They are also used to measure a wide range of frequencies with precision. Oscilloscopes or "scopes" are used to examine wave shapes. For electronic servicing, it is necessary to be able to observe the voltage waveform while troubleshooting. Oscilloscopes are discussed in more detail in Chap. 8.

An oscilloscope permits various voltage waveforms to be visually analyzed. It produces an image on its screen. The controls must be properly adjusted. The image, called a *trace,* is usually a line on the screen or cathode-ray tube (CRT). A stream of electrons strikes the phosphorescent coating on the inside of the screen, causing the screen to produce light.

The oscilloscope displays voltage waveforms on two axes, like a graph. The horizontal axis on the screen is the time axis. The vertical axis is the voltage axis. An ac waveform is displayed on the CRT as shown in Fig. 6–13. For the CRT to display a trace properly, the internal circuits of the scope must be properly adjusted. These adjustments are made by controls on the front of the oscilloscope. Oscilloscopes are slightly different, but most scopes have some of the following controls:

1. *Intensity:* Controls the brightness of the trace, sometimes the on-off control.
2. *Focus:* Adjusts the thickness of the trace so that it is clear and sharp.
3. *Vertical position:* Adjusts the entire trace up or down.

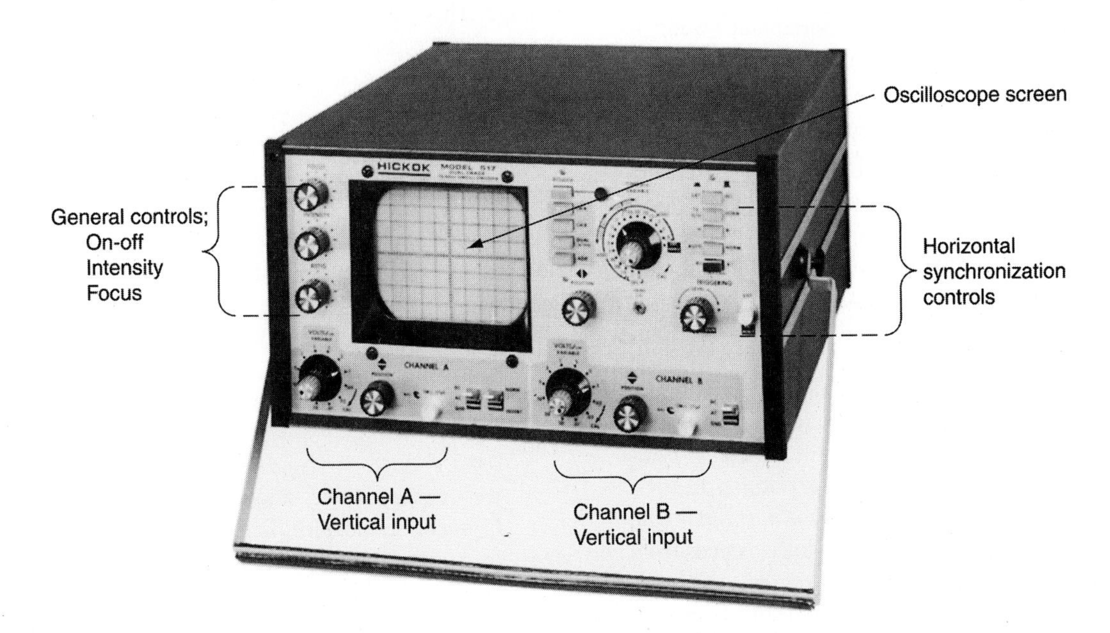

Figure 6–12. Oscilloscope used to measure ac voltage. (Courtesy of Hickok Electrical Co.)

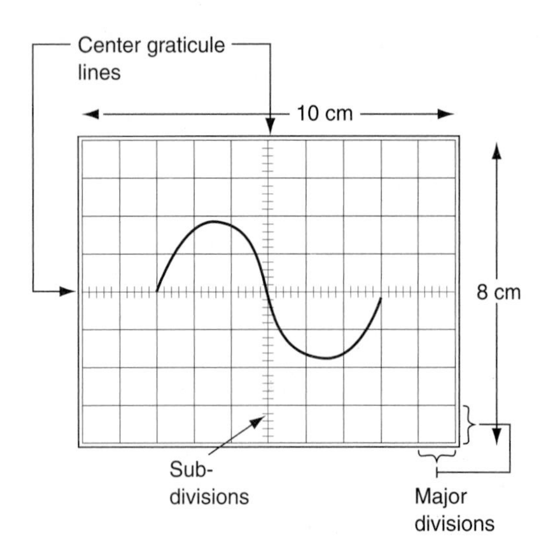

Figure 6–13. Ac waveform displayed on the screen of an oscilloscope.

4. *Horizontal position:* Adjusts the entire trace to the left or right.
5. *Vertical gain:* Controls the height of the trace.
6. *Horizontal gain:* Controls the horizontal size of the trace.
7. *Vertical attenuation* or *variable volts/cm:* Acts as a "coarse" adjustment to reduce the trace vertically.
8. *Horizontal sweep* or *variable time/cm:* Controls the speed at which the trace moves across (sweeps) the CRT horizontally. This control determines the number of waveforms displayed on the screen.
9. *Trigger:* Controls how the horizontal sweep is "locked in" with the circuitry of the scope.
10. *Vertical input:* External connections used to apply an input to the vertical circuits of the scope.
11. *Horizontal input:* External connections used to apply an input to the horizontal circuits of the scope.

The following procedure is used to adjust the oscilloscope controls to measure ac voltage. The names of some controls vary on different types of oscilloscopes.

1. Turn on the oscilloscope and adjust the *intensity* and *focus* controls until a bright, narrow, straight-line trace appears on the screen. Use the *horizontal position* and *vertical position* controls to position the trace in the center of the screen. Adjust the *horizontal gain* and the *variable time/cm* until the trace extends from the left of the screen to the right side of the screen. This allows the entire waveform to be displayed.
2. Connect the proper *test probes* into the oscilloscope's *vertical input* connections.
3. The scope is now ready to measure ac voltage.
4. After a waveform is displayed, adjust the *vertical attenuation* (volts/cm) and *vertical gain* controls until the height of the trace equals about 2 in., or 4 cm. Most scopes have scales that are marked in centimeters. Adjust the *vernier* or stability control until the trace becomes stable. One or more ac waveforms should appear on the screen of the scope.

6.5 OHM'S AND KIRCHHOFF'S LAWS FOR AC CIRCUITS

In Chap. 3, Ohm's law was used with dc circuits. As discussed there, Ohm's law can be used with ac circuits containing resistance *only.* Kirchhoff's voltage and current laws also apply to ac circuits containing resistance *only.* The use of Ohm's and Kirchhoff's laws for ac circuits is influenced by the effects of two circuit elements which are not present in dc circuits. Before learning to apply basic electrical laws to ac circuits, it is important to learn about inductance and capacitance.

6.6 INDUCTANCE

When energized with dc voltage, a magnetic field is produced around a coil. Dc current flow produces a constant magnetic field around a coil. When the same coil is supplied with ac voltage, the constantly changing ac current produces a constantly changing magnetic flux. This changing flux sets up a magnetic field with a constantly reversing polarity and changing strength. It also induces a counter EMF or counter voltage. This counter electromotive force (CEMF) opposes the source voltage. CEMF limits current flow from the source.

SAMPLE PROBLEM: ENERGY STORED IN AN INDUCTOR

A coil stores energy in its magnetic field because of current flow through it. The amount of energy is defined by the equation:

$$W + \frac{1}{2} \times L \times I^2$$

where:

W = energy stored in joules
L = coil inductance in henries
I = current flow through the coil in amperes

Given: a 10-henry coil has 5.8 amperes of current flowing through it.

Find: the amount of energy stored in the coil.

Solution:

$$W = \frac{1}{2} \times L \times I^2$$

$$= \frac{1}{2} \times 10 \times (5.8)^2$$

$$W = 168.2 \text{ joules}$$

The opposition to the flow of ac current by a magnetic field is due to a property called *inductance (L)*. The opposition to current flow of an inductive device depends on the resistance of the wire and the magnetic properties of the circuit. The opposition due to the magnetic effect is called *inductive reactance* (X_L), which varies with the applied frequency and is found by using the formula:

$$X_L = 2\pi \times f \times L$$

where $2\pi = 6.28$, f is the applied frequency in hertz, and L is the inductance in henries. The basic unit of inductance is the henry (H). X_L is measured in ohms.

At zero frequency (or dc), there is no opposition due to inductance—only a coil's resistance limits current flow. As frequency increases, the inductive effect becomes greater. Many ac machines use magnetic circuits in one form or another. The inductive reactance of an ac circuit usually has more effect on current flow than resistance. An ohmmeter measures dc resistance only. Inductive reactance must be calculated or determined experimentally by a specialized meter.

Ohm's law does not apply to ac circuits with inductance as it did with dc circuits. Resistance (R) *and* inductive reactance (X_L) are considered, however, and both values limit current flow. The total opposition to current flow is called *impedance (Z)*. Impedance is a combination of resistance and reactance. In ac circuits with both types of opposition, the Ohm's law relationship becomes $I = V/Z$.

Figure 6–14 shows some symbols used to indicate various types of inductors. Figure 6–15 shows some types of inductors. They are all coils of wire designed for specific functions. Often they are called "choke" coils. Choke coils are used to pass dc current and block ac current flow.

6.7 CAPACITANCE

When two conductors are separated by an insulator (dielectric), an electrostatic charge may be set up on the conductors. This charge becomes a source of stored energy. The strength of the charge depends on the applied voltage, the size of the conductors (which are called *plates*), and the quality or dielectric constant of the insulation. The closer the two plates are placed together, the more charge may be set up on them. This type of device is called a *capacitor*. The size of capacitors is measured in units called *farads, microfarads,* and *picofarads*. Some capacitor symbols are shown in Fig. 6–16. Various types of capacitors are shown in Fig. 6–17.

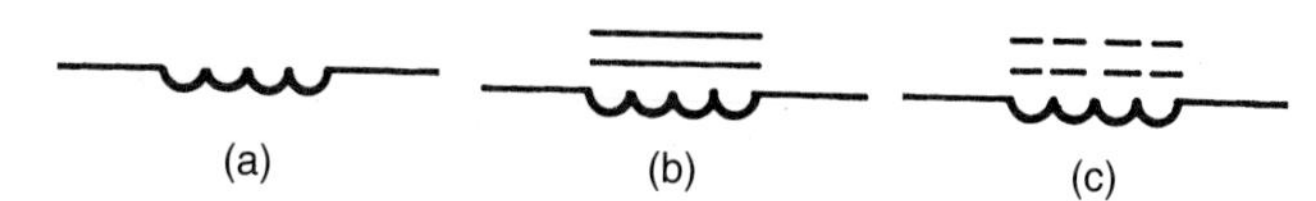

Figure 6–14. Symbols used for inductors: (a) air core; (b) iron core; (c) powdered metal core.

When a dc voltage is applied to the plates of a capacitor, it will charge to the value of the source voltage. The dielectric between the plates then stops current flow. The capacitor remains charged to the value of the dc voltage source. When the voltage source is removed, the capacitor charge will leak away. Dc current flows to or from a capacitor only when the source voltage is turned on or off.

If the same capacitor is connected to an ac voltage source, the voltage constantly changes. The capacitor receives energy from the source during one-quarter cycle and little current flows through the dielectric of a capacitor. Ac voltage applied to a capacitor causes it to constantly change its amount of charge. Capacitors have the ability to pass ac current because of their charging and discharging action. They can be used to pass ac current and block dc current flow. The opposition of a capacitor to a source voltage depends on frequency. The faster the applied voltage changes across a capacitor, the easier the capacitor passes current.

SAMPLE PROBLEM: ENERGY STORED IN A CAPACITOR

Energy is stored by a capacitor in its electrostatic field when voltage is applied to the capacitor. The amount of energy is defined by the equation:

$$W = \frac{1}{2} \times C \times V^2$$

where:

W = energy stored in a capacitor in joules
C = capacitance of the capacitor in farads
V = applied voltage in volts

Given: a 100-μF capacitor has 120 volts applied.
Find: the amount of energy stored in the capacitor.
Solution:

$$W = \frac{1}{2} \times C \times V^2$$

$$W = \frac{1}{2} \times 100^{-6} \times 120^2 = 72 \text{ joules}$$

are in developmental stages, a *liquid* capacitor and a *dry* type. The charge in an ultracapacitor is in the form of ions, which are trapped in tiny grooves within a coating of titanium. The ions can be released upon demand to produce a flow of current, similar to the discharging of a conventional capacitor.

The ultracapacitor has several advantages over other electrical energy sources. Unlike batteries, the ultracapacitor can be recharged any number of times. Also unlike batteries, it has an unlimited shelf life. Unlike the conventional capacitor, it has no leakage current.

6.13 CAPACITOR TESTING

Capacitor problems are of three types: opens, shorts, or leaks. An *open capacitor* is often caused by one of the connecting wires burning open where it connects to the plate. *Shorts* occur when current arcs through the dielectric, leaving a current path that allows excess current to flow. A *leaky capacitor* is one that is not shorted but does allow current through the dielectric. Thus, it never fully charges. Leaky capacitors are more likely to occur as equipment ages.

An analog scale ohmmeter can be used to test capacitors for opens or shorts. A normal capacitor, when connected across the ohmmeter, initially shows a low resistance. Then, as the capacitor charges, the needle moves toward infinity. If the capacitor is shorted or leaky, the resistance reading stays low and does not move toward infinity. The normal resistance should be between 500 kΩ and 1 MΩ for capacitors above 1 μF. Smaller capacitors charge too quickly for the ohmmeter to respond and should read infinite on all scales. When checking electrolytics, the polarity of the meter must match the polarity of the capacitor leads. A better check of a capacitor is made with a capacitor analyzer. This instrument checks the value of the capacitor and measures the leakage current under normal working voltage of the capacitor.

6.14 ALTERNATING-CURRENT CIRCUITS

Ac circuits are similar in many ways to dc circuits. They have a source, a load, a path, and usually controls and indicators. Ac circuits are classified by their electrical characteristics (resistive, inductive, or capacitive). All ac circuits are either resistive, inductive, capacitive, or a combination of these. The operation of each type of electrical circuit is different. The nature of alternating current causes certain electrical circuit properties to exist.

Resistive Circuits

The simplest type of ac circuit is a resistive circuit, such as that illustrated in Fig. 6–22. A resistive circuit is the same with ac applied as it is with dc applied. In dc circuits, the following formulas are used:

$$V = I \times R$$

$$I = \frac{V}{R}$$

$$R = \frac{V}{I}$$

These show that when the voltage applied to a circuit is increased, the current will increase. Also, when the resistance of a circuit is increased, the current decreases. The waveforms of Fig. 6–23 show the relationship of the voltage and current in a resistive ac circuit. Voltage and current are in phase. *In phase* means that the minimum and maximum values of voltage and current occur at the same time. The power converted by the resistance is found by multiplying voltage times current ($P = V \times I$). A power curve for a resistive ac circuit is shown in Fig. 6–24. When an ac circuit has only resistance, it is similar to a dc circuit.

Inductive Circuits

The property of inductance (L) is quite common. It adds more complexity to the relationship between voltage and current in an ac circuit. All motors, generators, and transformers have inductance. Inductance is due to the counter electromotive force (CEMF) produced when a magnetic field is developed around a coil of wire. The magnetic field produced around coils affects a circuit. The CEMF produced by a magnetic field offers opposition to change in the current of a circuit. In an inductive circuit, voltage leads the current. If the circuit was purely inductive (containing no resistance), the voltage would lead the current by 90° (see Fig. 6–25).

A purely inductive circuit does not convert any power in the load. All power is delivered back to the source. Refer

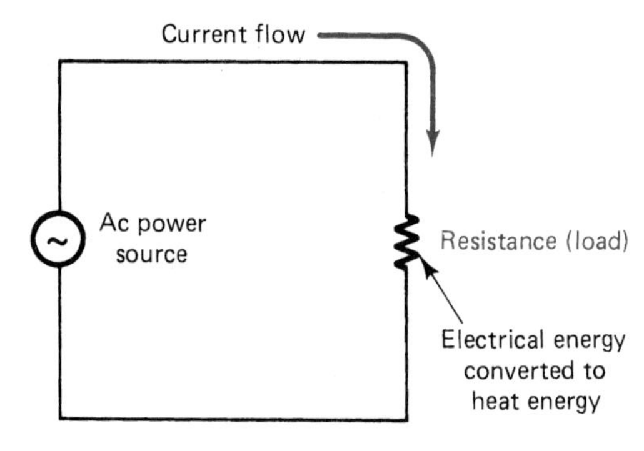

Figure 6–22. Resistive ac circuit.

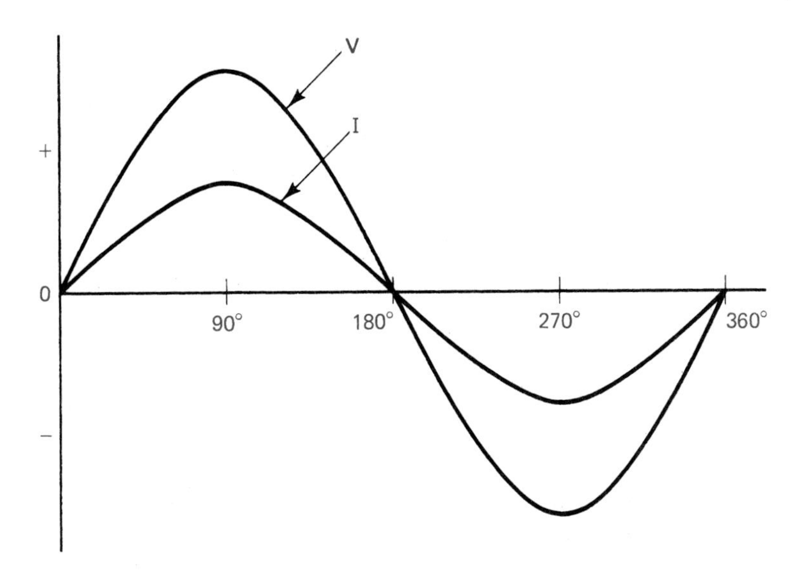

Figure 6–23. Voltage and current waveforms of a resistive ac circuit.

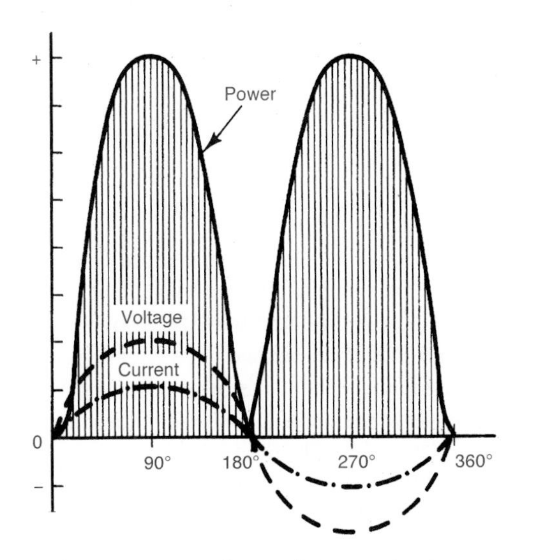

Figure 6–24. Power curve for a resistive ac circuit.

to points A and B on the waveforms of Fig. 6–26. At the peak of each waveform, the value of the other waveform is zero. The power curves are equal and opposite. They will cancel each other out. Where voltage and current are positive, the power is also positive, because the product of two positive values is positive. When voltage is positive and current is negative, the product of the two is negative. The power converted is also negative. Negative power means that electrical energy is returned from one load to the source without being converted to another form. The power converted in a purely inductive circuit is equal to zero.

Resistive-Inductive (*RL*) Circuits

Because all actual circuits have some resistance, the inductance in a circuit might typically cause the condition shown

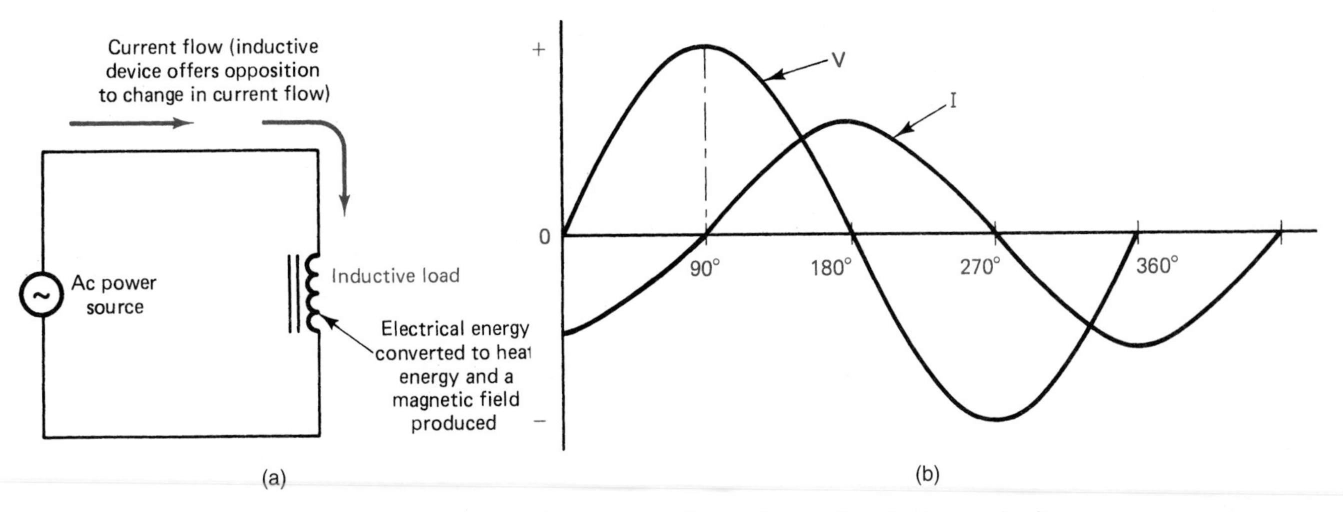

Figure 6–25. Voltage and current waveforms of a purely inductive ac circuit: (a) circuit; (b) waveforms.

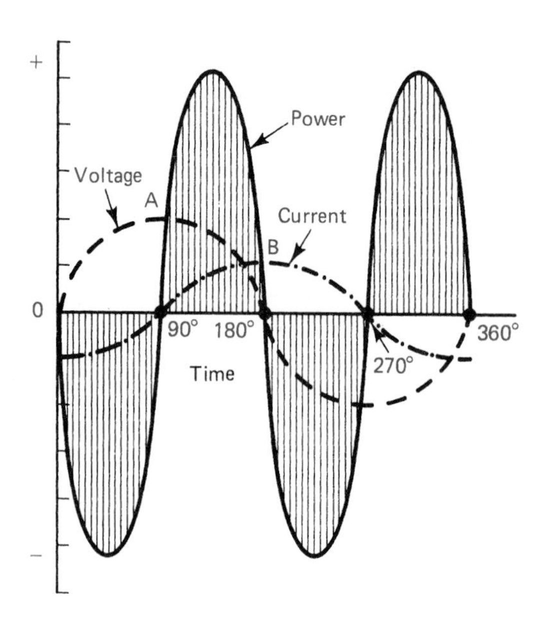

Figure 6–26. Power curve for a purely inductive circuit.

in Fig. 6–27. The voltage leads the current by 30°. The angular separation between voltage and current is called the *phase angle.* The phase angle increases as the inductance of the circuit increases. This type of circuit is called a resistive-inductive (*RL*) circuit.

Compare the purely inductive circuit's waveforms of Figs. 6–25 and 6–26 with those of Fig. 6–27. In a resistive-inductive (*RL*) circuit, part of the power supplied from the source is converted in the load. During the intervals from 0° to 30° and from 180° to 210°, negative power is produced. The remainder of the ac cycle produces positive power. Most of the electrical energy supplied by the source is converted to another form of energy in the circuit.

Capacitive Circuits

Figure 6–28 shows a capacitive circuit. Capacitors have the ability to store an electrical charge. They have many applications in electrical circuits. The operation of a capacitor in a circuit depends on its ability to charge and discharge.

If dc voltage is applied to a capacitor, the capacitor will charge to the value of that dc voltage. When the capacitor is fully charged, it blocks the flow of direct current. However, if ac is applied to a capacitor, the changing value of current will cause the capacitor to charge and discharge. The voltage and current waveforms of a purely capacitive circuit (no resistance) are shown in Fig. 6–29. The highest amount of current flows in a capacitive circuit when the voltage changes more rapidly. The most rapid change in voltage occurs at the 0° and 180° positions as polarity changes from (−) to (+). At these positions, maximum current is developed in the circuit. The rate of change of the voltage is slow near the 90° and 270° positions. A small amount of current flows at these positions. Remember that current leads voltage by 90° in a purely capacitive circuit. No power is converted in this circuit, just as no power is developed in the purely inductive circuit. Figure 6–30 shows that the positive and negative power waveforms cancel each other out.

Resistive-Capacitive (*RC*) Circuits

All circuits contain some resistance. A more practical circuit is the resistive-capacitive (*RC*) circuit shown in Fig. 6–31. In an *RC* circuit, the current leads the voltage by a phase angle between 0° and 90°. If capacitance in a circuit increases, the phase angle increases. The waveforms of Fig. 6–32 show an *RC* circuit in which the current leads the voltage by 30°. This circuit is similar to the *RL* circuit of Fig. 6–27(a). No power is converted in the circuit during the 0° to 30° and the 180° to 210° intervals. In this *RC* circuit, most of the electrical energy supplied by the source is converted to another form of energy in the circuit.

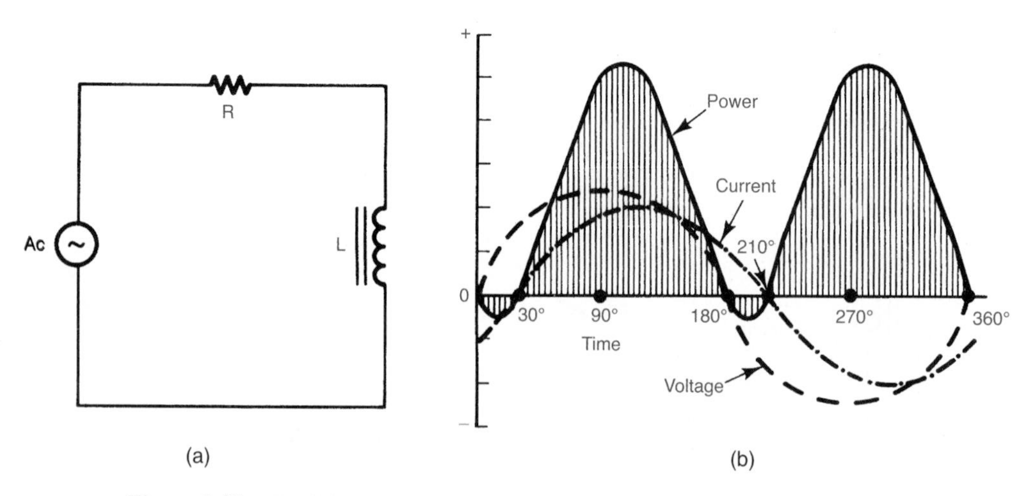

Figure 6–27. Resistive-inductive (*RL*) circuit and its waveforms: (a) *RL* ac circuit; (b) voltage and current waveforms.

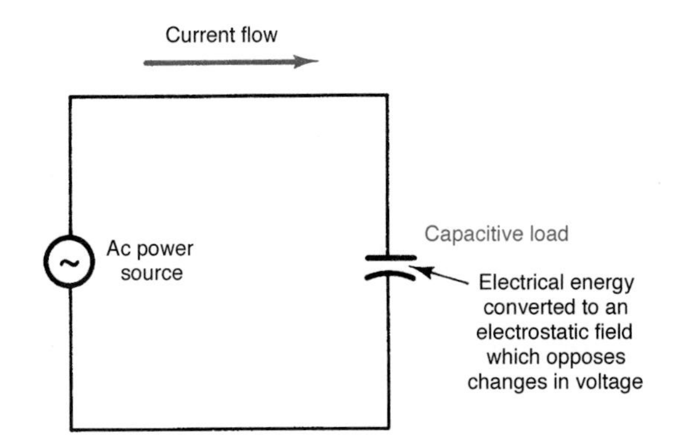

Figure 6–28. Capacitive ac circuit.

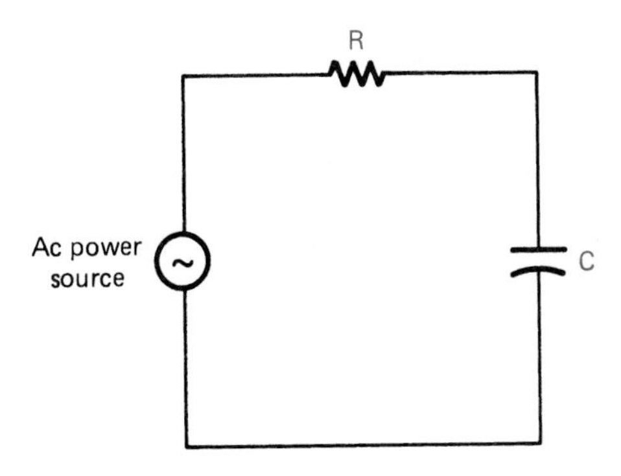

Figure 6–31. Resistive-capacitive (*RC*) circuit.

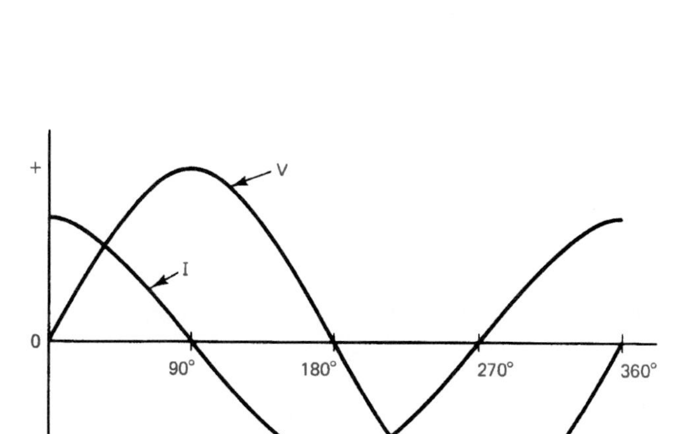

Figure 6–29. Voltage and current waveforms of a purely capacitive ac circuit.

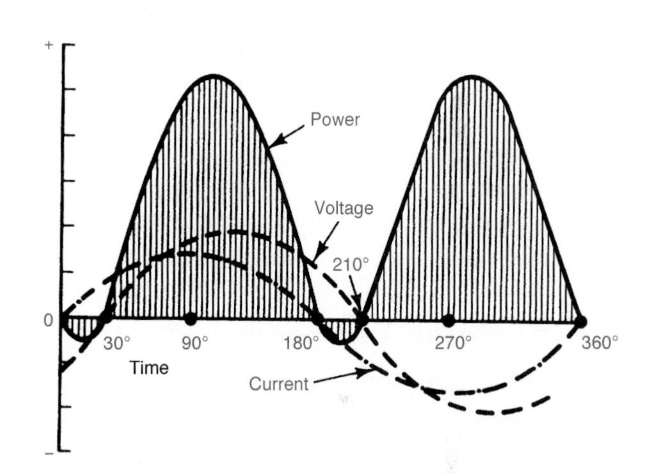

Figure 6–32. Waveforms of an *RC* circuit.

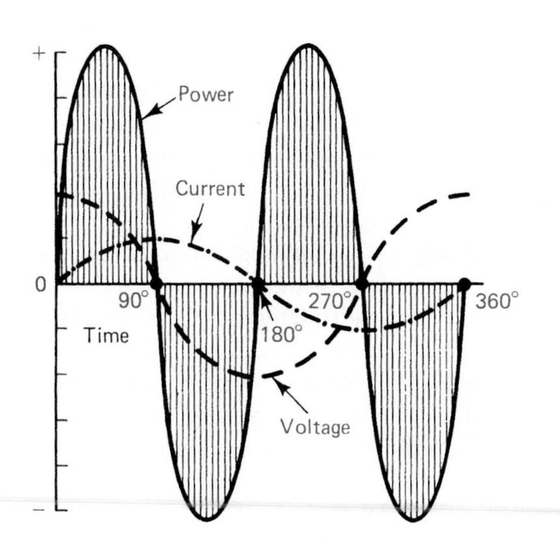

Figure 6–30. Power curves for a purely capacitive ac circuit.

6.15 VECTOR DIAGRAMS

A vector diagram for each ac circuit discussed in this chapter is shown in Fig. 6–33. The understanding of the vector diagram is helpful when working with ac circuits. Rather than using waveforms to show phase relationships, it is easier to use vector diagrams. Vectors are straight lines that have specific direction and length (magnitude). They are used to represent voltage and current values in ac circuits.

A horizontal line is drawn when beginning a vector diagram. Its left end is the reference point. In the diagrams of Fig. 6–33, the voltage vector line is the reference. For the inductive circuits, the current vectors are drawn in a clockwise direction from the voltage vector, which indicates that voltage leads current in an inductive circuit. For the capacitive circuits, the current vectors are drawn in a counterclockwise direction from the voltage vectors, which indicates that current leads voltage in a capacitive circuit.

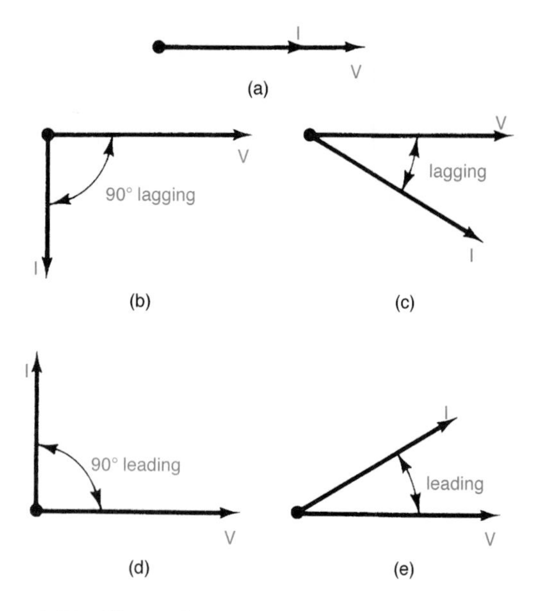

Figure 6–33. Vector diagrams showing voltage and current relationships of ac circuits: (a) resistive (R) circuit; (b) purely inductive (L) circuit; (c) RL circuit; (d) purely capacitive (C) circuit; (e) RC circuit.

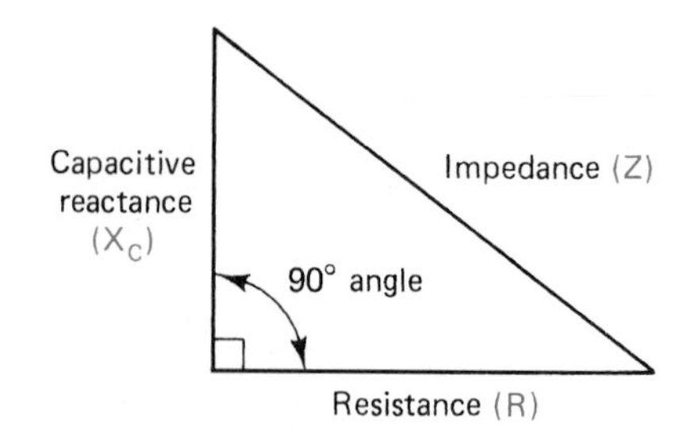

Figure 6–34. Right triangle for an ac series RC circuit.

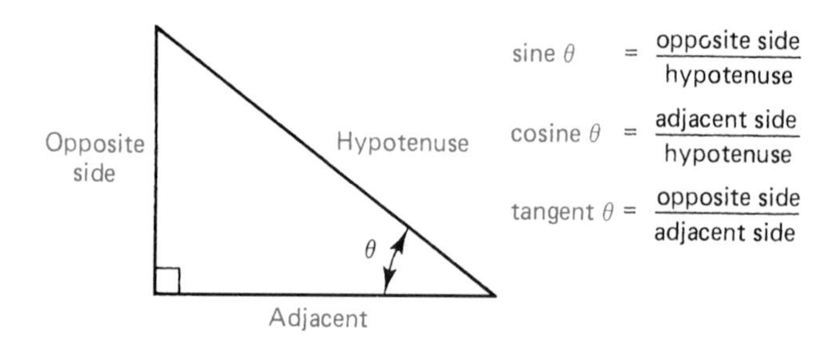

Figure 6–35. Sine, cosine, and tangent ratios.

6.16 MATHEMATICS FOR AC CIRCUITS

Right Triangles and Trigonometry

Right triangles and trigonometry are valuable in the study of electricity and electronics. Trigonometry, or "trig," is a form of mathematics that deals with angles and triangles, particularly the right triangle. A right triangle has one angle of 90°. An example of a right triangle is shown in Fig. 6–34. This example illustrates how resistance, capacitive reactance, and impedance are related in an ac series RC circuit. Resistance (R) and capacitive reactance (X_C) are 90° apart. Their angle of intersection forms a right angle. The law of right triangles, known as the Pythagorean theorem, can be used to solve circuit problems. The theorem states that "in a right triangle, the square of the hypotenuse is equal to the sum of the squares of the other two sides." With reference to Fig. 6–34, the Pythagorean theorem is used as:

$$Z^2 = R^2 + X_C^2 \text{ or } Z = \sqrt{R^2 + X_C^2}.$$

By using trig relationships, problems dealing with phase angles, power factor, and reactive power in ac circuits can be solved. The three most-used trig functions are the sine, the cosine, and the tangent. These functions show the ratios of the sides of a triangle. They determine the size of the angle. Figure 6–35 shows how these ratios are expressed. Their values are found by using a calculator. This process can be reversed to find the size of an angle when the value of the sides are known. The term *inverse* is used for this process. For example, "inverse sine 0.5 = 30°" means that 30° is the angle whose sine is 0.5. When given "inverse sine 0.866 = 60°," find the angle whose sine is 0.866 is equal to 60° (see Fig. F–18 in App. F on calculator use).

Trig ratios hold true for angles of any size. Angles in the first quadrant of a standard graph are from 0° to 90°. They are used as a reference. To solve for angles greater than 90° (second-, third-, and fourth-quadrant angles), an angle must be converted to a first-quadrant angle (refer to Fig. 6–36). All first-quadrant angles have positive values. Angles in the second, third, and fourth quadrants have two negative values and one positive value.

Rectangular Coordinates

A two-dimensional relationship can be established by drawing two perpendicular number lines that cross at their zero reference points, as shown in Fig. 6–37. The horizontal line is called the x axis and the vertical line is called the y axis. These coordinates are called *rectangular coordinates* because the axes are at right angles to each other. They are also called *Cartesian coordinates.*

For rectangular coordinates, two axes are placed in a plane along the surface of the paper, and any point on the plane can be located with reference to these axes. That is, any point on the plane must be a certain number of units to the left (negative) or to the right (positive) of the y axis. It must lie a certain number of units above (positive) or below (negative) the x axes. To locate a point with reference to this set of axes, the x value and the y value of the point must be known. These two values are called the *coordinates of the point.* The x value is called the abscissa, and the y value is called the ordinate. In describing a point in terms of its rectangular coordinates, the

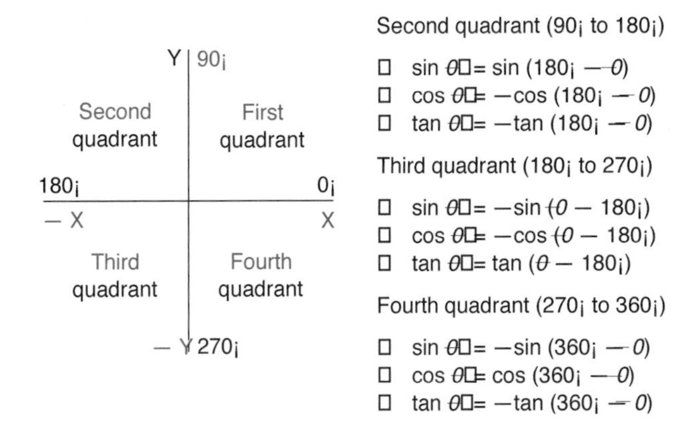

Figure 6–36. Standard graph and the method used to find sine, cosine, and tangent values.

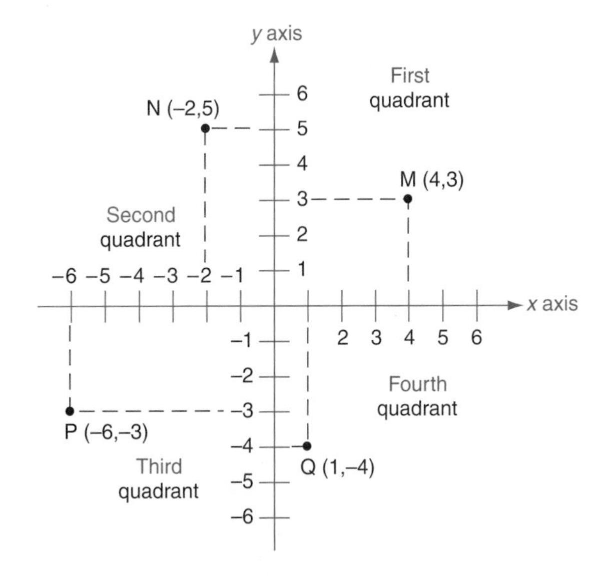

Figure 6–37. Rectangular coordinates.

abscissa is written first, followed by the *ordinate*. The two numbers are separated by a comma and are usually enclosed in parentheses. Thus, to describe the location of point M in Fig. 6–37, write M(4, 3). This means that to locate point M, count four divisions to the right from the origin along the x axis, and then count three divisions in the y direction above the x axis. By this method, the other points in the figure are described as N(-2, 5), P(-6, -3), and Q(1, -4).

Quadrants

The plane containing the x and y axes is called the x-y plane. The axes divide the plane into four sections, called *quadrants.* Figure 6–37 shows that in the first quadrant both coordinates are positive. In the second quadrant the abscissa is negative and the ordinate is positive. In the third quadrant both coordinates are negative. In the fourth quadrant the abscissa is positive and the ordinate is negative.

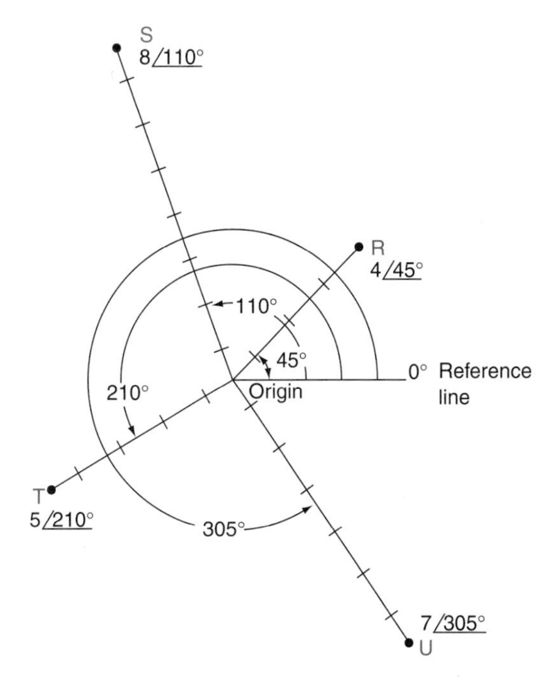

Figure 6–38. Polar coordinates.

Polar Coordinates

The position of any point on a plane can be determined by using a radius vector and a reference line that meet at a point called the origin. The length of the radius vector is the distance from the *origin* to the point. The direction of the radius vector is the angle it makes with the reference line. The length and angle of a vector used to locate a point on a plane are called *polar coordinates.* The term *polar* refers to the center (or polar) point of the graphed figure. The distance from the origin to the point is called the *modulus,* and the angle is called the *amplitude* (or argument).

In describing a point in terms of its polar coordinates, the modulus is written first, and then the argument. The general form $r/\underline{\theta}$ shows the polar coordinates for a point r units distance from the origin and θ angular units from the reference line. Thus $4\,/\underline{45^\circ}$ describes the location of point R in Fig. 6–38. Point R is located by constructing a vector that is four units in length at an angle of 45° to the reference line. By this method, the other points in the figure are found as S = $8/\underline{110^\circ}$, T = $5/\underline{210^\circ}$, and U = $7/\underline{305^\circ}$.

Angular Velocity

The rate at which an angle is generated by a vector radius is called the *angular velocity.* When the rotation is uniform, the unit of angular velocity is the angle generated per unit of time. Therefore, angular velocity can be measured in degrees per second, radians per second, revolutions per second, and

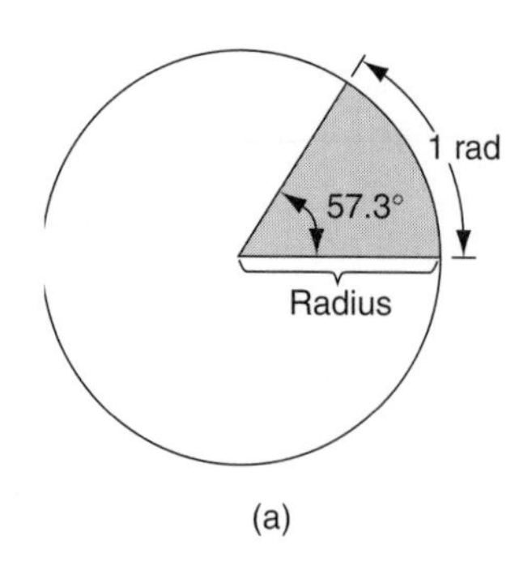

Radian/Degree Conversion

Radians can be converted to degrees using the equation:

$$\text{rad} = \left(\frac{\pi\ \text{rad}}{180^\circ}\right) \times \text{degrees}$$

Similarly, degrees can be converted to radians with the equation:

$$\text{degrees} = \left(\frac{180^\circ}{\pi\ \text{rad}}\right) \times \text{rad}$$

(b)

Degrees (°)	*Radians (rad)*
0	0
45	π/4
90	π/2
135	3π/4
180	π
225	5π/4
270	3π/2
315	7π/4
360	2π

(c)

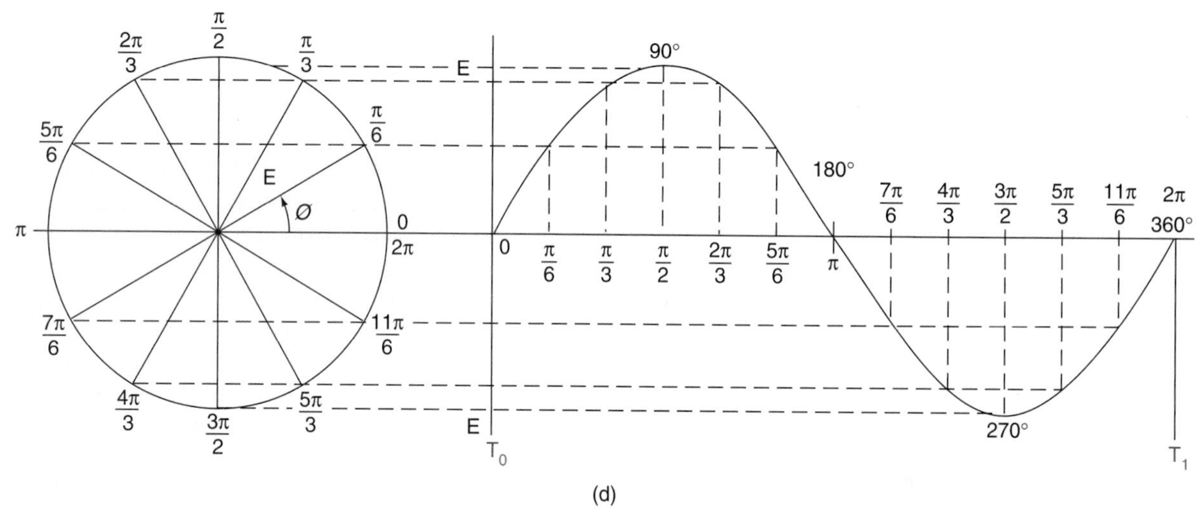

Figure 6–39. Radian values: (a) relation of radians to degrees; (b) radian/degree conversions; (c) table of radian/degree conversions; (d) sine-wave generation.

revolutions per minute. For ac waveforms, angular velocity is usually given in radians per second, and is denoted by the Greek lowercase letter omega (ω). The following equation relates ω to revolutions per second:

$$\omega = 2\pi f$$

where f is the number of revolutions per second. Hence the number of radians generated per second is $2\pi f$. For example, if a motor rotates at 3600 rpm, convert this to radians per second as follows:

$$\begin{aligned}\omega &= 2\pi f\\ &= 2\pi \frac{3600\ \text{rpm}}{60}\\ &= 120\pi\ \text{radians/second}\end{aligned}$$

Figure 6–39 is helpful in understanding the relationship of sine waves and the rotation of a vector quantity. The angular measurement of a sine wave is based on 360° or 2π radians for a complete cycle. A half-cycle is 180° or π radians; a quarter-cycle is 90° or π/2 radians; and so on. Figure 6–39 shows angles in degrees over a full cycle of a sine wave.

Complex Numbers

Electromotive forces, currents, and impedances are sometimes expressed in the form of complex numbers. There are two types of coordinates for locating a point on a plane surface: rectangular coordinates and polar coordinates. With either type you use a point and a line for reference; the point is called the *origin* and designated by the letter *O*, and the line is called the *axis.* With rectangular coordinates, a specific point is located on a plane by moving a measured distance from the origin along the real number axis in either the positive or negative direction, and then moving a measured distance perpendicular to the real number axis. The number written to represent these two distances and their directions is called the *rectangular form* of a complex number. With polar coordinates, a point on a plane is located by rotating a straight line about the origin as center, until it passes through the given point. You can then indicate the position of the point by the angle the rotating line makes with the real number axis, and the measured distance from the given point to the origin. The number representing these two items of information is called the *polar form* of a complex number. A *trigonometric form* of representing complex numbers is used to convert from the polar to the rectangular form. These three forms of complex-number representation can be directly related to graphic presentations. A more abstract expression is the *exponential form* of complex number, which cannot be shown.

Imaginary Numbers

The square root of a negative number is called an *imaginary number*. All other types of numbers, either positive or negative, such as 9, –5, and $\sqrt{3}$, are called *real numbers*.

The term $\sqrt{-1}$ can be considered to be an operator that indicates a rotation of 90°. Because the number $\sqrt{-1}$ is clumsy to write and is frequently used, it is represented by a letter. In pure mathematics, $\sqrt{-1}$ is represented by *i*. However, because *i* is the symbol for instantaneous current, the letter *j* is used to denote $\sqrt{-1}$ in electrical calculations. Thus the imaginary numbers $b\sqrt{-1}$ and $27\sqrt{-1}$ can be written as bi and $27i$ or jb and $j27$.

Rectangular Form of Complex Numbers

Combinations of real and imaginary numbers comprise the rectangular form of complex numbers. These can be written as follows: $4 + 3\sqrt{-1}$, $-3 + 5i$, $-6 - j2$, and $2 - j3$. These numbers are similar to algebraic binomials. Such numbers are not restricted to the real or imaginary number axes, but may have a magnitude in any direction on a plane surface. A complex number is represented by using rectangular coordinates with the horizontal axis for the real numbers and the vertical axis for the imaginary numbers. In Fig. 6–40, the complex number $4 + j3$ consists of a real part (4) and an imaginary part ($j3$). The point that represents this number is then four units in the positive direction and three units in the +*j* direction. The absolute value of the number is indicated by the distance from the resulting point to the zero point (origin) of the diagram. By this method, any complex number can be represented.

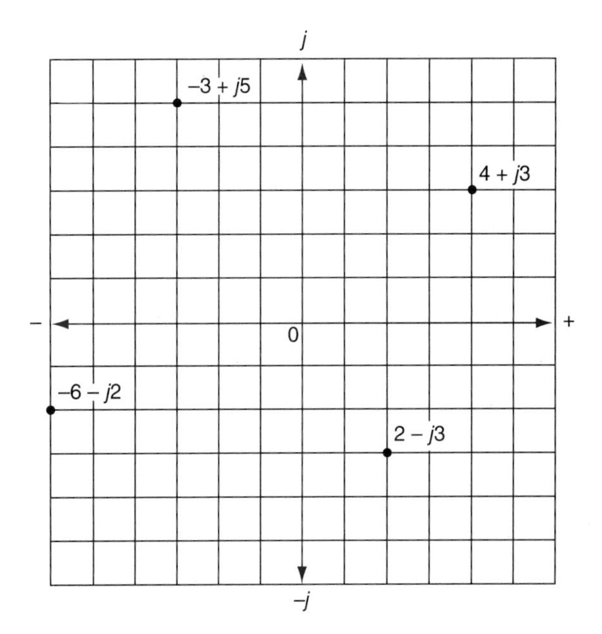

Figure 6–40. Coordinate diagram.

Addition of Complex Numbers

To find the sum of two complex numbers in the rectangular form, algebraically add the real parts and the imaginary parts in the same manner that algebraic binomials are added. The following two examples show the procedure:

$$(a + jb) + (c + jd) = (a + c) + j(b + d)$$
$$(4 + j3) + (-3 + j5) = (4 - 3) + j(3 + 5)$$
$$= 1 + j8$$

The sum of three or more complex numbers can be obtained by adding two of them, adding the sum to the third of the given numbers, and so on, until all the numbers have been added.

Impedances that are written in the form of complex numbers often have to be added. For example, Fig. 6–41 shows two impedances connected in series. Find the total impedance by adding them, as shown here.

$$Z_1 = R_1 - jX_C$$
$$Z_2 = R_2 + jX_L$$
$$Z_1 + Z_2 = (R_1 + R_2) + j(X_L - X_C)$$

Substituting numbers for resistances and reactances yields:

$$Z_1 = (2 - j750) \text{ ohms}$$
$$Z_2 = (280 + j430) \text{ ohms}$$
$$Z_1 + Z_2 = (2 + 280) + j(430 - 750)$$
$$= (282 - j320) \text{ ohms}$$

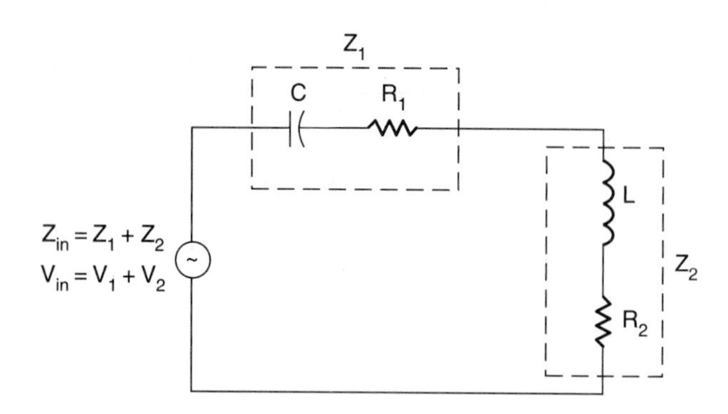

Figure 6–41. Series impedances.

Complex numbers can be subtracted as in algebra by changing the signs of the terms in the subtrahend and proceeding as in addition. Examples of subtraction are as follows:

$$(a + jb) - (c + jd) = (a - c) + j(b - d)$$
$$(4 + j3) - (-3 + j5) = 4 - (-3) + j(3 - 5)$$
$$= 7 - j2$$

For the circuit shown in Fig. 6–41, if $Z_{in} = (470 - j35)$ ohms and $Z_1 = (4 - j830)$ ohms, find Z_2 as follows:

$$Z_2 = Z_{in} - Z_1$$
$$= (470 - j35) - (4 - j830)$$
$$= (470 - 4) + j(-35 + 830)$$
$$= (466 + j795) \text{ ohms}$$

Polar and Trigonometric Forms

Complex numbers in the rectangular form can easily be handled for addition and subtraction operations. Two items of information must be given in both rectangular and polar coordinates to locate a point on a plane. In polar coordinates, one item is an angle and the other a distance. This information is written in the form $r/\underline{\theta}$, where r is the distance from the origin to the point that represents the complex number, and θ is the angle between the straight line r and the real number axis. The absolute value r is called the *modulus,* and the angle θ is called the *argument.* The imaginary numbers jr and $-jr$ can then be written as $r/\underline{90^\circ}$ and $r/\underline{-90^\circ}$, respectively. Figure 6–42 shows the position of complex numbers on a polar diagram.

6.17 SERIES AC CIRCUITS

In any series ac circuit, the current is the same in all parts of the circuit. The voltages must be added by using a voltage triangle. Impedance (Z) of a series ac circuit is found by using an impedance triangle. Power values are found by using a power triangle.

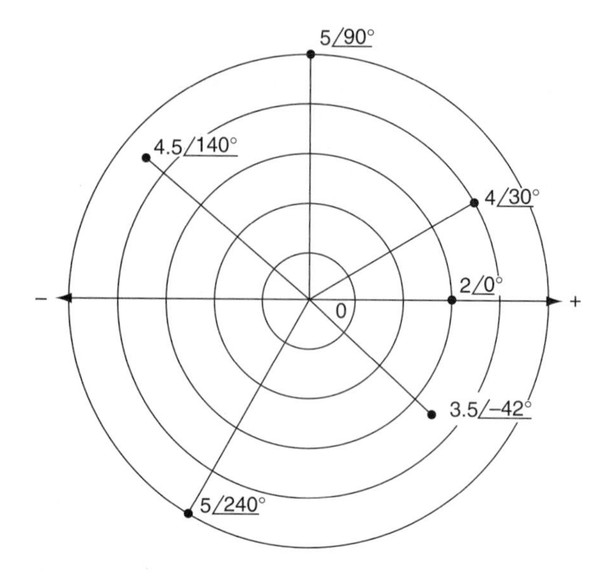

Figure 6–42. Polar diagram.

Series *RL* Circuits

Series *RL* circuits are often found in electronic equipment. When an ac voltage is applied to a series *RL* circuit, the current is the same through each part. The voltage drops across each component distribute according to the values of resistance (R) and inductive reactance (X_L) in the circuit.

The total opposition to current flow in any ac circuit is called impedance (Z). Both resistance and reactance in an ac circuit oppose current flow. The impedance of a series *RL* circuit is found by using either of these formulas:

$$Z = \frac{V}{I} \quad \text{or} \quad Z = \sqrt{R^2 + X_L^2}$$

The impedance of a series *RL* circuit is found by using an impedance triangle. This right triangle is formed by the three quantities that oppose ac. A triangle is also used to compare voltage drops in series *RL* circuits. Inductive voltage (V_L) leads resistive voltage (V_R) by 90°. V_A is the voltage applied to the circuit. Because these values form a right triangle, the value of V_A may be found by using the formula

$$V_A = \sqrt{V_R^2 + V_L^2}$$

An example of a series *RL* circuit is shown in Fig. 6–43. Note that the current is used as the reference in the voltage triangle. Refer to the calculator procedures for trigonometric functions in App. F (Fig. F–18).

Series *RC* Circuits

Series *RC* circuits also have many uses. This type of circuit is similar to the series *RL* circuit. In a capacitive circuit, current leads voltage. The reactive values of *RC* circuits act in

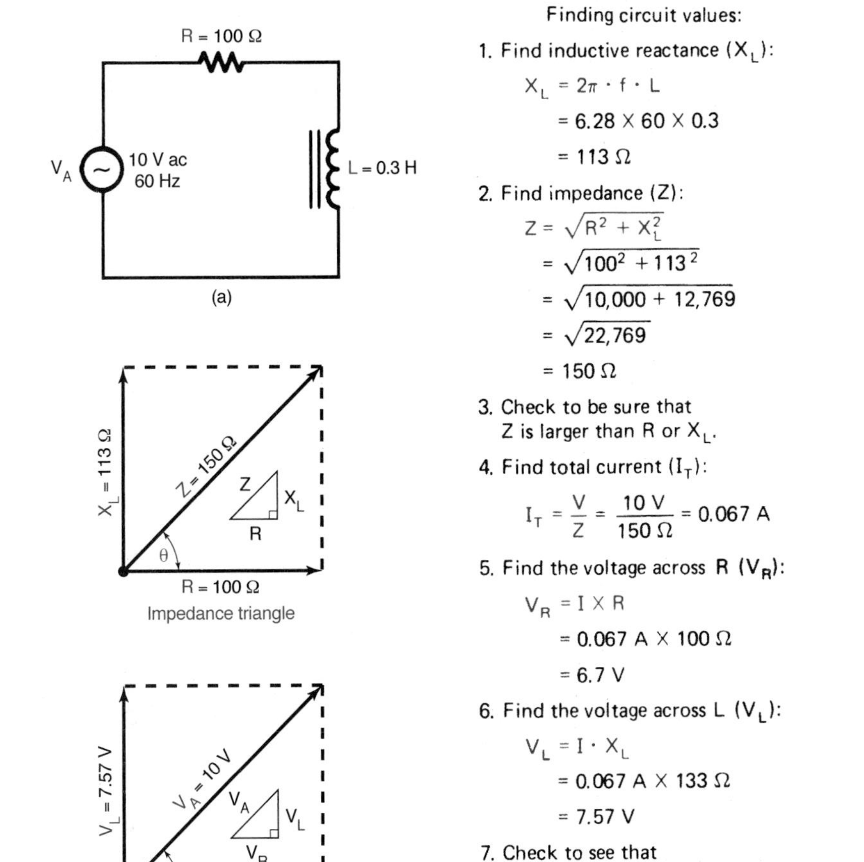

Figure 6–43. Series *RL* circuit example: (a) circuit; (b) procedure for finding circuit values; (c) circuit triangles.

the opposite directions to *RL* circuits. An example of a series *RC* circuit is shown in Fig. 6–44. Refer also to the calculator procedures for series *RC* circuits in App. F (Fig. F–19).

Series *RLC* Circuits

Series *RLC* circuits have resistance (*R*), inductance (*L*), and capacitance (*C*). The total reactance (X_T) is found by subtracting the smaller reactance (X_L or X_C) from the larger one. Reactive voltage (V_X) is found by obtaining the difference between V_L and V_C. The effects of the capacitance and inductance are 180° out of phase with each other. Right triangles are used to show the simplified relationships of the circuit values. An example of a series *RLC* circuit is shown in Fig. 6–45. The general procedure for solving series ac circuit problems is summarized next.

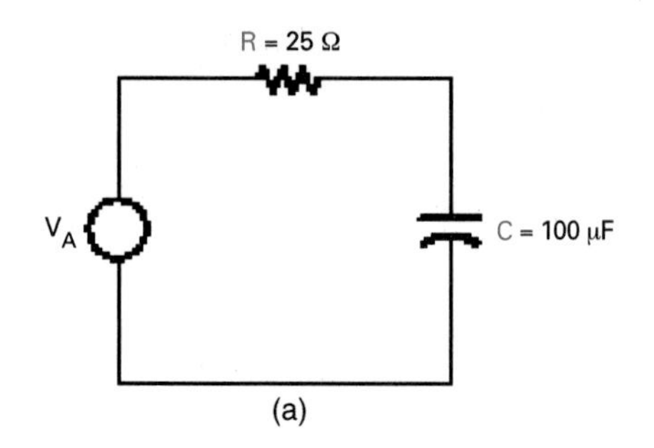

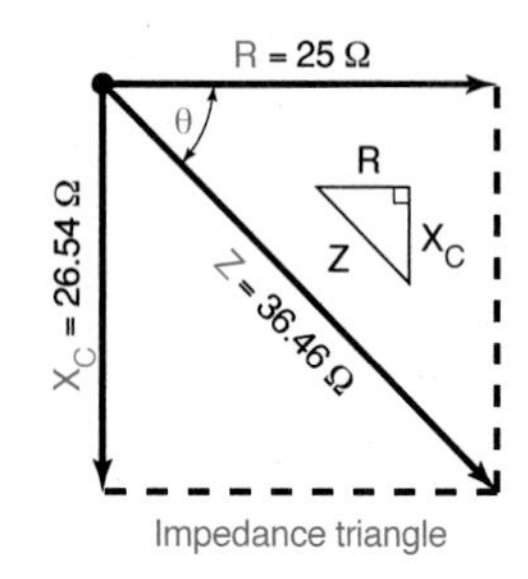

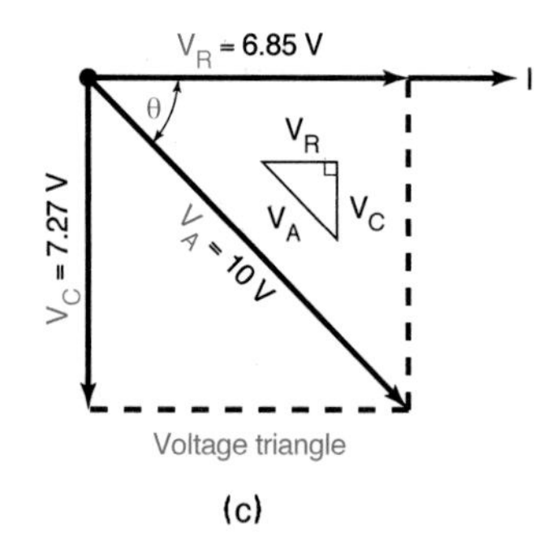

Finding circuit values:

1. Find capacitive reactance (X_C):

$$X_C = \frac{1}{2\pi \cdot f \cdot C} \text{ or } \frac{1{,}000{,}000}{2\pi \cdot f \cdot C}$$

(in farads) (in μF)

$$= \frac{1{,}000{,}000}{6.28 \times 60 \times 100}$$

$$= 26.54\ \Omega$$

2. Find impedance (Z):

$$Z = \sqrt{R^2 + X_C^2}$$
$$= \sqrt{25^2 + 26.54^2}$$
$$= \sqrt{625 + 704.37}$$
$$= \sqrt{1329.37}$$
$$= 36.46\ \Omega$$

3. Check to be sure that Z is larger than R or X_C.

4. Find total current (I_T):

$$I_T = \frac{V}{Z} = \frac{10\text{ V}}{36.46\ \Omega} = 0.274\text{ A}$$

5. Find the voltage across R (V_R):

$$V_R = I \times R$$
$$= 0.274\text{ A} \times 25\ \Omega$$
$$= 6.85\text{ V}$$

6. Find the voltage across C (V_C):

$$V_C = I \cdot X_C$$
$$= 0.274\text{ A} \times 26.54\ \Omega$$
$$= 7.27\text{ V}$$

7. Check to see that

$$V_A = \sqrt{V_R^2 + V_C^2}:$$
$$10\text{ V} = \sqrt{6.85\text{ V}^2 + 7.27\text{ V}^2}$$
$$= \sqrt{46.92 + 52.88}$$
$$= \sqrt{99.8}$$
$$\approx 9.99\text{ V}$$

8. Find the circuit phase angle (θ):

$${}^*\text{sine } \theta = \frac{\text{opposite}}{\text{hypotenuse}} = \frac{X_C}{Z} = \frac{26.54}{36.46}$$
$$= 0.727$$
$$\theta = \text{inverse sine } 0.727$$
$$= 47°$$

(b)

*Any trig function can be used with either triangle.

Figure 6–44. Series *RC* circuit example: (a) circuit; (b) procedure for finding circuit values; (c) circuit triangles.

General Procedure to Solve Series Ac Circuit Problems

1. Draw the circuit diagram. Label with known values.
2. If reactances are not given, solve for them using:

$$X_L = 2\pi fL \qquad X_C = \frac{1}{2\pi fC}$$

3. Draw an impedance triangle to calculate Z.
   a. Draw the *R* "vector" horizontally. From the tip of the *R* vector, draw the X_L vector vertically upward. Draw the X_C vector vertically downward.
   b. Vectorially add the vectors and then draw Z.
   c. Use the Pythagorean theorem to calculate the value of *Z* by using the formula

$$Z = \sqrt{R^2 + (X_L - X_C)^2}$$

4. Calculate the current in the circuit.

$$I = V/Z$$

5. Calculate the voltage drop across each component.

$$V_R = I \times R \qquad V_L = I \times X_L \qquad V_C = I \times X_C$$

6. Draw a vector diagram of the voltages. Use *I* as the horizontal reference. Draw V_R in phase with the current. Draw V_L vertically upward. Draw V_C vertically downward.
7. Vectorially add the voltages to obtain the applied voltage.

$$V_T = \sqrt{V_R^2 + (V_L - V_C)^2}$$

8. Use the Pythagorean theorem to calculate the value of the applied voltage as a check on the solution.

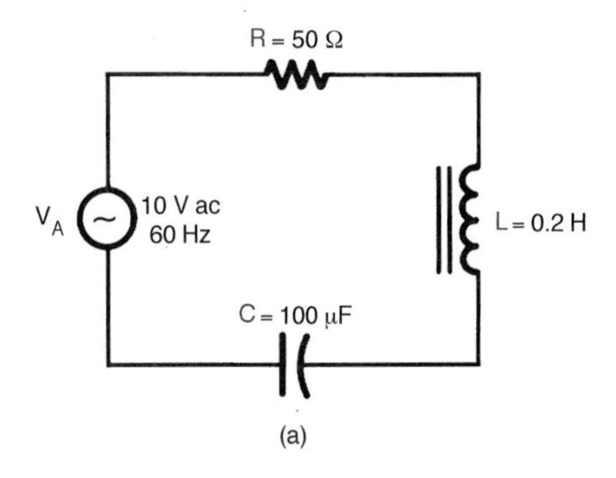

(a)

Finding circuit values:

1. Find inductive reactance (X_L):

$$X_L = 2\pi \cdot f \cdot L$$
$$= 6.28 \times 60 \times 0.2$$
$$= 75.36\ \Omega$$

2. Find capacitive reactance (X_C):

$$X_C = \frac{1{,}000{,}000}{2\pi \cdot f \cdot C\ (\mu F)}$$
$$= \frac{1{,}000{,}000}{6.28 \times 60 \times 100}$$
$$= 26.54\ \Omega$$

3. Find total reactance (X_T):

$$X_T = X_L - X_C$$
$$= 75.36\ \Omega - 26.54\ \Omega$$
$$= 48.82\ \Omega$$

4. Find impedance (Z):

$$Z = \sqrt{R^2 + X_T^2}$$
$$= \sqrt{50^2 + 48.82^2}$$
$$= \sqrt{2500 + 2383.4}$$
$$= \sqrt{4{,}883.4}$$
$$= 69.88\ \Omega$$

5. Check to be sure that Z is larger than X_T or R.

6. Find total current (I_T):

$$I_T = \frac{V}{Z} = \frac{10\ V}{69.88\ \Omega} = 0.143\ A$$

7. Find voltage across R (V_R):

$$V_R = I \times R$$
$$= 0.143\ A \times 50\ \Omega$$
$$= 7.15\ V$$

8. Find voltage across L (V_L):

$$V_L = I \cdot X_L$$
$$= 0.143\ A \times 75.36\ \Omega$$
$$= 10.78\ V$$

9. Find voltage across C (V_C):

$$V_C = I \cdot X_C$$
$$= 0.143\ A \times 26.54\ \Omega$$
$$= 3.8\ V$$

10. Find total reactive voltage (V_X):

$$V_X = V_L - V_C$$
$$= 10.78\ V - 3.8\ V$$
$$= 6.98\ V$$

11. Check to see that

$$V_A = \sqrt{V_R^2 + V_X^2}$$
$$= \sqrt{7.15\ V^2 + 6.98\ V^2}$$
$$10\ V = \sqrt{51.12 + 48.72}$$
$$= \sqrt{99.84}$$
$$\approx 9.99\ V$$

12. Find circuit phase angle (θ):

$$\text{tangent}\ \theta = \frac{\text{opposite}}{\text{adjacent}} = \frac{X_T}{R} = \frac{48.82}{50}$$
$$= 0.976$$
$$= \text{inverse tangent } 0.976$$
$$= 44°$$

(b)

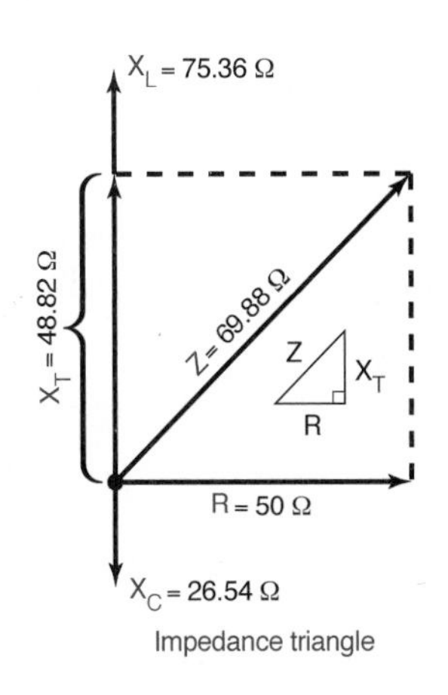

(c)

Figure 6–45. Series *RLC* circuit example: (a) circuit; (b) procedure for finding circuit values; (c) circuit triangles.

Drawing the vectors approximately to scale allows you to pick up any gross errors you may have made.

9. Calculate the phase angle.

$$\theta = \text{inv cos } V_R/V_A.$$

6.18 PARALLEL AC CIRCUITS

Parallel ac circuits have many applications. The basic formulas used with parallel ac circuits are different from those of series circuits. Impedance (Z) of a parallel circuit is less than individual branch values of resistance, inductive reactance, or capacitive reactance. There is no impedance triangle for parallel circuits because Z is smaller than R, X_L, or X_C. A right triangle is drawn to show the currents in the branches of a parallel circuit.

The voltage of a parallel ac circuit is the same across each branch. The currents through the branches of a parallel ac circuit are shown by a right triangle called a *current triangle.* The current through the capacitor (I_C) is shown leading the current through the resistor (I_R) by 90°. The current through the inductor (I_L) is shown lagging I_R by 90°. I_L and I_C are 180° out of phase. They are subtracted to find the total reactive current (I_X). Because these values form a right triangle, the total current may be found by using the formula

$$I_T = \sqrt{I_R^2 + I_X^2}$$

This method is used to find currents in parallel *RL, RC,* or *RLC* circuits.

When components are connected in parallel, finding impedance is more difficult. An impedance triangle cannot be used. A method that can be used to find impedance is to use an admittance triangle. The following quantities are plotted on the triangle: (1) admittance: $Y = 1/Z$; (2) conductance: $G = 1/R$; (3) inductive susceptance: $B_L = 1/XL$; and (4) capacitive susceptance: $B_C = 1/X_C$. These quantities are the inverse of each type of opposition to ac. Because total impedance (Z) is the smallest quantity in a parallel ac circuit, its reciprocal ($1/Z$) becomes the largest quantity on the admittance triangle (just as ½ is larger than ¼). The values are in siemens.

Parallel ac circuits are similar in several ways to series ac circuits except that the applied voltage is used as the reference. Study the examples of Figs. 6–46 to 6–48 on pages 213–215. The general procedure for solving parallel ac circuit problems is summarized next.

General Procedure to Solve Parallel Ac Circuit Problems

1. Draw the circuit diagram. Label the known values. Label the branch currents as I_R, I_C, I_L, and I_T depending on the kind of component in each branch.
2. If reactances are not given, solve for them using:

$$X_L = 2\pi fL \qquad X_C = \frac{1}{2\pi fC}$$

3. Use Ohm's law to calculate the current in each branch.
4. Draw a vector diagram of the currents. Use the applied voltage as the horizontal reference. Draw the total I_R in phase with the voltage. Draw the I_L vertically downward. Draw the I_C vertically upward.
5. Vectorially add the currents to obtain the total current, I_T, delivered by the source.
6. Use the Pythagorean theorem to calculate the value of I_T. Drawing the vectors approximately to scale will allow you to pick up any gross errors you may have made in calculations.
7. Solve for the total impedance of the parallel circuit using:

$$Z_T = \frac{V}{I_T}$$

Note: Impedance triangles are not applicable for parallel circuits.

6.19 POWER IN AC CIRCUITS

Power values in ac circuits are found by using power triangles, as shown in Fig. 6–49 on page 216. The volt-amperes (VA) delivered to an ac circuit are called the *apparent power.* The unit of measurement is the volt-ampere (VA) or kilovolt-ampere (kVA). Meters are used to measure the apparent power in ac circuits. The apparent power is the voltage applied to a circuit multiplied by the total current flow. The power converted to another form of energy by the load is measured with a wattmeter. This actual power converted is called *true power* and is equal to $VI \times$ PF. The ratio of true power converted in a circuit to apparent power delivered to an ac circuit is called the *power factor* (PF). The power factor is found by using the formula

$$\text{PF} = \frac{\text{true power } (W)}{\text{apparent power (VA)}}$$

or

$$\%\ \text{PF} = \frac{W}{\text{VA}} \times 100\%$$

SAMPLE PROBLEM:

Given: a 240-volt, 60-hertz, 30-ampere electric motor is rated at 6000 watts.

Find: power factor at which the motor operates.

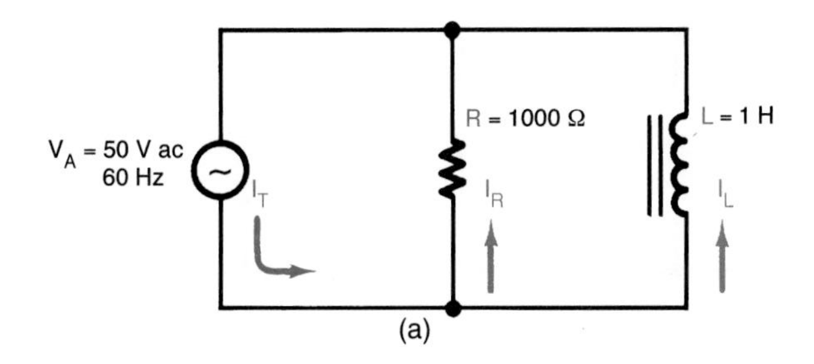

(a)

Finding circuit values:

1. Find inductive reactance (X_L):

$$X_L = 2\pi \cdot f \cdot L$$
$$= 6.28 \times 60 \times 1$$
$$= 376.8\ \Omega$$

2. Find current through R(I_R):

$$I_R = \frac{V}{R} = \frac{50\ \text{V}}{1000\ \Omega} = 0.05\ \text{A}$$

3. Find current through L (I_L):

$$I_L = \frac{V}{X_L} = \frac{50\ \text{V}}{376.8\ \Omega} = 0.133\ \text{A}$$

4. Find total current (I_T):

$$I_T = \sqrt{I_R^2 + I_L^2}$$
$$= \sqrt{0.05^2 + 0.133^2}$$
$$= \sqrt{0.0025 + 0.0177}$$
$$= \sqrt{0.0202}$$
$$= 0.142\ \text{A}$$

5. Check to see that I_T is larger than I_R or I_L.

6. Find impedance (Z):

$$Z = \frac{V}{I_T} = \frac{50\ \text{V}}{0.142\ \text{A}}$$
$$= 352.1\ \Omega$$

7. Check to see that Z is less than R or X_L.

8. Find circuit phase angle (θ):

$$\text{sine } \theta = \frac{\text{opposite}}{\text{hypotenuse}}$$
$$= \frac{I_L}{I_T} = \frac{0.133}{0.142}$$
$$= 0.937$$
$$\theta = \text{inverse sine } 0.937$$
$$= 70°$$

(b)

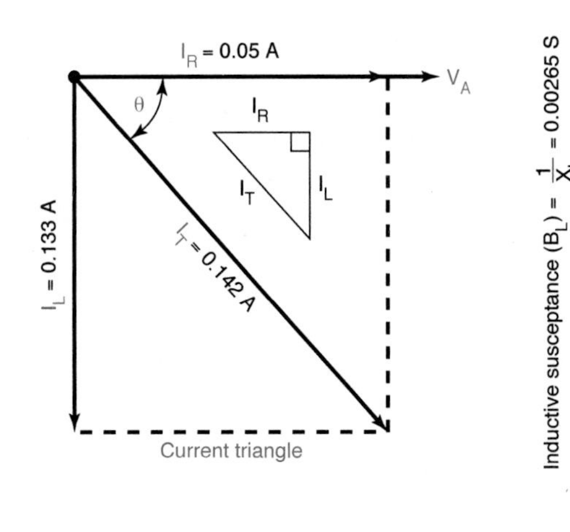

Inductive susceptance (B_L) = $\frac{1}{X_L}$ = 0.00265 S

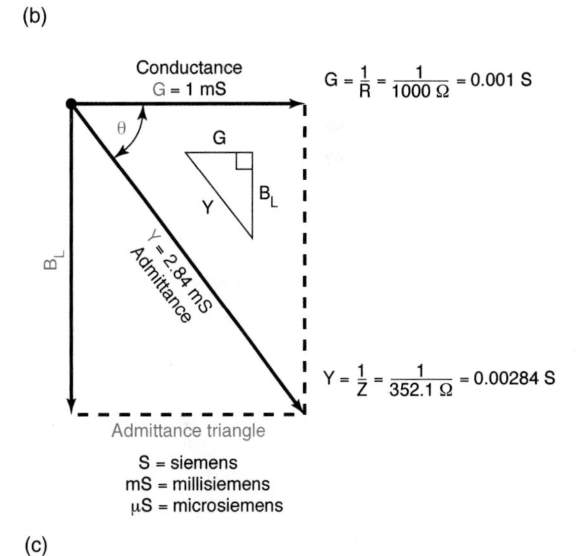

(c)

Figure 6–46. Parallel *RL* circuit example: (a) circuit; (b) procedure for finding circuit values; (c) circuit triangles.

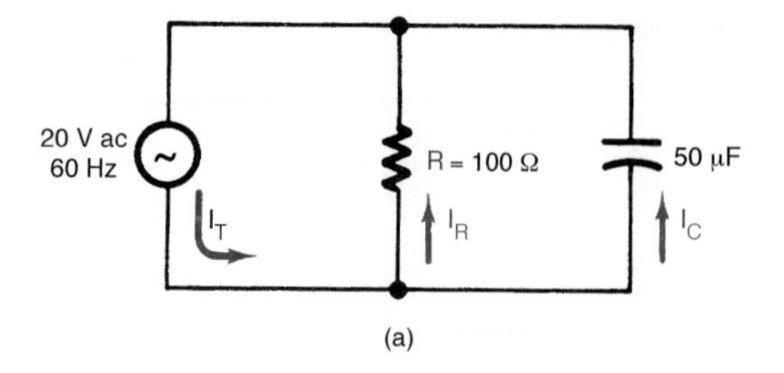

(a)

Finding circuit values:

1. Find capacitive reactance (X_C):

$$X_C = \frac{1{,}000{,}000}{2\pi \cdot f \cdot C\ (\mu F)}$$

$$= \frac{1{,}000{,}000}{6.28 \times 60 \times 50}$$

$$= 53\ \Omega$$

2. Find current through R (I_R):

$$I_R = \frac{V}{R} = \frac{20\ V}{100\ \Omega} = 0.2\ A$$

3. Find current through C (I_C):

$$I_C = \frac{V}{X_C} = \frac{20\ V}{53\ \Omega} = 0.377\ A$$

4. Find total current (I_T):

$$I_T = \sqrt{I_R^2 + I_C^2}$$

$$= \sqrt{0.2^2 + 0.377^2}$$

$$= \sqrt{0.04 + 0.142}$$

$$= \sqrt{0.182}$$

$$= 0.427\ A$$

5. Check to see that I_T is larger than I_R or I_C.

6. Find impedance (Z):

$$Z = \frac{V}{I_T} = \frac{20\ V}{0.427\ A} = 46.84\ \Omega$$

7. Check to see that Z is less than R or X_C.

8. Find circuit phase angle (θ):

$$\text{cosine}\ \theta = \frac{\text{adjacent}}{\text{hypotenuse}} = \frac{G}{Y} = \frac{10}{21.3}$$

$$= 0.469$$

$$\theta = \text{inverse cosine}\ 0.469$$

$$= 62^\circ$$

(b)

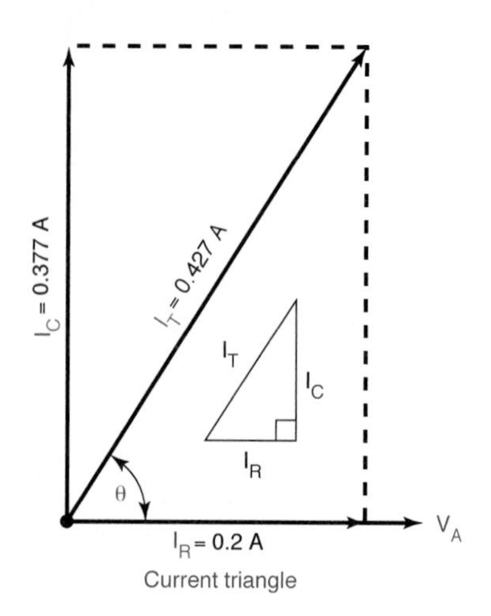

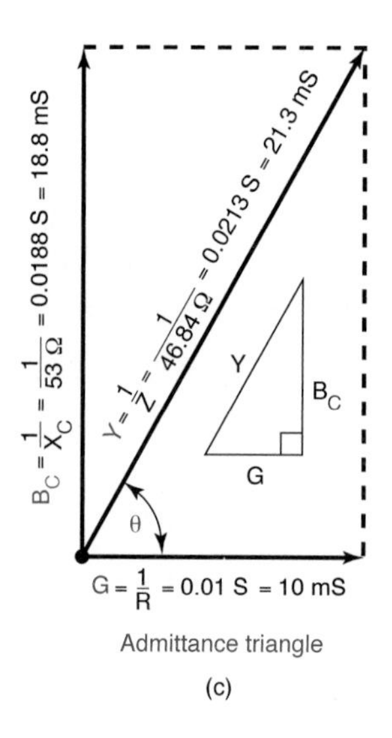

(c)

Figure 6–47. Parallel *RC* circuit example: (a) circuit; (b) procedure for finding circuit values; (c) circuit triangles.

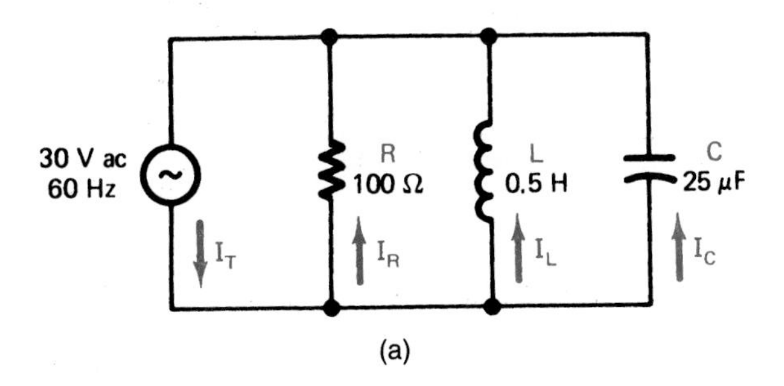

(a)

Finding circuit values:

1. Find inductive reactance (X_L):

$$X_L = 2\pi \cdot f \cdot L$$
$$= 6.28 \times 60 \times 0.5$$
$$= 188.4\ \Omega$$

2. Find capacitive reactance (X_C):

$$X_C = \frac{1{,}000{,}000}{6.28 \times 60 \times 25}$$
$$= 106\ \Omega$$

3. Find current through R (I_R):

$$I_R = \frac{V}{R} = \frac{30\ V}{100\ \Omega} = 0.3\ A$$

4. Find current through L (I_L):

$$I_L = \frac{V}{X_L} = \frac{30\ V}{188.4\ \Omega} = 0.159\ A$$

5. Find current through C (I_C):

$$I_C = \frac{V}{X_C} = \frac{30\ V}{106\ \Omega} = 0.283\ A$$

6. Find total reactive current (I_X):

$$I_X = I_C - I_L$$
$$= 0.283 - 0.159$$
$$= 0.124$$

7. Find total current (I_T):

$$I_T = \sqrt{I_R^C + I_X^2}$$
$$= \sqrt{0.3^2 + 0.124^2}$$
$$= \sqrt{0.9 + 0.0154}$$
$$= \sqrt{0.1054}$$
$$= 0.325\ A$$

8. Check to see that I_T is larger than I_R or I_X.

9. Find impedance (Z):

$$Z = \frac{V_T}{I_T} = \frac{30\ V}{0.325\ A} = 92.3\ \Omega$$

10. Check to see that Z is less than R or X_L or X_C.

11. Find circuit phase angle (θ):

$$\text{tangent}\ \theta = \frac{\text{opposite}}{\text{adjacent}} = \frac{I_X}{I_R} = \frac{0.124}{0.3}$$
$$= 0.413$$
$$\theta = \text{inverse tangent}\ 0.413$$
$$= 22°$$

(b)

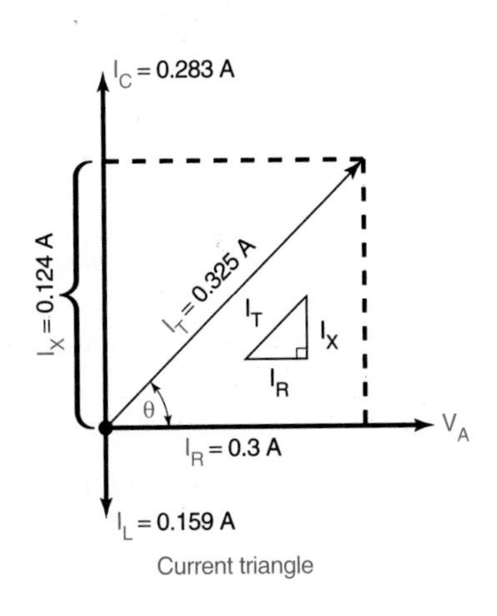

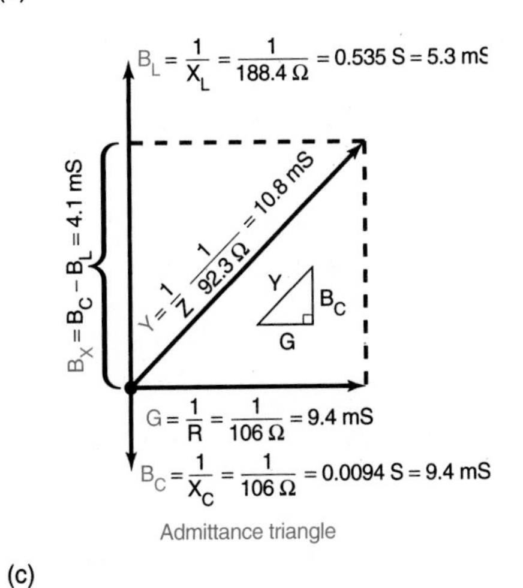

(c)

Figure 6–48. Parallel *RLC* circuit example: (a) circuit; (b) procedure for finding circuit values; (c) circuit triangles.

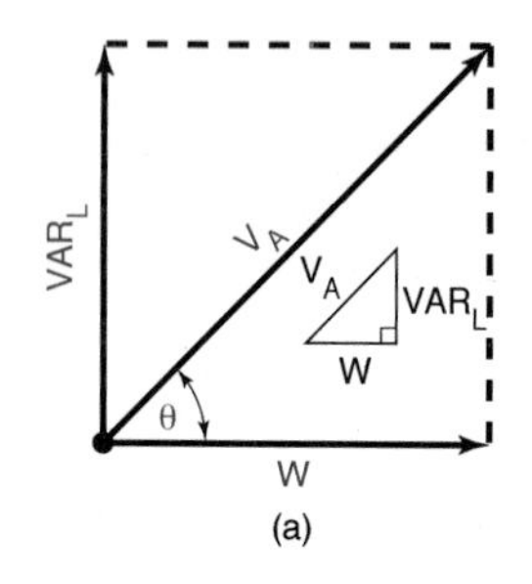

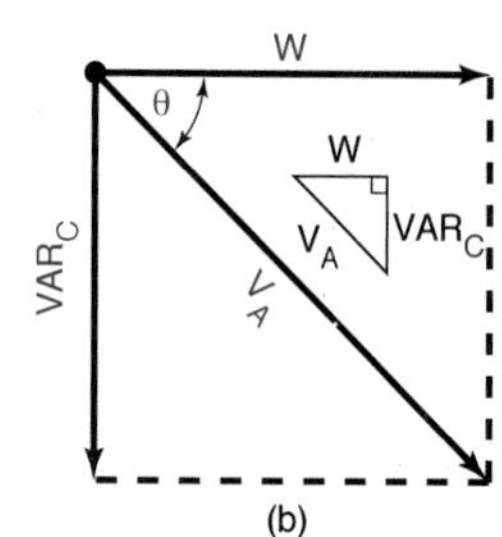

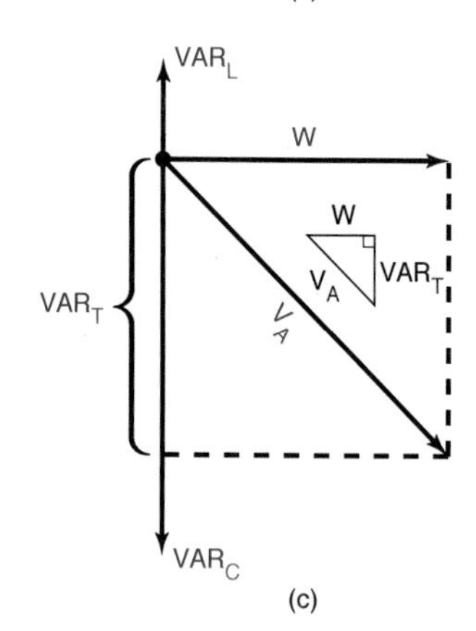

VA = volt-amperes (apparent power)
W = watts (true power)
VAR_L = volt-amperes reactive (inductive)
VAR_C = volt-amperes reactive (capacitive)
VAR_T = volt-amperes reactive (total)

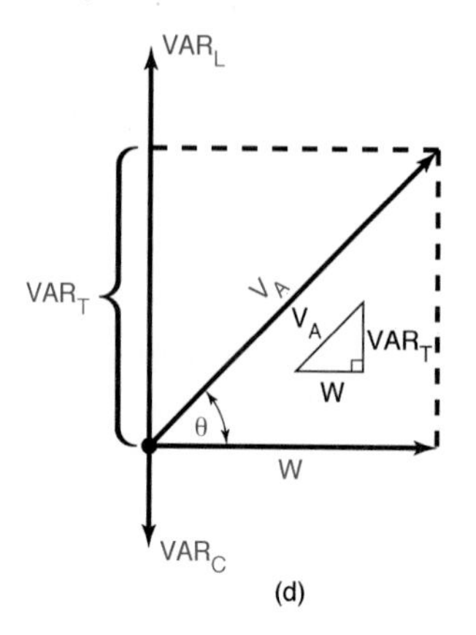

Calculating power values:

1. In dc circuits:

$$P = V \times I = I^2 \times R = \frac{V^2}{R}$$

2. In ac circuits:

a. $VA = V_A \times I_T = I_T^2 \times Z = \frac{V_A^2}{Z}$

b. $W = V_R \times I_R = I_R^2 \times R = \frac{V_R^2}{R}$

c. $VAR_L = V_L \cdot I_L = I_L^2 \cdot X_L = \frac{V_L^2}{X_L}$

d. $VAR_C = V_C \cdot I_C = I_C^2 \cdot X_C = \frac{V_C^2}{X_C}$

e. $VAR_T = VAR_L - VAR_C$ or $VAR_C - VAR_L$

(e)

Figure 6–49. Power triangles for ac circuits: (a) *RL* circuit; (b) *RC* circuit; (c) *RLC* circuit with X_L larger than X_C; (d) *RLC* circuit with X_C larger than X_L; (e) formulas for calculating power values.

Solution:

$$\% \text{ PF} = \frac{W}{VA} \times 100$$

$$\frac{6000W}{240V \times 30A} \times 100$$

$$\% \text{ PF} = 83.3\%$$

The term *W* is the true power in watts and VA is the apparent power in volt-amperes. The highest power factor of an ac circuit is 1.0 or 100%. This value is called *unity power factor.*

The phase angle (θ) of an ac circuit determines the power factor and is equal to PF = cos θ. Remember that the phase angle is the angular separation between voltage applied to an ac circuit and current flow through the circuit.

SAMPLE PROBLEM:

Given: a circuit has the following values: applied voltage = 240 volts, current = 12 amperes, power factor = 0.83.

Find: the true power of the circuit.

Solution:

$$W = \text{VA} \times \text{cosine } \theta$$
$$= 240 \text{ V} \times 12 \text{ A} \times 0.83$$
$$= 2390 \text{ watts}$$

Note that the expression

$$\frac{\text{true power}}{\text{apparent power}}$$

is the *power factor* of a circuit; therefore, the power factor is equal to the cosine of the phase angle (PF = cosine θ).

Right triangle relationships can be expressed as equations that determine the value of any of the sides of the power triangle when the other two values are known. These expressions are as follows:

$$W = \sqrt{VA^2 - var^2}$$
$$VA = \sqrt{W^2 + var^2}$$
$$var = \sqrt{VA^2 - W^2}$$

SAMPLE PROBLEM:

Given: total reactive power = 54 var, applied; voltage = 120 volts; current = 0.5 amperes.

Find: the true power of the circuit.

Solution:

$$W = \sqrt{VA^2 - var^2}$$
$$= \sqrt{(120 \times 0.5)^2 - 54^2}$$
$$= \sqrt{3600 - 2916}$$
$$= 26.15 \text{ watts}$$

More inductive or capacitive reactance causes a larger phase angle. In a purely inductive or capacitive circuit, a 90° phase angle causes a power factor of zero. The power factor varies according to the values of resistance and reactance in a circuit.

There are two types of power that affect power conversion in ac circuits. Power conversion in the resistive part of the circuit is called *active power* or true power. True power is measured in watts. The other type of power is caused by inductive or capacitive loads. It is 90° out of phase with the true power and is called *reactive VA*. It is a type of "unused" power. Reactive VA is measured in volt-amperes reactive (VAR).

The power triangle of Fig. 6–49 has true power (watts) marked as a horizontal line. Reactive power (VAR) is drawn at a 90° angle from the true power. Volt-amperes or apparent power (VA) is the longest side (hypotenuse) of the right triangle. This right triangle is similar to the impedance triangle and the voltage triangle for series ac circuits, and to the current triangle and admittance triangle for parallel ac circuits. Each type of right triangle has a horizontal line used to mark the resistive part of the circuit. The vertical lines are used to mark the reactive part of the circuit. The longest side of the triangle (hypotenuse) is used to mark total values of the circuit. The size of the hypotenuse depends on the amount of resistance and reactance in the circuit. Vector diagrams and right triangles are extremely important for understanding ac circuits.

Power in Three-Phase Circuits

In Chap. 5, discussion involved three-phase wye and delta connections. A review of these topics might be helpful. The formulas for three-phase power relationships are as follows:

Apparent power (volt-amperes) per phase:

$$VA_p = V_p \times I_p$$

Total apparent power (volt-amperes):

$$VA_T = 3 \times V_p \times I_p \text{ or } 1.73 \text{ (V3)} \times V_L \times I_L$$

Power factor:

$$PF = \frac{\text{true power } (W)}{1.73 \times V_L \times I_L} \quad \text{or} \quad \frac{\text{watts}}{3 \times V_p \times I_p}$$

True power per phase:

$$P_p = V_p \times I_p \times PF$$

Total true power:

$$P_T = 3 \times V_p \times I_p \times PF$$

or

$$1.73 \times V_L \times I_L \times PF$$

Remember that the symbols used previously are as follows: V_L is the line voltage in volts, I_L is the line current in amperes, V_p is the phase voltage in volts, and I_p is the phase current in amperes.

SAMPLE PROBLEM:

Given: phase voltage = 120 volts for a three-phase wye system.

Find: line voltage.

Solution:

$$V_L = V_p \times 1.73$$
$$= 120 \text{ V} \times 1.73$$
$$= 208 \text{ volts}$$

SAMPLE PROBLEM:

Given: a three-phase delta system has a phase voltage of 240 volts, a phase current of 20 amperes, and a power factor of 0.75.

Find: power per phase.

Solution:

$$P_P = V_p \times I_p \times \text{PF}$$
$$= 240 \text{ V} \times 20 \text{ A} \times 0.75$$
$$= 3600 \text{ watts}$$

where PF is the power factor of the load.

The total power (P_T) developed by all three phases of a three-phase system is expressed as:

$$P_T = 3 \times P_p$$
$$= 3 \times V_p \times I_p \times \text{PF}$$
$$= 1.73 \times V_L \times I_L \times \text{PF}$$

SAMPLE PROBLEM:

Given: a three-phase wye system has a phase voltage of 277 volts, a phase current of 10 amperes, and a power factor of 0.85.

Find: total three-phase power.

Solution:

$$P_T = 3 \times 277 \text{ V} \times 10 \text{ A} \times 0.85$$
$$= 7063.5 \text{ watts}$$

To summarize three-phase power relationships:

Volt-amperes per phase (VA_P) = $V_P I_P$

Total volt-amperes (VA_T) = $3\ V_P I_P$
$= 1.73\ V_L I_L$

SAMPLE PROBLEM:

Given: a three-phase delta system has a line voltage of 208 volts and a line current of 4.86 amperes.

Find: total three-phase volt-amperes.

Solution:

$$\text{VA} = 1.73 \times 208 \text{ V} \times 4.86 \text{ A}$$
$$= 1748.8 \text{ volt-amperes}$$

$$\text{power factor (PF)} = \frac{\text{true power } (W)}{1.73\ V_L I_L}$$

$$= \frac{W}{3\ V_P I_P}$$

$$\text{power per phase } (P_P) = 3\ V_P I_P \times \text{PF}$$
$$= 1.173\ V_L I_L \times \text{PF}$$

where:

V_L is the line voltage in volts
I_L is the line current in amperes
V_P is the phase voltage in volts
I_P is the phase current in amperes

SAMPLE PROBLEM:

Given: a three-phase wye system has the following values: phase voltage = 120 V, phase current = 18.5 A, and power factor = 0.95.

Find: total three-phase power of the circuit.

Solution:

$$P_T = 3 \times V_P \times I_P \times \text{PF}$$
$$= 3 \times 120 \text{ V} \times 18.5 \text{ A} \times 0.95$$
$$= P_T = 6.327\ W = 6.327 \text{ kW}$$

Calculations involving three-phase power are somewhat more complex than single-phase power calculations. Keep in mind the difference between phase values and line values to avoid making mistakes.

SAMPLE PROBLEM:

Given: the following 120-volt single-phase loads are connected to a 120/208-volt wye system. Phase A has 2000 watts at a 0.75 power factor, Phase B has 1000 watts at a 0.85 power factor, and Phase C has 3000 watts at a 1.0 power factor.

Find: the total power of the three-phase system, and the current flow through each line.

Solution: To find the phase currents, use the formula $P_P = V_P \times I_P \times$ power factor, and transpose. Thus, we have $I_P = P_P / V_P \times \text{PF}$. Substitution of the values for each leg of the system gives:

1.
$$*I_{P-A} = \frac{P_{P-A}}{V_P \times \text{PF}}$$
$$= \frac{2000 \text{ watts}}{120 \text{ volts} \times 0.75}$$
$$= 22.22 \text{ amperes}$$

*This notation means phase current (I_P) of "A" power line.

2.
$$I_{P-B} = \frac{P_{P-B}}{V_P \times \text{PF}}$$
$$= \frac{1000 \text{ watts}}{120 \text{ volts} \times 0.85}$$
$$= 9.8 \text{ amperes}$$

3.
$$I_{P-C} = \frac{P_{P-C}}{V_P \times \text{PF}}$$
$$= \frac{3000 \text{ watts}}{120 \text{ volts} \times 1.0}$$
$$= 25.0 \text{ amperes}$$

To determine the total power, use the total power formula for a three-phase unbalanced system:

$$\begin{aligned} P_T &= P_{P-A} + P_{P-B} + P_{P-C} \\ &= 2000 + 1000 + 3000 \\ &= 6000 \text{ watts (6 kW)} \end{aligned}$$

6.20 FILTERS AND RESONANT CIRCUITS

Some types of ac circuits are designed to respond to specific ac frequencies. Circuits that are used to pass some frequencies and block others are called filter circuits and resonant circuits. Each type of circuit uses reactive devices to respond to different ac frequencies. These circuits have frequency response curves which show their operation graphically. Frequency is graphed on the horizontal axis and voltage output on the vertical axis. Sample frequency response curves for each type of filter and resonant circuit are shown in the examples that follow.

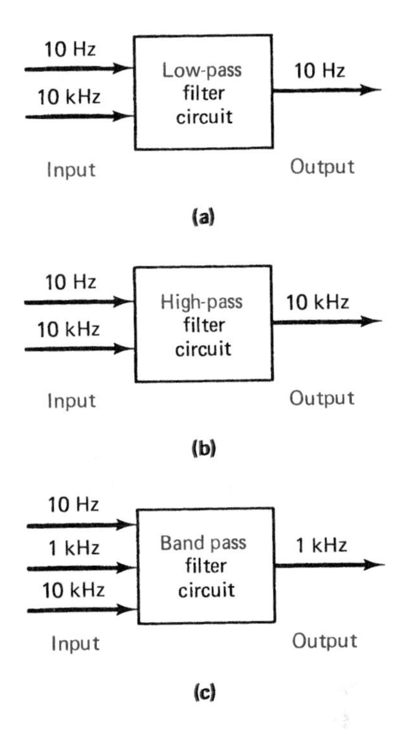

Figure 6–50. Three types of filter circuits: (a) low-pass filter—passes low frequencies and blocks high frequencies; (b) high-pass filter—passes high frequencies and blocks low frequencies; (c) bandpass filter—passes a midrange of frequencies and blocks high and low frequencies.

Filter Circuits

The three types of filter circuits are shown in Fig. 6–50. Filter circuits are used to separate one range of frequencies from another. Low-pass filters pass low ac frequencies and block higher frequencies. High-pass filters pass high frequencies and block lower frequencies. Bandpass filters pass a midrange of frequencies and block lower and higher frequencies. All filter circuits have resistance and capacitance *or* inductance.

Figure 6–51 shows the circuits used for low-pass, high-pass, and bandpass filters and their frequency response curves. Many low-pass filters are series *RC* circuits, as shown in Fig. 6–51(a). Output voltage (V_{out}) is taken across a capacitor. As frequency increases, capacitive reactance (X_C) decreases, because

$$X_C = \frac{1}{2\pi \times f \times C}$$

The voltage drop across the output is equal to $I \times X_C$. So as frequency increases, X_C decreases, and voltage output decreases. Series *RL* circuits may also be used as low-pass filters. As frequency increases, inductive reactance (X_L) increases, because $X_L = 2\pi \times f \times L$. Any increase in X_L reduces the circuit's current. The voltage output taken across the resistor is equal to $I \times R$. So when I decreases, V_{out} also decreases. As frequency increases, X_L increases, I decreases, and V_{out} decreases.

Figure 6–51(b) shows two types of high-pass filters. The series *RC* circuit is a common type. The voltage output (V_{out}) is taken across the resistor (R). As frequency increases, X_C decreases. A decrease in X_C causes current flow to increase. The voltage output across the resistor (V_{out}) is equal to $I \times R$. So as I increases, V_{out} increases. As frequency increases, X_C decreases, I increases, and V_{out} increases. A series *RL* circuit may also be used as a high-pass filter. The V_{out} is taken across the inductor. As frequency increases, X_L increases, and V_{out} is equal to $I \times X_L$. So as X_L increases, V_{out} also increases. In this circuit, as frequency increases, X_L increases and V_{out} increases.

The bandpass filter of Fig. 6–51(c) is a combination of low-pass and high-pass filter sections. It is designed to pass a midrange of frequencies and block low and high frequencies. R_1 and C_1 form a low-pass filter and R_2 and C_2 form a high-pass filter. The range of frequencies to be passed is determined by calculating the values of resistance and capacitance.

Resonant Circuits

Resonant circuits are designed to pass a range of frequencies and block all others. They have resistance, inductance, and capacitance. Figure 6–52 shows the two types of resonant circuits: (1) series resonant and (2) parallel resonant and their frequency response curves.

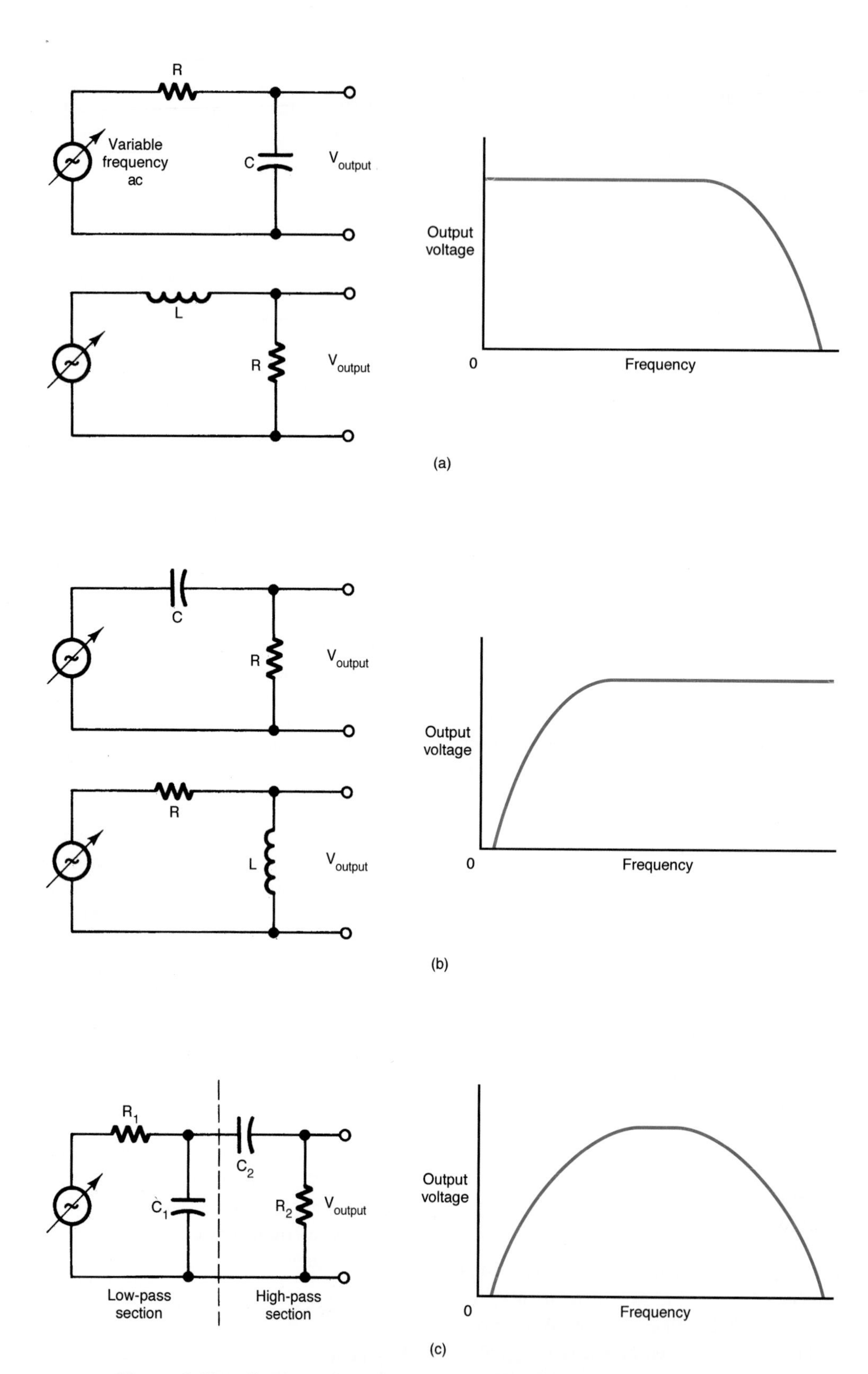

Figure 6–51. Circuits used to filter ac frequencies and their response curves: (a) low-pass filters; (b) high-pass filters; (c) bandpass filter.

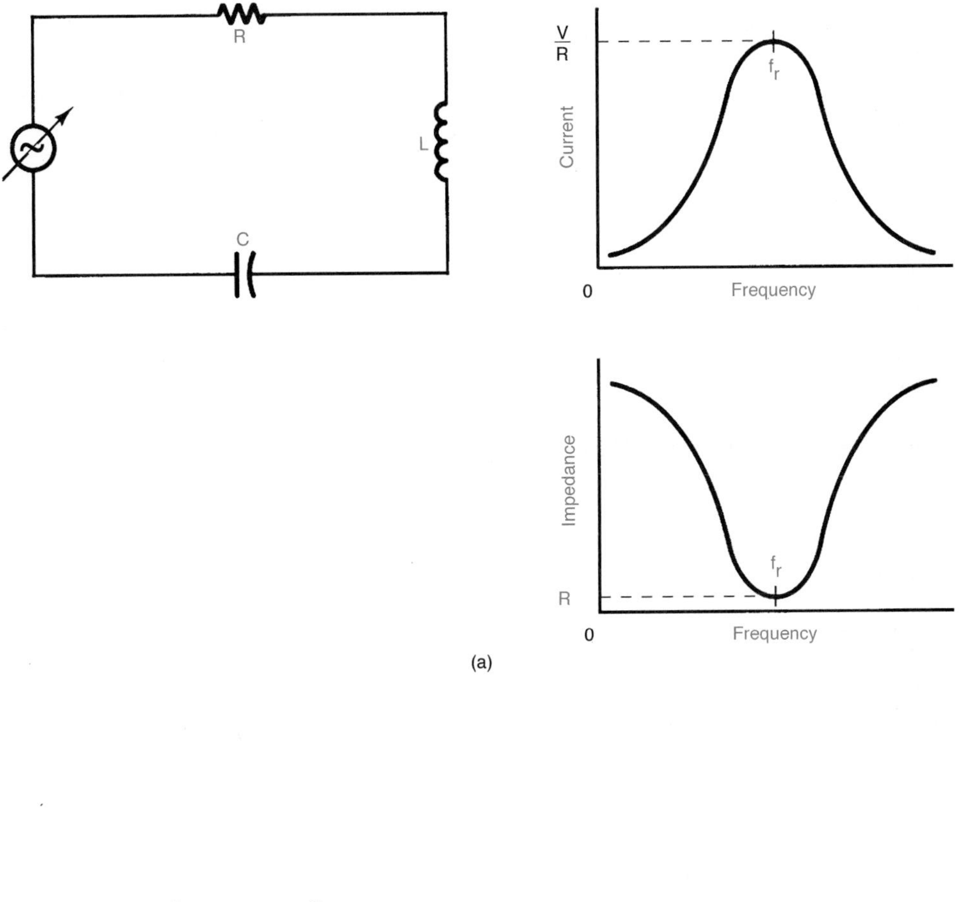

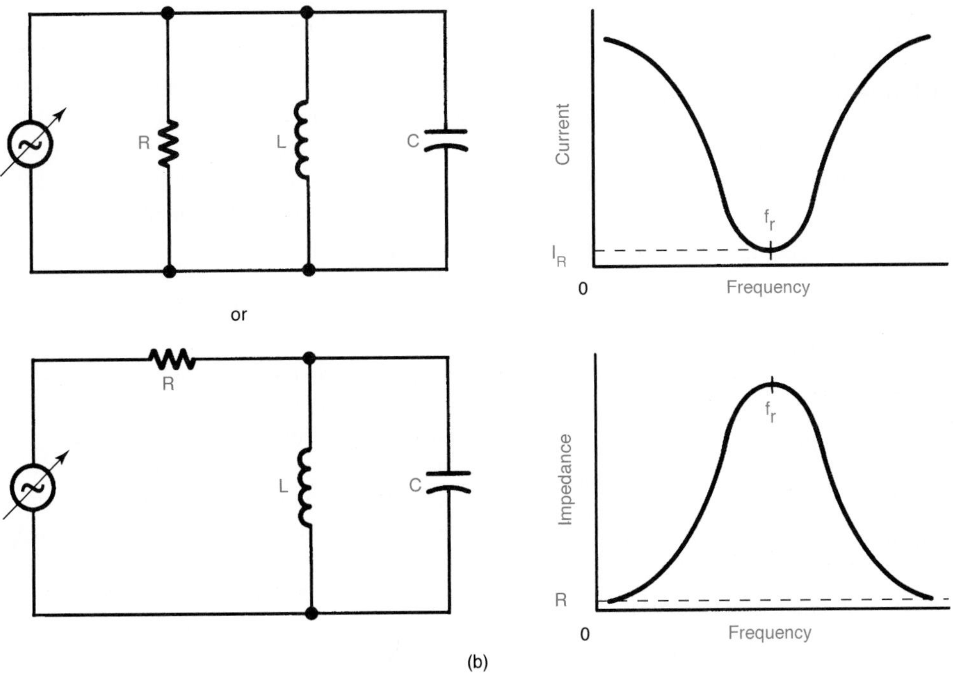

Figure 6–52. Resonant circuits: (a) series and (b) parallel resonant circuits and their frequency response curves.

Series Resonant Circuits. Series resonant circuits are a series arrangement of inductance, capacitance, and resistance. A series resonant circuit offers a small amount of opposition to some ac frequencies and much more opposition to other frequencies. They are important for selecting or rejecting frequencies.

The voltages across inductors and capacitors in ac series circuits are in direct opposition to each other (180° out of phase). They tend to cancel each other out. The frequency applied to a series resonant circuit affects inductive reactance and capacitive reactance. At a specific input frequency, X_L will equal X_C. The voltages across the inductor and capacitor are then equal. The total reactive voltage (V_X) is zero volts at this frequency. The opposition offered by the inductor and the capacitor cancel each other at this frequency. The total reactance (X_T) of the circuit ($X_L - X_C$) is zero. The impedance (Z) of the circuit is then equal to the resistance (R).

The frequency at which $X_L = X_C$ is called *resonant frequency*. To determine the resonant frequency (f_r) of the circuit, use the formula

$$f_r = \frac{1}{2\pi \times \sqrt{L \times C}}$$

In the formula, L is in henries, C is in farads, and f_r is in hertz. As either inductance or capacitance increases, resonant frequency decreases. When the resonant frequency is applied to a circuit, a condition called *resonance* exists. Resonance for a series circuit causes the following:

1. $X_L = X_C$.
2. X_T is equal to zero.
3. $V_L = V_C$.
4. Total reactive voltage (V_X) is equal to zero.
5. $Z = R$.
6. Total current (I_T) is maximum.
7. Phase angle (θ) is 0°.

Refer to the series *RLC* circuit of Fig. 6–45 to clarify these values.

The ratio of reactance (X_L or X_C) to resistance (R) at resonant frequency is called *quality factor (Q)*. This ratio is used to determine the range of frequencies or bandwidth (BW) a resonant circuit will pass. A sample resonant circuit problem is shown in Fig. 6–53. The frequency range that a resonant circuit will pass (BW) is found by using the procedure of steps 5 and 6 in Fig. 6–53.

The cutoff points are at about 70% of the maximum output voltage. These are called the low-frequency cutoff (f_{lc}) and high-frequency cutoff (f_{hc}). The bandwidth of a resonant circuit is determined by the Q. Q is determined by the ratio of X_L and X_C to R. Resistance mainly determines bandwidth. This effect is summarized as follows:

1. When R is increased, Q decreases, as $Q = X_L/R$.
2. When Q decreases, BW increases, as $\text{BW} = f_r/Q$.
3. When R is increased, BW increases.

Two curves shown in Fig. 6–54 illustrate the effect of resistance on bandwidth. The curve of Fig. 6–54(b) has high selectivity, which means that a resonant circuit with this response curve would select a small range of frequencies. This selectivity is important for radio and television tuning circuits.

Parallel Resonant Circuits. Parallel resonant circuits are similar to series resonant circuits. Their electrical characteristics are somewhat different, but they accomplish the same purpose. Another name for parallel resonant circuits is *tank circuits*. A tank circuit is a parallel combination of L and C used to select or reject ac frequencies.

With the resonant frequency applied to a parallel resonant circuit, the following occurs:

1. $X_L = X_C$.
2. X_T equals zero.
3. $I_L = I_C$.
4. I_X equals zero, so the total current (I_T) is minimum.
5. $Z = R$ and is maximum.
6. Phase angle (θ) = 0°.

Refer to the parallel *RLC* circuit of Fig. 6–48 to clarify these values.

The calculations used for parallel resonant circuits are similar to those for series circuits with one exception. The quality factor (Q) is found by using this formula for parallel circuits: $Q = R/X_L$. Otherwise, the example of Fig. 6–53 is the same for parallel resonant circuits.

6.21 TRANSFORMERS

Transformers are important electrical devices. They are used to either increase or decrease alternating-current (ac) voltage. Transformers are made by using two separate sets of wire windings which are wound on a core. These are called the primary and the secondary windings. A transformer that increases voltage is called a step-up transformer and one that decreases voltage is called a step-down transformer. Several types of transformers are shown in Fig. 6–55.

Transformer Operation

Transformers are electrical control devices used to either increase or decrease ac voltage. They do not operate with dc voltage applied. Notice in Fig. 6–56 that ac voltage is applied to the primary winding of the transformer. There is no con-

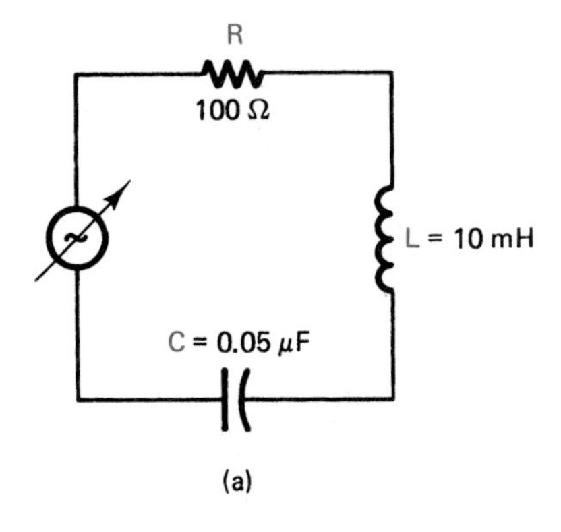

Finding circuit values:

1. Find resonant frequency (f_r):

$$f_r = \frac{1}{2\pi \times \sqrt{L \times C}} = \frac{1}{6.28 \times \sqrt{(10 \times 10^{-3}) \times (0.05 \times 10^{-6})}} =$$

$$= \frac{1}{6.28 \times \sqrt{0.5 \times 10^{-9}}} = \frac{1}{6.28 \times (2.23 \times 10^{-5})} = \frac{1}{1.4 \times 10^{-4}} = 7121 \text{ Hz}$$

2. Find X_L^* and X_C at resonant frequency:

$$X_L = 2\pi \cdot f \cdot L = 6.28 \times 7121 \times (10 \times 10^{-3}) = 447\ \Omega$$

*Easier to calculate.

3. Find quality factor (Q):

$$Q = \frac{X_L}{R} = \frac{447\ \Omega}{100\ \Omega} = 4.47$$

4. Find bandwidth (BW):

$$BW = \frac{f_r}{Q} = \frac{7121 \text{ Hz}}{4.47} = 1593 \text{ Hz}$$

5. Find low-frequency cutoff (f_{lc}):

$$f_{lc} = f_r - \tfrac{1}{2} BW = 7121 \text{ Hz} - 797 \text{ Hz} = 6324 \text{ Hz}$$

6. Find high-frequency cutoff (f_{hc})

$$f_{hc} = f_r + \tfrac{1}{2} BW = 7121 + 797 = 7918 \text{ Hz}$$

(b)

Figure 6–53. Sample resonant circuit problem: (a) circuit; (b) procedure to find circuit values.

nection of the primary and secondary windings. The transfer of energy from the primary to the secondary winding is due to magnetic coupling or mutual inductance. The transformer relies on electromagnetism to operate.

Although many different types and sizes of transformers exist, the same basic principles of operation apply to all. The operation of a transformer relies on the expanding and collapsing of the magnetic field around the primary winding. When current flows through a conductor, a magnetic field is developed around the conductor. When ac voltage is applied to the primary winding, it causes a constantly changing magnetic field around the winding. During times of increasing ac voltage, the magnetic field around the primary winding expands. After the peak value of the ac cycle is reached, the voltage decreases toward zero. When the ac voltage decreases, the magnetic field around the primary winding collapses. The collapsing magnetic field is transferred to the secondary winding and induces a voltage in it. Transformers will not operate with dc voltage applied, because dc voltage does not change in value.

Types of Transformers

The three basic transformer types are air-core, iron-core, and powered metal-core transformers. The core material used to construct transformers depends on the application of the transformers. Radios and televisions use several types of air-core and powered metal-core transformers. Iron-core transformers are sometimes called *power transformers.* They are often used in the power supplies of electrical equipment and at electrical power distribution substations.

Two other classifications of transformers are step-up transformers and step-down transformers. Step-up transformers have more turns of wire used for their secondary windings than for their primary windings. Step-down transformers have more turns of wire on their primary windings than on their secondary windings. An example of a step-up transformer is shown in Fig. 6–57(a). A step-up transformer has a higher voltage across the secondary winding than the input voltage applied to the primary winding. The step-up is caused by the turns ratio of the primary and secondary windings. The turns ratio is the ratio of primary turns of wire (N_P) to secondary turns of wire (N_S). If 100 turns of wire are used for the primary winding and 200 turns for the secondary winding, the turns ratio is 1:2. The secondary voltage will be two times as large as the primary voltage. If 10 V ac is applied to the primary, 20 V ac is developed across the load connected across the secondary winding. The voltage ratio (V_P/V_S) is the same as the turns ratio:

$$\frac{V_P}{V_S} = \frac{N_P}{N_S}$$

Figure 6–57(b) shows an example of a step-down transformer. The primary winding has 500 turns and the secondary has 100 turns of wire. The turns ratio (N_P/N_S) is 5:1.

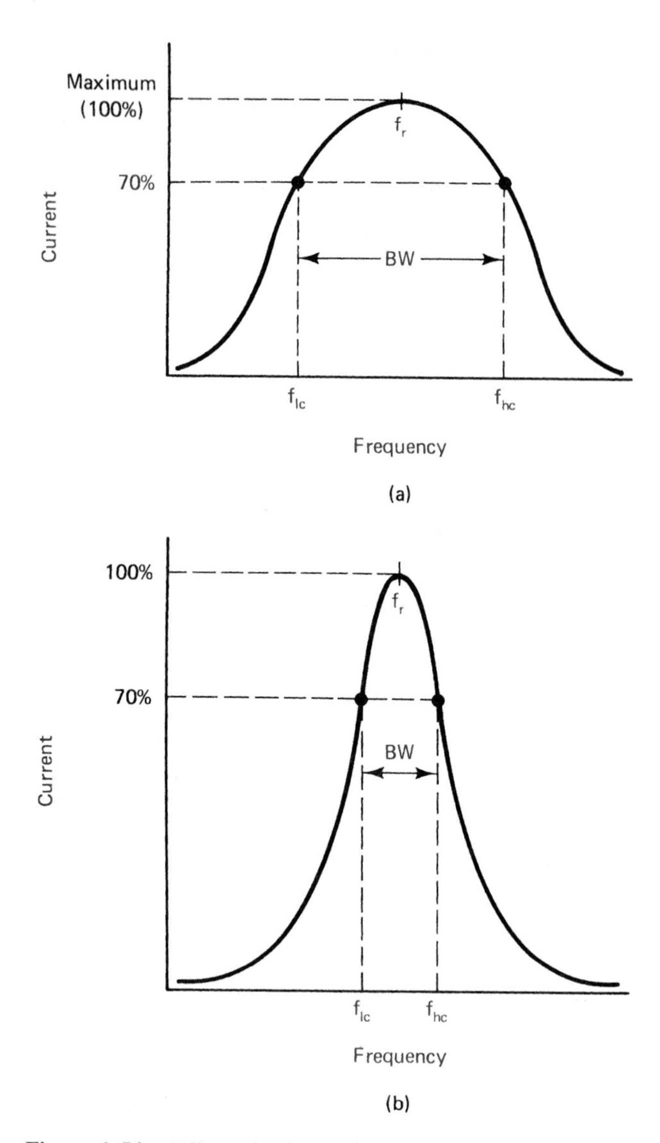

Figure 6–54. Effect of resistance on bandwidth of a series resonant circuit: (a) high resistance, low selectivity; (b) low resistance, high selectivity.

Figure 6–55. Transformers. (Courtesy of TRW/UTC Transformers.)

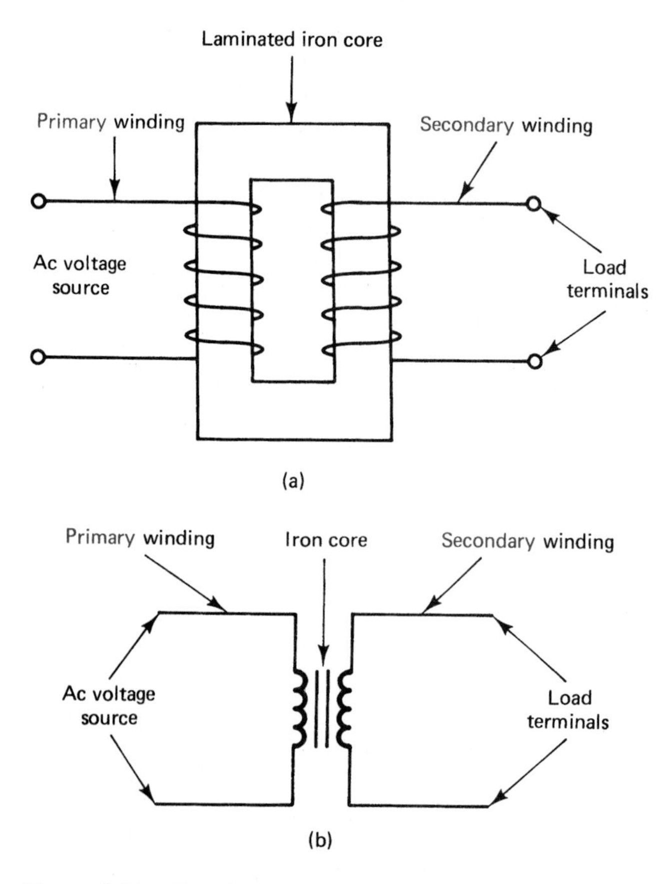

Figure 6–56. Transformer: (a) pictorial; (b) schematic symbol.

This means that the primary voltage (V_P) is five times greater than the secondary voltage (V_S). If 25 V is applied to the primary, 5 V is developed across the secondary. Remember that a step-up transformer increases voltage and a step-down decreases voltage. The voltage formula shown in Fig. 6–57 shows the relationship between the primary and secondary voltages and the turns ratio of the primary and secondary windings.

Power converted by a transformer is equal to voltage (V) multiplied by current (I). In a transformer, primary power (P_P) is equal to secondary power (P_S). Therefore, primary voltage (V_P) times primary current (I_P) is equal to secondary voltage (V_S) times secondary current (I_S). This is true if we

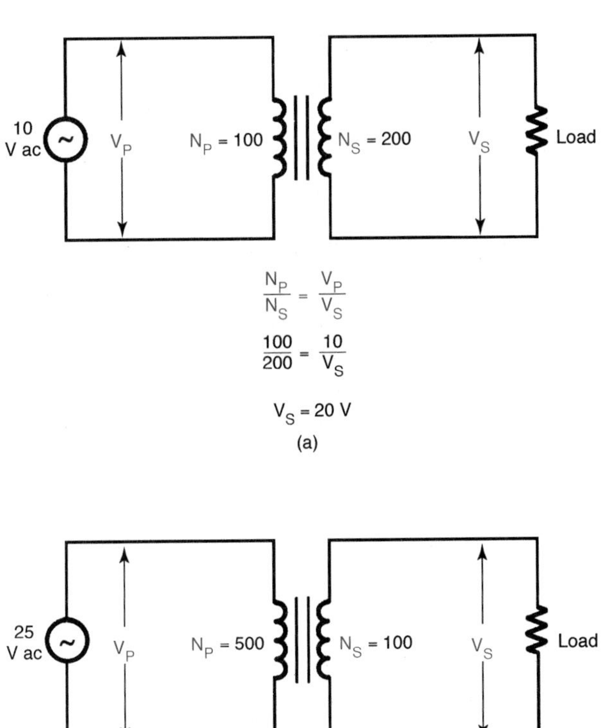

$$\frac{N_P}{N_S} = \frac{V_P}{V_S}$$

$$\frac{500}{100} = \frac{25}{V_S}$$

$$V_S = 5 \text{ V}$$

(b)

Figure 6–57. Transformers: (a) step-up; (b) step-down.

neglect circuit losses. This condition where $P_P = P_S$ is called an *ideal* transformer.

$$P_P = P_S$$

$$V_P \times I_P = V_S \times I_S$$

The current ratio of transformers is an inverse ratio:

$$\frac{N_P}{N_S} = \frac{I_S}{I_P}$$

An application of the current ratio is shown in Fig. 6–58. If voltage is increased, current is decreased in the same proportion. The secondary voltage is decreased by 10 times and secondary current is increased by 10 times in the example shown. This is a step-down transformer, because voltage decreased.

SAMPLE PROBLEM:

Given: an ideal transformer circuit has the following values:

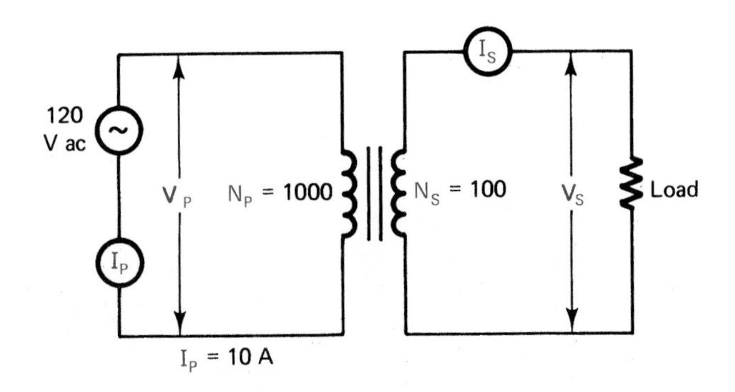

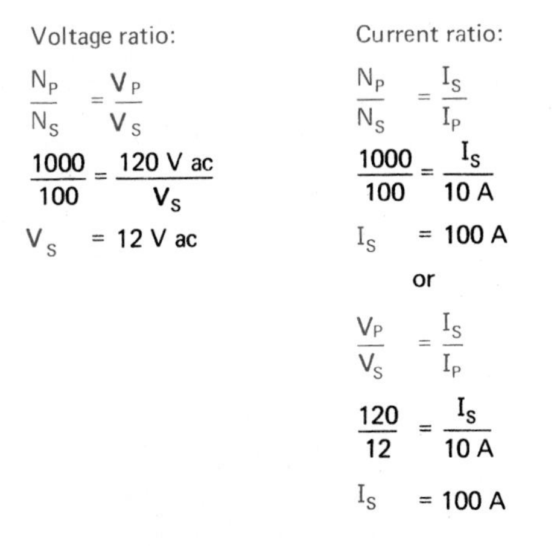

Figure 6–58. Transformer current ratio.

$$V_P = 600 \text{ V}$$

$$V_S = 2400 \text{ V}$$

$$I_S = 80 \text{ A}$$

Find: the primary current drawn by the step-up transformer.

Solution:

$$\frac{I_P}{I_S} = \frac{V_S}{V_P}$$

$$\frac{I_P}{80 \text{ A}} = \frac{2400 \text{ V}}{600 \text{ V}}$$

$$I_P = 3200 \text{ A}$$

Some transformers have more than one secondary winding. These are called *multiple secondary transformers.* Some are step-up and some are step-down windings, depending on the application. There are many other special types of transformers. *Autotransformers* have only one winding. They may be either step-up or step-down transformers, depending on how they are connected. *Variable autotransformers,* such as the one shown in Fig. 6–59, are used in power supplies to provide adjustable ac voltages.

4. Convert the following peak values to effective values:

   12 W peak = _____ W RMS

   100 mA peak = _____ mA RMS

   2 mW = _____ mW RMS

5. How does the power produced by 10 V dc across 10 Ω compare with the power produced by 10-V ac RMS across the same resistance?
6. Why is a cycle said to consist of 360°?
7. Compute the periods for the following ac frequencies:

   10 kHz = _____ s

   1.2 kHz = _____ s

   1 MHz = _____ s

   0.6 MHz = _____ s

   60 Hz = _____ s

8. What is the standard frequency of the alternating current used in the United States?
9. What is alternating current?
10. If ac is continually changing its direction, why do the lights in your home not blink on and off?

• *Measuring Ac Voltage with an Oscilloscope*

Complete Activity 6–2, page 119 in the manual.

• •

ANALYSIS

1. What is the most important application of the oscilloscope?
2. Briefly describe the purpose of each of the following controls:
   - **a.** Vertical gain
   - **b.** Horizontal gain
   - **c.** Intensity
   - **d.** Focus
   - **e.** Vertical position
   - **f.** Horizontal position
3. What effect does adjusting the *horizontal sweep selection* switch have upon the trace?
4. What does *synchronize* mean?

• *Alternating-Current Values (see Fig. 6–3)*

There are several ac values that are commonly used. These units are ac sine-wave relationships. The following ac sine-wave values are often converted from one value to the other.

$$\text{Effective value (rms)} = 0.707 \times \text{peak value}$$

$$\text{Average value} = 0.636 \times \text{peak value}$$

$$\text{Peak value} = 1.41 \times \text{rms value}$$

$$\text{Peak-to-peak value} = 2.82 \times \text{rms value}$$

Solve the following problems. Record all answers on a sheet of paper. Find the following *effective* values.

Peak Voltage	*Rms Voltage*	*Peak Voltage*	*Rms Voltage*
1. 4 V ac =	______ V rms	5. 1.5 V ac =	______ V rms
2. 12 V ac =	______ V rms	6. 2 V ac =	______ V rms
3. 6 V ac =	______ V rms	7. 5 V ac =	______ V rms
4. 10 V ac =	______ V rms	8. 9 V ac =	______ V rms

Find the following *peak* and *peak-to-peak* (*p-p*) values.

Rms Voltage Peak Voltage	*P-P Voltage*
9. 3 V ac = ______	V peak; 10. ______ V p-p
11. 8 V ac = ______	V peak; 12. ______ V p-p
13. 7 V ac = ______	V peak; 14. ______ V p-p
15. 9 V ac = ______	V peak; 16. ______ V p-p
17. 10 V ac = ______	V peak; 18. ______ V p-p
19. 15 V ac = ______	V peak; 20. ______ V p-p
21. 11 V ac = ______	V peak; 22. ______ V p-p
23. 18 V ac = ______	V peak; 24. ______ V p-p

• Ac Measurements

Refer to Fig. 6–63 and determine each of the ac voltage values that correspond to the pointer location.

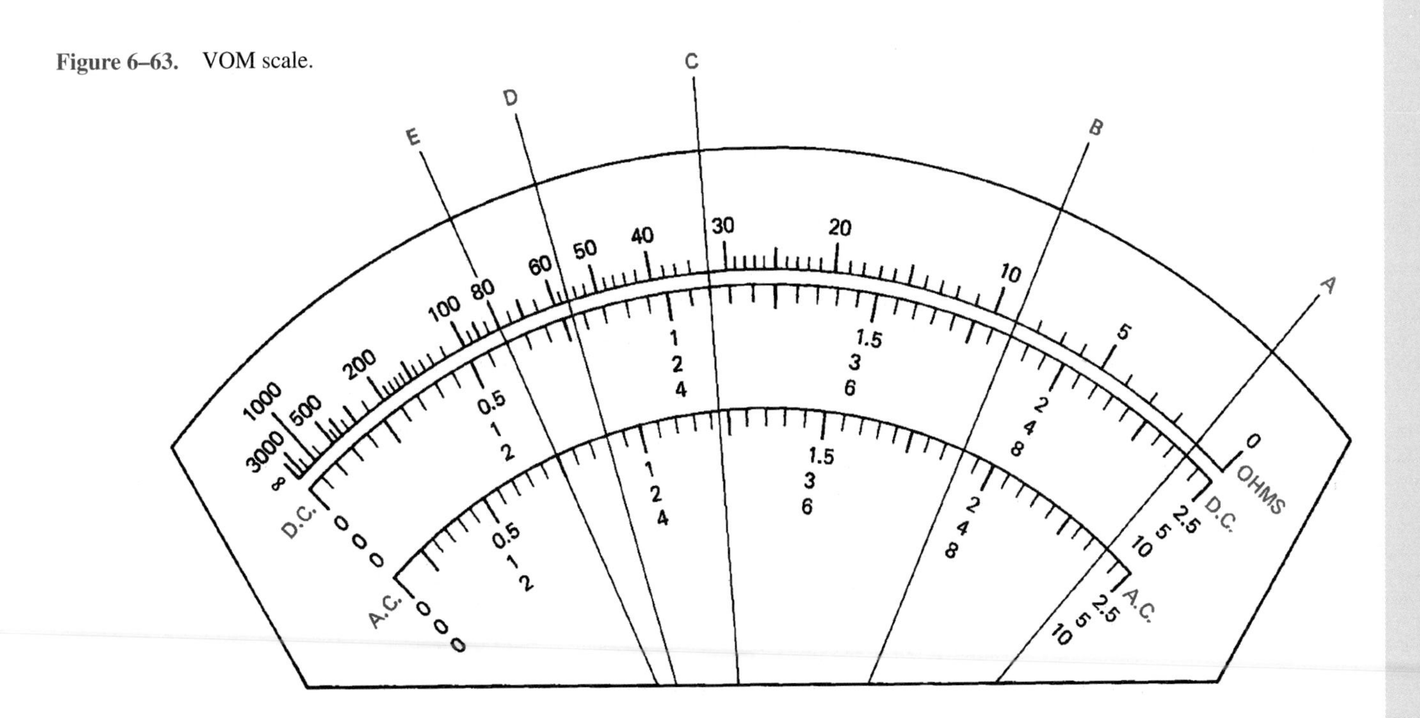

Figure 6–63. VOM scale.

Ac Volts Range	*A*	*B*	*C*	*D*	*E*
2.5 V ac	1. ____	2. ____	3. ____	4. ____	5. ____
10 V ac	6. ____	7. ____	8. ____	9. ____	10. ____
50 V ac	11. ____	12. ____	13. ____	14. ____	15. ____
250 V ac	16. ____	17. ____	18. ____	19. ____	20. ____
500 V ac	21. ____	22. ____	23. ____	24. ____	25. ____
1000 V ac	26. ____	27. ____	28. ____	29. ____	30. ____

• *Alternating-Current Circuit Problems*

For more practice in solving ac circuit problems, complete this activity. It is important to have an understanding of basic ac circuit problems. Examples of several types of ac circuit problems are shown.

1. Series inductance:

$L_1 = 2$ H, $L_2 = 5$ H, $L_3 = 6$ H

What is the inductance of this circuit? $L_T =$ ____ H.

2. Parallel inductance:

$L_1 = 2$ H, $L_2 = 4$ H, $L_3 = 8$ H

What is the inductance of this circuit? $L_T =$ ____ H.

3. Inductive reactance:

$L = 2$ H, $f = 60$ Hz

What is the inductive reactance? $X_L =$ ____ Ω.

4. Series capacitance:

$C_1 = 2\ \mu$F, $C_2 = 4\ \mu$F, $C_3 = 20\ \mu$F

What is the capacitance? $C_T =$ ____ μF.

5. Parallel capacitance:

$C_1 = 20\ \mu$F, $C_2 = 40\ \mu$F, $C_3 = 65\ \mu$F

What is the capacitance? C_T ____ μF.

6. Capacitive reactance:

$C = 20\ \mu$F, $f = 20$ Hz

What is the capacitive reactance? $X_C =$ ____ Ω.

7. *RL* time constant:

$R = 4\ \Omega$, $L = 20$ H

What is the time constant? $t =$ ____ s.

8. *RC* time constant:

$R = 50\ \Omega$, $C = 1\ \mu$F

What is the time constant? $t =$ ____ s.

9. Impedance in *RC* circuit:

$R = 10\ \Omega$, $X_C = 10\ \Omega$

What is the impedance? $Z =$ ____ Ω.

10. Impedance in *RLC* circuits:

$R = 30\ \Omega$, $X_L = 250\ \Omega$, $X_C = 210\ \Omega$

What is the impedance? $Z =$ ____ Ω.

☐ Inductance and Inductive Reactance Problems

Solve the following problems.

1. Total inductance in series: $L_1 = 2$ H, $L_2 = 3$ H, $L_3 = 2$ H. $L_T =$ ___ H.
2. Total inductance in parallel: $L_1 = 2$ H, $L_2 = 3$ H, $L_3 = 8$ H. $L_T =$ ___ H.

Inductive reactance (X_L) for the following:

3. 30 Hz, 8 H, $X_L =$ _____ Ω.
4. 50 Hz, 2 H, $X_L =$ _____ Ω.
5. 60 Hz, 1.0 H, $X_L =$ _____ Ω.
6. 40 Hz, 3 H, $X_L =$ _____ Ω.

Mutual inductance in series to increase inductance (fields aiding):

7. $L_1 = 2$ H, $L_2 = 5$ H, $M = 0.55$
   $L_T =$ _____ H.
8. $L_1 = 3$ H, $L_2 = 2$ H, $M = 0.35$
   $L_T =$ _____ H.

Mutual inductance in series to decrease inductance (fields opposing):

9. $L_1 = 4$ H, $L_2 = 3$ H, $M = 0.85$
   $L_T =$ _____ H.
10. $L_1 = 3$ H, $L_2 = 3$ H, $M = 0.4$
    $L_T =$ _____ H.

Compute the problems of 7, 8, 9, and 10 when the inductors are connected in parallel:

11. $L_T =$ _____ H. (for No. 7 values)
12. $L_T =$ _____ H. (for No. 8 values)
13. $L_T =$ _____ H. (for No. 9 values)
14. $L_T =$ _____ H. (for No. 10 values)

☐ Inductance and Inductive Reactance

Complete Activity 6–3, page 123 in the manual.

ANALYSIS

1. What factors determine the inductance of an inductor?
2. What factors determine the value of inductive reactance?
3. Why does inductance cause ac current to lag voltage?
4. What is the total inductance when inductors of 4 H and 3 H are connected in series with no mutual inductance factor?
5. What is the total inductance when inductors of 4 H and 3 H are connected in parallel with no mutual inductance factor?
6. Assume the inductors in question 4 were connected to aid with a mutual inductance of 0.6 H. What is the total inductance?
7. Assume the inductors in question 4 were connected to oppose with a mutual inductance of 0.86 H. What is the total inductance?
8. Assume the inductors in question 5 were connected to aid with a mutual inductance of 0.2 H. What is the total inductance?

9. Assume the inductors in question 5 were connected to oppose with a mutual inductance of 0.9 H. What is the total inductance?

10. What is mutual inductance?

11. If two inductors valued at 8 H and 10 H were connected in parallel, which would allow the most ac current if the ac frequency was 1000 Hz?

12. Why will an inductor oppose ac more than dc for any given voltage?

• *Capacitance and Capacitive Reactance Problems*

Solve the following problems.

- Compute the capacitance for the following capacitors, using the formula

$$C = \frac{0.0885KA}{D}$$

where C is capacitance in picofarads (pF), K is the dielectric constant, A is the plate area in square centimeters, and D is the dielectric thickness in centimeters.

1. Dielectric constant = 3
Size of plate = 2 × 200 cm
Thickness of dielectric = 0.03 cm

$C =$ _____ pF

2. Dielectric constant = 3
Size of plate = 3 × 200 cm
Thickness of dielectric = 0.04 cm

$C =$ _____ pF

3. Dielectric constant = 2
Size of plate = 0.25 × 275 cm
Thickness of dielectric = 0.02 cm

$C =$ _____ pF

- Compute the following for the total capacitance in a series circuit.

4. $C_1 = 10\ \mu F$
$C_2 = 10\ \mu F$
$C_3 = 30\ \mu F$

$C_T =$ _____ μF

5. $C_1 = 40\ \mu F$
$C_2 = 20\ \mu F$
$C_3 = 20\ \mu F$

$C_T =$ _____ μF

6. $C_1 = 80\ \mu F$
$C_2 = 60\ \mu F$
$C_3 = 80\ \mu F$

$C_T =$ ______ μF

7. Compute the capacitive reactance for an ac circuit of 60 Hz with a 30-μF capacitor:

$X_C =$ ______ Ω

- Compute the following for *total capacitance* in *parallel.*

8. $C_1 = 80\ \mu F$
$C_2 = 60\ \mu F$
$C_3 = 80\ \mu F$

$C_T =$ ______ μF

9. $C_1 = 50\ \mu F$
$C_2 = 20\ \mu F$
$C_3 = 30\ \mu F$

$C_T =$ ______ μF

10. $C_1 = 70\ \mu F$
$C_2 = 50\ \mu F$
$C_3 = 50\ \mu F$

$C_T =$ ______ μF

• Capacitance and Capacitive Reactance

Complete Activity 6–4, page 127 in the manual.

• • ANALYSIS

1. What is capacitive reactance?
2. What is meant by a capacitor's working voltage?
3. What variables determine the capacitance of a capacitor?
4. If you connected in series two capacitors, valued at 4 μF each, what would be their total capacitance?
5. If you connected the two capacitors of question 4 in parallel, what would be their total capacitance?
6. What is the relation between ac frequency and X_C?
7. How does capacitance affect the phase relation between alternating current and voltage?
8. How does the relation between ac frequency and X_C compare with the relation between ac frequency and X_L?

• Series and Parallel Ac Circuit Problems

Solve each of the following series and parallel ac circuit problems. Record all answers and calculations on a sheet of paper.

- A 20-μF capacitor and a 1000-Ω resistor are connected in series with a 120-V, 60-Hz ac source. Find:

1. X_C = _____ Ω.

2. Z = _____ Ω.

3. I = _____ A = _____ mA.

4. V_C = _____ V.

5. V_R = _____ V.

6. True power = _____ W = _____ mW.

7. Apparent power = _____ W = _____ mW.

8. Reactive power = _____ W = _____ mW.

9. Power factor = _____ = _____ %.

10. Draw an impedance triangle, a voltage triangle, and a power triangle for the previous circuit and label each value.

- A parallel ac circuit converts 12,000 W of power. The applied voltage is 240 V and the total current is 72 A. Find:

11. Apparent power = _____ W = _____ mW.

12. Power factor = _____ = _____ %.

13. Phase angle = _____ °.

14. Draw a power triangle for the previous circuit and label each value.

- A series circuit with 20 V, 60 Hz applied has a resistance of 100 Ω, a capacitance of 40 μF, and an inductance of 0.15 H. Find:

15. X_C = _____ Ω.

16. X_L = _____ Ω.

17. X_T = _____ Ω.

18. Z = _____ Ω.

19. I = _____ A = _____ mA.

20. V_C = _____ V.

21. V_L = _____ V.

22. V_R = _____ V.

23. Phase angle = _____ °.

24. Draw an impedance triangle and a voltage triangle for the previous circuit and label each value.

- Use the same values that were given for *R*, *C*, and *L* in the problem above. Connect them in a parallel circuit that has 10 V, 60 Hz applied to it. Find:

25. I_R = _____ A = _____ mA.

26. I_C = _____ A = _____ mA.

27. I_L = _____ A = _____ mA.

28. I_T = _____ A = _____ mA.

29. Z = _____ Ω.

30. Conductance (G) = _____ Siemens = _____ mS = _____ μS.

31. Susceptance (B_T) = _____ Siemens = _____ mS = _____ μS.

32. Admittance (Y) = _____ Siemens = _____ mS = _____ μS.

33. Phase angle (θ).

34. Draw a current triangle and an admittance triangle for the previous circuit and label each value.

☐ *Series RL Circuits*

Complete Activity 6–5, page 131 in the manual.

ANALYSIS

1. Why must the voltages in a series *RL* circuit be added by using a right triangle?
2. Determine the phase angle of Fig. 6–64 by using the following trigonometric methods.
   a. Phase angle (θ) = inverse tangent (X_L/R) = _____; θ = _____.
   b. Phase angle (θ) = inverse tangent (V_L/V_R) = _____; θ = _____.
   c. Phase angle (θ) = inverse tangent (var/watts) = _____; θ = _____.
3. Determine the power factor of the circuit of Fig. 6–64 as:

$$F = \frac{\text{true power}}{\text{apparent power}} = \_\_\_\_\_\_\_\_$$

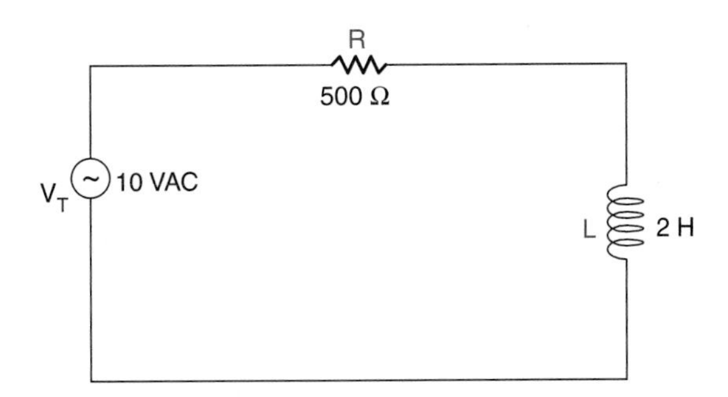

Figure 6–64. Series RL circuit.

4. Determine the cosine of the phase angle (θ) for the circuit of Fig. 6–64. This value should be equal to the circuit power factor PF = cosine θ: cosine θ = _____.
5. What is meant by power factor?
6. Make the necessary calculations and sketch the following triangles for the circuit of Fig. 6–64: (a) voltage triangle, (b) impedance triangle, (c) power triangle.

☐ *Series RC Circuits*

Complete Activity 6–6, page 133 in the manual.

ANALYSIS

1. Determine the phase angle of the circuit of Fig. 6–65 by using the following trigonometric methods:
   a. Phase angle (θ) = inverse tangent (X_C/R) = _____; θ = _____.
   b. Phase angle (θ) = inverse tangent (V_C/V_R) = _____; θ = _____.
   c. Phase angle (θ) = inverse tangent (watts/*VA*) = _____; θ = _____.

2. Determine the power factor of the circuit of Fig. 6–65 as:

$$\text{PF} = \frac{\text{true power}}{\text{apparent power}} = \_\_\_\_\_\_\_\_$$

3. Determine the cosine of the phase angle. This value should equal the power factor of the circuit of Fig. 6–65 cosine θ = _____.

4. Make the necessary calculations and sketch the following triangles for the circuit of Fig. 6–65: (a) voltage triangle, (b) impedance triangle, (c) power triangle.

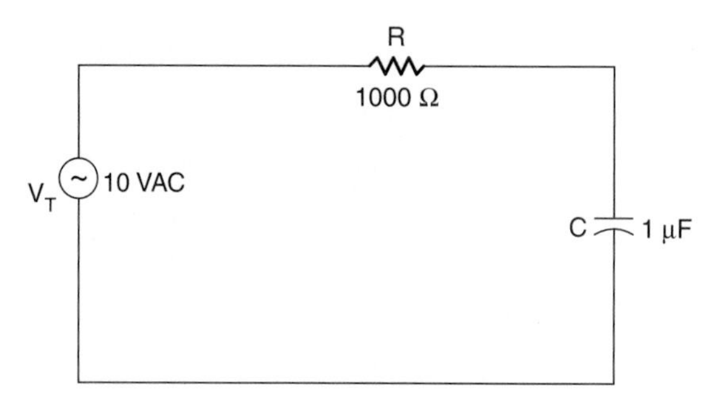

Figure 6–65. Series RC circuit.

Series RLC Circuits

Complete Activity 6–7, page 135 in the manual.

ANALYSIS

1. Determine the phase angle of the circuit of Fig. 6–66 by using the following trigonometric methods:

a. Phase angle (θ) = inverse tangent (X_T/R) = _____.

b. Phase angle (θ) = inverse tangent (V_X/V_R) = _____.

c. Phase angle (θ) = inverse tangent (VAR_T/watts) = _____.

2. Determine the power factor of the circuit of Fig. 6–66 as

$$\text{PF} = \frac{\text{true power}}{\text{apparent power}} = \_\_\_\_\_\_\_\_$$

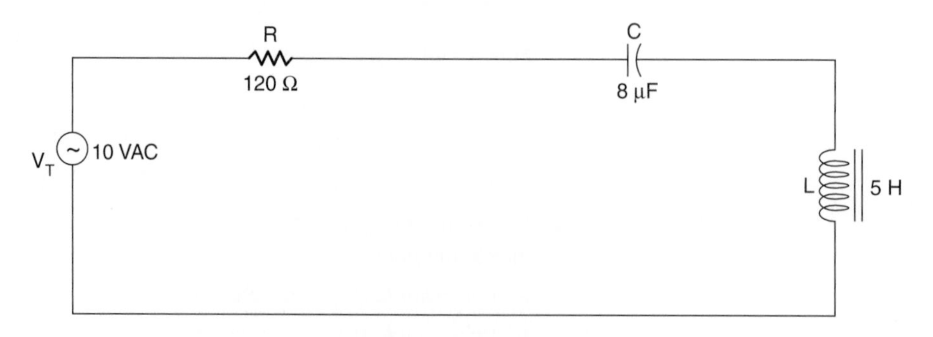

Figure 6–66. Series RLC circuit.

3. Determine the cosine of the phase angle of the circuit of Fig. 6–66. This value should equal the circuit power factor: cosine θ = _____.
4. How is power factor correction (reduction of phase angle by adding capacitance) accomplished in an inductive circuit?
5. As total reactive power (VAR_T) increases, the phase angle (θ) = _____.
6. Make the necessary calculations and sketch the following triangles for the circuit of Fig. 6–66: (a) voltage triangle, (b) impedance triangle, (c) power triangle.

• *Parallel RL Circuits*

Complete Activity 6–8, page 139 in the manual.

• • ANALYSIS

1. Determine the phase angle of the circuit of Fig. 6–67 by the following methods:
   a. Phase angle (θ) = inverse cosine (I_R/I_T) = _____; θ = _____.
   b. Phase angle (θ) = inverse sine (var/VA) = _____; θ = _____.
2. Determine the power factor of this circuit:

$$PF = \frac{\text{true power}}{\text{apparent power}} = \_\_\_\_\_\_\_\_$$

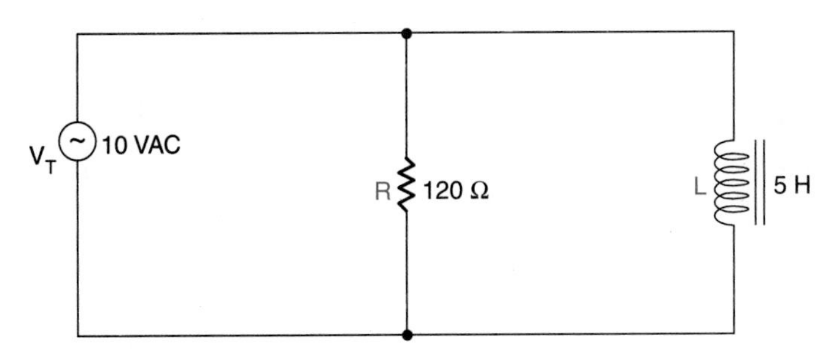

Figure 6–67. Parallel RL circuit.

3. Compare the power factor value of question 2 with the value of cosine θ from question 1(a).
4. Define the following terms associated with parallel ac circuits:
   a. Admittance
   b. Conductance
   c. Inductive susceptance
   d. Capacitive susceptance
5. Make the necessary calculations and sketch the following triangles for the circuit of Fig. 6–67: (a) current triangle, (b) admittance triangle, (c) power triangle.

• *Parallel RC Circuits*

Complete Activity 6–9, page 141 in the manual.

ANALYSIS

1. Determine the phase angle (θ) of the circuit of Fig. 6–68 using the following methods:
   a. Phase angle: θ = inverse sine (I_C/I_T) = _____; θ = _____.
   b. Phase angle: θ = inverse tangent (VAR/watts) = _____; θ = _____.
2. What is the value of:
   a. Cosine θ = _____ (use correct triangle values).
   b. Power factor = $\frac{\text{true power}}{\text{apparent power}}$ = _____.
3. Suppose that a 5-H inductor has been connected in parallel with the circuit of Fig. 6–68, and compute the following:
   a. $X_L = 2\pi fL$ = _____ Ω.
   b. $I_L = V_L/X_L$ = _____ mA.
   c. $I_X = I_L - I_C$ = _____ mA.
   d. $I_T = \sqrt{I_R^2 + I_X^2}$ = _____ mA.
   e. $Z = V_S/I_T$ = _____ Ω.
   f. θ = inverse sine (I_X/I_T) = _____ °.
   g. PF = cosine θ = _____.
4. How do the values of power factor for the original circuit and that of question 3 differ? Why?
5. How do the values of total current differ in the two circuits? Why?
6. Make the necessary calculations and sketch the following triangles for the circuit of Fig. 6–68: (a) current triangle, (b) admittance triangle, (c) power triangle.

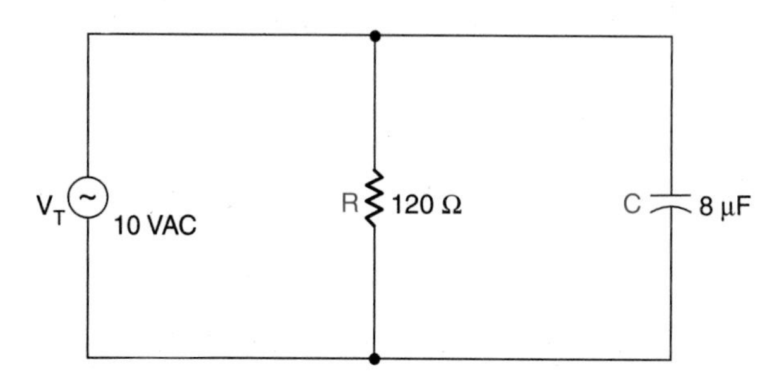

Figure 6–68. Parallel RC circuit.

Resonant Circuit Problems

1. Calculate the resonant frequency (f_r) for the circuit of Fig. 6–69.
2. Calculate the quality factor (Q) of the circuit.

$$Q = \frac{X_L}{R} = \_\_\_\_\_$$

3. Calculate the bandwidth of the circuit.

$$\text{BW} = \frac{f_r}{Q} = \_\_\_\_\_ \text{ Hz}$$

4. How do the values of V_R, V_L, and V_C vary as the frequency is moved to values *above* and *below* resonance?

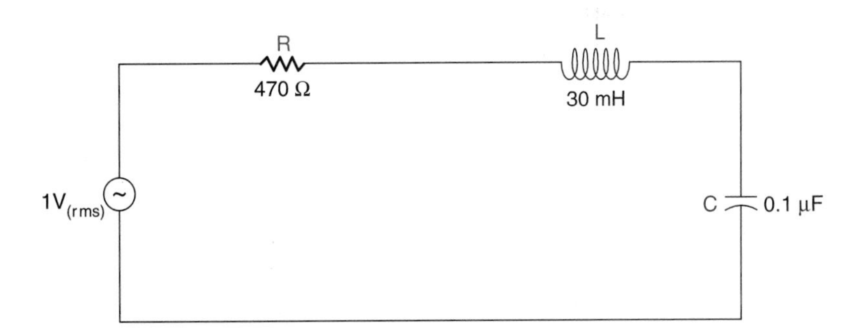

Figure 6–69. Series resonant circuit.

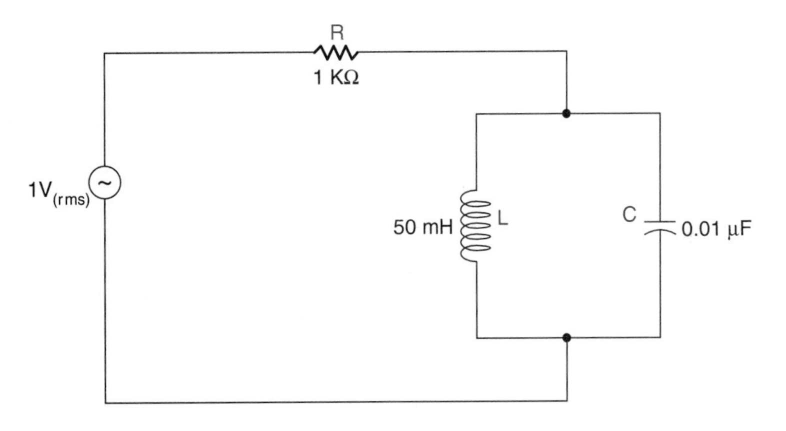

Figure 6–70. Parallel resonant circuit.

5. Complete the following statement: In a series resonant circuit with the resonant frequency applied to the circuit, impedance is _____, current is _____, phase angle is _____, $X_L =$ _____, and $R =$ _____.

6. What effect does changing resistance have on a series resonant circuit?

7. Calculate the resonant frequency (f_r) for the circuit of Fig. 6–70.

8. Calculate the impedance of the circuit at resonance by using the following formula:

$$Z_r = \frac{V_T}{I_T} = \_\_\_\_\_\ \Omega$$

9. Calculate the quality factor of the circuit:

$$Q = \frac{R}{X_L} = \_\_\_\_\_$$

10. Calculate the bandwidth of the circuit:

$$\text{BW} = \frac{f_r}{Q} = \_\_\_\_\_\ \text{Hz}$$

11. How do the values of I_R, I_L, and I_C vary as the frequency is moved to *above* and *below* resonance?

12. Complete the following statement: In a parallel resonant circuit at resonance, the impedance is _____, I_R is _____, the phase angle is _____, X_L is _____, and R is _____.

• Series Resonant Circuits

Complete Activity 6–10, page 143 in the manual.

• • ANALYSIS

1. Describe how to use a voltmeter to determine resonant frequency of a circuit.
2. Calculate the quality factor (Q) of the circuit of Fig. 6–71.

$$Q = \frac{X_L}{R} = \_\_\_\_\_\_$$

3. Calculate the bandwidth of the circuit of Fig. 6–71.

$$\text{BW} = \frac{f_r}{Q} = \_\_\_\_\_\_ \text{ Hz}$$

4. How do the values of V_R, V_L, and V_C vary as the frequency is moved to values *above* and *below* resonance?
5. Complete the following statement: In a series resonant circuit with the resonant frequency applied to the circuit, impedance is ______, current is ______, phase angle is ______, X_L = ______, and R = ______.
6. What effect does increasing resistance have on a series resonant circuit?

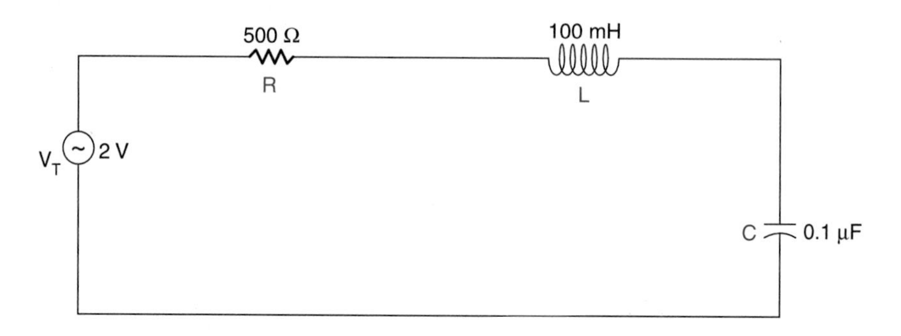

Figure 6–71. Series resonant circuit.

• Parallel Resonant Circuits

Complete Activity 6–11, page 145 in the manual.

• • ANALYSIS

1. Calculate the impedance of the circuit of Fig. 6–72 at resonance by using the following formula:

$$Z_r = \frac{E_T}{I_T} = \_\_\_\_\_\_ \ \Omega$$

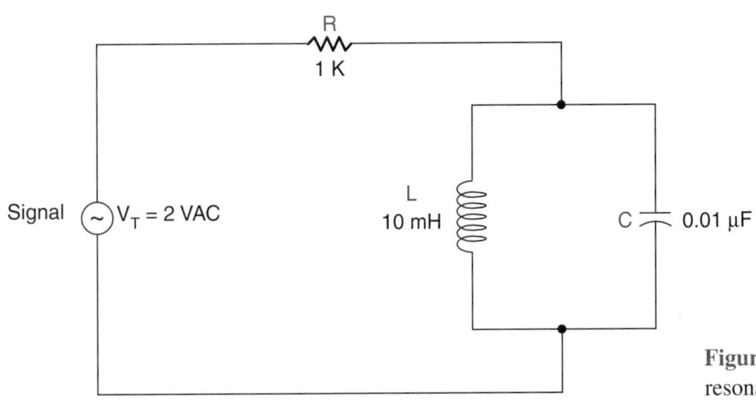

Figure 6–72. Parallel resonant circuit.

2. Calculate the quality factor of the circuit:

$$Q = \frac{R}{X_L} = \_\_\_\_\_$$

3. Calculate the bandwidth of the circuit:

$$\text{BW} = \frac{f_r}{Q} = \_\_\_\_\_ \text{ Hz}$$

4. How do the values of I_R, I_L, and I_C vary as the frequency is moved to *above* and *below* resonance?
5. Complete the following statement: In a parallel resonant circuit at resonance, the impedance is _____, I_R is _____, the phase angle is _____, X_L is _____, and R is _____.

• *Transformer Operation*

1. Obtain a transformer and a VOM. Record all data on a sheet of paper.
2. Identify the primary windings of the transformer. The primary will be connected to an ac voltage source. Plug it in when you are ready to make the necessary measurements. Remove the power cord when the transformer is not in use. Be aware of safety—it is best to make measurements with a low ac voltage applied to the primary (from 6 to 15 V). If a 120-V ac source is used, take extreme caution.
3. Some transformers have one secondary winding. Many transformers have three or more secondary windings. Count the number of secondary windings on the transformers. Number of secondary windings = _____.
4. Make sure that none of the wires from the secondary windings touch.
5. Prepare the VOM to measure ac voltage. Begin with the highest ac voltage range for each secondary voltage measurement.
6. Carefully connect the primary winding to an ac source. Measure the voltage across each secondary winding. Record these voltages.
7. Disconnect the ac voltage from the transformer primary. With no voltage applied, make the following resistance measurements with a VOM.
   a. Resistance of primary winding = _____ Ω.
   b. Resistance of each secondary winding = _____ Ω.
8. Turn in your data to the teacher for grading.

• *Transformer Problems*

Solve the following problems which deal with transformer operation. Record your answers on a sheet of paper.

1. A 1500-turn primary winding has 120 V applied. The secondary voltage is 20 V. How many turns of wire does the secondary winding have?
2. When a 200-Ω resistance is placed across a 240-V secondary winding, the primary current is 30 A. What is the value of the primary voltage of the transformer?
3. A transformer has a 20:1 turns ratio. The primary voltage is 4800 V. What is the secondary voltage?
4. A transformer has a 2400-V primary winding and a 120-V secondary. The primary winding has 1000 turns. How many turns of wire does the secondary winding have?
5. A 240- to 4800-V transformer has a primary current of 95 A and a secondary current of 4 A. What is its efficiency?
6. A single-phase transformer has a 20-kVA power rating. Its primary voltage is 120 V and its secondary voltage is 480 V. What are the values of its maximum primary current and secondary current?

• *Transformer Analysis*

Complete Activity 6–12, page 147 in the manual.

• • ANALYSIS

1. What conclusion may be drawn from a winding resistance increase concerning current capacity for a transformer secondary winding?
2. What is the relationship between primary power and secondary power in a multiple secondary transformer?
3. Discuss the following characteristics of transformers:
   a. Winding resistance
   b. Volt-amp rating
   c. Cove construction
   d. Magnetic coupling
   e. Turns ratio
   f. Voltage ratio

• *Self-Examination*

1. Complete the Self-Examination for Chap. 6, page 151 in the manual.
2. Check your answers on pages 412–414 in the manual.
3. Study the questions which were answered incorrectly.

• *Chapter 6 Examination*

Complete the Chap. 6 Examination on page 159 in the manual.

CHAPTER

7 ELECTRICAL ENERGY CONVERSION

OUTLINE

7.1 Lighting Systems
7.2 Heating Systems
7.3 Mechanical Loads (Motors)
7.4 Direct-Current Motors
7.5 Three-Phase Ac Motors
7.6 Single-Phase Ac Motors
7.7 Synchro Systems and Servo Systems
7.8 Motor Performance
7.9 Motor Control Basics

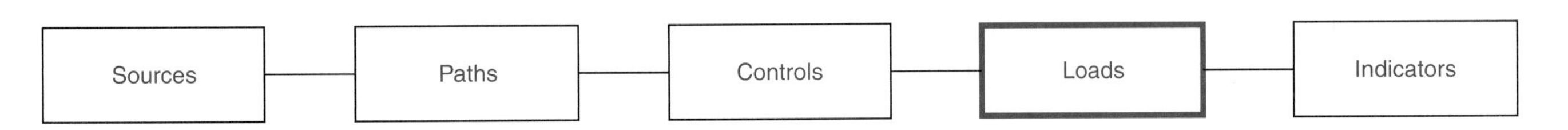

***Loads* convert electrical energy into another form.**

TABLE 7–1 Lighting Terms

Lighting Quantity	*Definition*	*Symbol*	*Unit* *SI*	*English*	*Definition of Unit*
Luminous intensity (candlepower)	Ability of source to produce light in a given direction	*I*	Candela (cd)		Approximately equal to the luminous intensity produced by a standard candle
Luminous flux	Total amount of light	ϕ	Lumen (lm)		Luminous flux emitted by a 1 candela uniform point source
Illumination	Amount of light received on a unit area of surface (density)	*E*	Lux (lx)	Footcandle (fc)	One lumen equally distributed over one unit area of surface
Luminance (brightness)	Intensity of light per unit of area reflected or transmitted from a surface	*L*	cd/m² (foot-lambert)	cd/in.²	A surface reflecting or emitting light at the rate of 1 candela per unit of area

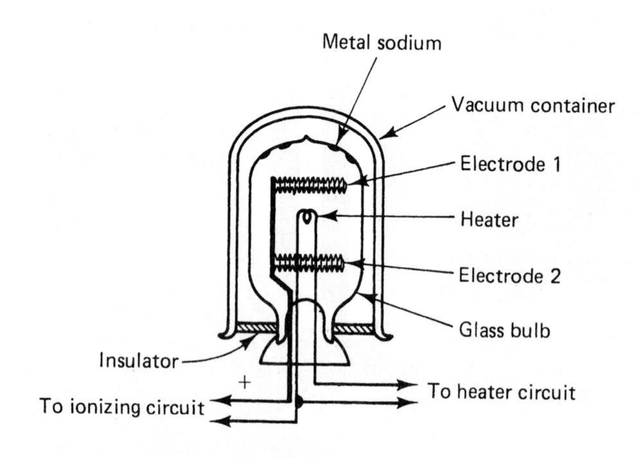

Figure 7–6. Sodium vapor lamp.

the mercury in the inner tube to ionize. When the mercury has ionized, greenish-blue light is produced.

Mercury vapor lamps are long lasting and easy to maintain. They are used in industry and commercial buildings to provide intense light outputs. Mercury vapor lamps may be used for outdoor lighting.

The sodium vapor lamp shown in Fig. 7–6 is popular for outdoor industrial and commercial lighting and for highway lighting. This lamp contains low-pressure neon gas and sodium. When electrical current passes through a heater, electrons are given off. An ionizing circuit causes positive charges to be placed on the positive electrodes. Electrons pass from the heater to the positive electrodes. This causes the neon gas to ionize. The neon produces enough heat to cause the sodium to ionize. A yellowish light is then produced by the sodium vapor. Sodium vapor lamps can produce about three times as much light as incandescent lamps of the same wattage.

TABLE 7–2 Comparison of Light Sources

Lamp Type	*Efficacy (Lumens/watt)*
200-watt incandescent lamp	20
400-watt mercury lamp	50
40-watt fluorescent lamp	70
400-watt metal halide lamp	75
400-watt high-pressure sodium lamp	110

The types of lighting loads discussed in this section are three of the most common—incandescent, fluorescent, and vapor lamps—used for many lighting applications. A few other specialized types of lighting are also used.

Units of light measurement are expressed in both English and metric units, which can be somewhat confusing. Table 7–1 summarizes some common terms used for light measurement.

The purpose of a light source is to convert electrical energy into light energy. A measure of how well this is done is called *efficacy.* Efficacy is the lumens of light produced per watt of electrical power converted. A comparison of some different types of light sources is shown in Table 7–2.

7.2 HEATING SYSTEMS

Most electrical loads produce a certain amount of heat. Heating is caused when current flows through a resistive device.

Figure 7–7. Resistance heating element. (Courtesy of Williamson Co.)

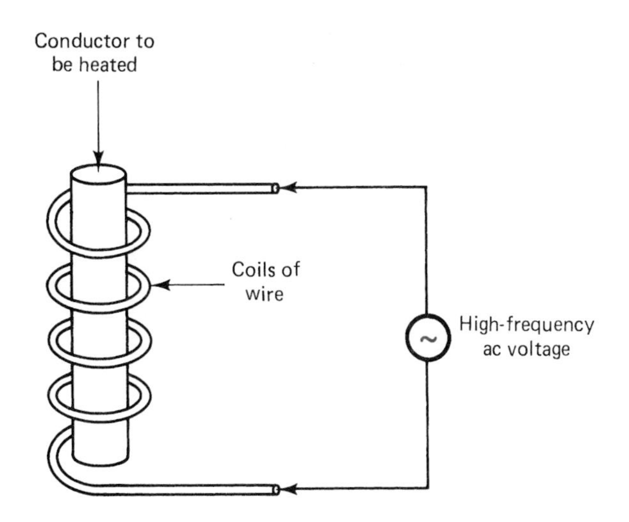

Figure 7–8. Principle of induction heating.

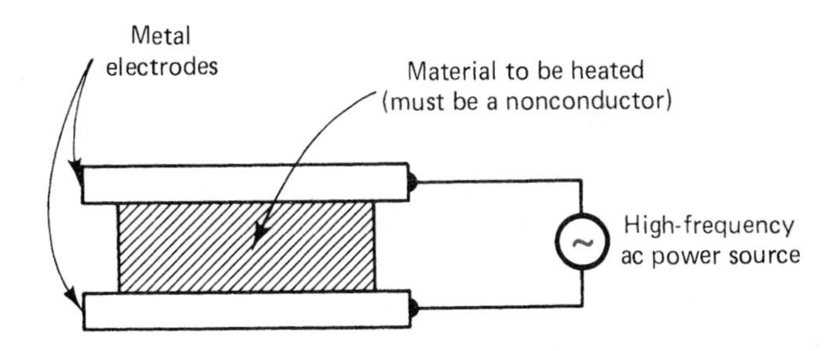

Figure 7–9. Dielectric (capacitive) heating.

In some equipment, heat causes a power loss. Lamps, for example, produce heat energy and light energy. Several types of loads are mainly heating loads—their main purpose is to convert electrical energy into heat energy. These load devices include resistance heating, inductive heating, and dielectric (capacitive) heating.

Resistance Heating

Heat energy is produced when electrical current flows through a resistive material. In some cases, the heat energy produced is undesirable. Resistance heating produces useful heat that can be transferred from a resistive material to where it is used. The heating element has coiled resistance wire designed as a heat-conducting material. The element is contained in a metal sheath. This principle can be used to heat water, oil, or a room. Resistive heaters may be used in open air or immersed in a material that is to be heated. The heat energy produced depends on the amount of current flow and the resistance of the element. A typical application is shown in Fig. 7–7, where resistance heating coils are used as supplemental elements for a heat pump unit.

Induction Heating

The principle of induction heating is shown in Fig. 7–8. Heat is produced in magnetic materials when alternating current is applied. In the example, current is induced into the material to be heated. Electromagnetic induction is caused by the application of alternating current to the heating coil. The material to be heated must be a conductor. Current is induced into the conductor. A high-frequency ac source is used to produce higher heat output. The high-speed magnetic field created by the high-frequency ac source moves across the material to be heated. The induced voltage causes circulating currents called *eddy currents* to flow in the material. Heat is produced due to the material's resistance to the flow of eddy currents. The heat produced by this method is used for many industrial processes, such as heating metals.

Dielectric (Capacitive) Heating

Induction heating is used only for conductive materials. Another method must be used to heat nonconductive materials. Dielectric or capacitive heating is shown in Fig. 7–9. Nonconductors may be heated by placing them in an electrostatic field between two metal electrodes. The electrodes are supplied by a high-frequency ac source. The material to be heated is like the dielectric of a capacitor. The metal electrodes are like two plates of a capacitor.

High-frequency ac is applied to the dielectric heater. The high-frequency ac voltage causes the atomic structure of the dielectric material to become distorted. As frequency increases, the amount of distortion also increases. Internal friction is caused which produces a large amount of heat in the nonconductive material. Dielectric heating produces rapid heat that spreads throughout the heated material. Common applications of dielectric heating are for making plywood and bonding plastic sheets.

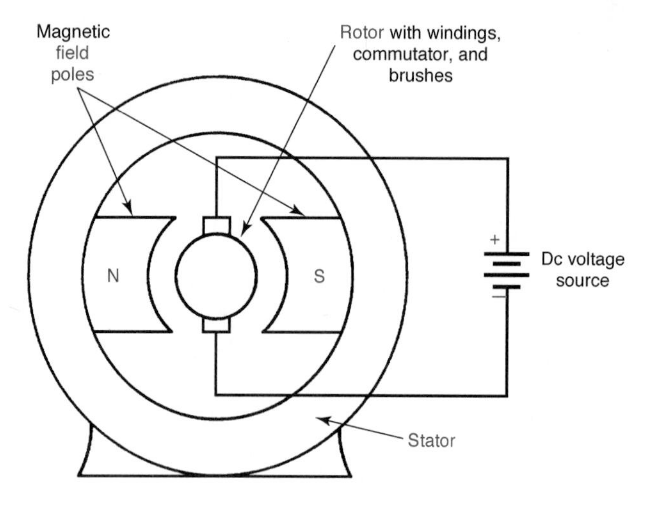

Figure 7–10. Basic parts of a dc motor.

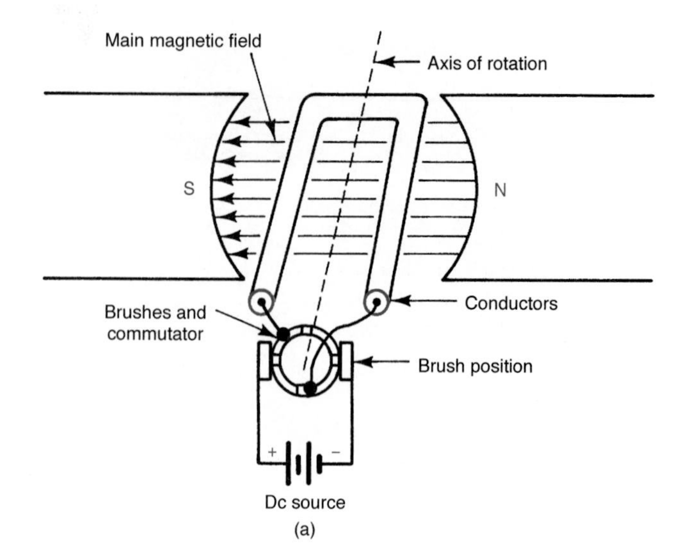

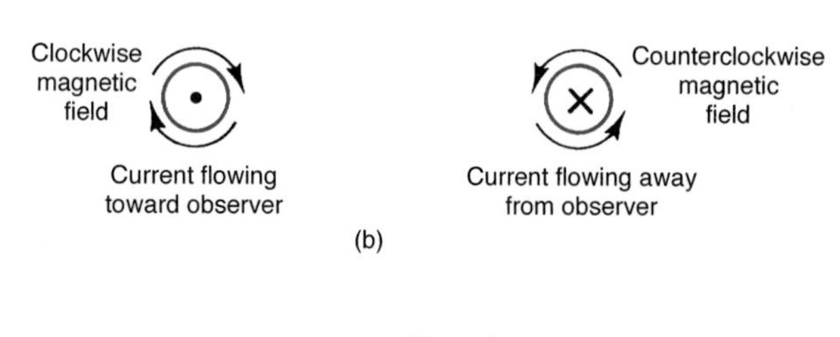

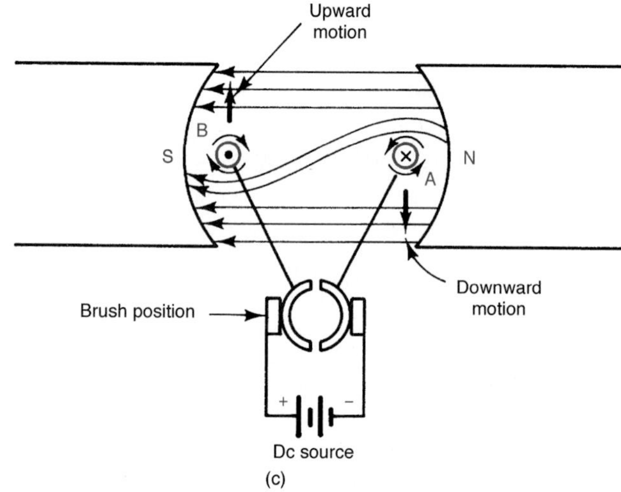

Figure 7–11. Motor principle: (a) no current flows through rotor conductor; (b) direction of circular magnetic fields around the motor conductors when current flows; (c) compression points produced at A and B when current flows through rotor conductors—causing rotation.

7.3 MECHANICAL LOADS (MOTORS)

Equipment that converts electrical energy into mechanical energy is another type of electrical load. Electric motors are the main type of mechanical load. There are many types of motors used today. Motors are our major energy-consuming load. Motors of various sizes are used for many applications.

All motors have several basic parts in common, including (1) a stator, which is the frame and other stationary parts; (2) a rotor, which is a rotating shaft and other parts that rotate; and (3) other equipment, such as a brush commutator assembly for dc motors and a starting circuit for certain ac motors. The basic parts of a dc motor are shown in Fig. 7–10. Dc motors are constructed in the same way as dc generators. Their basic parts are the same. The function of a motor is to convert electrical energy into mechanical energy. Motors produce rotary motion. A motor has an electrical input and a mechanical output. Mechanical energy is produced by a motor due to the interaction of a magnetic field and a set of conductors.

The motor principle is shown in Fig. 7–11. In Fig. 7–11(a), no current is flowing through the conductors due to the position of the brushes against the commutator, and no motion is produced at this time. When current flows through the conductors, a circular magnetic field is developed around the conductors. The direction of current flow determines the direction of the circular magnetic field, as shown in Fig. 7–11(b). The magnetic field around the conductors interacts with the main magnetic field. The interaction of these two magnetic fields results in rotary motion. The circular magnetic field around the conductors causes a compression of the main magnetic field at points A and B in Fig. 7–11(c). This compression causes the magnetic field to react in the opposite direction. Motion is produced away from points A and B. Rotary motion is produced in a clockwise direction. To change the direction of rotary motion, the direction of current flow through the conductors is reversed. This can be observed by using the right-hand rule shown in Fig. 7–12.

The rotary force produced by the interaction of two magnetic fields is called *torque* or *motor action.* The amount of torque produced by a motor depends on the strength of the main magnetic field and the amount of current flowing

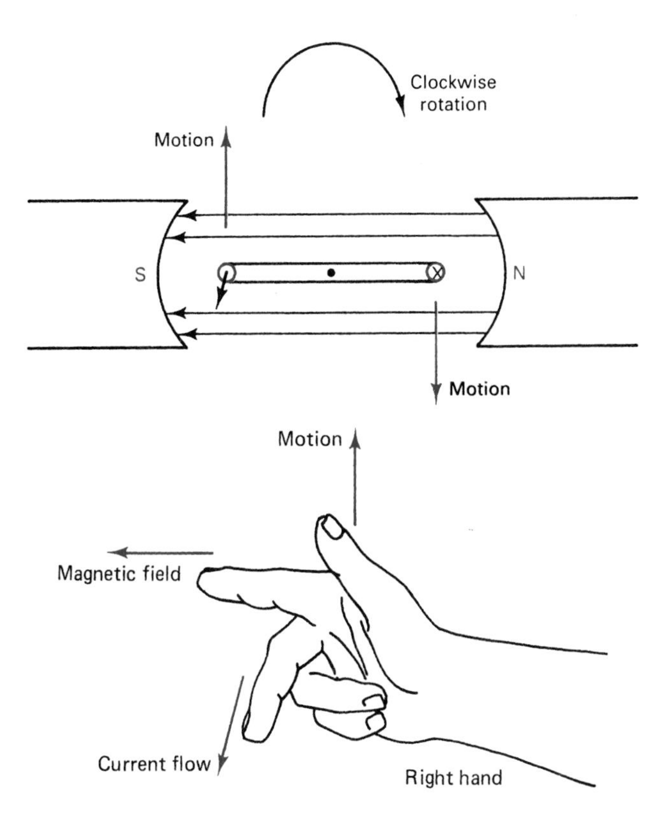

Figure 7–12. Right-hand motor rule.

through the rotor conductors. If the magnetic field or the current through the conductors increases, the amount of torque will increase.

7.4 DIRECT-CURRENT MOTORS

Motors that operate from direct-current power sources are often used. Direct-current motors are classified as series, shunt, or compound machines. The method of connecting the armature and field determines the type. Also, permanent magnet dc motors are used.

The operation of dc motors can be explained by referring to Fig. 7–13. Some important terms are used here. These terms are load, speed, counter electromotive force (CEMF), current, and torque. The amount of mechanical load applied to the shaft of a motor depends on how it is used. Load applied to the motor affects its operation. As the mechanical load increases, the speed of a motor tends to decrease. As speed decreases, CEMF induced into the rotor conductors decreases. This generated voltage (counter electromotive force) depends on the number of rotating conductors and the speed of rotation, just like a generator. The counter electromotive force produced by a motor opposes the applied voltage. The actual voltage applied to a motor increases as the CEMF decreases. When the working voltage increases, more current flows from the power source. The torque of a motor depends on current flow. Torque will increase as current increases. The opposite situation occurs when the mechanical load connected to the shaft of a motor decreases. The speed of the motor tends to increase. Increased speed causes increased CEMF. The CEMF is in opposition to the applied voltage, so the CEMF increases. Current decreases as CEMF increases. Decreased current causes a decrease in torque. Torque varies with changes in load. As the load on a motor is increased, its torque also increases. The current drawn from the power source also increases when the load is increased.

CEMF is important in motor operation. When a motor is started a large starting current flows. The starting current is much larger than the current when full speed is reached. Current flow at full speed is called running current. Maximum current flows when there is no CEMF (when the rotor is not turning). As the CEMF increases, the current flow decreases. A resistance in series with the armature circuit is often used to reduce the starting current of a motor. After the motor has reached full speed, this resistance is bypassed by automatic or manual switching systems. This allows the motor to produce maximum torque.

The torque of a motor is measured in foot-pounds. Motors are rated in terms of their horsepower. The horsepower rating of a motor is based on the amount of torque produced. Horsepower can be found by using the formula

$$\text{hp} = \frac{2\pi \times \text{speed} \times \text{torque}}{33{,}000} = \frac{\text{speed} \times \text{torque}}{5252}$$

The speed is in revolutions per minute (rpm) and the torque is in foot-pounds.

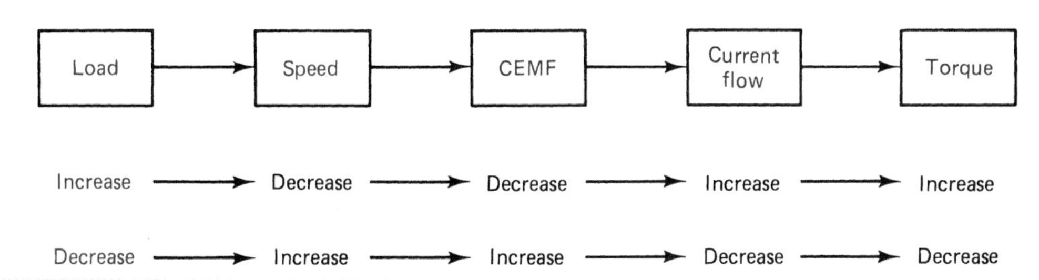

Figure 7–13. Dc motor operating characteristics.

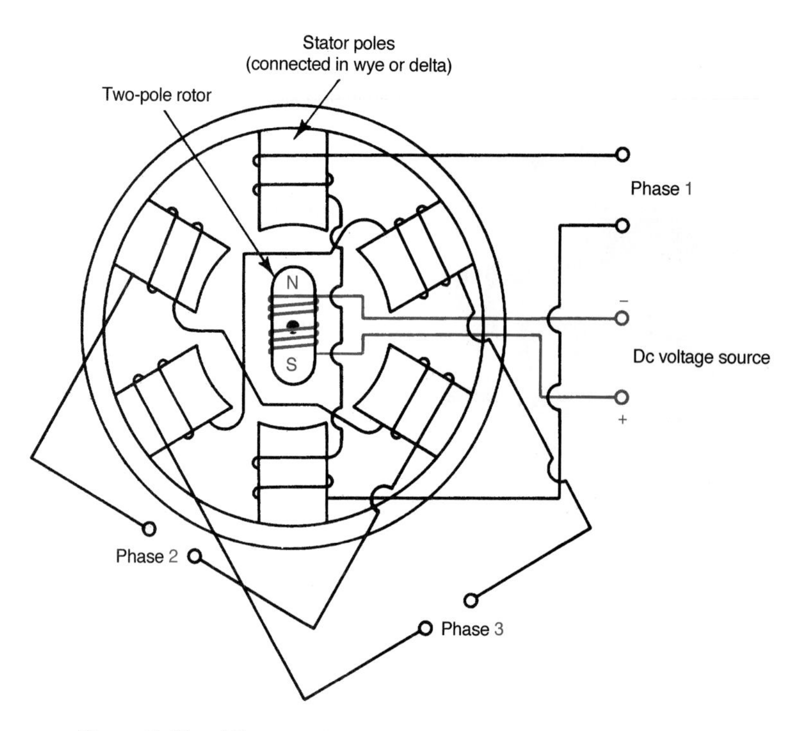

Figure 7–21. Diagram of a three-phase ac synchronous motor.

self-starting—some method must be used to start the motor. Synchronous motors are designed so that they will rotate at one speed regardless of the load. This speed is called *synchronous speed.* The speed of an ac synchronous motor may be found by using the formula

$$\text{speed (rpm)} = \frac{\text{ac frequency} \times 120}{\text{number of poles} \div 3}$$

The number of poles must be divided by 3 (the number of phases). For example, if a three-phase ac motor with 12 stator poles has 4 poles per phase ($12 \div 3 = 4$), its speed is 1800 rpm, as shown:

$$\begin{aligned} \text{Speed} &= \frac{60 \times 120}{12 \div 3} \\ &= \frac{7200}{4} \\ &= 1800 \text{ rpm} \end{aligned}$$

Synchronous ac motors have speeds based on the number of stator poles and the ac frequency.

One method of starting a three-phase ac synchronous motor is to use a small dc machine connected to its shaft, as shown in Fig. 7–22(a). It is one of the most complex types of motors in use. Listed below are the steps for starting this motor.

1. Dc power is applied to the auxiliary motor. This causes it to increase in speed.
2. Three-phase ac power is applied to the stator of the synchronous motor.
3. When motor speed nears the synchronous speed of the motor, the dc power circuit is opened. At the same time, the line terminals of the auxiliary machine are connected across the slip ring/brush assembly of the synchronous motor.
4. The auxiliary machine now converts to generator operation. It supplies dc current to the rotor of the synchronous motor. The synchronous motor now acts as a prime mover to turn the auxiliary machine.
5. When the rotor is magnetized, it will "lock in step" or "synchronize" with the revolving stator field.
6. The speed of the motor will remain the same even with changes in load as long as the rotor is magnetized.

Another starting method is shown in Fig. 7–22(b). This method uses "damper" windings. They are similar to the conductors of a squirrel-cage rotor. A damper winding is placed inside the laminated iron of the rotor. No auxiliary machine is required to start a synchronous motor when damper windings are used. The steps of operation are now simplified as follows:

1. Three-phase ac power is applied to the stator.
2. The motor operates like a three-phase induction motor due to the damper windings.
3. The rotor speed builds up to slightly less than the motor's synchronous speed. Remember that the

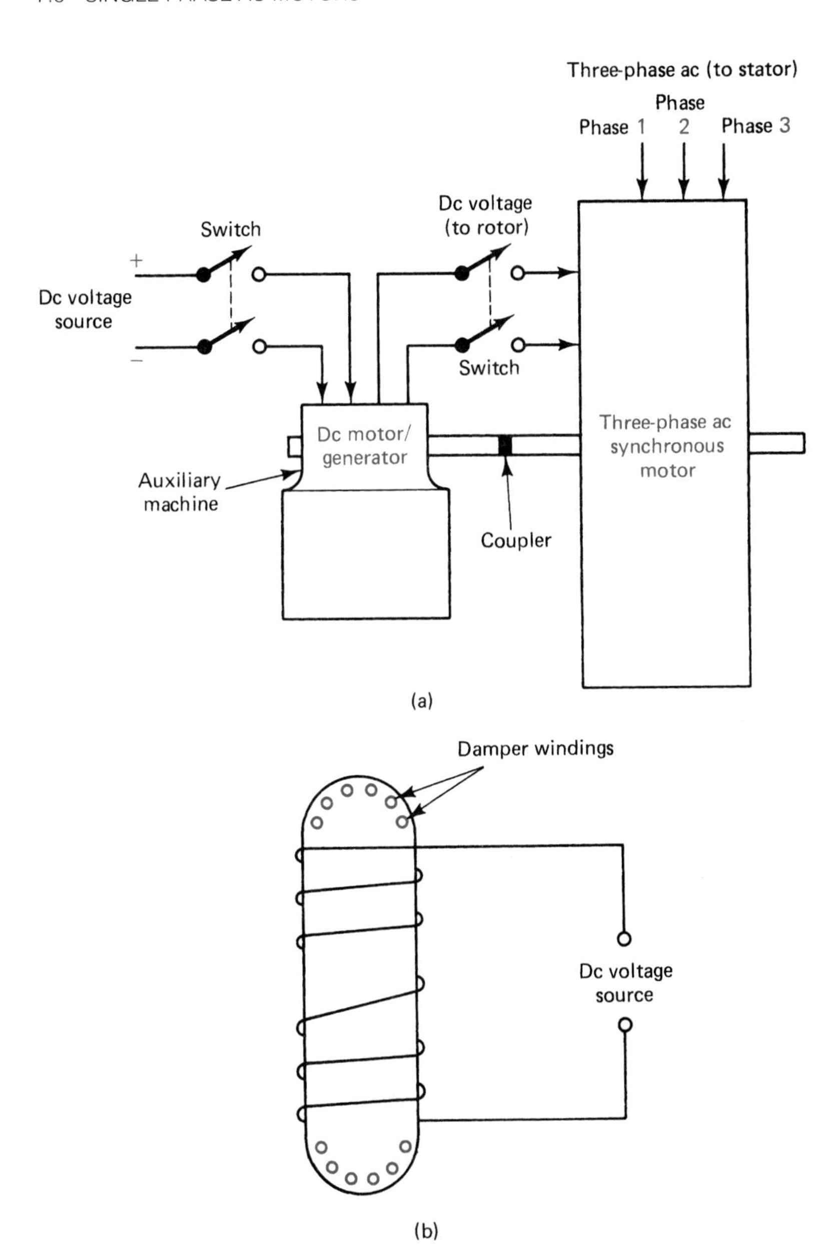

Figure 7–22. Methods of starting three-phase ac synchronous motors: (a) auxiliary dc motor/generator used to accelerate the motor; (b) squirrel-cage or damper windings placed in the rotor.

speed of an induction motor is always slower than its synchronous speed.

4. Dc power is applied to the slip ring/brush assembly.
5. The rotor becomes magnetized. The rotor then increases speed until it reaches synchronous speed.
6. The speed of the motor remains the same regardless of the load placed on the shaft of the motor.

An advantage of the three-phase ac synchronous motor is that it can be connected to a three-phase power system to increase power factor of ac circuits. Three-phase ac synchronous motors are sometimes used only to increase power factor. If no load is connected to the shaft of a three-phase ac synchronous motor, it is called a synchronous capacitor. In many cases, this motor is used for its constant-speed capability.

As load increases on a three-phase synchronous motor, the stator current will increase. The motor will remain "synchronized" (same speed) unless the load causes "pull out" to take place. The motor would then stop rotating due to excessive load. Many three-phase synchronous motors are larger than 100 hp. They are used mainly for industrial applications that require constant speed.

7.6 SINGLE-PHASE AC MOTORS

Another type of mechanical load is the single-phase ac motor. This type of motor is common for industrial, commercial,

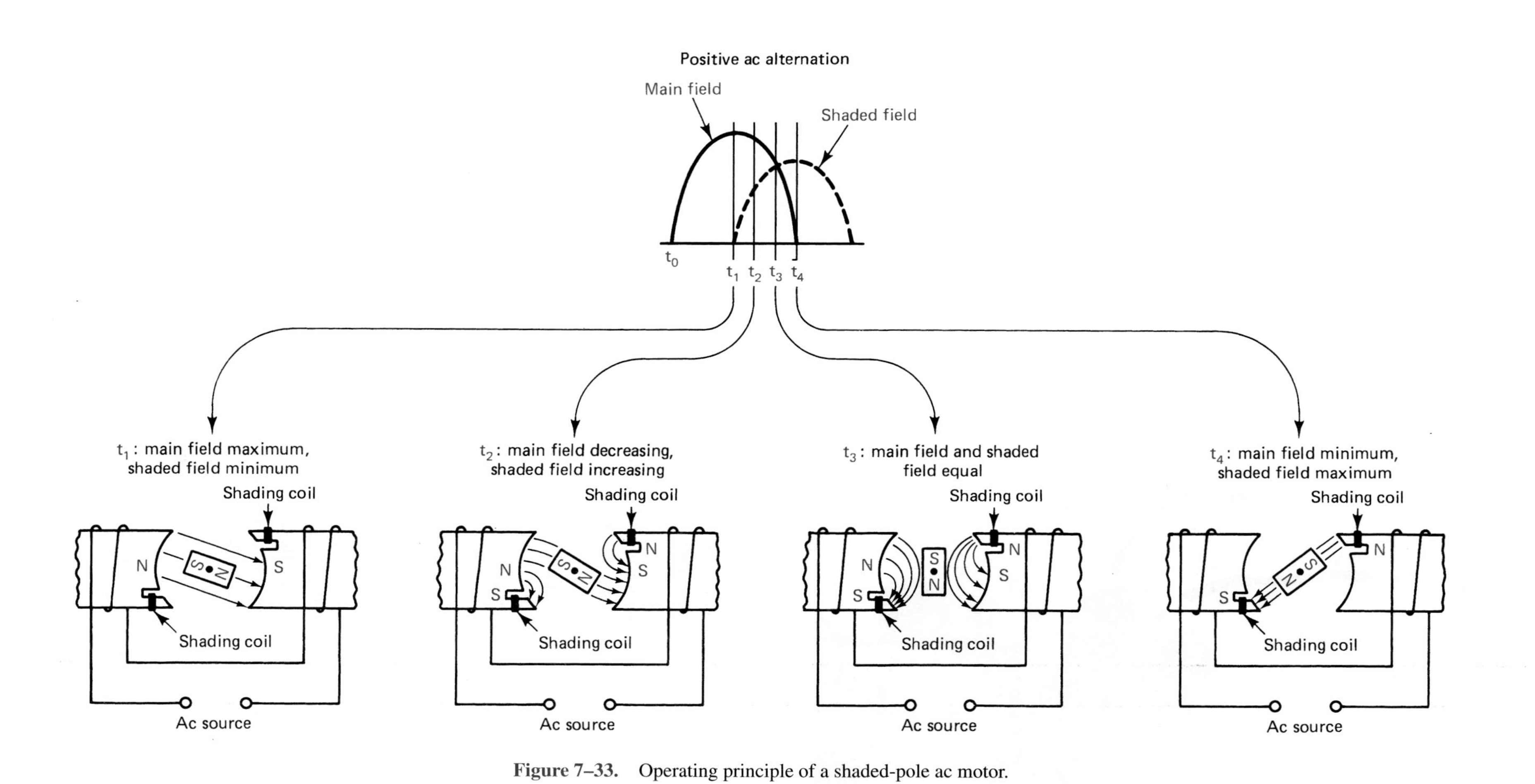

Figure 7–33. Operating principle of a shaded-pole ac motor.

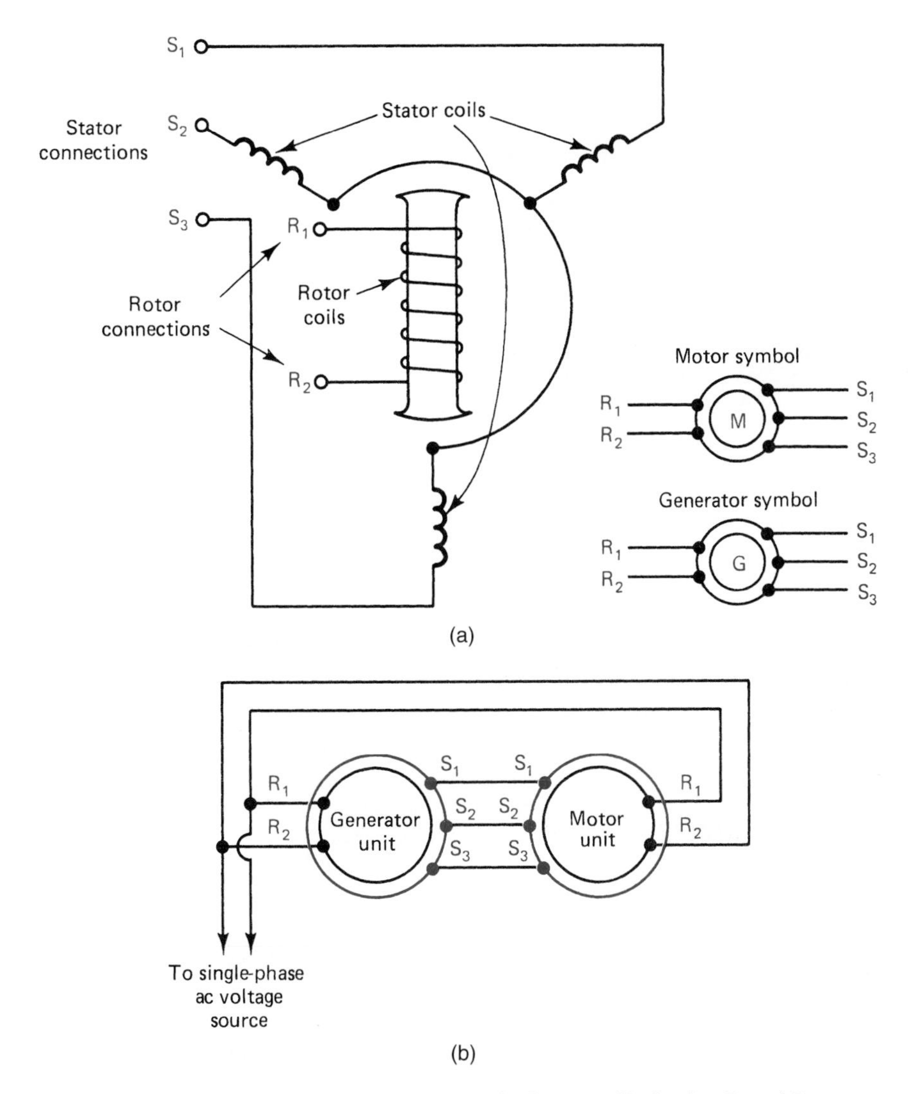

Figure 7–34. Synchro systems: (a) schematic diagram; (b) circuit—R_1 and R_2 are rotor connections and S_1, S_2, and S_3 are stator connections.

can be transferred to the motor through three small wires which connect the stators.

Synchro and servo systems are used to change the position or speed of an object. Position changes are needed for numerically controlled machinery and process control equipment in industry. Speed control is used for conveyor belt units, operating speed of machine tools, and disk or magnetic-tape drives for computers. A servomotor is shown in Fig. 7–35.

Dc Stepping Motors

Dc stepping motors are servomotors used to drive many types of computer-controlled machines. They convert electrical pulses into rotary motion. The shaft of a dc stepping motor rotates a specific number of degrees when a pulse is applied to it. The amount of movement caused by each pulse can be repeated. The shaft of a dc stepping motor is used to position industrial machinery accurately. The speed, distance, and direction of a machine can be controlled by a dc stepping motor. The stator construction and coil placement are shown in Fig. 7–36. The rotor is made of permanent magnet material. The permanent magnet rotor is placed between two series-connected stator coils as shown in Fig. 7–37. With power applied to the stator, the rotor can be repelled in either direction. With the stator polarities shown, the rotor aligns itself midway between two pairs of stator coils. The direction of rotation is determined by the polarities of the stator coils. Adding more stator coils to this motor makes positioning the rotor extremely accurate. Operation of a dc stepping motor is achieved by electrical pulses from a computerized power supply. The stepping motor rotor shown in Fig. 7–38 has 50 teeth. It takes 200 pulses or "steps" for each complete revolution. The amount of movement is determined by the number of teeth on the rotor and the electrical pulse sequence.

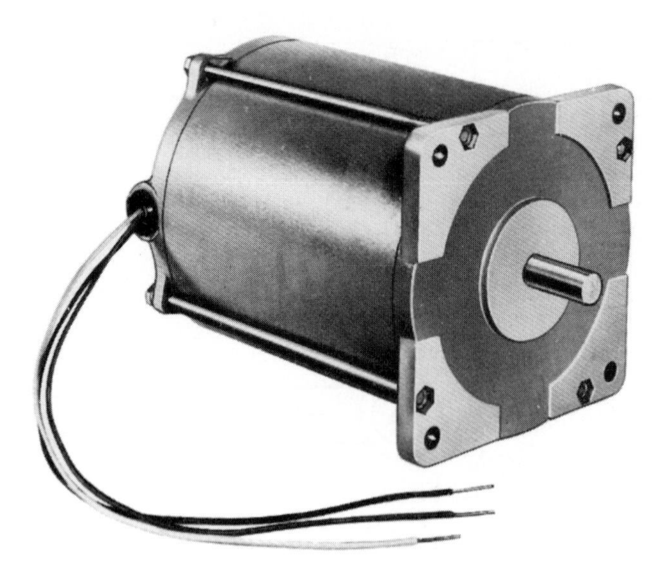

Figure 7–35. Servomotor. (Courtesy of Superior Electric Co.)

A dc stepping motor that takes 200 steps for one revolution will move 1.8° per step (360° divided by 200). Dc stepping motors are important for use in industry.

7.8 MOTOR PERFORMANCE

The major consumer of electrical power is the electric motor. It is estimated that electric motors account for 50% of the electrical power consumed in industrial usage, and that 35% of all electrical power is used by electrical motors. For these reasons, consider the efficient operation of motors to be a major part of energy conservation efforts.

Both efficiency and the power factor must be considered to determine the effect of a motor, in terms of efficient power conversion. Remember the following relationships. First,

$$\text{efficiency (\%)} = \frac{P_{out} \times 100}{P_{in}}$$

where:

P_{in} = the power input in horsepower
P_{out} = the power output in watts

(To convert horsepower to watts, remember that 1 horsepower = 746 watts.)

Then,

$$\text{PF} = \frac{P}{\text{VA}}$$

where:

PF = the power factor of the circuit
P = the true power in watts
VA = the apparent power in volt-amperes

Figure 7–36. Dc stepping motor construction. (Courtesy of Superior Electric Co.)

The maximum PF value is 1.0, or 100%, which would be obtained in a purely resistive circuit. This is referred to as *unity power factor.*

Effect of Load

Because electrical power will probably become more expensive and less abundant, the efficiency and power factor of electric motors will become increasingly important. The efficiency of a motor shows mathematically just how well a motor converts electrical energy into mechanical energy. A mechanical load placed on a motor affects its efficiency. Thus, it is particularly important for industrial users to load motors so that their maximum efficiency is maintained.

Power factor is also affected by the mechanical load placed on a motor. A higher power factor means that a motor requires less current to produce a given amount of torque or mechanical energy. Lower current levels mean that less energy is being wasted (converted to heat) in the equipment and circuits connected to the motor. Penalties are assessed on industrial users by the electrical utility companies for having low system power factors (usually less than 0.8 or 0.85 values). By operating at higher power factors, industrial users can save money on penalties and can help, on a larger scale,

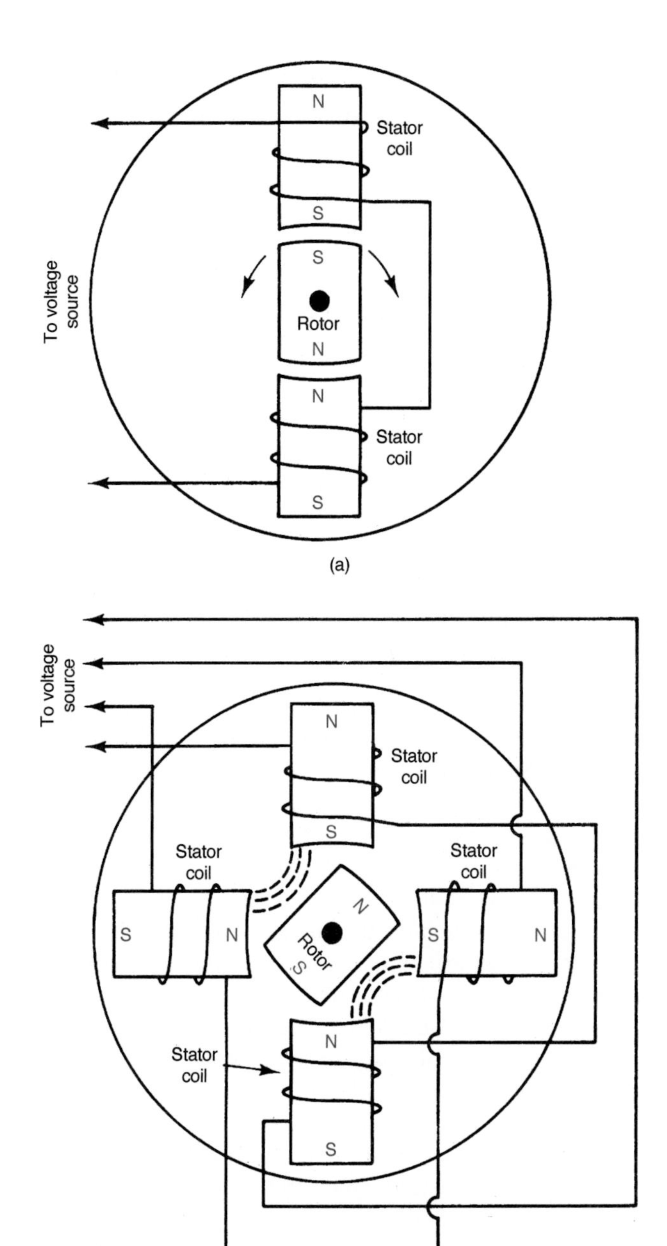

Figure 7–37. Stator coil placement of a dc stepping motor: (a) one set of stator coils—rotor can move in either direction; (b) two sets of stator coils—rotor will move in a clockwise direction.

Figure 7–38. Dc stepping motor rotor construction. (Courtesy of Superior Electric Co.)

with more efficient utilization of electrical power. Motor load affects the power factor to a much greater extent than it does the efficiency. Therefore, motor applications should be carefully studied to ensure that motors (particularly very large ones) are not overloaded or underloaded, so that the available electrical power will be used more effectively.

Effect of Voltage Variations

Voltage variation also has an effect on the power factor and efficiency. Even slight changes in voltage produce a distinct effect on the power factor; however, a less distinct effect results when the voltage causes a variation in the efficiency. Because proper power utilization is becoming more and more important, motor users should ensure that their motors do not operate at undervoltage or overvoltage conditions.

Considerations for Mechanical (Motor) Loads

The three basic types of mechanical (motor) loads connected to electrical power systems are dc, single-phase ac, and three-phase ac systems. Dc motors are ordinarily used for special applications, because they are more expensive than other types and require a dc power source. Typically, they are used for small, portable applications and powered by batteries, or for industrial or commercial applications with ac converted to dc by rectification systems. A major advantage of dc motors is the ease of speed control. The shunt-wound dc motor can be used for accurate speed control and good speed regulation. A disadvantage is the increased maintenance caused by the brushes and commutator of the machines. Dc shunt-wound motors are used for variable-speed drives on printing presses, rolling mills, elevators, hoists, and automated industrial machine tools.

Series-wound dc motors have a high starting torque. Their speed regulation is not as good as that of shunt-wound dc motors. The series-wound motor also requires periodic maintenance because of the brush/commutator assembly. Typical applications of series-wound dc motors are automobile starters, traction motors for trains and electric buses, and mobile equipment operated by batteries. Compound-wound dc motors have few applications today.

Single-phase ac motors are relatively inexpensive. Most types have good starting torque and are easily provided 120-volt and 240-volt electrical power. Disadvantages include maintenance problems due to centrifugal switches, pulsating torque, and rather noisy operation. They are used in fractional-horsepower sizes (less than 1 horsepower) for residential, commercial, and industrial applications. Some integral horsepower sizes are available in capacitor-start types. Uses include machine cooling system blowers and clothes dryer motors.

Specialized applications for single-phase motors include:

1. Shaded-pole motors used for portable fans, record players, dishwasher pumps, and electric typewriters. They are low cost and small, but inefficient.
2. Single-phase synchronous motors used for clocks, appliance timers, and recording instruments (compact disk players). They operate at a constant speed.
3. Universal (ac/dc) motors used for many types of portable tools and appliances such as electric drills, saws, office machines, mixers, blenders, sewing machines, and vacuum cleaners. They operate at speeds up to 20,000 r/min and have easy speed control. Remember that ac induction motors do not have speed control capability without the addition of expensive auxiliary equipment.

Three-phase ac motors of the induction type are simple in construction, rugged, and reliable in operation. They are less expensive (per horsepower) than other motors. Applications of three-phase induction motors include industrial and commercial equipment and machine tools. Three-phase ac synchronous motors run at constant speeds and may be used for power factor correction of electrical power systems. However, they are expensive, require maintenance of brushes/slip rings, and need a separate dc power supply.

7.9 MOTOR CONTROL BASICS

A block diagram of a motor control circuit is shown in Fig. 7–39. The components used to control motors include circuit breakers, switches, magnetic contactors, protective overloads, limit switches, and many other devices.

Motor Starting Control

On many motor control circuits, pushbutton switches such as those shown in Fig. 7–40 are used. A pushbutton may be pressed to cause a motor to start. Another pushbutton may be used to cause it to stop. Pushbuttons either apply power from a source or open the power circuit to stop a motor.

Limit switches may also be used to control a motor's on or off operation. Limit switches are used to limit the movement or travel of a piece of equipment; for example, they could be used in an elevator to cause it to stop at the desired floor. Some types of limit switches are shown in Fig. 7–41.

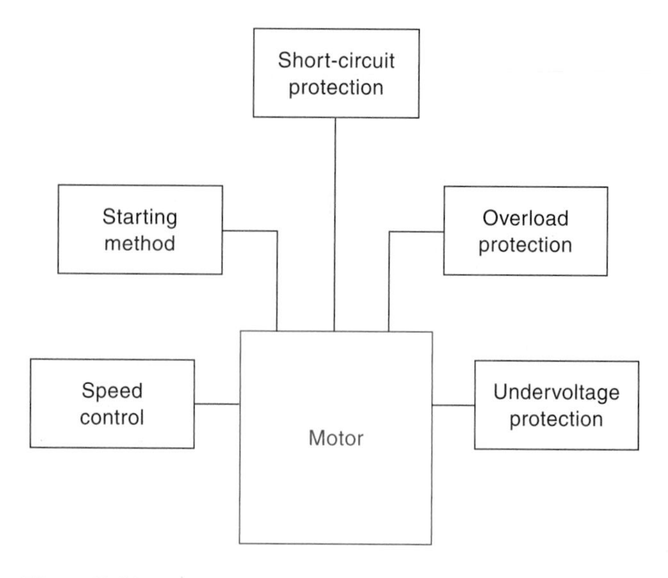

Figure 7–39. Block diagram of a motor control circuit.

Fuses or circuit breakers (see Fig. 7–42) are also part of a motor control circuit. Fuses are made so that they will open a circuit by melting. If the current flowing through a circuit exceeds a safe limit, the fuse element will melt. A fuse is used to prevent excessive current from flowing through a motor. It is a form of protective control. Circuit breakers have the same function. A circuit breaker must be reset after it is opened. If a circuit breaker is rated at 20 A, a current greater than that amount will disconnect the motor power source. By automatically disconnecting the power source, circuit breakers act as a protective control. They are usually preferred over fuses because they do not have to be replaced.

There are several types of electromechanical equipment used for the control of electric motors. The selection of control equipment will affect the efficiency of operation and the performance of the machinery. It is important to use the proper type of equipment for each control application.

A *motor-starting device* is a type of power control used to accelerate a motor from a stopped condition to its normal operating speed. There are many variations in motor-starter design, the simplest being a manually operated on-off switch connected in series with one or more power lines. This type of starter is used only for smaller motors that do not draw an excessive amount of current (see Fig. 7–43 on page 275).

Another type of motor starter is the *magnetic starter* (see Fig. 7–44 on page 275), which relies on an electromagnetic effect to open or close the power source circuit of the motor.

Often, motor starters are grouped together for the control of equipment in an industrial plant. Such groupings of motor starters and associated control equipment are called

Figure 7–40. Pushbutton switches. [(a) Courtesy of Cutler-Hammer Co.; (b) courtesy of Furnas Electric Co.]

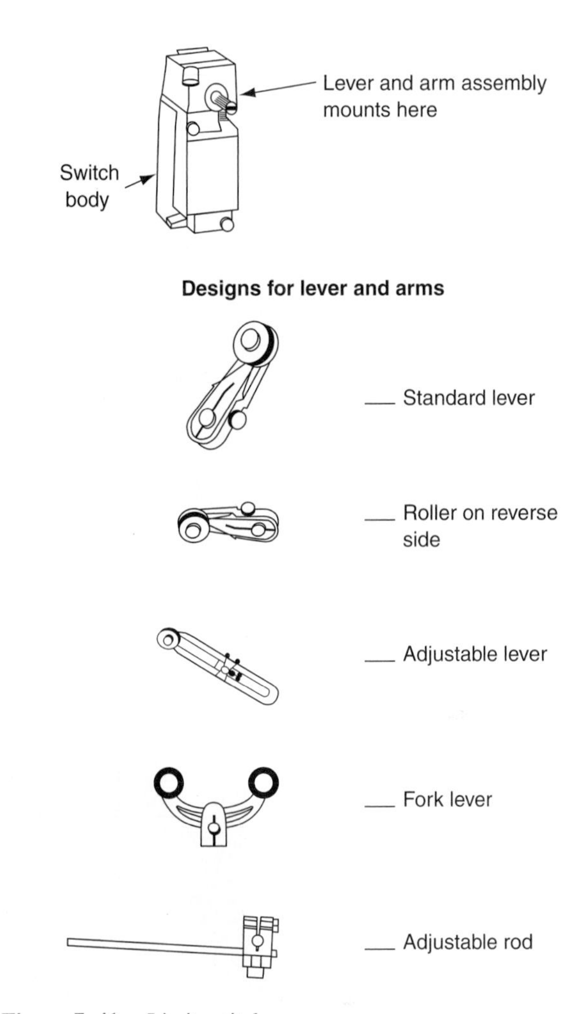

Figure 7–41. Limit switches.

control centers. Control centers (see Fig. 7–45) provide a relatively easy access to the power distribution system, because they are compact, and the control equipment is not scattered throughout a large area.

Various types of motor starters are used for control of motors. The functions of a starter vary in complexity; however, motor starters usually perform one or more of the following functions:

1. On and off control
2. Acceleration
3. Overload protection
4. Reversing the direction of rotation

Some starters control a motor by being connected directly across the power input lines. Other starters reduce the level of input voltage that is applied to the motor when it is started, so as to reduce the value of the starting current. Ordinarily, motor overload protection is contained in the same enclosure as the starter.

A popular type of motor starter used in control applications is the *combination starter.* These starters incorporate protective devices such as fused-disconnect switches, circuit breakers, or a system of fuses and circuit breakers mounted in a common enclosure.

Electronic controllers are capable of replacing electromagnetic circuit breakers and relays used for electrical control. The introduction of computerized control for equipment has brought about a new technology of machine control. Many industrial machines, such as automated manufacturing and robotic equipment, are now controlled by computerized circuits. An understanding of the basic principles of electrical control is, however, still important for technicians.

Figure 7–46 is a start-stop pushbutton control circuit (one-line diagram) with overload protection (OL). Notice that the start pushbutton is normally open (NO), and the stop pushbutton is normally closed (NC). Single-phase lines L1 and L2 are connected across the control circuit. When the start pushbutton is pushed, a momentary contact is made between points 2 and 3. This causes the NO contact (M) to

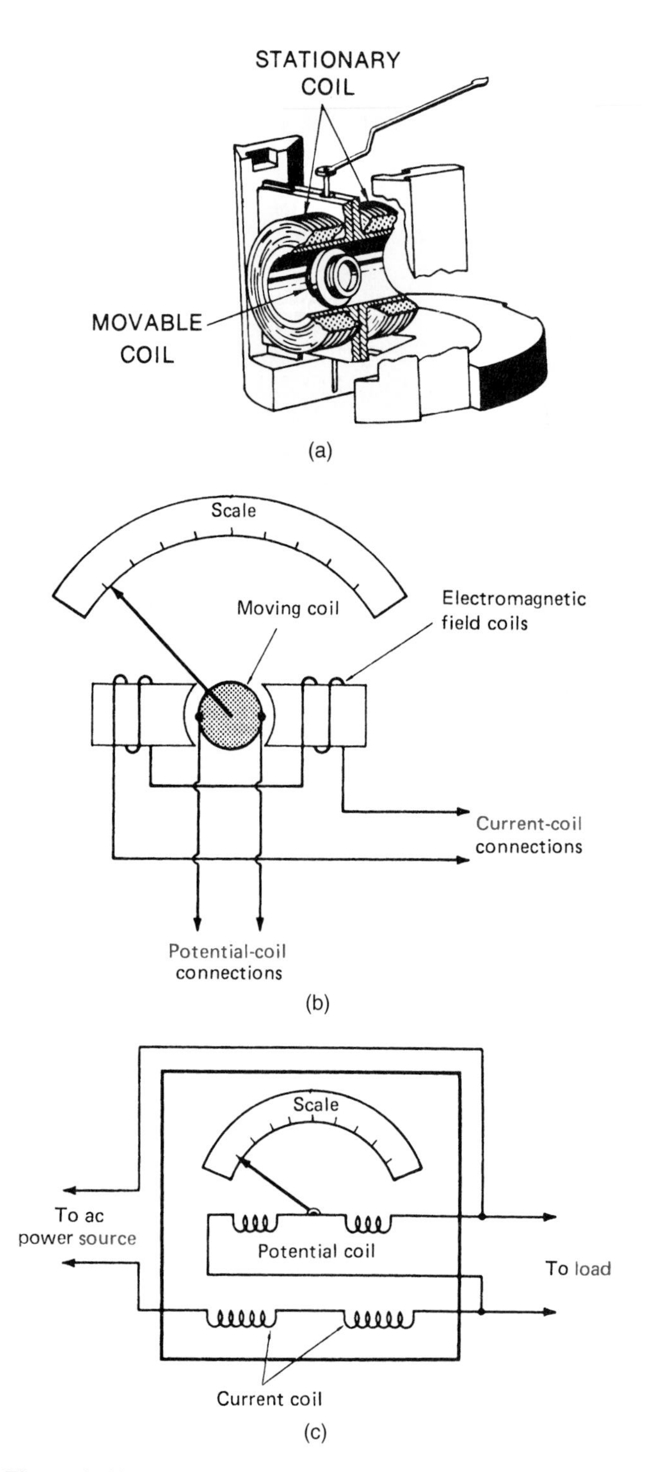

Figure 8–10. Meter used to measure electrical power: (a) dynamometer movement; (b) meter movement schematic; (c) circuit.

Figure 8–11. Watt-hour meter. (Courtesy of Sangamo-Schlumberger.)

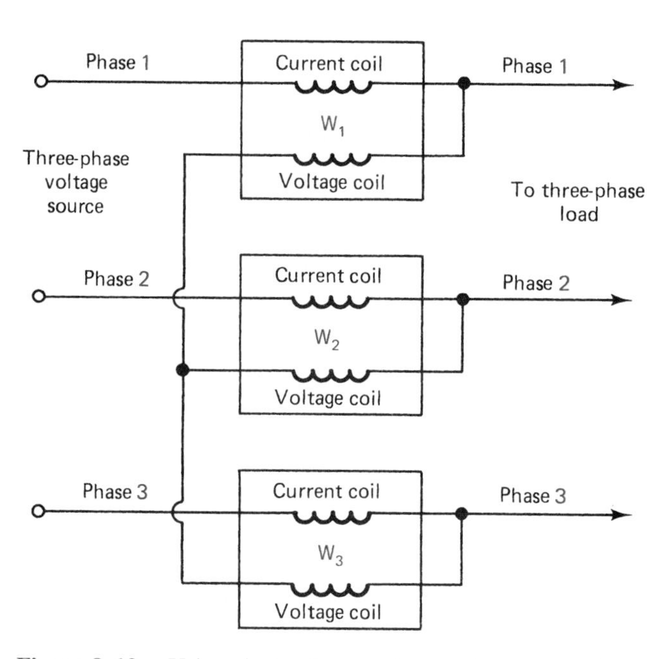

Figure 8–12. Using single-phase wattmeters to measure three-phase power.

8.8 MEASURING POWER FACTOR

Power factor is the ratio of the measured true power of a system to the apparent power (volts × amperes). To find the power factor of a system, a wattmeter, a voltmeter, and an ammeter may be used. The power factor could be found as PF = W/VA. It is more convenient to use a *power factor meter* when the power factor must be measured.

The circuit of a power factor meter is shown in Fig. 8–14. A power factor meter is similar to a wattmeter—it has two armature coils that rotate due to their electromagnetic field strengths. The armature coils are mounted on the same shaft. They are placed about 90° apart. One coil is connected across the power line in series with a resistance (*R*). The other coil is connected across the line in series with an inductance (*L*). The resistor in series with the coil produces a magnetic field due to the in-phase part of the power. The inductor in series with the other coil produces a magnetic field due to the out-of-phase part of the power. The scale of the meter is calibrated to measure power factors of 0 to 1.0, or 0% to 100%.

Figure 8–13. Chart—Recording three-phase power analyzer with clamp-on line monitors. (Courtesy of Esterline Angus Co.)

8.9 MEASURING POWER DEMAND

Figure 8–15 shows a power demand meter which is an important instrument for industries and large commercial buildings. Power demand is found by comparing the average electrical power used with the peak power used by an industry or commercial building. Power demand is important because it shows the ratio of average power used to the peak value that a utility company must supply. Power demand is usually calculated over 15-, 30-, or 60-min. intervals. It may be converted into longer periods of time. Ideally these values should be close.

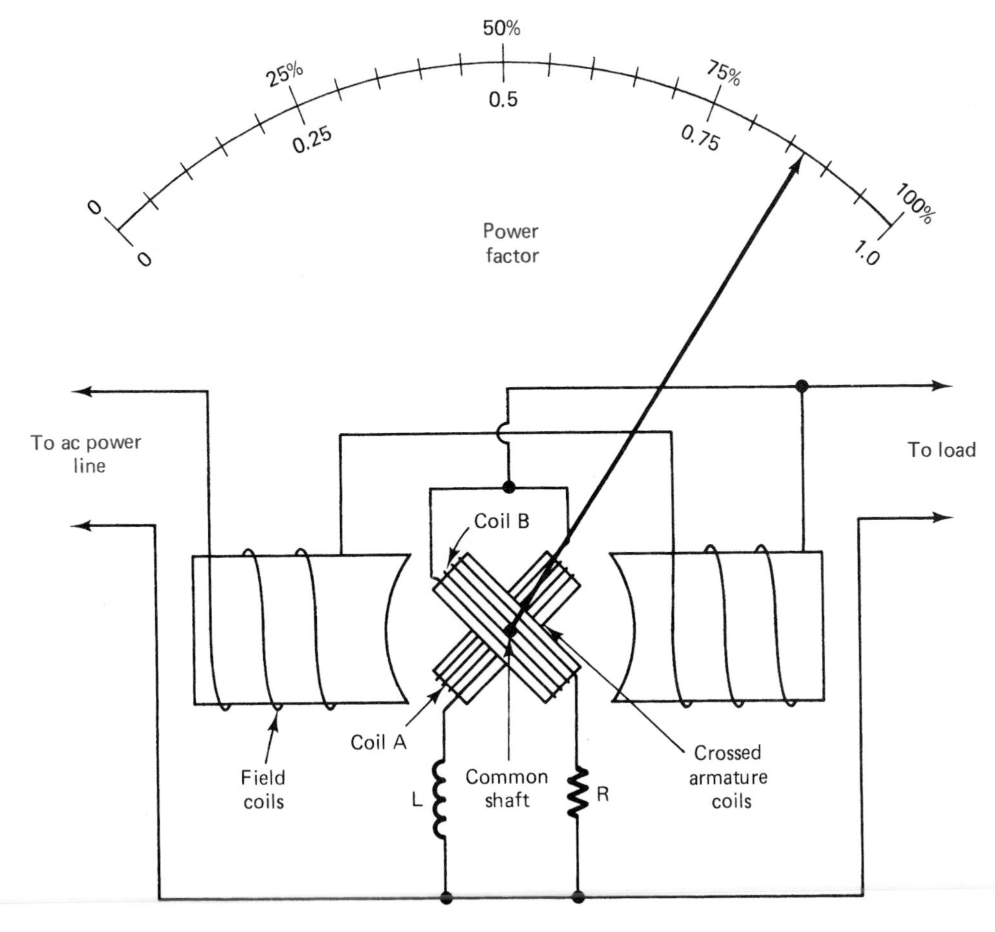

Figure 8–14. Circuit of a power factor meter.

Figure 8–15. Power demand meter. (Courtesy of Sangamo-Schlumberger.)

Industries and businesses may be penalized by the utility company if their peak power demand is far greater than their average demand. This is called a demand charge. Power demand meters help companies to better use their available electrical power. A high peak demand means that the equipment used for power distribution must be larger. The closer the peak demand is to the average demand, the more efficient the power system.

8.10 MEASURING FREQUENCY

Another important measurement is frequency. The frequency of the power system must remain the same at all times. *Frequency* refers to the number of cycles of voltage or current that occur in a given time. The international unit used to measure frequency is the hertz (Hz). One hertz equals one cycle per second. A list of frequencies is shown in Table 8–1. The standard power frequency in the United States is 60 Hz. Some countries use 50 Hz.

Frequency is measured with several different types of meters. An electronic counter is used to measure frequency. Vibrating-reed frequency meters are used for measuring frequencies. Figure 8–16 shows some frequency meters. An oscilloscope (see Fig. 8–17) is also used to measure frequency. Graphic recording instruments can be used to give a visual display of frequency over a given time. The power industry, for example, must at all times monitor frequency of its alternators.

TABLE 8–1 Frequency Values*

Band	*Frequency Range*
Extremely low frequency	30 Hz to 300 Hz
Voice frequency	300 Hz to 3 kHz
Very low frequency	3 kHz to 30 kHz
Low frequency	30 kHz to 3 MHz
Medium frequency	300 kHz to 3 MHz
High frequency	3 MHz to 3 MHz
Very high frequency	30 MHz to 300 MHz
Ultra-high frequency	300 MHz to 3 GHz
Super-high frequency	3 GHz to 30 GHz
Extremely high frequency	30 GHz to 300 GHz

*Power frequency = 60 Hz; AM radio band = 550–1600 kHz; FM radio band = 88–108 MHz.

8.11 GROUND-FAULT INDICATORS

A ground-fault indicator is used to locate faulty grounding of equipment or power systems. Equipment must be properly grounded. A ground-fault indicator, shown in Fig. 8–18, is used to check for faulty grounding. Several grounding conditions exist that might be dangerous. Faulty conditions include (1) hot and neutral wires reversed, (2) open equipment ground wires, (3) open neutral wires, (4) open hot wires, (5) hot and equipment grounds reversed, and (6) hot wires on

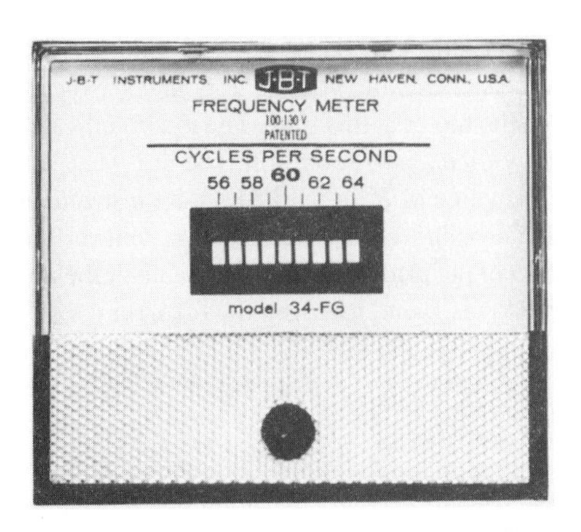

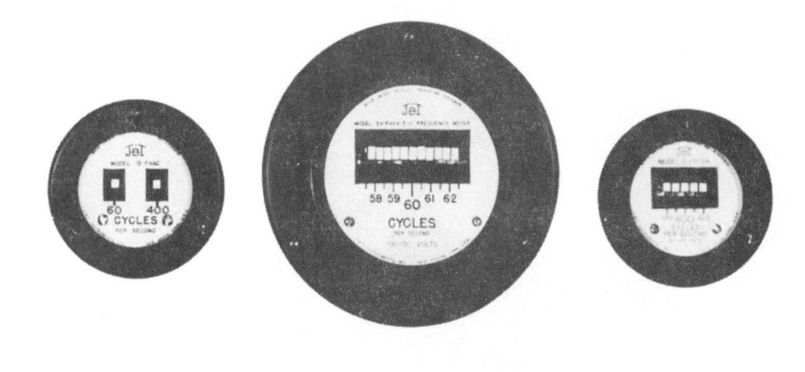

Figure 8–16. Frequency meters. (Courtesy of J-B-T Instruments.)

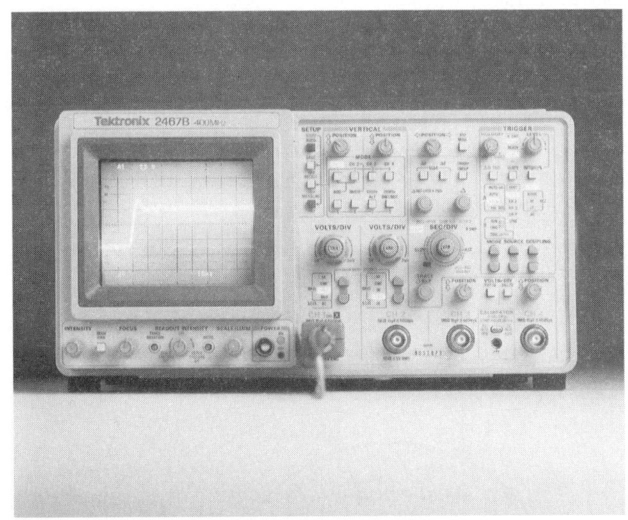

(a)

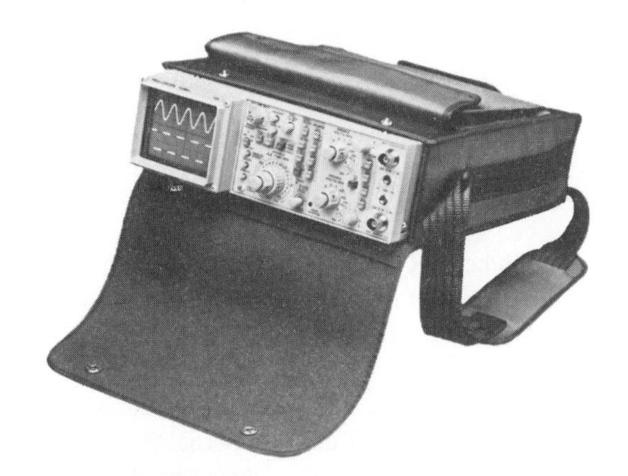

LEADER
FOR PROFESSIONALS WHO KNOW THE DIFFERENCE

Model 326
100-MHz Attache Case Oscilloscope

(b)

Figure 8–17. Oscilloscopes may be used to measure frequency. [(a) Courtesy of Tektronix Co; (b) courtesy of Leader Co.]

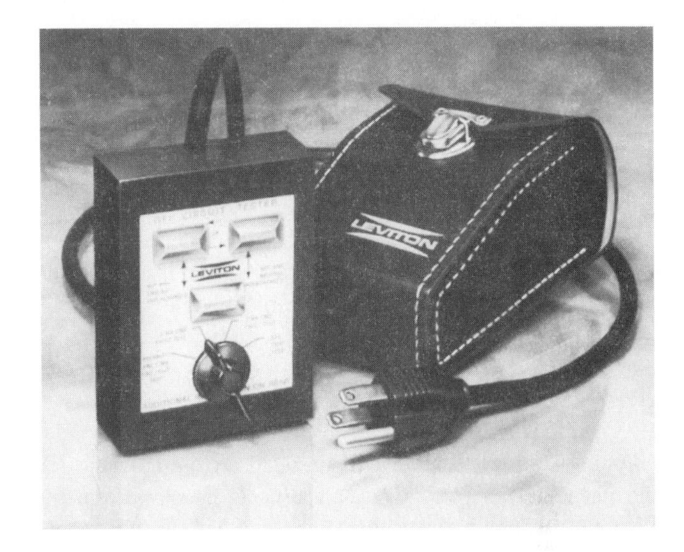

Figure 8–18. GFI circuit tester. (Courtesy of Leviton Co.)

neutral terminals. Each of these conditions presents a serious problem. Proper wiring eliminates most of these problems. A check with a ground-fault indicator ensures safe and efficient electrical wiring in a building.

8.12 MEASURING HIGH RESISTANCE

A megohmmeter, as shown in Fig. 8–19, is used to measure very high resistances. These resistances are beyond the range of most ohmmeters. Megohmmeters are used to check the quality of insulation on electrical equipment in industry. The quality of insulation of equipment varies with age, moisture content, and applied voltage. Megohmmeters are similar to most ohmmeters. They use a hand-crank, permanent magnet dc generator as a voltage source rather than a battery. The dc generator is cranked by the operator while making a test.

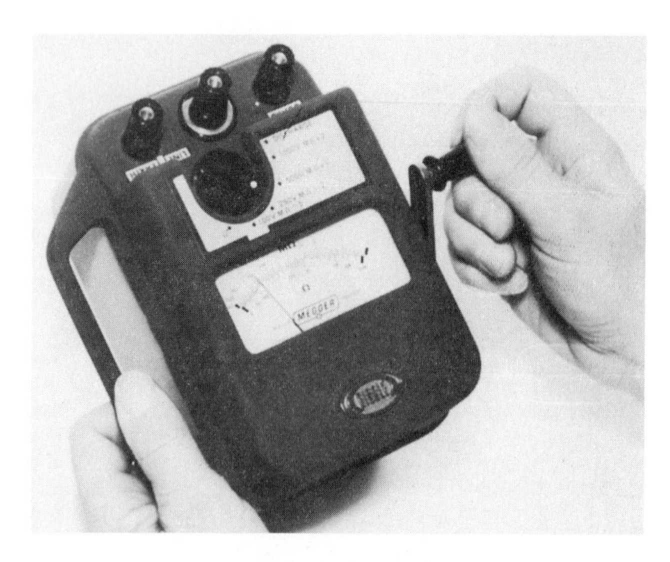

Figure 8–19. Megohmmeter used to measure high resistance. (Courtesy of James G. Biddle Instruments.)

Insulation tests should be made on all power equipment. Insulation breakdown causes equipment to fail. Insulation resistance value can be used to determine when equipment needs to be replaced or repaired. A decrease in insulation resistance over a time means that a problem could soon exist.

8.13 CLAMP-ON METERS

Clamp-on meters, as shown in Fig. 8–20, are used to measure current in power lines. They are used to check current by clamping around a power line, and are easy to use for maintenance and testing of equipment. The meter is simply clamped around a conductor. Current flow through a conductor creates a magnetic field around the conductor. The magnetic field induces a current into the iron core of the clamp-on part of the meter. The meter scale is calibrated to indicate current flow. An increase in the current flow through a power line causes the current induced into the clamp-on part of the meter to increase. Clamp-on meters usually have voltage and resistance ranges to make them more versatile. Because they operate as a transformer, clamp-on ammeters can only be used on ac circuits. Their primary advantage is that the circuit does not have to be broken, as with a VOM.

8.14 WHEATSTONE BRIDGE

The Wheatstone bridge is a comparison instrument (see Fig. 8–21). A voltage source is used with a sensitive zero-centered movement and a circuit called a *resistance bridge.* The resistance circuit has an unknown external resistance (R_x) which is the resistance to be measured. Resistor R_s is the known "standard" resistance. R_s is adjusted so that the voltage at point A is the same as the voltage at point B. No current flows through the meter at this time. The meter indicates zero current and is called a *null* condition. The bridge is said to be balanced. The value of R_s is marked on the instrument to compare with the value of R_x. Resistors R_1 and R_2 are called the *ratio arm* of the bridge circuit. The value of the unknown resistance (R_x) is found by using the formula

$$R_x = \frac{R_1}{R_2} \times R_s \text{ or } \frac{R_1}{R_x} = \frac{R_2}{R_s}$$

Wheatstone bridges measure resistance with precise accuracy. Other comparison instruments use the Wheatstone bridge principle. Comparison instruments compare unknown quantities with known quantities in the circuit of the instruments.

SAMPLE PROBLEM:

Given: in the bridge circuit of Fig. 8–21, $R_1 = 500$ ohms, $R_2 = 2000$ ohms, and $R_s = 1000$ ohms.

Find: the value of unknown resistance (R_x).

Solution:

$$R_x = \frac{R_1}{R_2} \times R_s$$

$$\frac{500\ \Omega}{2000\ \Omega} \times 1000\ \Omega$$

$$R_x = 250 \text{ ohms}$$

The Wheatstone bridge will measure most values of resistance with considerable accuracy. Several other types of comparative instruments also use the Wheatstone bridge principle.

8.15 CATHODE-RAY TUBE INSTRUMENTS

Cathode-ray tube (CRT) instruments are usually called oscilloscopes. The use of oscilloscopes was discussed in Chap. 6. Oscilloscopes monitor voltages of a circuit visually. The basic part of the oscilloscope is called a cathode-ray tube.

Figure 8–22 on page 299 shows the construction of an electrostatic-deflection CRT and its internal electron gun assembly. A beam of electrons is produced by the cathode of the tube. Electrons are given off when the filament is heated by a filament voltage. The electrons have a negative (−) charge. They are attracted to the positive (+) potential of anode 1. The number of electrons which pass to anode 1 is changed by the amount of negative (−) voltage applied to the control grid. Anode 2 has a higher positive voltage applied.

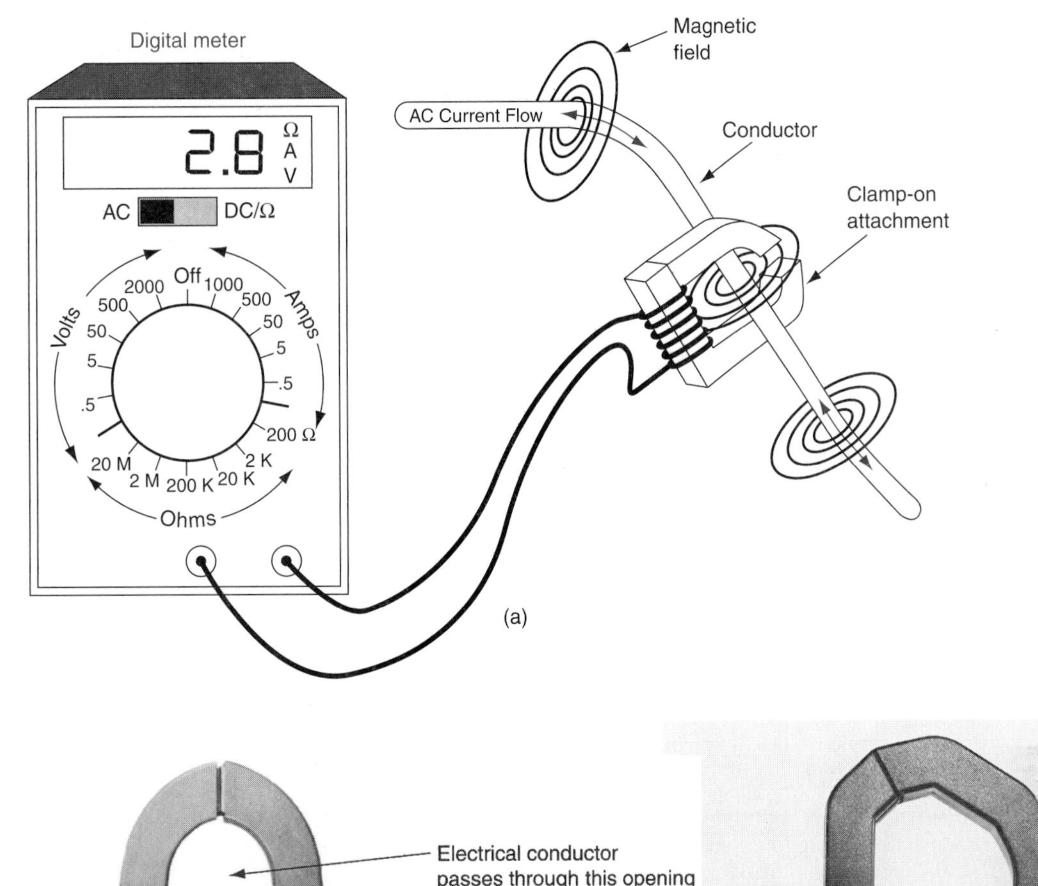

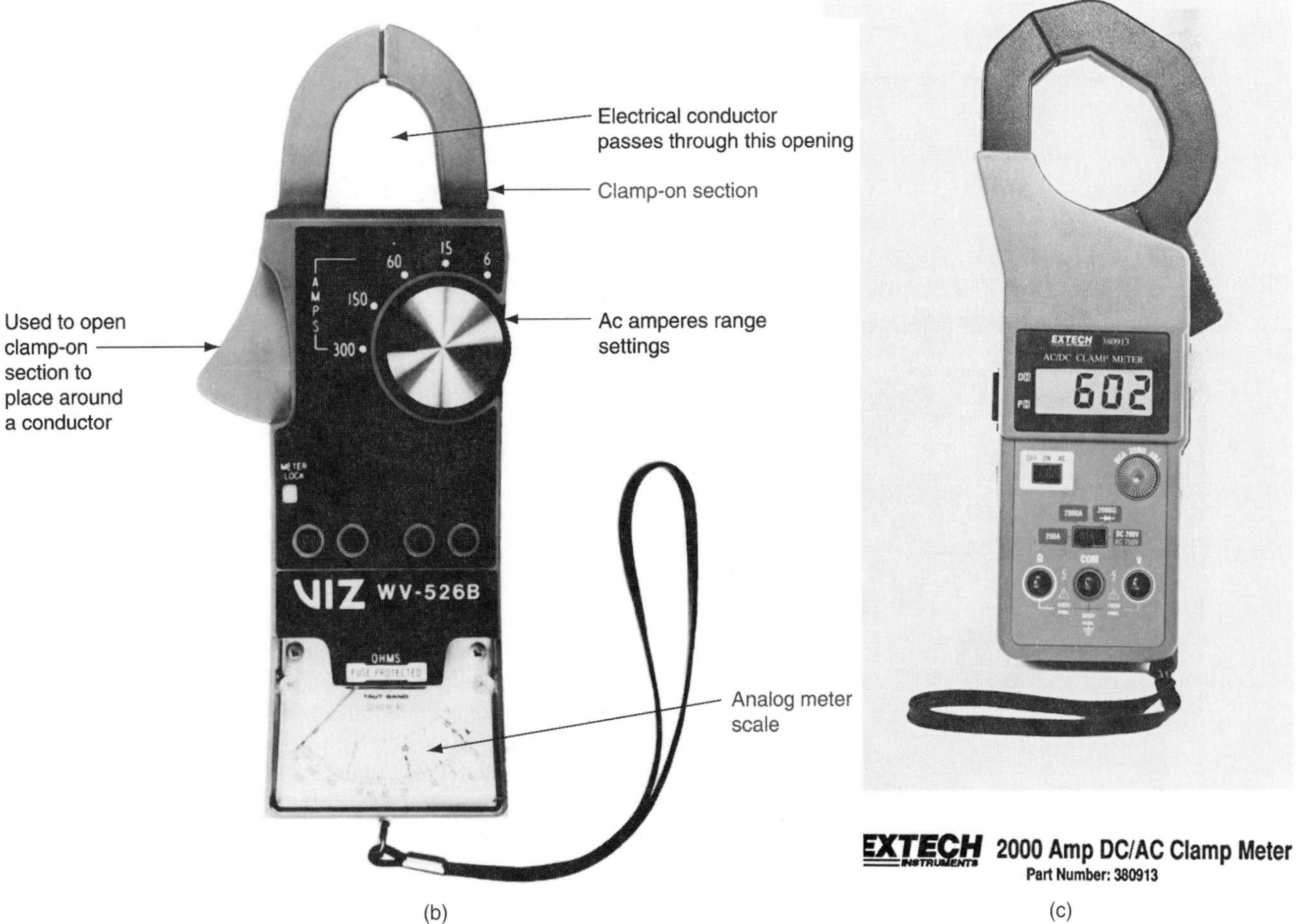

Figure 8–20. Clamp-on current meter: (a) operational principle; (b) analog meter; (c) digital ac/dc clamp meter. [(b) Courtesy of VIZ Manufacturing Co.; (c) courtesy of Extech Instruments Corp.]

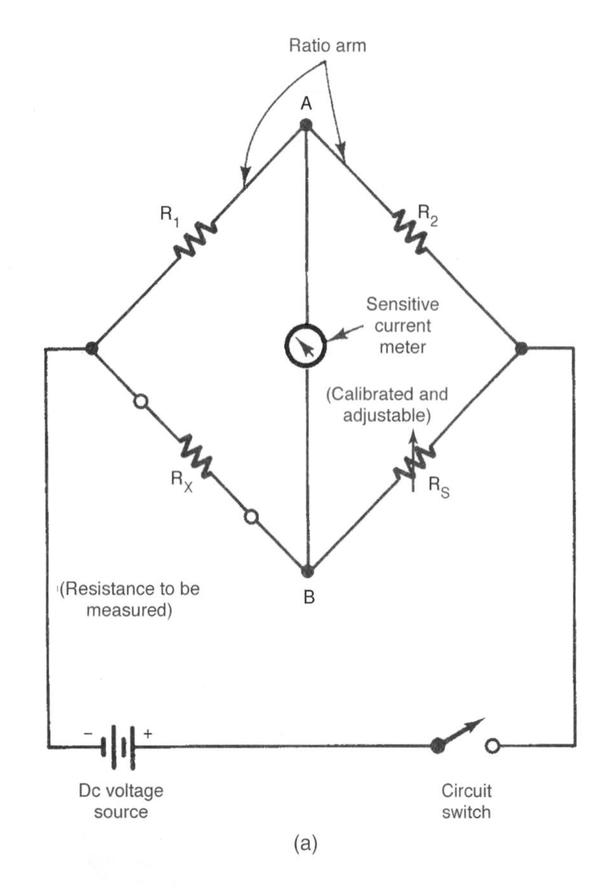

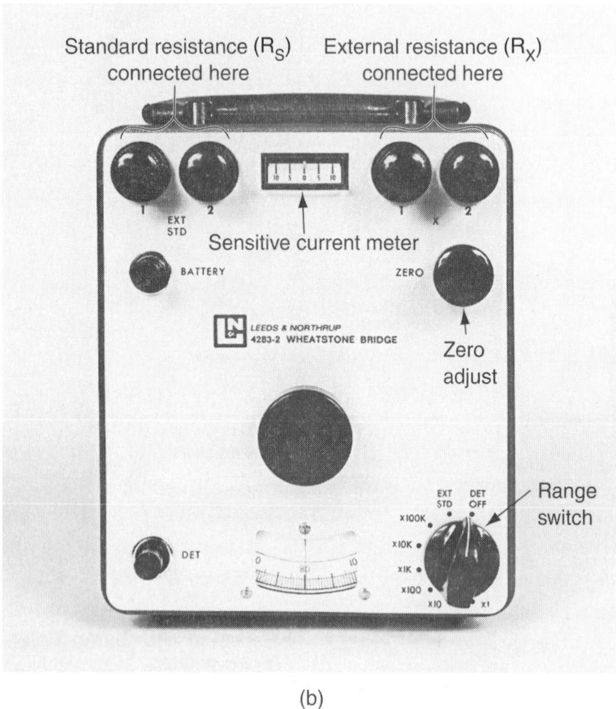

Figure 8–21. Wheatstone bridge: (a) circuit of a Wheatstone bridge; (b) comparative instrument. [(b) Courtesy of Leeds & Northrup, a unit of General Signal.]

The higher voltage accelerates the electron beam toward the screen of the CRT. The difference in voltage between anodes 1 and 2 sets the point where the beam strikes the CRT screen.

The screen has a phosphorescent coating. When electrons strike the CRT, light is produced. The phosphorescent coating of the screen allows an electrical charge to produce light. The horizontal and vertical movement of the electron beam is controlled by deflection plates. With no voltage applied to either set of plates, the electron beam appears as a dot in the center of the CRT screen. The movement of the electron beam caused by the change of the plates is called *electrostatic deflection.* When a potential is placed on the horizontal and vertical deflection plates, the electron beam is moved. Horizontal deflection is produced by a circuit called a *sweep oscillator circuit* inside the oscilloscope. A voltage "sweeps" the electron beam back and forth across the CRT screen. The horizontal setting of an oscilloscope is adjusted to match the frequency of the voltage being measured. Vertical deflection is caused by the voltage being measured. The voltage to be measured is applied to the vertical deflection of the oscilloscope.

General-Purpose Oscilloscopes

A block diagram of a general-purpose oscilloscope circuit is shown in Fig. 8–23. Oscilloscopes are used to measure ac and dc voltages, frequency, phase relationships, distortion in amplifiers, and various timing and special-purpose applications. They also have some important medical uses, such as monitoring heartbeat.

Digital Storage Oscilloscopes

An oscilloscope provides a time graph of a voltage amplitude versus time, which allows you to see the change in voltage as it happens. In most applications, a real-time display is adequate. At times this type of graph is not desirable—if a voltage change is short in duration and only occurs after a given time, as shown in Fig. 8–24, it would be difficult to study on a real-time display. As another example, if a voltage change is only a small part of a much longer complex waveform, getting the one small portion would be difficult on a real-time oscilloscope. At times it is best to store a waveform sampled by an oscilloscope for future analysis. Such storage can be accomplished using a digital storage oscilloscope (DSO).

With a DSO (see Fig. 8–25), the signal viewed can be stored in memory and later retrieved for viewing and analysis. This method of storage "digitizes" the waveform. Digitizing means to sample an analog signal and assign each sample a number. In other words, the analog (continuously changing) voltage is converted to a series of digital (binary 1 or 0) numbers. Each digital number is then stored for either immediate or future display. To display the waveform, the binary numbers of the digitized waveform are used to restore the points on the graph.

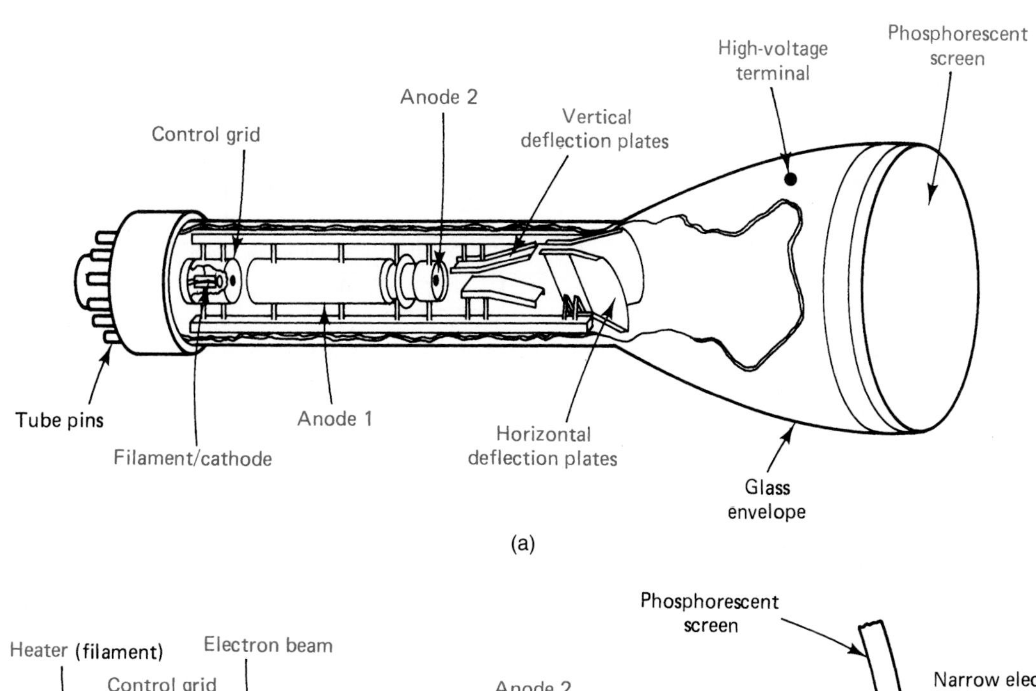

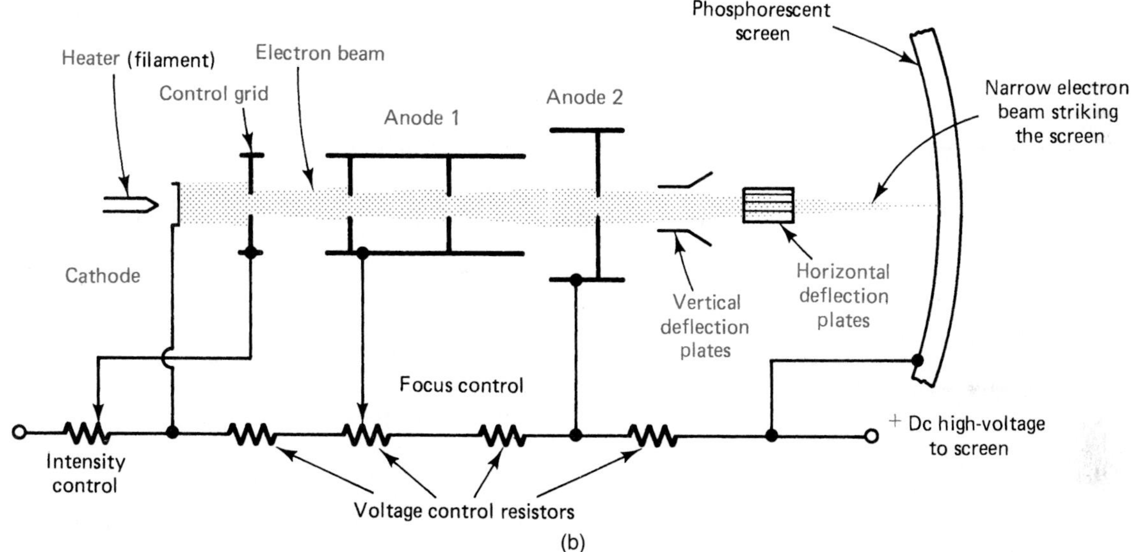

Figure 8–22. Construction of an electrostatic-deflection CRT: (a) cutaway view; (b) electron beam detail.

Figure 8–26 shows a block diagram of a typical digital storage oscilloscope. Because DSOs have basically the same circuitry and input probes as other scopes, most function as general-purpose oscilloscopes. If the scope displays the waveform at the same time it is being sampled, the real-time mode of operation is used. The DSO is a valuable troubleshooting instrument.

8.16 NUMERICAL READOUT INSTRUMENTS

Many instruments have numerical readouts to simplify and make accurate measurements. Instruments such as digital counters and digital multimeters (Fig. 8–27) are commonly used. Numerical readout instruments have internal circuits used to produce a digital display of the quantity being measured. These instruments are easy to read because a scale does not have to be interpreted.

8.17 CHART RECORDING INSTRUMENTS

Most instruments are used in applications where no permanent record of the measured quantity is needed. Sometimes instruments must provide a permanent record of some quantity over time. Typical chart recording instruments are shown in Fig. 8–28. Some chart recorders use pen and ink recorders and inkless recorders.

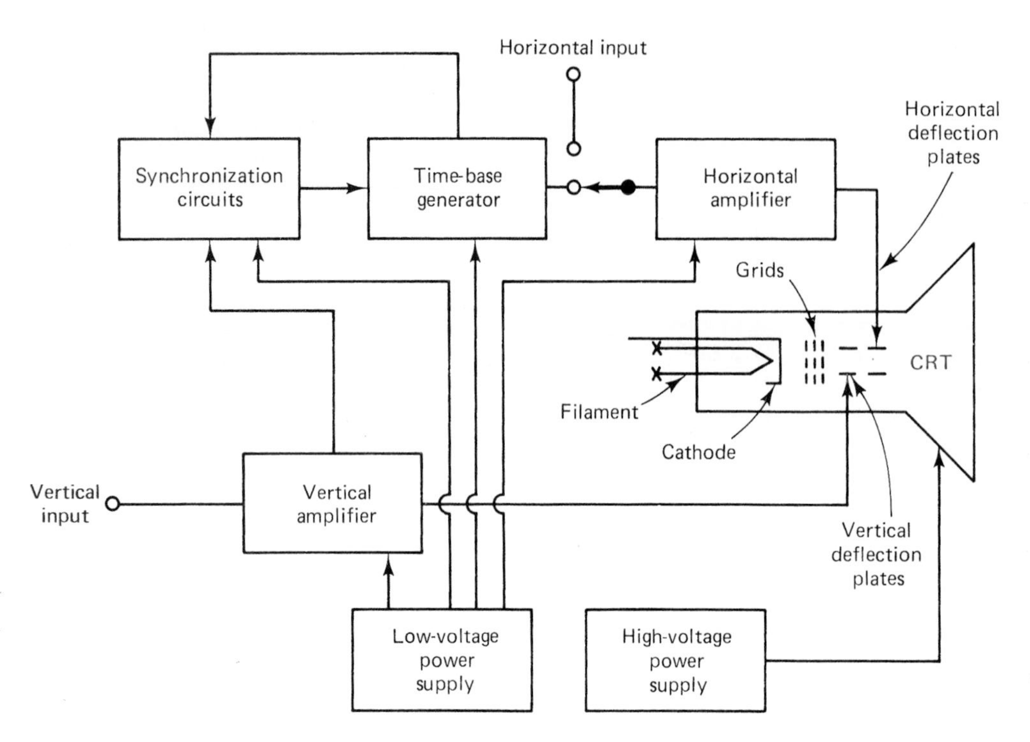

Figure 8–23. Block diagram of a general-purpose oscilloscope.

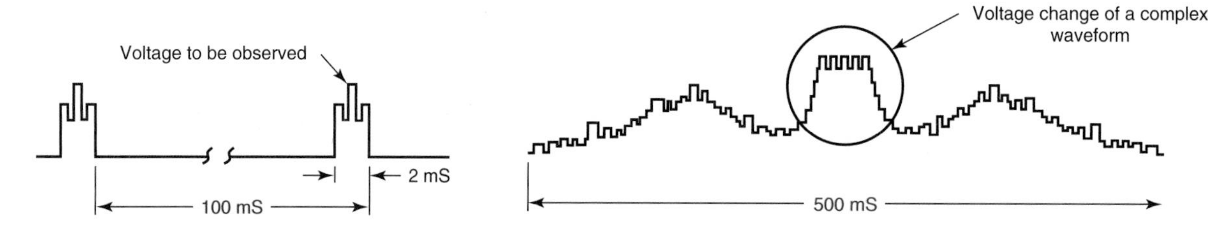

Figure 8–24. Waveforms that could easily be observed on a digital storage oscilloscope (DSO).

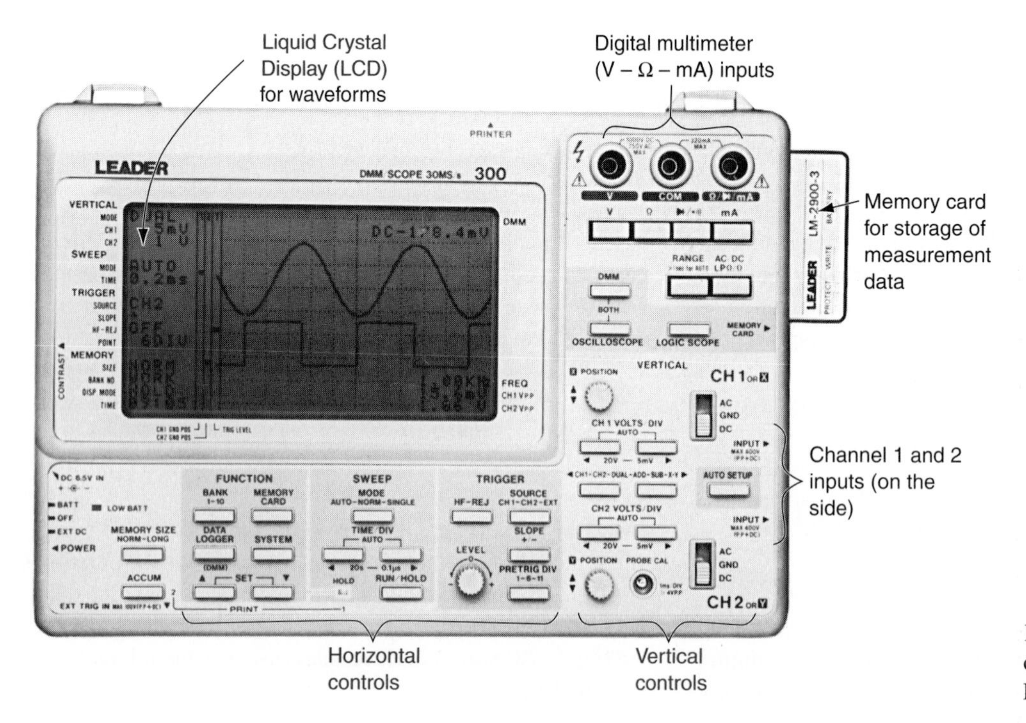

Figure 8–25. Digital storage oscilloscope. (Courtesy of Leader Instrument Corp.)

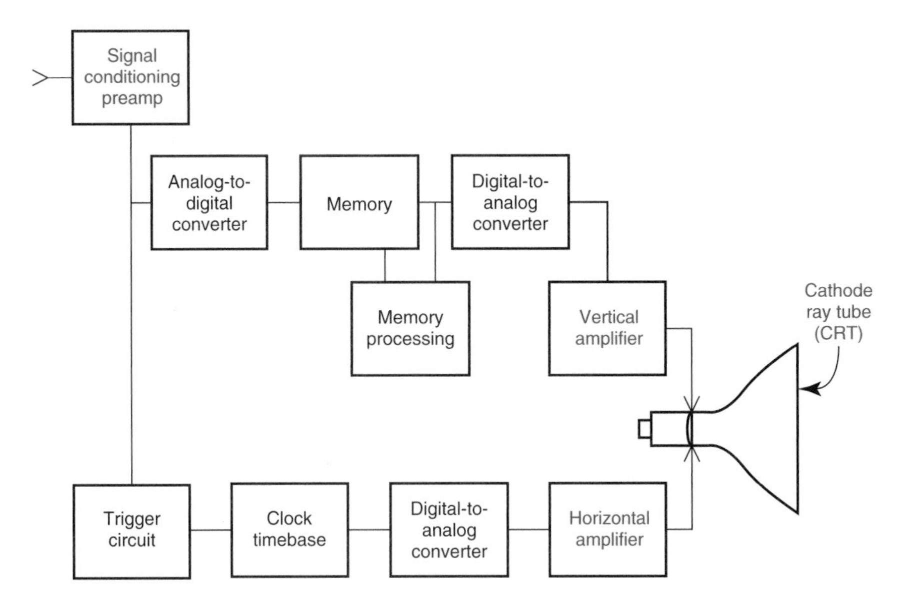

Figure 8–26. Block diagram of a typical digital storage oscilloscope (DSO).

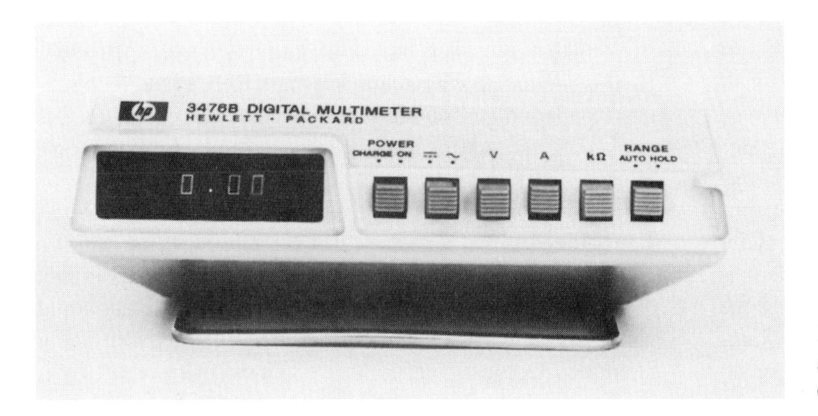

Figure 8–27. Autoranging digital multimeter—range adjusts automatically. (Courtesy of Hewlett-Packard Co.)

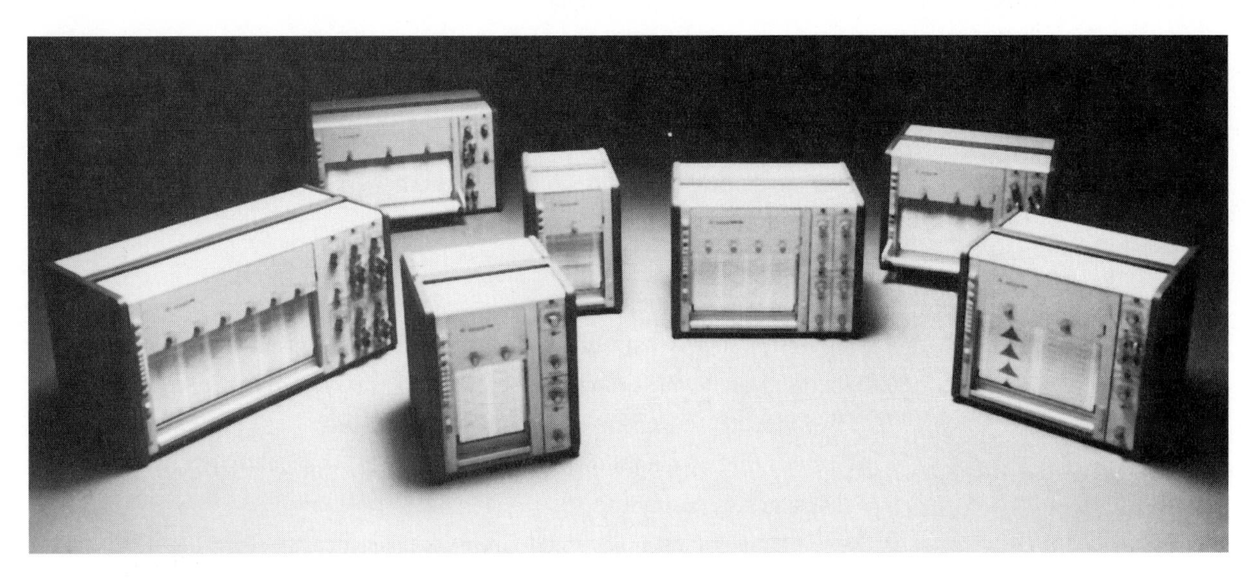

Figure 8–28. Chart recording instruments. (Courtesy of Gould, Inc., Instruments Division, Cleveland, Ohio.)

Figure 8–29. *X-Y* recorder. (Courtesy of Esterline Angus Instruments Division.)

Pen and ink recorders have a pen that touches a paper chart. The ink leaves a permanent record of the measured quantity on the chart. The charts are either roll charts that revolve on rollers under the pen mechanism or circular charts that rotate under the pen. Chart recorders sometimes use more than one pen to record several different quantities simultaneously.

The pen of a chart recorder is connected to the meter movement. The pen is supplied with a source of ink. It is moved by the meter movement in the same way as the pointer of a hand-deflection meter. The charts used ordinarily have lines that indicate the amount of pen movement. Spaces on the chart are marked according to time periods. Charts are moved under the pen at a constant speed. Spring-drive mechanisms, synchronous ac motors, or dc servomotors are used to drive charts.

Inkless recorders have voltage applied to the pen point. Heat is produced which causes a mark to be made on a sensitive paper chart. The advantage of inkless recorders is that ink is not required. Ink must be replaced and it is often messy to use.

A special type of chart recording instrument is the *X-Y* recorder shown in Fig. 8–29. *X-Y* recorders are used to plot the relationship of two quantities. Often some variable quantity is plotted versus time. One variable quantity could also be plotted versus another variable quantity. One input is applied to cause vertical (y axis) movement of the indicating pen. The other input causes horizontal (x axis) movement of the pen. The plot produced on the chart shows the relationship of the two quantities connected to the X and Y inputs.

REVIEW

1. What are the five basic types of instruments?
2. Discuss the construction of a meter movement.
3. What is a single-function meter?
4. What is a multifunction meter?
5. How is the range of a meter movement extended to measure higher current values?
6. What is meant by the following abbreviations? (a) I_{FS}, (b) R_m, (c) R_{sh}, and (d) I_{sh}
7. How is a meter movement used to measure voltages?
8. What is meant by the following abbreviations? (a) V_{FS} and (b) R_{mult}
9. What is meter sensitivity?
10. How is a meter movement used to measure resistances?
11. What is the difference between a linear and nonlinear scale?
12. What is the purpose of the following in an ohmmeter circuit? (a) R_{lim}, (b) voltage source, and (c) ohms adjust resistance

13. Discuss the method used to calibrate an Ohm's scale.
14. Describe the construction of an electrical power meter or wattmeter.
15. What is the difference between dc power and ac power measurement?
16. Describe a kilowatt-hour (kWh) meter.
17. How is three-phase ac power measured?
18. How could power factor of an ac circuit be determined if a power factor meter is not available?
19. Why is power demand important to industries?
20. What is the purpose of a ground-fault indicator?
21. What is the purpose of a megger?
22. What is the purpose of a Wheatstone bridge?
23. Discuss the operation of an oscilloscope.
24. What is an advantage of using a numerical readout instrument?
25. Why are chart recording instruments used?

STUDENT ACTIVITIES

• *Measuring Analog Meter Movement Resistance*

1. Obtain a meter movement of any value, two potentiometers (1000- to 25,000-Ω values will usually work), a VOM, and a 6-V lantern battery.
2. Meter movement resistance cannot be measured with a VOM. It would possibly damage the moving coil because it has a low current rating.
3. Connect the circuit shown in Fig. 8–30. Record all data on a sheet of paper.
4. Disconnect the wire at point A and adjust R_1 until the meter reads full scale. R_1 should first be set to maximum resistance.
5. Connect the wire at point A. Adjust R_2 until the meter reads half scale. The resistance of R_2 (from point A to point B) will now equal the resistance of the meter movement.

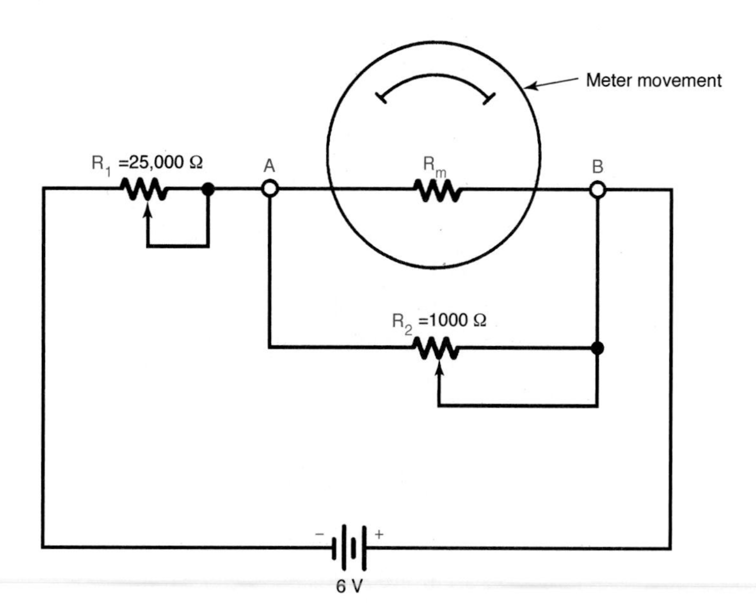

Figure 8–30. Circuit for measuring meter movement resistance.

6. Do not disturb the setting of R_2. Remove it from the circuit. Measure the resistance from point A to point B with a VOM. R = ______ Ω.
7. On a sheet of paper, discuss the construction and operation of a moving-coil meter movement.

• Current Meter Design (see Fig. 8–3)

In this activity, any meter movement with known values of I_{FS} and R_m can be used. If no meter movements are available, values of I_{FS} and R_m can be selected by the teacher. In either case, the following procedure is used to design a current meter to measure a given value of current flow.

1. Obtain a meter movement, a 6-V battery, and a resistor (50- to 250-Ω values will work well). Record all data on a sheet of paper.
2. Using any meter movement with known resistance (R_m) and current (I_{FS}), calculate the values of shunt resistance (R_{sh}) needed to extend its range to measure 200 mA of current.
3. Use the necessary resistor combinations connected together to construct the shunt.
4. Connect the shunt across the meter movement. Have the teacher check the R_{sh} value.
5. Connect the resistor (50 to 250 Ω) to the 6-V battery. Then put the meter that was designed in series with this circuit to test the meter.
6. Describe the construction of the meter movement.
7. If a 200-Ω, 1-mA meter movement is used, what are the values of shunt resistance required to extend its range to measure the following currents? Show your calculations on a sheet of paper.
   a. 5 mA—R_{sh} = ______ Ω.
   b. 10 mA—R_{sh} = ______ Ω.
   c. 50 mA—R_{sh} = ______ Ω.
   d. 100 mA—R_{sh} = ______ Ω.
   e. 9.5 mA—R_{sh} = Ω.
   f. 1 A—R_{sh} = ______ Ω.

• Current Measurement—Meter Shunts

Complete Activity 8–1, page 189 in the manual.

• • ANALYSIS

1. Describe the construction of an analog hand-deflection meter movement.
2. If a 200-ohm, 1.0-mA meter movement is employed, what are the values of shunt resistance required to extend its range to measure the following currents?
   a. 5 mA; R_{sh} = ______ Ω.
   b. 10 mA; R_{sh} = ______ Ω.
   c. 50 mA; R_{sh} = ______ Ω.
   d. 100 mA; R_{sh} = ______ Ω.
   e. 0.5 A; R_{sh} = ______ Ω.
   f. 1.0 A; R_{sh} = ______ Ω.

• Voltmeter Design (see Fig. 8–5)

In this activity, any meter movement with known values of I_{FS} and R_m can be used. If no meter movements are available, values of I_{FS} and R_m can be selected by the teacher. In either case, the following procedure is used to design a voltmeter to measure a given value of voltage. In this activity, 10 V will be used as the voltage range.

1. Obtain a meter movement and a 6-V battery. Record all data on a sheet of paper.
2. Find the voltage for full-scale deflection of the meter movement.

$$V_{FS} = I_{FS} \times R_m = \_\_\_\_\_\_ \text{ V}$$

3. Find the voltage that must be dropped across the multiplier resistor to measure 10 V.

$$V_{mult} = 10 \text{ V} - V_{FS} = \_\_\_\_\_\_ \text{ V}$$

4. Find the value of multiplier resistance needed in series with the meter movement to measure 10 V.

$$R_{mult} = \frac{V_{mult}}{I_m} = \_\_\_\_\_\_ \ \Omega$$

5. Use the necessary resistor combination connected together to construct the multiplier.
6. Connect the multiplier in series with the meter movement.
7. Have the teacher check the R_{mult} value.
8. Connect a 6-V battery across the meter which was designed to test the meter.
9. Calculate the values of multiplier resistance needed to extend the range of a 1-mA 100-Ω meter movement to measure the following voltages.
   - **a.** 1 V, R_{mult} = ______ Ω.
   - **b.** 5 V, R_{mult} = ______ Ω.
   - **c.** 15 V, R_{mult} = ______ Ω.
   - **d.** 150 V, R_{mult} = ______ Ω.

• Voltage Measurement—Meter Multipliers

Complete Activity 8–2, page 191 in the manual.

• • ANALYSIS

1. Compute the ohms-per-volt rating of a 200-μA meter movement.
2. What is meant by meter "loading" of a circuit?
3. Calculate the values of multiplier resistance needed to extend the range of a 1.0-mA, 100-ohm meter movement to measure the following voltages:
   - **a.** 1.0 volt; R_{mult} = ______ Ω.
   - **b.** 5 volts; R_{mult} = ______ Ω.
   - **c.** 15 volts; R_{mult} = ______ Ω.
   - **d.** 150 volts; R_{mult} = ______ Ω.

• Ohmmeter Design [see Fig. 8–7(a)]

In this activity, any meter movement with known values of I_m and R_m can be used. If no meter movements are available, values of I_m and R_m can be selected by the teacher. In either case, the following procedure is used to design an ohmmeter.

1. Obtain a meter movement and a 1.5-V battery. Record all data on a sheet of paper.
2. Find the total resistance needed in the ohmmeter circuit using a 1.5-V battery.

$$R_T = \frac{1.5 \text{ V}}{I_m}$$

3. Find the value of limiting resistor (R_{lim}) needed.

$$R_{\text{lim}} = R_T - R_m$$

4. Use the necessary resistor combination connected together to construct the limiting resistor (R_{lim}).
5. Connect R_{lim} in series with the meter movement and a 1.5-V battery.
6. Have the teacher check the R_{lim} value.
7. Draw a half-circle on a sheet of paper to represent the ohmmeter scale.
8. Mark the following points on the scale.
   a. Full scale on the right side (0 Ω)
   b. Infinite (∞) on the left side (∞Ω)
   c. Half scale ($R_{\text{lim}} + R_m$ value) ______ Ω
9. Take some known values of resistors (1 to 100 kΩ) and test the meter which was designed.
10. Find the value of R_{lim} needed for a 1-mA, 100-Ω meter movement using a 9-V battery.
11. Turn in the data to the teacher for grading.

• Resistance Measurement—Ohmmeters

Complete Activity 8–3, page 193 in the manual.

• • ANALYSIS

1. Explain why the scale of an ohmmeter is nonlinear.
2. Why would a higher voltage battery be used for an $R \times 10\text{K}$ range than for an $R \times 1$ range of a multirange ohmmeter?
3. What is the purpose of the "Ohms Adjust" control of an ohmmeter?
4. Why should power never be applied to a circuit when resistance is being measured?
5. Calculate the value of current-limiting resistor needed for a 0- to 1-mA, 200-Ω meter movement used to design a series ohmmeter circuit with a 30-volt battery used as the power source.

• Wheatstone Bridge Measurement

Complete Activity 8–4, page 193 in the manual.

• •

ANALYSIS

1. What is the relationship between the two branches of a bridge network?
2. How does the voltage drop across R_S of Fig. 8–18 compare with the voltage drop across R_X? How does the drop across R_A compare with the drop across R_B?
3. Briefly describe the operation of the Wheatstone bridge.

• Self-Examination

1. Complete the Self-Examination for Chap. 8, page 197 in the manual.
2. Check your answers on pages 414–415 in the manual.
3. Study the questions that were answered incorrectly.

• Chapter 8 Examination

Complete the Chap. 8 Examination on page 199 in the manual.

PART

II

ELECTRONICS

CHAPTER 9 Electronic Basics

CHAPTER 10 Electronic Diodes

CHAPTER 11 Electronic Power Supplies

CHAPTER 12 Transistors

CHAPTER 13 Amplification

CHAPTER 14 Amplifying Systems

CHAPTER 15 Oscillators

CHAPTER 16 Communications Systems

CHAPTER 17 Digital Electronic Systems

CHAPTER 18 Electronic Power Control

CHAPTER

9

ELECTRONIC BASICS

OUTLINE

9.1 Semiconductor Theory
9.2 Semiconductor Materials
9.3 Electron Emission

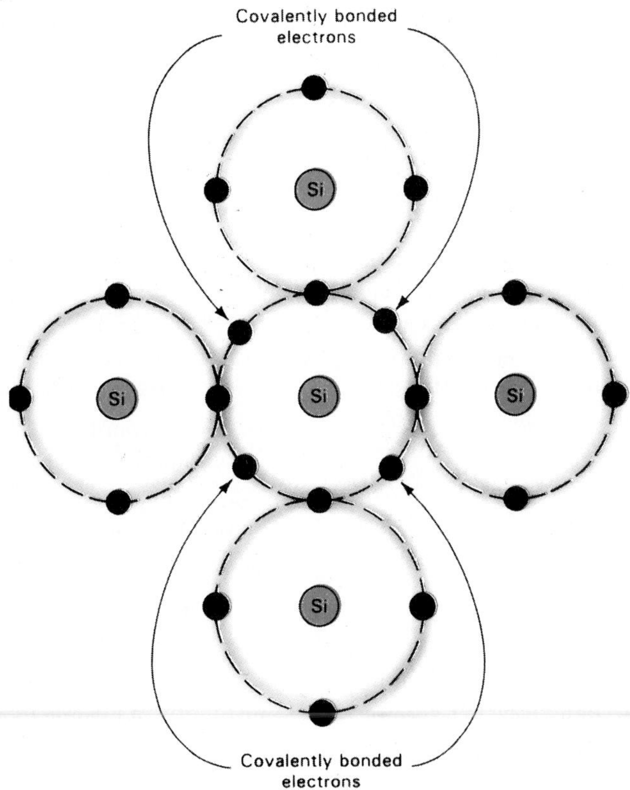

Covalently bonded silicon atoms.

OBJECTIVES

Upon completion of this chapter, you will be able to:

1. Explain the electrical differences among conductors, semiconductors, and insulators.
2. Describe how *N*-type and *P*-type semiconductors are produced.
3. Explain how current carriers move through semiconductors.
4. Explain the meaning of terms such as *intrinsic, doping, holes, covalent bonding, energy level,* and *emission.*

Electronics is a unique part of the electrical field. It has made possible such things as radio, television, computers, pocket calculators, and stereo amplifiers. The term *electronics* takes its name from the electron. Electrons are the tiny, negatively charged particles of an atom. Electronics deals with the flow of these particles through a number of rather specialized devices. One division of this field deals with devices that respond to electrons released from a metal surface. This is called electron emission. Vacuum tubes and photoelectric devices operate on this principle. Solid-state electronics is another division. It deals with the flow and control of electrons through semiconductor materials. Transistors, diodes, and integrated circuits are solid-state devices. Electronics deals with devices of both types.

In 1948, scientists at the Bell Telephone Laboratories were able to make pieces of solid material amplify an electric current. This development has brought a significant change to electronics. A number of unique solid-state devices have been developed. Solid-state electronic applications have surpassed the vacuum tube. Complete electronic systems are now built on a single integrated-circuit chip. These systems have caused a significant change in the electronics field.

Electronic basics deals with the elements that a person should know about electronic devices in order to use them. Solid-state basics will be discussed first, which deals with the structure of a semiconductor, crystal formation, covalent bonding, and current flow. Material theory of this type applies to all solid-state devices. Electron emission is then discussed. It is important today because of applications in communications and electronic control circuitry. Solid-state electronics and emission electronics both play an essential role in the field of electronics.

IMPORTANT TERMS

A number of new and, to some extent, unusual terms are used in this chapter. In general, these terms will assist the reader in understanding the material. Review each term carefully before proceeding with the chapter. Refer back to a specific term if its meaning is not clear in the text.

Acceptor impurity. An element with less than four valence electrons that is used as a dopant for semiconductors. Acceptor atoms mixed with silicon makes a *P* material.

Cathode. The negative terminal of an electronic device, such as a solid-state diode, which is responsible for electron emission.

Compound. The chemical combination of two or more elements to make an entirely different material.

Conduction band. The outermost energy level of an atom.

Covalent bonding. Atoms joined together by an electron-sharing process that causes a molecule to be stable.

Donor impurity. An element with more than four valence electrons that is used as a dopant for semiconductors. Donor atoms mixed with silicon makes an *N* material.

Dopant. An impurity added to silicon or germanium.

Doping. The process of adding impurities to a pure semiconductor.

Electrode. A specific part of a unit such as the cathode of a semiconductor.

Electron volt (eV). A measure of energy acquired by an electron passing through a potential of 1 V.

Electrostatic force. An interaction between negative and positive charged particles.

Emission. The release of electrons from the surface of a conductive material.

Energy level. A discrete amount of energy possessed by each electron in a specific shell or layer of an atom.

Extrinsic. A purified material that has a controlled amount of impurity added to it when manufactured.

Forbidden gap. An area between the valence and conduction band on an energy-level diagram.

Intrinsic. A semiconductor crystal with a high level of purification.

Ion. An atom that has lost or gained electrons, making it a negative or positive charge.

Ionic bond. A binding force that joins one atom to another in a way that fills the valence band of an atom.

Kinetic energy. Energy due to motion.

Light-dependent resistor (LDR). A photoconductive device that changes resistance with light energy.

Majority current carrier. Electrons of an *N* material and holes of a *P* material.

Metallic bond. A force such as a floating cloud of ions that holds atoms loosely together in a conductor.

Minority current carrier. A conduction vehicle opposite to the majority current carrier, such as holes in an *N* material and electrons in a *P* material.

Orbital. A mathematical probability where an electron will appear in the structure of an atom.

Photon. A small packet of energy or quantum of energy associated with light.

Quanta. A discrete amount of energy required to move an electron into a higher energy level.

Quantum theory. A theory based on the absorption and emission of discrete amounts of energy.

Secondary emission. The electrons released by bombardment of a conductive material with other electrons.

Solid state. A branch of electronics dealing with the conduction of current through semiconductors.

Stabilized atom. An atom that has a full complement of electrons in its valence band.

Thermionic emission. The release of electrons from a conductive material by the application of heat.

Valence band. The part of an energy-level diagram that deals with the outermost electrons of an atom.

Valence electrons. Electrons in the outer shell or layer of an atom.

9.1 SEMICONDUCTOR THEORY

Nearly every study of semiconductors begins with an investigation of atomic theory. The purpose of this investigation is to acquaint the reader with some important principles that show how a material responds electronically. In this regard, we need to become familiar with such things as energy levels, ionic bonding, covalent bonding, stable atoms, and electrical conductivity. Semiconductor materials respond to these ideas. By becoming familiar with this information, it is easier to understand how a particular device is formed and how it responds when energy is applied.

Elements are classified according to the number of particles they possess. Atoms of an element are distinguished by the number of protons in their nucleus, which is known as the *atomic number*. No two elements have the same atomic number. The nucleus of every atom is composed of one or more positively charged particles called protons. In addition to the protons, in all elements except the lightest, there are one or more neutrons. Hydrogen, which is the simplest atom, has one proton and no neutrons in its nucleus. The nucleus of uranium has 92 protons and 146 neutrons. The atomic weight of an atom is the sum of the protons and neutrons in the nucleus. The atomic weight of uranium is 238. The atomic weight of an element is rarely a whole number because it represents the average value of a large number of atoms. The atomic weight of a single atom is always a whole number.

Semiconductor electronics concerns primarily the electron content of an atom. An electron is a negatively charged particle that revolves around the nucleus of an atom. An electron is extremely small. It weighs 1/1850 as much as a proton. Practically all the weight of an atom is in the nucleus. Figure 9–1 shows the particle structure of an atom of copper. This structure is shown in a two-dimensional representation. Keep in mind that an atom is a three-dimensional object.

The electrons of an atom are not at equal distances from the nucleus. They revolve around the nucleus in shells or layers. Niels Bohr, a Danish physicist, identified these shells by the letters *K, L, M, N, O,* and *P.* The *K* shell is closest to the nucleus. Each shell or layer has a distinct energy level. Electrons do not exist in the space between these energy levels. The energy levels of a shell are identified by the letters *s, p, d, f, g,* and *h.* The *s* level is closest to the nucleus. A shell may have from one to six distinct energy levels in its structure. Energy levels represent two things: distance from the nucleus and the amount of energy possessed by an electron. Energy levels closest to the nucleus are lowest in value while those farther away are occupied by electrons with more energy. The structure of the atom determines this feature. The number of electrons in each energy level follows a unique pattern: 2, 6, 10, 14, 18, and 22. Table 9–1 shows the shells, the energy levels found in each shell, and the maximum number of electrons that exist in each level. Note that shell *P* has six energy levels. The maximum number of electrons at each level follows the 2, 6, 10, 14, 18, 22 pattern. This number sequence appears in the other shells but decreases in value with the last level. The complexity of an atom usually dictates its electron shell and energy-level assignment.

The exact path an electron follows in the structure of an atom is not known. This path is generally described as an orbital. An orbital shows the mathematical probability of where an electron will appear in the structure of an atom. Today we believe there are four types of orbitals in atoms. These orbitals are identified by the letters *s, p, d,* and *f.* Figure 9–2 on page 316 shows the representation of *s, p,* and *d* orbitals. The *f*-type orbital is not shown in this sequence because little evidence suggests it is as active as the others in atomic bonding, but rather is more meaningful in chemical analysis.

Atom Combinations

Hydrogen is an atom that contains one proton in its nucleus. This proton possesses a positive charge. The number of electrons that

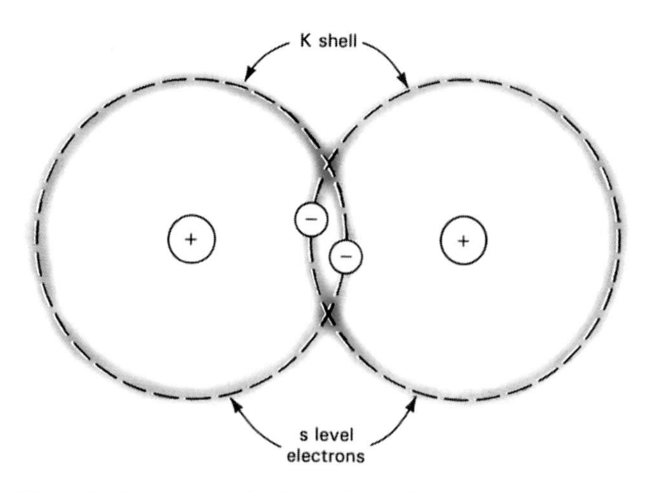

Figure 9–5. Two covalently bonded hydrogen atoms.

1.21 eV of energy at absolute zero (−273°C) to go into conduction. Energies of this magnitude are not acquired easily. As a result, the valence band remains full, the conduction band is empty, and these materials respond as insulators. However, an increase in temperature causes the conductivity of this material to change. At normal room temperature or 25°C, the valence electrons acquire thermal energy which is greater than the normal eV value. This essentially reduces the width of the forbidden gap and causes a semiconductor to become a conductor. This particular characteristic is extremely important in solid-state electronic devices.

The energy-level diagram of a conductor is quite unusual compared with other materials. In a sense, the valence band and conduction band are one and the same. Conductivity is explained as having an interaction of different energy

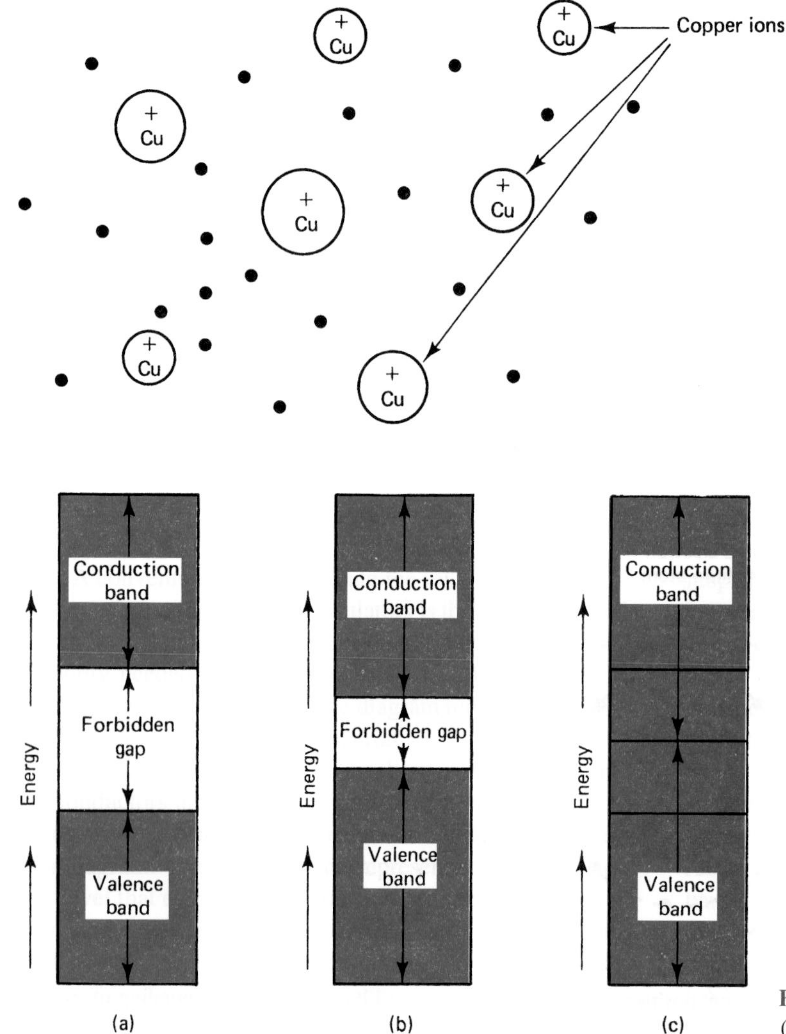

Figure 9–6. Metallic bonding of copper.

Figure 9–7. Energy-level diagrams for (a) insulators; (b) semiconductors; and (c) conductors.

levels of the valence band. Studies show that the atoms of most metals and semiconductors are in the form of a crystal lattice structure. A crystal consists of a space array of atoms or molecules built up by regular repetition in a three-dimensional pattern. The energy levels of electrons in the crystal do not respond in the same manner as those of an individual atom. When atoms form crystals, it is found that the energy levels of inner-shell electrons are not affected by the presence of neighboring atoms. However, the valence electrons of individual atoms are often shared by more than one atom. The new energy level of valence electrons is found in a distinct band. The spacing between the energy levels of this band is quite small compared with that of isolated atoms. Thus, electrons are free to absorb energy and to move from one point to another, therefore allowing conduction of heat and electricity. In good conductors, the energy-level bands of valence electrons tend to overlap, which lowers the energy level of valence electrons and increases the electrical conductivity of the material.

9.2 SEMICONDUCTOR MATERIALS

Materials such as silicon, germanium, and carbon are found naturally in crystalline form. Instead of being a random mass, these atoms are arranged in an orderly manner. A crystal of silicon or germanium forms a definite geometric pattern, namely, a cube. Figure 9–8 is a two-dimensional simplification of the silicon crystal showing only the nucleus and valence electrons. Note that the electrons of individual atoms are covalently bonded together. Germanium has a similar type of crystal structure.

Crystals of silicon and germanium can be manufactured by melting the natural elements. The process is somewhat complex and rather expensive to achieve. Manufactured crystals must be made extremely pure to be usable in semiconductor devices. A very pure semiconductor crystal is called an *intrinsic* material. Germanium is considered to be intrinsic when only one part of impurity exists in 10^8 parts of germanium. Silicon is intrinsic when the impurity ratio is $1{:}10^{13}$. In more sophisticated solid-state devices, the ratio may even be higher.

An intrinsic crystal of silicon would appear as the structure in Fig. 9–9. Covalent bonding changes the electrical conductivity of the material, causing each group of atoms to become stable; thus, only a limited number of free electrons are available for conduction. Intrinsic silicon and germanium crystals respond to some extent as insulators.

Intrinsic silicon at absolute zero (−273°C) is considered to be a perfect insulator. The valence electrons of each atom are firmly bound together in perfect covalent bonds. No free electrons are available for conduction. In actual circuit operation, the absolute-zero condition is not very meaningful. It cannot be attained readily. Any temperature above −273°C causes silicon to become somewhat conductive. The insulating quality of a semiconductor material is therefore dependent on its operating temperature.

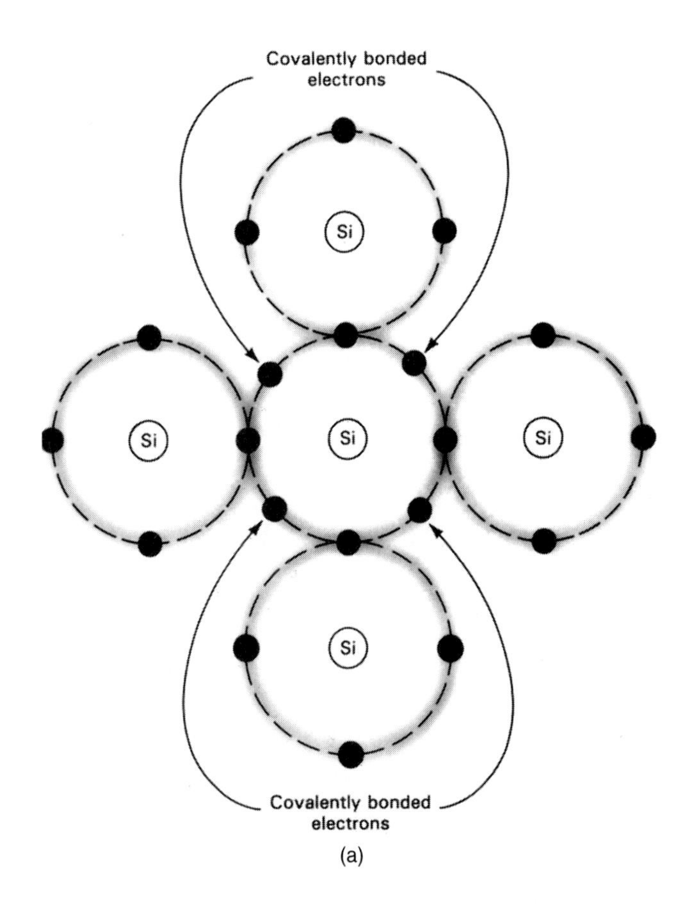

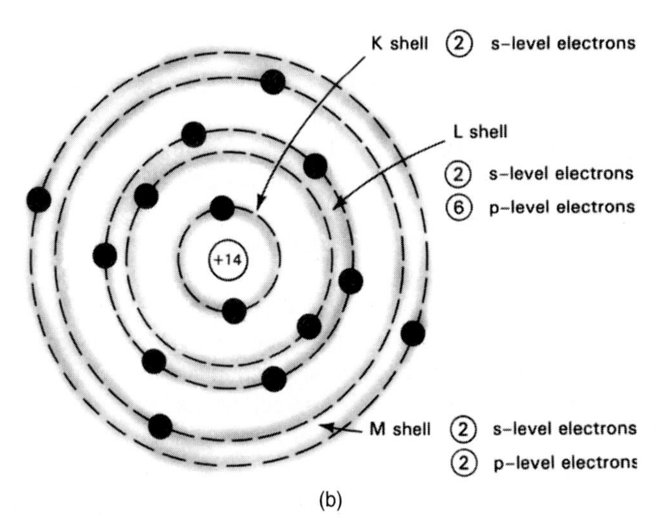

Figure 9–8. (a) Covalently bonded silicon atoms; (b) a single silicon atom (Si).

At room temperature, such as 25°C, silicon atoms receive enough energy from heat to break their bonding. A number of free electrons become available for conduction. At room temperature, intrinsic silicon becomes somewhat conductive.

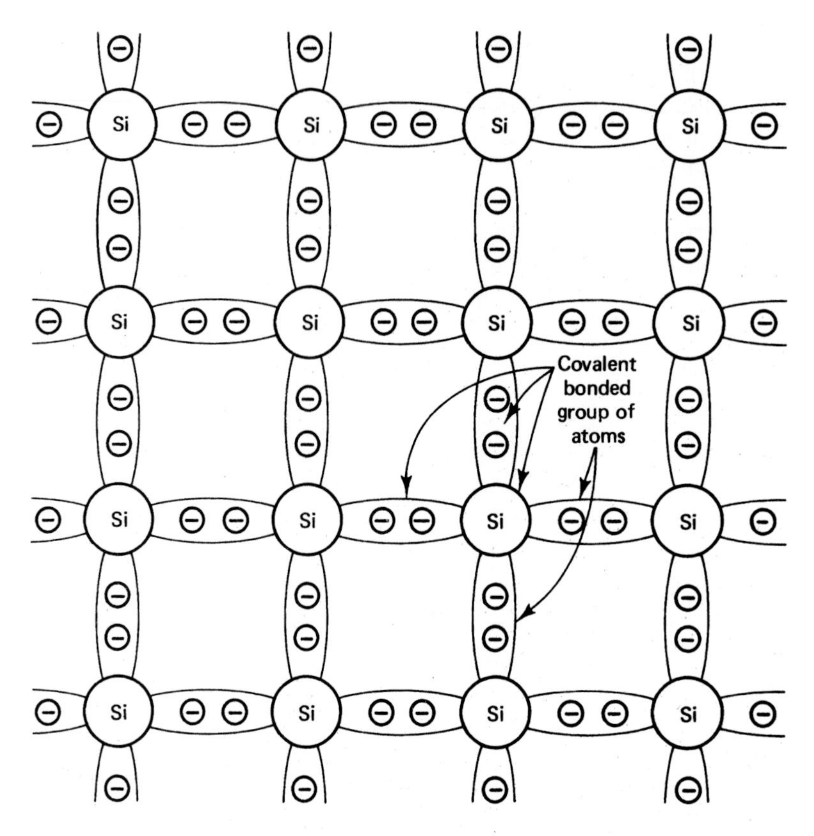

Figure 9–9. Intrinsic crystal of silicon.

An important solid-state concept takes place when intrinsic silicon goes into conduction. For every electron that is freed from its covalent bonding, a void spot is created. This spot, which is normally called a *hole,* represents an electron deficiency. A hole in a covalent bonding group occurs when an electron is released. An increase in the temperature of a piece of silicon causes the number of free electrons and holes to increase.

It should be remembered that when a neutral atom loses an electron, it acquires a positive charge. The atom then becomes a positive ion. Because an atom acquires a hole at the same time as the positive charge, the hole bears a positive charge. It should be noted, however, that the crystal remains electrically neutral. For every free electron, there is an equivalent hole. These two balance out the overall charge of the crystal.

When a valence electron leaves its covalent bond to become a free electron, a hole appears in its place. This hole can then attract a different electron from a nearby bonded group. Upon leaving its bonded group, this electron creates a new hole. The original hole is then filled and becomes void. Each electron that leaves its bonding to fill a hole creates a new hole in its original group. In a sense, this means that electrons move in one direction and holes in the opposite direction. Because electron movement is considered to be an electrical current, holes are also representative of current flow.

Electrons are called negative current carriers and holes are called positive current carriers. These current carriers move in opposite directions in a semiconductor material. When voltage is applied, holes move toward the negative side of the source. Electron current flows toward the positive side of the source. This condition only occurs in a semiconductor material. In a conductor such as copper there is no covalent bonding. Current carriers are restricted to only electrons.

Figure 9–10 shows how an intrinsic piece of silicon will respond at room temperature when voltage is applied. Note that the free electrons move toward the positive terminal of the battery. Electrons leaving the semiconductor flow into the copper connecting wire. Each electron that leaves the material creates a hole in its place. The holes appear to be moving by jumping between covalent bonded groups. Holes are attracted by the negative terminal connection. For each electron that flows out of the material, a new electron enters at the negative connection point. In effect, the new electron fills a hole at this point. The process of hole flow and electron flow through the material is continuous as long as energy is supplied. Current flow is the resulting carrier movement. An intrinsic semiconductor has an equal number of current carriers moving in each direction. The resultant current flow of a semiconductor is limited primarily to the applied voltage and the operating temperature of the material.

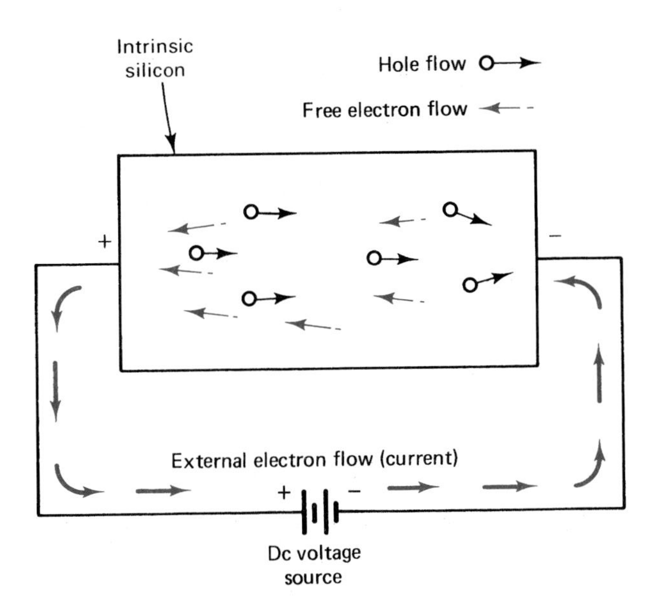

Figure 9–10. Current carrier movement in silicon.

Pure silicon or germanium in its intrinsic state is rarely used as a semiconductor. Usable semiconductors must have controlled amounts of impurities added to them. The added impurities change the conduction capabilities of a semiconductor. The process of adding an impurity to an intrinsic material is called *doping*. The impurity is called a *dopant*. Doping a semiconductor causes it to be an extrinsic material. Extrinsic semiconductors are the operational basis of nearly all solid-state devices.

N-Type Material

When a specific material or dopant is added to silicon or germanium which does not alter the structure of the basic crystal, an *N*-type material is produced. Atoms of arsenic (As) and antimony (Sb) have five electrons in their valence band (refer to the periodic table in Fig. 1–9). Adding this impurity to silicon must not alter the crystal structure or bonding process. Each impurity atom has an extra electron that does not take part in a covalent bonding group. These electrons are loosely held together by their parent atoms. Figure 9–11 shows how the crystal is altered with the addition of an impurity atom.

When arsenic is added to pure silicon, the crystal becomes an *N*-type material. It has extra electrons or negative (*N*) charges that do not take part in the covalent bonding process. Impurities that add electrons to a crystal are generally called *donor* atoms. An *N*-type material has more extra or free electrons than an intrinsic piece of material. A piece of *N* material is not negatively charged. Its atoms are all electrically neutral. There is, however, a number of extra electrons that do not take part in the covalent bonding process. These electrons are free to move about through the crystal structure.

An extrinsic silicon crystal of the *N* type will go into conduction with only 0.05 eV of energy applied. An intrinsic crystal requires 1.12 eV to move electrons from the valence band into the conduction band. Essentially, this means that an *N* material is a fairly good electrical conductor. In this type of crystal, electrons are considered to be the *majority current carriers*. Holes are the minority current carriers. The amount of donor material added to silicon determines the number of majority current carriers in its structure.

The number of electrons in a piece of *N*-type silicon is a million or more times greater than the electron-hole pairs of a piece of intrinsic silicon. At room temperature, there is a decided difference in the electrical conductivity of this material. Extrinsic silicon becomes a rather good electrical conductor. A larger number of current carriers take part in conduction. Current flow is achieved primarily by electrons in this material. Figure 9–12 shows how the current carriers respond in a piece of *N* material. A larger number of electrons are indicated than holes, which shows that electrons are the majority current carriers and holes are the minority carriers.

If the voltage source of Fig. 9–12 is reversed, the current flow would reverse its direction. Essentially, this means that *N*-type silicon will conduct equally as well in either direction. The flow of current carriers is simply reversed, and is an important consideration in the operation of a device that employs *N* material in its construction.

P-Type Material

P-type material is formed when indium (In) or gallium (Ga) is added to intrinsic silicon. Indium or gallium are type III elements according to the periodic table of Fig. 1–9. These elements are often called *acceptors*—they are readily seeking a fourth electron. In effect, this type of dopant material has three valence electrons. Each covalent bond that is formed with an indium atom will have an electron deficiency or hole, which represents a positive charge area in the covalent bonding structure. A positive charge area crystal or *P* material describes this type of structure. Each hole in the *P* material can be filled with an electron. Electrons from neighboring covalent bond groups require little energy to move in and fill a hole.

The ratio of doping material to silicon is typically in the range 1 to 10^6. This means that *P* material will have a million times more holes than the heat-generated electron-hole pairs of pure silicon. At room temperature, there is a decided difference in the electrical conductivity of this material. Figure 9–13 shows how the crystal structure of silicon is altered when doped with an acceptor element—in this case, indium.

A piece of *P* material is not positively charged. Its atoms are primarily all electrically neutral. There are, however, holes in the covalent structure of many atom groups. When an electron moves in and fills a hole, the hole becomes void. A new hole is created in the bonded group where the

gap and move into the conduction band. See the energy-level diagram of Fig. 9–7. Continued energy increases may cause electrons to escape the retaining forces of the parent atom. This effect is called electron emission.

Thermionic Emission

The process of freeing electrons from a metal or metal substance by heat is called *thermionic emission*. Ordinarily, electrons would not escape from the surface of a solid substance. In the structure of an atom, electrons are in a continuous state of motion. If the material is heated, individual electrons begin to move faster. Kinetic energy, which is a result of motion, increases with electron speed. Valence electrons eventually gain enough energy to escape from the surface of the material. The amount of heat needed to produce emission is usually quite high. For materials such as oxided barium, 1000°C is needed. Tungsten emits electrons at 2300°C. The emitting material is normally heated by electricity. Tungsten, which has a rather high resistance, produces heat when current flows through it.

When heat is applied directly to a piece of metal, electrons are emitted from its surface. The emitting electrode, in this case, is called a filament. Tubes with this type of emitter are called directly heated filaments. Battery-powered tube circuits and high-power transmitting tubes employ directly heated filaments. Figure 9–15 shows a directly heated filament and its schematic symbol.

If alternating current is used to energize the tube filament, then electron emission will vary according to the frequency of the source. These fluctuations will produce an ac hum in the stream of emitted electrons. In most applications, this causes a form of interference. Indirect heating of the emitting surface isolates it from the ac filament source. The emitting electrode of this tube is called a cathode. Essentially, the emitter consists of a cathode and a heater. The heater or filament is placed inside the cathode. Heat from the filament is transferred to the cathode. Electrons are emitted from the surface of the cathode after it has been heated.

Indirectly heated cathodes do not reach emission temperatures as quickly as directly heated surfaces. Heat takes time to be transferred to the cathode. The cathode usually has a rather large surface area compared with a directly heated filament. Once the surface is raised to its emitting temperature, it retains this heat for a short time. Its temperature will therefore not fluctuate with ac heater voltage. Indirectly heated tubes have a constant emission level while in operation. Figure 9–16 shows the structure of an indirectly heated tube and its schematic symbol.

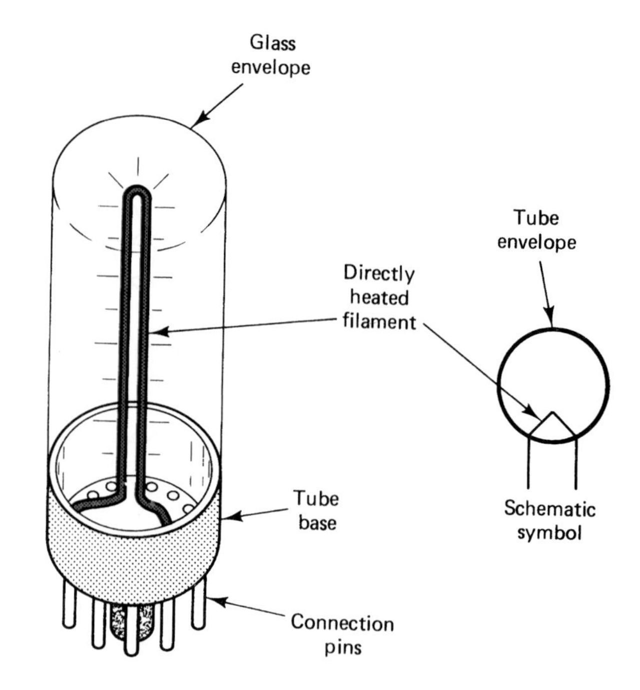

Figure 9–15. Directly heated filament.

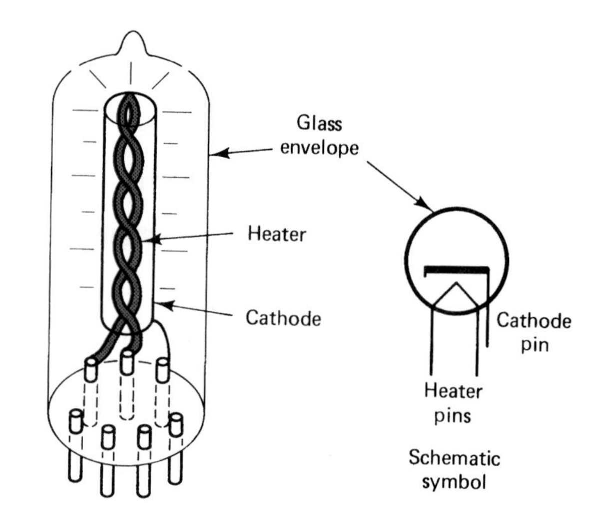

Figure 9–16. Indirectly heated tube.

Secondary Emission

After being emitted from the cathode surface, electrons have little velocity. A large metal electrode, with a high positive charge, is often placed inside the tube. This electrode is called a plate or anode. The positive charge on the plate attracts the electrons emitted from the cathode. A high positive charge will cause electrons to move to the plate with tremendous velocity. When striking the plate, these electrons will transfer some of their energy to the atoms of the plate material. If the energy level is great enough, it will cause electrons to be released from the surface of the plate. This effect is called *secondary emission.* Figure 9–17 shows an illustration of secondary emission.

In normal circuit operation, each primary electron may cause the release of several secondary electrons, meaning that a cloud of secondary electrons may form around the sur-

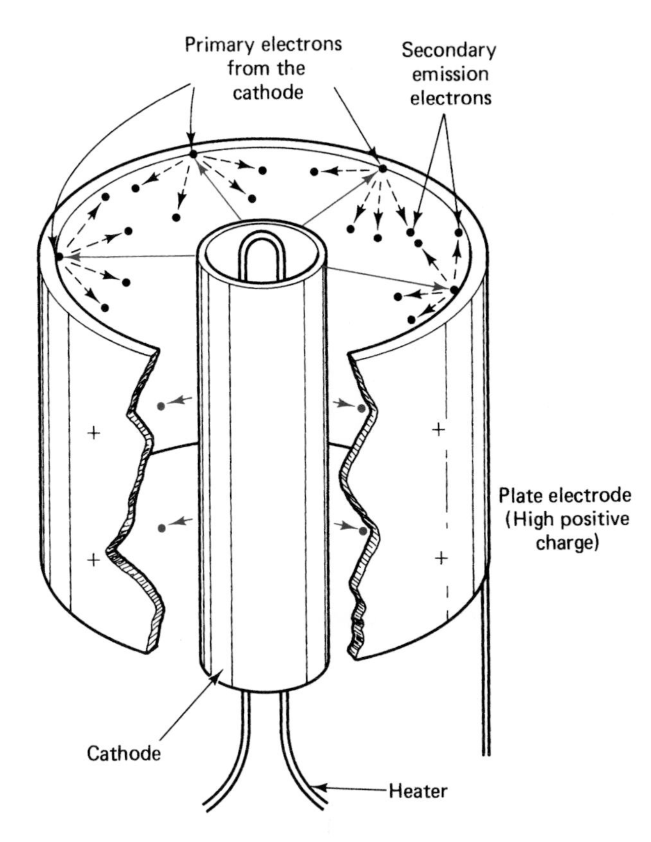

Figure 9–17. Secondary emission.

face of the plate. In effect, this can restrict the flow of primary electrons. As a general rule, secondary emission is not a desirable condition. In some tubes, special provisions are made to reduce or eliminate secondary emission. In some phototubes, secondary emission is used to achieve electron multiplication. This action permits the tube to have greater output capabilities.

Photoemission

Materials that emit or give off electrons when light strikes them are considered to be photoemissive. These materials are sensitive to light. Devices that make use of this effect are called photoelectric tubes. These devices are used to detect light in various control applications.

The operating principle of a photoemissive device is best explained by the quantum theory of light. According to this theory, light consists of tiny packages of energy called photons. A small packet of energy is called a *quanta,* hence the name quantum theory. These packages contain uncharged particles that possess certain levels of energy. This energy depends on the frequency of light.

The construction of a photoemissive cell and its schematic symbols are shown in Fig. 9–18. In this illustration, photons of light energy enter the window of the cell and strike the light-sensitive material deposited on the cathode.

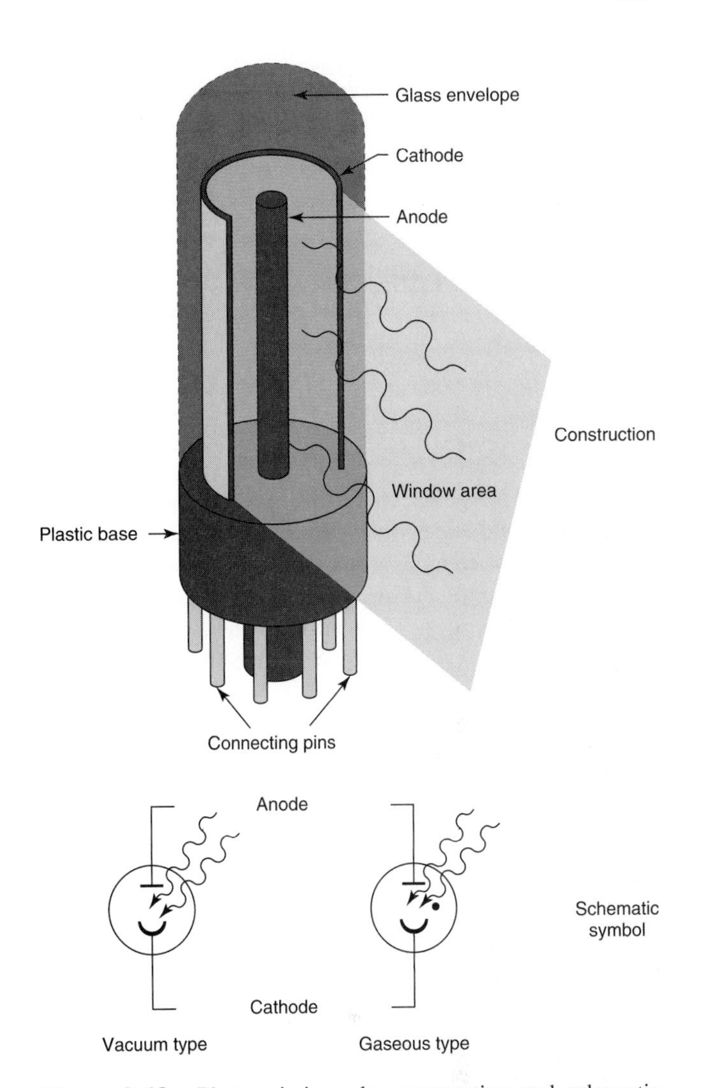

Figure 9–18. Photoemissive tube construction and schematic symbol.

When enough light energy is added to the kinetic energy of moving electrons in the cathode material, photoemission takes place. Electrons released from the cathode move away from the light-sensitive material. If the rod-shaped anode is positively charged, it attracts the emitted electrons. Photoemission is dependent on the shape and area of the two electrodes, the inside environment of the tube, and the voltage across the anode and cathode.

A light-detection circuit using a photoemissive cell is shown in Fig. 9–19. When light of the proper wavelength is focused on the cathode, electrons are emitted and travel to the positively charged anode. An anode-cathode current (V_{AK}) flows through the cell. This causes a voltage drop across the load resistor *RL.* The anode-cathode current caused by various combinations of light and anode voltage are determined by using characteristics supplied by the manufacturer.

At one time, all light detection was achieved by photoemissive tubes. The unusually high voltage needed to

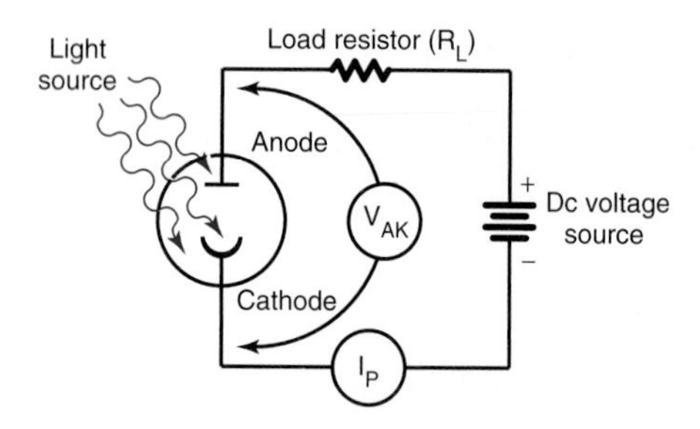

Figure 9–19. Phototube circuit operation.

energize a phototube is rather uncharacteristic of most electronic devices today. In general, this type of device has been replaced by solid-state photodevices. As a rule, circuit control can be achieved more efficiently with lower voltage values. The photoemissive principle is important, however, because it is the operational basis of the photomultiplier tube.

Photoconduction

A device that changes light intensity to electrical conduction is classified as a *photoconductive* device. Conduction or conductivity is the reciprocal of resistance. Conductivity is a measure of the ease with which current carriers pass through a material. Because of the relationship between conductance and resistance, these devices are usually called photoresistive cells, or light-dependent resistors (LDRs). These devices are essentially semiconductors. Light-sensitive materials such as cadmium sulfide (CdS), cadmium selenide (CdSe), and cadmium telluride (CdTe) are used in the construction of these devices. The material determines how it will respond to different levels of light and its wavelength.

Figure 9–20 shows the construction of a CdS photoresistive cell. A thick film of cadmium sulfide is deposited on a glass substrate. The film is in a zig-zag pattern that divides the top side of the glass substrate. Aluminum is deposited on each side of the pattern. Leads are attached to the aluminum on each side of the pattern. A transparent window is placed over the top of the cell. The entire assembly is then covered with a protective coating, which seals the cell and protects it from air and foreign material.

Operation of a CdS cell is based on the amount of light energy striking its working surface. Photons of light striking the cadmium sulfide surface transfer their energy to the valence electrons of each atom. If the transfer of energy is great enough, valence electrons will cross the forbidden gap of the material and go into the conduction band. These electrons then become free and respond as current carriers. Each electron that goes into the conduction band leaves a corresponding hole in the crystal structure. The amount of applied light energy therefore determines the electron-hole production of the material. An increase in the number of current carriers causes a decrease in material resistance. Electron and hole movement through the cell is from one side of the pattern to the other. An increase in the number of current carriers causes a decrease in material resistance across the cell.

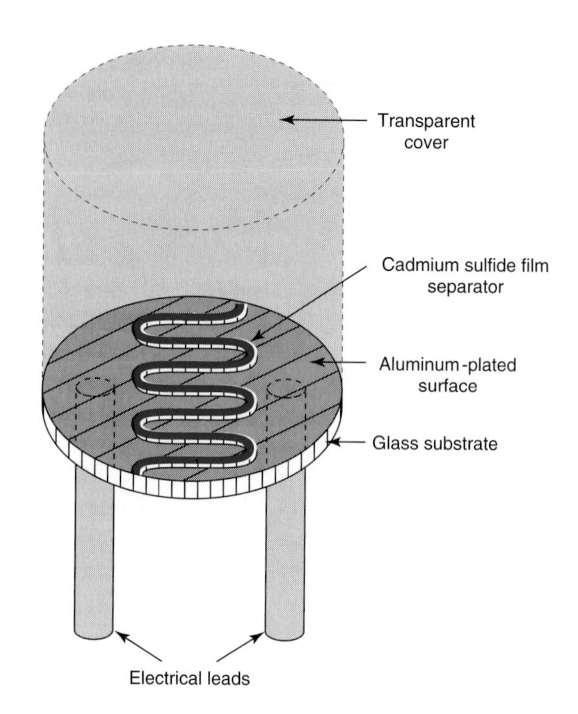

Figure 9–20. Cadmium sulfide cell construction.

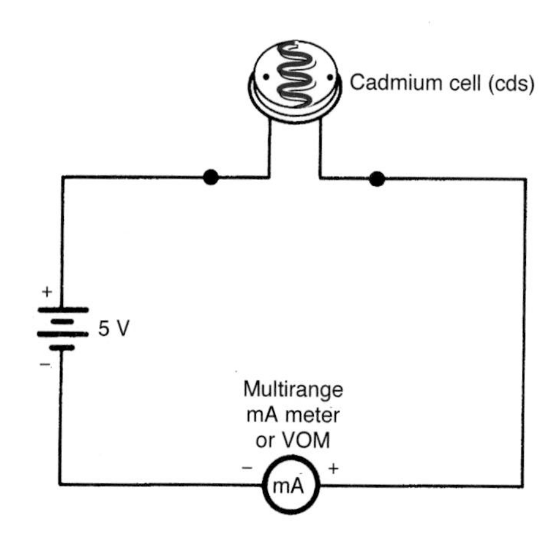

Figure 9–21. Light-detection circuit using a cadmium sulfide cell.

A light-detection circuit using a cadmium sulfide cell is shown in Fig. 9–21. The CdS cell responds in this circuit as a variable resistor. In a dark area the cell is highly resistant; little current will be indicated by the meter. An increase in light intensity will cause a decrease in cell resistance. This, in turn, will cause an increase in circuit current. The meter shows how the cell responds to different light levels.

REVIEW

1. Explain the differences among conductors, semiconductors, and insulators.
2. What is ionic bonding?
3. Explain the meaning of covalent bonding.
4. Explain the energy-level concept of solid material conduction.
5. What is an intrinsic material and an extrinsic material?
6. How does electrical conduction occur in a piece of intrinsic silicon?
7. Explain how the current carriers of a piece of *N* material respond when a voltage source is connected to it.
8. How is current conduction achieved in *P* material?
9. Explain how thermionic emission is achieved.
10. What is secondary emission?
11. What is a photon?
12. How is photoemission achieved?
13. How does the current conduction of a photoemissive cell change when exposed to light?
14. Explain the differences in energy-level diagrams for insulators, semiconductors, and conductors.

STUDENT ACTIVITIES

• *N-Material Conduction or Complete Activity 9–1, page 201 in the manual.*

1. Select either an *N*-channel or a *P*-channel junction field-effect transistor or JFET.
2. With an ohmmeter, measure the resistance between any two of the three leads. If a resistance value of a few hundred ohms is measured, reverse the two leads. A reading of the same resistance in the reverse direction indicates the *N* channel of the device. Find the two leads of the JFET where the resistance value is the same in each direction. Record the measured resistance.
3. Connect the two JFET leads into the circuit as indicated in Fig. 9–22.
4. Turn on the voltage source. Measure and record the value of current.
5. Turn off the voltage source. Reverse the two selected leads.
6. Turn on the voltage source. Measure and record the value of current. This shows how current passes through an *N* or *P* material.

• •

ANALYSIS

1. Explain the relationship between current and voltage applied to a semiconductor material.
2. What influence does an *N*-type semiconductor material have on direction of current flow?
3. How does the application of heat alter the current through a semiconductor material?

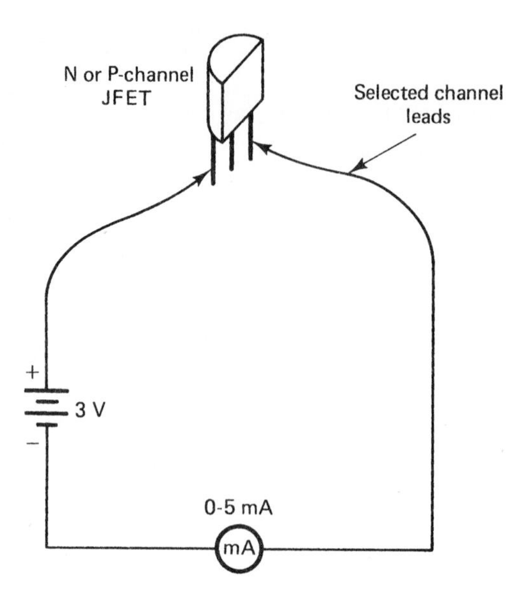

Figure 9–22. Semiconductor conduction circuit.

Light-Dependent Resistor Testing

1. Connect an ohmmeter to the leads of a CdS cell. Use an $R \times$ 1-kΩ range.
2. Cover the cell so that no light is on the lens. The dark resistance is _______ Ω.
3. In normal room light the resistance is _______ Ω.
4. Exposed to bright sunlight or near an incandescent lamp of 100 W, the resistance is _______ Ω.
5. Connect the light-detection circuit of Fig. 9–21.
6. Test the operation of the cell for dark, normal room light, and bright light. Measure and record the resultant current values for these light levels.

Self-Examination

1. Complete the Self-Examination for Chap. 9, page 205 in the manual.
2. Check your answers on page 415 in the manual.
3. Study the questions that were answered incorrectly.

Chapter 9 Examination

Complete the Chap. 9 Examination on page 207 in the manual.

CHAPTER

10

ELECTRONIC DIODES

OUTLINE

10.1 Junction Diode
10.2 Junction Biasing
10.3 Diode Characteristics
10.4 Diode Specifications
10.5 Diode Packaging
10.6 Diode Testing
10.7 Semiconductor Diode Devices
10.8 Vacuum Tubes

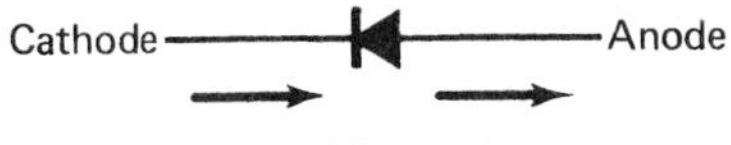

Crystal structure and symbol of a diode.

OBJECTIVES

Upon completion of this chapter, you will be able to:

1. Explain how a diode is constructed of *N* and *P* material.
2. Connect a diode so that it is either forward or reverse biased.
3. Distinguish between the forward and reverse characteristics of a diode.
4. Interpret the graphic relationship of plotted current and voltage characteristics for solid-state diodes.
5. Identify the schematic symbol of a diode while noting the name of the electrodes and the material of its construction.
6. Identify the electrodes of a diode with an ohmmeter and evaluate the condition of the device as good, shorted, or open.
7. Interpret the information from a manufacturer's data sheet for a silicon diode.
8. Identify several diodes by their package style or case structure.

Diodes are used rather extensively in nearly all phases of the electronics field today. Radio, television, industrial control circuits, home entertainment equipment, computers, and electrical appliances are only a few of the applications. Diodes are probably the simplest of all electronic components. Only two leads or electrodes are used in their construction. Functionally, this device has unidirectional conductivity, which means that it conducts well in only one direction. Ac can be converted easily into dc with a diode. In this application, a diode is used as a rectifier. In addition to this, other functions can be performed. Some of these are the operational bases of more complex electronic devices. Diode operation is therefore extremely important. A person working in electronics must be familiar with the characteristic operation of this device.

IMPORTANT TERMS

The study of diode devices frequently involves some rather unusual terms in common usage. As a rule, these terms play a key role in the presentation of new material. Review these terms carefully before proceeding with the text of this chapter.

Anode. The positive terminal or electrode of an electronic device such as a solid-state diode.

Barrier potential or voltage. The voltage that is developed across a *P-N* junction due to the diffusion of holes and electrons.

Bias, forward. Voltage applied across a *P-N* junction that causes the conduction of majority current carriers.

Bias, reverse. Voltage applied across a *P-N* junction that causes little or no current flow.

Covalent bonding. Atoms joined together by an electron-sharing process that causes a molecule to be stable.

Depletion zone. The area near a *P-N* junction that is void of current carriers.

Doping. The process of adding impurities to a pure semiconductor.

Efficiency. A ratio between output and input.

Electron. A negatively charged atomic particle; electrons cause the transfer of electrical energy from one place to another.

Energy. Something that is capable of producing work, such as heat, light, chemical, and mechanical action.

Filament. The heating element of a vacuum tube.

Hole. A charge area or void where an electron is missing.

Ion. An atom that has lost or gained electrons, making it a negative or positive charge.

Junction. The point where *P* and *N* semiconductor materials are joined.

Thermionic emission. The release of electrons from a conductive material by the application of heat.

Valence electrons. Electrons in the outer shell or layer of an atom.

10.1 JUNCTION DIODE

The term *junction diode* is often used to describe a crystal structure made of *P* and *N* materials. A diode is generally described as a two-terminal device. One terminal is attached to *P* material and the other to *N* material. The common connecting point where these materials are joined is called a *junction.* A junction diode permits current carriers to flow readily in one direction and blocks the flow of current in the opposite direction.

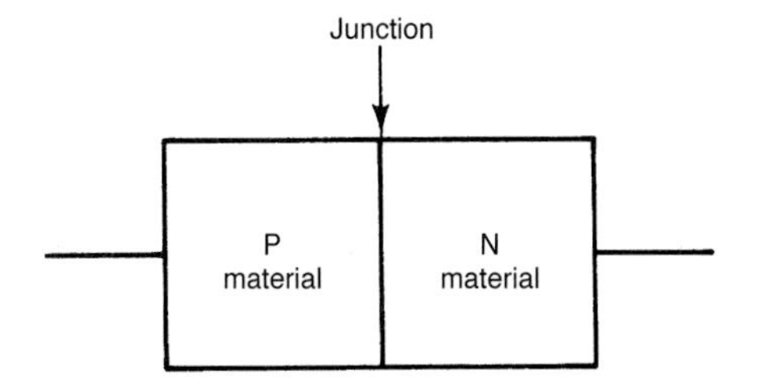

Figure 10–1. Crystal structure of a junction diode.

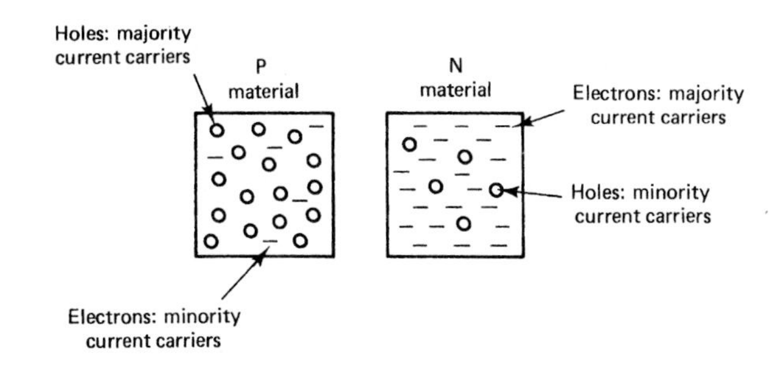

Figure 10–2. Semiconductor materials.

Figure 10–1 shows the crystal structure of a junction diode. Notice the location of the *P* and *N* materials with respect to the junction. The crystal structure is continuous from one end to the other. The junction serves only as a dividing line that marks the ending of one material and the beginning of the other. This type of structure permits electrons to move readily in one direction.

Figure 10–2 shows two pieces of semiconductor material before they are formed into a *P-N* junction. As indicated, each piece of material has majority and minority current carriers. The number of carrier symbols shown in each material indicates the minority or majority function. Electrons are the majority carriers in the *N* material and are the minority carriers in the *P* material. Holes are the majority carriers in the *P* material and are in the minority in the *N* material. Both holes and electrons have freedom to move about in their respective materials.

Depletion Zone

When a junction diode is first formed, a unique interaction occurs between current carriers. Electrons from the *N* material move readily across the junction to fill holes in the *P* material. This action is commonly called *diffusion.* Diffusion is the result of a high concentration of carriers in one material and a lower concentration in the other. Only those current carriers near the junction take part in the diffusion process.

The diffusion of current carriers across the junction of a diode causes a change in its structure. Electrons leaving the *N* material cause positive ions to be generated in their place. Upon entering the *P* material to fill holes, these same electrons create negative ions. The area on each side of the junction then contains a large number of positive and negative ions. The number of holes and electrons in this area becomes depleted. The term *depletion zone* is used to describe this area. It represents an area that is void of majority current carriers. All *P-N* junctions develop a depletion zone when they are formed. Figure 10–3 shows the depletion zone of a junction diode.

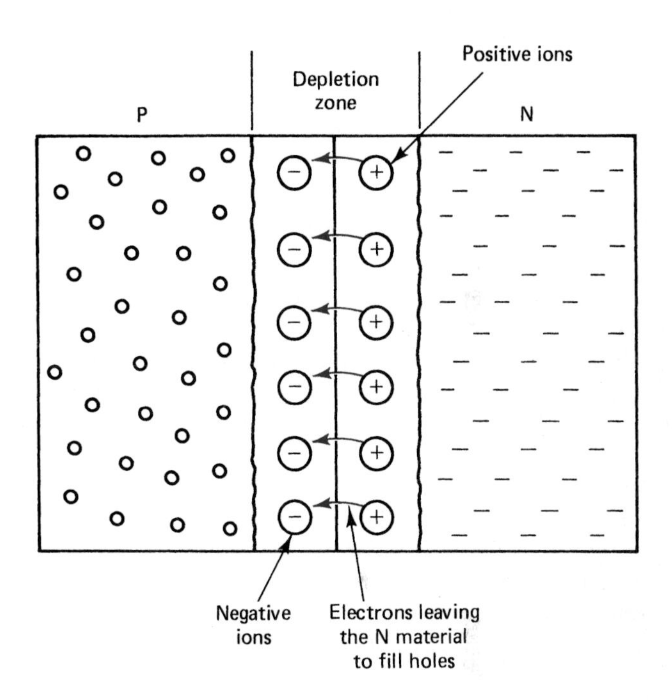

Figure 10–3. Depletion-zone formation.

Barrier Potential

Before *N* and *P* materials are joined together at a common junction, they are considered to be electrically neutral. After being joined, however, diffusion takes place immediately. Electrons crossing the junction to fill holes cause negative ions to appear in the *P* material. This action causes the area near the junction to take on a negative charge. Similarly, electrons leaving the *N* material cause it to produce positive ions. This, in turn, causes the *N* side of the junction to take on a net positive charge. These respective charge creations tend to drive remaining electrons and holes away from the junction. This action makes it somewhat difficult for additional charge carriers to diffuse across the junction. The end result is a charge buildup or *barrier potential* appearing across the junction.

Figure 10–4 shows the resulting barrier potential as a small battery connected across the *P-N* junction. Note the polarity of this potential with respect to *P* and *N* material. This voltage will exist even when the crystal is not connected to an outside source of energy. For germanium, the barrier

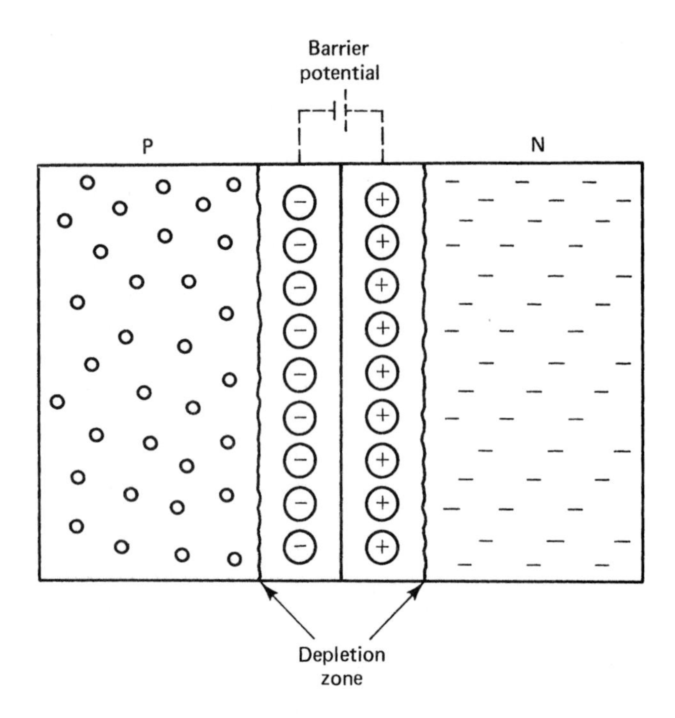

Figure 10–4. Barrier potential of a diode.

potential is approximately 0.3 V, and is 0.7 V for silicon. These voltage values cannot, however, be measured directly. They appear only across the space charge region of the junction. The barrier potential of a *P-N* junction must be overcome by an outside voltage source to produce current conduction.

10.2 JUNCTION BIASING

When an external source of energy is applied to a *P-N* junction, it is called a bias voltage, or simply biasing. This voltage either adds to or reduces the barrier potential of the junction. Reducing the barrier potential causes current carriers to return to the depletion zone. Forward biasing is used to describe this condition. Adding external voltage of the same polarity to the barrier potential causes an increase in the width of the depletion zone. This action hinders current carriers from entering the depletion zone. The term *reverse biasing* is used to describe this condition.

Reverse Biasing

When a battery is connected across a *P-N* junction as in Fig. 10–5, it causes reverse biasing. Note that the negative terminal of the battery is connected to the *P* material and the positive to the *N* material. Connection in this way causes the battery polarity to oppose the material polarity of the diode. Because unlike charges attract, the majority charge carriers of each material are pulled away from the junction. Reverse biasing of a diode normally causes it to be nonconductive.

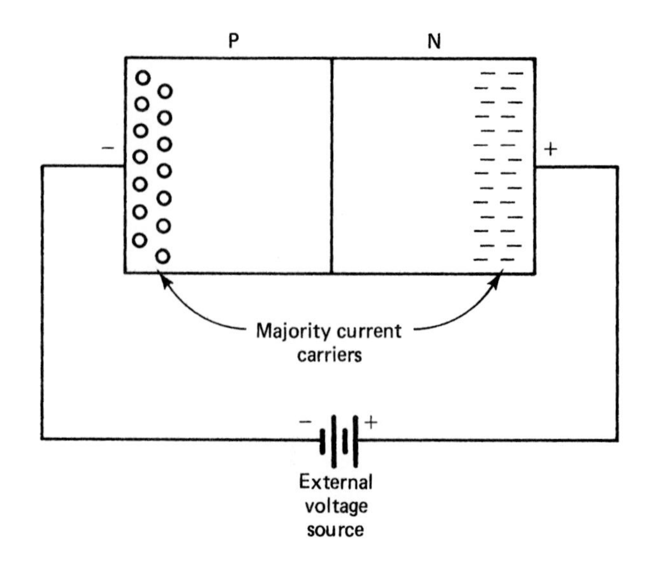

Figure 10–5. Reverse-biased diode.

Figure 10–6 shows how the majority current carriers are rearranged in a reverse-biased diode. As shown, electrons of the *N* material are pulled toward the positive battery terminal. Each electron that moves or leaves the diode causes a positive ion to appear in its place, which causes a corresponding increase in the width of the depletion zone on the *N* side of the junction.

The *P* side of the diode has a reaction similar to the *N* side. In this case, a number of electrons leave the negative battery terminal and enter the *P* material. These electrons immediately move in and fill a number of holes. Each filled hole then becomes a negative ion. These ions are then repelled by the negative battery terminal and driven toward the junction. As a result, the width of the depletion zone increases on the *P* side of the junction.

The overall depletion zone width of a reverse-biased diode is directly dependent on the value of the supply voltage. With a wide depletion zone, a diode cannot effectively support current flow. The charge buildup across the junction will increase until the barrier voltage equals the external bias voltage. When this occurs, a diode is effectively a nonconductor.

Leakage Current. When a diode is reverse biased, its depletion zone increases in width. Normally, this condition would restrict current carrier formation near the junction. The depletion zone, in effect, represents an area primarily void of majority current carriers. Therefore, the depletion zone will respond as an insulator. Ideally, current carriers do not pass through an insulator. In a reverse-biased diode some current actually flows through the depletion zone, which is called leakage current. Leakage current is dependent on minority current carriers.

Remember that minority carriers are electrons in the *P* material and holes in the *N* material. Figure 10–7 shows how

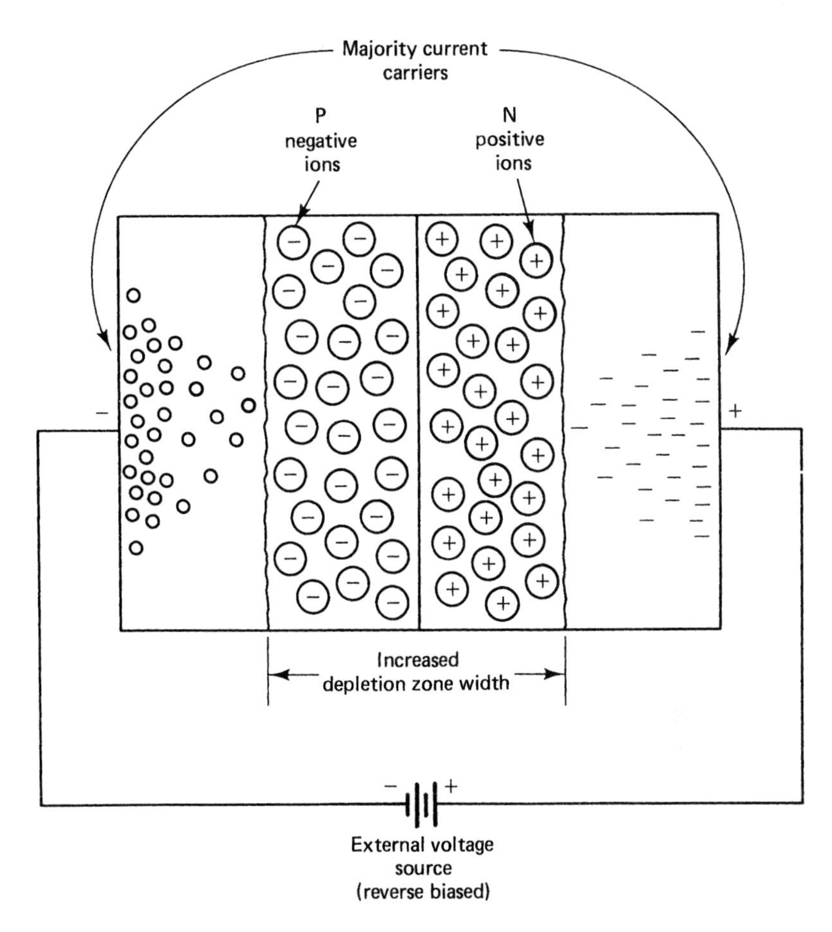

Figure 10–6. Reverse-biased *P-N* junction.

these carriers respond when a diode is reverse biased. It should be noted that the minority carriers of each material are pushed through the depletion zone to the junction. This action causes a very small leakage current to occur. Normally, leakage current is so small that it is often considered negligible.

The minority current carrier content of a semiconductor is primarily dependent on temperature. At normal room temperatures of 25°C or 77°F, a rather limited number of minority carriers are present in a semiconductor. However, when the surrounding temperature rises, it causes considerable increase in minority carrier production. This causes a corresponding increase in leakage current.

Leakage current occurs to some extent in all reverse-biased diodes. In germanium diodes, leakage current is only a few microamperes. Silicon diodes normally have a lower minority carrier content. This, in effect, means less leakage current. Typical leakage current values for silicon are a few nanoamperes. The construction material of a diode is an important consideration to remember. Germanium is much more sensitive to temperature than is silicon. Germanium therefore has a higher level of leakage current. This factor is largely responsible for the widespread use of silicon in modern semiconductor devices.

Forward Biasing

When voltage is applied to a diode as in Fig. 10–8, it causes forward biasing. In this example, the positive battery terminal is connected to the *P* material and the negative terminal to the *N* material. This voltage repels the majority current carriers of each material. A large number of holes and electrons therefore appear at the junction. On the *N* side of the junction, electrons move in to neutralize the positive ions in the depletion zone. In the *P* material, electrons are pulled from negative ions which causes them to become neutral again. This action means that forward biasing causes the depletion zone to collapse and the barrier potential to be removed. The *P-N* junction will therefore support a continuous current flow when it is forward biased.

Figure 10–9 shows how the current carriers of a forward-biased diode respond. Because the diode is connected to an external voltage source, it has a constant supply of electrons. Large arrows are used in the diagram to show the direction of current flow outside the diode. Inside the diode, smaller arrows show the movement of majority current carriers. Remember that electron flow and current are the same.

Starting at the negative battery terminal, assume now that electrons flow through a wire to the *N* material. Upon

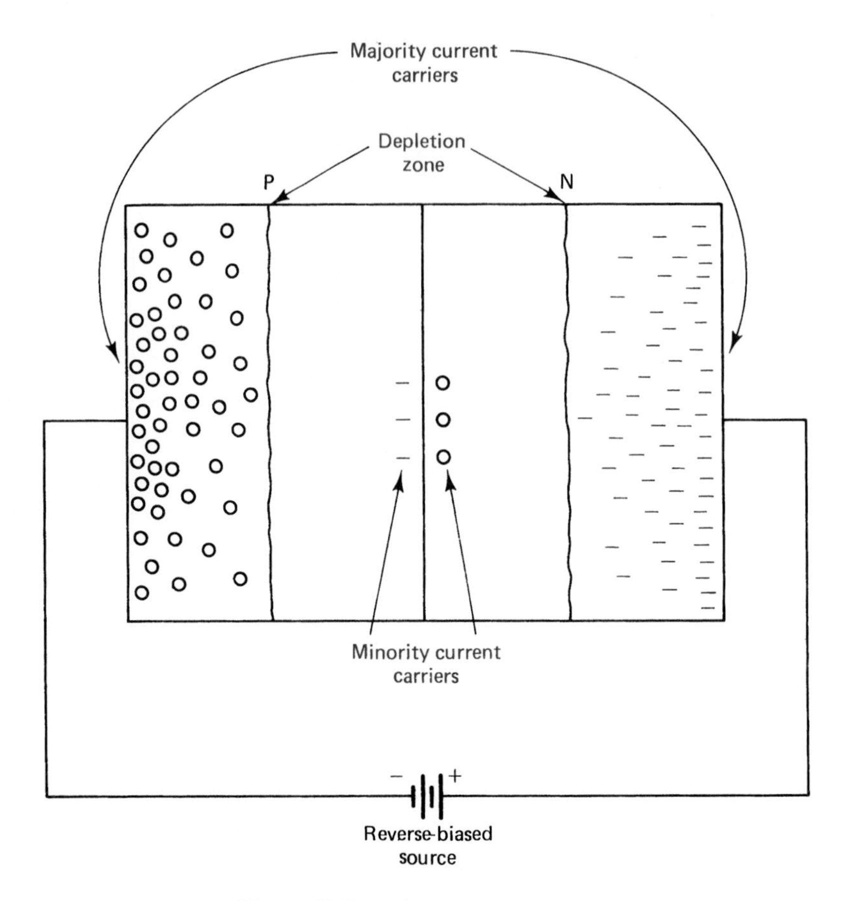

Figure 10–7. Minority current carriers.

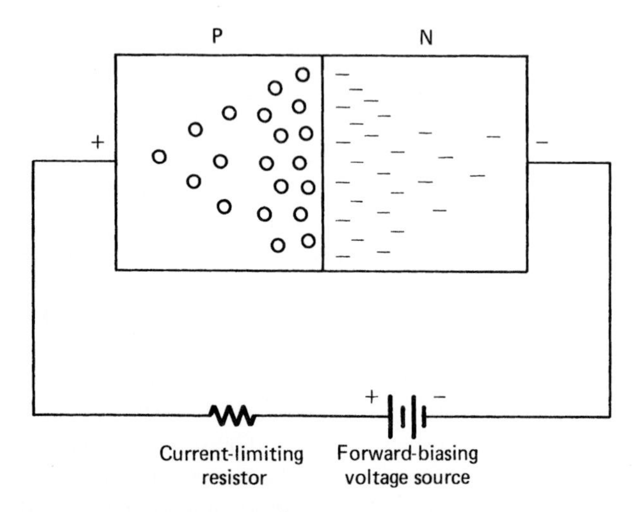

Figure 10–8. Forward-biased diode.

entering this material, they flow immediately to the junction. At the same time, an equal number of electrons are removed from the P material and are returned through a resistor to the positive battery terminal. This action generates new holes and causes them to move toward the junction. When these holes and electrons reach the junction, they combine and effectively disappear. At the same time, new holes and electrons appear at the outer ends of the diode. These majority carriers are generated on a continuous basis. This action continues as long as the external voltage source is applied.

It is important to realize that electrons flow through the entire diode when it is forward biased. In the N material this is quite obvious. In the P material, however, holes are the moving current carriers. Remember that hole movement in one direction must be initiated by electron movement in the opposite direction. Therefore, the combined flow of holes and electrons through a diode equals the total current flow.

The current-limiting resistor of Fig. 10–9 is essential in a forward-biasing diode circuit. This resistor is needed to keep the current flow at a safe operating level. The maximum current (I_{max}) rating of a diode is representative of this value. As long as diode current does not exceed this value, it can operate satisfactorily. Resistors are used to limit current flow to a reasonable operating level.

10.3 DIODE CHARACTERISTICS

Now that we have seen how a junction diode operates, it is time to examine some of its electrical characteristics, which takes into account such things as voltage, current,

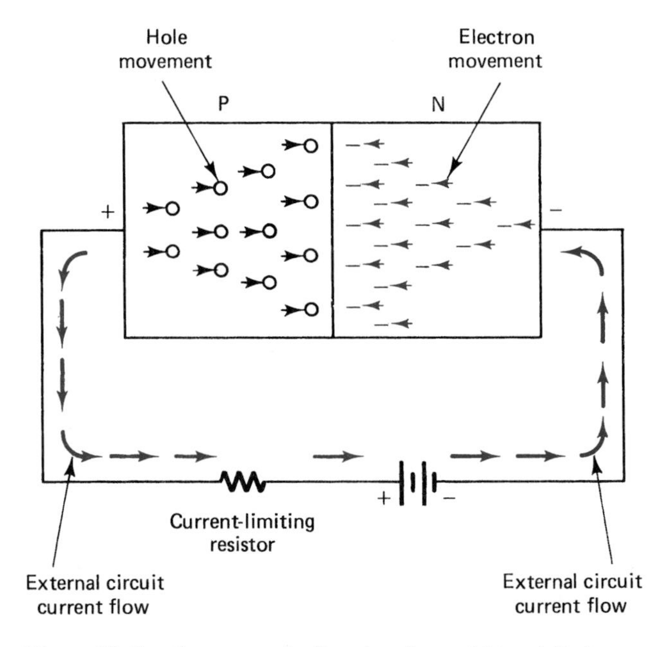

Figure 10–9. Current carrier flow in a forward-biased diode.

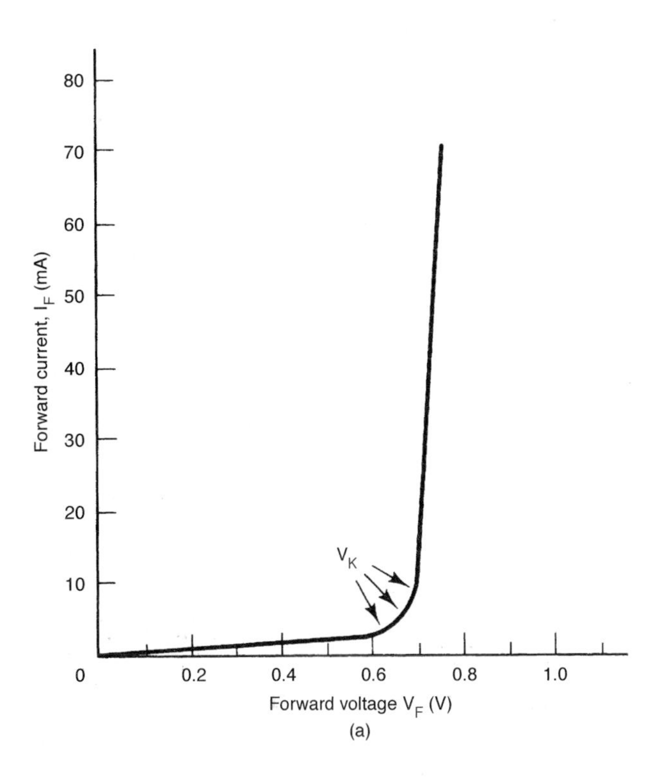

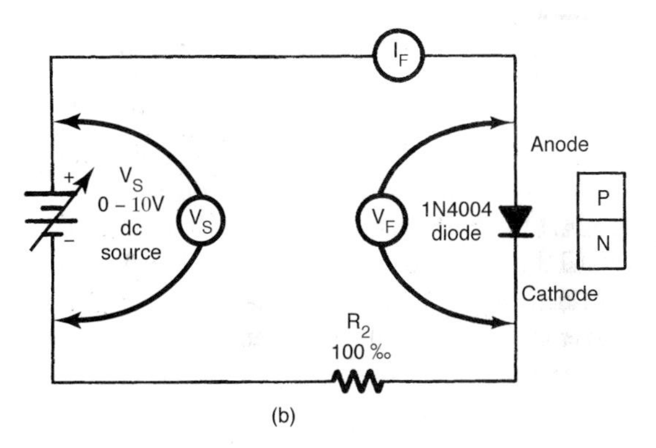

Figure 10–10. (a) *I-V* characteristics of a diode; (b) diode characteristic test circuit.

and temperature. Because these characteristics vary widely in an operating circuit, it is best to look at them graphically to see how the device will respond under different operating conditions.

Forward Characteristic

When a diode is connected in the forward-bias direction, it conducts forward current I_F. The value of I_F is directly dependent on the amount of forward voltage V_F. The relationship of forward voltage and forward current is called the ampere-volt or *I-V* characteristic of a diode. A typical diode forward *I-V* characteristic is shown in Fig. 10–10(a) with its test circuit in Fig. 10–10(b). Note, in particular, that V_F is measured across the diode and that I_F is a measure of what flows through the diode. The value of the source voltage V_S does not necessarily compare in value with V_F.

When the forward voltage of the diode equals 0 V, the I_F equals 0 mA. This value starts at the origin point (0) of the graph. If V_F is gradually increased in 0.1-V steps, I_F begins to increase. When the value of V_F is great enough to overcome the barrier potential of the *P-N* junction, a substantial increase in I_F occurs. The point at which this occurs is often called the knee voltage (V_K). For germanium diodes, V_K is approximately 0.3 V, and is 0.7 V for silicon.

If the value of V_F increases much beyond V_K, the forward current becomes quite large. This in effect causes heat to develop across the junction. Excessive junction heat can destroy a diode. To prevent this from happening, a protective resistor is connected in series with the diode. This resistor limits the I_F to its maximum-rated value. Diodes are rarely operated in the forward direction without a current-limiting resistor.

Reverse Characteristic

When a diode is connected in the reverse-bias direction, it has an I_R-V_R characteristic. Figure 10–11(a) shows the reverse *I-V* characteristic of a diode and its test circuit. This characteristic has different values of I_R and V_R. Reverse current is usually quite small. The vertical I_R line in this graph has current values graduated in microamperes. The number of minority current carriers that take part in I_R is quite small. In general, this means that I_R remains rather constant over a large part of V_R. Note also that V_R is graduated in 100-V

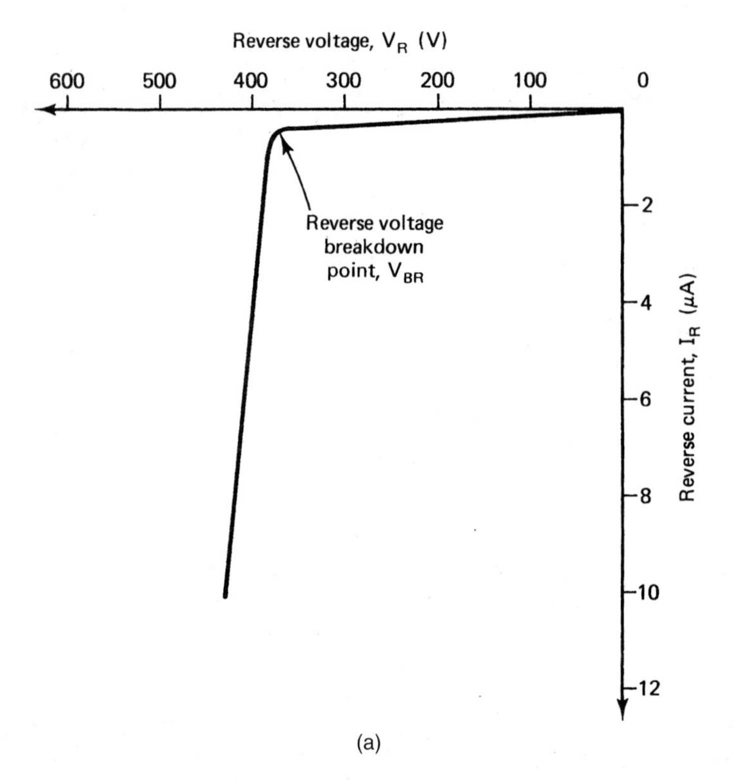

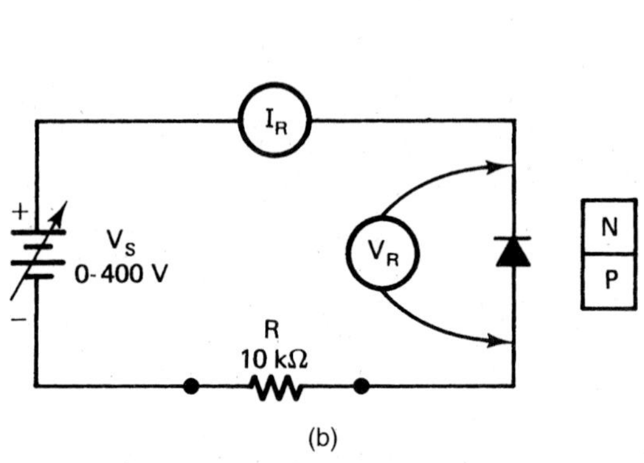

Figure 10–11. (a) Reverse *I-V* characteristics of a diode; (b) reverse characteristic test circuit.

increments. Starting at zero when the reverse voltage of a diode is increased, a slight change occurs in I_R. At the voltage breakdown (V_{BR}) point, current increases rapidly. The voltage across the diode remains fairly constant at this time. This constant-voltage characteristic leads to a number of reverse-bias diode applications. Normally, diodes are used in applications where the V_{BR} is not reached.

The physical processes responsible for current conduction in a reverse-biased diode are called *zener breakdown* and *avalanche breakdown.* Zener breakdown takes place when electrons are pulled from their covalent bonds in a strong electric field which occurs at a rather high value of V_R. When large numbers of covalent bonds are broken at the same time, a sudden increase in I_R occurs.

Avalanche breakdown is an energy-related condition of reverse biasing. At high values of V_R, minority carriers gain tremendous energy. This gain may be great enough to drive electrons out of their covalent bonding. This action creates new electron-hole pairs. These carriers then move across the junction and produce other ionizing collisions. Additional electrons are produced. The process continues to build until an avalanche of current carriers takes place. When this occurs, the process is irreversible.

Combined *I-V* Characteristics

The forward and reverse *I-V* characteristics of a diode are generally combined on a single characteristic curve. Figure 10–12 shows a rather standard method of displaying this curve. Forward- and reverse-bias voltages V_F and V_R are usually plotted on the horizontal axis of the graph. V_F extends to the right and V_R to the left. The point of origin or zero value is at the center of the horizontal line. Forward and reverse current values are shown vertically on the graph. I_F extends above the horizontal axis with I_R extending downward. The origin point serves as a zero indication for all four values. This means that combined V_F and I_F values are located in the upper right part of the graph and V_R and I_R in the lower left corner. Different scales are normally used to display forward and reverse values.

A rather interesting comparison of silicon and germanium characteristic curves is shown in Fig. 10–13. Careful examination shows that germanium requires less forward voltage to go into conduction than silicon. This characteristic is a distinct advantage in low-voltage circuits. Note also that a germanium diode requires less voltage drop across it for different values of current, which means that germanium has a lower resistance to forward current flow. Germanium therefore appears to be a better conductor than silicon. Silicon is more widely used, however, because of its low leakage current and cost.

The reverse-bias characteristics of silicon and germanium diodes can also be compared in Fig. 10–13. The I_R of a silicon diode is quite small compared with that of a germanium diode. Reverse current is determined primarily by the minority current content of the material. This condition is influenced primarily by temperature. For germanium diodes, I_R doubles for each 10°C rise in temperature. In a silicon diode, the change in I_R is practically negligible for the same rise in temperature. As a result, silicon diodes are preferred

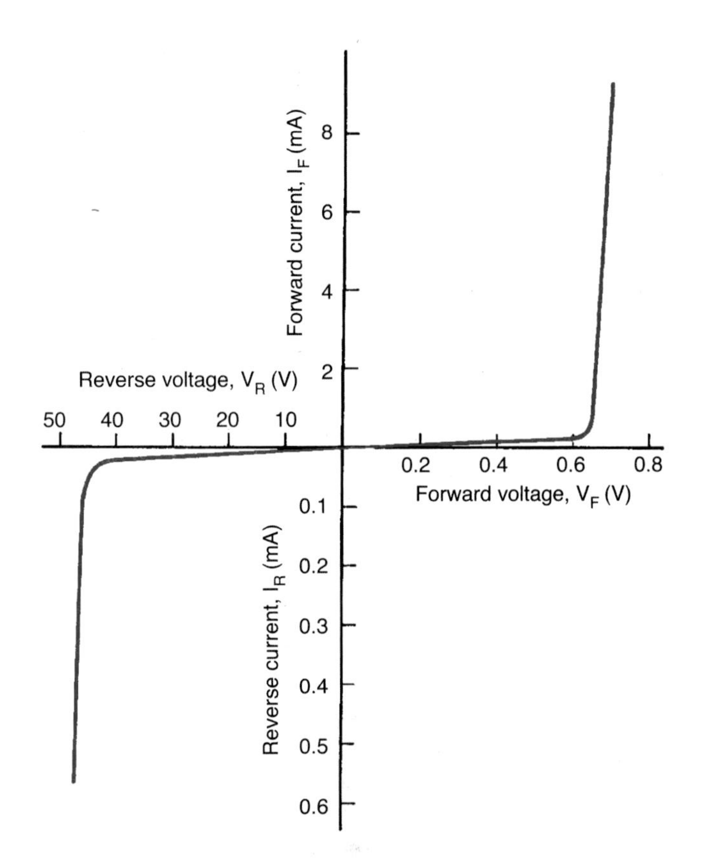

Figure 10–12. Diode current-voltage characteristics.

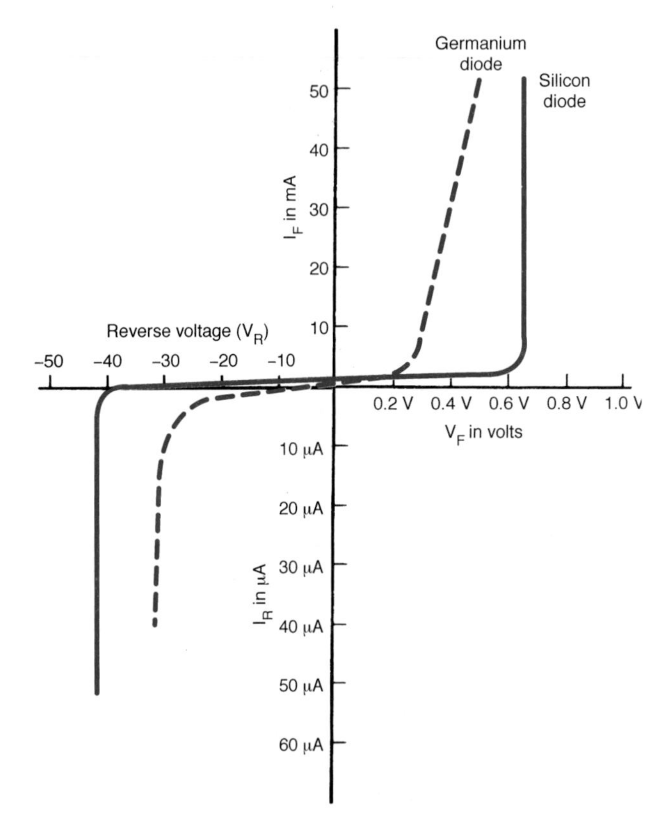

Figure 10–13. Silicon and germanium diode characteristics.

over germanium diodes in applications when large changes in temperature occur. Comparisons of this type are quite obvious through the study of characteristic curves.

10.4 DIODE SPECIFICATIONS

When selecting a diode for a specific application, some consideration must be given to specifications. This type of information is usually made available through the manufacturer. Diode specifications usually include such things as absolute maximum ratings, typical operating conditions, mechanical data, lead identification, mounting procedures, and characteristic curves. Figure 10–14 shows a representative diode data sheet. Following is an explanation of some important specifications.

1. *Maximum reverse voltage* V_{RM}: The absolute maximum or peak reverse-bias voltage that can be applied to a diode. This may also be called the peak inverse voltage (PIV) or peak reverse voltage (PRV).
2. *Reverse breakdown voltage* V_{BR}: The minimum steady-state reverse voltage at which breakdown will occur.
3. *Maximum forward current* I_{FM}: The absolute maximum repetitive forward current that can pass through a diode at 25°C (77°F). This is reduced for operation at higher temperatures.
4. *Maximum forward surge content* I_{FM} *(surge)*: The maximum current that can be tolerated for a short time. This current value is much greater than I_{FM} and represents the increase in current that occurs when a circuit is first turned on.
5. *Maximum reverse current* I_R: The absolute maximum reverse current that can be tolerated at device operating temperature.
6. *Forward voltage* V_F: Maximum forward voltage drop for a given forward current at device operating temperature.
7. *Power dissipation* P_D: The maximum power that the device can safely absorb on a continuous basis in free air at 25°C (77°F).
8. *Reverse recovery time* T_{rr}: The maximum time it takes the device to switch from its on to its off state.

Diode Temperature

The operation of a diode is influenced greatly by temperature. All semiconductor materials are similar in this respect.

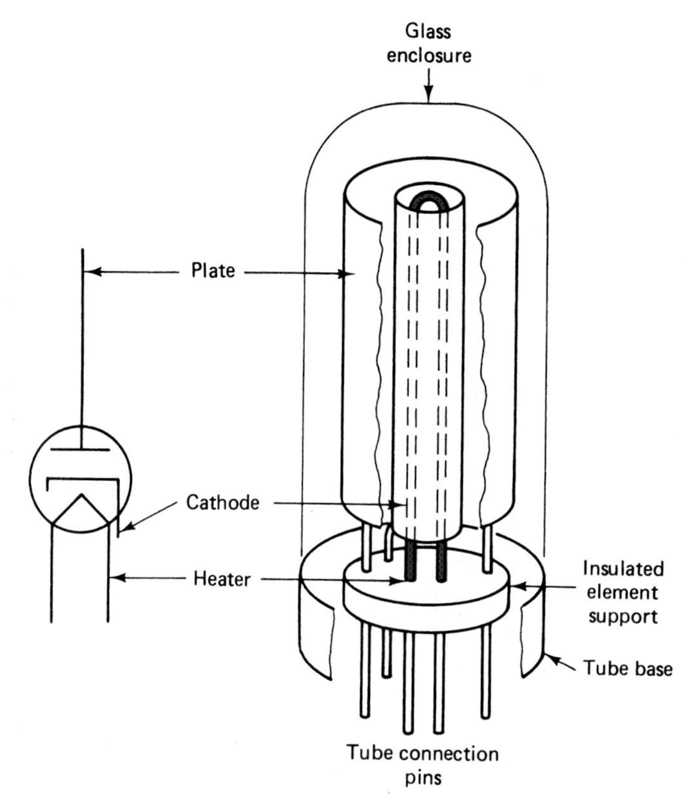

Figure 10–34. Indirectly heated tube.

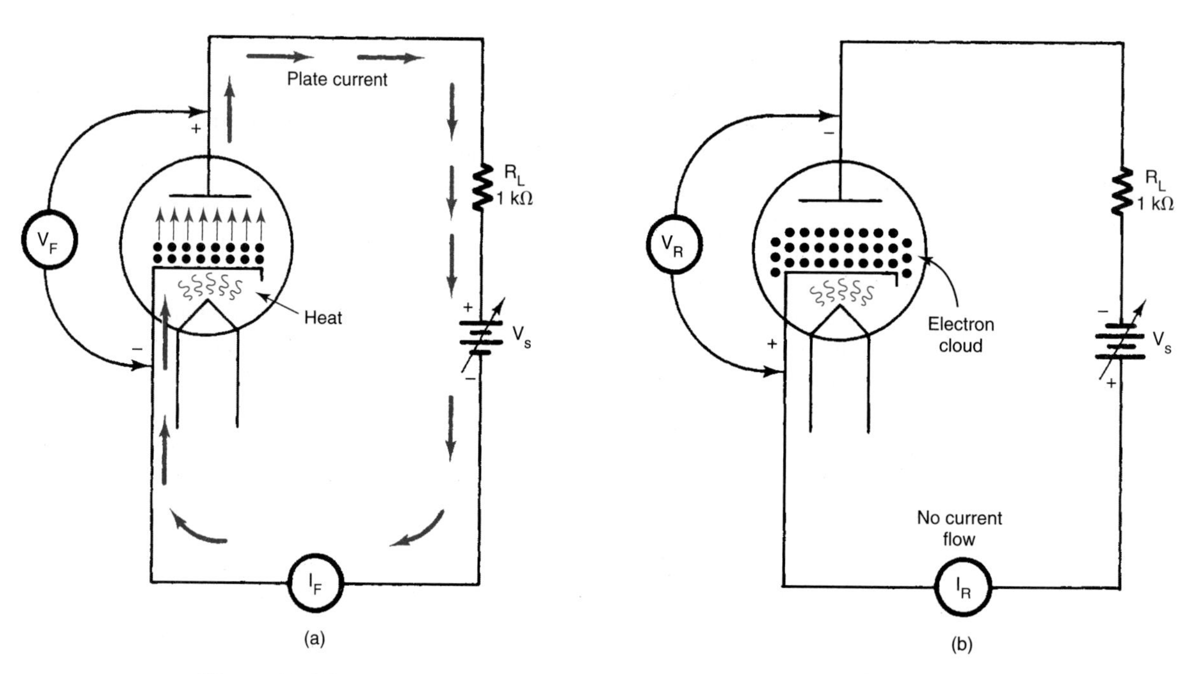

Figure 10–35. Operating conditions of a vacuum-tube diode: (a) forward bias; (b) reverse bias.

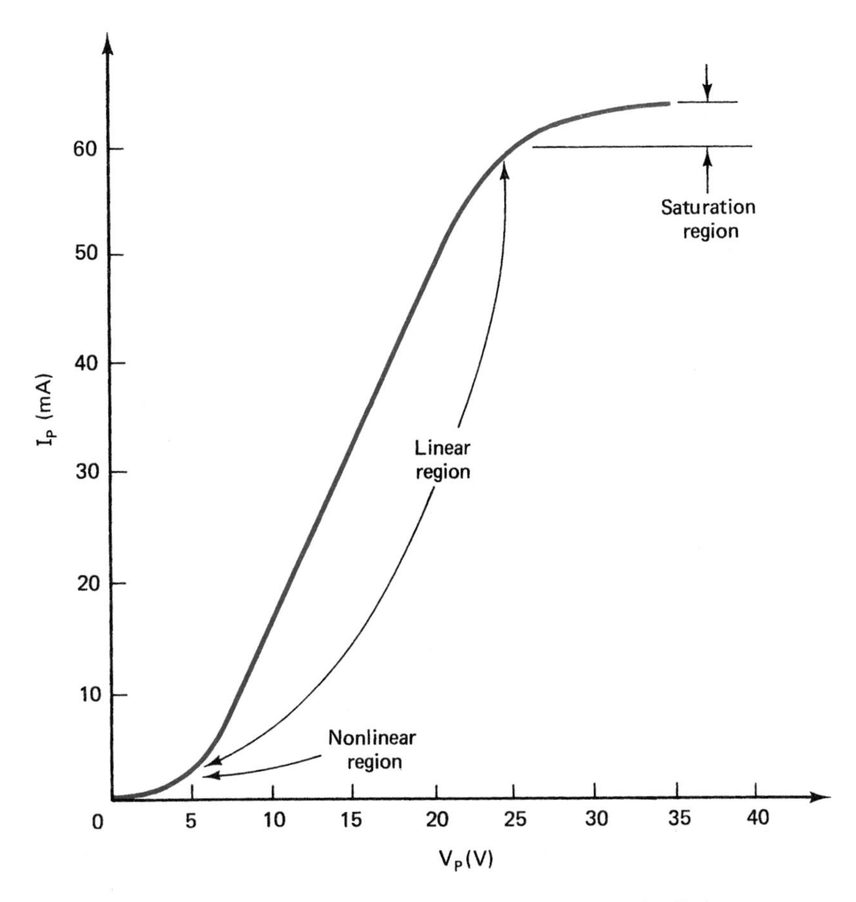

Figure 10–36. V_P-I_P characteristics of a vacuum-tube diode.

increase in V_P does not cause an increase in I_P. Essentially, the plate is attracting all the electrons emitted by the cathode. The I_P therefore levels off when saturation occurs.

As a general rule, the V_P-I_P characteristics of a diode vacuum tube are only concerned with forward-bias operation. The reverse characteristics are simply not shown. Ordinarily, no current flows in the reverse direction until the applied voltage is high enough to cause an arc between the electrodes. When this occurs there is permanent damage to the electrodes. The PRV rating usually indicates where this will occur.

When using a vacuum-tube diode, four ratings must be considered:

1. *Maximum plate current* (I_{max}): The maximum plate current that can flow through without causing electrode damage.
2. *Plate dissipation* (P_D): Power loss due to heat developed by electrons striking the plate.
3. *Peak reverse voltage* (*PRV*): The maximum reverse plate voltage that can be applied without causing electrode damage.
4. *Plate resistance* (R_P): The internal resistance that current encounters while passing between electrodes. It is calculated by the formula

$$R_P = \frac{\Delta V_P}{\Delta I_P}$$

Some representative vacuum-tube diodes are shown in Fig. 10–37. Connection to the electrodes is made through the bottom pins or the top cap. The enclosure or envelope is usually made of glass or glass-lined metal. The size of the tube is dependent primarily on its current and voltage capability. In general, these devices are rather fragile and should be handled with care.

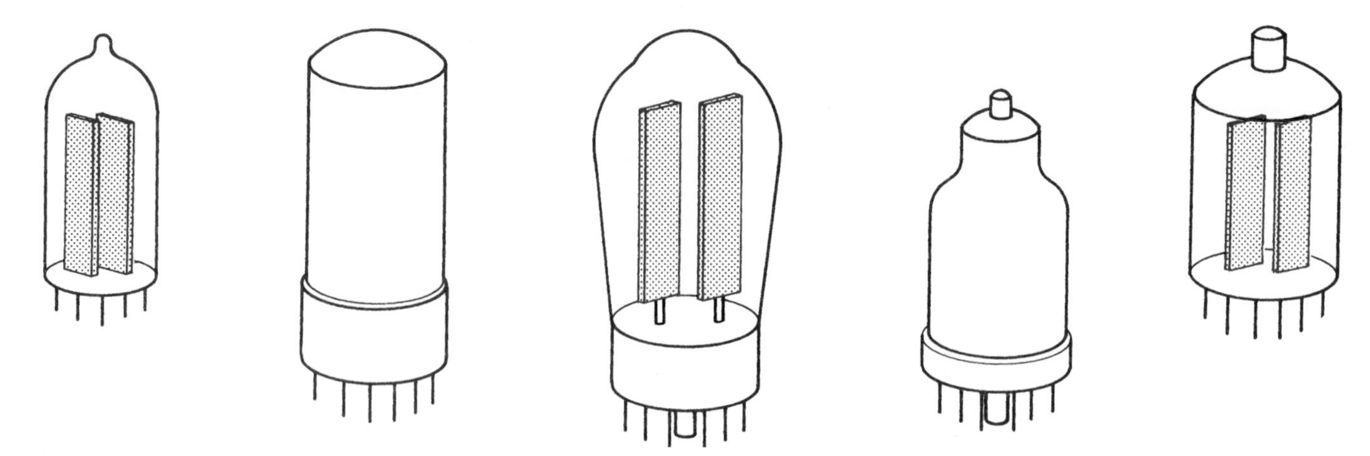

Figure 10–37. Representative vacuum-tube diodes.

REVIEW

1. When P and N materials are connected together, what causes the depletion zone to be formed?
2. What is barrier potential, and what causes it in a semiconductor diode?
3. Explain what is meant by forward biasing, and explain the influence that it has on the depletion zone and the barrier potential in a semiconductor diode.
4. Explain why conduction increases when a semiconductor diode is forward biased.
5. What causes leakage current in a semiconductor diode?

STUDENT ACTIVITIES

• *Diode Testing or Complete Activity 10–1, page 209 in the manual.*

1. With an ohmmeter or a VOM, test the forward and reverse resistance of several diodes.
2. If possible, test a shorted and an open diode. Compare the resistance differences.
3. Identify the voltage polarity of the ohmmeter leads. Record the voltage and lead polarity. Identify the leads of several diodes.

• •

ANALYSIS

1. Explain how an ohmmeter can be used to determine the status of a diode.
2. Explain how the anode and cathode of an unmarked diode can be identified with an ohmmeter.

• *Diode Characteristics or Complete Activity 10–2, page 211 in the manual.*

1. Connect the diode characteristic test circuit of Fig. 10–10(b). Adjust the V_S control in 0.2-V increments of V_F while observing I_F.
2. Plot an I_F -V_F characteristic curve on graph paper.

3. Test the forward characteristics of a silicon diode and a germanium diode, if possible.
4. To see the reverse characteristic, simply reverse the polarity of the diode and alter the voltage.

ANALYSIS

1. Explain what the forward characteristics of a silicon diode tell about its operation.
2. Explain what the reverse characteristics of a silicon diode tell about its operation.

Zener Diode Characteristics or Complete Activity 10–3, page 215 in the manual.

1. Connect the zener diode characteristic test circuit of Fig. 10–19. Adjust V_S in 1-V increments while observing V_Z and I_Z. The diode should be from 2 to 15 V_Z at 1 W.
2. Reverse the polarity of the diode and test the forward conduction of the zener diode.
3. Plot an I_Z -V_Z characteristic curve on graph paper.

ANALYSIS

1. Explain what causes a zener diode to maintain its V_Z at a constant value when the source voltage is increased.
2. Explain why the power dissipation rating of a zener diode is an important selection characteristic.

Zener Diode Voltage Regulation

Complete Activity 10–4, page 217 in the manual.

ANALYSIS

1. What determines the range of regulated voltage for Fig. 10–21, the Zener diode regulator circuit?
2. When the value of R_S in Fig. 10–21 is reduced, the load current ______ and zener current ______.

Light-Emitting Diode Characteristics

1. Connect the LED circuit of Fig. 10–38. Starting at 0 V_S, increase V_F in 0.5-V increments until illumination occurs.
2. Record the V_F and I_F that occur when the LED is lighting.
3. Reverse the polarity of the LED and repeat the procedure.

Photovoltaic Cell Characteristics

1. Connect the leads of a photovoltaic cell to a dc milliammeter.
2. Determine the output current of the cell at (a) normal room light, (b) near a window, (c) in bright sunlight, and (d) in total darkness.
3. Record the current values for each test condition.

OBJECTIVES

Upon completion of this chapter, you will be able to:

1. Explain the characteristic differences among half-wave, full-wave, and bridge rectifier circuits.
2. Explain the effect of a filter capacitor on the output voltage and ripple voltage of a rectifier circuit.
3. Describe the characteristics of capacitor, resistor-capacitor, and pi filters.
4. Explain zener regulator circuits.
5. Understand simple transistor and op-amp regulator circuits.
6. Explain the operation of series and shunt transistor regulators.
7. Calculate percentage of ripple of a power supply circuit.
8. Calculate percentage of regulation for a power supply.
9. Explain the operation of half-wave, full-wave, and bridge rectifier circuits.

All electronic systems require certain direct-current voltage and current values for operation. The energy source of the system is primarily responsible for this function. As a general rule, the source is more commonly called a power supply. In some systems, the power supply may be a simple battery or a single dry cell. Portable radios, tape recorders, and calculators are examples of systems that employ this type of supply. Television receivers, computer terminals, stereo amplifier systems, and electronic instruments usually derive their energy from the ac power line. This part of the system is called an electronic power supply. Electronic devices are used in the power supply to develop the required output energy.

In an electronic system the power supply serves as the primary source of electrical energy. As a rule, this part of the system is responsible for changing energy of one form into something that is more useful. The block diagram of Fig. 11–1 shows the location of the power supply in an electronic system.

An electronic power supply is often considered to be a smaller part or subsystem within the overall system. It has its own energy source, path, control, load, and indicator. Figure 11–2 shows a block diagram of the electronic power supply as a subsystem.

The energy source of an electronic power supply is alternating current. In most systems this is supplied by the ac power line. Most small electronic systems use 120-V single-phase 60-Hz ac as their energy source. This energy is readily available in homes, buildings, and industrial facilities.

The path of electrical energy in a power supply is similar to that of other systems. Copper wire and the metal foil of a printed circuit board serve as the conduction path. Conductors are generally insulated to prevent them from touching. Electrical energy flows to each system component through insulated conductors.

The control function of an electronic power supply is achieved by several components. Transformers, resistors, capacitors, inductors, and electronic devices are all used to achieve control. Each component has a unique role to play in the operation of the supply. The type of system, its operational capacity, and application have a great deal to do with the amount of control required.

The load of a power supply is somewhat unusual. A composite of all electrical and electronic components that

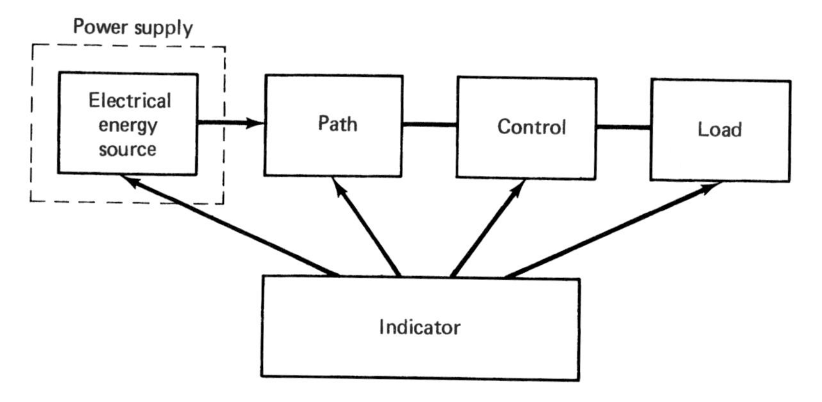

Figure 11–1. Electrical system parts.

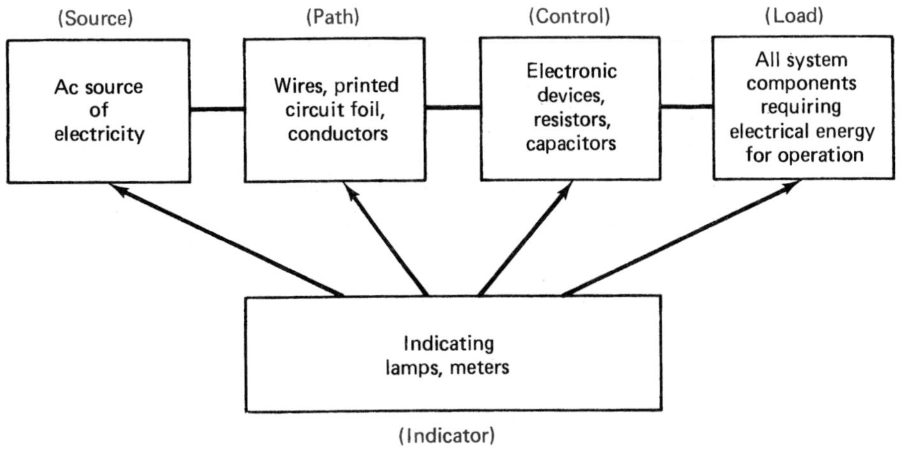

(Source)
Ac source of electricity

(Path)
Wires, printed circuit foil, conductors

(Control)
Electronic devices, resistors, capacitors

(Load)
All system components requiring electrical energy for operation

Indicating lamps, meters
(Indicator)

Figure 11–2. Power supply subsystem.

are supplied energy serves as the load. These components do some type of work when energized. Heat, light, sound, and mechanical motion are typical examples of the work produced by load components. A specific resistor value will serve as the load of a power supply in this discussion.

IMPORTANT TERMS

A discussion of electronic power supplies frequently involves some new and unusual terms. These terms generally make the discussion more meaningful if they are reviewed before studying the chapter. Some of the terms are discussed in detail in the chapter, whereas others are listed for reference purposes.

Alternation. Half of an ac sine wave. There is a positive alternation and a negative alternation for each ac cycle.

Anode. The positive terminal or electrode of an electronic device such as a solid-state diode.

Average value. A voltage or current value based on the average of all instantaneous values for one alternation.

$$V_A = 0.637 \times V_P$$

Capacitance input filter. A filter circuit employing a capacitor as the first component of its input.

Cathode. The negative terminal of an electronic device, such as a solid-state diode, which is responsible for electron emission.

Center tap. An electrical connection point at the center of a wire coil or transformer.

Counter electromotive force (*CEMF*). The voltage induced in a conductor as a result of a magnetic field; *CEMF* opposes the original source voltage.

Forward bias. A voltage connection method that produces current conduction in a *P-N* junction.

Inductance (*L*). The property of a circuit to oppose changes in current due to energy stored in a magnetic field.

Peak value. The maximum voltage or current value of an ac wave.

$$V_P = 1.414 \times \text{rms}$$

Pi filter. A filter with an input capacitor connected to an inductor-capacitor filter, forming the shape of the capital Greek letter pi (π).

Pulsating dc. A voltage or current value that rises and falls with current flow always in the same direction.

Rectification. The process of changing ac into pulsating dc.

Reverse bias. A voltage connection method producing little or no current through a *P-N* junction.

Schematic. A diagram of an electronic circuit made with symbols.

Time constant (*RC*). The time required for the voltage across a capacitor in an *RC* circuit to increase to 63% of its maximum value or decrease to 37% of its maximum;

$$\text{time} = RC.$$

Turns ratio. The ratio of the number of turns in the primary winding (N_P) of a transformer to the number of turns in the secondary winding (N_S).

11.1 POWER SUPPLY FUNCTIONS

All electronic power supplies have a number of functions that must be performed in order to produce operation. Some of these functions are achieved by all power supplies. Others

connection of this divider serves as the common point of the output. Note that the center tap of transformer T_1 and the common load resistor connection points are connected. This serves as a path for the ground current of the circuit. If the load resistors are equal or balanced, no ground current flows, and in turn causes all the current to pass through the equal-valued load resistors which divide the voltage. Ground current only flows when an imbalance occurs. This type of power supply will therefore produce an output when the load is either balanced or unbalanced.

Diodes D_1 and D_2 are rectifiers for the positive supply. They are connected to opposite ends of transformer T_1. Diodes D_3 and D_4 are rectifiers for the negative supply. They are connected in a reverse direction to the opposite ends of the transformer. The positive or negative output is with respect to the center tap of the transformer.

For one alternation, assume that the top of the transformer is positive and the bottom is negative. Current flows out of point B through D_4, R_{L2}, R_{L1}, D_1, and returns to terminal A of the transformer. The bottom of R_{L2} becomes negative and the top of R_{L1} becomes positive. This current conduction path is made complete by forward biasing diodes D_4 and D_1. Diodes D_2 and D_3 are reverse biased by this transformer voltage. Solid arrows show the path of current flow for this alternation.

The next alternation makes the top of the transformer negative and the bottom positive. Current flows out of point A through diode D_3, R_{L2}, R_{L1}, D_2, and returns to terminal B of the transformer. The top of R_{L1} continues to be positive and the bottom of R_{L2} is negative. This alternation forward biases D_3 and D_2 while reverse biasing D_4 and D_1. The direction of current flow through R_{L2} and R_{L1} is the same as the first alternation. The resultant output voltage of the supply appears at the top and bottom of R_{L2} and R_{L1}. Dashed arrows show the direction of current flow for the second alternation.

11.4 FILTERING

The output of a half- or full-wave rectifier is pulsating direct current. This type of output is generally not usable for most electronic circuits. A rather pure form of dc is usually required. The filter section of a power supply is designed to change pulsating dc into a rather pure form of dc. Filtering takes place between the output of the rectifier and the input to the load device. Power supplies discussed up to this point have not employed a filter circuit.

The pulsating dc output of a rectifier contains two components. One of these deals with the dc part of the output. This component is based on the combined average value of each pulse. The second part of the output refers to its ac component. Pulsating dc, for example, occurs at 60 Hz or 120 Hz, depending on the rectifier being employed. This part of the output has a definite ripple frequency. Ripple must be minimized before the output of a power supply can be used by most electronic devices.

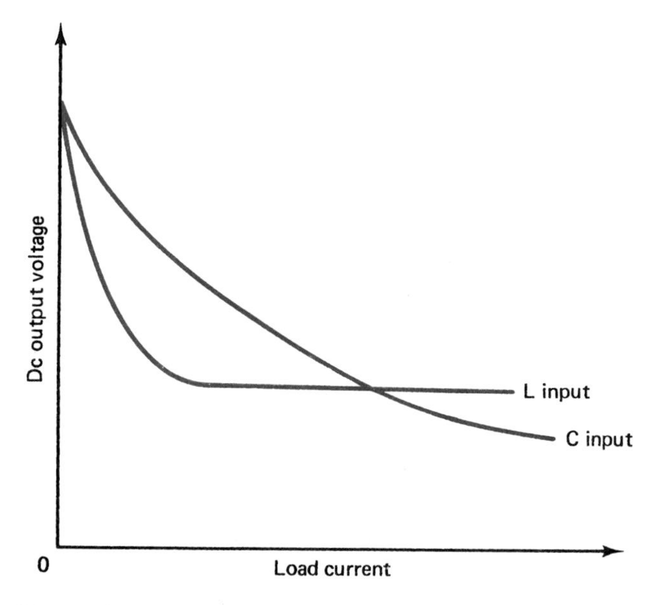

Figure 11–16. Load current versus output voltage comparisons for L and C filters.

Power supply filters fall into two general classes according to the type of component used in the input. If filtering is first achieved by a capacitor, it is classified as a capacitor or C-input filter. When a coil of wire or inductor is used as the first component, it is classified as an inductive or L-input filter. C-input filters develop a higher value of dc output voltage than does an L filter. The output voltage of a C filter usually drops in value when the load increases. L filters, by comparison, tend to keep the output voltage at a rather constant value, which is particularly important when large changes in the load occur. The output voltage of an L filter is, however, somewhat lower than that of a C filter. Figure 11–16 shows a graphic comparison of output voltage and load current for C and L filters.

C-INPUT FILTERS. The ac component of a power supply can be reduced effectively by a C-input filter. A single capacitor is simply placed across the load resistor as shown in Fig. 11–17. For alternation 1 of the circuit [Fig. 11–17(a)], the diode is forward biased. Current flows according to the arrows of the diagram. C charges quickly to the peak voltage value of the first pulse. At the same point in time, current is also supplied to R_L. The initial surge of current through a diode is usually quite large. This current is used to charge C and supply R_L at the same time. A large capacitor, however, responds somewhat like low resistance when it is first being charged. Notice the amplitude of the I_d waveform during alternation 1.

When alternation 2 of the input occurs, the diode is reverse biased. Figure 11–17(b) shows how the circuit re-

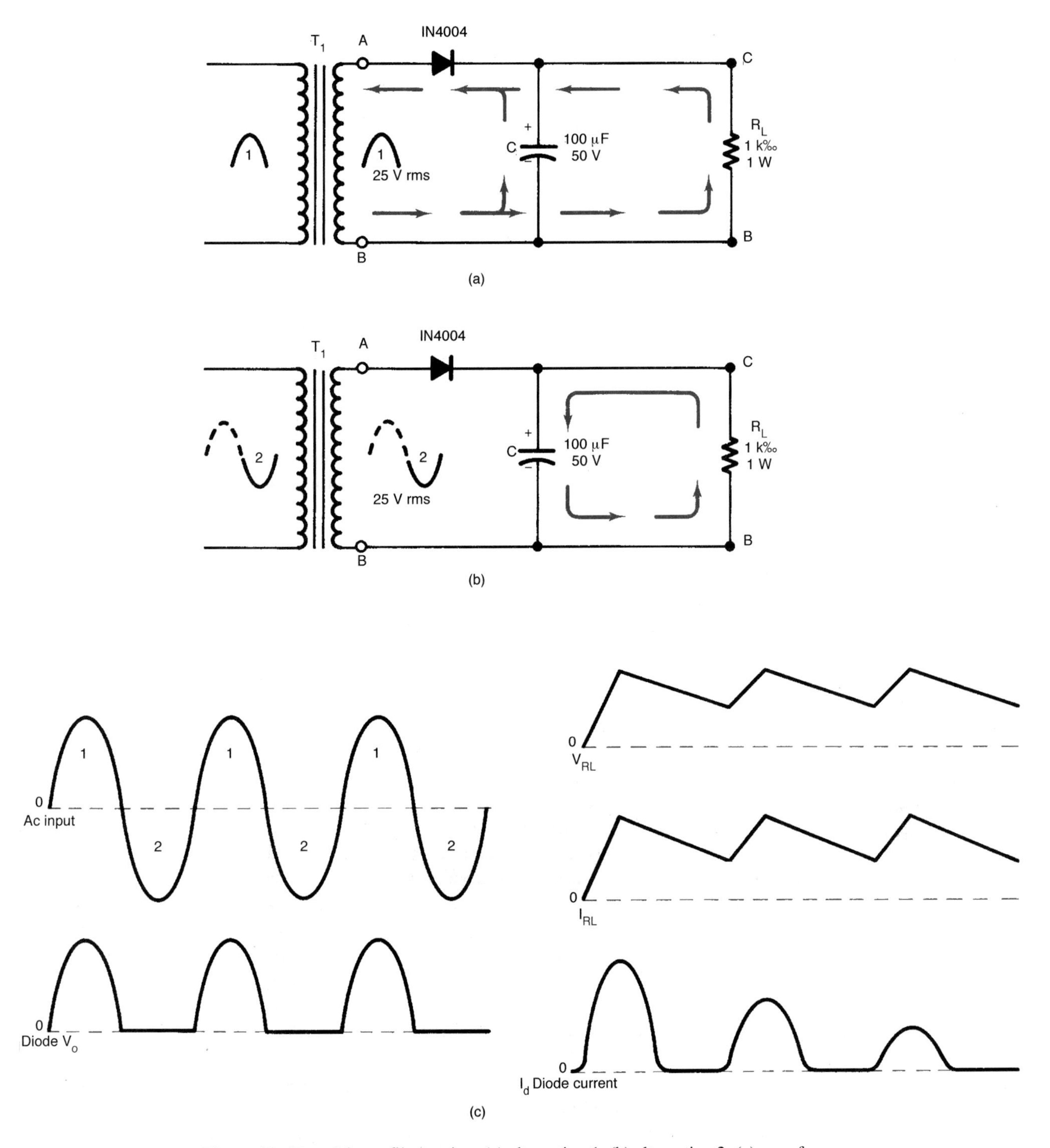

Figure 11–17. *C*-input filter action: (a) alternation 1; (b) alternation 2; (c) waveforms.

sponds for this alternation. Notice that no current flows from the source through the diode. The charge acquired by C during the first alternation now finds an easy discharge path through R_L. The resulting discharge current flow is indicated by the arrows between C and R_L. In effect, R_L is now being supplied current even when the diode is not conducting. The voltage across R_L is therefore maintained at a much higher value. See the V_{RL} waveform for alternation 2 in Fig. 11–17(c).

Discharge of C continues for the full time of alternation 2. Near the end of the alternation, there is somewhat of a drop in the value of V_{RL} due primarily to a depletion of

2. Calculate the dc output voltage for the circuit.
3. With a dc voltmeter, measure and record the dc output voltage at points CT and *D*. How do the measured and calculated values compare?
4. With an oscilloscope, observe the ac input at points A–CT and B–CT. Use line sync for the oscilloscope. Make a sketch of the observed waveforms. How do they compare?
5. Connect the oscilloscope at points CT and *D*. Make a sketch of the observed waveform.

ANALYSIS

1. What is the frequency of the ac ripple voltage across R_L of a full-wave rectifier?
2. Would using a lower value resistance for R_L in Fig. 11–10 have the same effect as on the single-phase half-wave rectifier? Why?
3. What is the ratio of dc voltage output to peak ac voltage input for the full-wave rectifier?
4. What is the formula for determining the value of dc voltage output for a single-phase full-wave rectifier?
5. What would happen if the polarities of both D_1 and D_2 of Fig. 11–10 were reversed?

Bridge Rectifiers or *Complete Activity 11–3, page 227 in the manual.*

1. Construct the bridge rectifier of Fig. 11–13. All four diodes are 1N4004 or the equivalent.
2. Calculate the dc output voltage for the circuit.
3. With a dc voltmeter, measure and record the dc output voltage. How does it compare with the calculated value?
4. With an oscilloscope, observe the ac input at points A–B and the dc output at points C–D. Make a sketch of the observed waveforms.

ANALYSIS

1. What is the frequency of the ac ripple voltage for a bridge rectifier?
2. What is the ratio of dc voltage output to the peak ac voltage input for a bridge rectifier circuit?
3. What would happen to the dc output of Fig. 11–13 if D_1 became short-circuited?
4. What would happen to the dc output of Fig. 11–13 if D_3 became an open-circuit path?
5. What would happen to the dc output of Fig. 11–13 if *all* diodes were reversed?

Dual Power Supply

1. Construct the dual full-wave power supply of Fig. 11–15. All diodes are IN4004 or the equivalent.
2. Calculate the dc output voltage for the top section and the bottom section.
3. With a dc voltmeter, measure and record the dc output voltage at points *D–G, E–G,* and *D–E.*
4. With an oscilloscope, observe the ac input at points *A*–CT, *B*–CT, *D–G, E–G,* and *D–E.* Make a sketch of the observed waveforms.

• *C-Input Filtering* or *Complete Activity 11–4, page 231 in the manual.*

1. Construct the half-wave rectifier with a *C*-input filter of Fig. 11–17.
2. Calculate the dc voltage value appearing across C_1. With a dc voltmeter, measure and record the dc voltage.
3. With an oscilloscope, observe the waveforms at points A–B and C–B. Make a sketch of the observed waveforms.
4. Turn off the circuit and change the value of R_L to 10 kΩ, then 470 Ω, 1 W. Repeat step 3 for each value of R_L. How does the value of R_L influence the dc ripple?

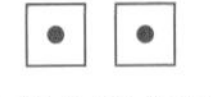

ANALYSIS

1. In Fig. 11–17, if V_{in} was increased, what effect would be observed on:
   a. V_{peak}?
   b. V_{dc}?
   c. Ripple factor?
2. What effect would the addition of a parallel capacitor have on load current of Fig. 11–17? Why?
3. What effect would the added capacitor have on:
   a. V_r (rms)?
   b. V_{dc}?
   c. Ripple factor?
4. What effect does increased load (R_L decreased) of Fig. 11–17 have on:
   a. V_r (rms)?
   b. V_{dc}?
   c. Ripple factor?
5. Calculate the percentage voltage regulation if the no-load resistance = 10 V and the full-load resistance = 9 V.

$$\% \text{ regulation} = \frac{V_{NL} - V_{FL}}{V_{FL}} \times 100 = \_\_\_\_\_\_ \%$$

• L-*Input Filter*

1. Construct the full-wave power supply with an *L*-input filter of Fig. 11–19.
2. Calculate the dc voltage output of the rectifier section. With a dc voltmeter, measure and record the dc voltage at CT–*D*, CT–*E*, and *D*–*E*.
3. With an oscilloscope, observe the waveforms at points *A*–CT, *B*–CT, *D*–CT, and *E*–CT. Make a sketch of the observed waveforms.
4. Turn off the circuit and change the value of R_L to 470 Ω. Turn on the circuit and observe the waveforms at *D*–CT and *E*–CT. What influence does increased load have on the output ripple?

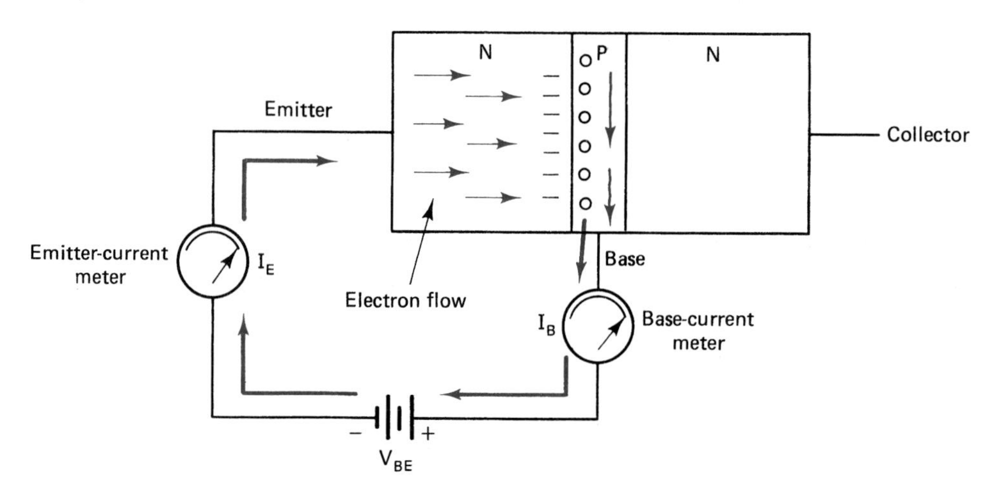

Figure 12–3. Emitter-base biasing.

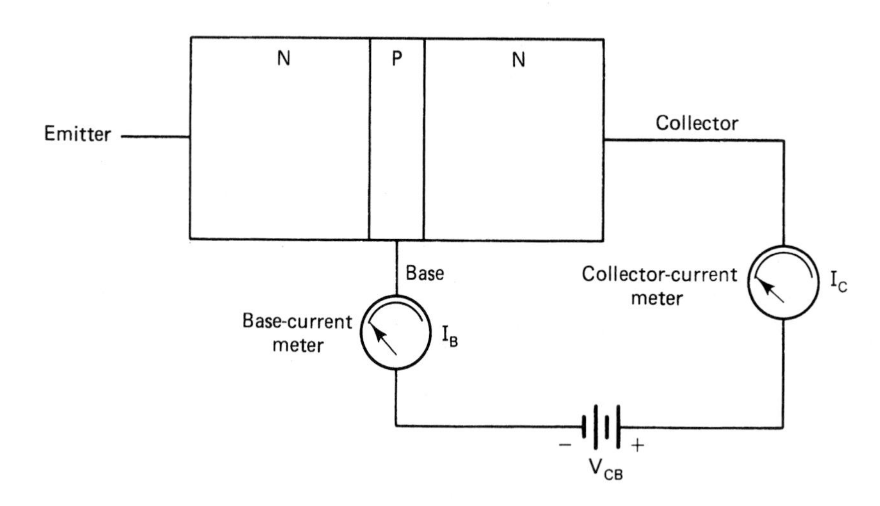

Figure 12–4. Base-collector biasing.

the base-collector junction. In this case, there is no indication of base current I_B or collector current I_C. In an actual reverse-biased junction, there could be a minute amount of current, which would be supported primarily by minority current carriers. As a general rule, this is called leakage current. In a silicon transistor, leakage current is usually considered to be negligible. In an actual circuit, V_{CB} is not normally applied to a transistor without the V_{BE} voltage source.

For a transistor to function properly, the emitter-base junction must be forward biased and the base-collector junction reverse biased. Both junctions must have bias voltage applied at the same time. In some circuits, this voltage may be achieved by separate V_{BE} and V_{CB} sources. Other circuits may utilize a single battery with specially connected bias resistors. In either case, the transistor responds quite differently when all its terminals are biased.

Consider now the action of a properly biased *NPN* transistor. Figure 12–5 shows separate V_{BE} and V_{CB} sources connected to the transistor. V_{BE} provides forward bias for the emitter-base junction while V_{CB} reverse biases the collector junction. Connected in this manner, the two junctions do not respond as independent diodes.

Figure 12–6 shows how current carriers pass through a properly biased *NPN* transistor. The emitter-base junction being forward biased causes a large amount of I_E to move into the emitter-base junction. Upon arriving at the junction, a large number of the electrons do not effectively combine with holes in the base. The base is usually made thin (0.0025 cm or 0.001 in.) and it is lightly doped, meaning that the majority current carriers of the emitter exceed the majority carriers of the base. Most of the electrons that cross the junction do not combine with holes. They are, however, immediately influenced by the positive V_{CB} voltage applied to the collector. A high percentage of the original emitter current enters the collector. Typically, 95% to 99% of I_E flows into the collector junction. It then becomes collector current. After passing through the collector region, I_C returns to V_{CB}, V_{BE}, and ultimately to I_E. The current flow inside the transistor is in-

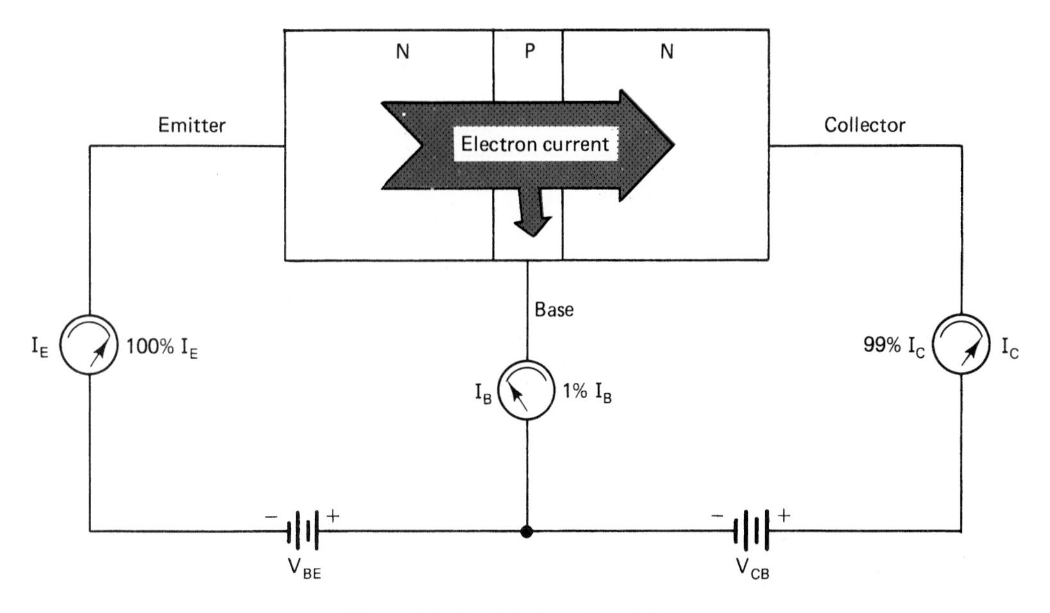

Figure 12–5. *NPN* transistor biasing.

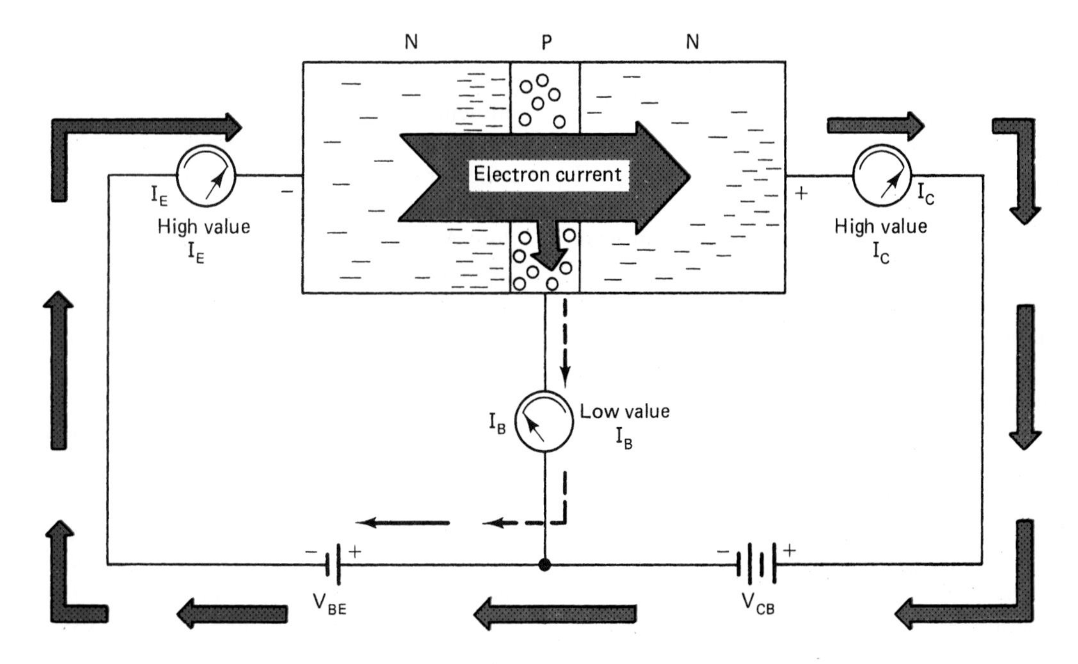

Figure 12–6. Current carriers passing through an *NPN* transistor.

dicated by a large arrow. Outside the transistor, current flow is indicated by small arrows.

The difference between the amount of I_E and I_C of Fig. 12–6 is base current. Essentially, base current is due to the combining of a small number of electrons and holes in the emitter-base junction. In a typical circuit, I_B is approximately 1% to 5% of I_E. The base region is narrow so it cannot support a large number of current carriers. A small amount of base current is needed, however, to make the transistor operational. Note the direction of I_B and its flow path in the diagram.

The relationship of I_E, I_B, and I_C in a transistor can be expressed by the equation

$$\text{emitter current} = \text{base current} + \text{collector current}$$

or

$$I_E = I_B + I_C$$

This equation shows that the emitter current is equal to the sum of I_B and I_C. The largest current flow in a transistor takes

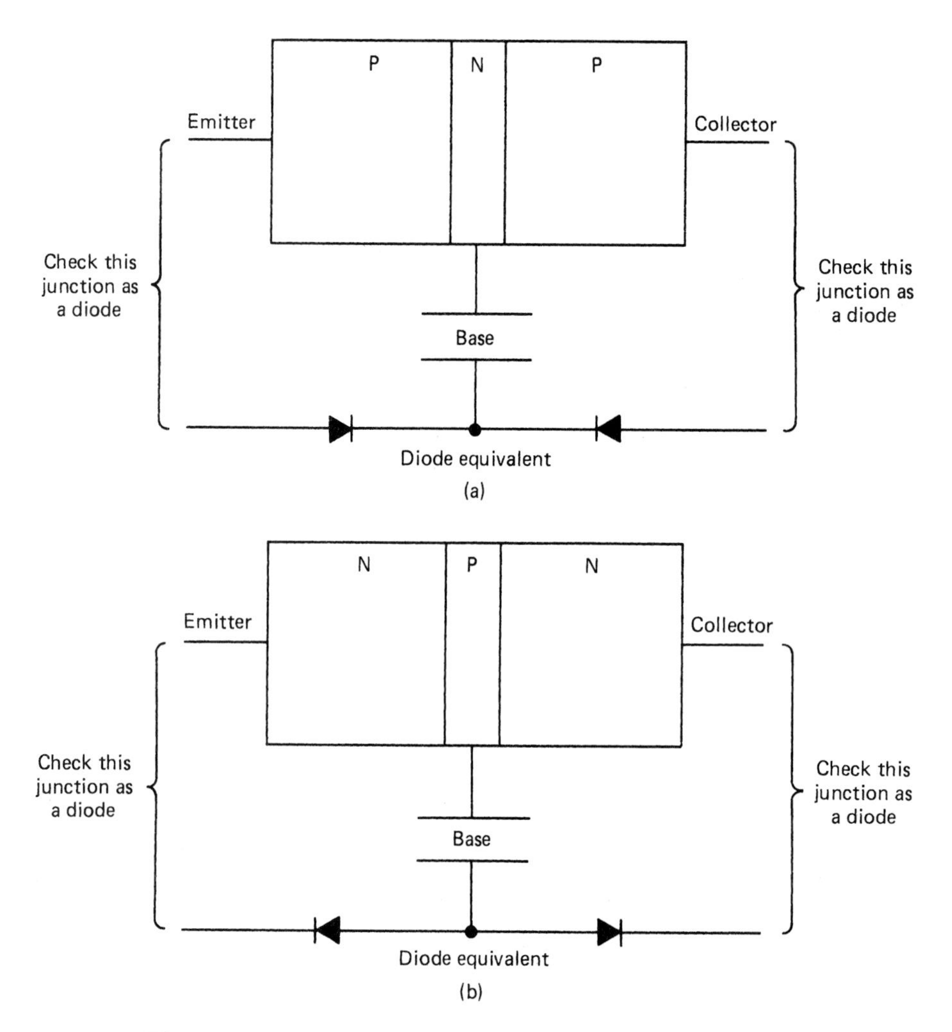

Figure 12–13. Junction polarity of (a) *PNP* and (b) *NPN* transistors.

polarity in this case is not important. This connection should cause an extremely high resistance reading. Then reverse the two ohmmeter leads. High resistance should also occur in this direction. A good silicon transistor will show an infinite resistance in either direction. Any measurable resistance in either direction indicates leakage. Germanium transistors will show some leakage in this test. Unless the leakage is excessive, it can be tolerated in a germanium transistor. A perfect transistor would have no indication of emitter-collector leakage.

Lead Identification

An ohmmeter can also be used to identify the leads of a transistor. The polarity of the ohmmeter voltage source must be known in order for this test to be meaningful. Straight-polarity ohmmeters have the black or common lead negative and the red or probe lead positive. Reverse-polarity ohmmeters are connected in the opposite direction. The polarity of an ohmmeter can be tested with a separate dc voltmeter if it is unknown. The ohmmeter used in this explanation has straight polarity.

To identify transistor leads, inspect the lead location of the device under test. Pick out the center lead of the transistor. Assume that it is the base lead. Connect the negative lead of the ohmmeter to it as shown in Fig. 12–14. Then alternately touch the positive ohmmeter lead to the two outside transistor leads. If a low resistance indication occurs for each lead, the center lead is actually the base. The test also indicates that the transistor is the *PNP* type. If the resistance is high between the center lead and the two outside leads, reverse the meter polarity. The positive lead should now be connected to the assumed base. Once again, alternately switch the negative ohmmeter lead between the two outside transistor leads. If a low resistance reading is obtained, the transistor is an *NPN* type.

If the center lead does not produce low resistance in either of the two conditions, it is not the base. Then select one of the outside leads as the assumed base. Try the same procedure again with the newly assumed base. If this does not

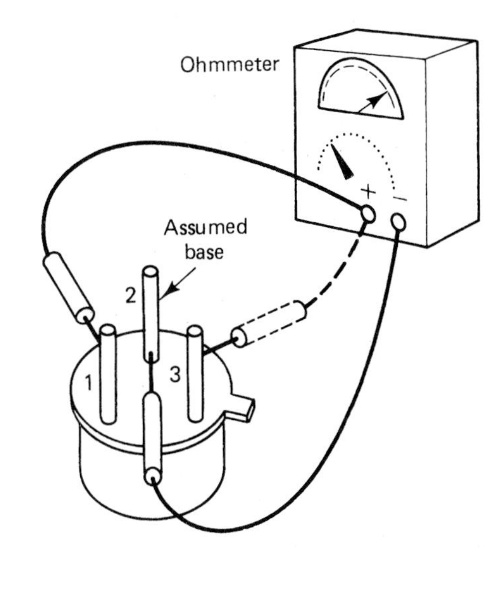

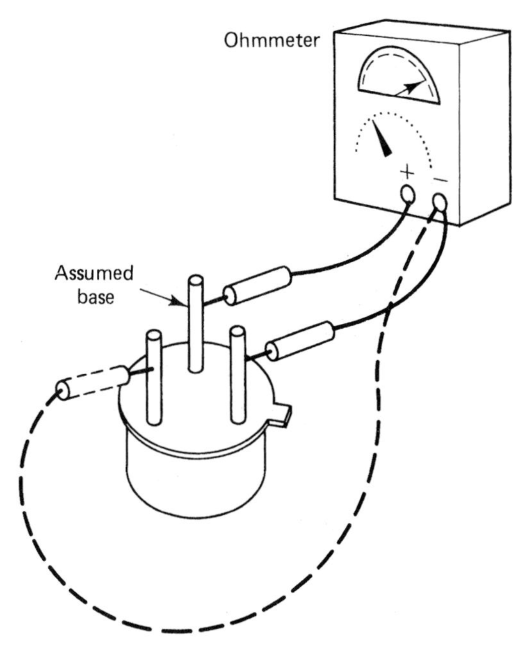

Figure 12–14. Ohmmeter transistor testing.

produce results, try the other outside lead. One of the three leads must respond as the base if the transistor is good. If a response cannot be obtained, the transistor must be open or shorted.

Thus far we have identified the base lead and determined the polarity of the transistor to be *NPN* or *PNP*. It is now possible to test the gain and to identify the remaining two leads. A 100,000-Ω resistor and an ohmmeter are needed for this part of the test. The resistor is used to provide base current from the ohmmeter. If a power transistor is being tested, use the $R \times 1$ meter range and a 1000-Ω base resistor.

The process of testing transistor gain will be more meaningful if we know how it works. Figure 12–15 shows how *PNP* and *NPN* transistors will respond to the ohmmeter test. The energy source of the ohmmeter is used to supply the bias voltage to each of the transistor elements. For the *PNP* transistor of part (a), the positive ohmmeter lead is connected to the emitter and the negative lead to the collector. When a resistor is connected between the collector and the base, it causes base current. With the emitter and base forward biased and the collector reverse biased, the transistor becomes low resistant. The ohmmeter will respond to this condition by showing a change in resistance.

The ohmmeter test of an *NPN* transistor is shown in Fig. 12–15(b). For this type of transistor, the emitter is connected to the negative ohmmeter lead and the collector to the positive. When a resistor is connected between the collector and the base, it will cause base current to flow. With the emitter and base forward biased and the base and collector reverse biased, the transistor becomes low resistant. An indication of this type on the ohmmeter shows that the transistor has gain and correct lead selection.

The procedure for testing an *NPN* transistor is shown in Fig. 12–16. In this case, the ohmmeter is connected to the two outside leads. We have previously identified the center lead as the base. If another lead were found to be the base, it would be used in place of the center lead. Note that one end of the resistor is connected to the base with the other lead open. We have in this case assumed that the negative ohmmeter lead is

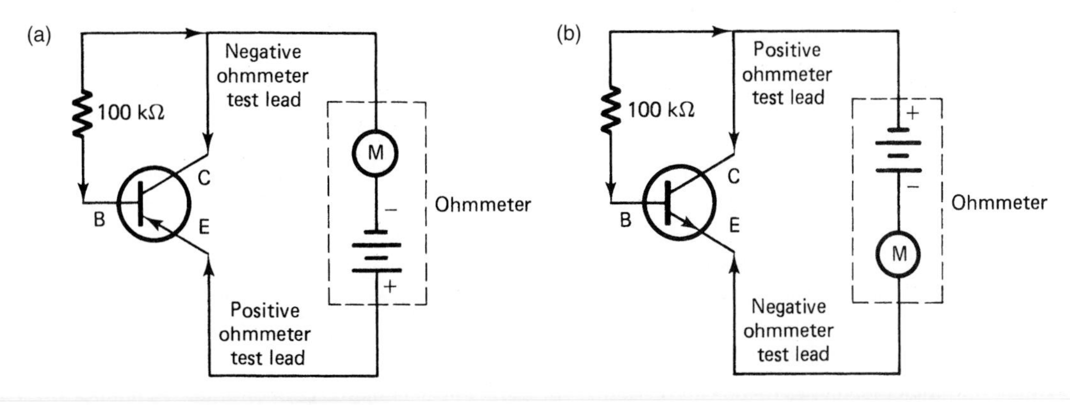

Figure 12–15. Transistor gain test: (a) *PNP* test circuit; (b) *NPN* test circuit.

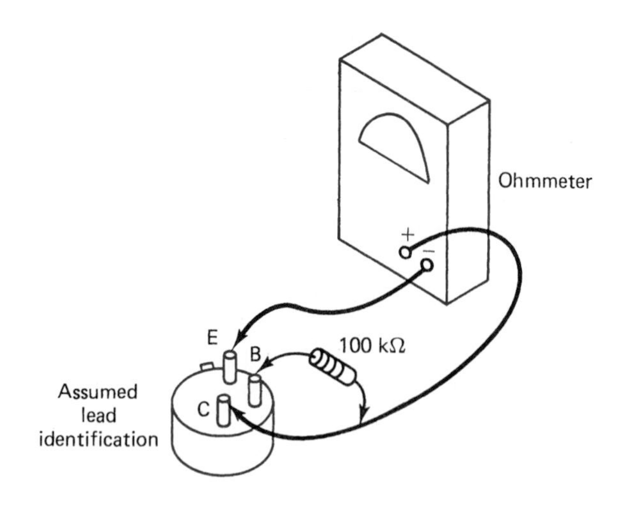

Figure 12–16. *NPN* gain test.

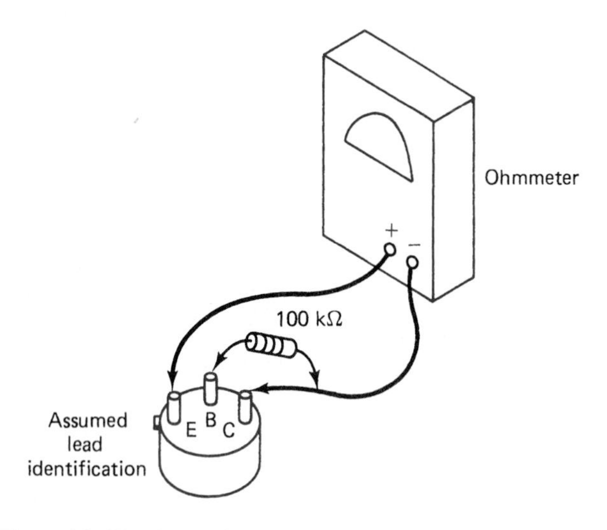

Figure 12–17. *PNP* gain test.

connected to the emitter and the positive lead to the collector. If the assumed leads are correct, the emitter will be forward biased and the collector reverse biased. Touching the open end of the base resistor to the positive lead will cause a base current flow. If it does, the ohmmeter will indicate a low resistance. In effect, the emitter-base junction is forward biased and the base-collector junction reverse biased. If no base current flows, the assumed emitter-collector leads are reversed. Simply reverse the ohmmeter leads and again touch the resistor lead to the positive ohmmeter lead. If the transistor is good and the assumed leads are correct, the ohmmeter will show a low resistance. If it does not, the transistor has low gain.

Figure 12–17 shows the procedure for testing a *PNP* transistor. In this case, the ohmmeter is connected to the two outside leads. We have identified the center lead as the base. The base resistor is now connected to this lead. For a *PNP* transistor, the positive ohmmeter lead goes to the emitter and the negative lead to the collector. If the assumed leads are correct, touching the open end of the base resistor to the negative lead will cause base current to flow. If it does, the ohmmeter will show a low resistance reading. This indicates that the emitter-base junction is forward biased and the base-collector junction reverse biased as in the circuit. If no current flows, the assumed emitter-collector leads are reversed. Reverse the two ohmmeter leads and again touch the base resistor to the negative ohmmeter lead. A good transistor with correct lead identification will indicate low resistance. A transistor with low gain will not cause much of a change in resistance.

12.6 UNIPOLAR TRANSISTORS

A transistor that has only one *P-N* junction in its construction is called a unipolar device. Unlike the bipolar transistor, a unipolar transistor conducts current through a single piece of semiconductor material. The current carriers of this device have only one polarity. The term *unipolar* is used to describe this characteristic.

Field-effect transistors and unijunction transistors are unipolar devices. These transistors are similar in construction but differ a great deal in operation and function. Field-effect transistors are normally used to achieve amplification. Unijunction transistors are used as wave-shaping devices. Both devices play an important role in solid-state electronics.

12.7 JUNCTION FIELD-EFFECT TRANSISTORS

The junction field-effect transistor (JFET) is a three-element electronic device. Its operation is based on the conduction of current carriers through a single piece of semiconductor material. This piece of material is called the *channel.* An additional piece of semiconductor material is diffused into the channel. This element is called the *gate.* The entire unit is built on a third piece of semiconductor known as the *substrate.* The assembled device is housed in a package. The physical appearance of a JFET in its housing is similar to that of a bipolar transistor.

N-Channel JFETs

Figure 12–18 shows the crystal structure, element names, and schematic symbol of an *N*-channel JFET. Its construction has a thin channel of *N* material formed on the *P substrate.* The *P* material of the gate is then formed on top of the *N*-channel. Lead wires are attached to each end of the channel and the gate. No connection is made to the substrate of this device.

The schematic symbol of an *N*-channel JFET is somewhat representative of its construction. The bar part of the

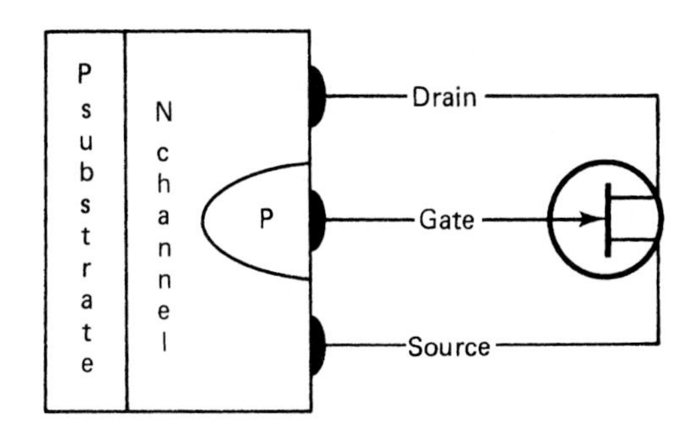

Figure 12–18. *N*-channel junction field-effect transistor.

symbol refers to the channel. The drain (*D*) and source (*S*) are attached to the channel. The gate (*G*) has an arrow, which shows that it forms a *P-N* junction. The arrowhead of an *N*-channel symbol "*P*oints i*N*" toward the channel. This indicates that it is a *P-N* junction. The arrowhead or gate is *P* material and the channel *N* material. This part of the device responds as a junction diode.

The operation of a JFET is somewhat unusual when compared with that of a bipolar transistor. Figure 12–19 shows the source and drain leads of a JFET connected to a dc voltage source. In this case, maximum current will flow through the channel. The I_S and I_D meters will show the same amount of current. The value of V_{DD} and the internal resistance of the channel determine the amount of channel current flow. Typical source-drain resistance values of a JFET are several hundred ohms. Essentially, this means that full conduction will take place in the channel even when the gate is open. A JFET is considered to be a normally-on device.

Current carriers passing through the channel of a JFET are controlled by the amount of bias voltage applied to the gate. In normal circuit operation, the gate is reverse biased with respect to the source. Reverse biasing of the gate and source of any *P-N* junction will increase the size of its depletion region. In effect, this will restrict or deplete the number of majority carriers that can pass through the channel. This means that drain current (I_D) is controlled by the value of V_{GS}. If V_{GS} becomes great enough, no I_D will be permitted to flow through the channel. The voltage that causes this condition is called the cutoff voltage. I_D can be controlled anywhere between full conduction and cutoff by a small change in gate voltage.

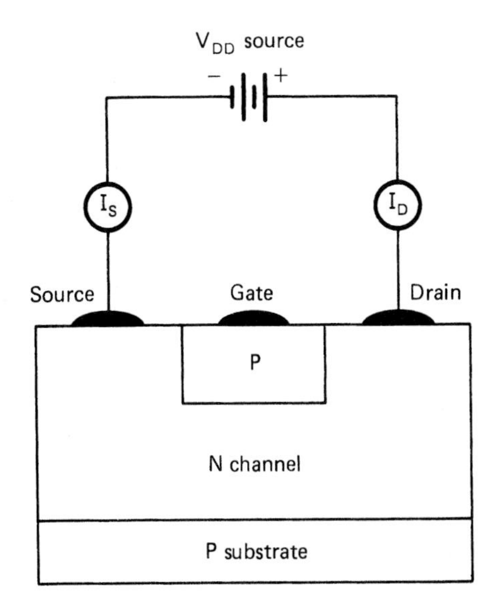

Figure 12–19. Source-drain conduction of a JFET.

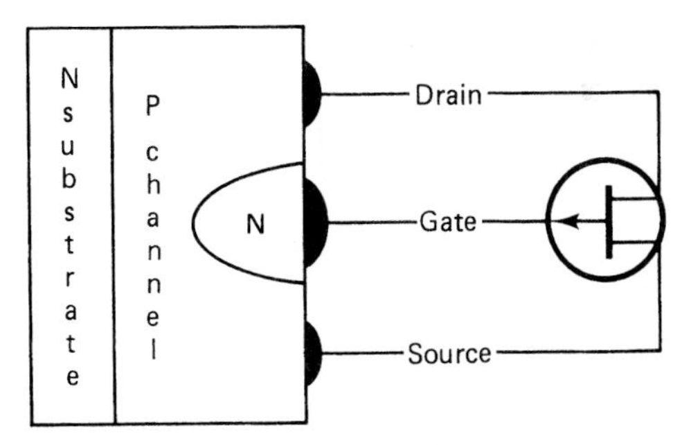

Figure 12–20. *P*-channel junction field-effect transistor.

P-Channel JFETs

The crystal structure, element names, and schematic symbol of a *P*-channel JFET are shown in Fig. 12–20. Construction of this device has a thin channel of *P* material formed on an *N* substrate. The *N* material of the gate is then formed on top of the *P*-channel. Lead wires are attached to each end of the channel and to the gate. Other construction details are the same as those of the *N*-channel device.

The schematic symbol of a JFET is different only in the gate element. In a *P*-channel device the arrow is "*N*ot *P*ointing" toward the channel, which means that the gate is *N* material and the channel is *P* material. In conventional operation, the gate is made positive with respect to the source. Varying values of reverse-bias gate voltage will change the size of the *P-N* junction depletion zone. Current flow through the channel can be altered between cutoff and full conduction. *P*-channel and *N*-channel JFETs cannot be used in a circuit without a modification in the source voltage polarity.

JFET Characteristic Curves

The JFET is a unique device compared with other solid-state components. A small change in gate voltage will, for example, cause a substantial change in drain current. The JFET is therefore classified as a voltage-sensitive device. By comparison, bipolar transistors are classified as current-sensitive devices. A JFET has a rather unusual set of characteristics compared with other solid-state devices.

A family of JFET characteristic curves is shown in Fig. 12–21. The horizontal part of the graph shows the voltage appearing across the source and drain as V_{DS}. The vertical axis shows the drain current (I_D) in milliamperes. Individual curves of the graph show different values of gate voltage (V_{GS}). The cutoff voltage of this device is approximately −7 V. The control range of the gate is from 0 to − 6 V.

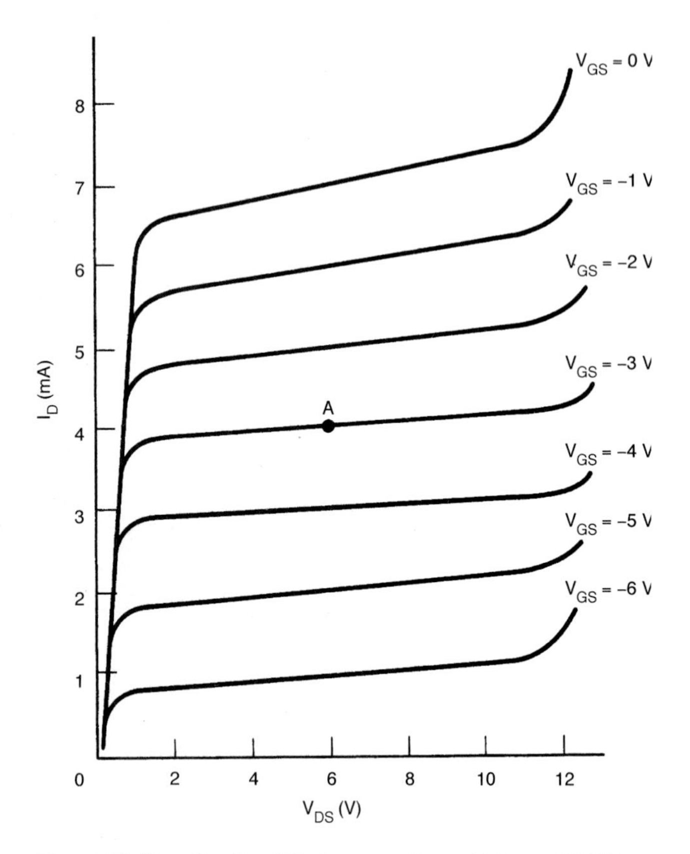

Figure 12–21. Family of drain curves for an *N*-channel JFET.

A drain family of characteristic curves tells much about the operation of a JFET. Refer to point A on the $V_{GS} = -3$ V curve. If the device has −3 V applied to its gate and a V_{DS} of 6 V, there is approximately 4 mA of I_D flowing through the channel. This is determined by projecting a line to the left of the intersection of −3 V_{GS} and 6 V_{DS}. Any combination of I_D, V_{GS}, and V_{DS} can be determined from the family of curves.

Developing JFET Characteristic Curves. A special circuit is used to develop the data for a drain family of characteristic curves. Figure 12–22 shows a circuit that is used to find the data points of a characteristic curve for an *N*-channel JFET. Three meters are used to monitor these data. Gate voltage is monitored by a V_{GS} meter connected between the gate-source leads. Drain current is measured by a milliampere meter connected in series with the drain. V_{DS} is measured with a voltmeter connected across the source and drain. V_{GS} and V_{DS} are adjusted to different values while monitoring I_D. Two variable dc power supplies are used in the test circuit.

The data points of a single curve are developed by first adjusting V_{GS} to 0 V. V_{DS} is then adjusted through its range starting at 0 V. Normally, V_{DS} is increased in 0.1-V steps up to 1 V. I_D increases quickly during this operational time. V_{DS} is then increased in 1-V steps recording the I_D values. Corresponding I_D and V_{GS} values are then plotted on a graph as the 0 V_{GS} curve.

To develop the second curve, V_{DS} must be returned to zero. V_{GS} is then adjusted to a new value—1 V would be a suitable value for most JFETs. V_{DS} is again adjusted through its range while monitoring I_D. Data for the second curve are then recorded on the graph.

To obtain a complete family of drain curves, the process would be repeated for several other V_{GS} values. A

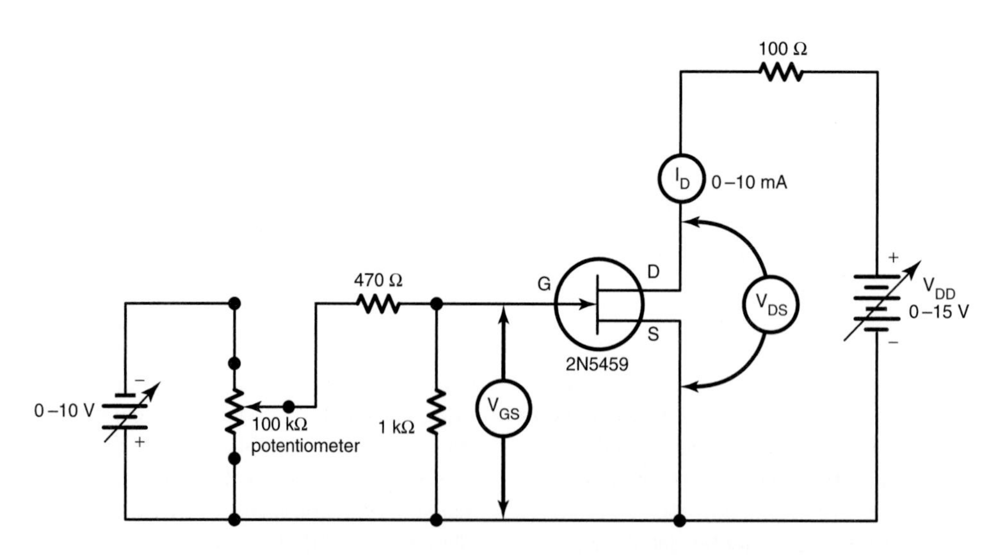

Figure 12–22. Drain curve test circuit for an *N*-channel JFET.

typical family of curves may have 8 to 10 different values. The step values of V_{DS}, V_{GS}, and I_D will obviously change with different devices. Full conduction is usually determined first. This gives an approximation of representative I_D and V_{DS} values. Normally, I_D will level off to a rather constant value when V_{DS} is increased. V_{DS} can be increased in value to a point where I_D starts a slight increase. Generally, this indicates the beginning of the breakdown region. JFETs are usually destroyed if conduction occurs in this area of operation. Maximum V_{DS} values for a specific device are available from the manufacturer. It is a good practice to avoid operation in or near the breakdown region of the device.

12.8 MOS FIELD-EFFECT TRANSISTORS

Metal-oxide semiconductor field-effect transistors (MOSFETs) are an addition to the unipolar transistor family. A distinguishing feature of this device is its gate construction. The gate is, for example, completely insulated from the channel. Voltage applied to the gate will cause it to develop an electrostatic charge. No current is permitted to flow in the gate area of the device. The gate is simply an area of the device coated with metal. Silicon dioxide is used as an insulating material between the gate and the channel. Construction of a MOSFET is rather easy to accomplish.

The channel of a MOSFET has a number of differences in its construction. One type of MOSFET responds to the depletion of current carriers in its channel. MOSFETs have an interconnecting channel built on a common substrate. Direct connection is made between the source and drain of this device. Channel construction is similar to that of the JFET. A second type of MOSFET is called an enhancement type of device. E-MOSFETs do not have an interconnecting channel. Current carriers pulled from the substrate form an induced channel. Gate voltage controls the size of the induced channel. Channel development is a direct function of the gate voltage. E- and D-type MOSFETs both have unique operating characteristics.

Enhancement-type MOSFET construction, element names, and schematic symbol are shown in Fig. 12–23. Note in the crystal structure that no channel appears between the source and drain. The gate is also designed to cover the entire span between the source and drain. The gate is actually a thin layer of metal oxide. Conductivity of this device is controlled by the voltage polarity of the gate. A *P-N* junction is not formed by the gate, source, and drain.

An E-MOSFET is considered to be a normally-off device, which means that without an applied gate voltage, there is no conduction of I_D. When the gate of an *N*-channel device is made positive, electrons are pulled from the substrate into the source-drain region. This action causes the induced channel to be developed. With the channel complete, conduction

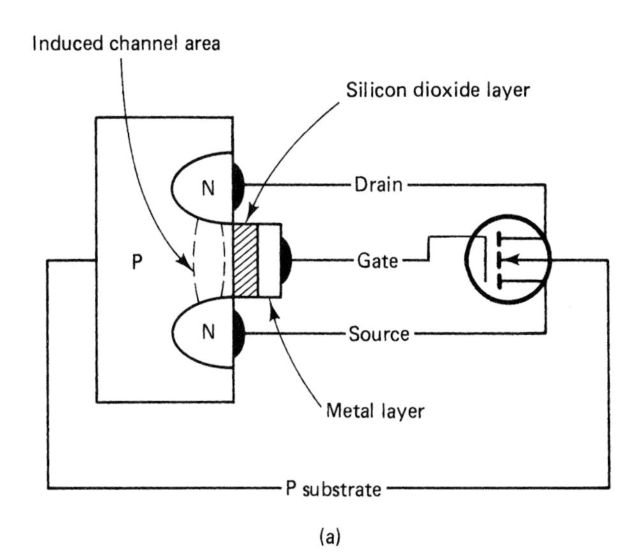

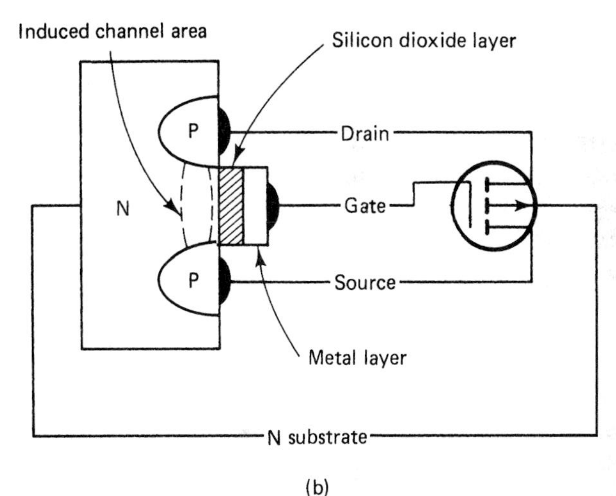

Figure 12–23. Enhancement mode MOSFET: (a) *N*-channel; (b) *P*-channel.

occurs between the source and drain. In effect, drain current is aided or enhanced by the gate voltage. Without V_G, the channel is not properly developed and no I_D will flow.

A family of characteristic curves for an E-type MOSFET is shown in Fig. 12–24. This particular set of curves is representative of an *N*-channel device. A *P*-channel device would have the polarity of the gate voltage reversed. Note that an increase of gate voltage causes a corresponding increase in I_D. The input impedance of this device is extremely high regardless of the crystal material used or the polarity of the gate voltage.

Depletion type MOSFET construction, element names, and schematic symbol are shown in Fig. 12–25. The *N*-channel device has a thin channel of *N* material formed on a *P* substrate. Source and gate connections are made directly to the channel. A thin layer of silicon dioxide (SiO_2) insulation covers the channel. The gate is a metal-plated area formed on the

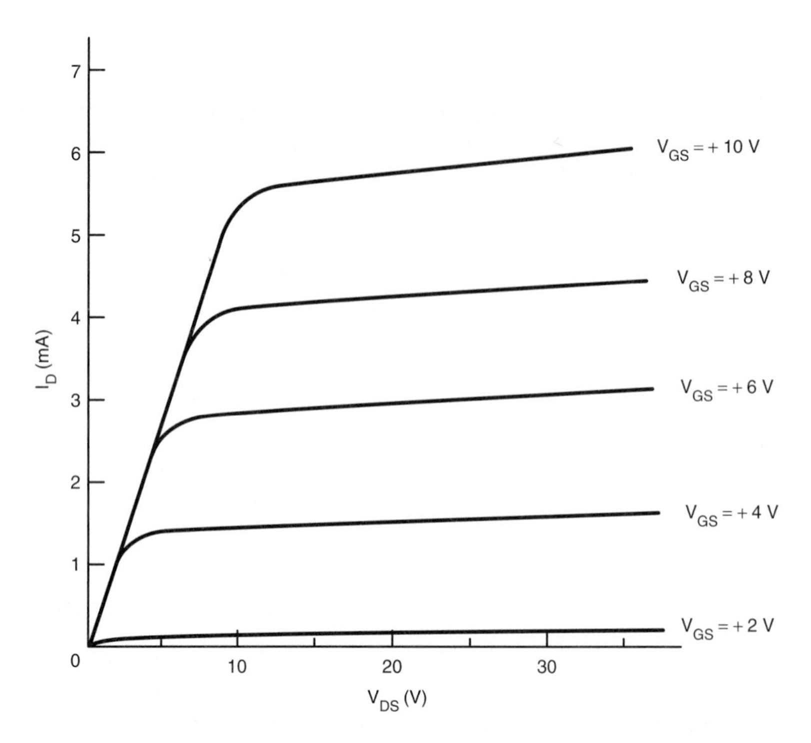

Figure 12–24. Family of drain curves for an E-type MOSFET (*N*-channel type).

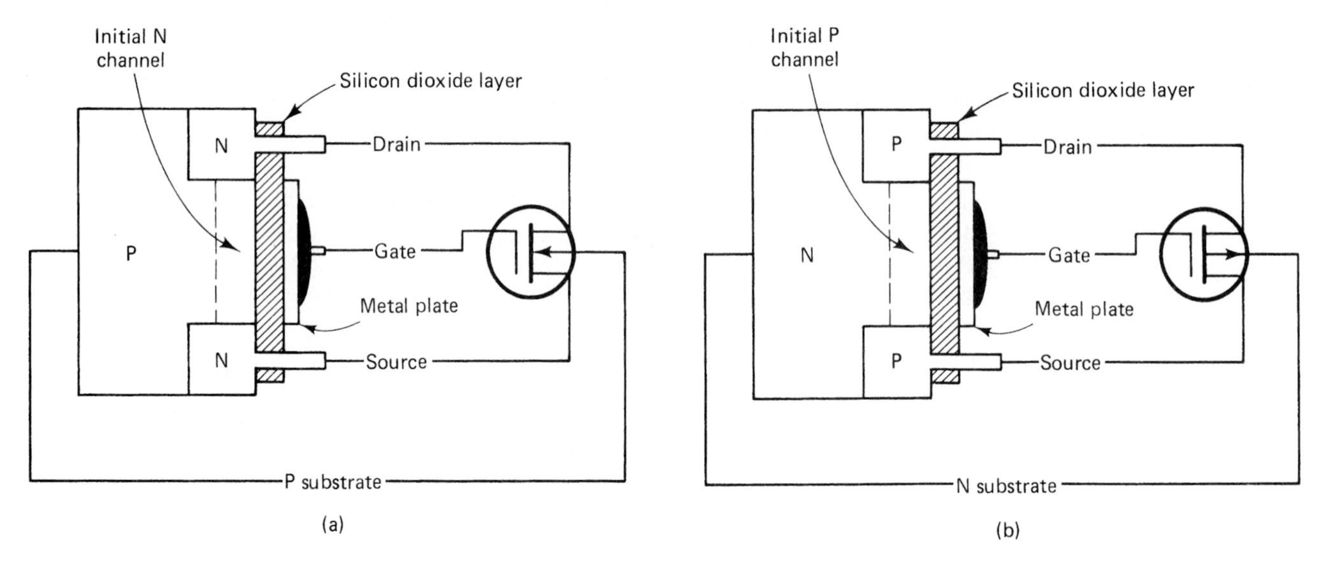

Figure 12–25. D-type MOSFET: (a) *N*-channel; (b) *P*-channel.

silicon dioxide layer. The entire unit is then built on a substrate of *P* material. The arrow of the schematic symbol refers to the material of the substrate. When it "*P*oints i*N*," this shows that the substrate is *P* material and the channel is *N* material.

The construction of *P*- and *N*-channel devices is essentially the same. The crystal material of the channel and substrate is the only difference. Current carriers are holes in the *P*-channel device, whereas electrons are carriers in the *N*-channel unit. The schematic symbol differs only in the direction of the substrate arrow. It does "*N*ot *P*oint" toward the substrate in a *P*-channel device. This means that the substrate is *N* material and the channel is *P* material.

The operation of a D-MOSFET is similar to that of a JFET. With voltage applied to the source and drain, current carriers pass through the channel. The gate does not need to be energized to produce conduction. In this regard, a D-MOSFET is considered to be a normally-on device. Control of I_D is, however, determined by the polarity of gate voltage.

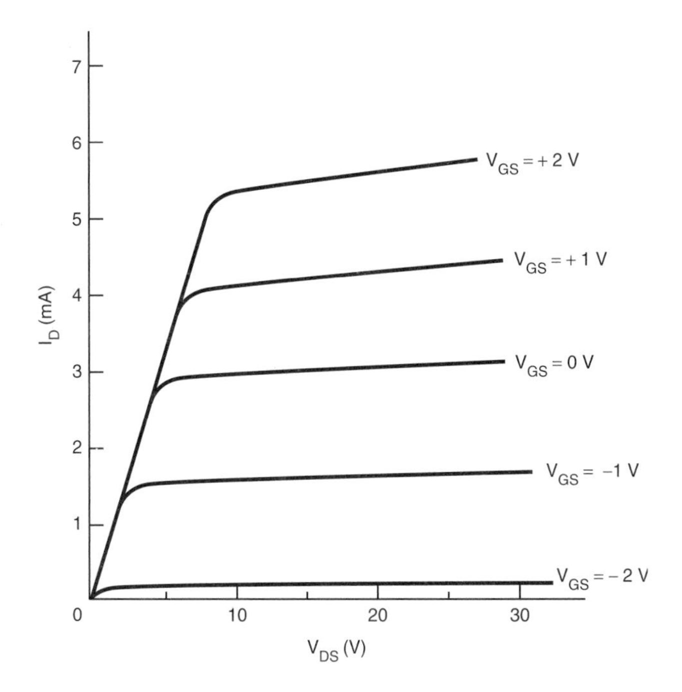

Figure 12–26. Family of drain curves for a D-type *N*-channel MOSFET.

When the polarity of V_G is the same as the channel, current carrier depletion occurs. A reverse in polarity causes I_D to increase due to the enhancement of the current carriers. In a sense, the D-MOSFET should be classified as a depletion-enhancement device. Its conduction normally responds to both conditions of operation.

A family of characteristic curves for an *N*-channel D-MOSFET is shown in Fig. 12–26. The horizontal part of the graph shows the source-drain voltage as V_{DS}. The vertical axis shows the drain current in milliamperes. Individual curves of the graph show different values of gate voltage. Note that zero V_G is near the center of the curve. This means that the gate can swing above or below zero. For an *N*-channel device, negative gate voltage reduces I_D. This voltage, in effect, pulls holes from the substrate into the initial channel area. Electrons normally in the channel move in to fill the holes. This action depletes the number of electrons in the initial channel. An increase in negative gate voltage causes a corresponding decrease in I_D.

When V_G swings positive, it causes the number of current carriers in the initial channel to be increased. A positive gate voltage attracts electrons from the *P* substrate. This action increases the width of the initial channel. As a result, more I_D flows, making V_G more positive. This causes an increase in I_D, which shows that the channel is aided or enhanced by a positive gate voltage. A D-MOSFET therefore responds to both the depletion and enhancement of its current carriers. Selection of an operating point is easy to accomplish when this device is used as an amplifier.

In a *P*-channel D-MOSFET, operation is similar to that of the *N*-channel device. In the *P*-channel device, holes are the current carriers. Current conduction is normally on when V_G is zero. A positive gate voltage will reduce I_D. Electrons are drawn out of the substrate to fill holes in the channel, causing a reduction in channel current. When the gate voltage swings negative, the channel current carriers are enhanced. Holes pulled into the initial channel cause an increase in current carriers. Current carriers are enhanced or depleted according to the polarity of V_G.

P-channel D-MOSFETs are not as readily available as the *N*-channel type of device. Fabrication of the *P*-channel device is somewhat more difficult to achieve. As a general rule, the *N*-channel unit is somewhat more costly to produce. Applications of D-MOSFETs tend to use the *N*-channel device more than the *P*-channel device.

VMOS Field-Effect Transistors

Another type of field-effect transistor is used mainly to replace the bipolar transistor. Power FETs are used in power supplies and solid-state switching applications. Collectively, these devices are called V-MOSFETs or V-MOSs. The *V* designation finds its origin in vertical-groove MOS technology. The device is constructed with a V-shaped groove etched into the substrate. Construction of this type requires less space than a horizontally assembled device. The construction of a V-MOS device permits better heat dissipation and a high density of material in the channel area.

Figure 12–27 shows a cross-sectional view of the crystal structure, the element names, and a schematic symbol of an *N*-channel V-MOS transistor. Notice that a V groove is etched into the surface of the structure. From the top, the V cut penetrates through the *N*+, *P*, and *N*− layers and stops near the *N*+ substrate. The two *N*+ layers are heavily doped, and the *N*− layer is lightly doped. A thin layer of silicon dioxide covers both the horizontal surface and the V groove. A metal film deposited on top of the groove serves as the gate. The gate is therefore insulated from the groove. The source leads on each side of the groove are connected internally. The bottom layer of *N*+ material serves as a combined substrate and drain. Current carriers move between the source and drain vertically.

A V-MOS transistor is classified as an enhancement type of MOSFET. No current carriers exist in the source and drain regions until the gate is energized. An *N*-channel device, as shown in Fig. 12–27, does not conduct until the gate is made positive with respect to the source. When this occurs, the *N*-channel is induced between the two *N*+ areas near the groove. Current carriers can then flow through the vertical channel from source to drain. When the gate of an *N*-channel device is made negative, no channel exists and the current carriers cease to flow.

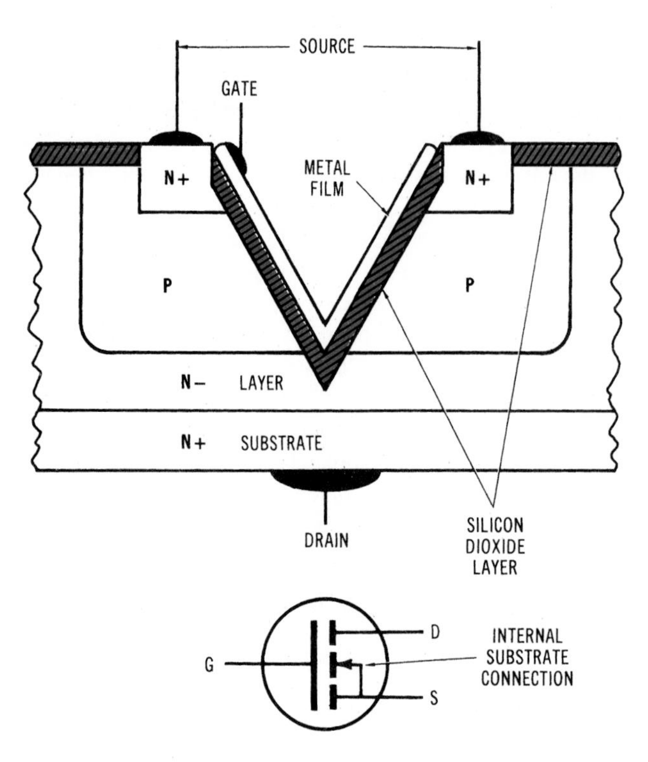

Figure 12–27. Cross-sectional view of an *N*-channel V-MOSFET.

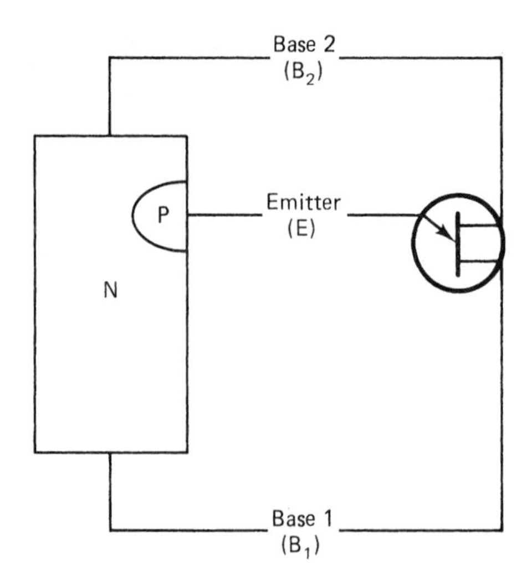

Figure 12–28. Unijunction transistor.

12.9 UNIJUNCTION TRANSISTORS

The unijunction transistor (UJT) is often described as a voltage-controlled diode. This device is not considered to be an amplifying device. It is, however, classified as a unipolar transistor. The UJT and an *N*-channel JFET are often confused because of the similarities in their schematic symbols and crystal structure. The operation and function of each device is entirely different.

Figure 12–28 shows the crystal structure, schematic symbol, and element names of a UJT. Note that the UJT is a three-terminal single-junction device. The crystal is an *N*-type bar of silicon with contacts at each end. The end connections are called base 1 (B_1) and base 2 (B_2). A small heavily doped *P* region is alloyed to one side of the silicon bar which serves as the emitter (*E*). A *P-N* junction is formed by the emitter and the silicon bar. The arrow of the symbol "*P*oints i*N*," which means that the emitter is *P* material and the silicon bar is *N* material. The arrow of the symbol is slanted, which distinguishes it from the *N*-channel JFET.

An interbase resistance (R_{BB}) exists between B_1 and B_2. Typically, R_{BB} is between 4 and 10 kΩ. This value can be measured easily with an ohmmeter. The resistance of the silicon bar is represented by R_{BB}. This resistance can be divided into two values. R_{B1} is between the emitter and B_1. R_{B2} is between the emitter and B_2. Normally, R_{B2} is somewhat less than R_{B1}. The emitter is usually closer to B_2 than B_1. When a UJT is made operational, the value of R_{B1} will change with different emitter voltages.

In circuit applications, B_1 is usually placed at circuit ground or the source voltage negative. The emitter then serves as the input to the device. B_2 provides circuit output. A change in E-B_1 voltage will cause a change in R_{BB}. The output current of the device will increase when E-B_1 turns on. A UJT is normally used as a trigger device, and it is used to control other solid-state devices. No amplification is achieved by a UJT.

A characteristic curve for a typical UJT is shown in Fig. 12–29. The vertical axis of this graph shows the emitter voltage as V_E. An increase in V_E causes the curve to rise vertically. The horizontal axis of the graph shows the emitter current (I_E). Note that the curve has *peak voltage* (V_P) and *valley voltage* (V_V) points. An increase in I_E causes V_E to rise until it reaches the peak point. A further increase in I_E will cause the emitter voltage to drop to the valley point. This condition is called the negative resistance region. A device that has a negative resistance characteristic is capable of regeneration or oscillation.

In operation, when the emitter voltage reaches the peak voltage point, E-B_1 becomes forward biased. This condition draws holes and electrons to the *P-N* junction. Holes are injected into the *B*1 region. This action causes B_1 to be more conductive. As a result, the E-B_1 region became low resistant. A sudden drop in R_{B1} will cause a corresponding increase in B_1-B_2 current. This change in current can be used to trigger or turn on other devices. The trigger voltage of a UJT is a predictable value. Knowing this value will permit the device to be used as a control element.

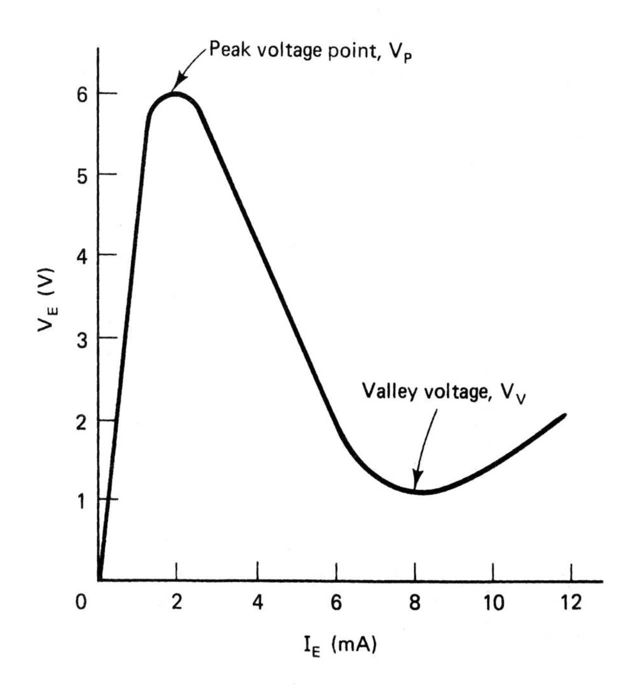

Figure 12–29. Characteristic curve for a UJT.

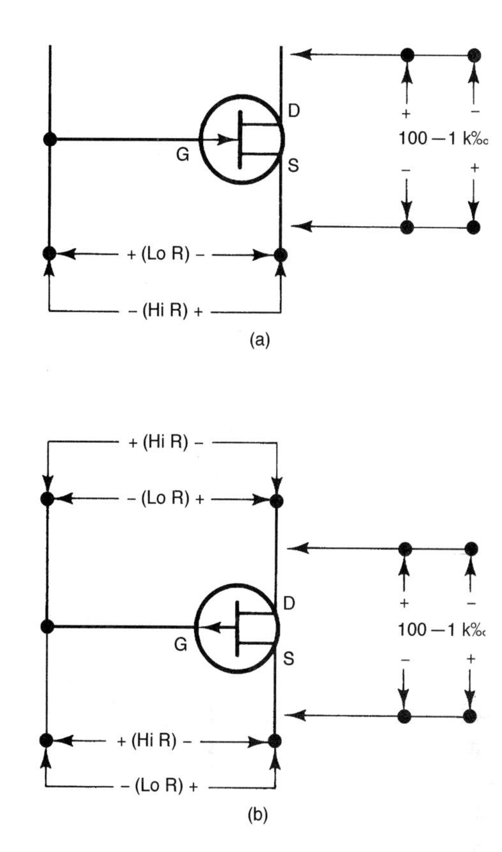

Figure 12–30. Resistance values of a JFET: (a) *N*-channel; (b) *P*-1 channel.

12.10 UNIPOLAR TRANSISTOR TESTING

Unipolar transistors, like bipolar transistors, periodically require testing. The condition of the device can be tested with a curve tracer oscilloscope, sound-producing instruments, or in-circuit/out-of-circuit testers. This equipment is normally used to evaluate a device and to identify leads. Test procedures are rather easy to accomplish and are a reliable indicator of device condition. If equipment of this type is available, become familiar with its operation by reviewing the operational manual. Use it to test devices and identify leads.

The ohmmeter section of a multimeter can also be used to evaluate unipolar devices. This instrument is more readily available than specialized testers. Lead identification, shorted devices, open conditions, and gain can be evaluated. Before using the ohmmeter, it is important that the polarity of its voltage source be known. The following procedure describes an ohmmeter with straight polarity. The black or common lead is negative and the probe or red lead is positive. If the meter polarity is unknown, it may be tested by connecting a voltmeter across its leads.

JFET testing with an ohmmeter is relatively easy to accomplish. Remember that a JFET is a single bar of silicon with one *P-N* junction. To identify the leads, first select any two of the three leads. Connect the ohmmeter to these leads. Note the resistance. If it is 100 to 1000 Ω, the two leads are the source and drain. Reverse the ohmmeter and connect it to the same leads. If the two leads are actually the source and drain, the same resistance value will be indicated. The third lead is the gate. If the two selected leads show a different resistance in each direction, one must be the gate. Select the third lead and one of the previous leads. Repeat the procedure. A good JFET must show the same resistance in each direction between two leads. This represents the source-drain channel connections.

The gate of a JFET responds as a diode. It will be low resistant in one direction and high resistant in the reverse direction. If the device shows low resistance when the positive probe is connected to the gate and negative to the source or drain, the JFET is an *N*-channel device. A *P*-channel device will show low resistance when the gate is made negative and the source or drain positive. The resistance ratio in the forward and reverse direction must be at least 1:1000 for a good device. Figure 12–30 indicates some representative resistance values for good JFETs.

A test of JFET amplification is indicated by connecting the ohmmeter between the source and drain. The polarity of the meter is not significant. Touch the gate with your finger. This action should cause a decrease in resistance. A more significant indication will take place when a 200-kΩ resistor is used. Touching the resistor between the gate and either the source or drain will cause reduced resistance. If little or no resistance change occurs during this test, it indicates an open or faulty gate.

REVIEW

1. Explain what is meant by the term *bipolar transistor.*
2. Why is it important that the base of a transistor be thin and lightly doped?
3. Why must the emitter-base junction and base-collector junction of a bipolar transistor both be biased at the same time to produce operation?
4. Why does current flow through the reverse-biased base-collector junction of a bipolar transistor?
5. Why can *PNP* and *NPN* transistors not be used interchangeably in a circuit?
6. Explain how current carriers flow through the forward-biased emitter-base junction of a bipolar transistor.
7. What is the relationship of I_B, I_C, and I_E in a bipolar transistor?
8. What is dc beta?
9. What is ac beta?
10. What does a collector family of characteristic curves show about a bipolar transistor?
11. How does a transistor respond when operating in the saturation region?
12. How do I_C and V_{CE} respond when a bipolar transistor is in the cutoff region?
13. What is the difference between a unipolar and a bipolar transistor?
14. Describe the physical construction of JFET.
15. Why is a JFET described as a voltage-controlled device?
16. How do a MOSFET and JFET compare?
17. What is a depletion type MOSFET?
18. What is an enhancement type MOSFET?
19. Describe the physical construction of a UJT.
20. Why does the interbase resistance of a UJT change when the emitter becomes forward biased?
21. What is the negative resistance region of a UJT characteristic curve?
22. Describe the physical construction of a V-MOSFET.

STUDENT ACTIVITIES

• *Bipolar Transistor Type Identification*

1. With an ohmmeter, identify the type, *NPN* or *PNP,* of several bipolar transistors.
2. Record the number of the device and indicate its type according to your findings.

• *Bipolar Transistor Lead Identification*

1. Select a known *NPN* transistor type.
2. Make a sketch of the lead placement.
3. Identify the leads of the device.
4. Add the labels *E, B,* and *C* to the sketch.
5. Repeat the procedure of steps 1 to 4 for a *PNP* device.
6. If the manufacturer's data are available, compare your lead identification with the known information.
7. What type of packaging is used for each device?

• Bipolar Gain Test

1. Using the known lead data of a specific transistor, determine the gain using an ohmmeter.
2. Use a 100-kΩ resistor for the base connection.
3. Record the resistance change observed for each device.

• Bipolar Transistor Testing

Complete Activity 12–1, page 245 in the manual.

• • ANALYSIS

1. Explain how an ohmmeter is used to determine if a transistor is an *NPN* or a *PNP* type.
2. How is the leakage of a transistor evaluated with an ohmmeter?

• Bipolar Transistor Characteristic Curves

1. Construct the I_C–V_{CE} data circuit of Fig. 12–11.
2. Collect data for three or four individual I_B curves.
3. On a sheet of graph paper, plot curves using the data collected.

• Transistor Operation Regions or Complete Activity 12–2, page 247 in the manual.

1. Construct the circuit of Fig. 12–10(a).
2. Use the small-signal transistor indicated or an equivalent device.
3. Turn on the source voltage V_{CC}.
4. Measure and record V_C and V_R.
5. Turn off V_{CC} and modify the circuit as in Fig. 12–10(b).
6. Turn on V_{CC}. Measure and record V_C and V_R.
7. Turn off V_{CC} and modify the circuit to conform with Fig. 12–10(c).
8. Turn on V_{CC}. Measure and record V_C and V_R.
9. Note the operational differences of the transistor in its active, cutoff, and saturation regions.

• • ANALYSIS

1. Explain what conditions of operation must be met by a bipolar transistor for it to operate in the saturation region.
2. Explain what conditions of operation must be met by a bipolar transistor for it to operate in the cutoff region.
3. What are the conditions of operation that indicate when a bipolar transistor is operating in the active region?

• JFET Testing or Complete Activity 12-3, page 249 in the manual.

1. With an ohmmeter, determine the type of device being tested. Use both *N*- and *P*-channel devices.
2. Identify the leads.
3. Make a sketch showing the lead placement.
4. Test the device for gain.
5. Note any faulty devices.

• •

ANALYSIS

1. Explain how an ohmmeter can be used to evaluate the status of a JFET.
2. How would an ohmmeter test distinguish between an *N*-channel and a *P*-channel JFET?
3. What are some of the differences that an ohmmeter will show between a JFET and a MOSFET?

• JFET Characteristic Curves or Complete Activity 12–4, page 251 in the manual.

1. Construct the drain curve test circuit of Fig. 12–22.
2. Adjust V_{GS} to zero, then alter V_{DD} from 0 to 15 V while recording I_D values for each 0.5- or 1-V increment.
3. Adjust V_{GS} to 0.5 V, 1 V, and 1.5 V values while recording I_D values for the same V_{DD} values.
4. On a sheet of graph paper, plot an I_D family of curves.

• •

ANALYSIS

1. How would an increase in the resistance of the drain resistor change the operation of a JFET amplifier?
2. Explain how a JFET achieves amplification.
3. How does the operation of a JFET differ from that of a bipolar transistor?

• MOSFET Testing

1. With an ohmmeter, determine the type of MOSFET device being tested. Indicate E or D type, *N*- or *P*-channel.
2. Identify the leads. Note if it is a three- or four-lead device.
3. Make a sketch of the lead location.

• MOSFET Characteristics

Complete Activity 12–5, page 255 in the manual.

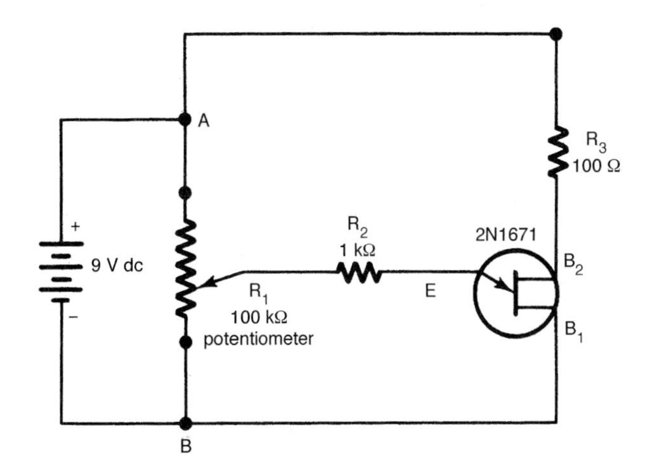

Figure 12–34. UJT test circuit.

ANALYSIS

1. Explain how an *N*-channel E-MOSFET achieves amplification.
2. How would an increase in the resistance of the drain resistor of a MOSFET amplifier change the operation of a MOSFET?
3. What are the operational differences between an E-MOSFET and a JFET?

• *UJT Characteristic Curves*

1. Construct the UJT characteristic test circuit of Fig. 12–34.
2. Turn on the power supply and adjust it to 9 V dc as measured at point A–B.
3. Connect the voltmeter across E and B_1 of the UJT.
4. Adjust the 100-kΩ potentiometer to an indication of 0 V.
5. Gradually adjust the potentiometer while observing V_E-B_1. Note where the peak voltage (V_P) and the valley voltage (V_V) are located.
6. Turn off the power supply and disconnect the circuit.

• *Self-Examination*

1. Complete the Self-Examination for Chap. 12, page 259 in the manual.
2. Check your answers on page 417 in the manual.
3. Study the questions that you answered incorrectly.

• *Chapter 12 Examination*

Complete the Chap. 12 Examination on page 263 in the manual.

CHAPTER

13

AMPLIFICATION

OUTLINE

13.1 Amplification Principles
13.2 Bipolar Transistor Amplifiers
13.3 Basic Amplifiers
13.4 Classes of Amplification
13.5 Transistor Circuit Configurations
13.6 Field-Effect Transistor Amplifiers
13.7 Integrated-Circuit Amplifiers

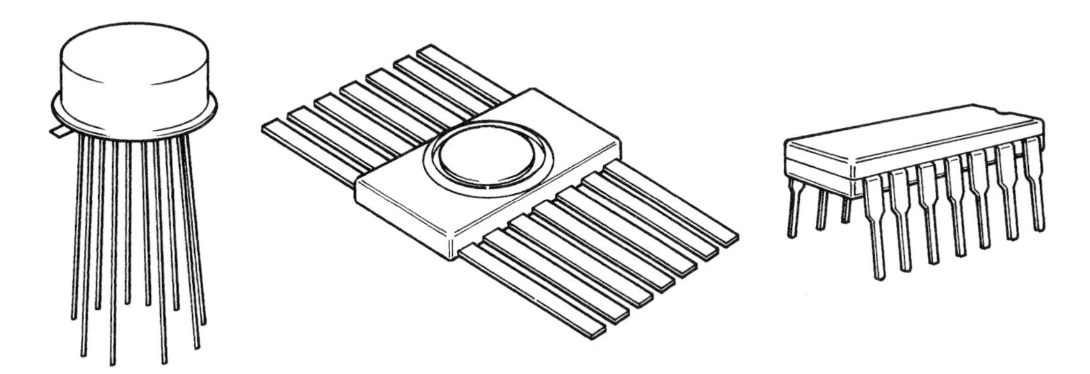

Integrated circuit amplifier packages.

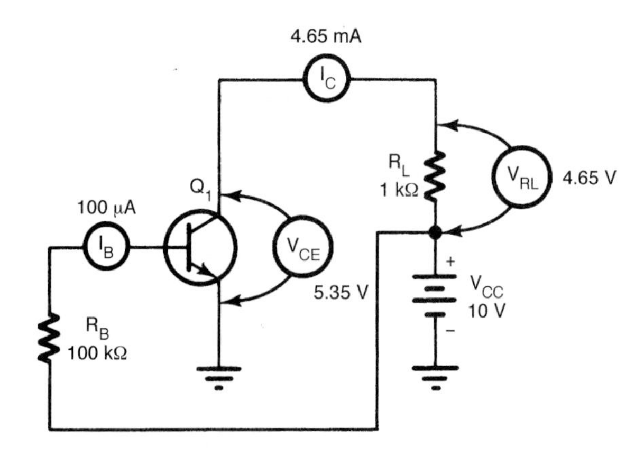

Figure 13–3. Static operating condition of a basic amplifier.

Signal Amplification

For the basic transistor circuit of Fig. 13–3 to respond as an amplifier, it must have a signal applied. The signal may be either voltage or current. An applied signal causes the transistor to change from its static state to a dynamic condition. Dynamic conditions involve changing values. All ac amplifiers respond in the dynamic state. The output of this should develop an ac signal.

Figure 13–4 shows a basic transistor amplifier with an ac signal applied. A capacitor is used, in this case, to couple the ac signal source to the amplifier. Remember that ac passes easily through a capacitor while dc is blocked. As a result, the ac signal is injected into the base-emitter junction. Dc does not flow back into the signal source. The ac signal is therefore added to the dc operating voltage. The emitter-base voltage is a dc value that changes at an ac rate.

Consider now how the applied ac signal alters the emitter-base junction voltage. Figure 13–5 shows some representative voltage and current values that appear at the emitter-base junction of the transistor. They can be measured with a voltmeter or observed with an oscilloscope. Figure 13–5(a) shows the dc operating voltage and base current. This occurs when the amplifier is in its steady or static state. Figure 13–5(b) shows the ac signal that is applied to the emitter-base junction. Note that the amplitude change of this signal is a minute value. This also shows how the resultant base current and voltage change with ac applied. The ac signal is essentially riding on the dc voltage and current.

Refer now to the schematic diagram of the basic amplifier in Fig. 13–6. Note in particular the waveform inserts that appear in the diagram. These show how the current and voltage values respond when an ac signal is applied.

An ac signal applied to the input of our basic amplifier rises in value during the positive alternation and falls during the negative alternation. Initially, this causes an increase and decrease in the value of V_{BE}. This voltage has a dc level with an ac signal riding on it. The indicated I_B is developed as a result of this voltage. The changing value of I_B causes a corresponding change in I_C. Note that I_B and I_C both appear to be the same. There is, however, a noticeable difference in values—I_B is in microamperes whereas I_C is in milliamperes. The resultant I_C passing through R_L causes a corresponding voltage drop (V_{RL}) across R_L. V_{CE} appearing across the transistor is the reverse of V_{RL}. These signals are both ac values riding on a dc level. The output, V_O, is changed to an ac value. Capacitor C_2 blocks the dc component and passes only the ac signal.

It is interesting to note in Fig. 13–6 that the input and output signals of the amplifier are reversed. When the ac input signal rises in value, it causes the output to fall in value. The negative alternation of the input causes the output to rise in value. This condition of operation is called *phase inversion.* Phase inversion is a distinguishing characteristic of the common-emitter amplifier.

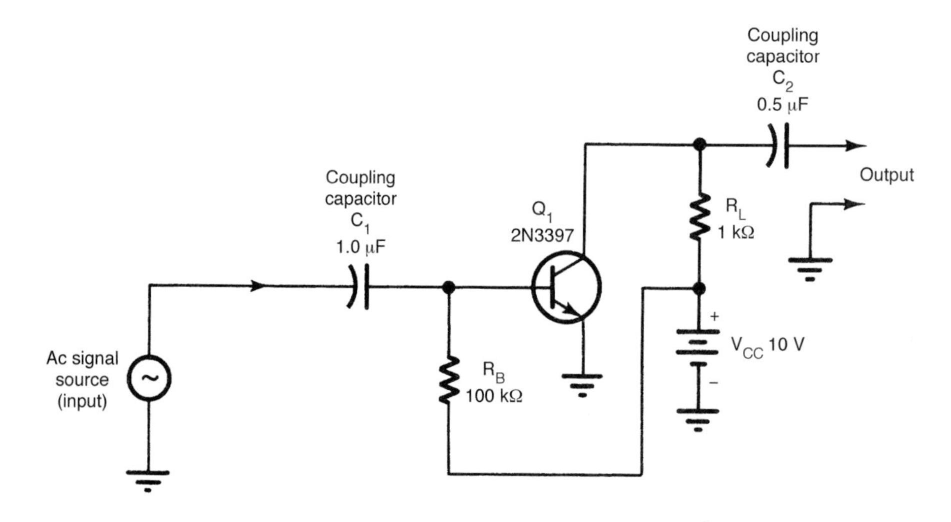

Figure 13–4. Amplifier with ac signal applied.

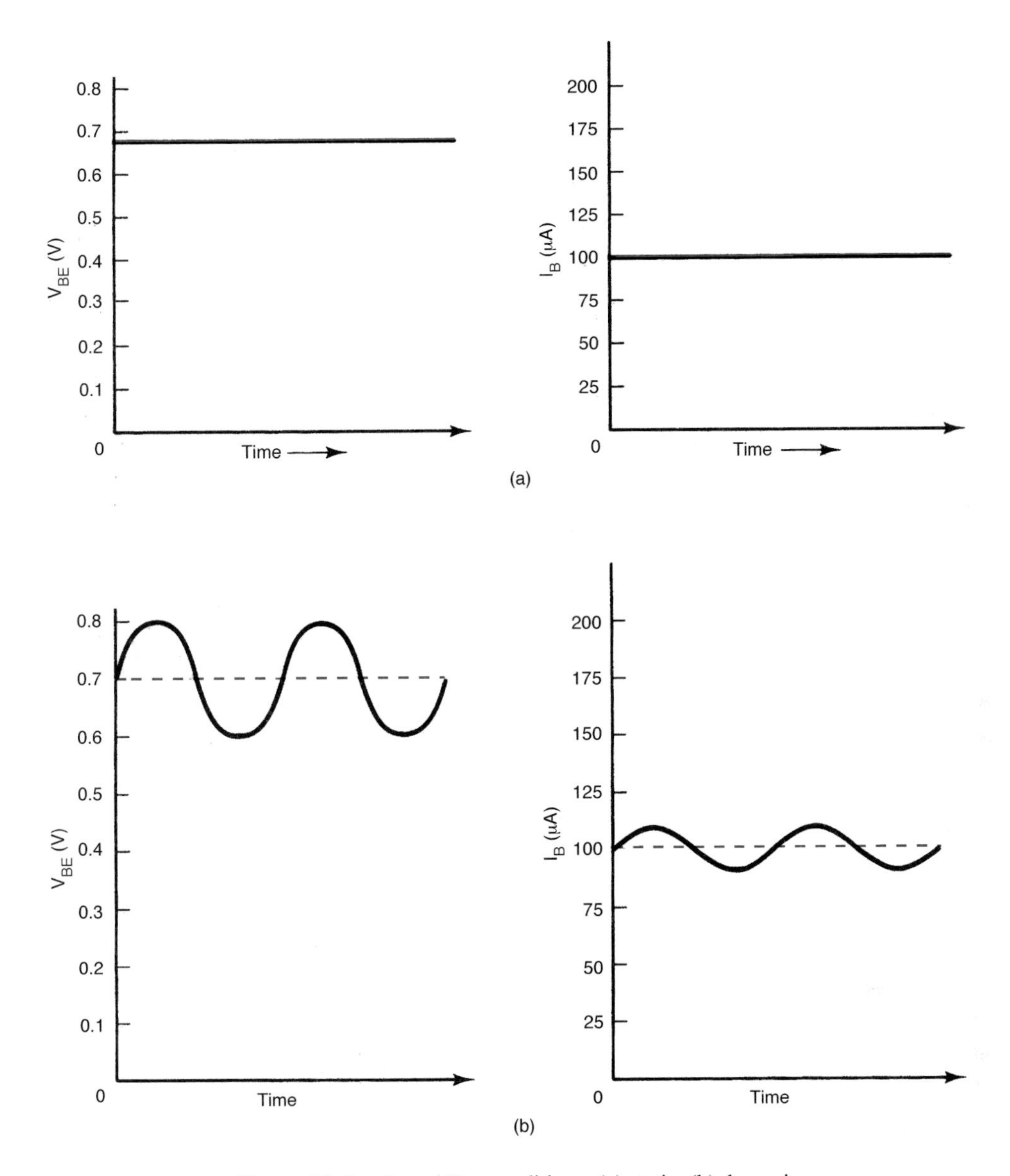

Figure 13–5. I_B and V_{BE} conditions: (a) static; (b) dynamic.

Amplifier Bias

If a transistor is to operate as a linear amplifier, it must be properly biased and have a suitable operating point. Its steady state of operation depends greatly on base current, collector voltage, and collector current. Establishing a desired operating point requires proper selection of bias resistors and a load resistor to provide proper input current and collector voltage.

Stability of operation is an important consideration. If the operating point of a transistor is permitted to shift with temperature changes, unwanted distortion may be introduced. In addition to distortion, operating point changes may also cause damage to a transistor. Excessive collector current, for example, may cause the device to develop too much heat.

The method of biasing a transistor has much to do with its thermal stability. Several different methods of achieving bias are in common use today. One method of biasing is considered to be beta dependent. Bias voltages are largely dependent on transistor beta. A problem with this method of biasing is transistor response. Transistor beta is rarely the same for a specific device. Biasing setup for one transistor will not necessarily be the same for another transistor.

Biasing that is independent of beta is important. This type of biasing responds to fixed voltage values. Beta does not alter these voltage values. As a general rule, this form of biasing is more reliable. The input and output of a transistor are stable and the results are predictable.

BETA-DEPENDENT BIASING. Four common methods of beta-dependent biasing are shown in Fig. 13–7. The

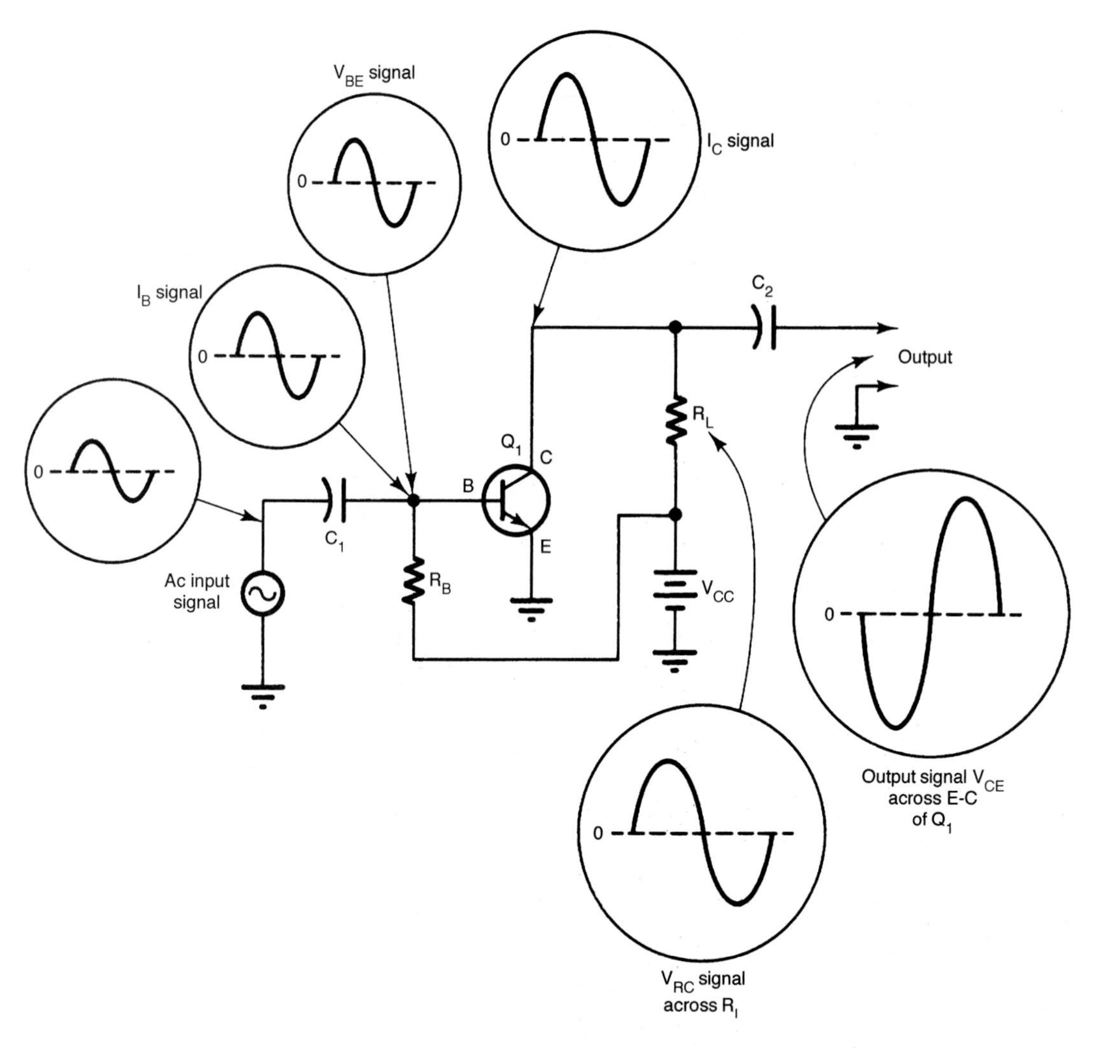

Figure 13–6. Ac amplifier operation.

steady-state or static conditions of operation are a function of the transistor's beta. These methods are easy to achieve with a minimum of parts. They are not used widely today.

The circuit in Fig. 13–7(a) is a simple form of biasing for a common-emitter amplifier. This method of biasing was used for the basic transistor amplifier. The base is simply connected to V_{CC} through R_B, which causes the emitter-base junction to be forward biased. The resulting collector current is beta times the value of I_B. The value of I_B is V_{CC}/R_B. As a general rule, biasing of this type is sensitive to changes in temperature. The resultant output of the circuit is rather difficult to predict. This method of biasing is often called *fixed biasing* and is primarily used because of its simplicity.

The circuit of Fig. 13–7(b) was developed to compensate for the temperature sensitivity of the circuit in Fig. 13–7(a). Bias current is used to counteract changes in temperature. In a sense, this method of biasing has negative feedback. R_B is connected to the collector rather than V_{CC}. The voltage available for base biasing is that remaining after a voltage drop across the load resistor. If the temperature rises, it causes an increase in I_B and I_C. With more I_C, there is more voltage drop across R_L. Reduced V_C voltage will cause a corresponding drop in I_B. This, in turn, brings I_C back to normal. The opposite reaction will occur when transistor temperature becomes less. This method of biasing is called *self-biasing.*

The circuit in Fig. 13–7(c) is an example of emitter biasing. Thermal stability is improved with this type construction. I_B is again determined by the value of R_B and V_{CC}. An additional resistor is placed in series with the emitter of this circuit. Emitter current passing through R_E produces emitter voltage (V_E). This voltage opposes the base voltage developed by R_B. Proper values of R_B and R_E are selected so that I_B and I_E will flow under ordinary operating conditions. If a change in temperature should occur, V_E will increase in value. This action will oppose the base bias. As a result, collector current will drop to its normal value. The capacitor connected across R_E is called an *emitter bypass capacitor.* It provides an ac path for signal voltages around R_E. With C in the circuit,

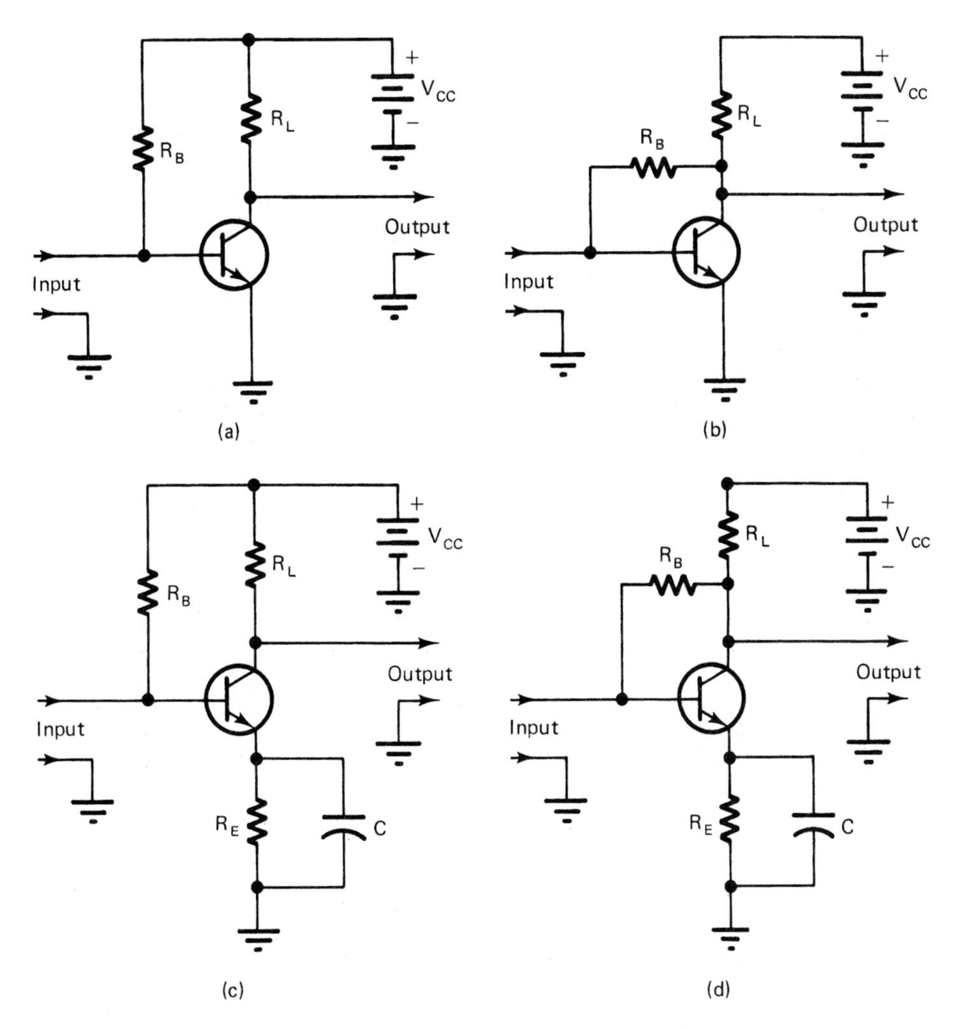

Figure 13–7. Methods of beta-dependent biasing: (a) fixed biasing; (b) self-biasing; (c) fixed-emitter biasing; (d) self-emitter biasing.

an average dc level is maintained at the emitter. Without C in the circuit, amplifier gain would be reduced. The value of C is dependent on the frequencies being amplified. At the lowest possible frequency being amplified, the capacitive reactance (X_C) must be 10 times as small as the resistance of R_E.

The circuit of Fig. 13–7(d) is a combination of the circuits in Fig. 13–7(b) and (c). It is often called *self-emitter bias.* In the same regard, the circuit in Fig. 13–7(c) could be called *fixed-emitter bias.* As a general rule, emitter biasing is not effective when used independently. The circuit in Fig. 13–7(d) has good thermal stability. The output has reduced gain because of the base resistor connection.

INDEPENDENT BETA BIASING. Two methods of biasing a transistor that is independent of beta are shown in Fig. 13–8. These circuits are extremely important because they do not change operation with beta. As a general rule, these circuits have reliable operating characteristics. The output is predictable and stability is excellent.

The circuit in Fig. 13–8(a) is described as the divider method of biasing. It is widely used today. The base voltage (V_B) is developed by a voltage-divider network made up of R_B and R_1. This network makes the circuit independent of beta changes. Voltages at the base, emitter, and collector all depend on external circuit values. By proper selection of components, the emitter-base junction is forward biased, with the collector being reverse biased. Normally, these bias voltages are all referenced to ground. The base, in this case, is made slightly positive with respect to ground. This voltage is somewhat critical. A voltage that is too positive, for example, will drive the transistor into saturation. With proper selection of bias voltage, however, the transistor can be made to operate in any part of the active region. The temperature stability of the circuit is excellent. With proper R_E bypass capacitor selection, this method of biasing produces high gain. The divider method of biasing is often a universal biasing circuit. It can be used to bias all transistor amplifier circuit configurations.

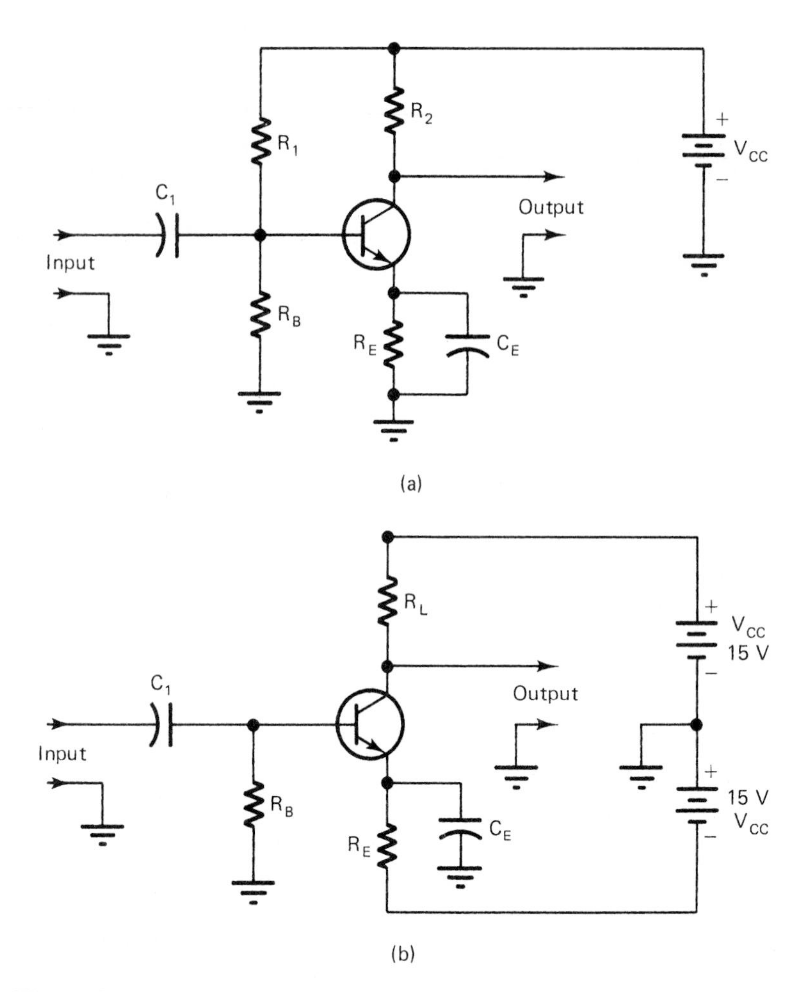

Figure 13–8. Independent beta biasing: (a) divider base; (b) divider base with split supply.

The circuit in Fig. 13–8(b) is similar in construction and operation to that shown in Fig. 13–8(a). One less resistor is used. The power supply requires two voltage values with reference to ground. A split power supply is used as an energy source for this circuit. Note the indication of $+V_{CC}$ and $-V_{CC}$. R_B is connected to ground. In this case, the value of R_B would determine the value of I_B with only half of total supply voltage. The values of R_L and R_E are usually larger to accommodate the increased supply voltage. If R_E is bypassed properly, the gain of this circuit is quite high. Thermal stability is excellent.

Load-Line Analysis

Earlier discussion in the chapter involved the operation of an amplifier with respect to its beta. Current and voltage were calculated and operation was related to these values. This method of analysis is important. Amplifier operation can also be determined graphically by employing a collector family of characteristic curves. A load line is developed for the graph. With the load-line method of analysis, it is possible to predict how the circuit will respond graphically. Much about the operation of an amplifier can be observed through this method.

The load-line method of circuit analysis can be used by engineers to design circuits. The operation of a specific circuit can also be visualized by this method. In circuit design, a specific transistor is selected for an amplifier. Source voltage, load resistance, and input signal levels may be given values in the design of the circuit. The transistor is made to fit the limitation of the circuit.

For our application of the load line, assume that a circuit is to be analyzed. Figure 13–9(a) shows the circuit being analyzed. A collector family of characteristic curves for the transistor is shown in Fig. 13–9(b). Note that the power dissipation rating of the transistor is included in the diagram.

Power Dissipation Curve. A common practice in load-line analysis is to first develop a power dissipation curve. This gives some indication of the maximum operating limits of the transistor. *Power dissipation* (P_D) refers to max-

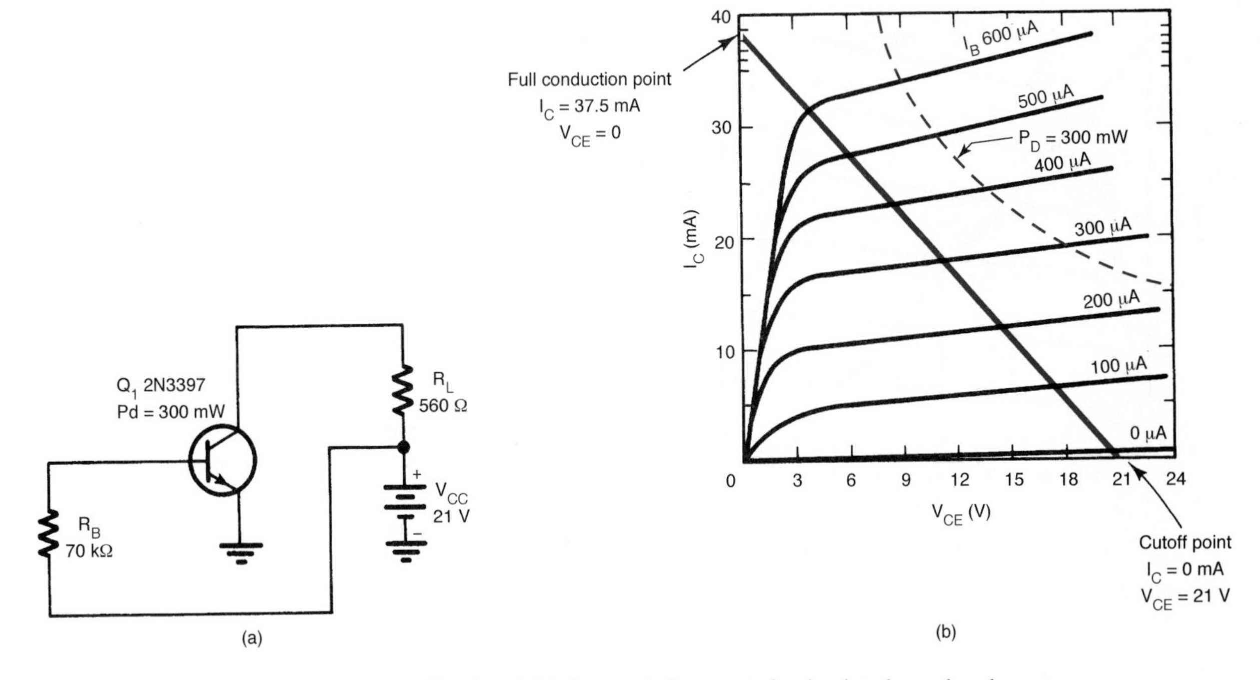

Figure 13–9. (a) Circuit and (b) characteristic curves of a circuit to be analyzed.

imum heat that can be given off by the base-collector junction. Usually, this value is rated at 25°C. P_D is the product of I_C and V_{CE}. In our circuit, the P_D rating for the transistor is 300 mW.

To develop a P_D curve, each value of V_{CE} is used with the P_D rating to determine an I_C value. The formula is

$$\text{collector current} = \frac{\text{power dissipation}}{\text{collector-emitter voltage}}$$

or

$$I_C = \frac{P_D}{V_{CE}}$$

Using this formula, calculate the I_C value for each of the V_{CE} values of the family of curves. Using the V_{CE} value and the corresponding calculated I_C value, note the location of the curve. These connected points are representative of the 300-mW power dissipation curve. In practice, the load line must be located to the left of the established P_D curve. Satisfactory operation without excessive heat generation can be ensured in that area of the curve.

STATIC LOAD-LINE ANALYSIS. The load line of a transistor amplifier represents two extreme conditions of operation. One of these is in the cutoff region. When the transistor is cut off, there is no I_C flowing through the device. V_{CE} equals the source voltage with zero I_C. The second load-line point is in the saturation region. This point assumes full conduction of I_C. Ideally, when a transistor is fully conductive, $V_{CE} = 0$ and $I_C = V_{CC}/R_L$.

The two load-line construction points for the analysis circuit are shown on the curves of Fig. 13–9(b). The cutoff point is located at the 0 I_C, 21 = V_{CE} point. The value of V_{CC} determines this point. At cutoff, $V_{CE} = V_{CC}$. Saturation is located at 37.5 mA of I_C and 0 V_{CE}. The I_C value is calculated using V_{CC} and the value of R_L. The formula is

$$I_C = \frac{V_{CC}}{R_L} = \frac{21\ \text{V}}{560\ \Omega}$$
$$= 0.0375\ \text{A or } 37.5\ \text{mA}$$

These two points are connected with a straight line.

Figure 13–10 shows a family of characteristic curves with a load line for the circuit of Fig. 13–9. Development of the load line makes it possible to determine the operating conditions of the amplifier. For linear amplification, the operating point should be located near the center of the load line. In our circuit, an operating point of 300 μA is used. In the circuit diagram, the value of R_B determines I_B. It is calculated by the equation $I_B = V_{CC} - V_{BE}/R_B$. The value is

$$I_B = \frac{V_{CC} - V_{BE}}{R_B} = \frac{21\ \text{V} - 0.7\ \text{V}}{70\ \text{k}\Omega}$$
$$= 0.00029\ \text{A or } 290\ \mu\text{A}$$

The operating point for this value is located below the intersection of the load line and the 300 μA I_B curve. It is indicated as point *Q*. Knowing this much about an amplifier shows how it will respond in its steady or static state. The

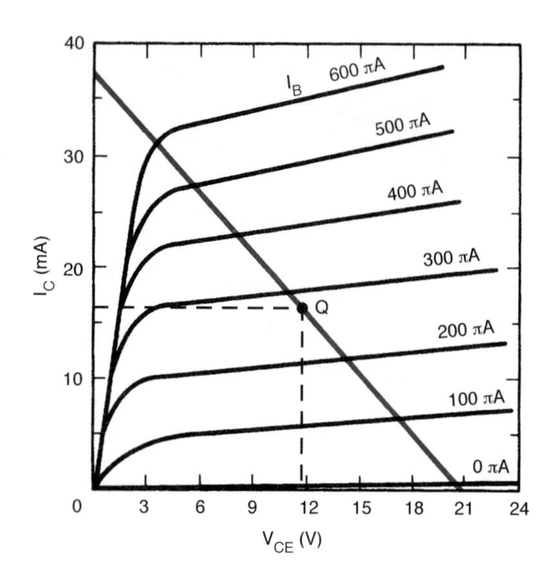

Figure 13–10. Family of characteristic curves for Fig. 13–9 for locating the Q point.

Q point shows how the amplifier will respond without a signal applied.

Operation of our amplifier in its static state is displayed by the family of curves in Fig. 13–10. Projecting a line from the Q point to the I_C scale shows the resultant collector current. In this case, the I_C is 17 mA. The dc beta for the transistor at this point is determined by the formula $\beta = I_C/I_B$. This value is

$$\beta = \frac{I_C}{I_B}$$

$$= \frac{17 \text{ mA}}{300 \ \mu\text{A}} \quad \text{or} \quad \frac{0.017 \text{ A}}{0.0003 \text{ A}}$$

$$= 56.67$$

The resultant V_{CE} that will occur for the amplifier can also be determined graphically. Projecting a line directly down from the Q point will show the value of V_{CE}. In our circuit, the V_{CE} is approximately 11 V, which means that 10 V will appear across R_L when the transistor is in its static state.

Dynamic Load-Line Analysis. Dynamic load-line analysis shows how an amplifier will respond to an ac signal. In this case, the collector curves and circuit of Fig. 13–11 are used. A load line and Q point for the circuit have been developed on the curves which establishes the static operation of the amplifier. Note the values of V_{CE} and I_C in the static state.

Assume now that the 0.1-V peak-to-peak ac signal is applied to the input of the amplifier and causes a 20-μA peak-to-peak change in I_B. During the positive alternation of the input, I_B will change from 30 μA to 40 μA. This is shown as point P on the load line. During the negative alternation, I_B will drop from 30 μA to 20 μA. This is indicated as point N on the load line. In effect, this means that 0.1-V p-p causes I_B to change 20-μA p-p. The I_B signal extends to the right of the load line. Its value is shown as ΔI_B.

To show how a change in I_B influences I_C, lines are projected to the left of the load line. Note the projection of lines P, Q, and N toward the I_C values. The changing value of I_C is indicated as ΔI_C. An increase and decrease in I_B causes a corresponding increase and decrease in I_C. This shows that I_B and I_C are in phase.

The ac current gain of the amplifier can be determined by ΔI_C and ΔI_B. First determine the peak-to-peak I_C and I_B values. Then divide ΔI_C by ΔI_B. Determine the ac current gain of the amplifier circuit. Using the same procedure, determine the dc current gain of the transistor at point Q. How do the ac and dc current gains of this circuit compare?

Projecting points P, Q, and N downward from the load line shows how V_{CE} changes with I_B. The value of V_{CE} is indicated as ΔV_{CE}. Note that an increase in I_B causes a decrease in the value of V_{CE}. A decrease in I_B causes V_{CE} to increase. This shows that I_B and V_{CE} are 180° out of phase. The difference in V_{CE} at any point appears across R_L.

The ac voltage gain of the amplifier can be determined from the dynamic load line. Remember that 0.1-V p-p input caused a change of 20-μA p-p in the I_B signal. The ac voltage gain can be determined by dividing ΔV_{CE} by ΔV_B. The ΔV_B value is 0.1-V p-p. Using the ΔV_{CE} value from the graph and ΔV_B, determine the ac voltage gain of the amplifier circuit.

Linear and Nonlinear Operation

The V_{CE} or output of an amplifier should be a duplicate of the input with some gain. When a sine wave is applied, the output should develop a sine wave. When an amplifier operates in this manner, it is considered to be *linear.* For linear operation to be achieved, the amplifier must operate in or near the center or linear area of the collector curves. As a rule, nonlinearity occurs near the saturation and cutoff regions. If the operating point is adjusted near these regions, it usually causes the output to be distorted. Normally, this is called *nonlinear distortion.*

Figure 13–12 shows how an amplifier will respond at three different operating points. Figure 13–12(a) shows linear operation. The input and output are duplicates in this case. Figure 13–12(b) shows operation in the saturation region. Note that the top of the input wave distorts the negative alternation of the output voltage. Figure 13–12(c) shows operation near the cutoff region. The lower part of the input wave causes distortion of the positive alternation. As a general rule, nonlinear distortion can be tolerated in some applications. In other applications, nonlinear distortion is quite obvious.

An input signal that is too large can also produce distortion. Figure 13–13 on page 423 shows how an amplifier

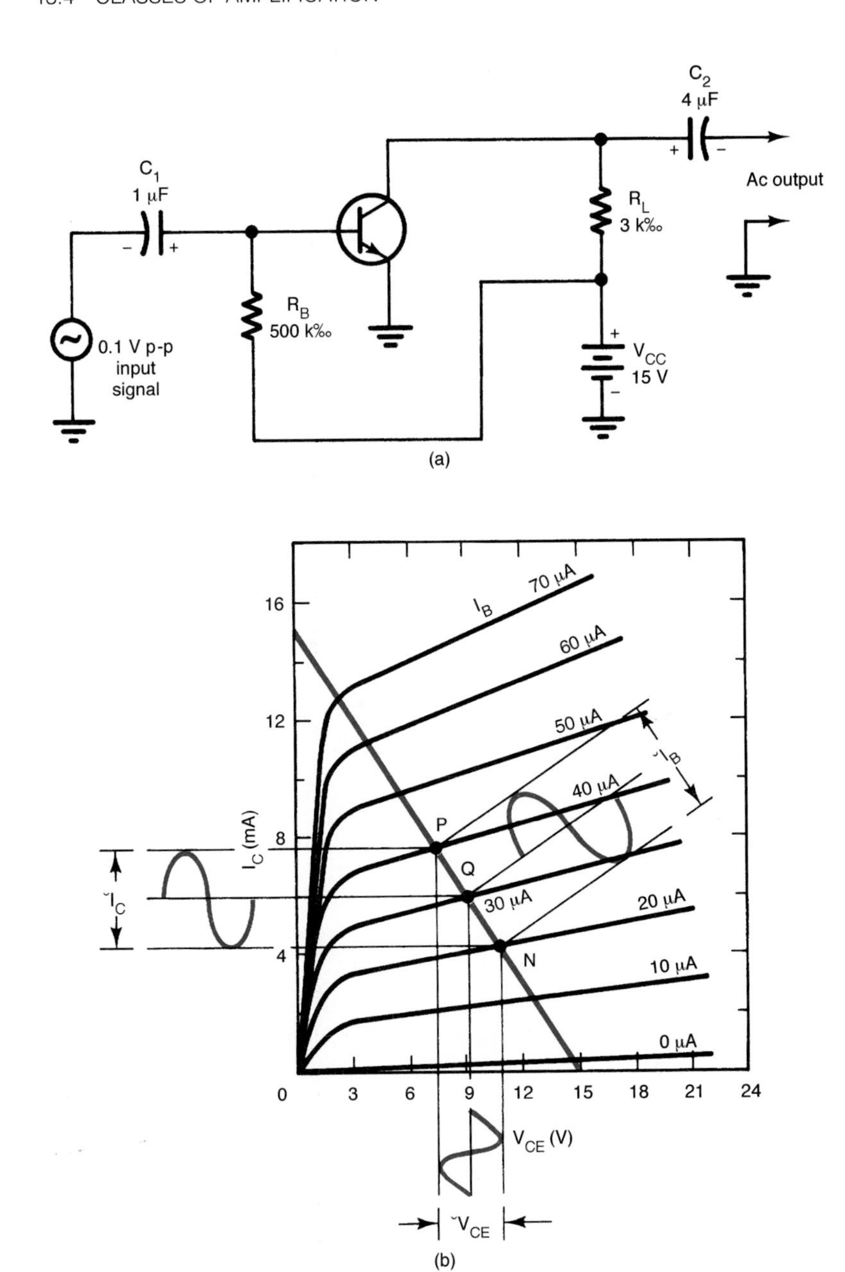

Figure 13–11. Dynamic load line: (a) circuit; (b) characteristic curves.

will respond to a large signal. Note that the input signal swings into the saturation region during the positive alternation and into the cutoff region during the negative alternation. This condition causes distortion of both alternations of the output. Normally, this condition is described as "overdriving." In a radio receiver or stereo amplifier, overdriving causes poor quality of speech or music and usually occurs when the volume control is turned too high. The audio or sound signal has its peaks clipped. The volume level may be higher, but the quality of reproduction is usually poor when this occurs. It is interesting to note that the operating point is near the center of the active region. Overdriving can occur even with a properly selected operating point.

13.4 CLASSES OF AMPLIFICATION

Transistor amplifiers are frequently classified according to their bias operating point. This means of classification describes the shape of the output wave. Three general groups of amplifiers are class A, class B, and class C. Figure 13–14 on page 424 shows a graphical display of these three amplifier classes.

Class A amplifiers generally have linear operation. The bias operating point is set near the center of the active region. With a sine wave applied to the input, the output is a complete sine wave. Figure 13–14(a) shows the input-output waveforms of a class A amplifier. This type of

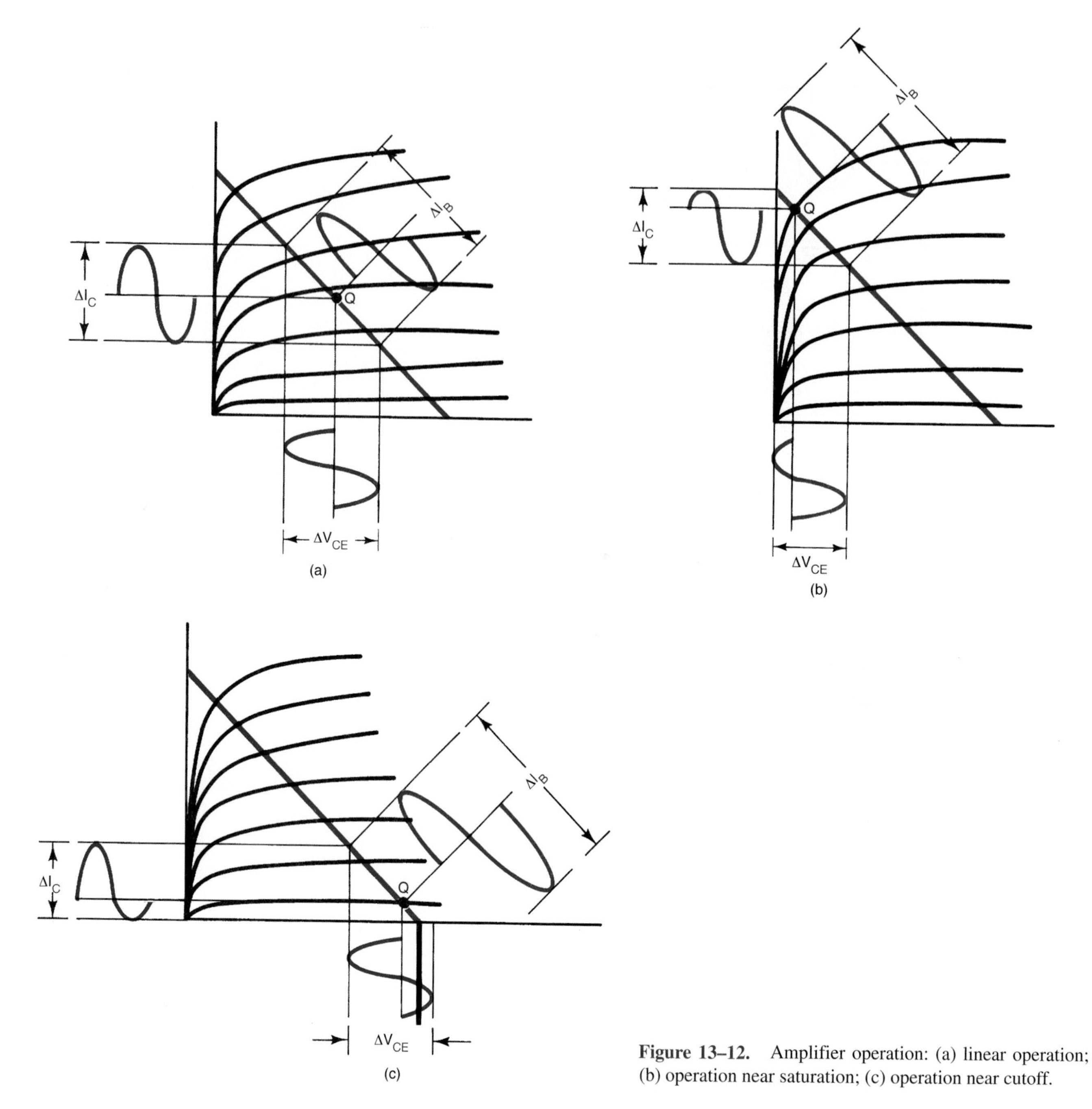

Figure 13–12. Amplifier operation: (a) linear operation; (b) operation near saturation; (c) operation near cutoff.

amplifier is used when a true reproduction of the input signal is required.

Figure 13–14(b) shows the input-output waveforms of a class B amplifier. The operating point of this amplifier is adjusted near the cutoff point. With a sine wave applied to the input, only one alternation of the signal is reproduced. When using class B amplifiers, it is possible to obtain a large change in output for one alternation. Two class B amplifiers working together can each amplify one alternation of a sine wave. When the wave is restored, a complete sine wave is developed. Class B amplifiers are commonly used in push-pull audio output circuits. These amplifiers are usually concerned with power amplification.

A class C amplifier is shown in Fig. 13–14(c). In this amplifier the bias operating point is below cutoff. With a sine wave applied to the input, the output is less than half of one alternation. Radio-frequency circuits are the primary application of class C amplifiers today. The operational efficiency of this amplifier is quite high. It consumes energy only for a small portion of the applied sine wave.

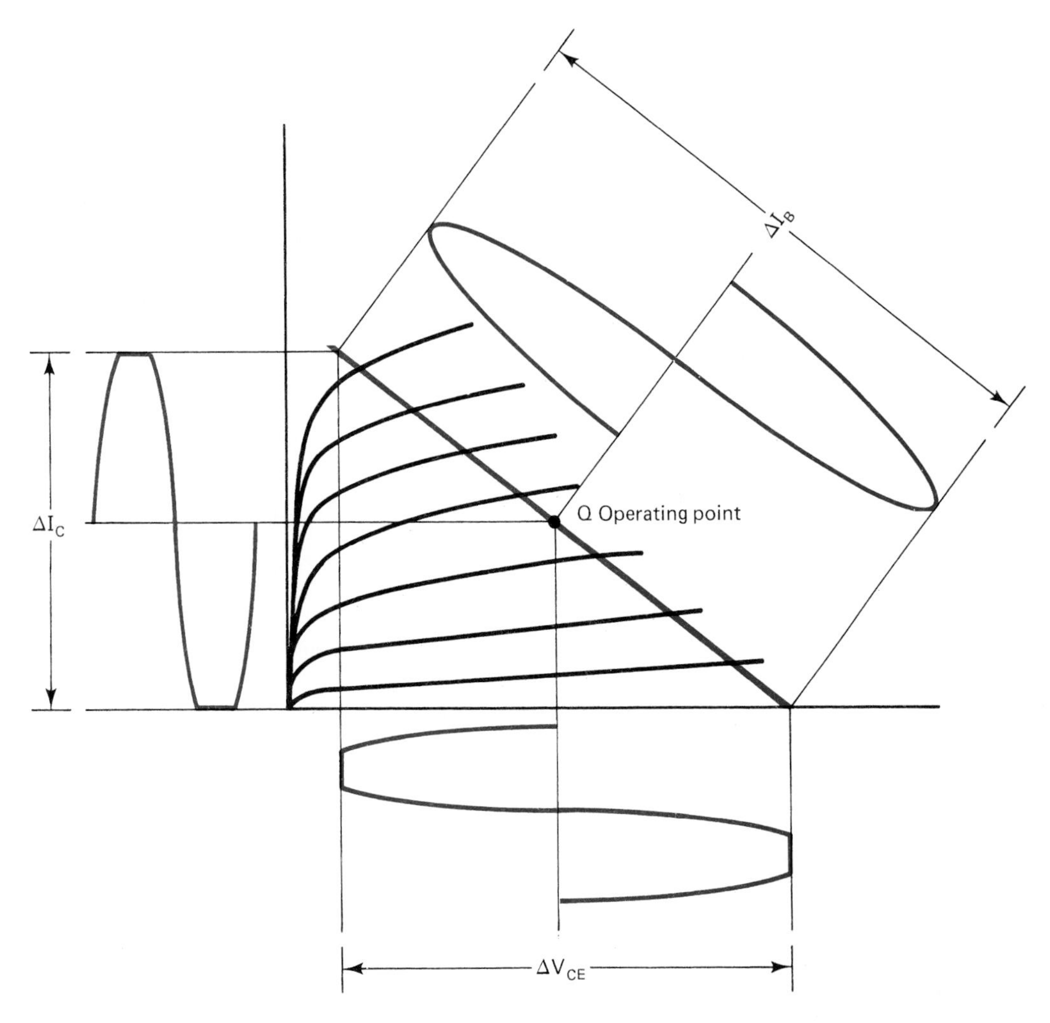

Figure 13–13. Overdriven amplifier.

13.5 TRANSISTOR CIRCUIT CONFIGURATIONS

The elements of a transistor can be connected in one of three different circuit configurations and are usually described as common-emitter, common-base, and common-collector circuits. Of the three transistor leads, one is connected to the input and one to the output. The third lead is commonly connected to both the input and output. The common lead is generally used as a reference point for the circuit. It is usually connected to the circuit ground or common point. This configuration has brought about the terms *grounded emitter, grounded base,* and *grounded collector.* The terms *common* and *ground* have the same meaning. In some circuit configurations, the emitter, base, or collector may be connected directly to ground. When this occurs, the lead is at both dc and ac ground potential. When the lead goes to ground through a battery or resistor that is bypassed by a capacitor, it is at an ac ground.

Common-Emitter Amplifiers

The common-emitter amplifier is an important transistor circuit. A high percentage of all amplifiers in use today are of the common-emitter type. The input signal of this amplifier is applied to the base and the output is taken from the collector. The emitter is the common or grounded element.

Figure 13–15 shows a circuit diagram of a common-emitter amplifier, which is similar to the basic amplifier used in the first part of the chapter. In effect, this circuit is used to acquaint you with amplifiers in general. It is presented here to make a comparison with the other circuit configurations.

The signal being amplified by the common-emitter amplifier is applied to the emitter-base junction. This signal is superimposed on the dc bias of the emitter and base. The base current then varies at an ac rate. This action causes a corresponding change in collector current. The output voltage developed across the collector and emitter is inverted 180°. The current gain of the circuit is determined by beta.

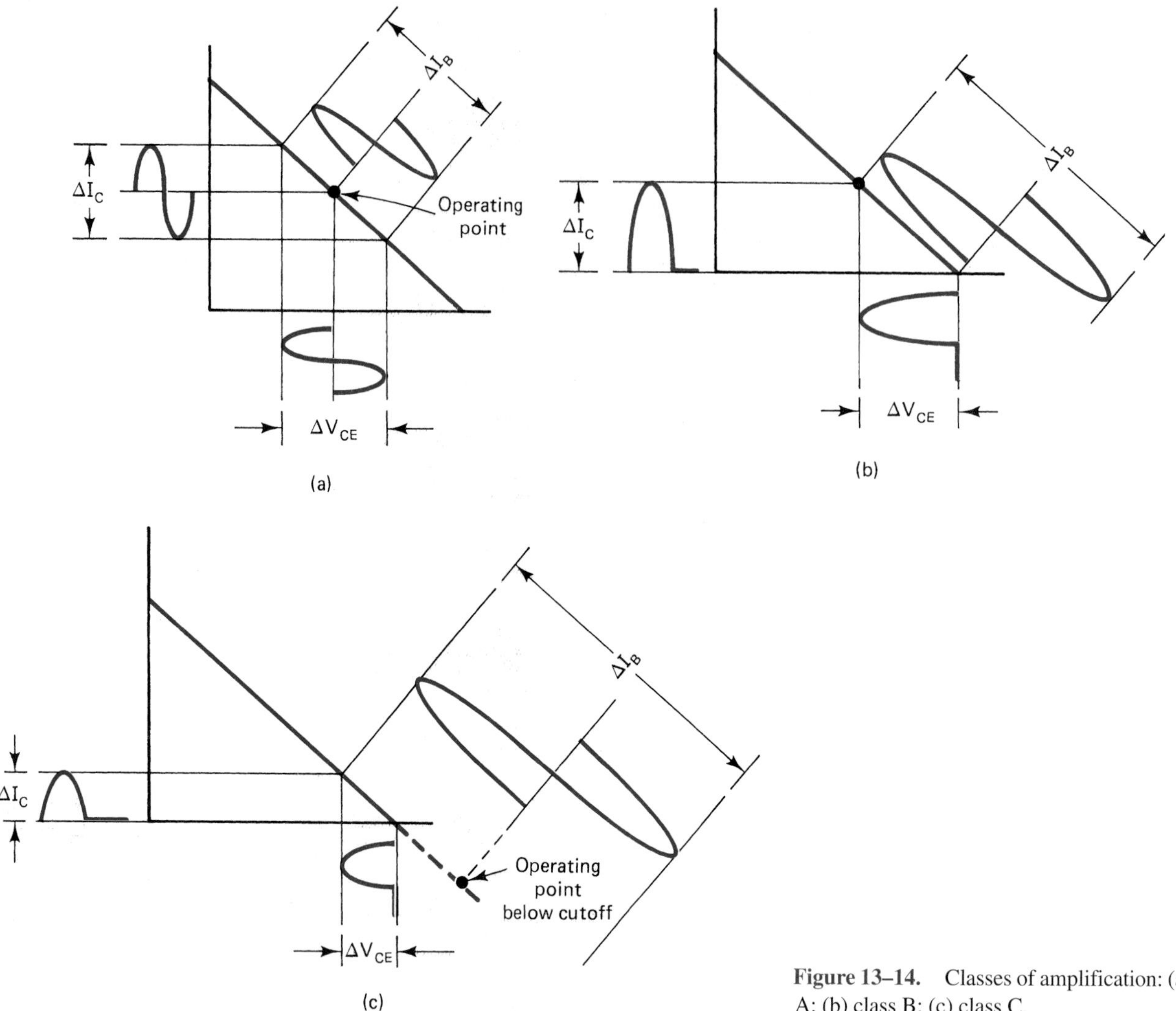

Figure 13–14. Classes of amplification: (a) class A; (b) class B; (c) class C.

Typical beta values are in the range of 50 to 200. The voltage gain of the circuit ranges from 250 to 500. The power gain is in the range of 20,000. Input impedance is moderately high, typical values being 100 Ω. Output impedance is moderate, typical values being 2000 Ω. In general, common-emitter amplifiers are used in small-signal applications or as voltage amplifiers.

Common-Base Amplifiers

A common-base amplifier is shown in Fig. 13–16. In this type of amplifier, the emitter-base junction is forward biased and the collector-base junction is reverse biased. In this circuit configuration, the emitter is the input. An applied input signal changes the circuit value of I_E. The output signal is developed across R_L by changes in collector current. For each value change in I_E, there is a corresponding change in I_C.

In a common-base amplifier, the current gain is called *alpha*. Alpha is determined by the formula I_C/I_E. In a common-base amplifier, the gain is always less than a value of 1. Remember that $I_E = I_B + I_C$. The I_C will therefore always be slightly less than I_E by the value of I_B. Typical values of alpha are 0.98 to 0.99.

In Fig. 13–16, V_{EE} forward biases the emitter and base while V_{CC} reverse biases the collector and base. Resistor R_E is an emitter-current-limiting resistor and R_L is the load resistor. Note that the transistor is a *PNP* type.

When an ac signal is applied to the input, it is added to the dc operating value of V_E. The positive alternation therefore adds to the forward-bias voltage of V_E. This condition causes an increase in I_E. A corresponding increase in I_C also takes place. With more I_C through R_L there is an increase in its voltage drop. The collector therefore becomes less negative or swings positive. In effect, the positive alternation of the input produces a positive alternation in the output. The input and output of this amplifier are in phase. See the waveform inserts in Fig. 13–16. The negative alternation causes the same reaction, the only difference being a reverse in polarity.

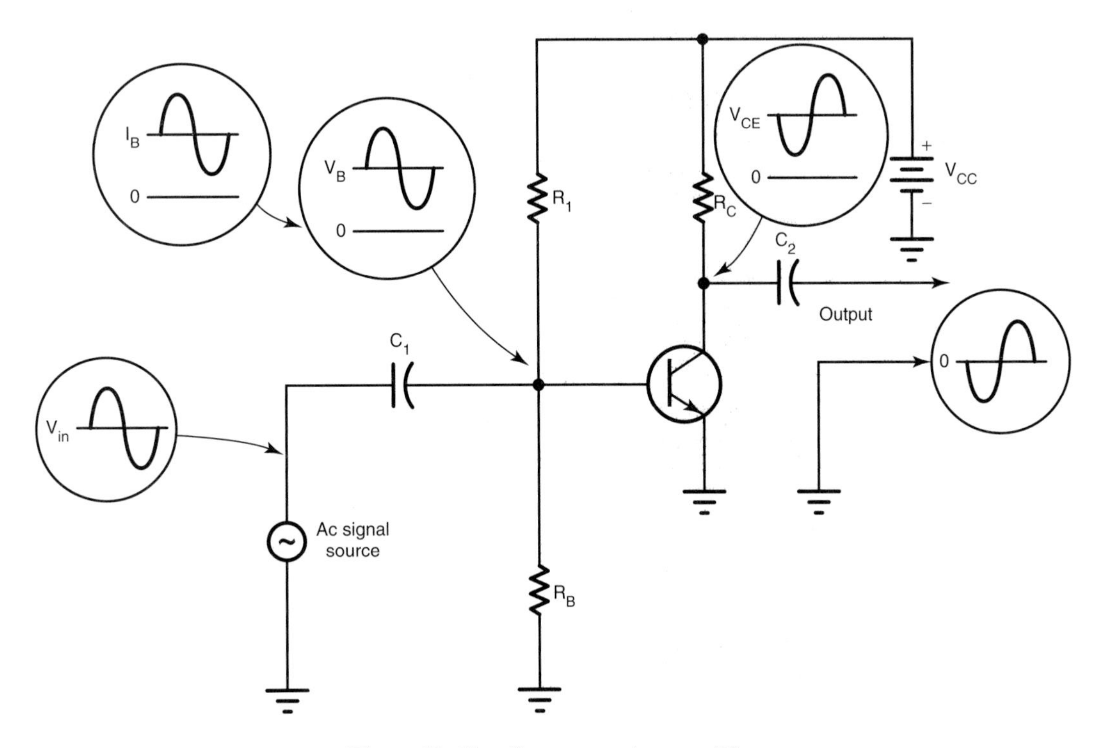

Figure 13–15. Common-emitter amplifier.

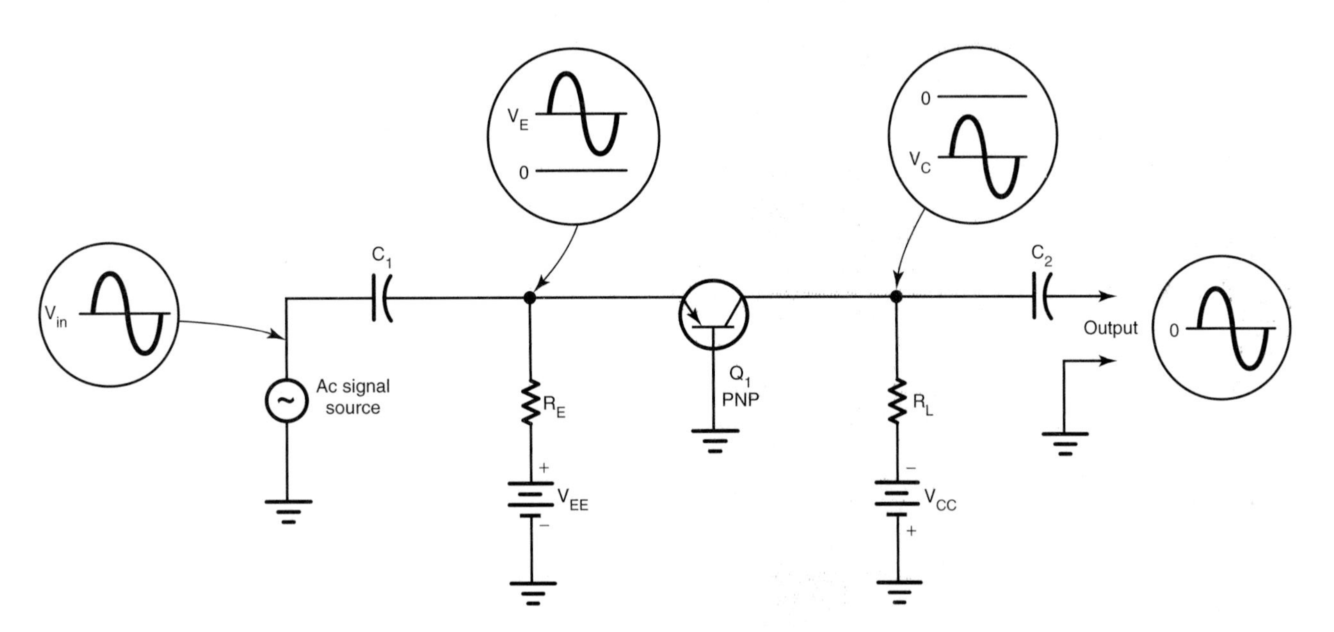

Figure 13–16. Common-base amplifier.

A common-base amplifier has a number of rather unique characteristics that should be considered. The current gain of a transistor connected in a common-base configuration is called *alpha.* This current gain is always less than 1. Typical values are 0.98 to 0.99. The voltage gain is usually quite high. Typical values range from 100 to 2500 depending on the value of R_L. The power gain is the same as the voltage gain. The input impedance of the circuit is quite low. Values of 10 to 200 Ω are common. The output impedance of the amplifier is somewhat moderate. Values range from 10 to 40 kΩ. This amplifier does not invert the applied signal.

Common-base amplifiers are used primarily to match a low-impedance input device to a circuit. This type of circuit configuration is also used in radio-frequency amplifier applications. As a general rule, the common-base amplifier is not widely used today.

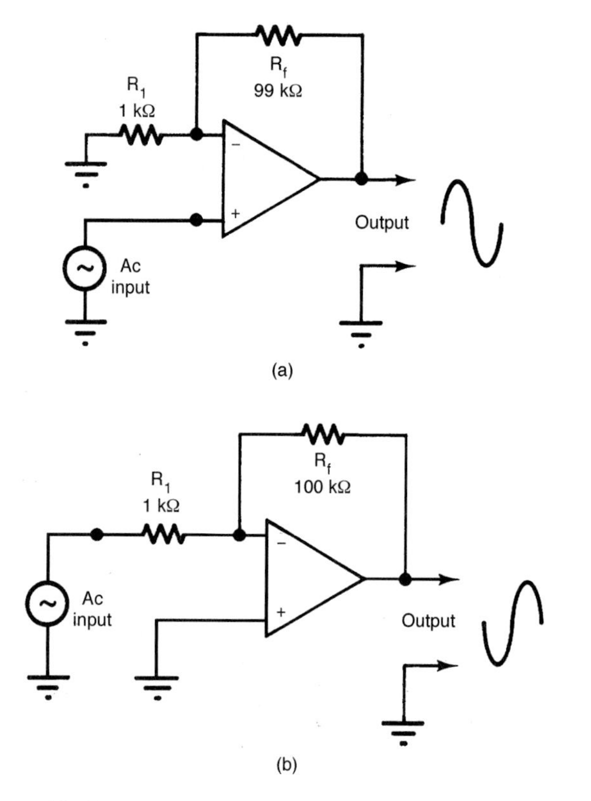

Figure 13–33. Closed-loop op-amp simplification: (a) noninverting amplifier; (b) inverting amplifier.

If R_f is changed to 9 kΩ and R_1 to 1 kΩ, would the voltage gain change?

Figure 13–33(b) shows an op-amp connected in an inverting circuit. The ac signal is applied to the negative (−) or inverting input. This amplifier has its output polarity inverted 180°. Voltage amplification is based on the formula

$$A_V = \frac{R_f}{R_1}$$

or

$$A_V = \frac{100\ \text{k}\Omega}{1\ \text{k}\Omega}$$

$$= 100$$

If R_f is changed to 50 kΩ and R_1 to 2.5 kΩ, what would be the voltage gain?

PRACTICAL OP-AMP CIRCUITS. Figure 13–34 shows a functioning op-amp circuit connected for noninverting amplification. The numbers near each lead are the pin numbers. The LM741 is housed in an eight-pin dual-in-line package. Notice that the operating voltage is supplied from a split or divided power supply. Common leads on each side of the supply are connected to ground. The ground is also con-

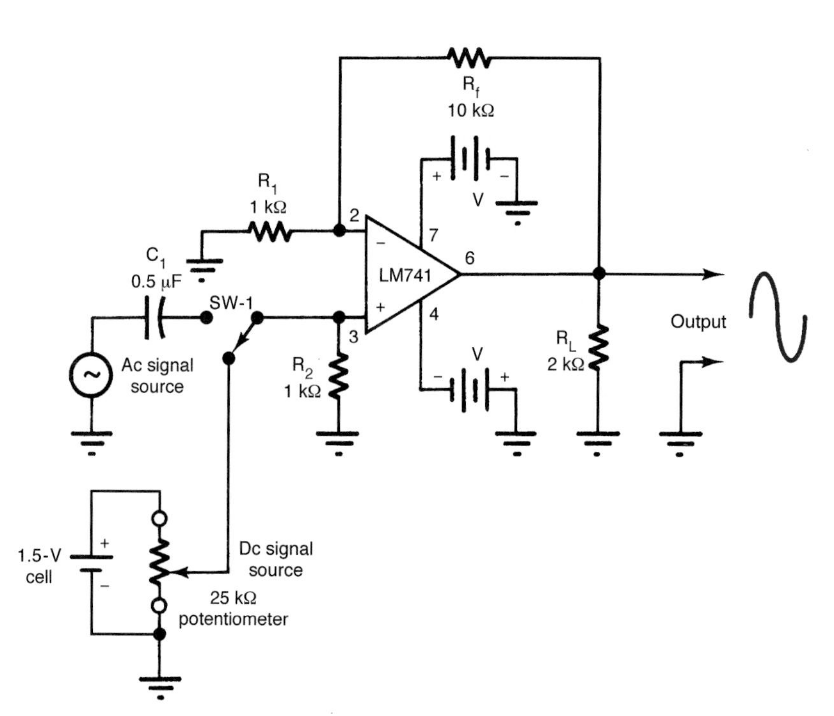

Figure 13–34. Noninverting op-amp circuit.

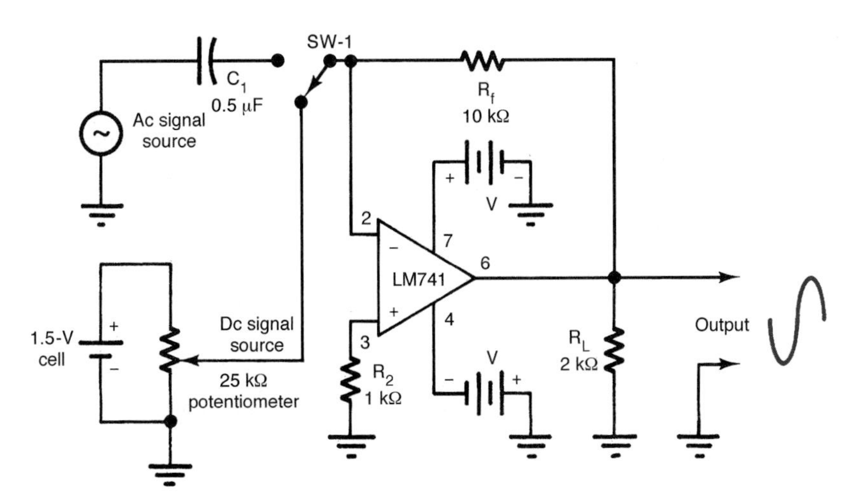

Figure 13–35. Inverting op-amp circuit.

nected commonly to a number of circuit points. The output of this circuit cannot exceed 10-V dc or 20-V p-p ac. The gain achieved by this circuit closely follows the calculated values of the feedback network. When the circuit is made operational, gain and signal phase can be observed at the input and output with an oscilloscope. Input and output voltages can also be measured with a voltmeter.

A practical inverting op-amp circuit is shown in Fig. 13–35. This amplifier has the same gain capabilities as the noninverting circuit. Note the connection differences on the input circuit. The operational waveforms, signal phase, and voltage values can be observed on an oscilloscope and measured with a voltmeter. Op-amp circuits of this type are widely used today.

REVIEW

1. What is the difference between amplification and reproduction?
2. If the output voltage of an amplifier is 5-V p-p and the input voltage is 0.025-V p-p, what is the voltage gain?
3. If the output current of an amplifier is 8 mA when the input current is 100 μA, what is the current gain?
4. How much base current occurs in a transistor amplifier when a 120-kΩ resistor is connected to 9 V?
5. What is the current gain of a transistor amplifier having an I_C of 6 mA, an I_B of 120 μA, and an I_E of 6.12 mA?
6. Explain how an ac signal is amplified by an *NPN* bipolar transistor.
7. What is beta-dependent biasing?
8. Why is independent biasing better than beta-dependent biasing?
9. Explain what is meant by the static state of a bipolar amplifier.
10. What are the two extreme conditions of operation of a bipolar transistor used to develop a load line?
11. What is meant by the term *dynamic characteristics of an amplifier?*
12. Explain how the input signal of a bipolar amplifier causes the output to be inverted 180°.
13. Explain the type of distortion that will occur when a bipolar amplifier is near cutoff or near saturation.
14. With a sine wave applied to the input of a bipolar amplifier, where must the operating point be adjusted to achieve class A, B, and C amplification?

15. What are the characteristic differences in common-emitter, common-base, and common-collector amplifiers with respect to A_V, current gain, phase, input impedance, and output impedance?
16. Describe the physical differences between a JFET, D-MOSFET, and an E-MOSFET.
17. Explain how a JFET achieves amplification.
18. What is the difference in biasing an *N*-channel JFET compared with a bipolar-transistor *NPN* type?
19. What are the distinguishing features of fixed biasing, self-biasing, and divider biasing for a JFET?
20. Why are op-amps rarely used in the open-loop gain configuration?
21. If the feedback resistor for a noninverting circuit is 78 kΩ and the input resistor is 2 kΩ, what is the A_V?
22. An inverting op-amp has a 10-kΩ R_f and a 2-kΩ R_1. What is the A_V?

STUDENT ACTIVITIES

• *Amplifier Static Operation*

1. Construct the amplifier of Fig. 13–3.
2. Turn on the energy source.
3. Measure and record the static operating conditions for I_B, I_C, V_{CE}, and V_{RL}. These values will differ widely among transistors.

• *Amplifier Dynamic Operation*

1. Modify the circuit of Fig. 13–3 to conform with the ac amplifier of Fig. 13–4.
2. With an oscilloscope connected to the output, adjust the ac input signal level to produce the greatest output with a minimum of distortion. Record the peak-to-peak output signal level.
3. Move the oscilloscope leads to the signal input points. Measure and record the signal level.
4. Calculate the ac voltage gain.

• *Load-Line Analysis of an Amplifier* or *Complete Activity 13–1, page 265 in the manual.*

1. Using the collector family of characteristic curves in Fig. 13–36(a), develop a dc load line for the circuit of Fig. 13–36(b).
2. The two locating points are _____ and _____.
3. With a straightedge, connect the locating points with a lead pencil.
4. Determine the *Q* point for the circuit. Locate this point on the load line.
5. Determine the static operation points for V_{CE} and I_C.
6. If a 0.5-V p-p signal is applied, it causes 20-μA p-p of I_B.
7. Locate these points on the load line. By the projection of lines, determine the ΔI_C, ΔV_{CE}, and ΔI_B.
8. Calculate the dc beta, ac beta, and ac voltage gain.

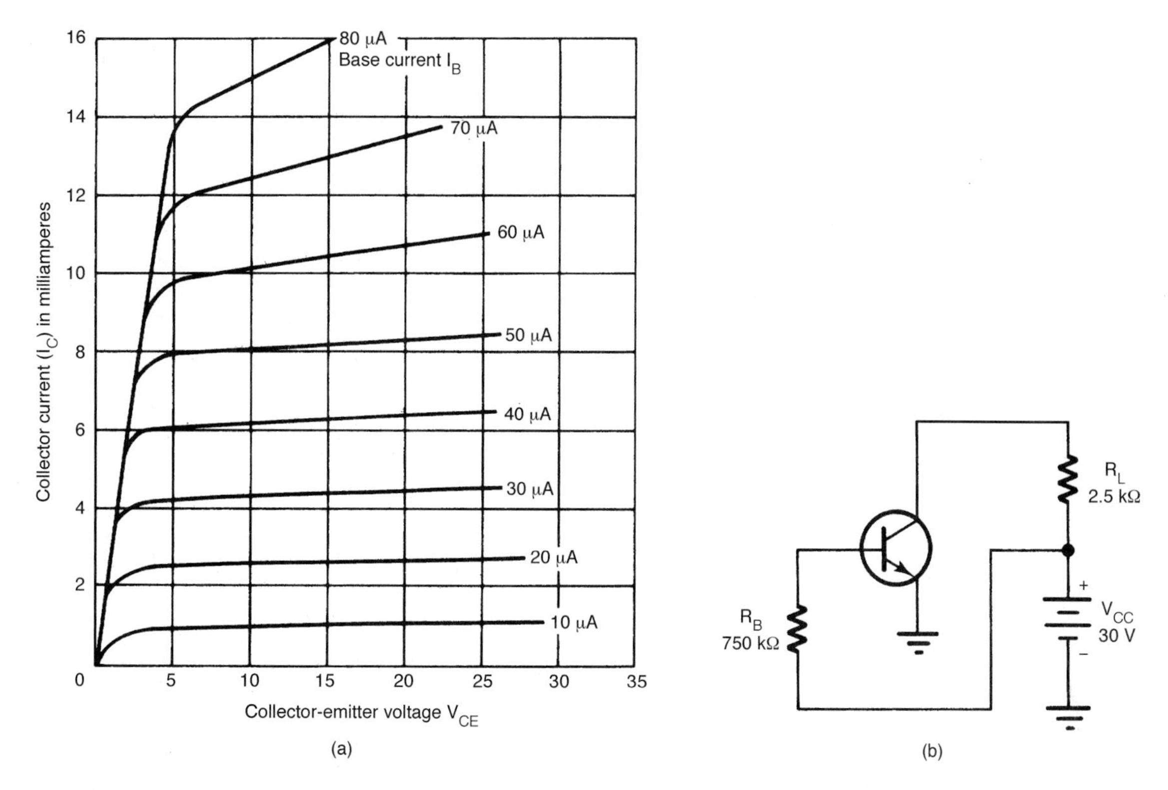

Figure 13–36. Transistor load-line analysis exercise: (a) $I_C - V_{CE}$ curves; (b) transistor circuit.

ANALYSIS

1. What are the two extreme operating conditions represented by a load line?
2. If a smaller V_{CC} value were used in a circuit, how would it alter the load line?
3. Define Q point.
4. What does the power dissipation curve of a transistor show?
5. If the base current of Fig. 13–36(b) were increased by a stronger signal, how would it alter circuit operation?
6. What is the phase relationship in Fig. 13–36(b) of I_B, I_C, and V_{CE}?

DC Transistor Amplifiers

Complete Activity 13–2, page 269 in the manual.

ANALYSIS

1. What determines the beta of a transistor amplifier?
2. If the value of R_L of a transistor amplifier was increased, how would the output respond?
3. Why does an increase in I_B of an amplifier cause an increase in I_C and a reduction in V_{CE}?
4. What is meant by the term *common-emitter* amplifier?

• *AC Transistor Amplifiers*

Complete Activity 13–3, page 271 in the manual.

• • ANALYSIS

Explain how the amplifier of Fig. 13–4 achieves ac amplification with respect to changes in V_{BE}, I_C, I_B, I_E, and V_{CE}.

• *JFET Amplifier—Common-Source* or *Complete Activity 13–4, page 275 in the manual.*

1. Construct the JFET amplifier of Fig. 13–26.
2. Apply energy to the circuit.
3. Measure and record the static operating conditions without a signal applied.
4. Apply the signal source to the amplifier.
5. Connect an oscilloscope to the output of the amplifier.
6. Adjust the signal source output to produce the greatest signal with a minimum of distortion. Note the peak-to-peak voltage signal level.
7. Move the oscilloscope leads to the amplifier input across R_1. Record the peak-to-peak voltage input signal level.
8. Calculate the voltage gain.
9. Connect the external sync input of the oscilloscope to the output. Observe the phase of the input and output signals. Describe the phase relationship.

• • ANALYSIS

1. In Fig. 13–26, what is the bias voltage polarity of the gate with respect to the source voltage?
2. What method of biasing is used by the JFET amplifier of Fig. 13–26?
3. What type of amplifier configuration is used in Fig. 13–26?
4. What does transconductance tell about a JFET?
5. What is dynamic resistance of a JFET?
6. Why is the JFET considered a voltage-operated device?
7. Why is the input impedance of the JFET high?

• *JFET Amplifier—Common-Gate*

1. Construct the common-gate JFET amplifier of Fig. 13–27.
2. Repeat steps 2 through 9 of the common-source amplifier for the common-gate amplifier.

• *JFET Amplifier—Common Drain*

1. Construct the common-drain JFET amplifier of Fig. 13–28.
2. Repeat steps 2 through 9 of the common-source amplifier for the common-drain amplifier.

• ***Op-Amp Circuit—Inverting*** or ***Complete Activity 13–5, page 277 in the manual.***

1. Construct the inverting op-amp circuit of Fig. 13–35.
2. Repeat steps 2 through 10 of the noninverting circuit for the inverting amplifier.

• •

ANALYSIS

1. If the source resistance (R_S) of Fig. 13–35 were 500 Ω, how would it alter I_{in}?
2. If R_F of Fig. 13–35 were increased to 250 kΩ, how would it alter the voltage gain (A_V)?
3. If R_F of Fig. 13–35 were changed to 50 kΩ, how would it alter the voltage gain (A_V)?

• ***Op-Amp Circuit—Noninverting*** or ***Complete Activity 13–6, page 279 in the manual.***

1. Construct the noninverting op-amp circuit of Fig. 13–34. Be certain to use a split or divider power supply.
2. Place SW-1 in the dc input position.
3. Adjust the dc voltage input to 0.1 V.
4. Measure and record the dc voltage across R_L.
5. Calculate the dc voltage gain. Compare measured and calculated values.
6. Place SW-1 in the ac input position.
7. Turn on the ac signal source and adjust it to 0.1-V p-p.
8. With an oscilloscope or voltmeter, measure and record the peak-to-peak output voltage across R_L.
9. Calculate A_V. Compare calculated and measured values.
10. Observe the phase of the input and output signals.

• •

ANALYSIS

1. What is the primary difference between an ac and a dc noninverting op-amp?
2. Why is a capacitor used on the input of the ac amplifier?
3. How would the value of capacitor C_1 of Fig. 13–34 limit the amplifying frequency of the amplifier?
4. Why does R_f of Fig. 13–34 influence voltage gain?

• ***Self-Examination***

1. Complete the Self-Examination for Chap. 13, page 281 in the manual.
2. Check your answers on pages 417 to 418 in the manual.
3. Study the questions that you answered incorrectly.

• ***Chapter 13 Examination***

Complete the Chap. 13 Examination on page 283 in the manual.

CHAPTER

14

AMPLIFYING SYSTEMS

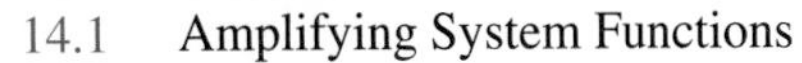

OUTLINE

14.1 Amplifying System Functions
14.2 Amplifier Gain
14.3 Decibels
14.4 Amplifier Coupling
14.5 Power Amplifiers
14.6 Integrated-Circuit Amplifying Systems
14.7 Speakers
14.8 Input Transducers

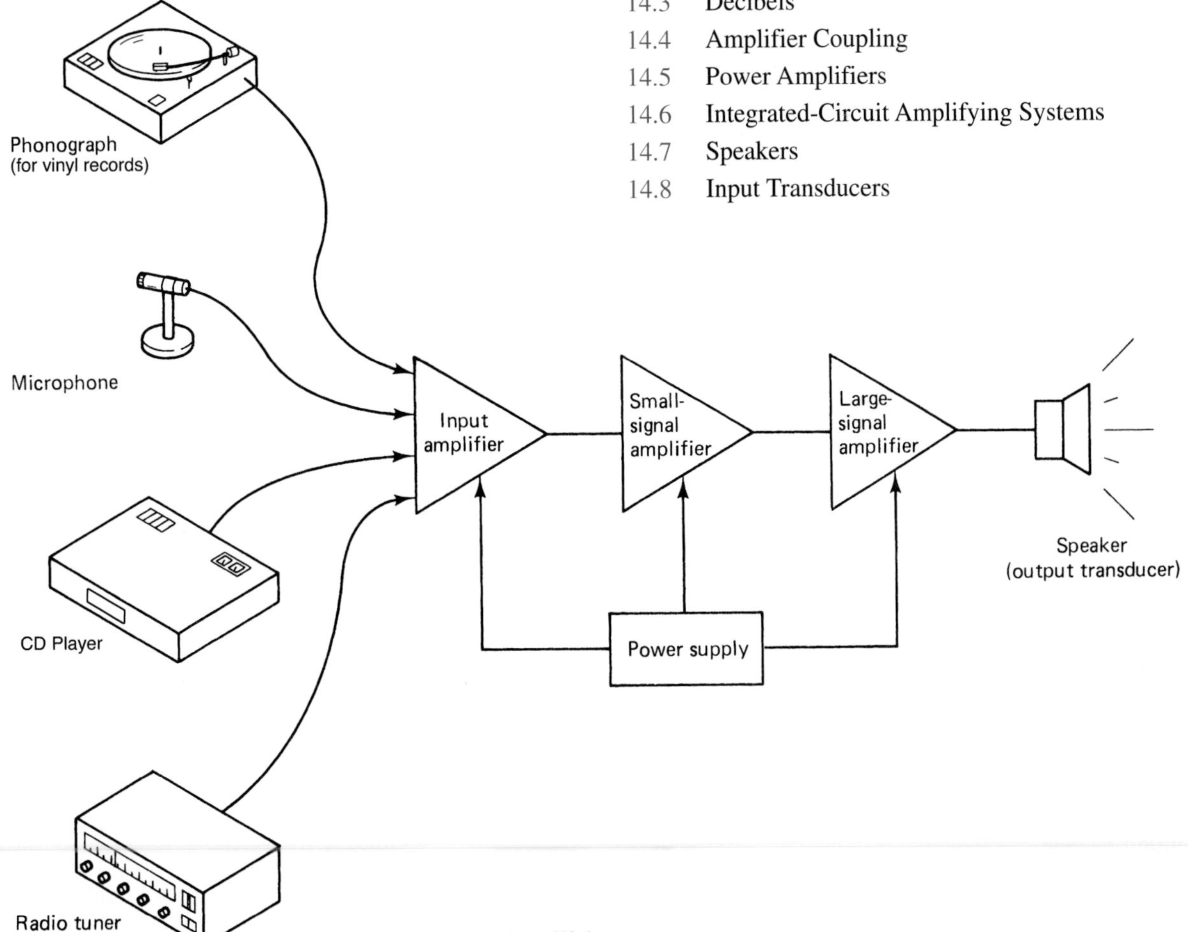

Amplifying system.

OBJECTIVES

Upon completion of this chapter, you will be able to:

1. List the basic functions of an amplifier system.
2. Explain what is meant by the term *gain.*
3. Identify basic coupling techniques and explain the advantages and disadvantages of each.
4. Determine the operational class of an amplifier with respect to its input and output.
5. Explain how dc, audio, power, RF, and IF amplification are accomplished.
6. Identify different power amplification circuits.
7. Explain how a complementary-symmetry amplifier works.
8. Define the meaning of input and output transducers and give examples of each.
9. Explain the operation of a Darlington amplifier.
10. Determine the power, voltage, or current gain of an amplifier in decibels.

Amplifying systems are used widely in the electronics field today. This particular type of system may be only one part or section of a rather large and complex system. A television receiver is a good example of this application. Several individual amplifying systems are included in a TV receiver. The amplifier's primary function is to process a sound signal and build it to a level that will drive a speaker. The function of an amplifier is basically the same regardless of its application.

This chapter will investigate how amplifying devices are used in an operating system. Transistors, integrated circuits, and vacuum tubes can all be used for this type of operation. A number of specific circuit functions are also discussed, such as gain, coupling, load devices, output circuits, transducers, and signal processing. The amplifying devices and their circuit components play an important role in the operation of a system.

IMPORTANT TERMS

Several new terms will be encountered in this chapter. As a rule, these terms play an important part in the understanding of amplifier systems. Review each term carefully before proceeding with the chapter. Refer back to a specific term when its meaning is not fully understood.

Amplitude modulation (AM). A method of transmitter modulation achieved by varying the strength of the RF carrier at an audio signal rate.

Audio frequency (AF). A range of frequency from 15 Hz to 15 kHz, the human range of hearing.

Bel (B). A measurement unit of gain that is equivalent to a 10:1 ratio of power levels.

Cascade. A method of amplifier connection in which the output of one stage is connected to the input of the next amplifier.

Characteristic. The magnitude range of a logarithmic number.

Compact disc (CD). A form of optical memory that stores audio and video signals digitally.

Complementary transistors. *PNP* and *NPN* transistors with identical characteristics.

Coupling. A process of connecting active components.

Crossover distortion. The distortion of an output signal at the point where one transistor stops conducting and another conducts, in class B amplifiers.

Crystal microphone. Input transducer that changes sound into electrical energy by the piezoelectric effect.

Darlington amplifier. Two transistor amplifiers cascaded together.

Decibel (dB). A unit used to express an increase or decrease in power, voltage, or current in a circuit; one-tenth of a bel.

Dynamic microphone. An input transducer that changes sound into electrical energy by moving a coil through a magnetic field.

Impedance ratio. A transformer value that is the square of the turns ratio.

Impedance (Z). The total opposition to current flow in an ac circuit which is a combination of resistance (R) and reactance (X) in a circuit; measured in ohms.

$$Z = \sqrt{R^2 + X^2}.$$

Logarithm. A mathematical expression dealing with exponents showing the power to which a value, called the base, must be raised to produce a given number.

Mantissa. Decimal part of a logarithm.

Midrange. A speaker designed to respond to audio frequencies in the middle part of the AF range.

Peak to peak (p-p). An ac value measured from the peak positive to the peak negative alternations.

Piezoelectric effect. The property of certain crystal materials to produce a voltage when pressure is applied to them.

Push-pull. An amplifier circuit configuration using two active devices with the input signal being of equal amplitude and 180° out of phase.

Read head. A signal pickup transducer that generates a voltage in a tape recorder.

Root mean square (rms). Effective value of ac or 0.707 × peak value.

Speaker. A transducer that converts electrical current and/or voltage into sound waves. Also called a loudspeaker.

Stage of amplification. A transistor, IC, or vacuum tube and all the components needed to achieve amplification.

Stereophonic (stereo). Signals developed from two sources that give the listener the impression of sound coming from different directions.

Stylus. A needlelike point that rides in a phonograph record groove.

Transducer. A device that changes energy from one form to another. A transducer can be an input or output device.

Tweeter. A speaker designed to reproduce only high frequencies.

Voice coil. The moving coil of a speaker.

Voltage divider. A combination of resistors that produces different or multiple voltage values.

Woofer. A speaker designed to reproduce low audio frequencies with high quality.

14.1 AMPLIFYING SYSTEM FUNCTIONS

Regardless of its application, an amplifying system has a number of primary functions that must be performed. An understanding of these functions is an extremely important part of operational theory. A block diagram of an amplifying system is shown in Fig. 14–1. The triangular-shaped items of the diagram show where amplification is performed. A stage of amplification is represented by each triangle. Three amplifiers are

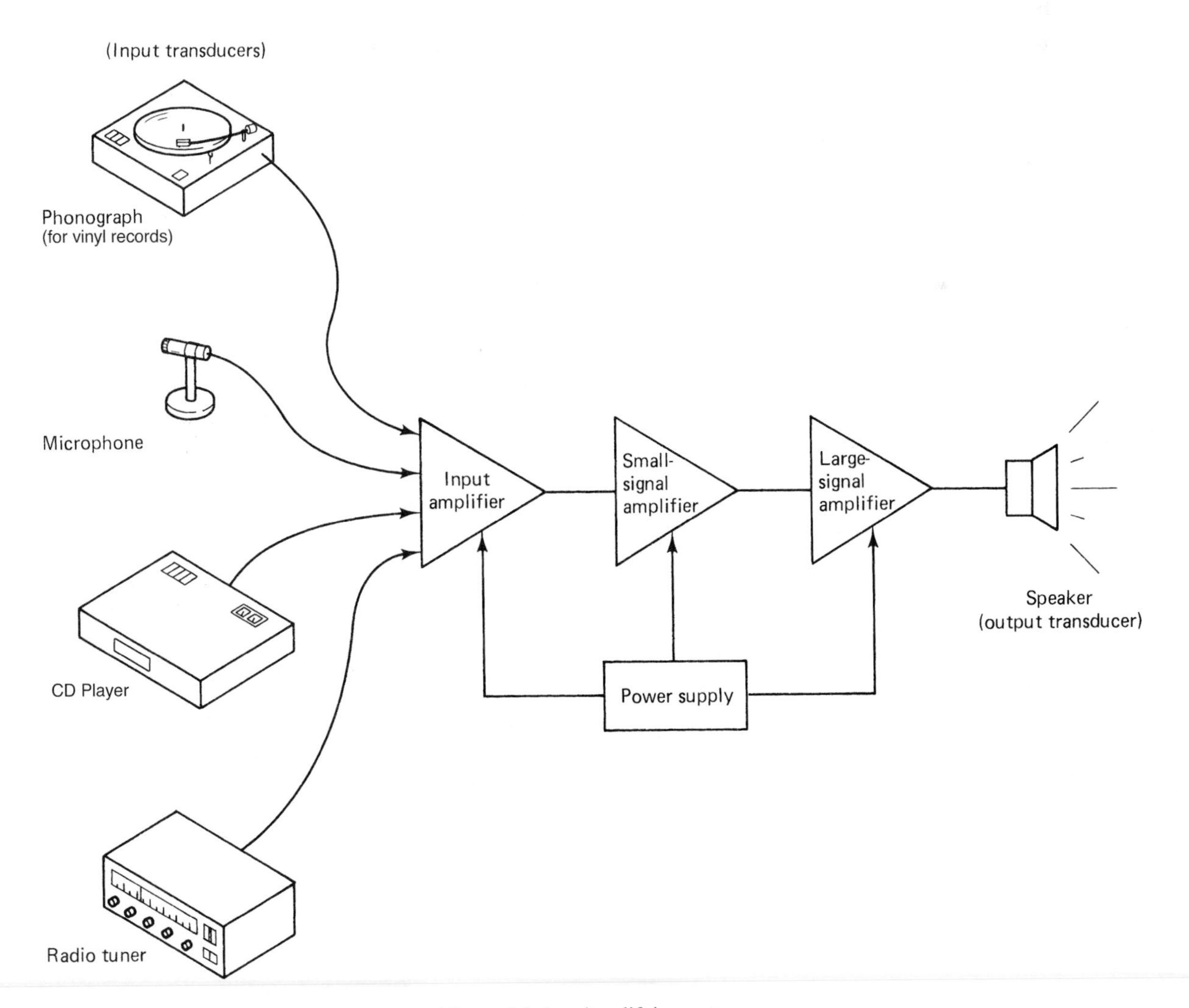

Figure 14–1. Amplifying system.

included in this particular system. A stage of amplification consists of an active device and all its associated components. Small-signal amplifiers are used in the first three stages of this system. The amplifier on the right side of the diagram is an output stage. A rather large signal is needed to control the output amplifier. An output stage is generally called a large-signal amplifier or a power amplifier. In effect, this amplifier is used to control a rather large amount of current and voltage. Remember that power is the product of current and voltage.

In an amplifying system, a signal must be developed and applied to the input. The source of this signal varies with different systems. In a *stereophonic* (stereo) phonograph, the signal is generated by a phonograph cartridge. Variations in the groove of a record are changed into electrical energy. This signal is then applied to the input of the system. The amplitude level of the signal is increased to a suitable value by the amplifying devices. A variety of different input signal sources may be applied to the input of an amplifying system. In other systems, the signal may also be developed by the input. Transducers are responsible for this function. An input transducer changes energy of one form into energy of a different type. Microphones, phonograph pickup cartridges for vinyl records and CD player heads are typical input transducers. Input signals may also be received through the air. Antenna coils and networks serve as the input transducer for this type of system. An antenna changes electromagnetic waves into radio-frequency (RF) voltage signals. The signal is then processed through the remainder of the system.

Signals processed by an amplifying system are ultimately applied to an output transducer. This type of transducer changes electrical energy into another form of energy. In a sound system, the speaker is an output transducer. It changes electrical energy into sound energy. Work is performed by the speaker when it achieves this function. Lamps, motors, relays, transformers, and inductors are frequently considered to be transducers. An output transducer is also considered to be the load of a system.

For an amplifying system to be operational, it must be supplied with electrical energy. A dc power supply performs this function. A relatively pure form of dc must be supplied to each amplifying device. In most amplifying systems, ac is the primary energy source. Ac is changed into dc, filtered, and, in some systems, regulated before being applied to the amplifiers. The reproduction quality of the amplifier is dependent on the dc supply voltage. Some portable stereo amplifiers may be energized by batteries.

14.2 AMPLIFIER GAIN

The gain of an amplifier system can be expressed in various ways such as voltage, current, power, and, in some systems, decibels. Nearly all input amplifier stages are voltage amplifiers. These amplifiers are designed to increase the voltage level of the signal. Several voltage amplifiers may be used in the front end of an amplifier system. The voltage of the input signal usually determines the level of amplification being achieved.

A three-stage voltage amplifier is shown in Fig. 14–2. These three amplifiers are connected in *cascade,* which refers to a series of amplifiers where the output of one stage is connected to the input of the next amplifier. The voltage gain of each stage could be observed with an oscilloscope. The waveform would show representative signal levels. Note the voltage-level change in the signal and the amplification factor of each stage.

The first stage has a voltage gain of 5 V. With 0.25-V p-p input, the output is 1.25-V p-p. The second stage also has a voltage gain of 5. With 1.25-V p-p input, the output is 6.25-V p-p. The output stage has a gain factor of 4. With 6.25-V p-p input, the output is 25-V p-p.

The total gain of the amplifier is 5 × 5 × 4, or 100. With 0.25-V p-p input, the output is 25-V p-p. Note that output is the product of the individual amplifier gains and not simply the addition of 5 + 5 + 4. Voltage gain (A_V) is an expression of output voltage (V_O) divided by input voltage (V_{in}). For the amplifier system, A_V is expressed by the formula

$$A_V = \frac{V_O}{V_{in}} = \frac{25\text{-V p-p}}{0.25\text{-V p-p}} = 100$$

Note that the units of voltage cancel each other in the problem. Voltage gain is therefore expressed as a pure number. It is not good practice to say that the voltage gain of an amplifier is 100-V p-p. The voltage gain is best described as a value such as 100.

Another consideration of an amplifier is its current gain. The current gain of a single amplifier was previously described as beta (β). The combined current amplification of the three-stage circuit of Fig. 14–2 is identified as A_I. It can be calculated with measured values of output current (I_O) and input current (I_{in}) and can be determined by the formula

$$A_I = \frac{I_O}{I_{in}}$$

Because the current of an amplifier depends on different voltage and impedance values, A_I is best determined by the formula

$$A_I = \frac{V_O/Z_O}{V_{in}/Z_{in}}$$

where

V_O = output voltage
Z_O = output impedance
V_{in} = input voltage
Z_{in} = input impedance

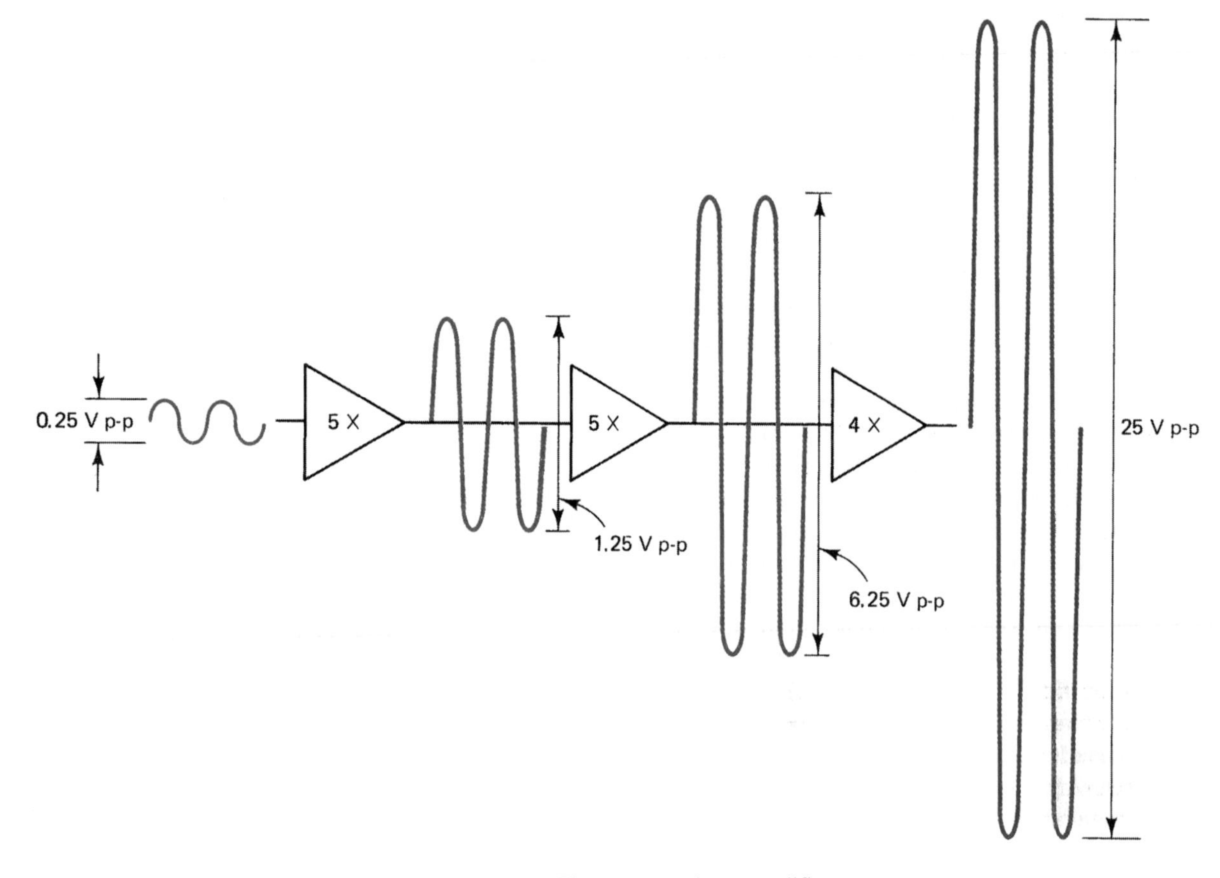

Figure 14–2. Three-stage voltage amplifier.

The power gain of an amplifier is another factor that needs to be considered. The combined power amplification (A_P) can be determined with measured values of output power (P_O) and input power (P_{in}) and can be expressed by the formula

$$A_P = \frac{P_O}{P_{in}}$$

Because power is the product of current and voltage, power amplification can also be determined by current amplification and voltage amplification factors and can be expressed by the formula

$$A_P = A_I A_V$$

Power amplification can also be determined by the square of the output voltage/output impedance divided by the square of the input voltage/input impedance and can be expressed by the formula

$$A_P = \frac{V_O^2/Z_O}{V_{in}^2/Z_{in}}$$

14.3 DECIBELS

The human ear does not respond to sound levels in the same manner as an amplifying system. An amplifier, for example, has a linear rise in signal level. An input signal level of 1 V could produce an output of 10 V. The voltage amplification would be 10:1 or 10. The human ear, however, does not respond in a linear manner. It is essentially a nonlinear device. As a result, sound amplifying systems are usually evaluated on a logarithmic scale that indicates how our ears will actually respond to specific signal levels. Gain expressed in logarithms is much more meaningful than linear gain relationships.

The logarithm of a given number is the power to which another number, called the base, must be raised to equal the given number. Basically, a logarithm has the same meaning as an exponent. A *common logarithm* is expressed in powers of 10 and is illustrated by the following:

$$10^3 = 1000$$
$$10^2 = 100$$
$$10^1 = 10$$
$$10^0 = 1$$

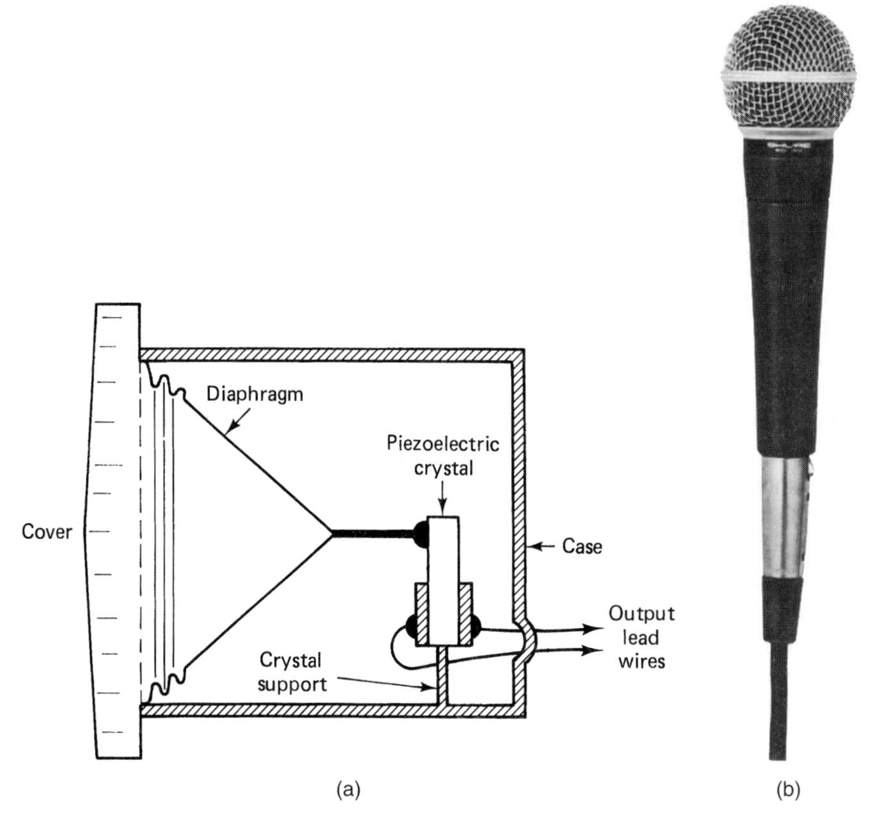

Figure 14–22. Crystal microphone: (a) construction; (b) assembled. (Courtesy of Shure Brothers, Inc.)

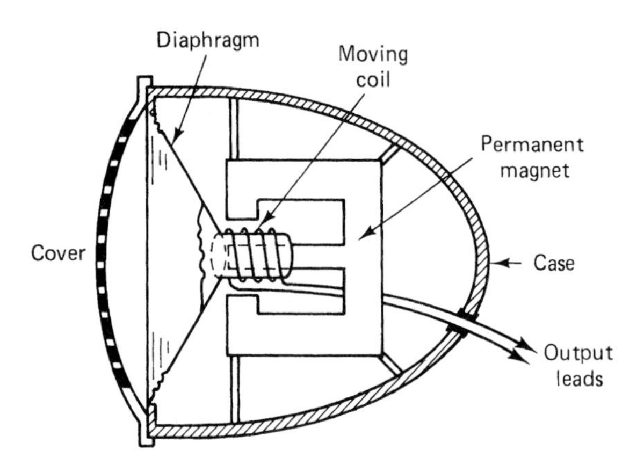

Figure 14–23. Structure of a dynamic microphone.

which is achieved by pushing the record button of a tape machine and speaking into a microphone. The signal is then processed by the amplifier. The output of the system supplies signal current to the electromagnetic recording head. A motor-drive mechanism is needed to move the tape at a selected recording speed. The tape head responds as a load device when the system responds as a recording device.

The recovery of taped information is achieved by a *read head.* This part of the system is much like a dynamic microphone; however, the electromagnetic coil is stationary in this type of device. Voltage is induced in the coil by a moving magnetic field. As the magnetic tape moves under the read head, voltage is induced in the coil. Changes in the field will cause variations in voltage values. Figure 14–25 shows a simplification of the tape reading function. The tape head responds as an input transducer when the system responds as a playback device.

The head of a cassette tape system is usually designed to perform both record and playback functions. Playback is shown in Fig. 14–26. The air gap of a tape head is 6 to 9 mils (0.006 to 0.009 in.) for audio tape recorders. This gap is provided by a thin layer of nonmagnetic material between the poles of an electromagnetic coil. The head must be kept clean and free of foreign material to function properly.

Phonograph Pickup Cartridges

The phonograph pickup cartridge is another type of amplifying system input transducer. This device develops an audio signal from the grooves of a vinyl record. Phono cartridges

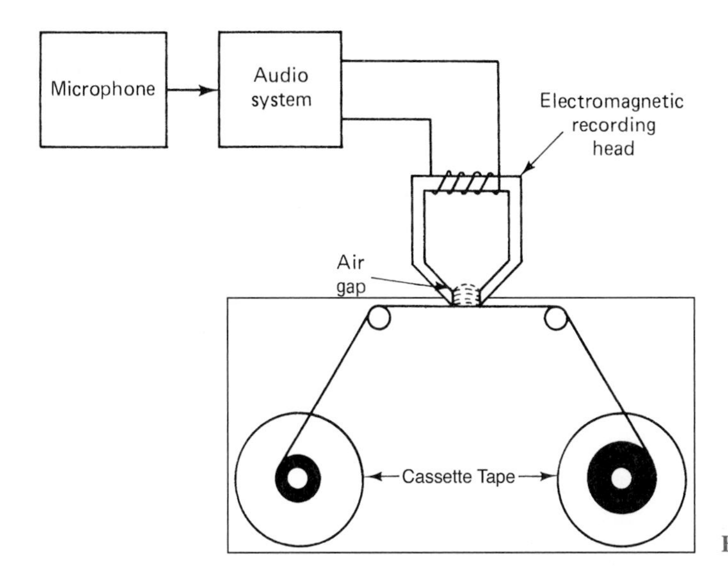

Figure 14–24. Tape recording function.

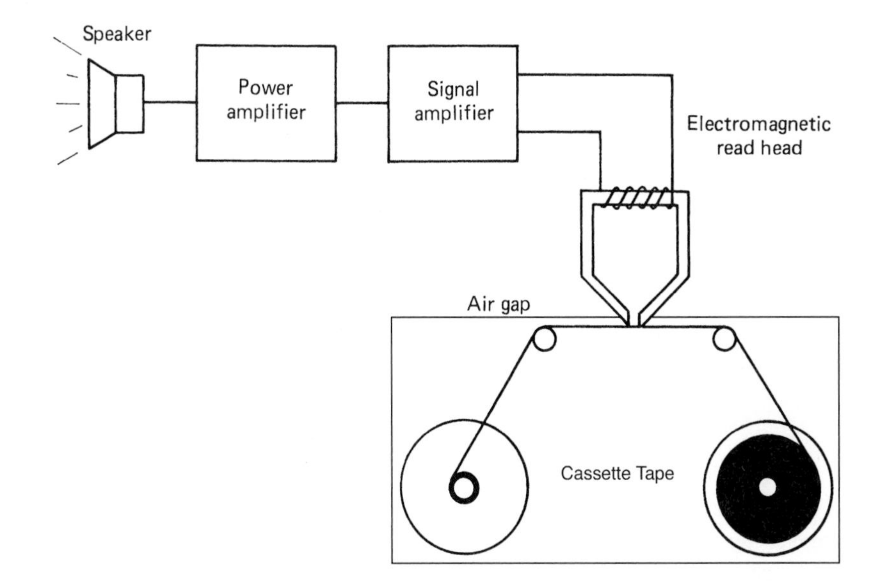

Figure 14–25. Tape reading or playback function.

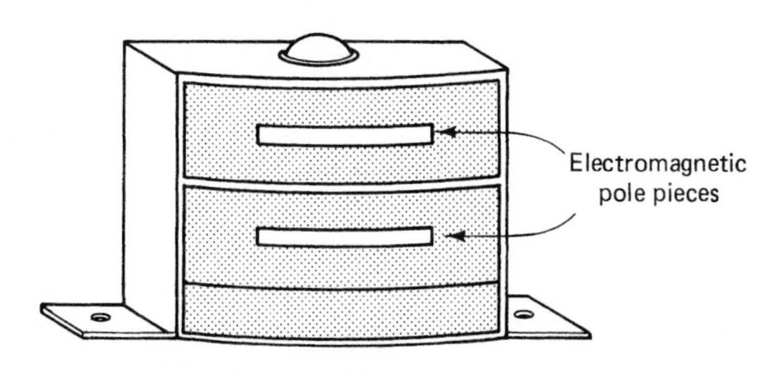

Figure 14–26. Recording-playback tape head.

are either of the electromagnetic or piezoelectric crystal type. The output of a phono pickup is much greater than that of a microphone. Typical voltage levels are 5 to 10 mV for magnetic devices and 200 to 1000 mV for crystal types. The resultant frequency output can be 15 to 16,000 Hz.

The construction of an electromagnetic phono pickup is shown in Fig. 14–27. This particular cartridge is designed for stereophonic reproduction. The grooves of a record are cut with two audio signals on opposite sides of a 45° angle. As the stylus moves the yoke, voltage is induced in each coil. The yoke is a small permanent magnet core. In other cartridges,

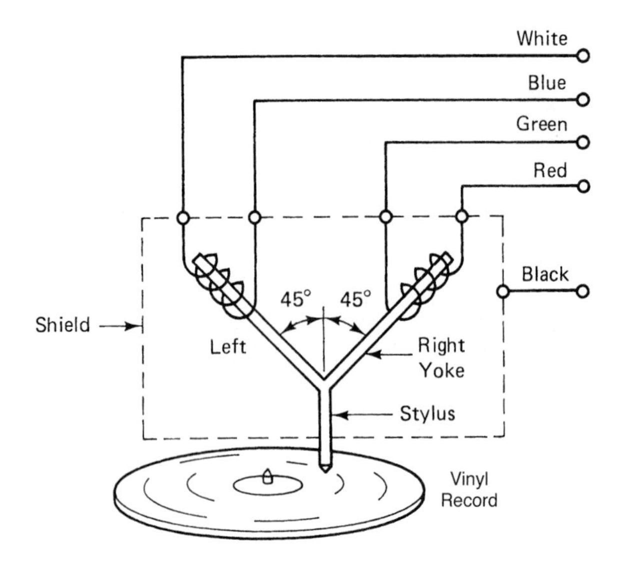

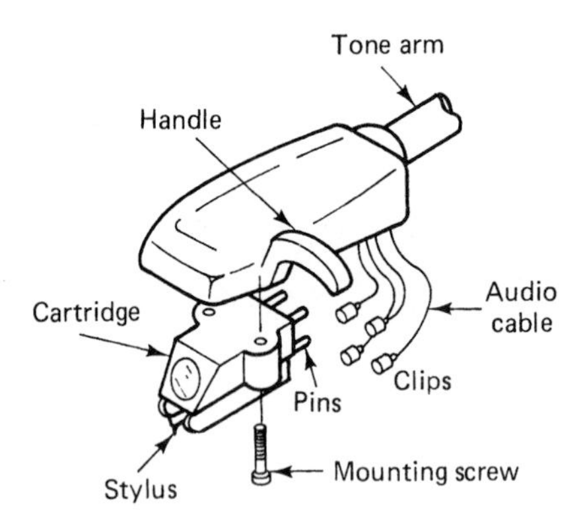

Figure 14–27. Electromagnetic cartridges.

the magnet may be stationary with the coil being moved by the stylus. In either type of construction, signal voltage is generated by electromagnetic induction.

A crystal or ceramic stereo cartridge is shown in Fig. 14–28. This particular design has a Y-shaped yoke structure. A piece of crystal or ceramic material is attached to each side of the yoke. When the stylus moves in the record groove, it causes the yoke to twist. This force on the crystal causes a corresponding voltage to be developed. Audio-frequency signal voltage is generated due to the piezoelectric effect. The output from each crystal is developed by wire connections at one end.

Compact Discs

Compact disc (CD) technology is a form of input for audio and video systems. CDs are considered to be optically actuated devices. Information is embedded on the surface of a disc in digital form. The encoded information is represented as submicroscopic pits and lands on the surface of the disc. The presence or absence of pits and lands is then read by a laser beam. Lands reflect the laser beam and pits cause it to be diffused or not reflected.

Two-channel stereophonic sound signals are stored as digital information on CDs. Stereophonic sound is digitally sampled at a rate of 44,100 times per second. The signal of each channel is then expressed as a 16-bit binary number. The generated digital signals, correction data, tracking codes, and cueing data are all placed on a compression-molded plastic disc that is 12 cm (4.72 in.) in diameter. The disc is coated with a thin reflective metal layer that is protected by a clear plastic coating.

Compact discs are generally classified as memory devices. Stereophonic sound can be retrieved from a CD by placing it into a playback or reading device. CD players have a low-powered laser beam mounted on a movable assembly that responds to the recorded information as the disc rotates. The laser beam distinguishes between the presence or absence of pits beneath the clear, transparent surface of the CD. Reflections of the laser beam are detected by a photodiode. The photodiode develops an output voltage value that is representative of the original stereophonic signal placed on the disc. This signal is then amplified and used to drive loudspeakers that produce sound.

Lasers

The development of the laser has had a significant continuing impact on electronic control systems. The major advantage of the laser is its tightly focused beam. This beam can travel long distances with relatively little spreading. The term *laser* stands for *l*ight *a*mplification by *s*timulated *e*mission of *r*adiation.

A ruby laser is shown in Fig. 14–29. When the xenon tube flashes, the chromium atoms in the ruby rod absorb photons of light. The chromium atoms then emit their own photons. Many of these travel along the axis of the ruby rod, and are reflected by the mirrors on each end. This causes additional photons to be released, which amplifies the light. Eventually, the photons fall into step with one another so that they produce only one wavelength (color) of light. The result is an intense beam of light energy.

Gas Lasers. Gas lasers are often used with control systems. A popular type is the helium-neon laser, shown in Fig. 14–30. Many other types are available, and all use basically the same operating principle, as follows: A high dc potential is applied to the plasma tube by means of a voltage-multiplier circuit and a pulse transformer. The filament within the tube is heated by a 6.3-V ac source. As

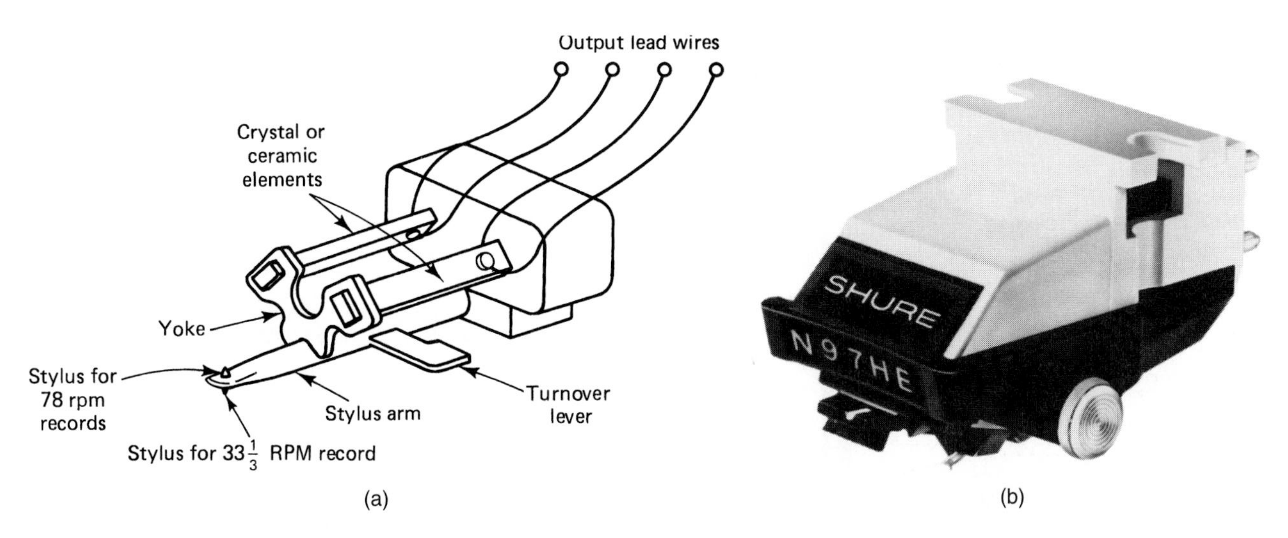

Figure 14–28. Stereo cartridge: (a) crystal or ceramic cartridge structure; (b) assembled phonograph cartridge. [(b) Courtesy of Shure Brothers, Inc.]

electrons from the filament are accelerated, they strike helium-neon gas atoms in the tube and cause them to ionize. The ionized gas emits light. This action is similar to that of a fluorescent tube used for household lighting. The light reflects from a flat, fully reflective mirror at the top. The plasma tube is cut at a precise angle to control reflection. The light is reflected back and forth several times to the partially reflective spherical mirror at the bottom, where it is concentrated into a laser beam that is emitted through the spherical mirror.

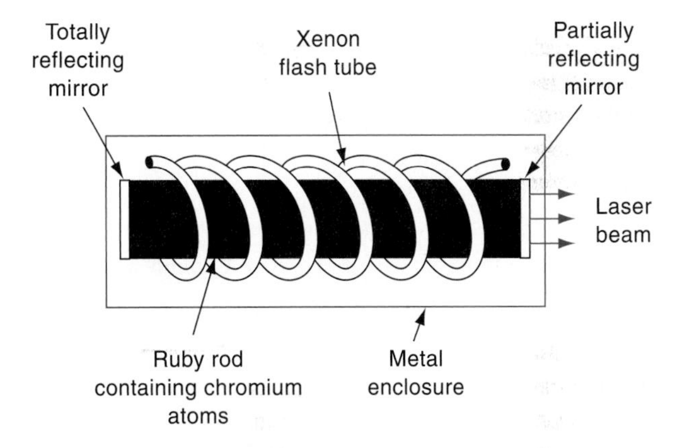

Figure 14–29. The ruby laser shown in this simplified drawing produces a beam of photons that do not spread out and scatter as does ordinary light.

Semiconductor Lasers. It is also possible to generate laser beams by means of semiconductors. These lasers have a resonant cavity similar to other lasers, except that this cavity is formed on a chip of semiconductor material.

Semiconductor injection lasers (Fig. 14–31) are efficient and extremely tiny in size compared with other lasers. The end faces of a gallium-arsenide chip are made parallel and flat. Since gallium-arsenide is a reflective material, no mirrors are needed. This is a distinct advantage in terms of complexity and cost.

As current flows through the chip, light is emitted from the gallium-arsenide. The atoms collide near the *P-N* junction of the material and cause the release of additional photons. Due to the reflective properties of the gallium-arsenide, a wave of photons is developed between its flat surfaces. The back-and-forth movement of this wave creates the resonant action required to produce a laser beam.

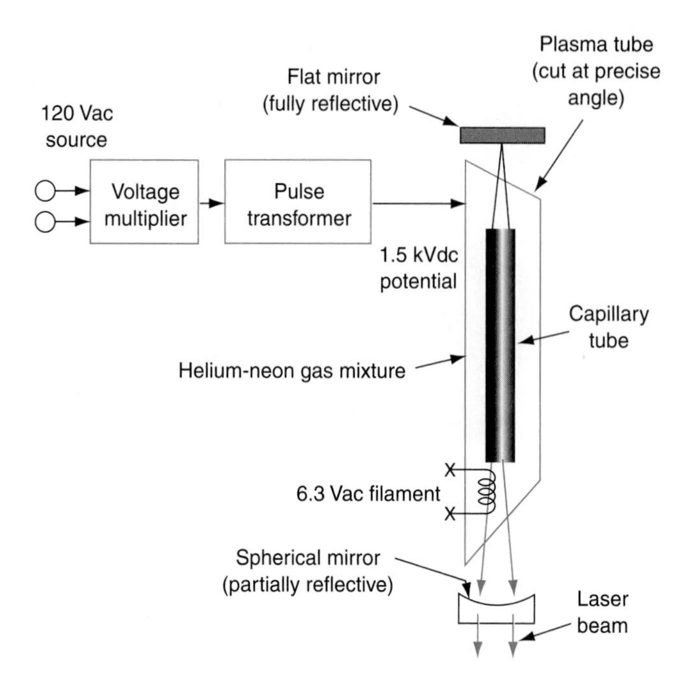

Figure 14–30. A laser using helium-neon gas is one of the types used for sensing and control applications.

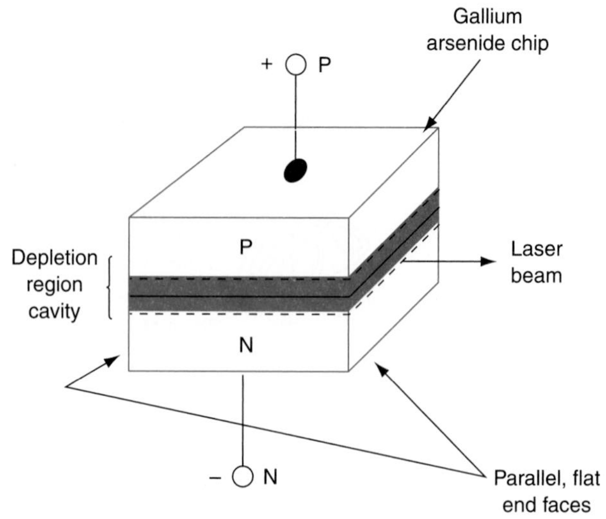

Figure 14–31. This semiconductor injection laser requires no mirrors. It is tiny and very efficient.

REVIEW

1. Explain what is included in an amplifying system.
2. If the input of an amplifier is 0.1 V and the output voltage is 10 V, what is the dB gain?
3. If an amplifying system has 0.02 W of input power and develops 20 W of output power, what is the dB gain?
4. If an amplifying system has 0.025-V input and 2.25-V output, how much dB gain is achieved?
5. Why is the value of the capacitance important in capacitance coupling?
6. What is a Darlington amplifier?
7. What is meant by impedance matching in transformer coupling?
8. Why is impedance matching important in transformer coupling?
9. Explain why the operational efficiency of a single-ended amplifier is low.
10. What is meant by the term *audio frequency?*
11. Explain how a sine-wave input causes output in a class B push-pull amplifier.
12. What is crossover distortion in a class B push-pull amplifier?
13. Why is a push-pull amplifier more efficient than a single-ended amplifier?
14. How does a class AB amplifier reduce crossover distortion?
15. What is a complementary-symmetry amplifier?
16. Why is a speaker considered to be an output transducer?
17. Explain how a speaker functions.
18. Explain how a dynamic microphone changes sound energy into electrical energy.
19. How is the piezoelectric effect used in a microphone to generate an electrical signal?
20. What is the primary function of a tape head?

21. How does a dynamic phonograph cartridge change a record groove vibration into electrical signal energy?
22. How is audio stored on a compact disc?
23. Briefly describe a gas laser and a semiconductor laser.

STUDENT ACTIVITIES

• *Capacitor-Coupled Amplifier* or *Complete Activity 14–1, page 287 in the manual.*

1. Construct the two-stage capacitor-coupled amplifier of Fig. 14–3. The transistors should be of the silicon *NPN* type with at least 250 mW power dissipation. A type 2N3397 works fairly well.
2. Measure the dc bias voltages at the emitter, base, and collector of each transistor.
3. Apply an ac signal of 400 Hz to the input. Adjust the signal level to produce the greatest output with a minimum of distortion. Observe this on an oscilloscope.
4. Determine the voltage gain of each stage and the combined system gain.

• •

ANALYSIS

1. Why is a capacitor used to couple signals?
2. What is the major advantage of *RC* coupling?
3. What limits the frequency passing through an *RC*-coupled amplifier?

• *Direct-Coupled Amplifier*

1. Construct the two-stage direct-coupled amplifier of Fig. 14–4. The transistors should be of the silicon *NPN* type with at least 250-mW power dissipation. A type 2N3397 could be used.
2. Measure and record the dc emitter, base, and collector voltages of each transistor.
3. Apply an ac signal of 1 kHz to the input. With an oscilloscope, observe the output and input waveforms. Adjust the signal level to produce the greatest output with a minimum of distortion.
4. Determine the ac gain using the measured input and output voltages. Determine the gain of each transistor and the overall system gain.

• *Darlington Transistor Amplifier*

1. Construct the Darlington amplifier of Fig. 14–5. A GE-62 amplifier can be used.
2. Apply a small signal to the input. Something less than 1 mV should be used.
3. With an oscilloscope, observe the output signal. Adjust the signal level to produce the greatest output with a minimum of distortion.
4. Measure the input and output signal levels. Calculate the ac voltage gain. Calculate the dB gain.
5. What is the phase relationship of the input-output signals?

• *Single-Ended Power Amplifier* or *Complete Activity 14–2, page 289 in the manual.*

1. Construct the single-ended power amplifier of Fig. 14–10. Use a 2N3053 transistor or a suitable equivalent. A 10:1-ratio filament transformer could be used for T_1 if a suitable audio output transformer is not available.
2. Measure and record the emitter, base, and collector bias voltages.
3. Apply an ac signal to the input. Adjust the output across the speaker to the greatest signal level with a minimum of distortion.
4. If a crystal microphone or phonograph input is available, connect it to the amplifier input.
5. Test the operation of the amplifier.
6. Observe the input and output signals with an oscilloscope.
7. Determine the signal voltage gain and dB gain.

• •

ANALYSIS

1. What is the primary function of a single-ended power amplifier?
2. What is the function of an output transformer?
3. What does the efficiency of an amplifier indicate?

• *Push-Pull Amplifier*

1. Construct the push-pull amplifier of Fig. 14–16. Transistors Q_1 and Q_2 are of the *NPN* silicon type. A 2N3053 or its equivalent can be used. R_{E1} and R_{E2} are 1-Ω resistors. R_{B1} and R_{B2} are selected to produce +0.6 V at the base of Q_1 and Q_2. Experiment with different values. An R_{B1} of 500 Ω and R_{B2} of 15 Ω are some representative values. T_2 should have a 2:1 impedance ratio.
2. After proper bias voltage selection for Q_1 and Q_2, test circuit operation. Measure and record the voltages of the emitter, base, and collector for both transistors.
3. Apply an ac signal to the input of T_1. Adjust the signal level to produce an undistorted output across the secondary winding of T_2. Proper base bias voltage is essential for undistorted output.
4. If undistorted output is achieved, observe the input signals to Q_1 and Q_2. Note the signal level and phase of each input. Observe the output signal at each collector. Note the amplitude and phase of each signal. Does this amplifier have crossover distortion?
5. Calculate the voltage gain in dB. Assume that the input and output impedances are equal.

• *IC Power Amplifier*

1. Construct the IC power amplifier of Fig. 14–19.
2. Connect a 400-Hz ac signal to the input. Use an audio signal generator.
3. With an oscilloscope, observe the output signal across the speaker.
4. Adjust R_1 through its range. Does the output become distorted?
5. Measure the input signal at pin 3. Then measure the output signal at pin 5. Calculate the voltage gain in dB for low-, medium-, and high-output signal levels.
6. Test the frequency response of the amplifier. What is the low-frequency limit where the signal output drops 30% in level?

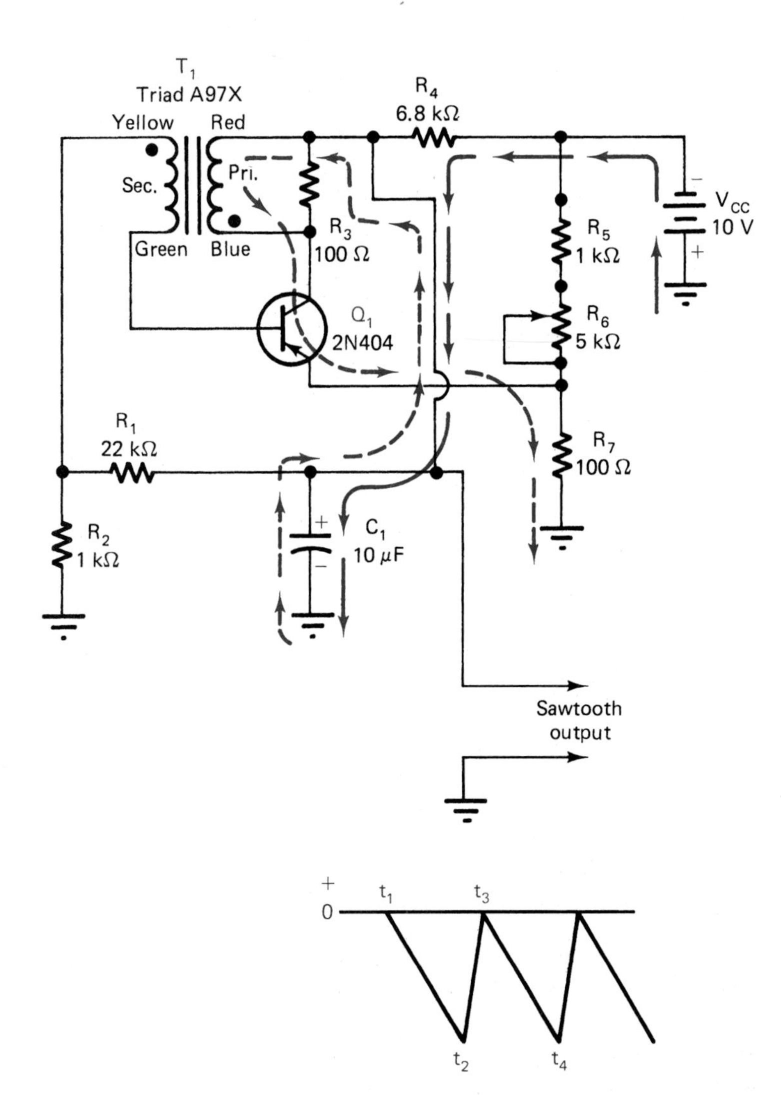

Figure 15–24. Transistor vertical-blocking oscillator. Solid arrows denote C_1 charge path; dashed arrows denote C_1 discharge path.

Blocking Oscillators

Blocking oscillators are an example of the relaxation principle. The active device, which is a transistor, is cut off during most of the operational cycle. It turns on for a short operational time to discharge an *RC* network. In appearance this circuit closely resembles the Armstrong oscillator. Feedback from the output to the input is needed to achieve the blocking function. The output of the oscillator is used to change the shape of a waveform.

A transistor blocking oscillator is shown in Fig. 15–24. A sawtooth-forming capacitor (C_1) charges when the source voltage is initially applied. It discharges through the transistor when it becomes conductive. The output across the capacitor is a sawtooth waveform.

When the supply voltage is initially applied, the charge of C_1 is zero, which is represented by T_1 on the output waveform. The transistor is cut off by the reverse-bias voltage developed by emitter resistor R_7. C_1 begins to charge to $-V_{CC}$ as indicated by the solid arrows. As C_1 charges, voltage is applied to the base of Q_1 through R_1 and the secondary winding of T_1. This voltage forward biases Q_1. The time of the wave from T_1 to T_2 represents the charging action of C_1. At T_2, the base voltage overcomes the reverse-bias emitter voltage. Q_1 goes into conduction and current flows through the primary winding of T_1. Feedback from primary to secondary drives the base more negatively. The process continues until Q_1 saturates.

When the transistor is conductive, it provides a low-resistance discharge path for C_1. See the dashed-line arrows of the discharge path. C_1 discharges quickly because of the low-impedance path. This is represented by the space between T_2 and T_3 of the waveform. When C_1 is fully discharged, Q_1 is cut off again by R_7. The process then repeats itself with C_1 charging again.

The frequency of the sawtooth output is dependent on the reverse-bias voltage of Q_1. An increase in negative voltage from R_6 will increase the cutoff time for Q_1, which will increase the charge time that C_1 needs to develop voltage to bring Q_1 into conduction. Value changes in R_6 alter the sawtooth frequency of the oscillator.

REVIEW

1. Explain what is meant by the term *feedback oscillator.*
2. What is a relaxation oscillator?
3. What are the fundamental parts of an oscillator?
4. Explain how the *LC* circuit responds when dc is applied to it.
5. What is the frequency of an *LC* circuit using a 0.5-μF capacitor and 100-mH inductor?
6. What causes the wave of an *LC* circuit to be damped?
7. Explain how the Armstrong oscillator operates.
8. What distinguishes a Hartley oscillator from other oscillator circuits?
9. What function does the capacitor divider of a Colpitts oscillator perform?
10. Explain how a crystal oscillator achieves stabilized oscillation.
11. How much charge voltage does a capacitor develop in one time constant? Two time constants? Five time constants?
12. Why does the E-B_1 junction of a UJT oscillator appear as a sawtooth wave?
13. Explain how an astable multivibrator changes states.
14. How does a trigger pulse cause a state change in a monostable multivibrator?
15. What is the polarity of the trigger pulse needed to cause a state change in a bistable multivibrator?
16. What are the internal functions of a 555 timer IC?
17. What is the frequency of an LM555 IC using 200-kΩ R_A and 200-kΩ R_B resistors and a 1-μF capacitor?
18. When does a blocking oscillator have its relaxing time?

STUDENT ACTIVITIES

• Armstrong Oscillator

1. Construct the Armstrong oscillator of Fig. 15–7.
2. Q_1 should be a transistor with a gain of 100. A 2N3397 or a 2N2405 transistor are typical. T_1 can be a 12-V step-down transformer.
3. Apply power to the circuit.
4. Connect an oscilloscope to the output to observe the generated waveform.
5. Adjust R_3 to produce the best sine wave.
6. Measure and record the dc voltages at the emitter, base, and collector.
7. If a frequency counter is available, determine the generated output. Some oscilloscopes may be used to determine frequency.

Hartley Oscillator or Complete Activity 15–1, page 303 in the manual.

1. Construct the Hartley oscillator of Fig. 15–8.
2. Turn on the power source. If the circuit is operating properly, test point *B* and ground should indicate a negative dc voltage.
3. Prepare an oscilloscope for operation. Connect the vertical probe to point *B* and the ground probe to circuit ground. Make a sketch of the observed waveform.
4. Observe and record the waveforms observed at test points *C, D,* and *A.*
5. Turn off the circuit power. Remove C_1 and replace it with a 365-pF variable capacitor.
6. Turn on the circuit power. Adjust the variable capacitor through its range while observing the waveform at test point *A*. Which position of the capacitor produces the highest and lowest frequency?
7. If an AM radio is available, turn it on and place it near the oscillator circuit. Set the receiver near the center of the dial that is not occupied by a station.
8. Adjust the variable capacitor of the oscillator through its range while listening for a whistling sound on the receiver. What frequency does the receiver indicate?

ANALYSIS

1. What components determine the frequency of a Hartley oscillator?
2. What achieves amplification in a Hartley oscillator circuit?
3. How does feedback from the output of the transistor occur?
4. What does feedback actually achieve in the operation of an oscillator?

Colpitts Oscillator or Complete Activity 15–2, page 305 in the manual.

1. Construct the Colpitts oscillator of Fig. 15–9.
2. Test the circuit for operation.
3. Observe the waveforms with an oscilloscope at test points *A, B,* and *C* with respect to ground. Make a sketch of the observed waves.
4. With a VOM, measure and record the dc voltages at the transistor's emitter, base, and collector.

ANALYSIS

1. What is the major difference between a Hartley and a Colpitts oscillator?
2. How does the Colpitts oscillator achieve feedback?
3. What are the frequency-determining components of the Colpitts oscillator?

Crystal Oscillator or Complete Activity 15–3, page 307 in the manual.

1. Construct the Pierce crystal oscillator of Fig. 15–12.
2. Test the circuit for operation.

3. Observe the waveforms with an oscilloscope at test points *A* and *B* with respect to ground.
4. With a VOM, measure and record the dc voltages at *B* and *C* of the transistor.
5. Determine the frequency of the oscillator with an oscilloscope or frequency counter.

ANALYSIS

1. How is the piezoelectric effect used in a crystal oscillator?
2. Why are crystals used in oscillators for communication circuits?
3. Why does the frequency of a crystal oscillator produce a number of different oscillations on a radio receiver?

UJT Oscillator

1. Construct the UJT oscillator of Fig. 15–17.
2. Test the circuit for operation.
3. Observe the waveform with an oscilloscope at test point *E* with respect to ground.
4. Exchange R_1 (2.7 kΩ) for a 10-kΩ resistor. How does this alter the sound output?
5. Remove the speaker and replace it with a 120-Ω resistor.
6. With an oscilloscope, observe the waveforms at test points *E* and B_1 with respect to ground. Make a sketch of the observed waveforms.

Astable Multivibrator

1. Construct the astable multivibrator of Fig. 15–18.
2. Turn on the circuit.
3. Observe the waveforms with an oscilloscope at points *Q, A, B,* and $\bar{Q}$ with respect to ground. Make a sketch of the waveforms.
4. Turn off the circuit. Exchange C_1 and C_2 for two large capacitors of 10 to 100 μF.
5. Turn on the circuit and measure the voltages at *Q, A, B,* and $\bar{Q}$ for an operational cycle. Explain your findings.

Monostable Multivibrator

1. Construct the monostable multivibrator of Fig. 15–19.
2. Turn on the circuit.
3. Test the dc voltages at *Q, A, B,* and $\bar{Q}$.
4. Connect a voltmeter to point $\bar{Q}$. Momentarily touch the base (*B*) of Q_2 to ground. How does the voltage at $\bar{Q}$ respond?
5. Feed a square-wave generator trigger pulse into the trigger input. (*Note:* The astable multivibrator of Fig. 15–18 or 15–23 may be used for the trigger input.)
6. With an oscilloscope, observe the waveforms at points *Q, A, B,* and $\bar{Q}$.

Bistable Multivibrator

1. Construct the bistable multivibrator of Fig. 15–21.
2. Apply power to the circuit.

3. Measure the dc voltages at Q and $\bar{Q}$. What is the initial status of Q_1 and Q_2? If Q_1 is set, Q will be the value of V_{CC} and Q will be 0.6 V. If Q_2 is set, Q will be the value of V_{CC} and Q will be 0.6 V.
4. Touch a jumper ground wire to the input of the transistor, which has a low output. This will turn the transistor off. Its output will go to V_{CC}.
5. Alternately test the set and reset of Q_1 and Q_2 while measuring the respective output voltage values.
6. Touch the ground lead to the set input. Measure and record the voltages at Q and $\bar{Q}$.
7. Touch the ground lead to the reset input. Measure and record the voltages at Q and $\bar{Q}$.
8. Explain how the circuit responds when set and reset.

• 555 Astable Multivibrator

1. Construct the IC multivibrator of Fig. 15–23.
2. Turn on the circuit power.
3. With an oscilloscope, observe the waveform at the input (pin 3) with respect to ground.
4. Calculate the frequency of the output.
5. Measure the frequency with an oscilloscope or frequency counter.
6. Observe the waveform across C.
7. Turn off the power.
8. Make the following changes: $R_A = 200\ \text{k}\Omega$; $C = 50\ \mu\text{F}$. Turn on the power and test the circuit.
9. Measure and record the dc voltage at the output through an operational cycle. How does it respond?
10. Measure the voltage at pins 2 and 7. How does it respond?

• Blocking Oscillators

1. Construct the blocking oscillator circuit of Fig. 15–24. The indicated transformer must have a 1:414 turns ratio.
2. Turn on the power source.
3. Measure dc voltages at the base and collector or plate and control grid.
4. With an oscilloscope, observe the base and collector waveforms with respect to ground. In the vacuum-tube circuit observe the grid and plate waveforms.
5. Adjust R_6 and observe the change in the waveforms. Make a sketch of the waveform change that occurs.

• Self-Examination

1. Complete the Self-Examination for Chap. 15, page 309 in the manual.
2. Check your answers on page 418 in the manual.
3. Study the questions that you answered incorrectly.

• Chapter 15 Examination

Complete the Chap. 15 Examination on page 311 in the manual.

CHAPTER

16

COMMUNICATIONS SYSTEMS

OUTLINE

16.1 Systems
16.2 Electromagnetic Waves
16.3 Continuous-Wave Communication
16.4 Amplitude Modulation Communication
16.5 Frequency Modulation Communication
16.6 Television Communication
16.7 Fiber Optics

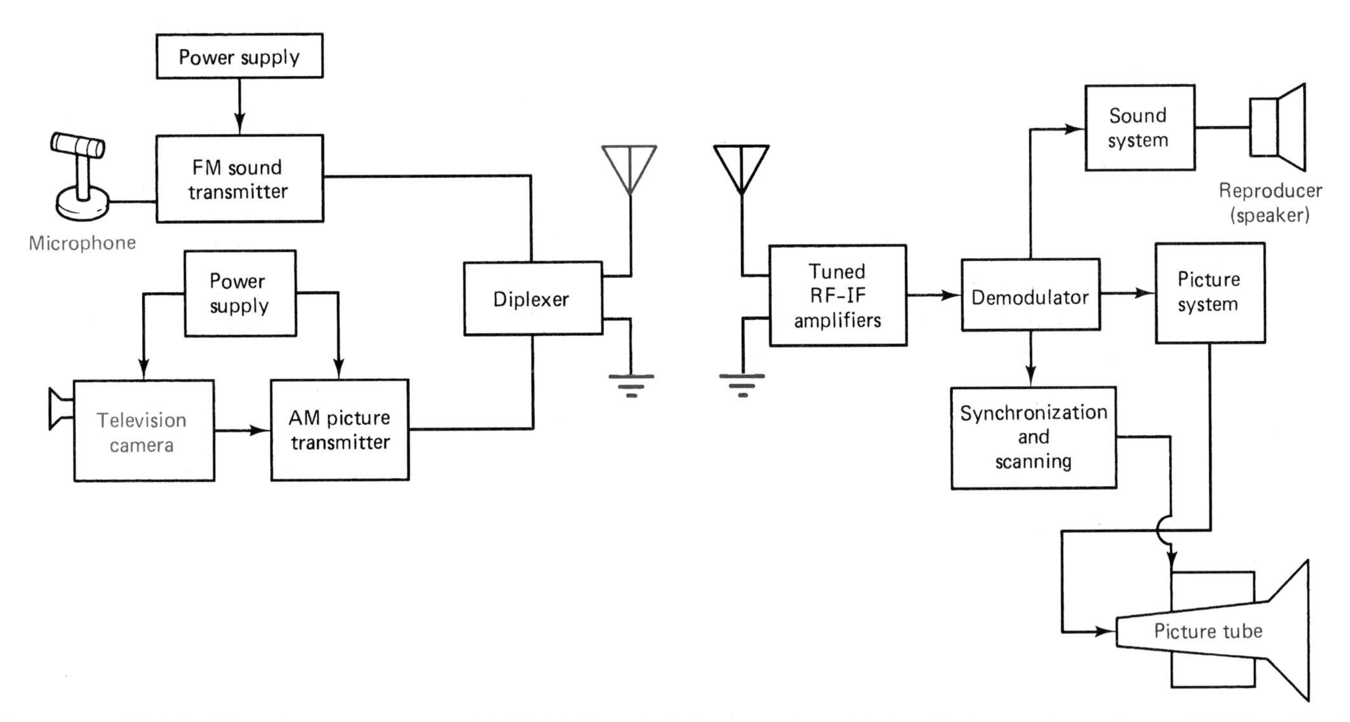

Television communication system.

OBJECTIVES

Upon completion of this chapter, you will be able to:

1. Explain the advantages, disadvantages, and characteristics of amplitude modulation and frequency modulation.
2. Explain the operation of AM and FM detectors.
3. Draw block diagrams of basic AM, CW, and FM transmitters and receivers.
4. Discuss the effects of heterodyning.
5. Discuss how intelligence is added to and removed from an RF carrier.
6. Calculate percentage modulation.
7. Explain the operation of a cathode-ray tube (CRT).
8. Describe how a picture is produced on the face of a picture tube.
9. Draw block diagrams of a black-and-white and a color TV set, and explain the function of each block.
10. Explain the functions of sweep signals and how they are synchronized.
11. Explain how high voltage is produced in a TV receiver.
12. Describe how video amplifiers obtain wide bandwidth, and list ways of increasing bandwidth.
13. How are fiber optics systems used in electronics?

In a previous chapter we discussed how to make sound louder by using electronic amplifiers. High-level sound amplification made it possible to communicate over a rather substantial distance. A public address amplifier system, for example, permits an announcer to communicate with a large number of people in a stadium or an arena. Even the most sophisticated sound system, however, has some limitations. Sound waves moving away from the source have a tendency to become somewhat weaker the farther they travel. An increase in signal strength does not, therefore, necessarily solve this problem. People near large speakers usually become uncomfortable when the sound is increased to a high level. It is possible, however, to communicate with people over long distances without increasing sound levels. Electromagnetic waves make this type of communication possible. Radio, television, long-distance telephone, and cellular phone communication are achieved by this process.

Electromagnetic wave communications systems play an important role in our daily lives. We listen to radio receivers, watch television, talk to friends on cellular telephones, and even communicate with astronauts by electromagnetic waves. Because this type of communication is widely used today, it is important that we have some basic understanding of it.

In this chapter we investigate how sound is transmitted through the air by electromagnetic waves. This type of communication uses high-frequency alternating current or radio-frequency energy for its operation. These signals travel through the air at 186,000 miles per second. Electromagnetic communication permits sound to travel long distances almost instantaneously.

IMPORTANT TERMS

The study of communications systems frequently involves several unique terms. These terms play an important role in the presentation of this material. As a rule, it is helpful to review these terms before proceeding with the chapter.

Beat frequency. A resultant frequency that develops when two frequencies are combined in a nonlinear device.

Beat frequency oscillator. An oscillator of a CW receiver. Its output beats with the incoming CW signal to produce an audio signal.

Blanking pulse. A part of the TV signal where the electron beam is turned off during the retrace period. There is both vertical and horizontal blanking.

Buffer amplifier. An RF amplifier that follows the oscillator of a transmitter. It isolates the oscillator from the load.

Carrier wave. An RF wave to which modulation is applied.

Center frequency. The carrier wave of an FM system without modulation applied.

Chroma. An abbreviation for "chrominance." Refers to color in general.

Compatible. A TV system characteristic in which broadcasts in color may be received in black and white on sets not adapted for color.

Deflection. Electron beam movement of a TV system that scans the camera tube or picture tube.

Deflection yoke. A coil fixture that electromagnetically deflects the CRT electron beam vertically and horizontally.

Diplexer. A special TV transmitter coupling device that isolates the audio carrier and the picture carrier signals from each other.

Field, even lined. The even-numbered scanning lines of one TV picture or frame.

Field, odd lined. The odd-numbered scanning lines of one TV picture or frame.

Frame. A complete electronically produced TV picture of 525 horizontally scanned lines.

Ganged. Two or more components connected together by a common shaft, such as a three-ganged variable capacitor.

Heterodyning. The process of combining signals of independent frequencies to obtain a different frequency.

High-level modulation. Where the modulating component is added to an RF carrier in the final power output of the transmitter.

Hue. A color, such as red, green, or blue.

Interlace scanning. An electronic picture production process in which the odd lines are all scanned, then the even lines, to make a complete 525-line picture.

Intermediate frequency (IF). A single frequency that is developed by heterodyning two input signals together in a superheterodyne receiver.

***I* signal.** A color signal of a TV system that is in phase with the 3.58-MHz color subcarrier.

Line-of-sight transmission. An RF signal transmission that radiates out in straight lines because of its short wavelength.

Modulating component. A specific signal or energy used to change the characteristic of an RF carrier. Audio and picture modulation are common.

Modulation. The process of changing some characteristic of an RF carrier so that intelligence can be transmitted.

Monochrome. Black-and-white television.

Negative picture phase. A video signal characteristic where the darkest part of a picture causes the greatest change in signal amplitude.

Optical fiber. A transmission medium made of glass or plastic.

Pixel. A discrete picture element on the viewing surface of a cathode-ray tube.

***Q* signal.** A color signal of a TV system that is out of phase with the 3.58-MHz color subcarrier.

Retrace. The process of returning the scanning beam of a camera or picture tube to its starting position.

Saturation. The strength or intensity of a color used in a TV system.

Scanning. In a TV system, the process of moving an electron beam vertically and horizontally.

Selectivity. A receiver function of picking out a desired radio-frequency signal. Tuning achieves selectivity.

Sidebands. The frequencies above and below the carrier frequency that are developed because of modulation.

Skip. An RF transmission signal pattern that is the result of signals being reflected from the ionosphere or the earth.

Sky wave. An RF signal radiated from an antenna into the ionosphere.

Sync. An abbreviation for "synchronization."

Synchronization. A control process that keeps electronic signals in step. The sweep of a TV receiver is synchronized by the transmitted picture signal.

Telegraphy. The process of conveying messages by coded telegraph signals.

Trace time. A period of the scanning process where picture information is reproduced or developed.

Vestigial sideband. A transmission procedure where part of one sideband is removed to reduce bandwidth.

Videcon. A TV camera tube that operates on the photoconductive principle.

Wavelength. The distance between two corresponding points that represents one complete wave.

***Y* signal.** The brightness or luminance signal of a TV system.

Zero beating. The resultant difference in frequency that occurs when two signals of the same frequency are heterodyned.

16.1 SYSTEMS

A communication system is similar to any other electrical system. It has an energy source, transmission path, control, load device, and one or more indicators. These individual parts are all essential to the operation of the system. The physical layout of the communication system is one of its most distinguishing features.

Figure 16–1 shows a block diagram of a radio-frequency communication system. The signal source of the system is a radio-frequency (RF) transmitter. The transmitter is the center or focal point of the system. The RF signal is sent to the remaining parts of the system through space. Air is the transmission path of the system. RF finds air to be an excellent signal path. The control function of the system is directly related to the signal path. The distance that the RF signal must travel has much to do with its strength. The load of the system is an infinite number of radio receivers. Each receiver picks the signal out of the air and uses it to do work. Any number of receivers can be used without directly influencing the output of the transmitter. System indicators may be found at several locations. Meters, indicating lamps, and waveform monitoring oscilloscopes are typical indicators. These basic functions apply in general to all RF communication systems.

Many different RF communication systems are in operation today. As a general rule, these systems are classified according to the method by which signal information or intelligence is applied to the transmitted signal. Three common communication systems are continuous wave (CW), amplitude modulation (AM), and frequency modulation (FM). Each system has several unique features that distinguish it

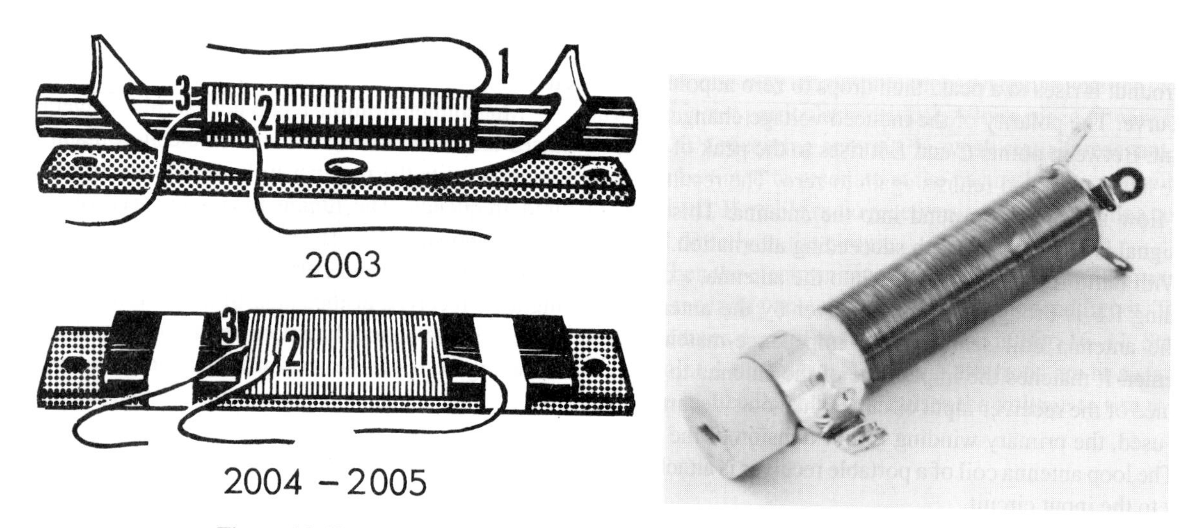

Figure 16–13. Antenna coils. (Courtesy of J.W. Miller Division/Bell Industries.)

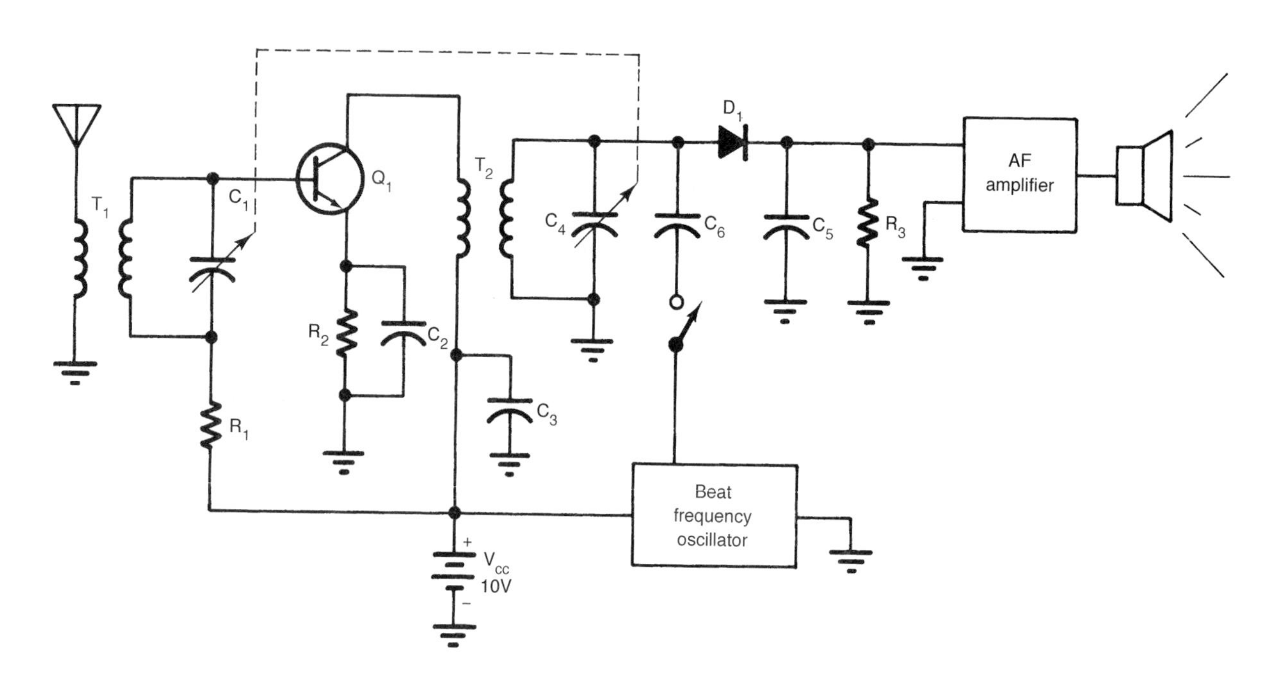

Figure 16–14. Tuned radio-frequency receiver.

Heterodyne Detection. Heterodyne detection is an essential function of the CW receiver. Heterodyning is simply a process of mixing two ac signals in a nonlinear device. In the CW receiver, the nonlinear device is a diode. The applied signals are called the *fundamental frequencies.* The resultant output of this circuit has four frequencies. Two of these are the original fundamental frequencies. The other two are beat frequencies. Beat frequencies are the addition and the difference in the fundamental frequencies. Heterodyne detectors used in other receiver circuits are commonly called *mixers.*

Figure 16–15 shows the signal frequencies applied to a heterodyne detector. Fundamental frequency F_1 is representative of the incoming RF signal. This signal has been keyed on and off with a telegraphic code. Fundamental frequency F_2 comes from the beat frequency oscillator (BFO) of the receiver. It is a CW signal of 1.001 MHz. Both RF signals are applied to the receiver's diode.

The resulting output of a heterodyne detector is F_1, F_2, $F_1 + F_2$, and $F_2 - F_1$. F_1 is 1.0 MHz and F_2 is 1.001 MHz. Beat frequency B_1 is 2.001 MHz and beat frequency B_2 is 1000 Hz. The output of the detector contains all four frequencies. The difference beat frequency of 1000 Hz is in the range of human hearing. F_1, F_2, and F_3 are RF signals.

When two RF signals are applied to a diode, the resultant output will be the sum and difference signals. If the two

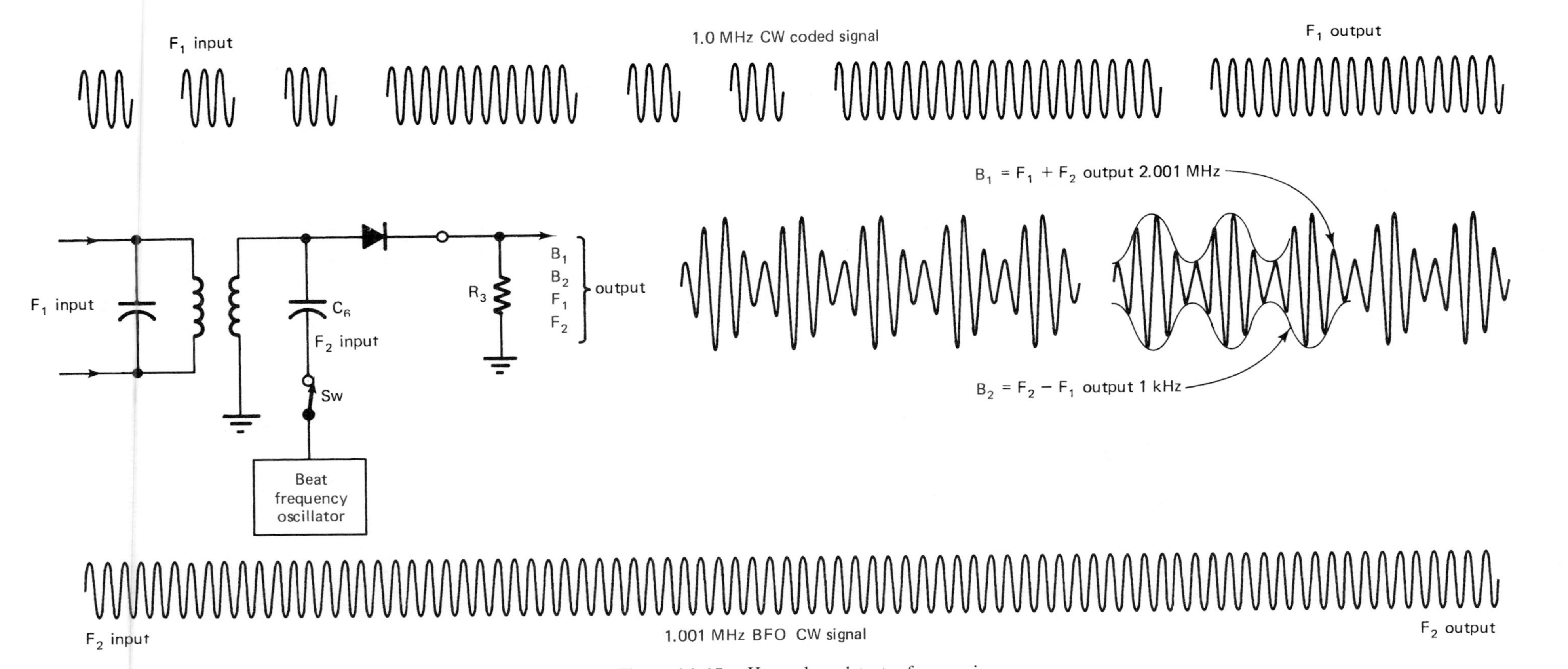

Figure 16–15. Heterodyne detector frequencies.

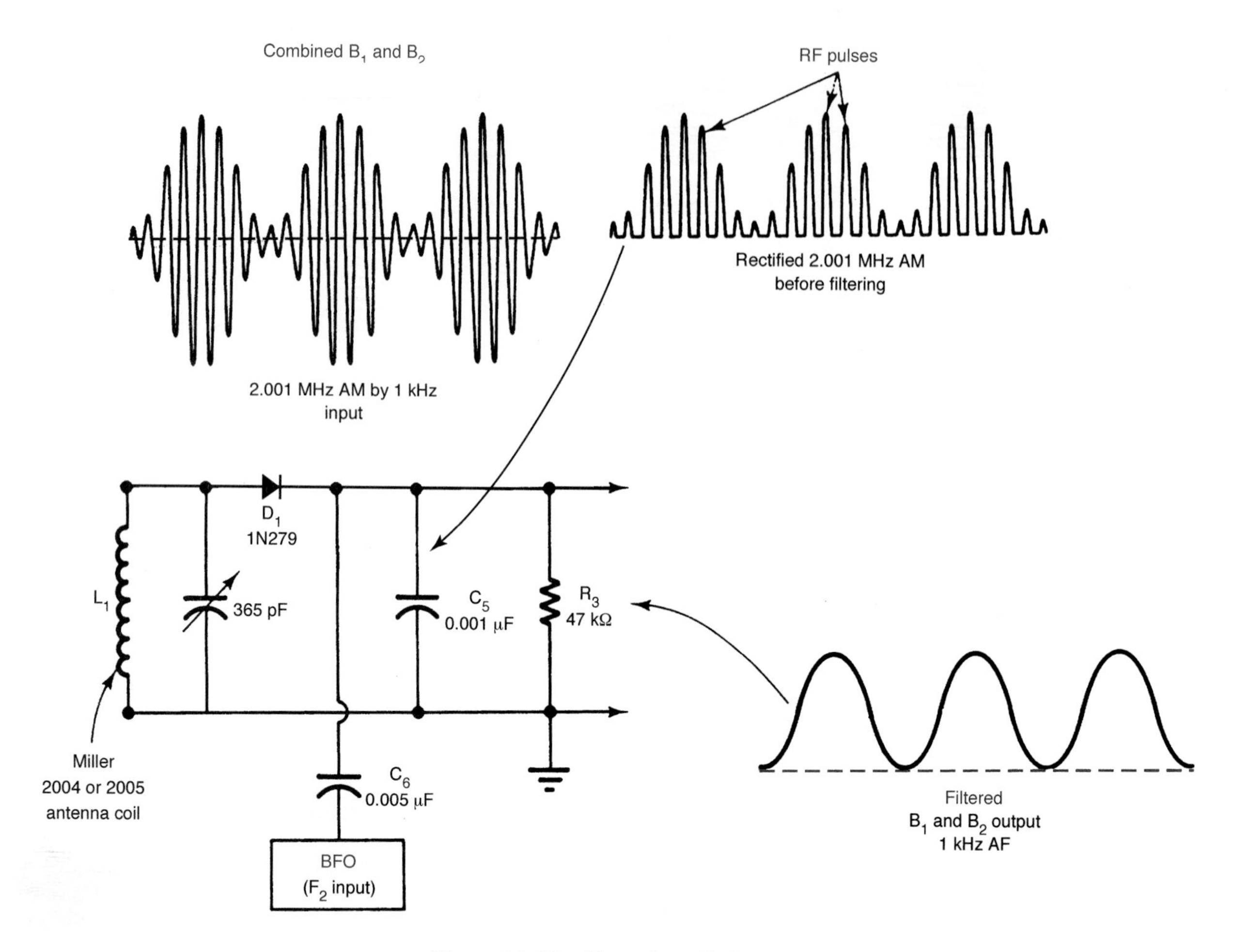

Figure 16–16. Heterodyne diode output.

signals are identical, the output frequency will be the same as the input, and is generally called *zero beating*. There is no developed beat frequency when this occurs.

When the two RF input signals are slightly different, beat frequencies will occur. At one instant, the signals will move in the same direction. The resultant output will be the sum of the two frequencies. When one signal is rising, and one is falling at a different rate, at times the signals will cancel each other. The sum and difference of the two signals are therefore frequency dependent. B_1 and B_2 are combined in a single RF wave. The amplitude of this wave varies according to frequency difference in the two signals. In a sense, the mixing process causes the amplitude of the RF wave to vary at an audio rate.

The diode of a heterodyne detector is responsible for rectification of the applied signal frequencies. The rectified output of the added beat frequency is of particular importance. It contains both RF and AF. Figure 16–16 shows how the diode responds. Initially, the RF signals are rectified. The output of the diode would be a series of RF pulses varying in amplitude at an AF rate. C_5 added to the output of the diode responds as a *C*-input filter. Each RF pulse will charge C_5 to its peak value. During the off period C_5 will discharge through R_3. The output will be a low-frequency AF signal of 1000 Hz, which will occur only when the incoming CW signal has been keyed. The AF signal is representative of the keyed information imposed on the RF signal at the transmitter.

Beat Frequency Oscillator. Nearly any basic oscillator circuit could be used as the BFO of a CW receiver. Figure 16–17 shows a Hartley oscillator. Feedback is provided by a tapped coil in the base circuit. C_1 and T_1 are the frequency-determining components. Note that this particular oscillator has variable-frequency capabilities. C_1 is usually connected to the front panel of the receiver. Adjustment of C_1 is used to alter the tone of the beat frequency signal. When the BFO signal is equal to the coded CW signal, no sound output will occur. This is where zero beating takes place. When the frequency of the BFO is slightly above or below the incoming signal frequency, a low AF tone is produced. Increasing the BFO frequency causes the pitch of the AF tone to increase. Adjustment of the AF output is a matter of personal preference. In practice, an AF tone of 400 Hz to 1 kHz is common.

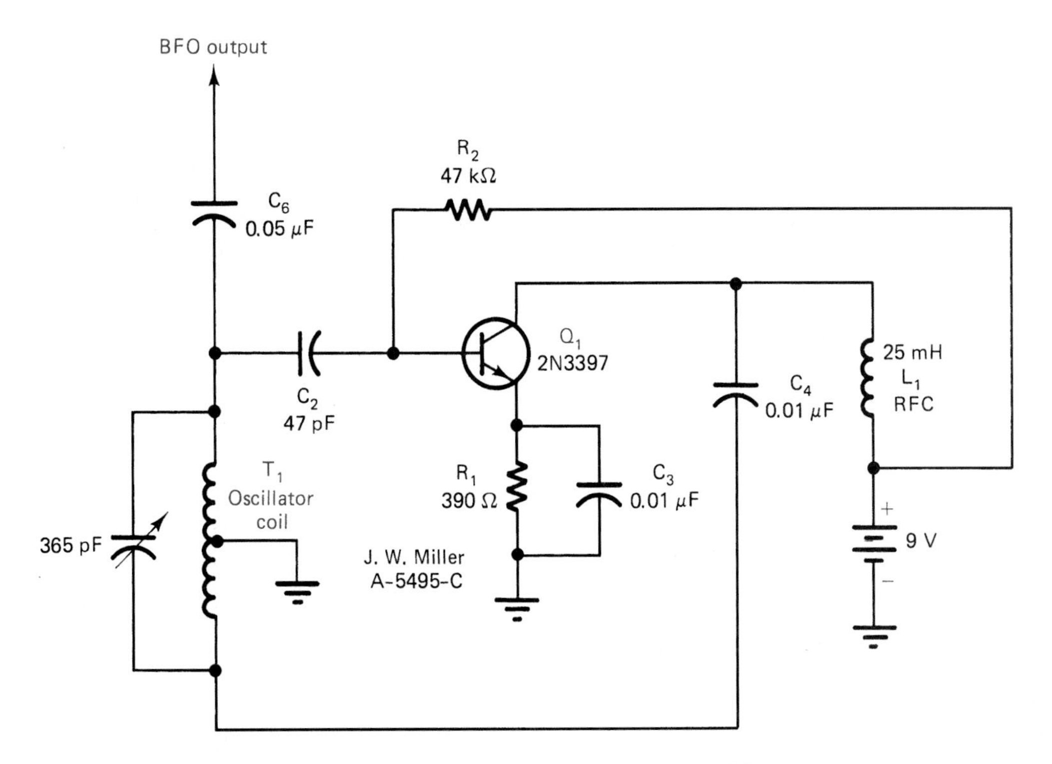

Figure 16–17. Hartley oscillator used as a BFO.

Output of the BFO is coupled to the diode through capacitor C_6. In a communication receiver, BFO output is controlled by a switch. With the switch on, the receiver will produce an output for CW signals. With the switch off, the receiver will respond to AM and possibly FM signals. The BFO is only needed to receive coded CW signals.

Audio-Frequency Amplification. The audio-frequency (AF) amplifier of a CW receiver is responsible for increasing the level of the developed sound signal. The type and amount of signal amplification varies widely among different receivers. Typically, a small-signal amplifier and a power amplifier are used. The small-signal amplifier responds as a voltage amplifier. This amplifier is designed to increase the signal voltage to a level that will drive the power amplifier. Power amplification is needed to drive the speaker. The power output of a communication receiver rarely ever exceeds 5 W. A number of the AF amplifier circuits in Chap. 14 could be used in a CW receiver.

16.4 AMPLITUDE MODULATION COMMUNICATION

Amplitude modulation or AM is an extremely important form of communication. It is achieved by changing the physical size or amplitude of the RF wave by the intelligence signal. Voice, music, data, and picture intelligence can be transmitted by this method. The intelligence signal must first be changed into electrical energy. Transducers such as microphones, phonograph cartridges, tape heads, and photoelectric devices are designed to achieve this function. The developed signal is called the *modulating component.* In an AM communication system, the RF transmitted signal is much higher in frequency than the modulating component. The RF component is an uninterrupted CW wave. In practice, this part of the radiated signal is called the *carrier wave.*

An example of the signal components of an AM system is shown in Fig. 16–18. The unmodulated RF carrier is a CW signal that is generated by an oscillator. In this example, the carrier is 1000 kHz or 1.0 MHz. A signal of this frequency would be in the standard AM broadcast band of 535 to 1605 kHz. When listening to this station, a receiver would be tuned to the carrier frequency. Assume now that the RF signal is modulated by the indicated 1000-Hz tone. The RF component will change 1000 cycles for each AF sine wave. The amplitude of the RF signal will vary according to the frequency of the modulating signal. The resulting wave is called an *amplitude-modulated RF carrier.*

To achieve amplitude modulation, the RF and AF components are applied to a nonlinear device. A vacuum-tube or solid-state device operating in its nonlinear region can be used to produce modulation. In a sense, the two signals are mixed or heterodyned together. This operation causes beat frequencies to be developed. The signals of Fig. 16–18 will have two beat frequencies. Beat frequency 1 is the sum of the

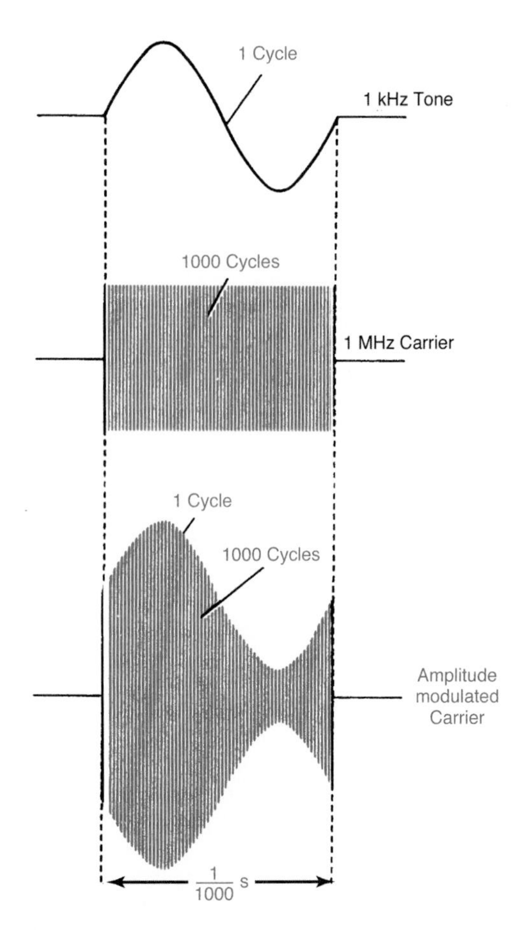

Figure 16–18. AM signal components.

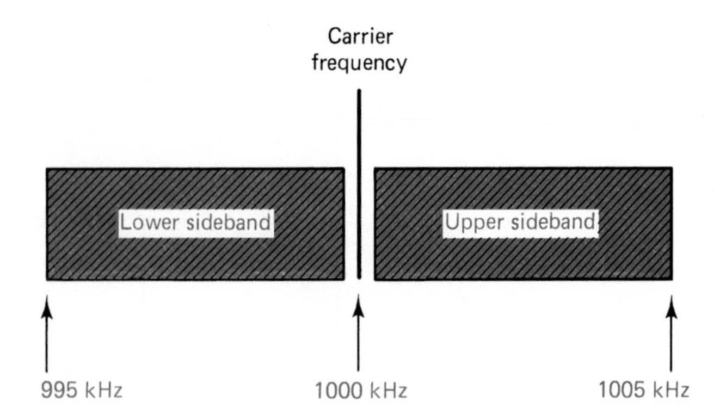

Figure 16–19. Sideband frequencies of an AM signal.

two signal frequencies. B_1 is 1,000,000 Hz plus 1000 Hz or 1,001,000 Hz. B_2 is 1,000,000 Hz minus 1000 Hz or 999,000 Hz. The resulting AM signal therefore contains three RF signals. These are the 1.0-MHz carrier, 0.999 MHz, and 1.001 MHz.

When the modulating component of an AM signal is music, the resultant beat frequencies become quite complex. As a rule, this involves a range or a band of frequencies. In AM systems, these are called *sidebands*. B_1 is the upper sideband and B_2 is the lower sideband.

The space that an AM signal occupies with its frequency is called a *channel*. The bandwidth of an AM channel is twice the highest modulating frequency. For our 1-kHz modulating component, a 2-kHz bandwidth is needed, which would be 1 kHz above and below the carrier frequency of 1 MHz. In commercial AM broadcasting, a station is assigned a 10-kHz channel which limits the AM modulation component to a frequency of 5 kHz. Figure 16–19 shows the sidebands produced by a standard AM station.

Interestingly, the carrier wave of an AM signal contains no modulation. All the modulation appears in the sidebands. If the modulating component is removed, the sidebands will disappear. Only the carrier will be transmitted. The sidebands are directly related to the carrier and the modulating component. The carrier has a constant frequency and amplitude. The sidebands vary in frequency and amplitude according to the modulation component. In AM radio, the receiver is tuned to the carrier wave.

Percentage of Modulation

In AM radio, it is not permitted by law to exceed 100% modulation. Essentially, this means that the modulating component cannot cause the RF component to vary over 100% of its unmodulated value. Figure 16–20 shows an AM signal with three different levels of modulation.

When the peak amplitude of a modulating signal is less than the peak amplitude of the carrier, modulation is less than 100%. If the modulation component and carrier amplitudes are equal, 100% modulation is achieved. A modulating component greater than the carrier will cause overmodulation. An overmodulated wave has an interrupted spot in the carrier wave. *Overmodulation* causes increased signal bandwidth and additional sidebands to be generated, thus causing interference with adjacent channels.

Modulation percentage can be calculated or observed on an indicator. When operating voltage values are known, the percentage of modulation can be calculated. The formula is

$$\% \text{ modulation} = \frac{V_{max} - V_{min}}{2\,V_{carrier}} \times 100\%$$

Using the values of Fig. 16–20(a), compute the percentage of modulation. Compute the percentage of modulation for Fig. 16–20(b). If the V_{min} value of the overmodulated signal is considered to be 0 V, compute the modulation percentage.

AM Communication System

A block diagram of an AM communication system is shown in Fig. 16–21. The transmitter and receiver respond as independent systems. In a one-way communication system, there is one transmitter and an infinite number of receivers. Commercial AM radio is an example of one-way communication.

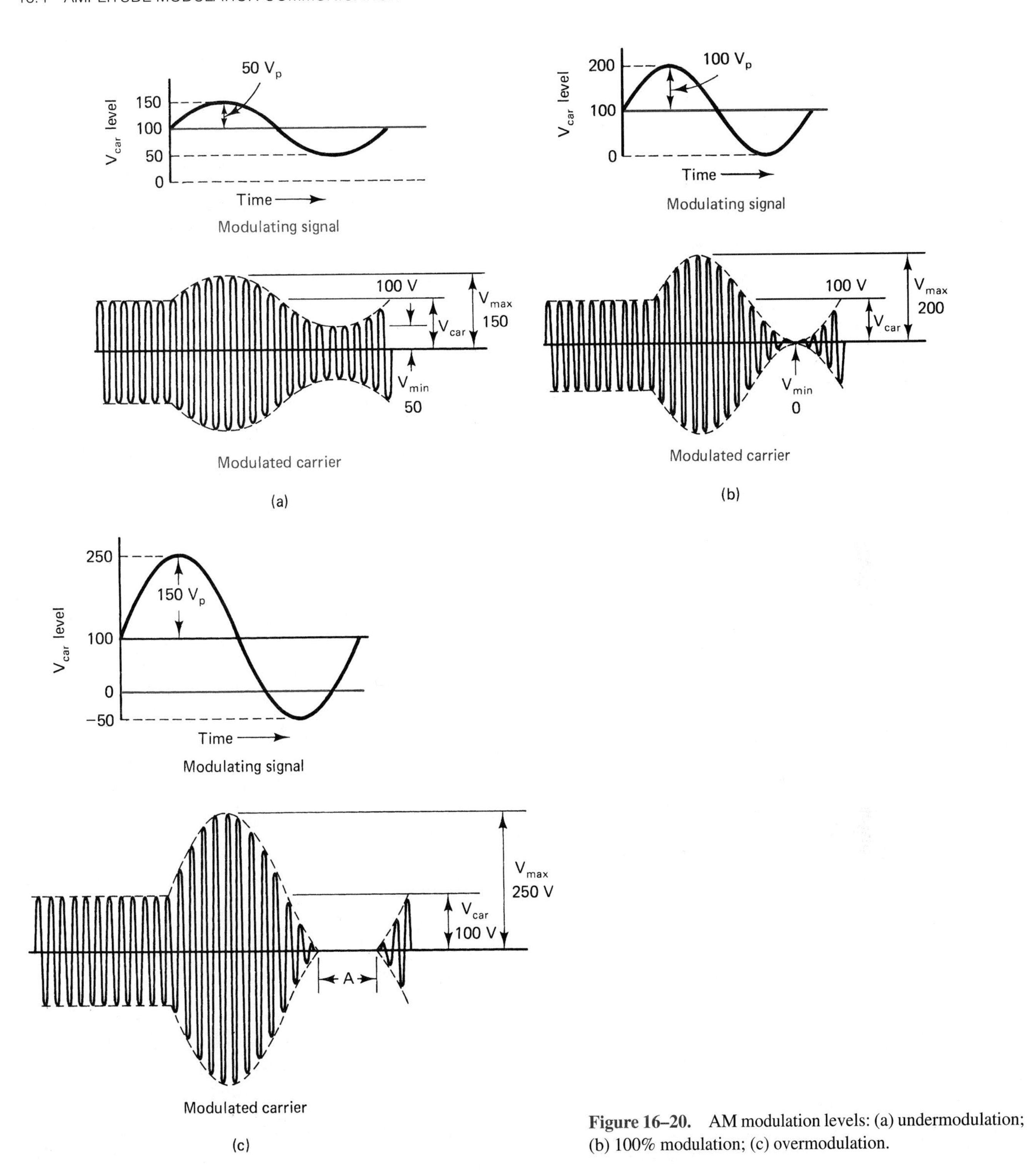

Figure 16–20. AM modulation levels: (a) undermodulation; (b) 100% modulation; (c) overmodulation.

Two-way communication systems have a transmitter and a receiver at each location. CB radio systems are of the two-way type. The operating principles are basically the same for each system.

The *transmitter* of an AM system is responsible for signal generation. The RF section of the transmitter is primarily the same as that of a CW system. The modulating signal component is, however, a unique part of the transmitter. Essentially, this function is achieved by an AF amplifier. A variety of amplifier circuits can be used. Typically, one or two small-signal amplifiers and a power amplifier are suitable for low-power transmitters. The developed modulation component power must equal the power level of the RF output. Modulation of the two signals can be achieved at several

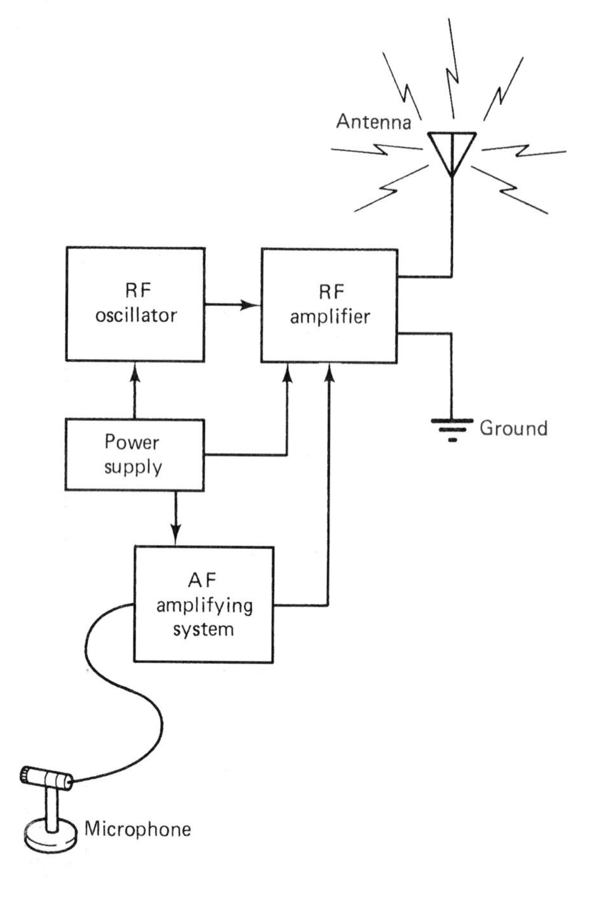

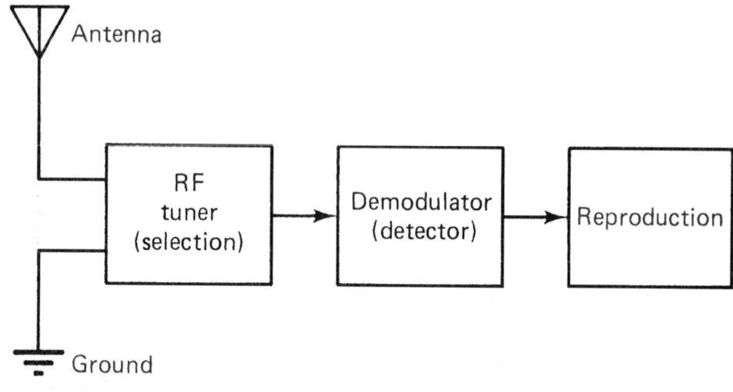

Figure 16–21. AM communication system.

places. High-level modulation occurs in the RF power amplifier. Low-level modulation is achieved after the RF oscillator. All amplification following the point of modulation must be linear. Only high-level modulation will be discussed in this chapter.

The *receiver* of an AM system is responsible for signal interception, selection, demodulation, and reproduction. Most AM receivers employ the heterodyne principle in their operation. This type of receiver is known as a *superheterodyne circuit.* It is somewhat different from the heterodyne detector of the CW receiver. A large part of the circuit is the same as the CW receiver. We will only discuss the new functions of the AM receiver.

AM Transmitters. Many types of AM transmitters are available today. Toy "walkie-talkie" units with an output of less than 100 mW are popular. Citizens band (CB) transmitters are designed to operate at 27 MHz with a 5-W output. AM amateur radio transmitters are available with a power output of up to 1 kW of power. Commercial AM transmitters are assigned power output levels from 250 W to 50 kW. These stations operate between 535 and 1605 kHz. In addition, there are 2-meter mobile communication systems, military transmitters, and public service radio systems that all use AM. As a general rule, frequency allocations and power levels are assigned by the Federal Communications Commission.

An AM transmitter has a number of fundamental parts regardless of its operational frequency or power rating. It must have an oscillator, an audio signal component, an antenna, and a power supply. The oscillator is responsible for the RF carrier signal. The audio component is responsible for the intelligence being transmitted. The modulation function can be achieved in a variety of ways. The signal is radiated from the antenna.

A simplified AM transmitter with a minimum of components is shown in Fig. 16–22. Transistor Q_1 is an AF signal amplifier. The sound signal being amplified is developed by a crystal microphone. It is then applied to transistor Q_2. This transistor is a Colpitts oscillator. L_1, C_4, and C_5 determine the RF frequency of the oscillator. The amplitude of the oscillator varies according to the AF component. Resistor R_3 is used to adjust the amplitude level of the AF signal. The developed AM output signal is applied to the transmitting antenna. With proper design of the transmitter, it can be tuned to the standard AM broadcast band. The frequency of the system is adjusted by capacitor C_4. The operating range of this unit is several hundred feet.

A schematic of an improved low-power AM transmitter is shown in Fig. 16–23. Transistors Q_1, Q_2, and Q_3 of this circuit are responsible for the RF signal component. The AF component is developed by Q_4, Q_5, and Q_6. High-level modulation is achieved by this transmitter. The AF signal modulates the RF power amplifier Q_3.

Transistor Q_1 is the active device of a Hartley oscillator. This particular oscillator has a variable-frequency output of 1 to 3.5 MHz. The frequency-determining components are C_1 and L_{T1}. The output of the oscillator is coupled to the base of Q_2 by capacitor C_3. The emitter-follower output of the oscillator has a low impedance.

Transistor Q_2 is an RF signal amplifier. This transistor is primarily responsible for increasing the signal level of the oscillator. It is also used to isolate the RF load from the oscillator, which is needed to improve oscillator frequency stability. When Q_2 is used in this regard, it is called a *buffer amplifier.*

The signal output of Q_2 must be capable of driving the power amplifier. In high-power transmitters, several RF signal amplifiers may be found between the oscillator and the

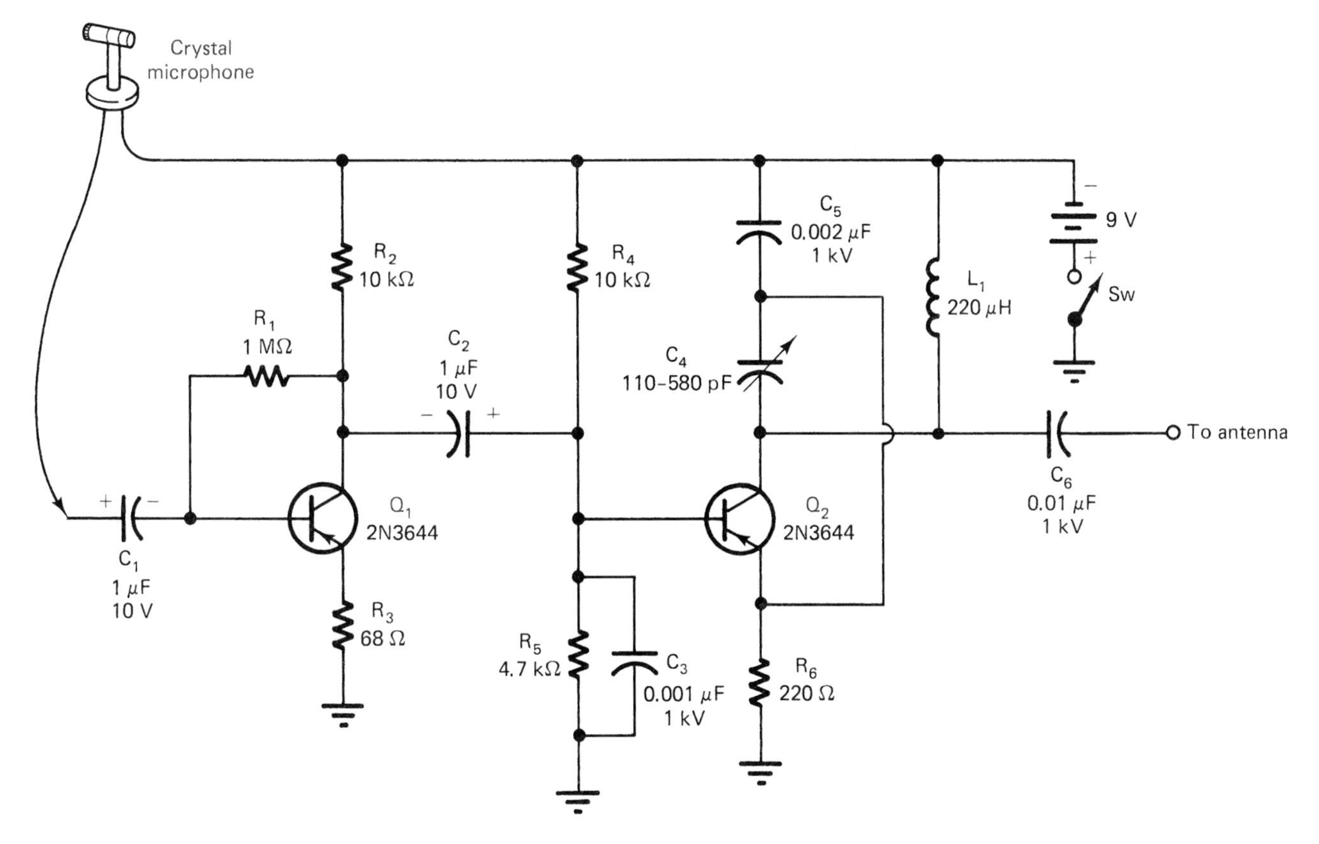

Figure 16–22. Simplified AM transmitter.

power amplifier. Each stage is responsible for increasing the signal level to a suitable level. RF amplifiers of this type are often called *drivers.* Q_3 is an RF power amplifier. It is designed to increase the power level of the RF signal applied to its input. The output is used to drive the transmitting antenna. The load of Q_3 is a tuned circuit composed of C_{10}, L_5, and C_{11}. This is a pi-section filter. A filter of this type is used to remove signals other than those of the resonant frequency. C_{11} is the output capacitor of the filter. It is adjusted to match the impedance of the antenna. C_{10} is the input capacitor. It is used to resonate the filter to the applied carrier frequency. Resonance of the tuning circuit occurs when the collector current meter dips to its lowest value. In the broadcasting field an adjustment of this type is called "dipping the final." Q_3 is operated as a class C power amplifier.

The modulating component of the transmitter is developed by an AF amplifier. Q_4 and Q_5 are push-pull AF power amplifiers. The developed AF signal is applied to the modulation transformer T_2. This signal causes the collector voltage of Q_3 to vary at an AF rate instead of being dc, which causes the RF output to vary in amplitude according to the AF component. The output signal has a carrier and two sidebands. The power output is approximately 5 W.

Do not connect the output of this transmitter to an outside antenna unless you hold a valid radio-telephone operator's license. A load lamp is used for operational testing. The intensity of the load lamp is a good indication of the RF power developed by the transmitter.

Simple AM Receiver. An AM receiver has four primary functions that must be achieved for it to be operational. No matter how complex or involved the receiver, it must accomplish these functions: signal interception, selection, detection, and reproduction. Figure 16–24 shows a diagram of an AM receiver that accomplishes these functions. Note that the circuit does not employ an amplifying device. No electrical power source is needed to make this receiver operational. The signal source is intercepted by the antenna-ground network. The receiver is energized by the intercepted RF signal energy. A receiver of this type is generally called a *crystal radio.* The detector is a crystal diode.

The functional operation of our crystal diode radio receiver is similar to the CW receiver. This particular circuit does not employ an RF amplifier, a beat frequency oscillator, or an AF amplifier. A strong AM signal is needed to make this receiver operational. Signal interception and selection are achieved in the same way as in the CW receiver.

The detection or demodulation function of an AM receiver is responsible for removing the AF component from the RF signal. The detector is essentially a half-wave rectifier for the RF signal. Germanium diodes are commonly used as detectors. They are more sensitive to RF signals than a silicon diode.

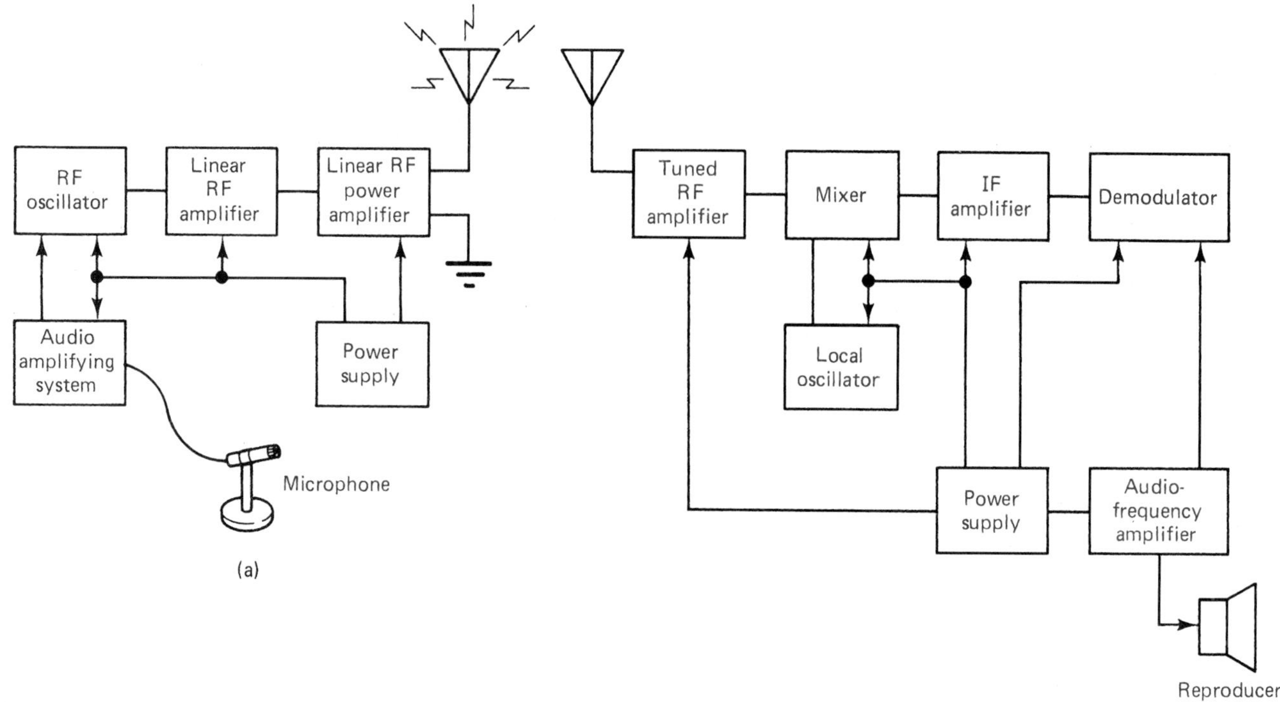

Figure 16–31. FM communication system: (a) direct FM transmitter; (b) superheterodyne FM receiver.

mobile FM communications systems operate with a few watts of power. Two-way land mobile communications employs low-power transmitters. Educational FM transmitters can operate with a power of several hundred thousand watts. Commercial FM transmitters have a similar power rating. In addition to these types are public service radio and military FM. The frequency allocation and power capabilities of an FM transmitter are allocated and regulated by the FCC.

An FM transmitter regardless of its power rating or operational frequency has a number of fundamental parts. The oscillator is responsible for the development of the RF carrier. Many FM transmitters employ direct oscillator modulation, meaning that the frequency of the oscillator is shifted by direct application of the modulation component. The RF signal is then processed by a number of amplifiers. The power amplifier ultimately feeds the FM carrier into the antenna for radiation.

A simplified direct FM transmitter is shown in Fig. 16–32. A basic Hartley oscillator is used as the RF signal source. With power applied, the oscillator will generate the RF carrier. The frequency of the oscillator is determined by the value of C_4 and T_1. Without modulation, the oscillator would generate a constant frequency, which is representative of the unmodulated carrier wave.

Modulation of the oscillator is achieved by voltage changes across a varicap diode. The capacitance of this diode changes with the value of the reverse-bias voltage. Bias voltage is used to establish a dc operating level for this diode. A change in audio signal voltage causes the capacitance of the diode to change at an audio rate. Note that the series network of C_1, D_1, and C_2 is connected in parallel with capacitor C_4 of the oscillator tank circuit. A change in audio signal level will cause a corresponding change in tank circuit capacitance. This, in turn, causes the oscillator frequency to change according to the modulating component. The modulating component is applied directly to the oscillator's frequency-determining components.

Direct FM can be achieved in a number of ways. The varicap diode method is commonly found in new transmitter equipment. This application is only one of a large number of circuit variations of the varicap diode. Circuits of this type are particularly well suited for voice communication in narrowband FM.

The power output of an FM transmitter is dependent primarily on its application. In some systems the oscillator output can be applied directly to the antenna for radiation. In other systems the power level must be raised to a higher level. The output of the oscillator would be applied to linear amplifiers for processing. The amplification function would be primarily the same as that achieved by CW and AM systems. We will not repeat the discussion of RF power amplification for FM transmitters. The primary difference in FM power amplification is the resonant frequency of the tuned circuits. FM usually operates in the VHF band which generally calls for smaller capacitance and inductance values in the resonant circuits.

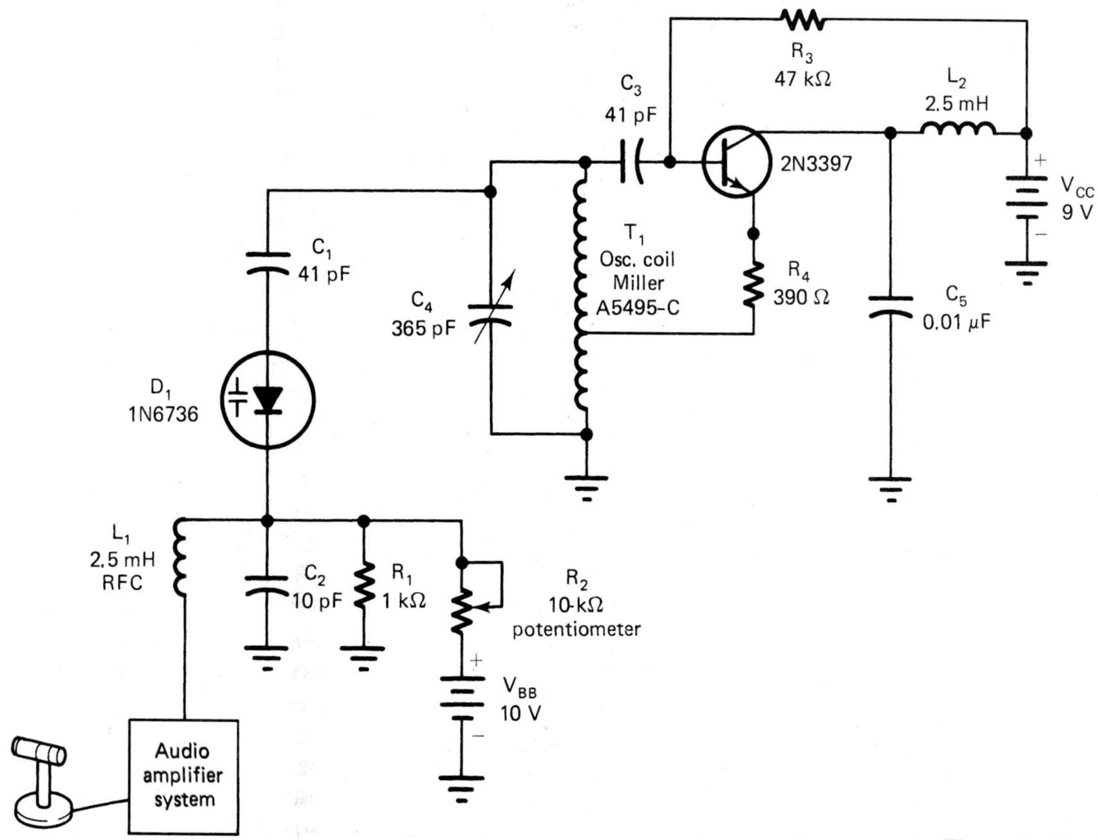

Figure 16–32. Direct FM transmitter.

FM RECEIVERS. FM signal reception is achieved in basically the same way as AM and CW. An FM superheterodyne circuit is designed to respond to frequencies in the VHF band. Commercial FM signal reception is in the 88- to 108-MHz range. Higher-frequency operation generally necessitates some change in the design of antenna, RF amplifier, and mixer circuits. These differences are due primarily to the increased frequency rather than the FM signal. The RF and IF sections of an FM receiver are somewhat different. They must be capable of passing a 200-kHz bandwidth signal instead of the 10-kHz AM signal. The most significant difference in FM reception is in the demodulator. This part of the receiver must pick out the modulating component from a signal that changes in frequency. In general, this circuit is more complicated than the AM detector. The AF amplifier section of an FM receiver is generally better than an AM receiver. It must be capable of amplifying frequencies of 30 Hz to 15 kHz. Figure 16–33 shows a block diagram of an FM superheterodyne receiver. Some differences in FM reception are indicated by each functional block.

FM DEMODULATION. The demodulator of an FM receiver is the primary difference between AM and FM superheterodyne receivers. A number of demodulators have been used to achieve this receiver function. Today the ratio detector seems to be the most widely used. The circuit is an outgrowth of earlier FM discriminators. Solid-state diodes and transformer design have made the ratio detector a practical circuit.

A basic ratio detector is shown in Fig. 16–34. The operation of this detector is based on the phase relationship of an FM signal that is applied to the input. This signal comes from the IF amplifier section of the receiver (see Fig. 16–33). It is 10.7 MHz and deviates in frequency. Design of the circuit causes two IF signal components to appear in the secondary winding of the transformer. One part of this component is coupled by capacitor C_3 to the center connection of the transformer. This resulting voltage is considered to be V_1. Note the location of V_1 in the circuit and voltage-line diagrams.

The second IF signal component is developed inductively by transformer action. Design of the secondary winding causes two voltage values to be developed. These voltages are labeled V_2 and V_3 in the circuit and voltage-line diagrams—they are of equal amplitude and 180° out of phase with each other. The center-tap connection of the secondary winding is used as the common reference point for these voltage values.

The resultant secondary voltages are based on the combined component voltage values of V_1, V_2, and V_3. V_{D1} and

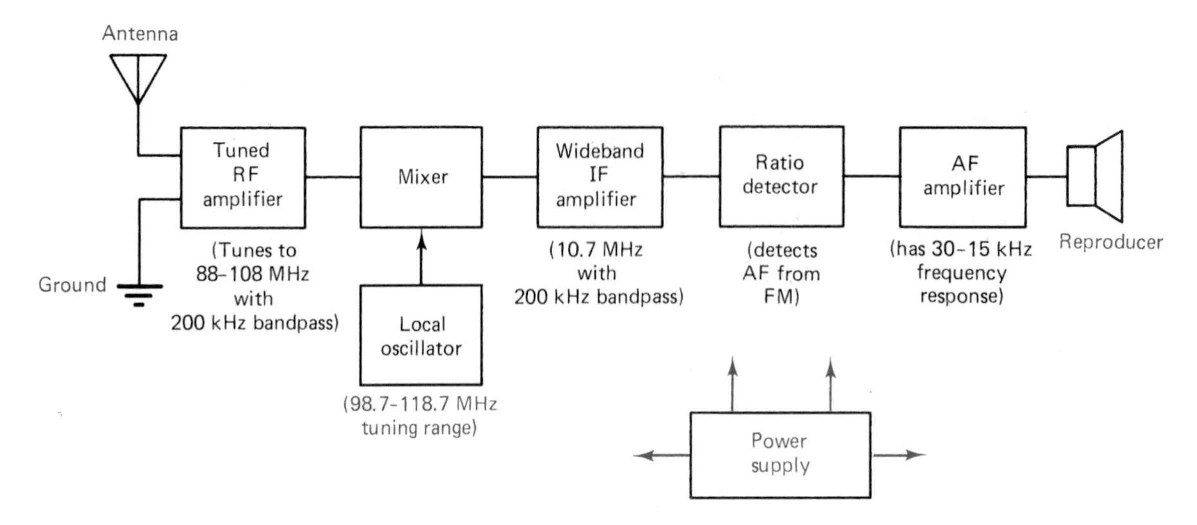

Figure 16–33. Block diagram of an FM receiver.

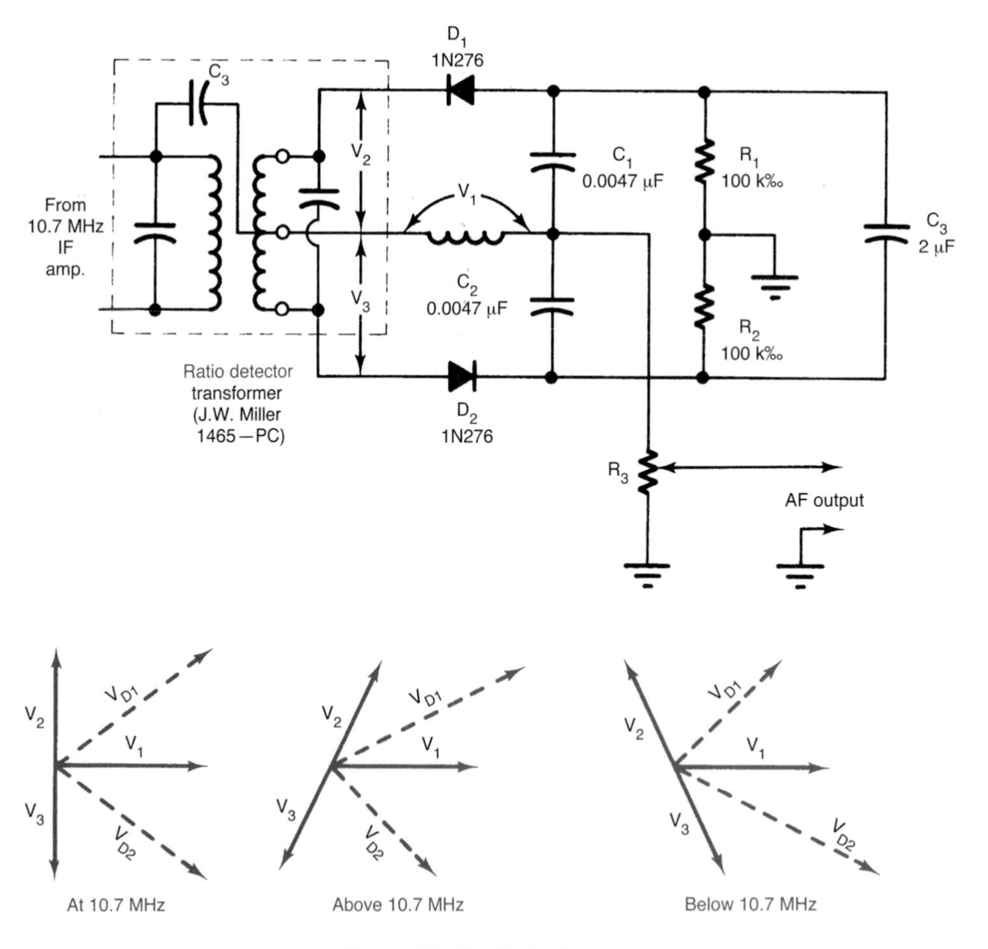

Figure 16–34. Ratio detector.

V_{D2} are the resulting secondary voltages. These two voltage values will appear across diodes D_1 and D_2. The developed voltage for each diode is based on the phase relationship of the two IF components. Note the location of V_{D1} and V_{D2} in the circuit and the voltage line diagram.

Operation of the ratio detector is based on voltage developed by the transformer for diodes D_1 and D_2. With no modulation applied to the FM carrier, the transformer will have a 10.7-MHz IF applied to the transformer. The developed voltage values are shown by the center-frequency voltage-line diagram. This method of voltage display is generally called a *vector diagram.* A vector shows the relationship of voltage values (line length) and their phase relationship (direction). At the center frequency, note that the resultant diode voltage values (V_{D1} and V_{D2}) are of the same length, meaning that each diode will receive the same voltage value. D_1 and D_2 will conduct an equal amount of current.

When the carrier swings above the center frequency, it causes the IF to swing above its resonant frequency. Note the resultant vector for this change in frequency. V_{D1} is longer than V_{D2}, meaning that D_1 will conduct more current than D_2. In the same regard, note how the vector changes when the frequency swings below resonance. This condition will cause D_1 to be less conductive and D_2 to be more conductive. Essentially, this means that the IF signal is translated into different diode voltage values.

Let us now see how the input RF voltage is used to develop an AF signal. The two diodes of the ratio detector are connected in series with the secondary winding and capacitors C_1 and C_2. For one alternation of the input signal both diodes are reverse biased. No conduction occurs during this alternation [see Fig. 16–35(a)]. For the next alternation both diodes are forward biased. The input signal voltage is then rectified [see Fig. 16–35(b)]. Essentially, this means that the incoming signal is changed into a pulsating waveform for one alternation.

Figure 16–36 shows how the ratio detector responds when the input signal deviates above and below the center frequency. Keep in mind that conduction occurs only for one alternation of the input. Figure 16–36(a) shows how the circuit will respond when the input is at its 10.7-MHz center frequency. Each diode has the same input voltage value for this condition of operation. Capacitors C_1 and C_2 will charge to equal voltages as indicated. With respect to ground, the output voltage will be 0 V. Note this point on the output voltage waveform in Fig. 16–36(d).

Assume now that the input IF signal swings to 10.8 MHz as shown in Fig. 16–36(b). This condition causes D_1 to receive more voltage than D_2. C_1 charges to 12 V while C_2 charges to -1 V. With respect to ground, the output voltage rises to $+1$ V. Note this point on the output voltage waveform in Fig. 16–36(d).

The input IF signal then swings to 10.6 MHz as shown in Fig. 16–36(c). This condition causes D_2 to receive more voltage than D_1. C_2 charges to -2 V while C_1 charges to $+1$ V. With respect to ground, the output voltage drops to -1 V. See this point on the output voltage waveform of Fig. 16–36(d).

The output of the ratio detector is an AF signal of 2-V p-p. This signal corresponds to the frequency changes placed on the carrier at the transmitter. In effect, we have recovered the AF component from the FM carrier signal. This signal can then be amplified by the AF section for reproduction.

16.6 TELEVISION COMMUNICATION

Nearly everyone has had an opportunity to view a television communication system in operation. This communication process plays an important role in our lives. Few people spend a day without watching television. It is probably the most significant application of electronics today.

The signal of a television system is quite complex. It is made up of a number of unique parts or components. Basically, the transmitted signal has a picture carrier and a sound carrier. These two signals are transmitted simultaneously from a single antenna. The picture carrier is amplitude modulated. The sound carrier is frequency modulated. A television receiver intercepts these two signals from the air. They are tuned, amplified, demodulated, and ultimately reproduced. Sound is reproduced by a speaker. Color and picture information are reproduced by a picture tube. Television is primarily a one-way communication system with a central transmitting station and an infinite number of receivers. A simplification of the television communication system is shown in Fig. 16–37 on page 534.

The television camera of our system is basically a transducer. It changes the light energy of a televised scene into electrical signal energy. Light energy falling on a highly sensitive surface varies the conduction of current through a resistive material. The resultant current flow is proportional to the brightness of the scene. An electron beam scans horizontally and vertically across the light-sensitive surface. Traditionally, the camera tube of a TV system has been a videcon. Videcons are now being replaced with solid-state tubes called *charge-coupled devices* or CCDs. These devices are smaller, use less power, and are more sensitive to low light levels. CCDs are also used in portable video recorders or camcorders.

Figure 16–38 on page 535 shows a simplification of the television camera tube. Note that a complete circuit exists between the cathode and power supply. Electrons are emitted from the heated cathode. These are formed into a thin beam and directed toward the back side of the photoconductive

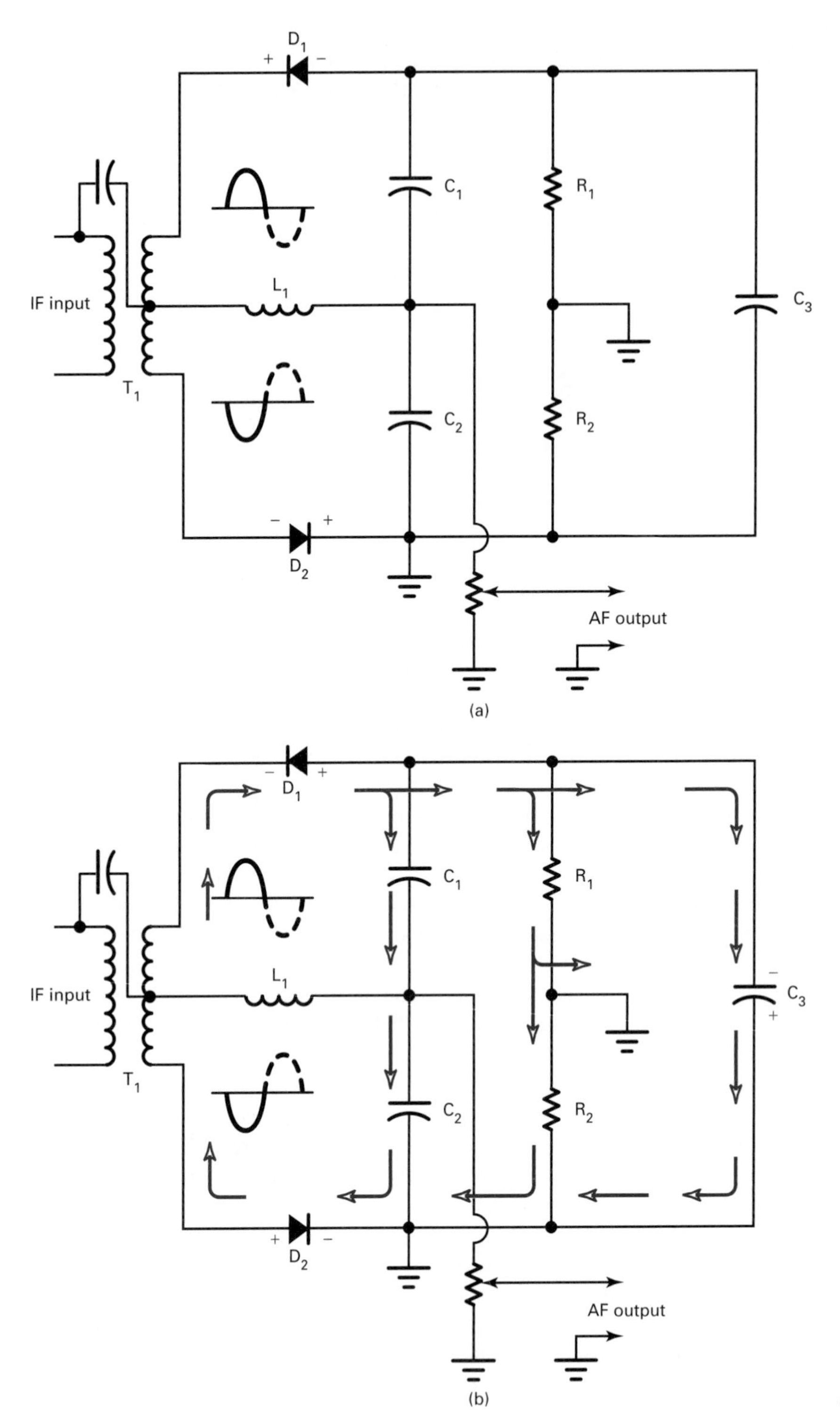

Figure 16–35. Diode conduction of a ratio detector: (a) nonconduction; (b) conduction.

layer. Conduction through the layer is based on the intensity of light from the scene being televised. A bright or intense light area becomes low resistant. Dark areas have a higher resistance. As the electron beam scans across the back of the photoconductive layer, it sees different resistance values. Conduction through the layer is based on this resistance. A discrete area with low resistance causes a large current through the layer. Dark areas cause less current flow. Current flow is directly related to the light intensity of the televised scene. Output current flow appears across the load resistor (R_L). Voltage developed across R_L is amplified and ultimately used to modulate the picture carrier. In practice, the devel-

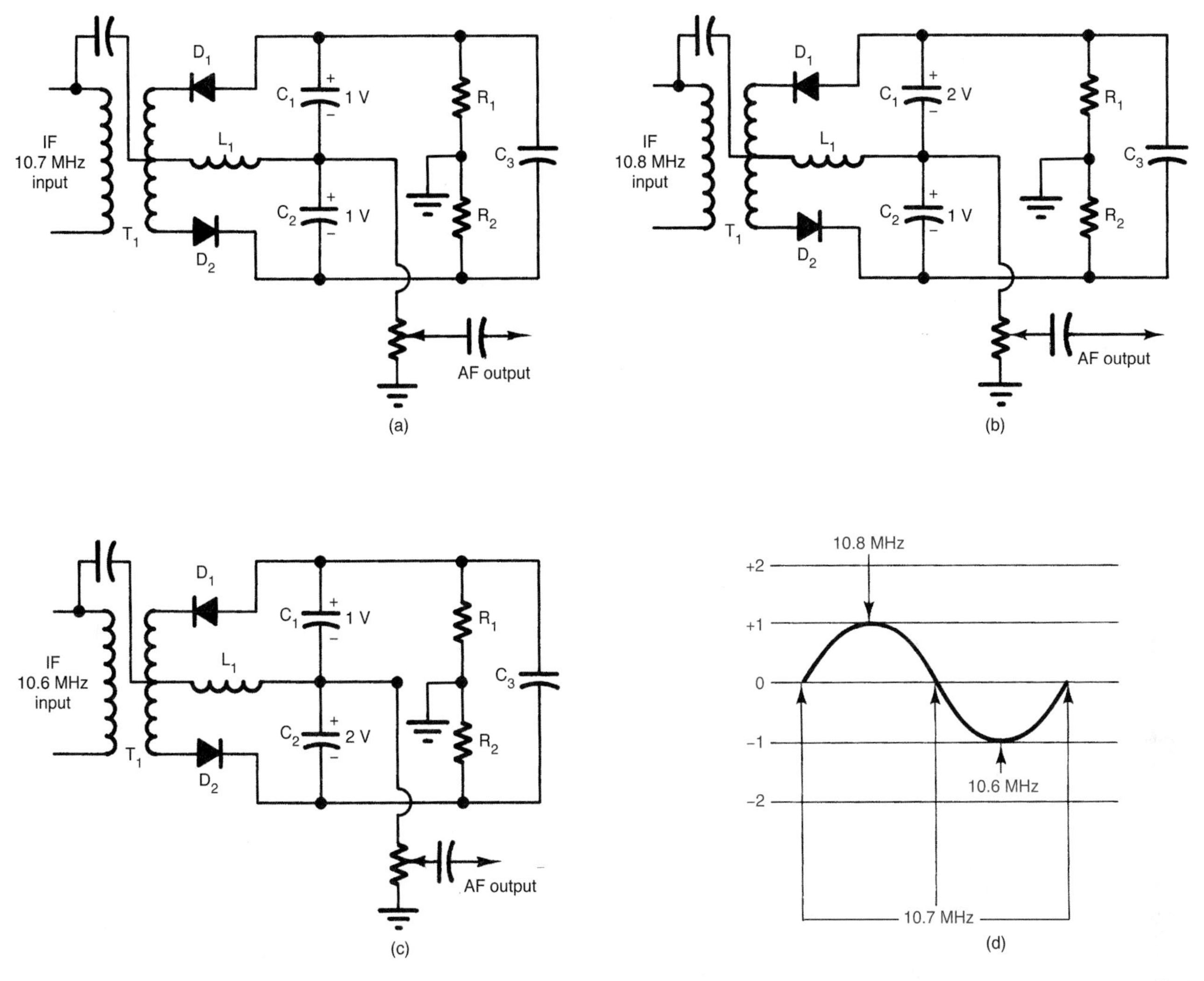

Figure 16–36. Ratio detector response to frequency: (a) center frequency; (b) above center; (c) below center; (d) output waveform.

oped camera tube voltage is called a video signal. *Vide* means "to see" in Latin. A camera tube sees things electronically.

The scene being televised by a camera tube must be broken into small parts called *picture elements* or *pixels*. For this to occur, it is necessary to scan the light-sensitive surface of a camera tube with a stream of electrons, similar to reading a printed page. Letters, words, and sentences are placed on the page by printing. We do not determine what is on a printed page at one instant. Our eyes must scan the page starting at the upper left-hand corner one line at a time. They move left to right, drop down one line and quickly return to the left, then scan right again for the next line. The process continues until all lines are scanned.

In a similar way the electron beam of a camera tube scans the back surface of the photoconductive layer. The electron beam is deflected horizontally and vertically by an electromagnetic field. A coil fixture known as the *deflection yoke* is placed around the neck of the camera tube. This coil deflects the electron beam. Current flow in the deflection yoke is varied so that the field rises to a peak value, then drops to its starting value. Figure 16–39 shows the deflection yoke current and the resultant electron beam scanning action. Notice that each line has a trace and a retrace time. During the trace period, the line is scanned from left to right, which takes a rather large portion of the complete sawtooth wave. Retrace occurs when the beam returns from right to left. Notice that this takes only a small portion of the total waveform. The same condition applies to the vertical sweep waveform.

Figure 16–39 shows another unique difference in the scanning lines. During the trace time, the scanning line is solid which indicates that the electron beam is conducting during this time. It also shows a broken line during the retrace period which indicates that the electron beam is nonconductive during this period. In effect, conduction occurs during the trace time and no conduction occurs during retrace. This same condition applies to the vertical trace and retrace time.

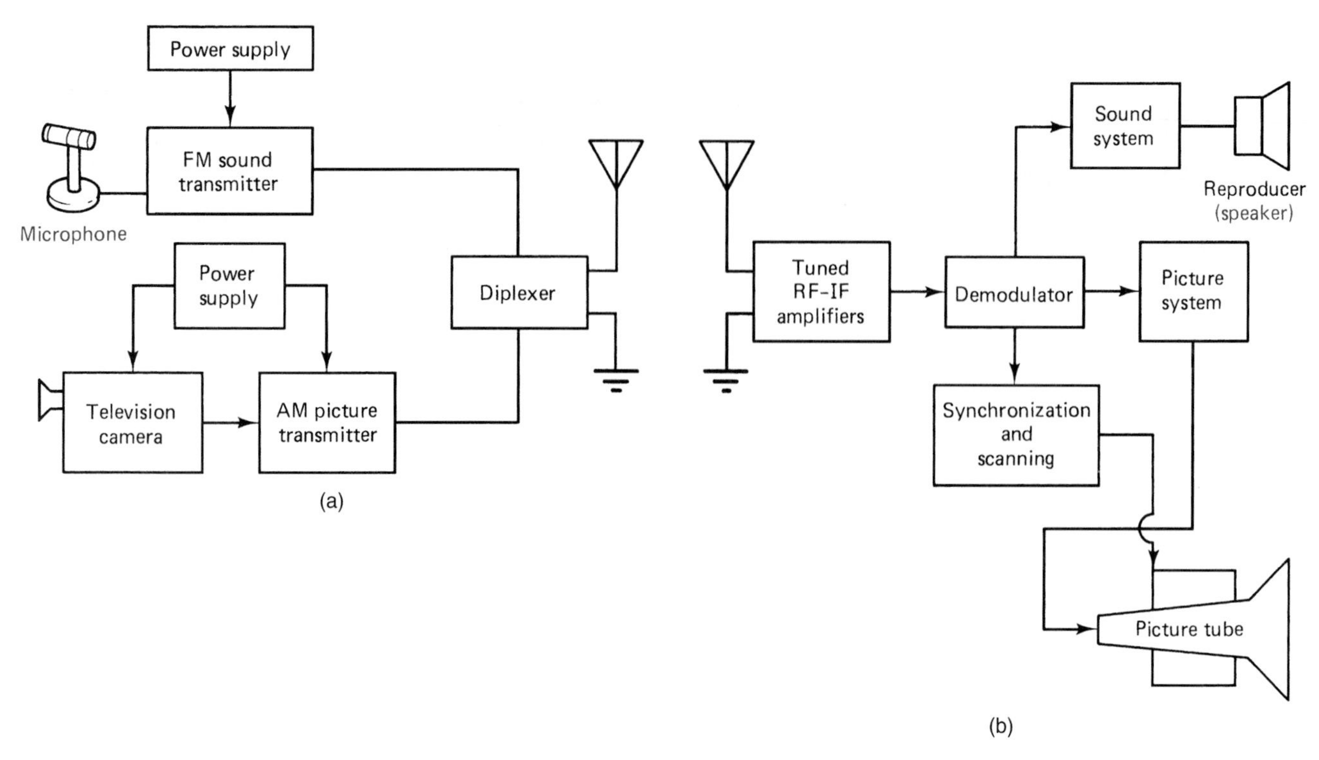

Figure 16–37. TV communication system: (a) transmitter; (b) receiver.

The scanning operation of a camera tube requires two complete sets of deflection coils. One set is for horizontal deflection and one is for vertical. The current needed to produce deflection comes from two sawtooth oscillators—a vertical blocking oscillator and a horizontal multivibrator could be used for this operation. In U.S. television, the horizontal sweep frequency is 15,750 Hz and the vertical frequency is 60 Hz.

To produce a moving television scene, there must be at least 30 complete pictures produced per second. In a TV system a complete picture is called a *frame*. The U.S. television system has 525 horizontal lines in a frame. These lines are not scanned progressively, such as 1, 2, 3, 4, 5, and so on. They are, however, divided into two fields. One field contains 262.5 odd-numbered lines. Lines 1, 3, 5, 7, 9, 11, and so on, make up the odd field. The other field has 262.5 even-numbered lines. Lines 2, 4, 6, 8, 10, and so on, are included in the even field. A complete picture or frame has one odd-lined field and one even-lined field.

Picture production in a TV system employs interlace scanning. To produce one frame, the odd-line field is scanned first. After scanning all the odd-numbered lines, the electron beam is deflected from the bottom position to the top. The even-numbered-line field is then scanned. A complete frame has 262.5 odd-numbered lines and 262.5 even-numbered lines. The odd-line field starts with a complete line and ends with a half line. The even field starts with a half line and ends with a complete line.

The picture repetition rate of U.S. television is 30 frames per second. Because two fields are needed to produce one frame, the vertical frequency is 2×30, or 60 Hz. This particular frequency was chosen to coincide with the ac power line frequency. In some foreign countries the vertical frequency is 50 Hz.

In television signal production, the vertical and horizontal sweep circuits must be synchronized properly to produce a picture. The signal sent out by the transmitter must contain synchronization or sync information. This signal is used to keep the oscillator of the receiver in step with the correct signal frequency. Separate generators are used to develop the sync signal which is added to the video signal developed by the camera.

Picture Signal

The picture signal of a TV transmitter contains several important parts. Each part of the signal plays a specific role in the operation of the system. The video signal, for example, is developed by the camera tube. It represents instantaneous variations in scene brightness. Figure 16–40(a) shows the video signal for one horizontal line.

The video signal of a television system has negative picture phase, which means that the highest amplitude part of the signal corresponds to the darkest picture area. Bright picture areas have the lowest amplitude level. Signal levels that are 75% of the total amplitude range are considered to be in the black region. All light disappears in this region. Signal-level amplitude percentages are shown in Fig. 16–40(a).

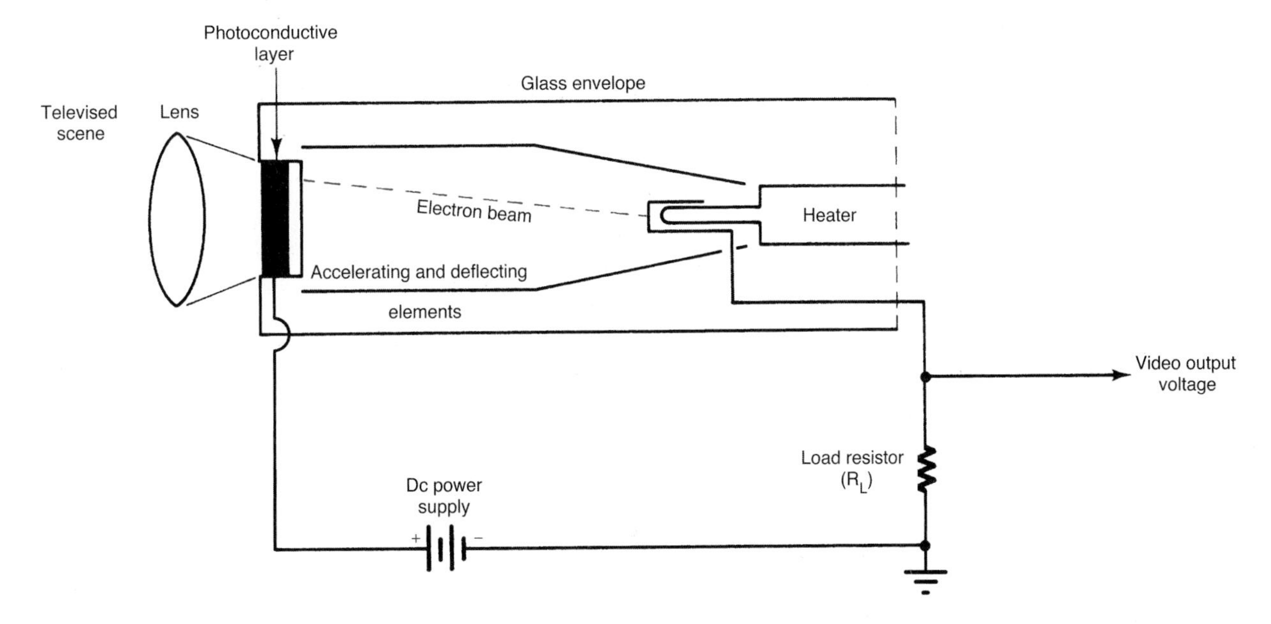

Figure 16–38. Simplification of a TV camera tube.

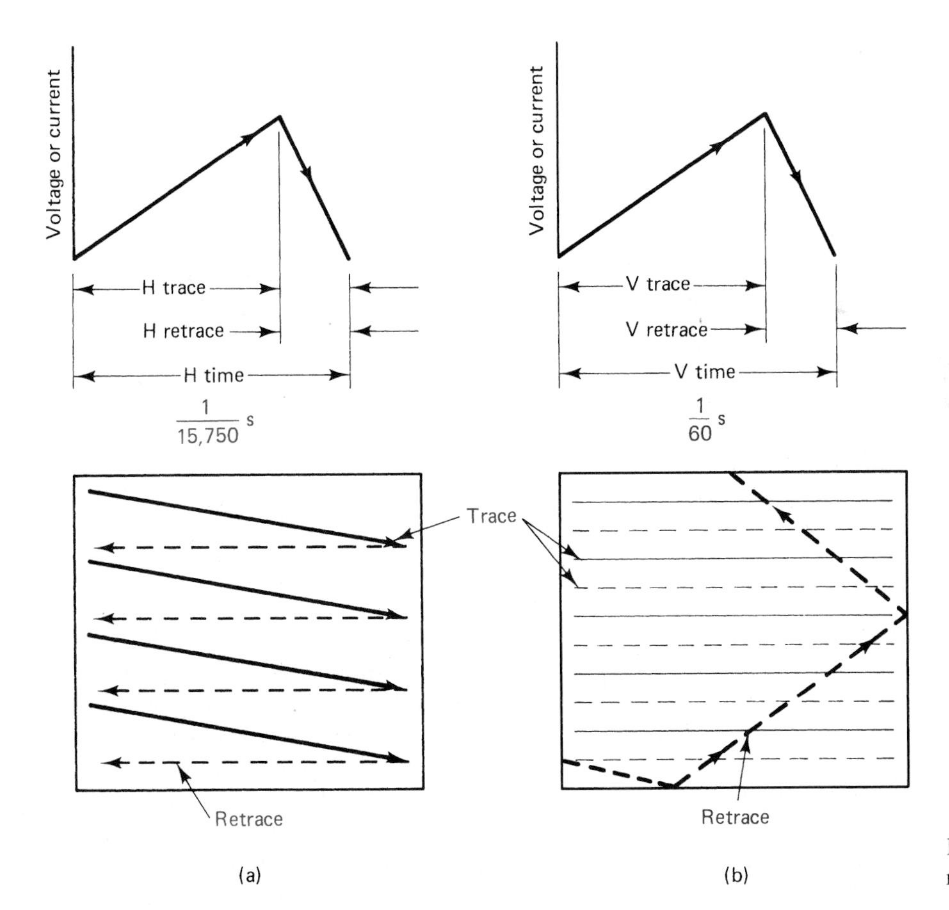

Figure 16–39. Scanning and sweep signals: (a) horizontal; (b) vertical.

In addition to the video information, the picture signal must also provide some way of cutting off the electron beam at certain times. When scanning occurs, the electron beam is driven to cut off during the retrace period. Horizontal retrace occurs at the end of one line and vertical retrace occurs at the end of each field. A rectangular pulse of sufficient amplitude is needed to reach cutoff. This condition permits the electron beam to retrace without producing unwanted lines. This part of the signal is called *blanking*. A composite signal has both horizontal and vertical blanking pulses. Figure 16–40(a)

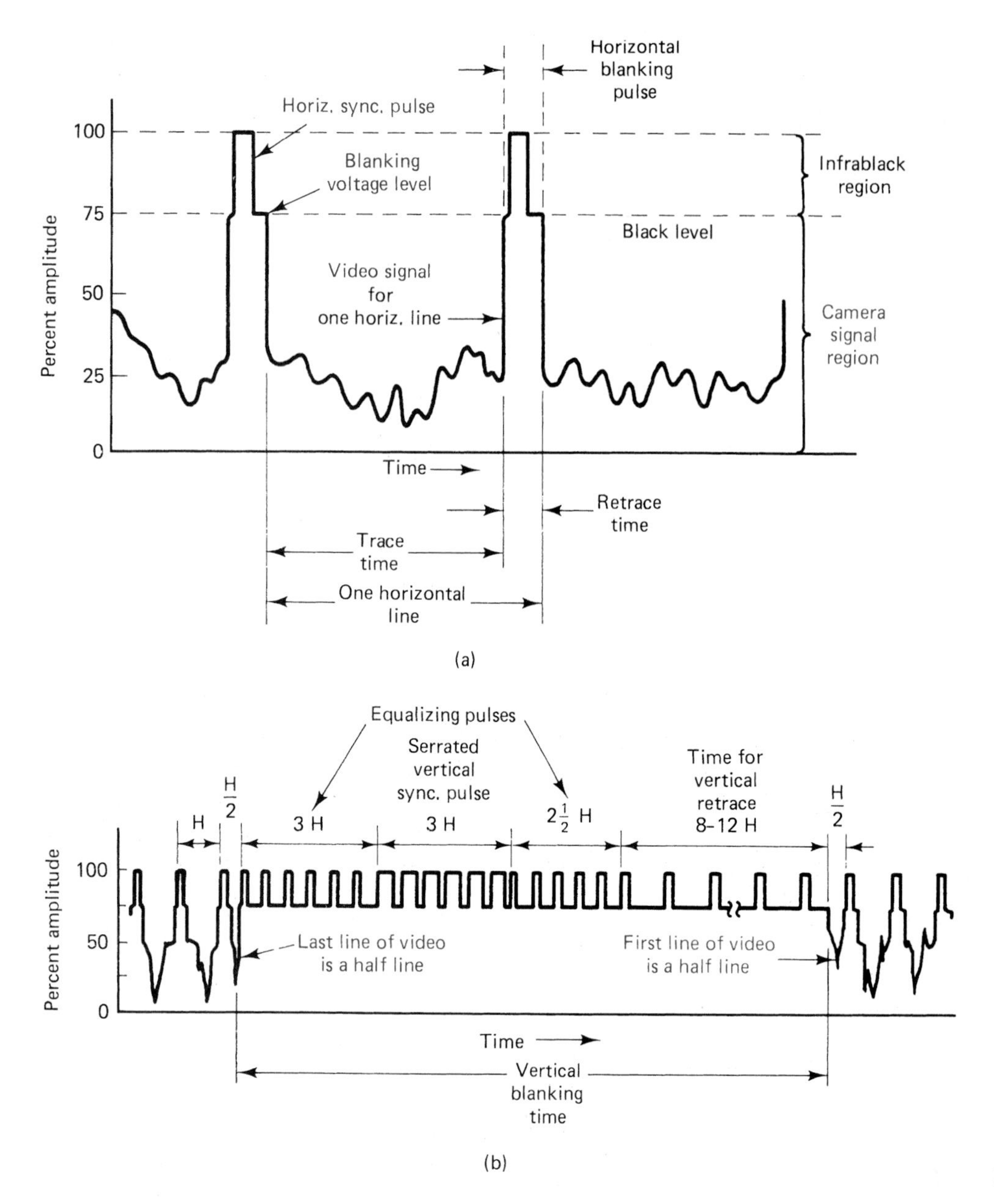

Figure 16–40. Composite TV picture signal.

shows the location of a horizontal blanking pulse. Vertical blanking occurs after 262.5 horizontal lines. Figure 16–40(b) shows the vertical blanking time.

A composite TV picture signal also has vertical and horizontal synchronization pulses. These pulses ride on the top of the blanking pulses. Horizontal sync pulses are shown in both parts of Fig. 16–40. The serrated pulses of Fig. 16–40(b) provide continuous horizontal sync during the vertical retrace time. The vertical sync pulse is made up of six rectangular pulses near the center of the vertical blanking time. The width of a vertical sync pulse is much greater than that of the horizontal sync pulse. All these pulses, plus blanking and the video signal, are described as a compositive picture signal.

Television Transmitter

A television transmitter is divided into two separate sections or divisions that feed outputs into a common antenna. The video section is responsible for the picture part of the signal. A crystal oscillator is used for carrier-wave generation. As a general rule, the frequency is multiplied to bring it up to an allocated channel in the VHF or UHF band. An intermediate power amplifier and a final power amplifier follow the last multiplier. The modulating component is a composite picture signal. It contains video, blanking pulses, sync, and equalizing pulses. The composite signal is amplified and ultimately applied to the final power amplifier. The final power amplifier is amplitude modulated by the composite picture signal. This section of the transmitter

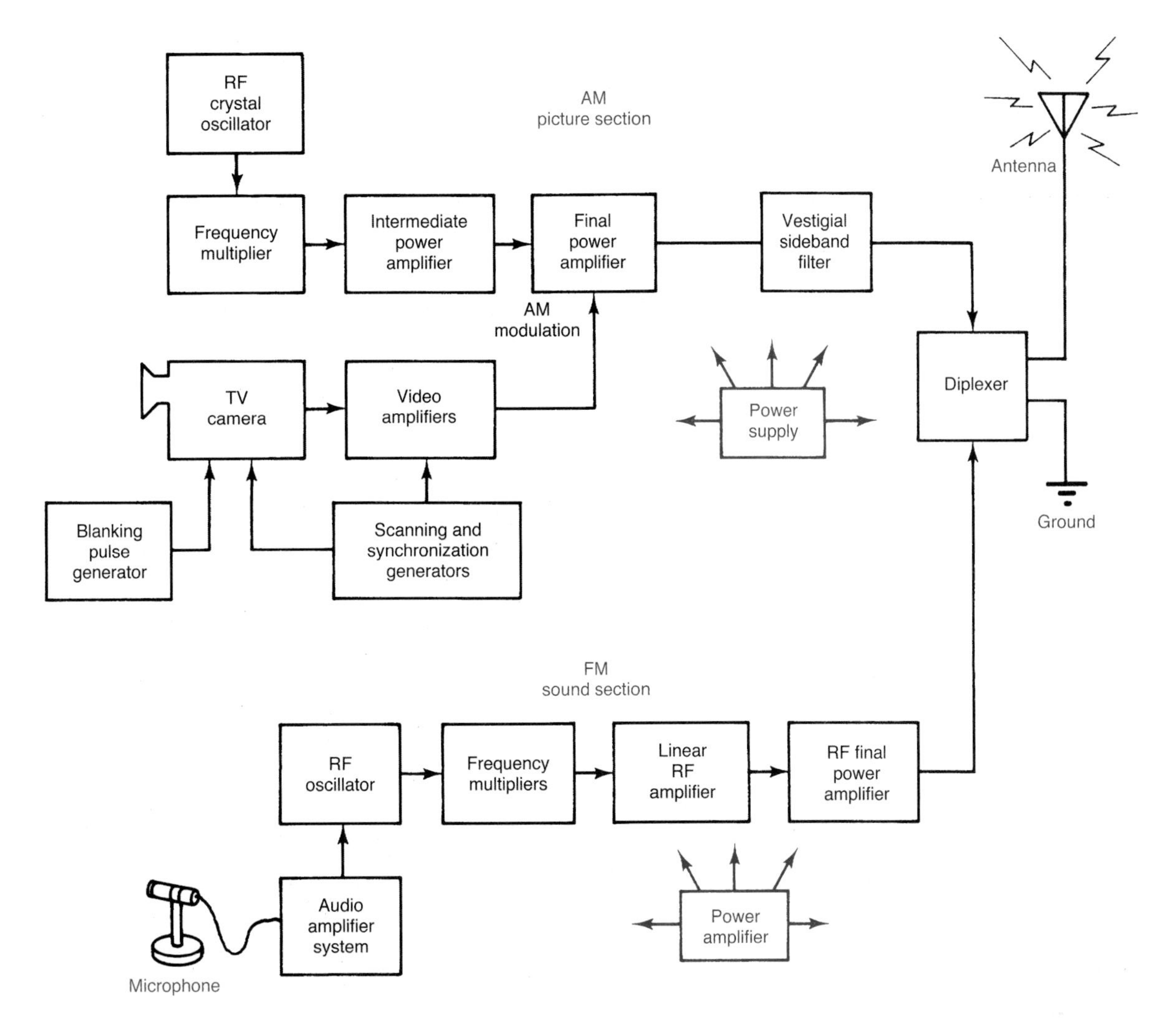

Figure 16–41. Monochrome TV transmitter.

is essentially the same as that of a commercial AM station. The obvious differences are frequency and power output. Figure 16–41 shows a block diagram of a black-and-white TV transmitter.

A unique difference in TV and commercial AM transmitter circuitry appears after the final power amplifier. The TV output signal is applied to a vestigial sideband filter. This filter is designed to remove all lower sideband frequencies 1.25 MHz below the carrier frequency. The entire upper sideband of the signal is transmitted. A large portion of the lower sideband is purposely suppressed to reduce the frequency occupied by a channel. With the lower sideband suppressed, the bandwidth of a TV channel is 6 MHz.

The vestigial sideband signal is then applied to a diplexer, which is a filter circuit that isolates the picture and sound carriers. Essentially, the sound carrier will not pass into the picture section and the picture carrier will not pass into the sound section, thus preventing undesirable interaction between the two carrier signals.

The sound section of a TV system is primarily an FM transmitter similar to a commercial FM transmitter. The center frequency of the FM carrier is always 4.5 MHz above the picture carrier. Carrier deviation is ±25 kHz in a TV system. *Modulation* is normally applied to the oscillator of the FM sound system. The remainder of the sound section is similar to that of a commercial FM transmitter. See the FM sound section of the transmitter in Fig. 16–41.

Television Receiver

A television receiver is designed to intercept the electromagnetic waves sent out by the transmitter and use them to develop sound and a picture. The received signals are in the VHF or UHF band. The FCC has allocated a 6-MHz bandwidth for each TV channel. Channels 2 through 13 are in the VHF band. These frequencies are from 54 to 216 MHz. Channels 14 through 83 are in the UHF band which ranges from 470 to 890 MHz. All channels in the immediate area

21. Describe the negative picture phase concept.
22. Why is horizontal and vertical blanking important in picture production?
23. What is the luminance signal?
24. How does a color transmitter differ from a monochrome transmitter?
25. Explain how a color signal modulates the RF carrier.
26. Describe how a color signal develops a picture in the receiver.
27. How are fiber optic systems used in communications?

STUDENT ACTIVITIES

• *CW Transmitter* or *Complete Activity 16–1, page 317 in the manual.*

1. Construct the CW transmitter of Fig. 16–8.
2. Wrap a few turns of wire around T_1. Connect an oscilloscope to the coil.
3. Close the code key and observe the waveform. Open and close the key while observing the waveform.
4. Place an AM radio receiver near the antenna coil. Turn it on and tune the receiver for an audio tone at the speaker.
5. Key the transmitter to see if it is producing the tone.
6. If the signal is close to a local station, adjust C_1 to a different frequency.
7. Send several messages in Morse code.
8. Test the transmitting range of the system.
9. Retain the CW transmitter circuit for additional activities.

• •

ANALYSIS

1. How is coded information transmitted by a CW transmitter?
2. What are the fundamental stages or parts of a CW transmitter?
3. What type of signal is radiated from a CW transmitter?

• *Beat Frequency Oscillator*

1. Construct the BFO of Fig. 16–17.
2. Apply power to the circuit. Measure and record the dc voltages at the emitter, base, and collector of Q_1.
3. With an oscilloscope, observe the output waveforms.
4. Make a sketch of the observed waveforms.
5. Retain the BFO oscillator circuit.

• *Heterodyne Detection*

1. Construct the heterodyne detector circuit of Fig. 16–16.
2. Use the BFO of Fig. 16–17 for the F_2 input.
3. Use the CW transmitter of Fig. 16–8 for the CW input. Only one CW transmitter is needed for the entire class.

4. Place the antenna coil near the CW transmitter. Key the transmitter.
5. Tune the detector input and BFO for an AF output signal. An oscilloscope may be used to observe the signal.
6. It may be possible to hear the detected signal with a set of high-impedance earphones. Remove R_3 and connect the earphones in its place.
7. Adjust the BFO for a zero beat and for frequencies above and below the beat frequency.

• AM Transmitter or Complete Activity 16–2, page 319 in the manual.

1. Construct the AM transmitter of Fig. 16–22.
2. Turn on the switch and test the circuit for operation.
3. Connect 4 to 5 ft of wire to the antenna connection.
4. Tune an AM receiver to the transmitted signal.
5. The transmitter can be tuned to a different frequency by adjusting C_4.
6. Record the operating frequency.
7. Retain an operating AM transmitter for the next activity.

• • ANALYSIS

1. What signal of a modulator circuit represents the carrier wave?
2. What signal represents the modulating wave?
3. If a 50-V peak-to-peak RF carrier changes in amplitude from 75-V p-p to 25-V p-p when modulated, what is the percent of modulation?
4. What causes an AM signal to be overmodulated?
5. If the modulated signal is stronger than the RF signal of the transmitter, what occurs?
6. How does a crystal microphone produce electrical energy from sound waves?
7. Make a sketch showing (a) an unmodulated RF wave, (b) a 50% modulated wave, (c) a 100% modulated wave, and (d) an overmodulated RF wave.

• AM Crystal Radio Receiver

1. Construct the crystal radio receiver of Fig. 16–24. *Note:* T_1 may be constructed by winding 15 turns and 90 turns of No. 28 wire on a 1-in.-diameter paper tube.
2. Place the receiving antenna close to the AM transmitting antenna of Fig. 16–22.
3. Ask another person to speak into the transmitter microphone.
4. Tune the receiver to the transmitted signal.
5. If an outside antenna is available, connect the receiver to the outside antenna and a ground terminal.
6. Tune in a local AM station.

• AM Receiver Circuitry or Complete Activity 16–3, page 323 in the manual.

1. If a superheterodyne AM receiver is available, you may want to investigate its circuit.
2. Locate such things as the antenna coil, tuning circuit, IF transformers, volume control, and audio amplifier circuit.

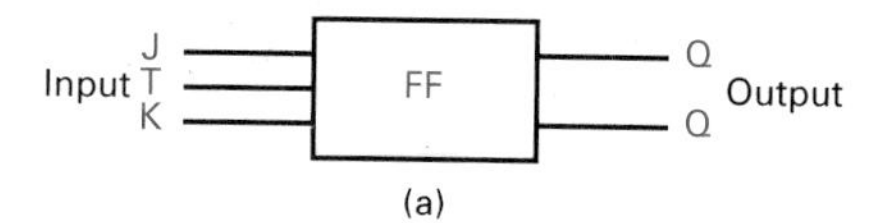

(a)

Applied inputs			Previous outputs		Resulting outputs		
J	T	K	Q	$\overline{Q}$	Q	$\overline{Q}$	
0	0	0	0	1	0	1	
0	0	1	0	1	0	1	
0	1	1	0	1	0	1	
1	0	0	0	1	0	1	
1	1	0	0	1	1	0	
1	1	1	0	1	1	0	← Toggle state
0	0	0	1	0	1	0	
0	0	1	1	0	1	0	
0	1	1	1	0	0	1	
1	0	0	1	0	1	0	
1	1	0	1	0	1	0	
1	1	1	1	0	0	1	← Toggle state

(b)

Figure 17–11. *J-K* flip-flop: (a) symbol; (b) truth table.

applied to the input of a counter. These pulses must initiate a state change in the circuit when they are received. The circuit must also be able to recognize where it is in the counting sequence at any particular time, which requires some form of memory. The counter must also be able to respond to the next number in the sequence. In digital electronic systems, flip-flops are primarily used to achieve counting. This type of device is capable of changing states when a pulse is applied, has memory, and will generate an output pulse.

Several types of counters are used in digital circuitry today. Probably the most common of these is the binary counter. This particular counter is designed to process two-state information. *J-K* flip-flops are commonly used in binary counters.

Refer now to the single *J-K* flip-flop of Fig. 17–11. In its toggle state, this flip-flop is capable of achieving counting. First, assume that the flip-flop is in its reset state, which would cause Q to be 0 and $\overline{Q}$ to be 1. Normally, we are only concerned with Q output in counting operations. The flip-flop is now connected for operation in the toggle mode of line 6. J and K must both be made high or in the "1" state. When a pulse is applied to the T, or clock input, Q changes to 1, meaning that with one pulse applied, a 1 is generated in the output. The flip-flop has therefore counted one time. When the next pulse arrives, Q resets or changes to 0. Essentially, this means that two input pulses produce only one output pulse—a divide-by-two function. For binary numbers, counting is achieved by several divide-by-two flip-flops.

To count more than one pulse, additional flip-flops must be employed. For each flip-flop added to the counter, its capacity is increased by the power of 2. With one flip-flop the maximum count was 2^0, or 1. For two flip-flops it would count two places, such as 2^0 and 2^1, and would reach a count of 3 or a binary number of 11. The count would be 00, 01, 10, and 11. The counter would then clear and return to 00, and in effect, count four pulses. Three flip-flops would count three places, or 2^0, 2^1, and 2^2, and would permit a total count of eight pulses. The binary values are 000, 001, 010, 011, 100, 101, 110, and 111. The maximum count is 7, or 111. Four flip-flops would count four places, or 2^0, 2^1, 2^2, and 2^3. The total count would make 16 pulses. The maximum count would be 15, or the binary number 1111. Each additional flip-flop would cause this number to increase one binary place.

Figure 17–12 shows a 4-bit binary counter. Four *J-K* flip-flops are included in this circuit. Sixteen counts can be made with this circuit. The counts are listed under the diagram. A light-emitting diode (LED) is used as an output display for each flip-flop. When a particular LED is on, it indicates the "1" state. An off LED indicates the "0" state.

Operation of the counter is based on the number of clock pulses applied to its input. The flip-flops will only trigger or change states on the negative-going part of the clock pulse. Note this on the input waveform. The output of flip-flop 1 (FF_1) will change back and forth between 1 and 0 for each pulse. One pulse applied to the input of FF_1 will cause its Q output to be 1. The 2^0 LED will turn on, indicating a "1" count. The display will be 0001. The second pulse will turn off FF_1. The negative-going output of Q will turn on FF_2. The Q output of FF_2 will then become a 1, which will light LED-2 or the 2^1 display. The entire display will now be 0010, or the binary 2 count. The next pulse will trigger FF_1. LED-1 will light again because the Q output of FF_1 is 1 again. The display will show 0011, or the binary 3 count. The fourth pulse will reset FF_1, and FF_2 and set FF_3. The Q output of FF_3 will go positive and turn on LED-3. The display will show 0100, or the binary 4 count. The operation is then repeated, the counts being 0101, 0110, and 0111. The next pulse will clear FF_1, FF_2, and FF_3. The state change of FF_3 will set FF_4. The count will now be 1000, or the binary 8 count. The operation is again repeated. The count display will be 1001, 1010, 1011, 1100, 1101, 1110, and 1111. This reaches the maximum count that can be achieved with four flip-flops. The next pulse will clear all four flip-flops. The LED display will then be 0000. The entire process will repeat itself starting with the first pulse.

A 4-bit IC binary counter is shown in Fig. 17–13. This particular circuit achieves the same thing as the counter of Fig. 17–12. Four *J-K* flip-flops are built into this chip. The Q output of each flip-flop appears at terminals *A, B, C,* and *D.* These serve as the 2^0, 2^1, 2^2, and 2^3 or 1, 2, 4, and 8 outputs. LEDs connected to each output will respond as the display.

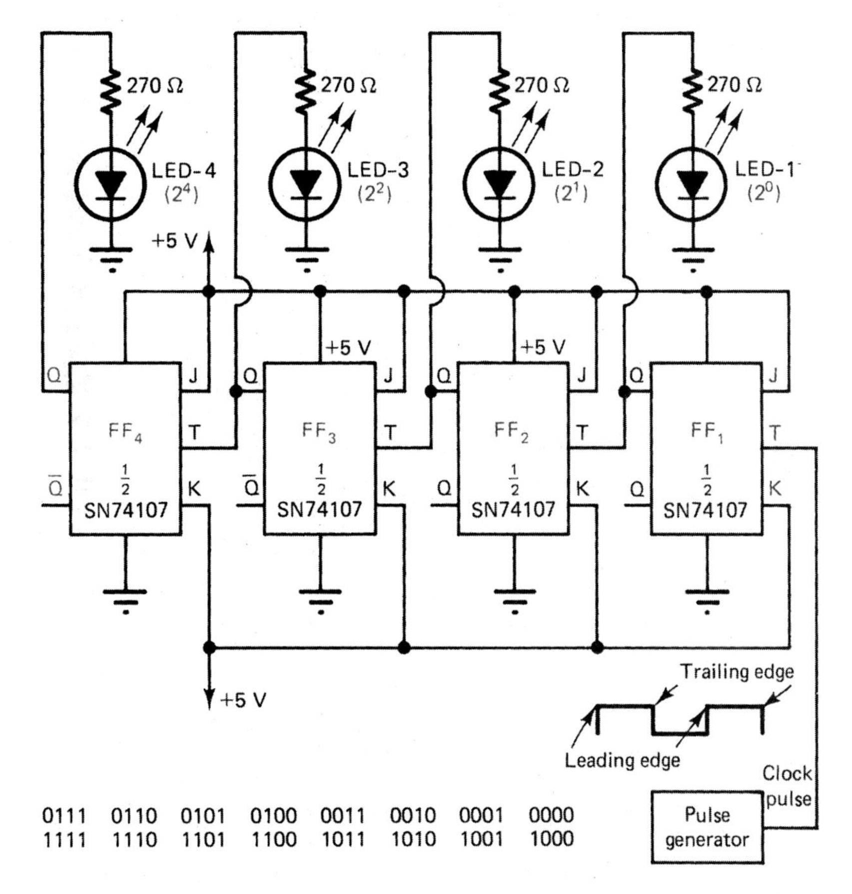

Figure 17–12. 4-bit binary counter.

This chip essentially has one input and four outputs. This chip employs interconnected transistors in its operations. It belongs to the transistor–transistor logic family of ICs.

17.7 DECADE COUNTERS

Decade counters are used widely in digital electronic systems. This type of counter has 10 counting states in its output. The output is in binary numbers. In practice, this is more commonly called a binary-coded decimal (BCD) counter. Its output is 0000, 0001, 0010, 0011, 0100, 0101, 0110, 0111, 1000, and 1001. It is used to change binary numbers into decimal form. One BCD counter is needed for each place of a decimal number. One counter of this type can only count to 9. Two would be required to count to 99. Three are needed for 999. The place value increases with each additional counter.

BCD counters are primarily binary counters that clear all flip-flops after the 9 count, or 1001. The first seven counts appear in the normal binary sequence. Figure 17–14 shows how the *J-K* flip-flops of the BCD counter are connected. Notice that FF_A, FF_B, and FF_C are connected in the same way as those of Fig. 17–12. The resultant binary count is shown under the diagram.

At count 7, or binary 0111, a 1 is applied to the AND gate from the Q outputs of FF_C and FF_B. Two 1s applied to an AND gate cause the J of FF_D to go to "1." The next pulse clears FF_A, FF_B, and FF_C and sets FF_D. The counter now reads 8, or a binary 1000. Note that the $\bar{Q}$ output of FF_D is returned to FF_B. With FF_D set to a "1," $\bar{Q}$ is 0, which prevents FF_B from further triggering until FF_D is cleared.

Arrival of the next clock pulse causes FF_A to be set. The count is now 9, or binary 1001. The next pulse clears FF_A and FF_D simultaneously. The count is now 0, or binary 0000. With FF_B and FF_C already clear, the counter is reset to 0 and is ready for the next count.

Figure 17–15 shows a BCD counter constructed from an IC chip. The ABCD output will be in binary code for numbers 0 through 9. The input is applied to pin 14. As a general rule, the output of a BCD counter is applied to a decoding device. The decoder changes binary data into an output that drives a display device.

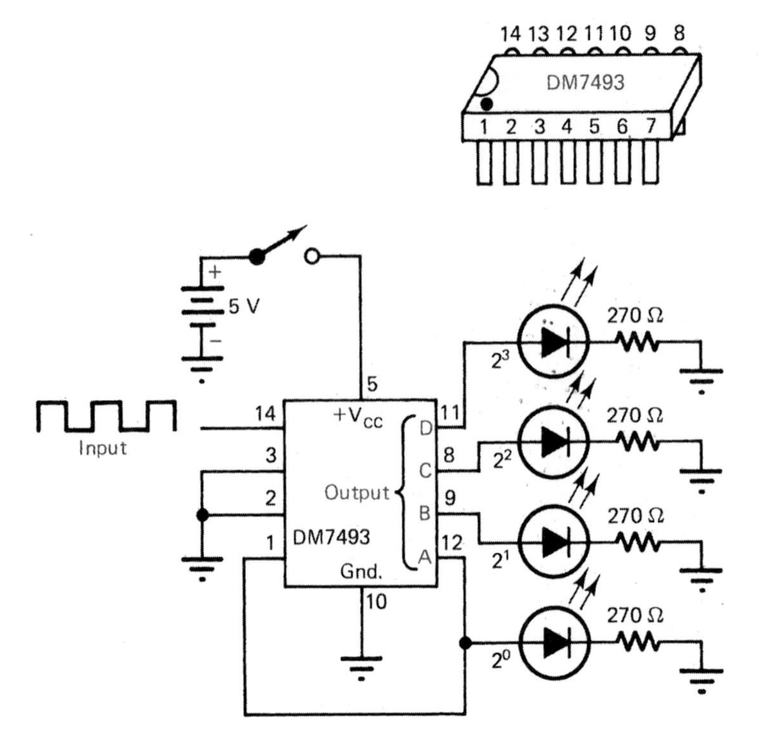

Figure 17–13. 4-bit IC binary counter.

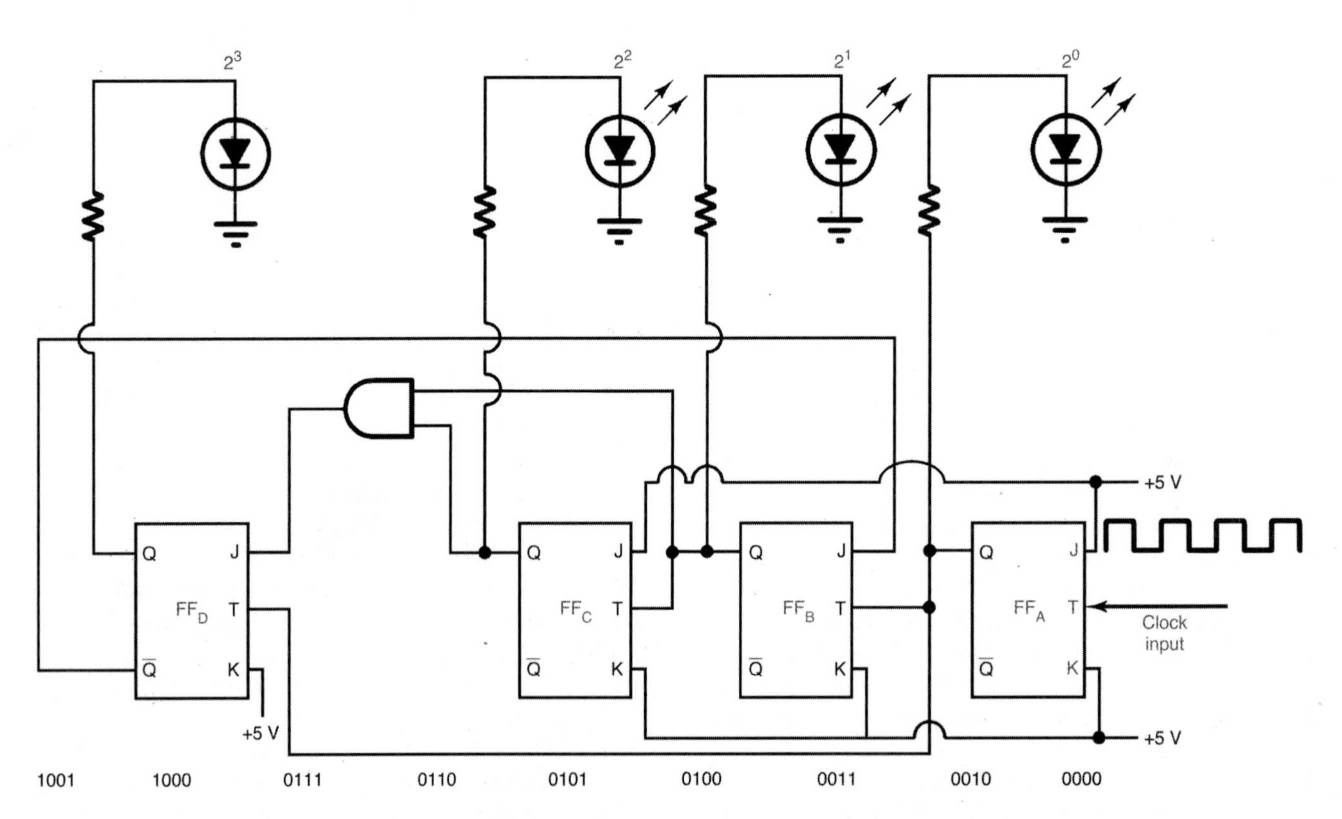

Figure 17–14. BCD flip-flop circuit.

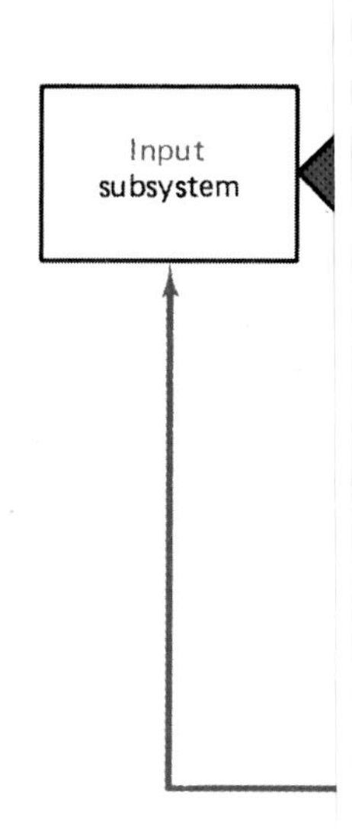

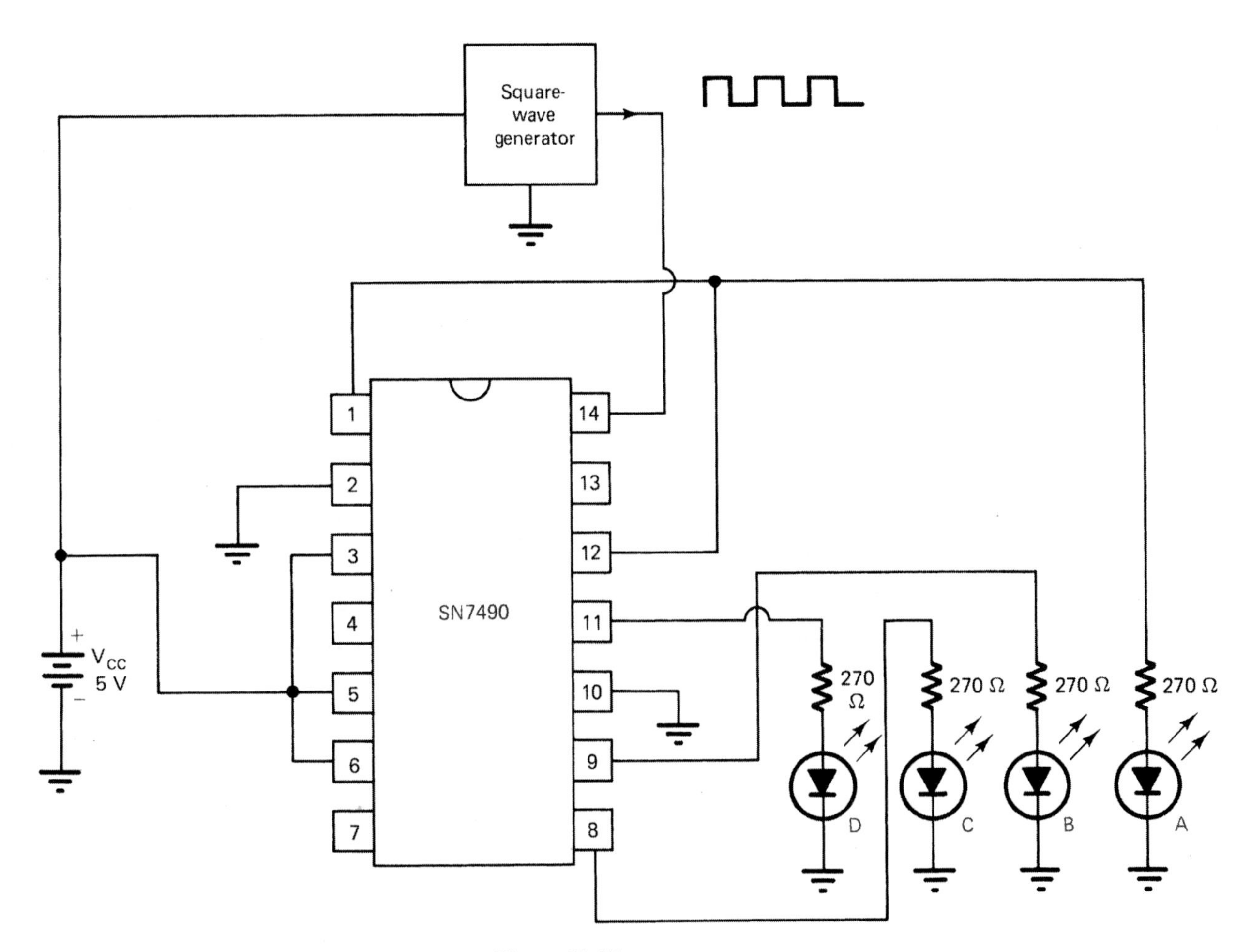

Figure 17–15. BCD counter.

17.8 DIGITAL SYSTEM DISPLAYS

Many people who need to read the output of a digital system may not be familiar with the BCD method of number display. The system must therefore change its output to a more practical method of display. A variety of display devices have been developed to perform this function. One that is used primarily for numbers is called a *seven-segment display.* Calculators, watches, and test instruments employ this type of display. It is widely used in nearly all digital electronic systems today.

A seven-segment number display divides the decimal numbers 0 through 9 into lines. Only four vertical lines and three horizontal lines are needed to display a specific number. Each line or segment is designed to be illuminated separately. Figure 17–16(a) shows a typical seven-segment display. Each segment may contain several LEDs in a common bar. The size of the display generally determines the number of LEDs in a specific segment. All the LEDs in a segment are energized at the same time.

Individual lines or bars of the display are labeled *a, b, c, d, e, f,* and *g*. A 7-bit binary number is applied to the display. Each segment is fed by a specific part of the binary number. In a common-anode type of display, when the input goes to ground or 0, it illuminates the segments. In a common-cathode display, the input must be 1 or +5 V to illuminate a segment. The truth table of Fig. 17–16(b) is for a common-anode display. For a common-cathode display, the polarity of each segment would be inverted.

A common-anode seven-segment display test circuit is shown in Fig. 17–17. Pins 14, 9, and 3 are the anode connections and are connected to the positive side of the power source. Pins 1, 2, 6, 7, 8, 10, 11, and 13 are the segment connections. Note that each segment is connected to a resistor, thus limiting the current to the LEDs in each segment to a reasonable value. Connecting one resistor to the ground will energize a specific segment. Connecting them all to ground will produce an 8 display. Seven segments and a decimal point are included in this display.

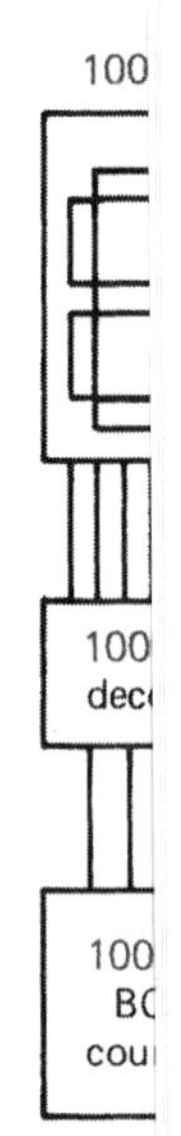

8. What is the mathematical equivalent of AND, OR, NOT, NAND, and NOR gates?
9. How does a flip-flop store data?
10. Explain how one flip-flop counts.
11. Explain how four flip-flops count to 1111.
12. Explain how decade or BCD counting is achieved with flip-flops.
13. Explain how a number, such as 5, is indicated on a seven-segment display.
14. What are the letter designations of individual segments in a seven-segment display?
15. Why is decoding needed for a binary counter with a seven-segment display?
16. What are the fundamental parts of a computer?
17. Explain the differences between software, hardware, and firmware.
18. What are encoding and decoding?
19. What are a *bit,* a *byte,* and a *word* in binary terms?
20. What is achieved during the fetch and execute cycles of a computer?
21. What is the ASCII code?
22. Describe the differences between volatile and nonvolatile memory.
23. What is meant by read-only memory?
24. What is a random access memory?
25. What is the function of an ALU?
26. How does a microcomputer differ from a regular computer?

STUDENT ACTIVITIES

• *Switch-Lamp Gates*

1. Connect the AND gate of Fig. 17–4.
2. Prepare a truth table showing the relationship of the two inputs and the output.
3. Connect the OR gate of Fig. 17–5. Test the circuit and prepare a truth table.
4. Connect the NOT gate of Fig. 17–6. Prepare a truth table of its operation.
5. Connect the NAND gate of Fig. 17–7. Test the circuit and prepare a truth table.
6. Connect the NOR gate of Fig. 17–8. Test the circuit and prepare a truth table.

• *The NAND Gate*

Complete Activity 17–1, page 343 in the manual.

ANALYSIS

1. What is the *mathematical* expression of a NAND gate?
2. Show how the *mathematical* expression of a NAND gate is modified to achieve the AND function.
3. Show how the *mathematical* expression of a NAND gate is modified to achieve the OR function.

• *The NOR Gate*

Complete Activity 17–2, page 347 in the manual.

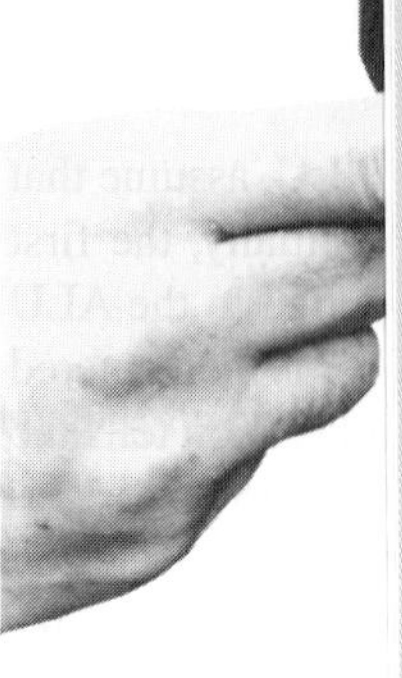

ANALYSIS

1. What is the *mathematical* expression of a NOR gate?
2. Show how the *mathematical* expression of a NOR gate is modified to achieve the OR function.
3. Show how the *mathematical* expression of a NOR gate is modified to achieve the AND function.

IC Logic Gate Test Circuit

1. In evaluating an IC logic gate, we will use the following test circuit. It can be used to evaluate any of the logic gates. Switch *A* and LED *A* are for one input. Switch *B* and LED *B* are for the other input. LED *Y* is used to observe the state of the output.
2. Connect the test circuit of Fig. 17–26.
3. To test the operation of a logic gate, look up the connection diagram in a manufacturer's data manual.
4. Connect an input *A* or *B* to the gate input terminals. Connect the *Y* output to the gate output.
5. Apply 5 V dc to the test circuit and the IC.
6. When switch *A* is 1, LED *A* should light. Switch *B* and LED *B* should respond in the same manner. In the "0" or off state, the respective LEDs will be off.
7. When the output of the gate is 1, LED *Y* will light; 0 causes LED *Y* to be off.

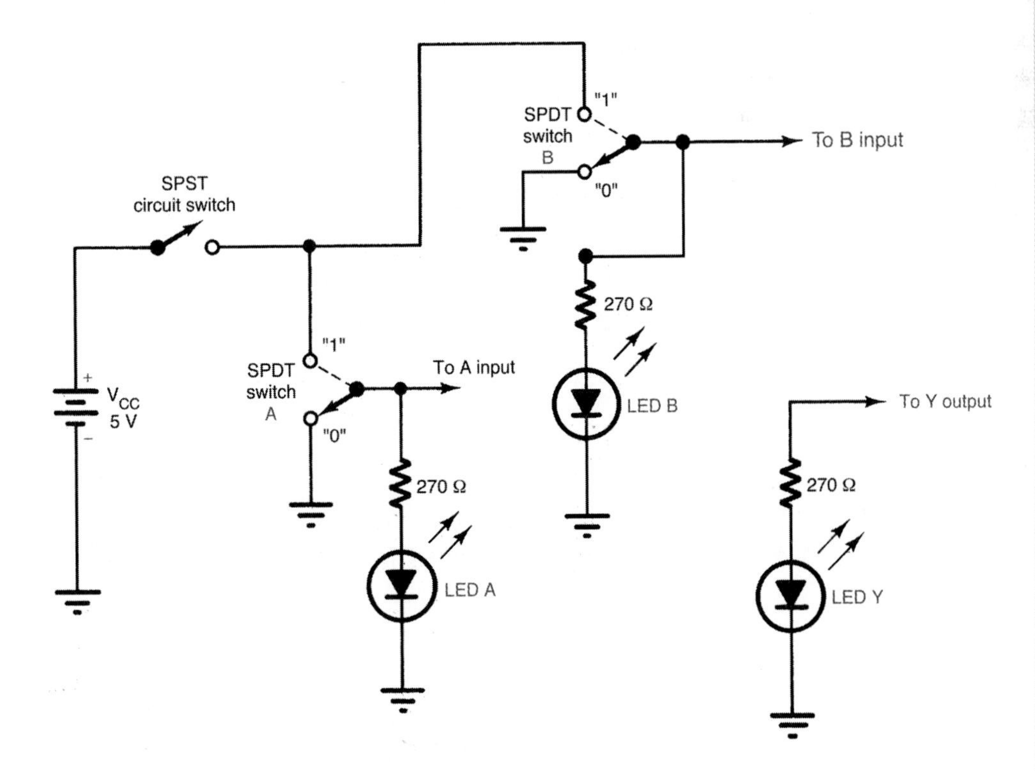

Figure 17–26. IC logic gate test circuit.

Logic Gate Testing

1. Look up the pin connections for an SN7404, an SN7408, a DM7400, an SN7402, and an SN7432.
2. Make a drawing of the pin connections for each chip being used in the test procedure.
3. Connect the *A* and *B* inputs and the *Y* output to one gate. By a switching sequence determine if the gate is good or faulty. Four gates are included on each chip except the SN7404. It has six gates.
4. Test each gate on the chip. Note the status of each gate.

R-S *Flip-Flop* or *Complete Activity 17–3, page 351 in the manual.*

1. Using the logic gate test circuit of Fig. 17–26, connect the *R-S* flip-flop circuit of Fig. 17–27(a). Note that one additional LED is needed. The inputs are labeled *R* and *S* and the outputs are *Q* and $\overline{Q}$.

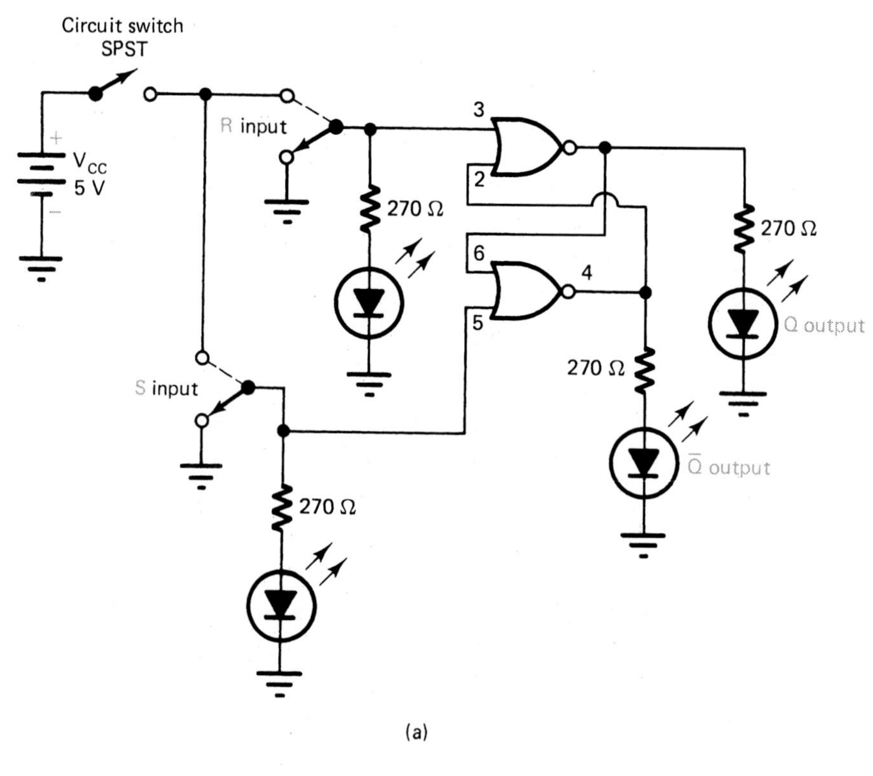

(a)

Step	Input		Output	
	R	S	Q	$\overline{Q}$
0	0	0		
1	0	1		
2	1	0		
3	1	1		

(b)

Figure 17–27. (a) *R-S* flip-flop circuits; (b) truth table.

2. Any two of the four gates on the SN7402 chip can be used.
3. Turn on the power supply. An "on" LED indicates the "1" state and off is "0." Be certain to apply power to the $+V_{CC}$ and ground pins of the IC.
4. What is the initial state of Q and $\bar{Q}$?
5. Alter the *R-S* switches so that both inputs are 0.
6. Record the Q and $\bar{Q}$ outputs in a truth table.
7. Alter the *R-S* switches to conform with steps 1, 2, and 3 of the truth table [Fig. 17–27(b)]. Record the Q and $\bar{Q}$ outputs for each step.

ANALYSIS

1. Explain what it takes to SET an *R-S* flip-flop constructed with NOR gates.
2. Explain what it takes to SET an *R-S* flip-flop constructed with NAND gates.

555 Astable Multivibrator

Complete Activity 17–4, page 355 in the manual.

ANALYSIS

1. If the clock circuit of an astable multivibrator (see Fig. 15–18) capacitor (C_1) value has changed, how would it alter the frequency of operation?
2. If resistors R_1 and R_4 of Fig. 15–18 were changed to 10 kΩ, how would it alter the operation of the circuit?

J-K *Flip-Flop*

1. Using the modified logic gate tester of Fig. 17–26, test one of the *J-K* flip-flops in SN74107. Look up the chip in a data manual to find the pin connections.
2. Initially, connect the clear (CLR) input at +5 V.
3. Place the *J-K* inputs both at 0. Note the state of Q and $\bar{Q}$.
4. Then connect the T input to +5 V. Does the resultant output change?
5. Set the *J-K* inputs as in Fig. 17–11(b) and check the previous and resultant outputs of Q and $\bar{Q}$. Then apply the designated T input and note the outputs.
6. Make a truth table of your data.

4-Bit Binary Counter or Complete Activity 17–5, page 359 in the manual.

1. Construct the 4-bit binary counter of Fig. 17–13. FF_1, FF_2, FF_3, and FF_4 are all built on the DM7493 chip.
2. The counting pulse may be achieved with an SPDT switch. If the count is irregular, it is probably due to contact bouncing of the switch.
3. A square-wave generator or a pulse generator may be used in place of the switch. The NE555 square-wave generator of Fig. 17–19 may also be constructed for this purpose.
4. Cycle the counter through its range following the binary count. An "on" LED indicates "1" and an "off" LED represents the "0" state.

ANALYSIS

1. What does it take to change the counting direction of a DM7493 IC?
2. Discuss the operation of the circuit of Fig. 17–13.

BCD Counter or Complete Activity 17–6, page 303 in the manual.

1. Construct the BCD counter of Fig. 17–15. The SN7490 has an AND gate and four flip-flops built on a chip.
2. The counting pulse may be achieved with an SPDT switch. If the count is irregular, it is probably due to contact bouncing of the switch.
3. A square-wave generator or a pulse generator should be used in place of the switch. The NE555 square-wave generator of Fig. 17–19 can be used for this purpose.
4. Cycle the counter through its range, noting the counting sequence.

ANALYSIS

1. For a binary counter to be converted to a BCD counter, how many counts must it lose?
2. Why is BCD counting an important operation for a digital system?

Seven-Segment Display

1. Construct the seven-segment display test circuit of Fig. 17–17.
2. Turn on the power supply and adjust it to 5 V dc.
3. Turn on the circuit switch.
4. Alternately connect each of the 270-Ω resistors to the ground side of the power supply.
5. Make a sketch of the segment illuminated and note the pin number that is energized.
6. Record the current needed to produce illumination.
7. Using the appropriate segment combinations, form the numbers 0, 1, 2, 3, 4, 5, 6, 7, 8, and 9. Note the segment combination used for each number.

Counting Timer

1. Construct the counting timer of Fig. 17–19.
2. Turn on the power supply and input switch.
3. The speed of the count is based on the setting of R_A and the capacitor selection of C. Adjust R_A to a rather fast timing rate.
4. Observe the counting cycle of the timer.
5. Calibrate R_A for a 1-s or a 1-min count.
6. Test the operation of the circuit for its count duration.

Seven-Segment Decoding and Driving

Complete Activity 17–7, page 367 in the manual.

ANALYSIS

1. What is the active status of the input and output of the 7447 of Fig. 17–19?
2. If the input to the 7447 of Fig. 17–19 is 0110, what is the status of the output?

Seven-Segment Displays and Decoding

Complete Activity 17–8, page 371 in the manual.

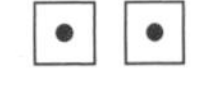

ANALYSIS

1. What does it mean for the output of the 7447 of Fig. 17–19 to be active low?
2. When the output of the 7447 of Fig. 17–19 is active low, what does this do to the segments of the display to produce illumination?

Self-Examination

1. Complete the Self-Examination for Chap. 17, page 375 in the manual.
2. Check your answers on pages 419–420 in the manual.
3. Study the questions that you answered incorrectly.

Chapter 17 Examination

Complete the Chap. 17 Examination on page 377 in the manual.

A rather interesting change in the lamp circuit occurs when the total resistance increases. Some of the source voltage will appear across the rheostat and will be deducted from the lamp voltage. With the lamp and the rheostat both being 100 Ω, there will now be 50 V across each. The lamp will obviously be operating with reduced power which is determined to be

power (P) = current (I) × voltage (V)

or

$$P = I \times V = 0.5\text{ A} \times 50\text{ V}$$
$$= 25\text{ W}$$

The lamp will therefore receive only 25 W of its original power. Its brightness will be reduced significantly. The rheostat in this case is consuming as much power as the load device. Power consumed by the rheostat appears as heat. Because the purpose of our circuit is to control light, heat is not a desirable feature. In effect, any heat dissipated by R is considered to be wasted energy. We must pay for this power to control lamp brightness. Control of a power system by changes in current is therefore not efficient.

As an alternative, electrical power can be controlled by variations in circuit voltage. Figure 18–2 shows a voltage-controlled lamp dimmer circuit. In this circuit, lamp brightness is controlled by a variable transformer. A reduction in voltage will produce a corresponding change in brightness. This type of control is more efficient than the rheostat circuit. In effect, the lamp is the primary resistive device. Power consumed by the circuit will appear only at the lamp. Some power is consumed by the primary winding of the transformer. This power is usually rather nominal compared with the rheostat circuit.

Power control by a variable transformer is more efficient than the rheostat method. A variable transformer is, however, a rather expensive item. When large amounts of power are being controlled, this device also becomes quite large. Control of this type is more efficient but has a number of limitations.

One of the most efficient methods of electrical power control that we have today is through circuit switching. When a switch is turned on, power is consumed by the load. When the switch is turned off, no power is consumed. The switching method of power control is shown in Fig. 18–3. If the switch is low resistant, it consumes little power. When the switch is open, it consumes no power. By switching the circuit on and off rapidly, the average current flow can be reduced. The brightness of the lamp will also be reduced. In effect, lamp brightness is controlled by the switching speed. Power in this method of control is not consumed by the control device. This type of control is both efficient and effective.

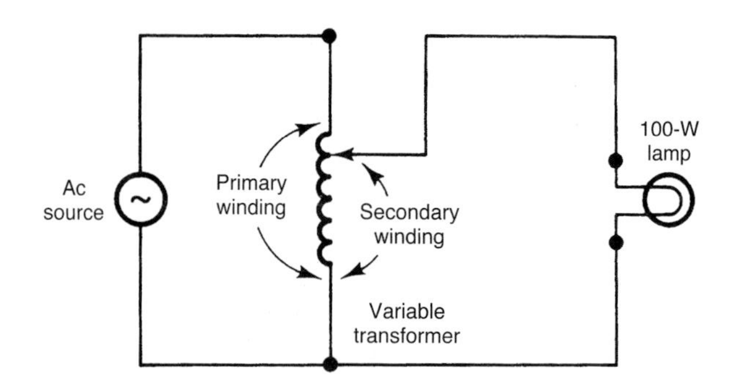

Figure 18–2. Voltage-controlled lamp circuit.

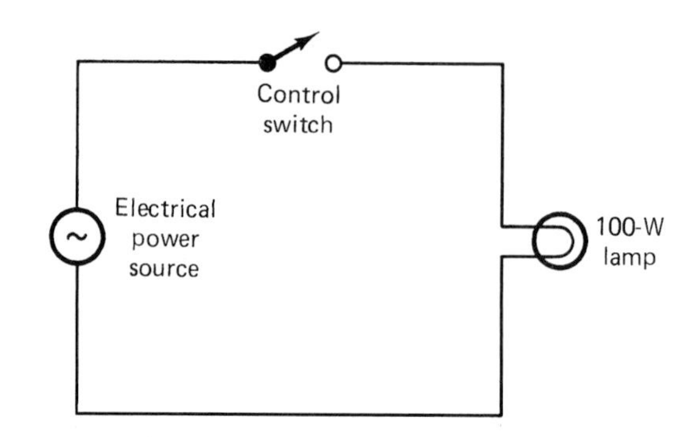

Figure 18–3. Switch-controlled lamp circuit.

The switching method of electrical power control cannot be achieved effectively with a manual switch. The mechanical action of the switch will not permit it to be turned on and off quickly. A switch would soon wear out if used in this manner. Electronic switching can be used to achieve the same result. Switching action of this type can be accomplished in a lamp circuit without a noticeable flicker. Electronic power control deals with the switching method of controlling power.

18.2 SILICON-CONTROLLED RECTIFIERS

A silicon-controlled rectifier or SCR is probably the most popular electronic power control device today. The SCR is used primarily as a switching device. Power control is achieved by switching the SCR on and off during one alternation of the ac source voltage. For 60 Hz ac the SCR would be switched on and off 60 times per second. Control of electrical power is achieved by altering or delaying the turn-on time of an alternation.

An SCR, as the name implies, is a solid-state rectifier device. It conducts current in only one direction. It is similar in size to a comparable silicon power diode. SCRs are usually small, rather inexpensive, waste little power, and require practically no maintenance. The SCR is available today in a full range of types and sizes to meet nearly any power control application. Presently, they are available in current ratings from less than 1 A to over 1400 A. Voltage values range from 15 to 2600 V.

SCR Construction

An SCR is a solid-state device made of four alternate layers of *P*- and *N*-type silicon. Three *P-N* junctions are formed by the structure. Each SCR has three leads or terminals. The anode and cathode terminals are similar to those of a regular silicon diode. The third lead is called the *gate.* This lead determines when the device switches from its off to on state. An SCR will usually not go into conduction by simply forward biasing the anode and cathode. The gate must be forward biased at the same time. When these conditions occur, the SCR becomes conductive. The internal resistance of a conductive SCR is less than 1 Ω. Its reverse or off-state resistance is generally in excess of 1 MΩ.

A schematic symbol and the crystal structure of an SCR are shown in Fig. 18–4. Note that the device has a *PNPN* structure from anode to cathode. Three distinct *P-N* junctions are formed. When the anode is made positive and the cathode negative, junctions 1 and 3 are forward biased. J_2 is reverse biased. Reversing the polarity of the source alters this condition. J_1 and J_3 would be reverse biased and J_2 would be forward biased and would not permit conduction. Conduction will occur only when the anode, cathode, and gate are all forward biased at the same time.

Some representative SCRs are shown in Fig. 18–5. Only a few of the more popular packages are shown here. As a general rule, the anode is connected to the largest electrode if there is a difference in their physical size. The gate is usually smaller than the other electrodes. Only a small gate current is needed to achieve control. In some packages, the SCR symbol is used for lead identification.

Anode
Gate
Cathode
(a)
Anode
P
N
P
N
J3
J2
J1
Gate
Cathode
(b)

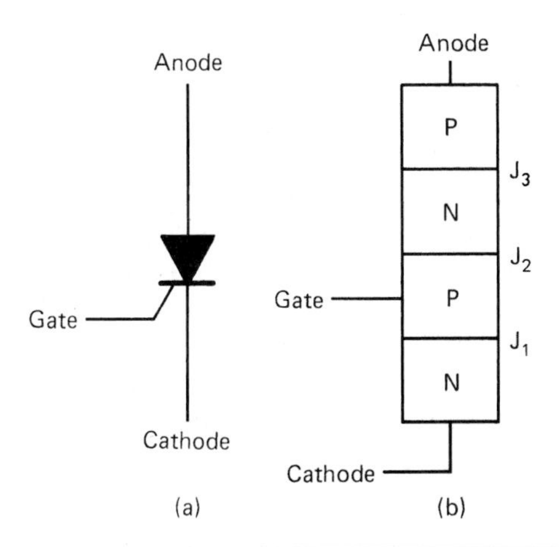

Figure 18–4. SCR crystal: (a) symbol; (b) structure.

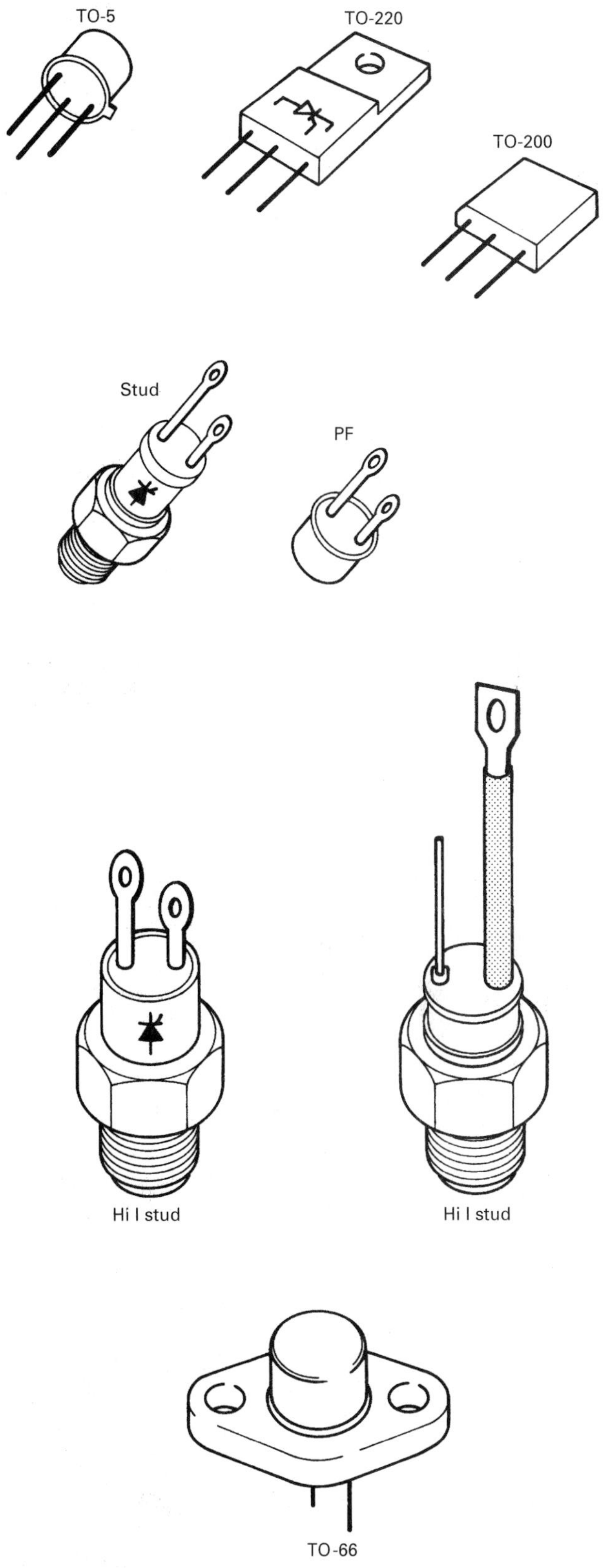

TO-5
TO-220
TO-200
Stud
PF
Hi I stud
Hi I stud
TO-66

Figure 18–5. Representative SCR packages.

SCR Operation

Operation of an SCR is not easily explained by using the four-layer *PNPN* structure of Fig. 18–4. A rather simplified method has been developed that describes the crystal structure as two interconnected transistors. Figure 18–6(a) shows this imaginary division of the crystal. Notice that the top three crystals form a *PNP* transistor. The lower three crystals form an *NPN* device. Two parts of each transistor are interconnected. The two-transistor circuit diagram of Fig. 18–6(b) is an equivalent of the crystal structure division.

Figure 18–7 shows how the two-transistor equivalent circuit of an SCR will respond when voltage is applied. Note that the anode is made positive and the cathode negative. This condition forward biases the emitter of each transistor and reverse biases the collector. Without base voltage applied to the *NPN* transistor, the equivalent circuit is nonconductive. No current will flow through the load resistor. Reversing the polarity of the voltage source will not alter the conduction of the circuit. The emitter of each transistor will be reverse biased by this condition, meaning that no conduction will take place when the anode and cathode are reverse biased. With only anode-cathode voltage applied, no conduction will occur in either direction.

Assume now that the anode and cathode of an SCR are forward biased, as indicated in Fig. 18–7. If the gate is momentarily made positive with respect to the cathode, the emitter-base junction of the *NPN* transistor becomes forward biased. This action will immediately cause collector current to flow in the *NPN* transistor. As a result, base current will also flow into the *PNP* transistor, which in turn will cause a corresponding collector current to flow. *PNP* collector current now causes base current to flow into the *NPN* transistor. The two transistors therefore latch or hold when in the conductive state. This action continues even when the gate is disconnected. In effect, when conduction occurs, the device will latch in its "on" state. Current will continue to flow through the SCR as long as the anode-cathode voltage is of the correct polarity.

To turn off a conductive SCR, it is necessary to momentarily remove or reduce the anode-cathode voltage. This action will turn off the two transistors. The device will then remain in this state until the anode, cathode, and gate are forward biased again. With ac applied to an SCR, it will automatically turn off during one alternation of the input. Control is achieved by altering the turn-on time during the conductive or "on" alternation.

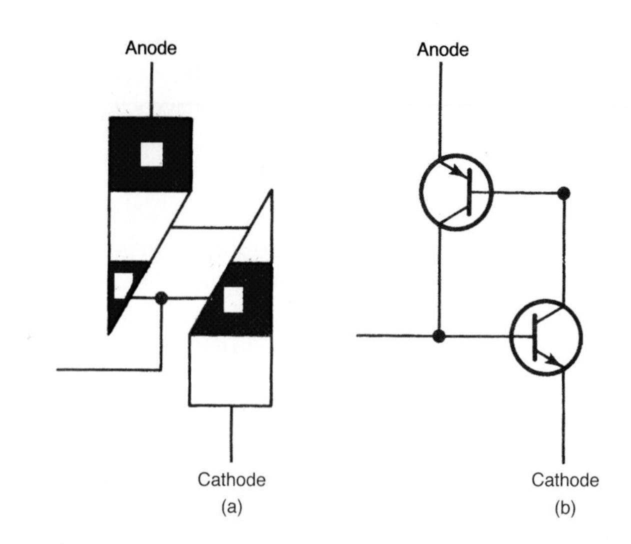

Figure 18–6. Equivalent SCR.

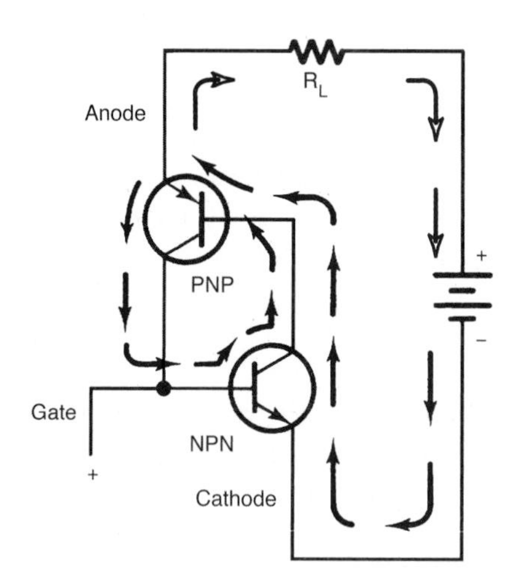

Figure 18–7. SCR response.

SCR *I-V* Characteristics

The current-voltage characteristics of an SCR tell much about its operation. The *I-V* characteristic curve of Fig. 18–8 shows that an SCR has two conduction states. Quadrant I shows conduction in the forward direction, which shows how conduction occurs when the forward breakover voltage (V_{BO}) is exceeded. Note that the curve returns to approximately zero after the V_{BO} has been exceeded. When conduction occurs, the internal resistance of the SCR drops to a minute value similar to that of a forward-biased silicon diode. The conduction current (I_{AK}) must be limited by an external resistor. This current, however, must be great enough to maintain conduction when it starts. The holding current or I_H level must be exceeded for this to take place. Note that the I_H level is just above the knee of the I_{AK} curve after it returns to the center.

Quadrant III of the *I-V* characteristic curve shows the reverse breakdown condition of operation. This characteristic of an SCR is similar to that of a silicon diode. Conduction occurs when the peak reverse voltage (PRV) value is reached. Normally, an SCR would be permanently damaged if the PRV is exceeded. Today, SCRs have PRV ratings of 25 to 2000 V.

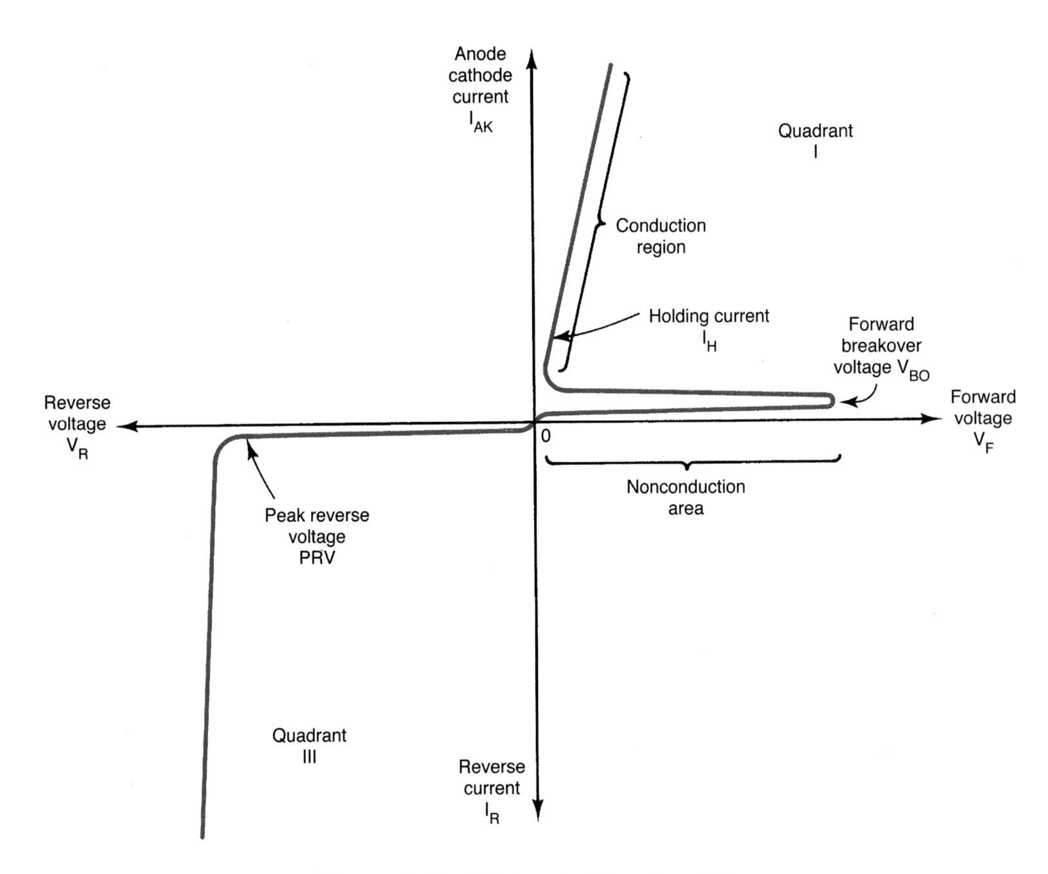

Figure 18–8. *I-V* characteristics of an SCR.

For an SCR to be used as a power control device, the forward V_{BO} must be altered. Changes in gate current will cause a decrease in the V_{BO}. This occurs when the gate is forward biased. An increase in I_G will cause a large reduction in the forward V_{BO}. An enlargement of quadrant I of the *I-V* characteristics is shown in Fig. 18–9, which also shows how different values of I_G change the V_{BO}. With 0 I_G it takes a V_{BO} of 400 V to produce conduction. An increase in I_G reduces this quite significantly. With 7 mA of I_G the SCR conducts as a forward-biased silicon diode. Lesser values of I_G will cause an increase in the V_{BO} needed to produce conduction.

The gate current characteristic of an SCR shows an important electrical operating condition. For any value of I_G there is a specific V_{BO} that must be reached before conduction can occur, which means that an SCR can be turned on when a proper combination of I_G and V_{BO} is achieved. This characteristic is used to control conduction when the SCR is used as a power control device.

DC Power Control with SCRs

When an SCR is used as a power control device, it responds primarily as a switch. When the applied source voltage is below the forward breakdown voltage, control is achieved by increasing the gate current. Gate current is usually made large enough to ensure that the SCR will turn on at the proper time. Gate current is generally applied for only a short time. In many applications this may be in the form of a short-duration pulse. Continuous I_G is not needed to trigger an SCR into conduction. After conduction occurs, the SCR will not turn off until the I_{AK} drops to zero.

Figure 18–10 shows an SCR used as a dc power control switch. In this type of circuit, a rather high load current is controlled by a small gate current. Note that the electrical power source (V_S) is controlled by the SCR. The polarity of V_S must forward bias the SCR, which is achieved by making the anode positive and the cathode negative.

When the circuit switch is turned on initially, the load is not energized. In this situation the V_{BO} is in excess of the V_S voltage. Power control is achieved by turning on SW-1 which forward biases the gate. If a suitable value of I_G occurs, it will lower the V_{BO} and turn on the SCR. The I_G can be removed and the SCR will remain in conduction. To turn the circuit off, momentarily open the circuit switch. With the circuit switch on again, the SCR will remain in the off state. It will go into conduction again by closing SW-1.

Dc power control applications of the SCR require two switches to achieve control, but this application of the SCR is not practical. The circuit switch would need to be capable

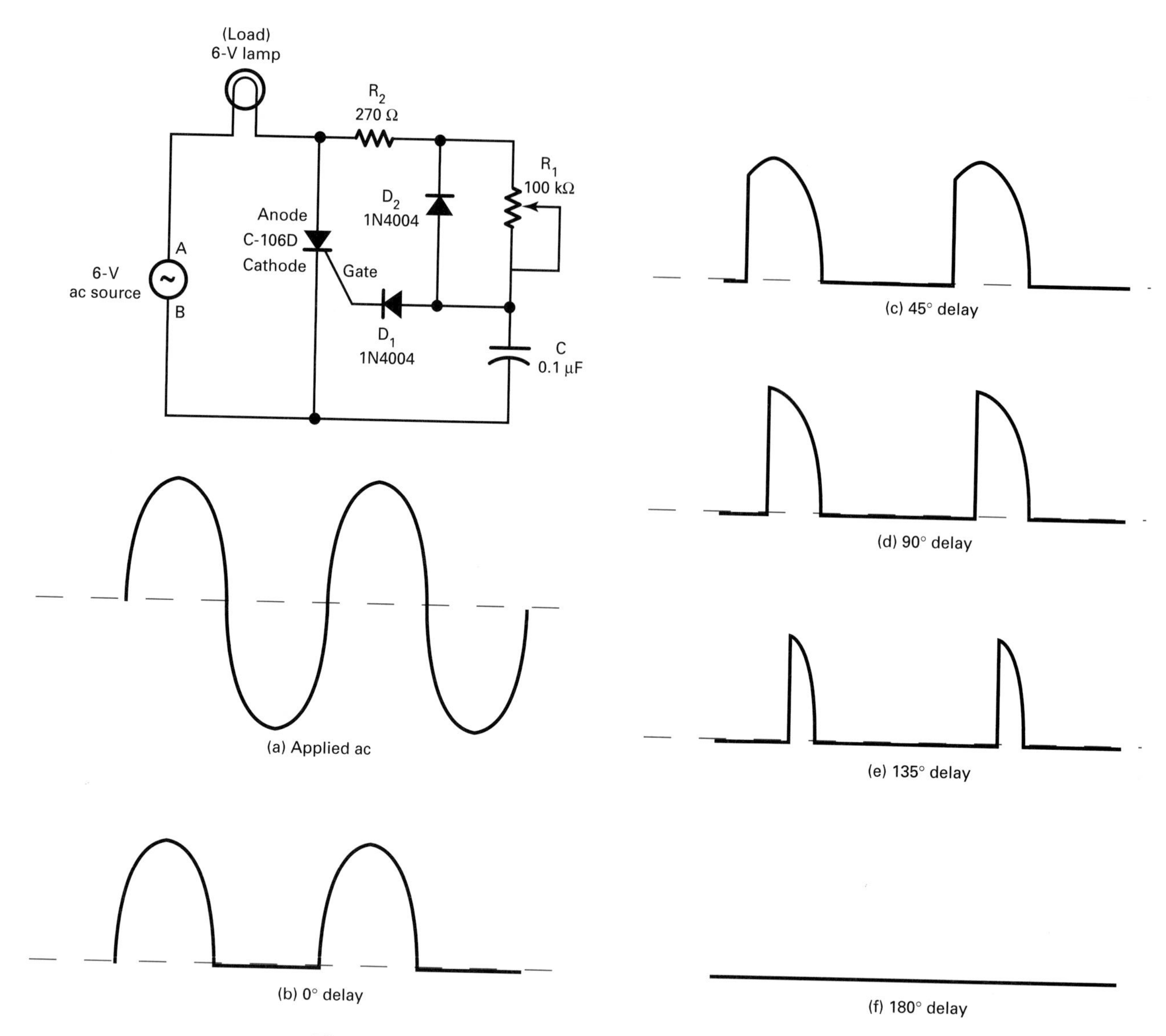

Figure 18–13. Phase control SCR circuit and waveforms.

A change in the resistance of R_1 will alter the conduction time of the SCR power control circuit. When R_1 is increased, C_1 cannot charge as quickly during the positive alternation. It therefore takes more time for C_1 to build its charge voltage, which means that the turn-on of I_G will be delayed for a longer time. As a result, conduction of the SCR will be delayed at the beginning of the alternation. See the waveform of Fig. 18–13(c). The resultant load current for this condition will be somewhat less than that of the waveform in Fig. 18–13(b). Variable control of the load is achieved by altering the conduction time of the SCR.

A further increase in the resistance of R_1 will cause more delay in the turn-on of the SCR during the positive alternation. Waveforms of Figs. 18–13(d), (e), and (f) show the result of this change. Waveform (d) has a 90° delay of SCR conduction time. Waveform (e) shows a 135° delay in the conduction time. If the value of R_1 is great enough, the SCR will delay its turn-on for the full alternation. The waveform in Fig. 18–13(f) shows a 180° turn-on delay. Note that the load does not receive any current for this condition. Control of this circuit is from full conduction as in waveform (b) to 180° delay as in waveform (f).

Power control with an SCR is highly efficient. If no delay of the conduction time occurs, the load develops full power. In this case, full power occurs only for one alternation. If conduction is delayed 90°, the load develops only half of its potential power. If the conduction delay is 135°, the load develops only 25% of its potential power. It is important to note that nearly all the power controlled by the circuit is applied to the load device—little is consumed by the SCR. In variable power control applications, it is extremely important that the load be supplied to its full value of power. With an SCR control circuit, no power is consumed by other components. By delaying the turn-on time of conduction, power can be controlled with the highest level of efficiency.

18.3 TRIAC POWER CONTROL

Ac power control can be achieved with a device that switches on and off during each alternation. Control of this type is accomplished with a special solid-state device known as a *triac*. This device is described as a three-terminal ac switch. Gate current is used to control the conduction time of either alternation of the ac waveform. In a sense, the triac is the equivalent of two reverse-connected SCRs feeding a common load device.

A triac is classified as a gate-controlled ac switch. For the positive alternation it responds as a *PNPN* device. An alternate crystal structure of the *NPNP* type is used for the negative alternation. Each crystal structure is triggered into conduction by the same gate connection. The gate has a dual-polarity triggering capability.

Triac Construction

A triac is a solid-state device made of two different four-layer crystal structures connected between two terminals. We do not generally use the terms *anode* and *cathode* to describe these terminals. For one alternation they would be the anode and cathode. For the other alternation they would respond as the cathode and anode. It is common practice therefore to use the terms *main 1* and *main 2* or *terminal 1* and *terminal 2* to describe these leads. The third connection is the gate. This lead determines when the device switches from its off to its on state. The gate G will normally go into conduction when it is forward biased, and is usually based on the polarity of terminal 1. If T_1 is negative, G must be positive. When T_1 is positive, the gate must be negative. This means that ac voltage must be applied to the gate to cause conduction during each alternation of the T_1-T_2 voltage.

The schematic symbol and the crystal structure of a triac are shown in Fig. 18–14. Notice the junction of the crystal structure simplification. Looking from T_1 to T_2, the structure involves crystals N_1, P_1, N_2, and P_2. The gate is used to bias P_1. This is primarily the same as an SCR with T_1 serving as the cathode and T_2 the anode.

Looking at the crystal structure from T_2 to T_1, it is N_3, P_2, N_2, and P_1. The gate is used to bias N_4 for control in this direction, which is similar to the structure of an SCR in this direction. Notice that T_1, T_2, and G are all connected to two pieces of crystal. Conduction will take place only through the crystal polarity that is forward biased. When T_1 is negative, for example, N_1 is forward biased and P_1 is reverse biased. Terminal selection by bias polarity is the same for all three terminals.

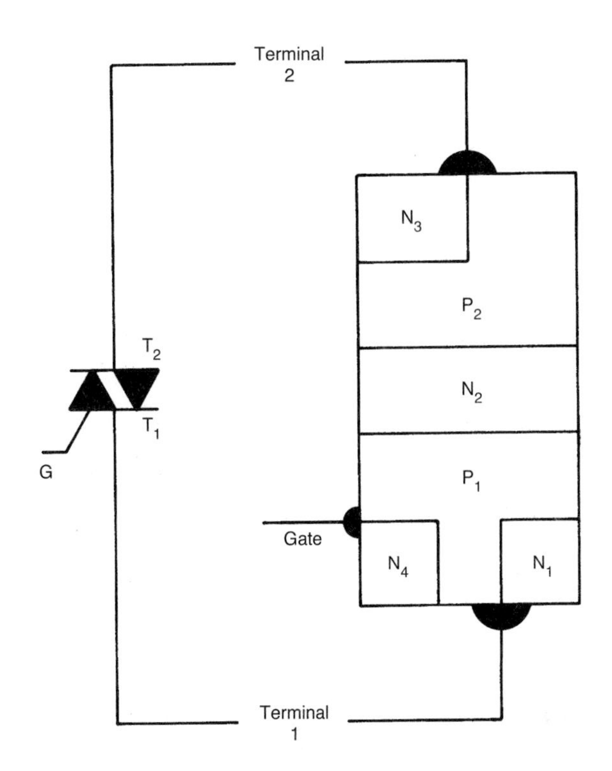

Figure 18–14. Triac symbol and crystal structure.

The schematic symbol of the triac is representative of reverse-connected diodes. The gate is connected to the same end as T_1, which is an important consideration when connecting the triac into a circuit. The gate is normally forward biased with respect to T_1.

Triac Operation

When ac is applied to a triac, conduction can occur for each alternation, but T_1 and T_2 must be properly biased with respect to the gate. Forward conduction occurs when T_1 is negative, the gate (G) is positive, and T_2 is positive. Reverse conduction occurs when T_1 is positive, G is negative, and T_2 is negative. Conduction in either direction is similar to that of the SCR. The two-transistor operation used with the SCR applies to the triac in both directions. Figure 18–15 shows an example of the operational conduction polarities of a triac. Note that dc is used as a power source. It is used here only to denote operational polarities. In practice, dc would not be used as a power source for a triac. It is an ac power control device.

The operation of a triac is primarily the same as that of the SCR. When T_1 and T_2 are of the correct polarity, gate current must occur to reduce the breakover voltage. Small values of I_G will cause the breakover voltage to be quite large. An increase in I_G decreases the breakover voltage. This applies equally to both forward and reverse conduction. After conduction occurs, the gate loses its effectiveness.

When a triac goes into conduction, its internal resistance drops to an extremely low value. Typical resistance

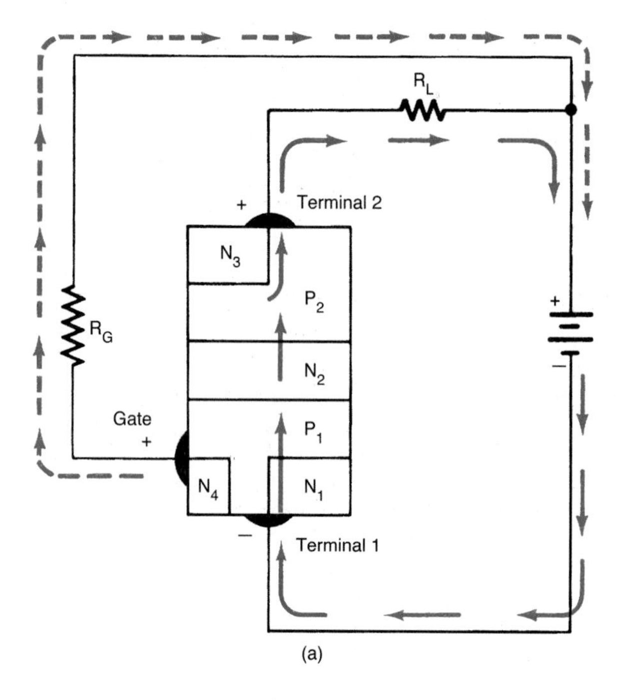

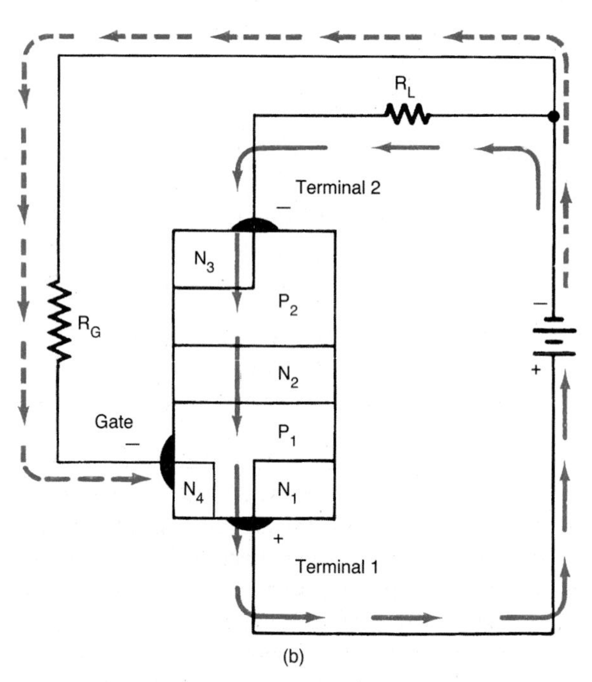

Figure 18–15. Triac condition polarities: (a) forward conduction, T_1-, T_2+, $G+$; (b) reverse condition.

values are less than 1 Ω. The device remains in this state for the remainder of the alternation. It is essential that current flowing through the triac be great enough to maintain conduction, which is generally called the *holding current* or I_H. The resistance of the load device and the value of the source voltage determine conduction current. This current must be slightly greater than I_H to maintain conduction.

Operation of a triac has one rather unique problem that does not occur in the SCR which refers to its conduction commutation. Commutation is used to describe the turn-off of conduction. In triac operation, commutation time can be critical. With a resistive load, turn-off starts when the conduction current drops below the I_H level. It usually continues into the next alternation until the breakover voltage is reached. In most applications the triac has sufficient time to reach turn-off. With an inductive load, conduction time is extended somewhat due to the inductive voltage caused by the collapsing magnetic field. A special *RC* circuit called a *snubber* is placed across the triac to reduce this problem. A 0.1-μF capacitor with a 100-Ω series resistor is placed across the triac. The capacitor charges during the collapsing of the inductive field and generally reduces the commutation problem.

Triac *I-V* Characteristics

The *I-V* characteristic of a triac shows how it responds to forward and reverse voltages. Figure 18–16 is a typical triac *I-V* characteristic. Note that conduction occurs in quadrants I and III. The conduction in each quadrant is primarily the same. With 0 I_G, the breakover voltage is usually quite high. When breakover occurs, the curve quickly returns to the center. This shows a drop in the internal resistance of the device when conduction occurs. Conduction current must be limited by an external resistor. The holding current or I_H of a triac occurs just above the knee of the I_T curve. I_H must be attained or the device will not latch during a specific alternation.

Quadrant III is normally the same as quadrant I and thus ensures that operation will be the same for each alternation. Because the triac is conductive during quadrant III, it does not have a peak reverse voltage rating. It does, however, have a maximum reverse conduction current value the same as the maximum forward conduction value. The conduction characteristics of quadrant III are mirror images of quadrant I.

Triac Applications

Triacs are used primarily to achieve ac power control. In this application the triac responds primarily as a switch. Through normal switching action, it is possible to control the ac energy source for a portion of each alternation. If conduction occurs for both alternations of a complete sine wave, 100% of the power is delivered to the load device. Conduction for half of each alternation permits 50% control. If conduction is for one-fourth of each alternation, the load receives less than 25% of its normal power. It is possible through this device to control conduction for the entire sine wave, which means that a triac is capable of controlling from 0% to 100% of the electrical power supplied to a load device. Control of this type is efficient as practically no power is consumed by the triac while performing its control function.

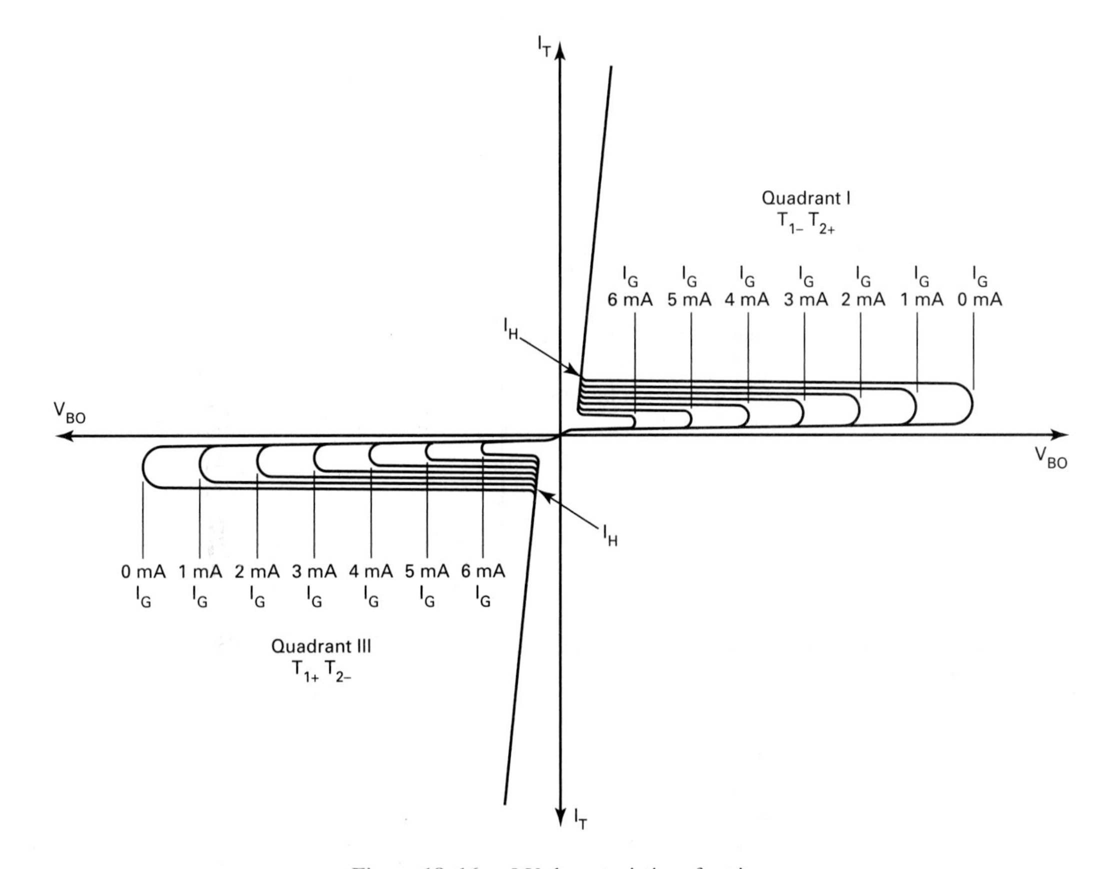

Figure 18–16. *I-V* characteristics of a triac.

STATIC SWITCHING. The use of a triac as a static switch is primarily an on-off function. Control of this type has a number of advantages over mechanical load switching. A high-current energy source can be controlled with a very small switch. No contact bounce occurs with solid-state switching which generally reduces arcing and switch contact destruction. Control of this type is rather easy to achieve. Only a small number of parts are needed for a triac switch.

Two rather simple triac switching applications are shown in Fig. 18–17. The circuit in Fig. 18–17(a) shows the load being controlled by an SPST switch. When the switch is closed, ac is applied to the gate. Resistor R_1 limits the gate current to a reasonable operating value. With ac applied to the gate, conduction occurs for the entire sine wave. The gate of this circuit requires only a few milliamperes of current to turn on the triac. Practically any small switch could be used to control a rather large load current.

The circuit of Fig. 18–17(b) is considered to be a three-position switch. In position 1, the gate is open and the power is off. In position 2, gate current flows for only one alternation. The load receives power during one alternation, which is the half-power operating position. In position 3, gate current flows for both alternations. The load receives full ac power in this position.

Start-Stop Triac Control

Some electrical power circuits are controlled by two pushbuttons or start-stop switches. Control of this type begins by momentarily pushing the start button. Operation then continues after releasing the depressed button. To turn off the circuit, a stop button is momentarily pushed. The circuit then resets itself in preparation for the next starting operation. Control of this type is widely used in motor control applications and for lighting circuits. A triac can be adapted for this type of power control.

A start-stop triac control circuit is shown in Fig. 18–18. When electrical power is first applied to this circuit, the triac is in its nonconductive state. The load does not receive any power for operation. All the supply voltage appears across the triac because of its high resistance. No voltage appears across the *RC* circuit, the gate, or the load device initially.

To energize the load device, the start pushbutton is momentarily pressed. C_1 charges immediately through R_1 and R_2, which in turn causes I_G to flow into the gate. The V_{BO} of the triac is lowered and it goes into conduction. Voltage now appears across the load, R_2-C_1, and the gate. The charging current of R_2-C_1 and the gate continue and are at peak value when the source voltage alternation changes. The gate then

Extrinsic. A purified material that has a controlled amount of impurity added to it when manufactured.

Farad (F). The unit of measurement of capacitance that is required to contain a charge of 1 C when a potential of 1 V is applied.

Feedback. Transferring some of the output signal of an amplifier back to its input.

Feedback network. A resistor-capacitor or resistor-inductor combination that returns some of the output signal to the input of an amplifier.

Ferrite core. An inductor core made of molded iron particles.

Fetch. A computer instruction that retrieves data from memory.

Field coils. Electromagnetic coils that develop the magnetic fields of electrical machines.

Field, even lined. The even-numbered scanning lines of one TV picture or frame.

Field, odd lined. The odd-numbered scanning lines of one TV picture or frame.

Field pole. Laminated metal that serves as the core material for field coils.

Filament. The heating element of a vacuum tube.

Filter. A circuit used to pass certain frequencies and attenuate all other frequencies.

Firmware. Permanently installed instructions in computer hardware that control operation.

Flip-flop. A multivibrator circuit having two stable states that can be switched back and forth.

Fluorescent lighting. A lighting method that uses phosphor-coated lamps which are tubes filled with mercury vapor to produce light.

Flux (Φ). Invisible lines of force that extend around a magnetic material.

Flux density. The number of lines of force per unit area of a magnetic material or circuit.

Forbidden gap. An area between the valence and conduction band on an energy-level diagram.

Forward bias. A voltage connection method that produces current conduction in a *P-N* junction.

Forward breakover voltage. The voltage whereby an SCR or triac goes into conduction in quadrant I of its *I-V* characteristics.

Fossil fuel system. A power system that produces electrical energy due to the conversion of heat from coal, oil, or natural gas.

Frame. A complete electronically produced TV picture of 525 horizontally scanned lines.

Free electrons. Electrons located in the outer orbit of an atom which are easily removed and result in electrical current flow.

Free running. An oscillator circuit that develops a continuous waveform that is not stable, such as an astable multivibrator.

Frequency. The number of ac cycles per second, measured in hertz (Hz).

Frequency response. A circuit's ability to operate over a range of frequencies.

Fuse. An electrical overcurrent device that opens a circuit when it melts due to excess current flow through it.

Gain. A ratio of voltage or current output to voltage or current input.

Galvanometer. A meter that measures minute current values.

Ganged. Two or more components connected by a common shaft, such as a three-ganged variable capacitor.

Gate. (1) The control electrode of a field-effect transistor, SCR, or triac. (2) A circuit that performs special logic operations such as AND, OR, NOT, NAND, and NOR.

Gate current (I_G). The current that flows in a gate when it is forward biased.

Gauss. A unit of measurement of magnetic flux density.

Generator. A rotating electrical machine that converts mechanical energy into electrical energy.

Gilbert. A unit of measurement of magnetomotive force (MMF).

Ground. Of two types: *system grounds,* current-carrying conductors used for electrical power distribution; and *safety grounds,* not intended to carry electrical current but to protect individuals from electrical shock hazards.

Ground fault. An accidental connection to a grounded conductor.

Ground-fault circuit interrupter (GFCI). A device used in electrical wiring for hazardous locations; it detects fault conditions and responds rapidly to open a circuit before shock occurs to an individual or equipment is damaged.

Hardware. Electronic circuits, parts, and physical parts of a computer.

Henry (H). The unit of measurement of inductance that is produced when a voltage of 1 V is induced when the current through a coil is changing at a rate of 1 A per second.

Hertz (Hz). The international unit of measurement of frequency equal to one cycle per second.

Heterodyning. The process of combining signals of independent frequencies to obtain a different frequency.

High-level language. A form of computer programming that uses words and statements.

High-level modulation. Where the modulating component is added to an RF carrier in the final power output of the transmitter.

Holding current (I_H). A current level that must be maintained in an SCR or triac when it is in the conductive state.

Hole. A charge area or void where an electron is missing.

Horsepower. One horsepower (hp) = 33,000 ft-lb of work per minute or 550 ft-lb per second; it is the unit of measuring mechanical power; 1 hp = 746 W.

Hot conductor. A wire that is electrically energized, or "live," and is not grounded.

Hue. A color, such as red, green, or blue.

Hydroelectric system. A power system that produces electrical energy due to the energy of flowing water.

Hydrometer. An instrument used to measure the specific gravity or "charge" of the electrolyte of a storage battery.

Hysteresis. The property of a magnetic material that causes actual magnetizing action to lag behind the force that produces it.

ICE. A term used to help remember that current (I) leads voltage (E) in a capacitive (C) circuit.

Impedance (Z). The total opposition to current flow in an ac circuit which is a combination of resistance (R) and reactance (X) in a circuit; measured in ohms.

$$Z = \sqrt{R^2 + X^2}.$$

Impedance ratio. A transformer value that is the square of the turns ratio.

Incandescent lighting. A method of lighting that uses bulbs that have tungsten filaments which, when heated by electrical current, produce light.

Indicator. The part of an electrical system that shows if it is on or off or indicates a specific quantity.

Induced channel. The channel of an E-MOSFET that is developed by gate voltage.

Induced current. The current that flows through a conductor due to magnetic transfer of energy.

Induced voltage. The potential that causes induced current to flow through a conductor which passes through a magnetic field.

Inductance (L). The property of a circuit to oppose changes in current due to energy stored in a magnetic field.

Induction motor. An ac motor that has a squirrel-cage rotor and operates on the induction (transformer action) principle.

Inductive circuit. A circuit that has one or more inductors or has the property of inductance, such as an electric motor circuit.

Inductive heating. A method of heating conductors in which the material to be heated is placed inside a coil of wire and high-frequency ac voltage is applied.

Inductive reactance (X_L). The opposition to current flow in an ac circuit caused by an inductance (L), measured in ohms. $X_L = 2\pi fL$.

Inductor. A coil of wire that has the property of inductance and is used in a circuit for that purpose.

In phase. Two waveforms of the same frequency which pass through their minimum and maximum values at the same time and polarity.

Instantaneous voltage (V_i). A value of ac voltage at any instant (time) along a waveform.

Insulator. A material that offers a high resistance to electrical current flow.

Interbase resistance. The resistance between B_1 and B_2 of a unijunction transistor.

Interface. Equipment that connects two circuits together in a computer.

Interlace scanning. An electronic picture production process in which the odd lines are all scanned, then the even lines, to make a complete 525-line picture.

Intermediate frequency (IF). A single frequency that is developed by heterodyning two input signals together in a superheterodyne receiver.

Internal resistance. The resistance between the anode and cathode of an SCR or T_1 and T_2 of a triac. Also called the *dynamic resistance.*

Intrinsic. A semiconductor crystal with a high level of purification.

Inverse. The value of 1 divided by some quantity, such as $1/R_T$ for finding parallel resistance.

Inversely proportional. When one quantity increases or decreases, causing another quantity to do the opposite.

Ion. An atom that has lost or gained electrons, making it a negative or positive charge.

Ionic bond. A binding force that joins one atom to another in a way that fills the valence band of an atom.

***I* signal.** A color signal of a TV system that is in phase with the 3.58-MHz color subcarrier.

Isolation transformer. A transformer with a 1:1 turns ratio used to isolate an ac power line from equipment with a chassis ground.

Junction. The point where P and N semiconductor materials are joined.

Open circuit. A circuit that has a broken path so that no electrical current can flow through it.

Open loop. An operational amplifier connection without a feedback network.

Orbit. The circular path along which electrons travel around the nucleus of an atom.

Orbital. A mathematical probability where an electron will appear in the structure of an atom.

Oscilloscope. An instrument that has a cathode-ray tube to allow a visual display of voltages.

Overcurrent device. A device such as a fuse or circuit breaker which is used to open a circuit when an excess current flows in the circuit.

Overload. A condition that results when more current flows in a circuit than it is designed to carry.

Parallel circuit. A circuit that has two or more current paths.

Parallel resonant circuit. A circuit that has an inductor and capacitor connected in parallel to cause response to frequencies applied to the circuit.

Path. The part of an electrical system through which electrons travel from a source to a load, such as the electrical wiring used in a building.

Peak-to-peak (p-p). An ac value measured from the peak positive to the peak negative alternations.

Peak-to-peak voltage ($V_{p\text{-}p}$). The value of ac sine-wave voltage from positive peak to negative peak.

Peak value. The maximum voltage or current value of an ac wave.

$$V_P = 1.414 \times \text{rms.}$$

Peak voltage (V_{peak}). The maximum positive or negative value of ac sine-wave voltage.

$$V_{peak} = V_{eff} \times 1.41.$$

Peak voltage point (P_V). A characteristic point of the unijunction transistor where the emitter voltage rises to a peak value.

Period (time). The time required to complete one ac cycle;

$$\text{time} = 1/\text{frequency.}$$

Permanent magnet. Bars or other shapes of materials that retain their magnetic properties.

Permeability (μ). The ability of a material to conduct magnetic lines of force as compared to air.

Phase. A position or part of an ac wave that is compared with another wave. A wave can be in phase or out of phase with respect to another.

Phase angle (θ). The angular displacement between applied voltage and current flow in an ac circuit.

Phase shifter. A circuit that changes the phase of an ac waveform.

Photon. A small packet of energy or quantum of energy associated with light.

Photovoltaic cell. A cell that produces dc voltage when light shines onto its surface.

Piezoelectric effect. The property of certain crystal materials to produce a voltage when pressure is applied to them.

Pi filter. A filter with an input capacitor connected to an inductor-capacitor filter, forming the shape of the capital Greek letter pi (π).

Pixel. A discrete picture element on the viewing surface of a cathode-ray tube.

Polarities. *See* Magnetic poles.

Polarity. The direction of an electrical potential (– or +) or a magnetic charge (north or south).

Positive logic. A system where low or 0 has no voltage and high or 1 has voltage.

Potential energy. Energy due to position.

Potentiometer. A variable-resistance component used as a control device in electrical circuits.

Power (P). The rate of doing work in electrical circuits, found by using the equation $P = I \times V$.

Power dissipation. An electronic device characteristic that indicates the ability of the device to give off heat.

Power factor (PF). The ratio of true power in an ac circuit and apparent power:

$$\text{PF} = \frac{\text{true power (W)}}{\text{apparent power (VA)}}$$

Precision resistor. A resistor used when a high degree of accuracy is needed.

Primary cell. A cell that cannot be recharged.

Primary winding. The coil of a transformer to which ac source voltage is applied.

Prime mover. A system that supplies the mechanical energy to rotate an electrical generator.

Programmable read-only memory (PROM). A memory device that can be programmed by the user.

Proton. A particle in the center of an atom which has a positive (+) electrical charge.

Pulsating dc. A voltage or current value that rises and falls with current flow always in the same direction.

Push-pull. An amplifier circuit configuration using two active devices with the input signal being of equal amplitude and 180° out of phase.

***Q* point.** An operating point of an electronic device that indicates its dc or static operation with no ac signal applied.

***Q* signal.** A color signal of a TV system that is out of phase with the 3.58-MHz color subcarrier.

Quadrant. One-fourth or a quarter part of a circle or a graph.

Quality factor (Q). The "figure of merit" or ratio of inductive reactance and resistance in a frequency-sensitive circuit.

Quanta. A discrete amount of energy required to move an electron into a higher energy level.

Quantum theory. A theory based on the absorption and emission of discrete amounts of energy.

Radio frequency (RF). An alternating-current (ac) frequency that is high enough to radiate electromagnetic waves, generally above 30 kHz.

Radio-frequency choke (RFC). An inductor or coil that offers high impedance to radio-frequency ac.

Radix. The base of a numbering system.

Random access memory (RAM). Memory used to run individual programs in a computer.

Reactance (*X*). The opposition to ac current flow due to inductance (X_L) or capacitance (X_C).

Reactive circuit. An ac circuit that has the property of inductance or capacitance.

Reactive power (VAR). The "unused" power of an ac circuit that has inductance or capacitance, which is absorbed by the magnetic or electrostatic field of a reactive circuit.

Read head. A signal pickup transducer that generates a voltage in a tape recorder.

Reading. Sensing binary information that has been stored on a magnetic tape, disk, or IC chip.

Read-only memory (ROM). A form of stored data that can be sensed or read. It is not altered by the sensing process.

Reciprocal. *See* Inverse.

Rectification. The process of changing ac into pulsating dc.

Regenerative feedback. Feedback from the output to the input that is in phase so that it is additive.

Regulation. A measure of the amount of voltage change that occurs in the output of a generator due to changes in load.

Relaxation oscillator. A nonsinusoidal oscillator that has a resting or nonconductive period during its operation.

Relay. An electromagnetically operated switch.

Reluctance (R). The opposition of a material to the flow of magnetic flux.

Repulsion motor. A single-phase ac motor that has a wound rotor and brushes which are shorted together.

Residual magnetism. The magnetism that remains around a material after the magnetizing force has been removed.

Resistance ($\Re$). Opposition to the flow of current in an electrical circuit; its unit of measurement is the ohm (Ω).

Resistive circuit. A circuit whose only opposition to current flow is resistance; a nonreactive circuit.

Resistive heating. A method of heating that relies on the heat produced when electrical current moves through a conductor.

Resistor. A component used to control either the amount of current flow or the voltage distribution in a circuit.

Resonant circuit. *See* Parallel resonant circuit *and* Series resonant circuit.

Resonant frequency (f_r). The frequency that passes most easily through a frequency-sensitive circuit when $X_L = X_C$ in the circuit:

$$f_r = \frac{1}{2\pi\sqrt{L \times C}}$$

Retentivity. The ability of a material to retain magnetism after a magnetizing force has been removed.

Retrace. The process of returning the scanning beam of a camera or picture tube to its starting position.

Reverse bias. A voltage connection method producing little or no current through a *P-N* junction.

Reverse breakdown. The conduction that occurs in quadrant III of the *I-V* characteristics of a triac.

Rheostat. A two-terminal variable resistor.

Right-hand motor rule. The rule applied to find the direction of motion of the rotor conductors of a motor.

Root mean square (rms). Effective value of ac or 0.707 × peak value.

Root mean square (rms) voltage. *See* Effective voltage.

Rotating-armature method. The method used when a generator has dc voltage applied to produce a field to the stationary part (stator) of the machine and voltage is induced into the rotating part (rotor).

Rotating-field method. The method used when a generator has dc voltage applied to produce a field to the rotor of the machine and voltage is induced into the stator coils.

Rotor. The rotating part of an electrical generator or motor.

Running neutral plane. The actual switching position of the commutator and brushes of a dc generator or motor which shifts the theoretical neutral plane due to armature reaction.

Saturation. (1) The strength or intensity of a color used in a TV system. (2) An active device operational region where the output current levels off to a constant value.

Saturation region. A condition of operation where a device is conducting to its full capacity.

Sawtooth waveform. An ac waveform shaped like the teeth of a saw.

Scanning. In a TV system, the process of moving an electron beam vertically and horizontally.

Schematic. A diagram of an electronic circuit made with symbols.

Schematic diagram. A diagram used to show how the components of electrical circuits are wired together.

Scientific notation. The use of "powers of 10" to simplify large and small numbers.

Secondary cell. A cell that can be recharged by applying dc voltage from a battery charger.

Secondary emission. The electrons released by bombardment of a conductive material with other electrons.

Secondary winding. The coil of a transformer into which voltage is induced; energy is delivered to the load circuit by the secondary winding.

Selectivity. A receiver function of picking out a desired radio-frequency signal. Tuning achieves selectivity.

Self-starting. The ability of a motor to begin rotation when electrical power is applied to it.

Semiconductor. A material that has electrical resistance somewhere between a conductor and an insulator.

Sensitivity. *See* Ohms-per-volt rating.

Series circuit. A circuit that has one path for current flow.

Series resonant circuit. A circuit that has an inductor and capacitor connected in series to cause response to frequencies applied to the circuit.

Shaded-pole motor. A single-phase ac induction motor that uses copper shading coils around its poles to produce rotation.

Short circuit. A circuit that forms a direct path across a voltage source so that a high and possibly unsafe electrical current flows.

Shunt. A resistance connected in parallel with a meter movement to extend the range of current it will measure.

Sidebands. The frequencies above and below the carrier frequency that are developed because of modulation.

Siemens (S). *See* Mho.

Signal. An electrical waveform of varying value which is applied to a circuit.

Sine wave. The waveform of ac voltage.

Single phase. The type of electrical power that is supplied to homes and is in the form of a sine wave when it is produced by a generator.

Single-phase ac generator. A generator that produces single-phase ac voltage in the form of a sine wave.

Single-phase motor. Any motor that operates with single-phase ac voltage applied to it.

Skip. An RF transmission signal pattern that is the result of signals being reflected from the ionosphere or the earth.

Sky wave. An RF signal radiated from an antenna into the ionosphere.

Slip rings. Copper rings mounted on the end of a rotor shaft and connected to the brushes and the rotor windings.

Snubber. A resistor-capacitor network placed across a triac to reduce the effect of an inductive voltage.

Software. Instructions, such as a program that controls the operation of a computer.

Solenoid. An electromagnetic coil with a metal core that moves when current passes through the coil.

Solid state. A branch of electronics dealing with the conduction of current through semiconductors.

Source. (1) The part of an electrical system that supplies energy to other parts of the system, such as a battery that supplies energy for a flashlight. (2) The common or energy input lead of a field-effect transistor.

Speaker. A transducer that converts electrical current and/or voltage into sound waves. Also called a *loudspeaker.*

Specific gravity. The weight of a liquid as compared with the weight of water, which has a value of 1.0.

Speed regulation. The ability of a motor to maintain a steady speed with changes in load.

Split-phase motor. A single-phase ac induction motor that has start windings and run windings.

Split-ring commutator. A group of copper bars mounted on the end of a rotor shaft of a dc motor and connected to the brushes and the rotor windings.

Squirrel-cage rotor. A solid rotor used in ac induction motors which has round metal conductors joined at the ends and placed inside the laminated metal motor core.

Stability. The ability of an oscillator to stay at a given frequency or given condition of operation without variation.

Stabilized atom. An atom that has a full complement of electrons in its valence band.

Stable atom. An atom that will not release electrons under normal conditions.

Stage of amplification. A transistor or IC and other components needed to achieve amplification.

Starter. A resistive network used to limit starting current in motors.

Static charge. A charge on a material which is said to be either positive or negative.

Static electricity. Electricity with positive and negative electrical charges which is "at rest."

Static state. A dc operating condition of an electronic device with operating energy but no signal applied.

Stator. The stationary part of an electrical generator or motor.

Steam turbine. A machine that uses the pressure of steam to cause rotation, which is used to turn an electrical generator.

Step-down transformer. A transformer that has a secondary voltage lower than its primary voltage.

Step-up transformer. A transformer that has a secondary voltage higher than its primary voltage.

Stereophonic (stereo). Signals developed from two sources that give the listener the impression of sound coming from different directions.

Storage battery. *See* Secondary cell.

Stylus. A needlelike point that rides in a phonograph record groove.

Substrate. A piece of *N* or *P* semiconductor material which serves as a foundation for other parts of a MOSFET.

Susceptance (*B*). The ability of an inductance (B_L) or a capacitance (B_C) to pass ac current; measured in siemens:

$$B_L = \frac{1}{X_L} \quad \text{and} \quad B_C = \frac{1}{X_C}$$

Switch. A control device used to turn a circuit on or off.

Switching device. An electronic device that can be turned on and off rapidly. This type of device is capable of achieving control by switching.

Symbol. Used as a simple way to represent a component on a diagram or an electrical quantity in a formula.

Symmetrical. A condition of balance where the parts, shapes, and sizes of two or more items are the same.

Sync. An abbreviation for *synchronization.*

Synchronization. A control process that keeps electronic signals in step. The sweep of a TV receiver is synchronized by the transmitted picture signal.

Synchronized. A condition of operation that occurs simultaneously. The vertical and horizontal sweeps of an oscilloscope are often synchronized or placed in sync.

Synchronous motor. An ac motor that operates at a constant speed regardless of the load applied.

Tank circuit. A parallel resonant *LC* circuit.

Telegraphy. The process of conveying messages by coded telegraph signals.

Terminal (T_1 or T_2). The connecting points where load current passes through a triac. These two connection points are also called *main 1* or *2*.

Thermal stability. The condition of an electronic device that indicates its ability to remain at an operating point without variation due to temperature.

Thermionic emission. The release of electrons from a conductive material by the application of heat.

Thermocouple. A device that has two pieces of metal joined together so that when its junction is heated, a voltage is produced.

Theta (θ). The Greek letter used to represent the phase angle of an ac circuit.

Three-phase ac. The type of electrical power that is generated at power plants and transmitted over long distances.

Three-phase ac generator. A generator that produces three ac sine-wave voltages that are separated in phase by 120°.

Threshold. The beginning or entering point of an operating condition. A terminal connection of the LM555 IC timer.

Time constant (*RC*). The time required for the voltage across a capacitor in an *RC* circuit to increase to 63% of its maximum value or decrease to 37% of its maximum value;

$$\text{time} = RC.$$

Time constant (*RL*). The time required for the current through an inductor in an *RL* circuit to increase to 63% of its maximum value or decrease to 37% of its maximum value;

$$\text{time} = L/R.$$

Toggle. A switching mode that changes between two states, such as on and off.

Torque. The turning force delivered by a motor.

Total current. The current that flows from the voltage source of a circuit.

Total resistance. The total opposition to current flow of a circuit, which may be found by removing the voltage source and connecting an ohmmeter across the points where the source was connected.

Total voltage. The voltage supplied by a source.

Trace time. A period of the scanning process where picture information is reproduced or developed.

Transducer. A device that changes energy from one form to another. A transducer can be an input or output device.

Transformer. An ac power control device that transfers energy from its primary winding to its secondary winding by mutual inductance and is ordinarily used to increase or decrease voltage.

Triggered. A control technique that causes a device to change its operational state, such as a triggered monostable multivibrator.

Triggering. The process of causing an *NPNP* device to switch states from off to on.

True power (*W*). The power actually *converted* by an ac circuit, as measured with a wattmeter.

Truth table. A graph or table that displays the operation of a logic circuit with respect to its input and output data.

Turn-on time. The time of an ac waveform when conduction starts.

Turns ratio. The ratio of the number of turns of the primary winding (N_P) of a transformer to the number of turns of the secondary winding (N_S).

Tweeter. A speaker designed to reproduce only high frequencies.

Unipolar. A semiconductor device that has only one *P-N* junction and one type of current carrier. FETs and UJTs are unipolar devices.

Universal motor. A motor that operates with either ac or dc voltage applied.

Valence band. The part of an energy-level diagram that deals with the outermost electrons of an atom.

Valence electrons. Electrons in the outer shell or layer of an atom.

Valley voltage (V_V). A characteristic voltage point of the UJT where the voltage value drops to its lowest level.

Vapor lighting. A method of lighting that uses lamps which are filled with certain gases that produce light when electrical current is applied.

Vector. A straight line whose length indicates magnitude and position indicates direction.

Vertical blocking oscillator. A TV circuit that generates the vertical sweep signal for deflection of the cathode-ray tube or picture tube.

Vestigial sideband. A transmission procedure where part of one sideband is removed to reduce bandwidth.

Videcon. A TV camera tube that operates on the photoconductive principle.

Voice coil. The moving coil of a speaker.

Volatile memory. Stored binary data that are lost or destroyed when the electrical power is removed.

Volt (V). The unit of measurement of electrical potential.

Voltage. The electrical force or "pressure" that causes current to flow in a circuit.

Voltage divider. A combination of resistors that produce different or multiple voltage values.

Voltage drop. The reduction in voltage caused by the resistance of conductors of an electrical distribution system; it causes a voltage less than the supply voltage at points near the end of an electrical circuit that is farthest from the source.

Voltage follower. An amplifying condition where the input and output voltages are of the same value across different impedances.

Voltage regulation. A measure of the amount of voltage change which occurs in the output of a generator due to changes in load.

Voltaic cell. A cell that produces voltage due to two metal electrodes that are suspended in an electrolyte.

Volt-ampere (VA). The unit of measurement of apparent power.

Volt-amperes reactive (VAR). The unit of measurement of reactive power.

Voltmeter. A meter used to measure voltage.

Volt-ohm-milliammeter (VOM). A multifunction, multirange meter which is usually designed to measure voltage, current, and resistance. Same as multimeter.

Watt (W). The unit of measurement of electrical power; the amount of power converted when 1 A of current flows under a pressure of 1 V.

Watt-hour (Wh). A unit of electrical energy measurement equal to watts times hours.

Watt-hour meter. A meter that measures accumulated energy conversion.

Wattmeter. A meter used to measure power conversion.

Waveform. The pattern of an ac frequency derived by looking at instantaneous voltage values that occur over time; on a graph, a waveform is plotted with instantaneous voltages on the vertical axis and time on the horizontal axis.

Wavelength. The distance between two corresponding points that represents one complete wave.

Wheatstone bridge. A bridge circuit that makes precision resistance measurements by comparing an "unknown" resistance with a known or standard resistance value.

Woofer. A speaker designed to reproduce low audio frequencies with high quality.

Work. The transforming or transferring of energy.

Working voltage. A rating of capacitors which is the maximum voltage that can be placed across the plates of a capacitor without damage occurring.

Writing. Placing binary data into the storage cells of a memory device.

Wye connection. A method of connecting three-phase alternator stator windings in which the beginnings *or* ends of each phase winding are connected to form a common or neutral point; power lines extend from the remaining beginnings or ends.

***X* axis.** A horizontal line.

***Y* axis.** A vertical line.

***Y* signal.** The brightness or luminance signal of a TV system.

Zero beating. The resultant difference in frequency that occurs when two signals of the same frequency are heterodyned.

APPENDIX

B

SOLDERING

Soldering is an important skill for electronics technicians. Good soldering is important for the proper operation of equipment.

Solder is an alloy of tin and lead. The solder that is most used is 60/40 solder, which means that it is made from 60% tin and 40% lead. Solder melts at a temperature of about 400°F.

For solder to adhere to a joint, the parts must be hot enough to melt the solder. The parts must be kept clean to allow the solder to flow evenly. Rosin flux is contained inside the solder. It is called *rosin-core solder.*

A good mechanical joint must be made when soldering. Heat is then applied until the materials are hot. When they are hot, solder is applied to the joint. The heat of the metal parts (not the soldering tool) is used to melt the solder. Only a small amount of heat should be used. Solder should be used sparingly. The joint should appear smooth and shiny. If it does not, it could be a "cold" solder joint. Be careful not to move the parts when the joint is cooling or it could also cause a cold joint.

When parts that could be damaged by heat are soldered, be careful not to overheat them. Semiconductor components such as diodes and transistors are heat sensitive. One way to prevent heat damage is to use a *heat sink,* such as a pair of pliers. A heat sink is clamped to a wire between the joint and the device being soldered. A heat sink absorbs heat and protects delicate devices. Printed circuit boards, such as the one shown in Fig. B–1, are also sensitive to heat. Care should be taken not to damage PCBs when soldering parts onto them.

Several types of soldering irons and soldering guns are shown in Figs. B–2 to B–7. Small, low-wattage irons should be used with PCBs and semiconductor devices.

Below are some rules for good soldering:

1. Be sure that the tip of the soldering iron is clean and tinned.
2. Be sure that all the parts to be soldered are heated. Place the tip of the soldering iron so that the wires and the soldering terminal are heated evenly.
3. Be sure not to overheat the parts.
4. Do not melt the solder onto the joint. Let the solder flow onto the joint.
5. Use the right kind and size of solder and soldering tools.
6. Use the right amount of solder to do the job but not enough to leave a "blob."
7. Be sure not to allow parts to move before the solder joint cools.

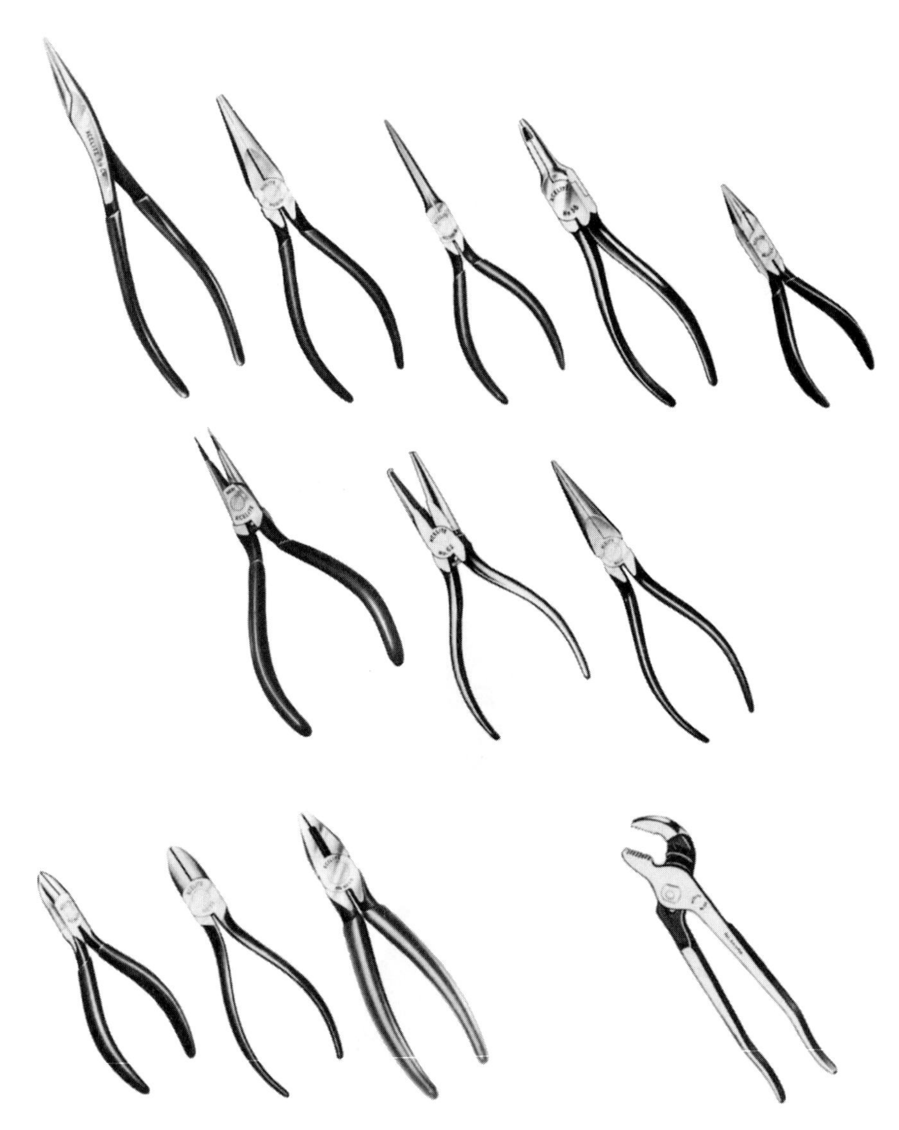

Figure C–8. (top) Long-nose or needle-nose pliers; (middle) diagonal cutting pliers; (bottom) channel lock pliers. (Courtesy of The Cooper Group.)

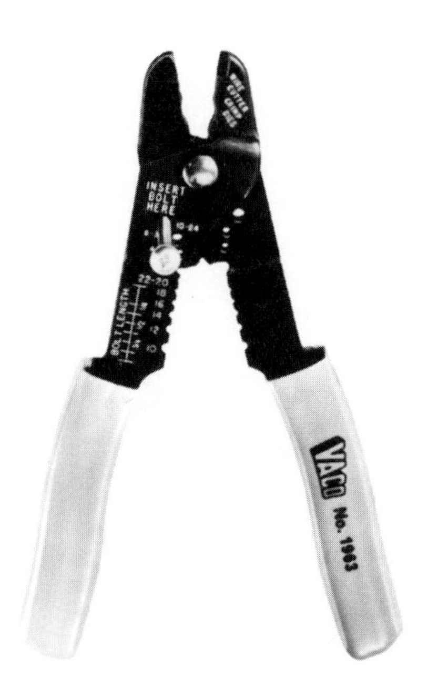

Figure C–9. Wire crimping tool with stripping and bolt-cutting capability. (Courtesy of Vaco Products Co.)

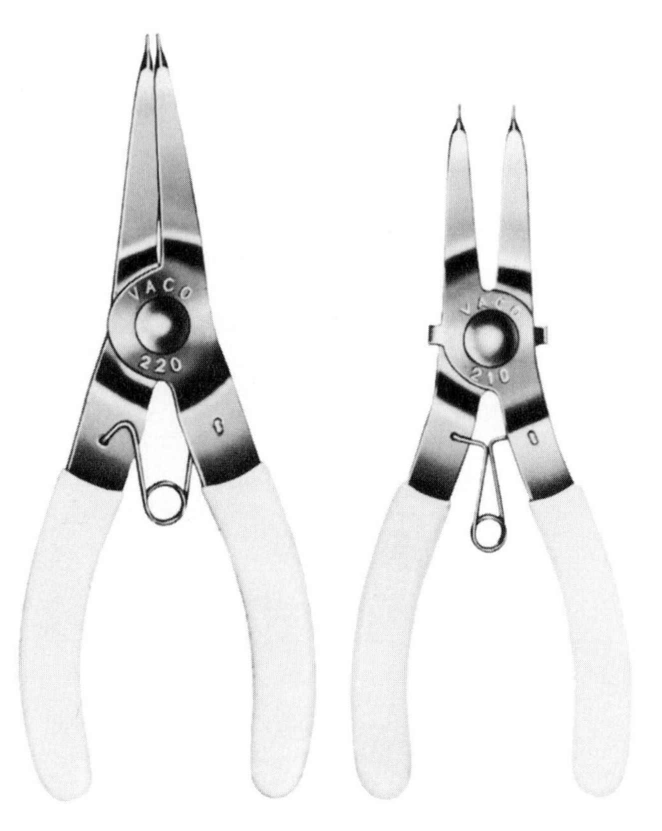

Figure C–10. C-ring pliers. (Courtesy of Vaco Products Co.)

Figure C–11. Cheater cord for electronic equipment repair. (Courtesy of GC Electronics.)

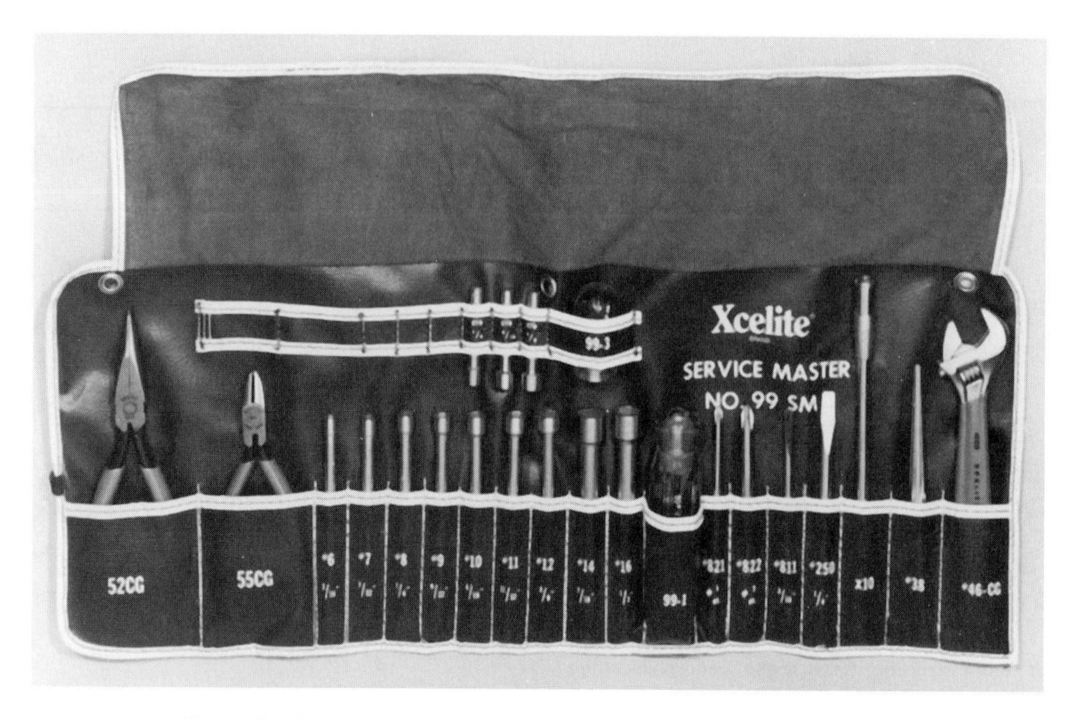

Figure C–12. Roll of tools for servicing. (Courtesy of The Cooper Group.)

Figure C–13. Electronic tool kit. (Courtesy of The Cooper Group.)

Figure C–14. Portable multipurpose tool. (Courtesy of The Cooper Group.)

SYMBOLS FOR ELECTRICITY AND ELECTRONICS

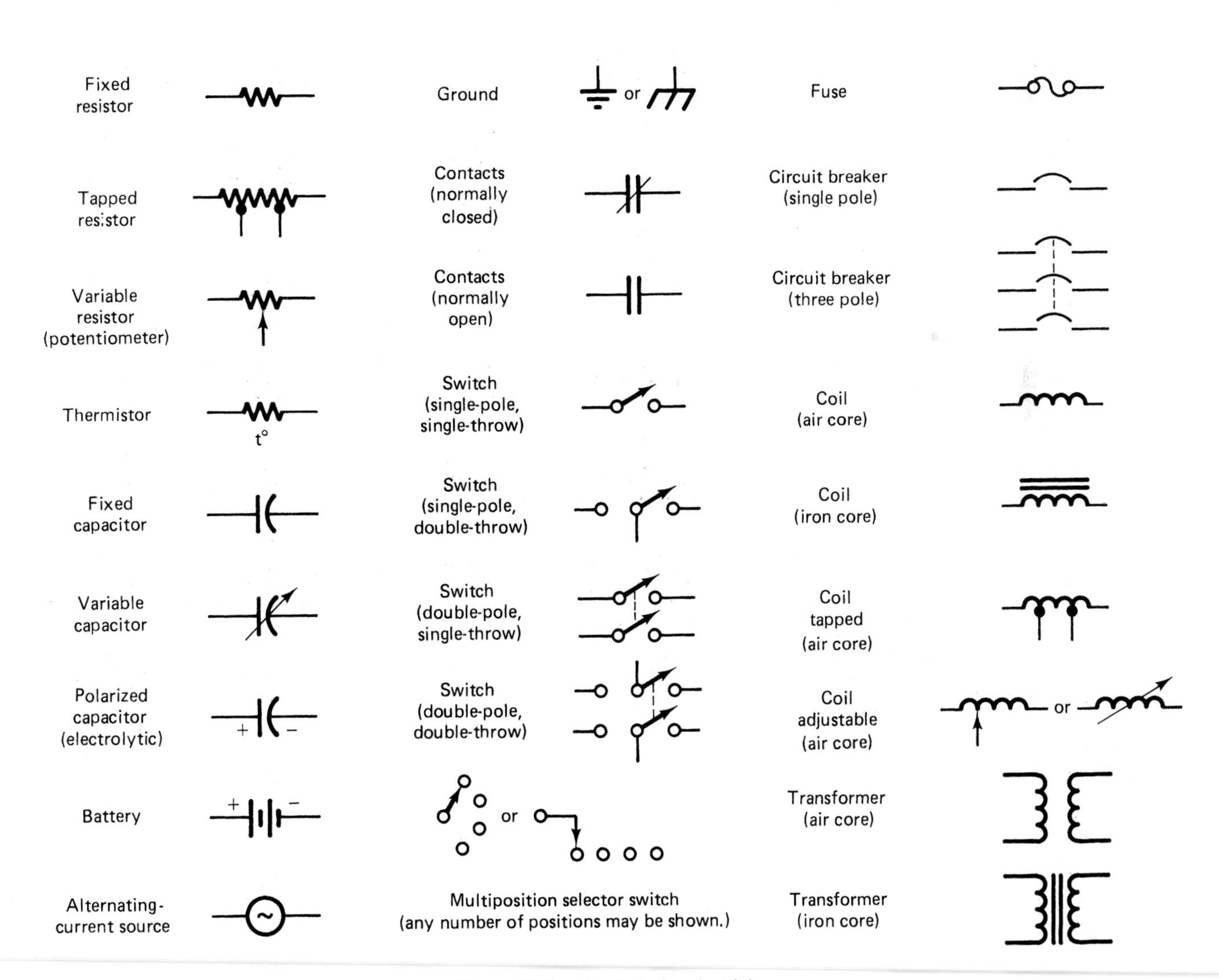

Figure D–1. Symbols for electricity.

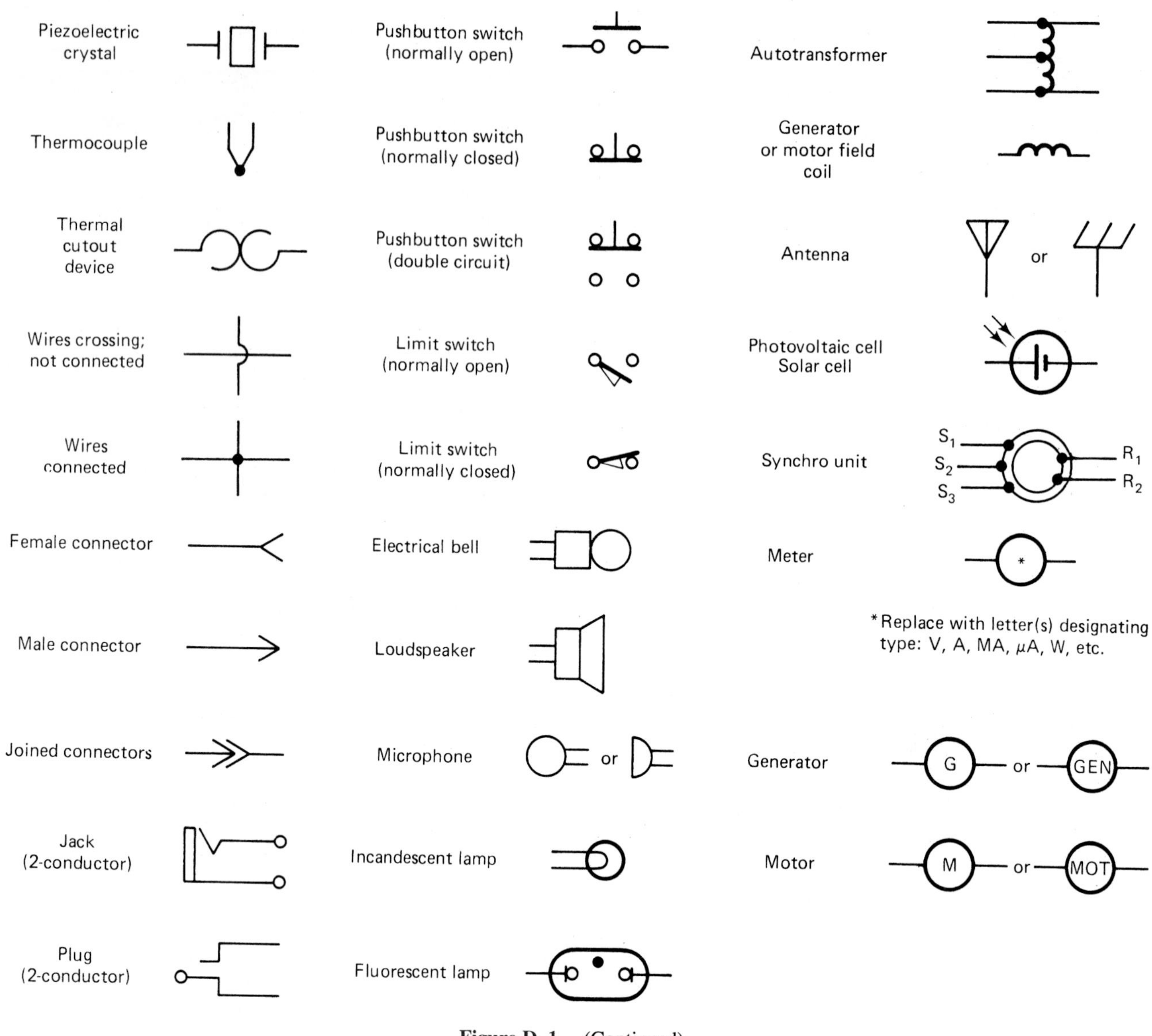

Figure D–1. (Continued)

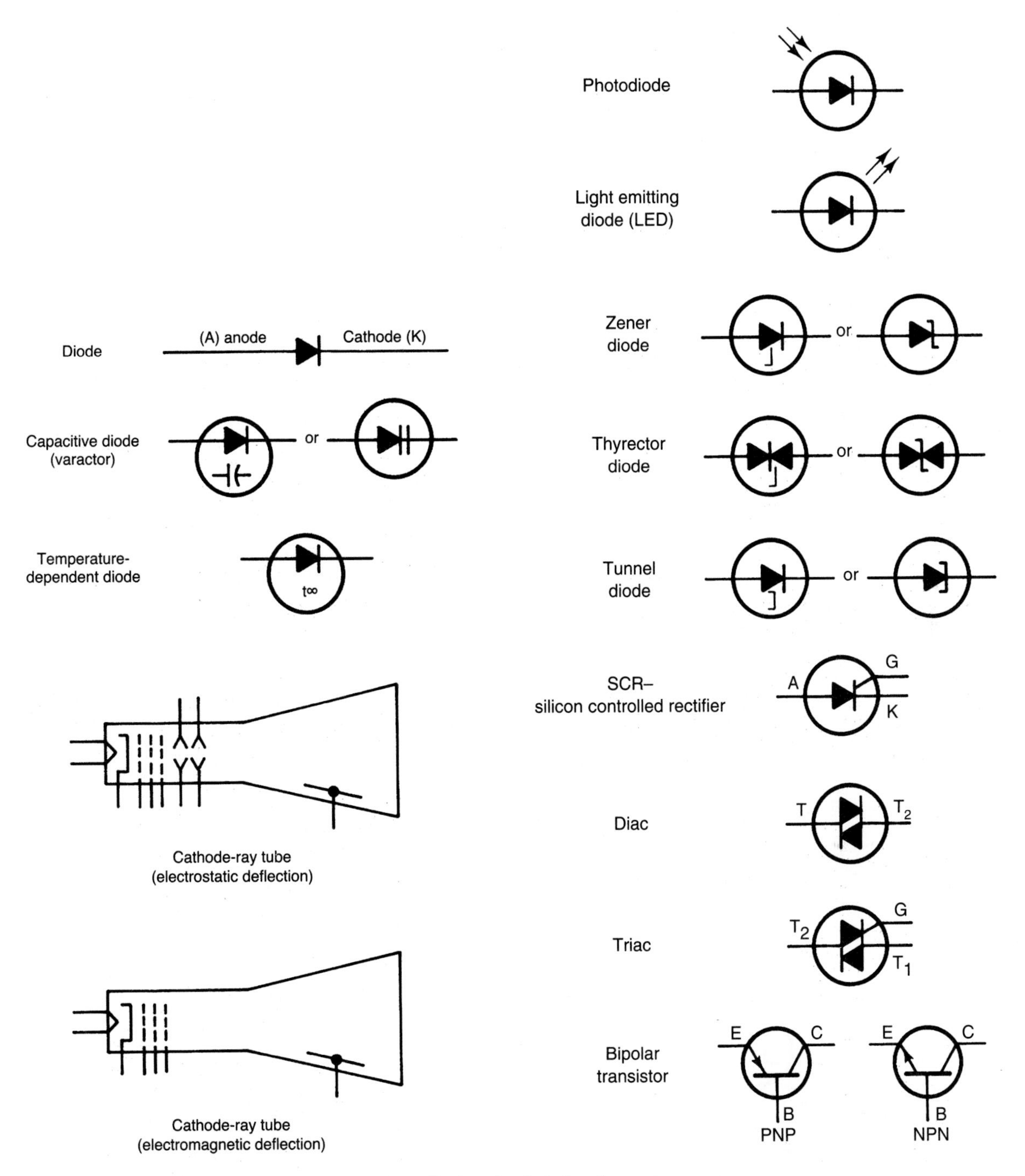

Figure D–2. Symbols for electronics.

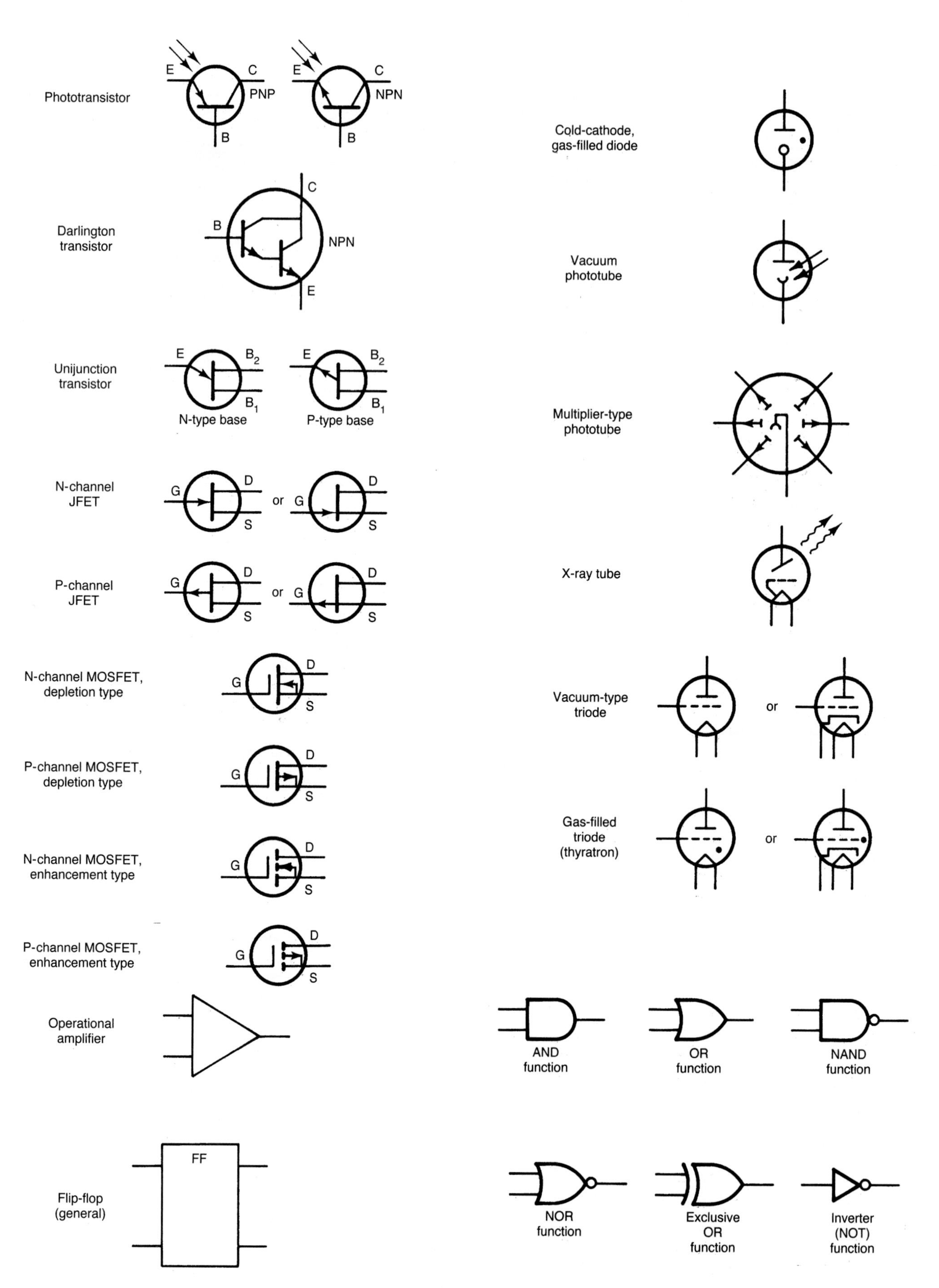

Figure D–2. (Continued)

COMMON LOGARITHMS AND NATURAL LOGARITHMS

LAWS OF EXPONENTS

Logarithms are based on the algebraic laws of exponents, as shown by the following equations:

$$b^m \cdot b^n = b^{m+n}$$
$$\frac{b^m}{b^n} = b^{m-n}$$
$$(b^m)^n = b^{mn}$$

where b is the base and m and n are exponents. When the bases of two quantities are the same, the first equation shows that multiplication is performed by adding the exponents. The second equation shows that when division is indicated, the exponents are subtracted. The third equation shows that to raise a number with an exponent to a power, you must multiply the exponent by the power. The product then becomes the exponent of the same base. It is usually easier and quicker when handling two numbers to add rather than to multiply, to subtract rather than divide, and to multiply rather than raise a number to a power. Therefore, when numbers are converted to exponent form, the mathematical operations of multiplication, division, and powers are reduced to addition, subtraction, and multiplication, respectively.

If a real number is replaced by the form b^x, the laws of exponents can be used for the mathematical operations of multiplication, division, powers, and roots. This is shown by the following equation:

$$N = b^x$$

where

N = any positive real number
b = base
x = logarithm (log) of N

The equation for the logarithm is

$$x = \log N$$

Any positive number except 1 can be used as the base b for a system of logarithms; 10 is used as the base for the common system of logarithms and ϵ, which is equal to approximately 2.718, is used as the base for the natural system of logarithms. The system of logarithms that is used can be indicated in the equation by a subscript following the abbreviation. Thus $x = \log_{10}N$ is for the base 10, $x = \log_2 N$ is for the base 2, and $x = \log_\epsilon N$ is for the base ϵ. When the base is not indicated, it is understood to be 10 and $x = 1\text{n}N$ is another method used for indicating the base ϵ.

COMMON LOGARITHMS

To every positive real number you can equate an exponential quantity to the base 10. Numbers can easily be written in common logarithmic form. The logarithm for 1 is 0, the logarithm for 10 is 1, the logarithm for 100 is 2, and so on. Therefore, the logarithm for any number between 1 and 10 must be between 0 and 1. A number between 10 and 100 has a logarithm between 1 and 2 (1 plus a decimal). A number between 100 and 1000 has a logarithm between 2 and 3. Notice that logarithms for positive numbers less than 1 are negative. A number between 1 and 0.1 has a logarithm between 0 and –1, a number between 0.1 and 0.01 has a logarithm between –1 and –2, and so on.

The common logarithm of a number consists of two parts, the characteristic, which is the integral part, and the mantissa, which is the decimal part. Table E–1 shows that a number with two digits to the left of the decimal point has a characteristic of 1, a number with three digits to the left of the decimal point has a characteristic of 2, and so on. Therefore, the characteristic of the logarithm of any number greater than 1 is always one less than the number of digits preceding the decimal point.

The characteristic of the logarithm of any number less than 1 is always negative, and its value is determined by the number of zeros between the decimal point and the first significant digit. Table E–1 shows that a number with no zeros between the decimal point and the first significant digit has a characteristic of –1. A number with one zero between the decimal point and the first significant digit has a characteristic of –2, and so on. It is usually more convenient to write the logarithm of a number in positive form; therefore, negative logarithms are usually transformed as follows:

$$-1 = 9.0000 - 10$$
$$-7 = 3.0000 - 10$$
$$-14 = 6.0000 - 20$$

When using the base 10, the mantissa of a number is independent of the position of the decimal point. Thus 14.83, 14.830, and 0.001483 all have the same mantissa.

INVERSE LOGARITHMS

The process of finding the number corresponding to a given logarithm is called *finding the antilogarithm* or *inverse logarithm.* To find the inverse logarithm, use the $\log^{-1}$ key on the calculator. The characteristic of the logarithm places the decimal point in the resulting number. Refer to Fig. F–20 for calculator procedures.

NATURAL LOGARITHMS

Many mathematical operations can be simplified by using logarithms with the base ϵ. The value to which the expression

$$(1 + X)^{1/Z}$$

approaches, as Z approaches zero, is called *epsilon.* This value is denoted by the Greek symbol ϵ and is equal to 2.71828. A sample problem follows to illustrate the use of natural logarithms.

TABLE E–1 Comparison of Exponential and Common Logarithmic Equations.

Exponential Equations	*Common Logarithmic Equations*
$10^6 = 1{,}000{,}000$	$\log 1{,}000{,}000 = 6$
$10^5 = 100{,}000$	$\log 100{,}000 = 5$
$10^4 = 10{,}000$	$\log 10{,}000 = 4$
$10^3 = 1000$	$\log 1000 = 3$
$10^2 = 100$	$\log 100 = 2$
$10^1 = 10$	$\log 10 = 1$
$10^0 = 1$	$\log 1 = 0$
$10^{-1} = 0.1$	$\log 0.1 = -1$
$10^{-2} = 0.01$	$\log 0.01 = -2$
$10^{-3} = 0.001$	$\log 0.001 = -3$
$10^{-4} = 0.0001$	$\log 0.0001 = -4$
$10^{-5} = 0.00001$	$\log 0.00001 = -5$
$10^{-6} = 0.000001$	$\log 0.000001 = -6$

SAMPLE PROBLEM

The instantaneous current in the series *RL* circuit can be found by using the formula

$$i = \frac{V}{R}(1 - \epsilon^{-Rt/L})$$

where $\epsilon = 2.718$. This formula shows that initially the current is zero. It then increases and approaches the value of V/R. If $V = 15$ V, $R = 130\ \Omega$, and $L = 0.03$ H, calculate the time it takes the current to reach 0.1 A. To find the time, use logarithms as follows:

$$\frac{iR}{V} - 1 = -\frac{1}{\epsilon Rt/L}$$

$$\epsilon^{Rt/L}\left(1 - \frac{iR}{V}\right) = 1$$

$$\frac{Rt}{L}\log \epsilon + \log\left(1 - \frac{iR}{V}\right) = \log 1$$

Because $\log 1 = 0$.

$$t = \frac{L}{R}\,\frac{-\log(1 - iR/V)}{\log \epsilon}$$
$$\log \epsilon = 0.4343$$

To find $\log(1 - iR/V)$, substitute values for i, R, and V as follows:

$$\log\left(1 - \frac{iR}{V}\right) = \log\left(1 - \frac{0.1 \times 130}{15}\right)$$
$$= \log(1 - 0.867)$$
$$= \log 0.133$$
$$= 9.1239 - 10$$
$$= -0.8761$$

t can then be found as follows:

$$
\begin{aligned}
t &= \frac{L}{R}\,\frac{-(-0.8761)}{0.4343} \\
&= \frac{0.03 \times 0.8761}{130 \times 0.4343} \\
&= \frac{0.0263}{56.5} \\
&= 0.000465 \text{ s} = 465\ \mu\text{s}
\end{aligned}
$$

EXPONENTIAL EQUATIONS

An exponential curve is the graph of an equation in which a variable appears as an exponent. Figure E–1 shows two exponential curves where the exponent is used as the independent variable. For the equation $y = 10^x$, y increases exponentially as x is increased linearly. In the equation $y = 10^{-x}$, y decreases exponentially as x is increased linearly. The two curves cross at the point where the exponent is 0 and y is 1.

Exponential curves are often useful when analyzing electrical circuit operation. A graph for equations that relate capacitor charge and discharge voltage to time is shown in Fig. E–2. The following equation is for the voltage across a capacitor in a series RC circuit as it is being charged from a dc source:

$$e = E(1 - \epsilon^{-t/RC})$$

where the variables are e, the instantaneous voltage across the capacitor; and t, the time in seconds that the capacitor has been charging. All other symbols represent constants. E is the voltage of the dc source, R is the resistance in ohms, C is the capacitance in farads, and ϵ is 2.718. When t is zero, t/RC is zero and e is zero, as follows:

$$
\begin{aligned}
e &= E(1 - \epsilon^{-0}) \\
&= E(1 - 1) \\
&= 0
\end{aligned}
$$

As t is increased, the value of $\epsilon^{-t/RC}$ decreases and e approaches the value of E.

The voltage across a capacitor in a series RC circuit as it is being discharged can be found by the following equation:

$$e = E(\epsilon^{-t/RC})$$

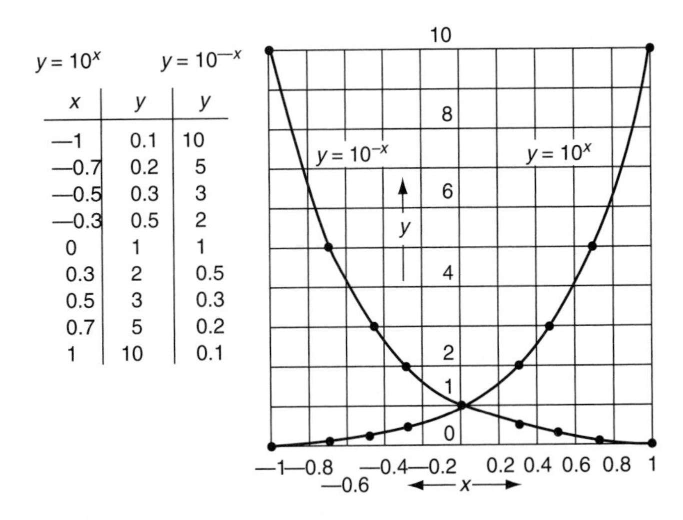

$y = 10^x$		$y = 10^{-x}$
x	y	y
−1	0.1	10
−0.7	0.2	5
−0.5	0.3	3
−0.3	0.5	2
0	1	1
0.3	2	0.5
0.5	3	0.3
0.7	5	0.2
1	10	0.1

Figure E–1. Exponential curves: equations and graph.

When t is 0, e equals E:

$$
\begin{aligned}
e &= E\epsilon^{-0} \\
&= E(1)
\end{aligned}
$$

where E is the voltage to which the capacitor has been charged. As t is increased, the value of $\epsilon^{-t/RC}$ decreases and e approaches 0.

LOGARITHMIC EQUATIONS AND GRAPHS

A logarithmic curve is the graph of an equation in which one variable is the logarithm of another variable. The general equation for a logarithm curve is $y = \log x$. Graphs for this general equation for logarithms to the base 10 ($\log x$) and the base ϵ ($\ln x$) are shown in Fig. E–3. For these graphs x is made the independent variable. For both curves the logarithm of 1 is zero; when x is greater than 1, y is positive, and when x is less than 1, y is negative.

A graph that has logarithmic scales for the two axes is called a *logarithmic graph.* Nonlinear equations that have only two terms and have exponents that are not variables can be converted to a logarithmic form that is a linear equation. Straight-line graphs are more easily constructed and are usually more accurate than curved-line graphs. Therefore, it is often desirable to convert a nonlinear equation into a logarithmic form for graphical presentation.

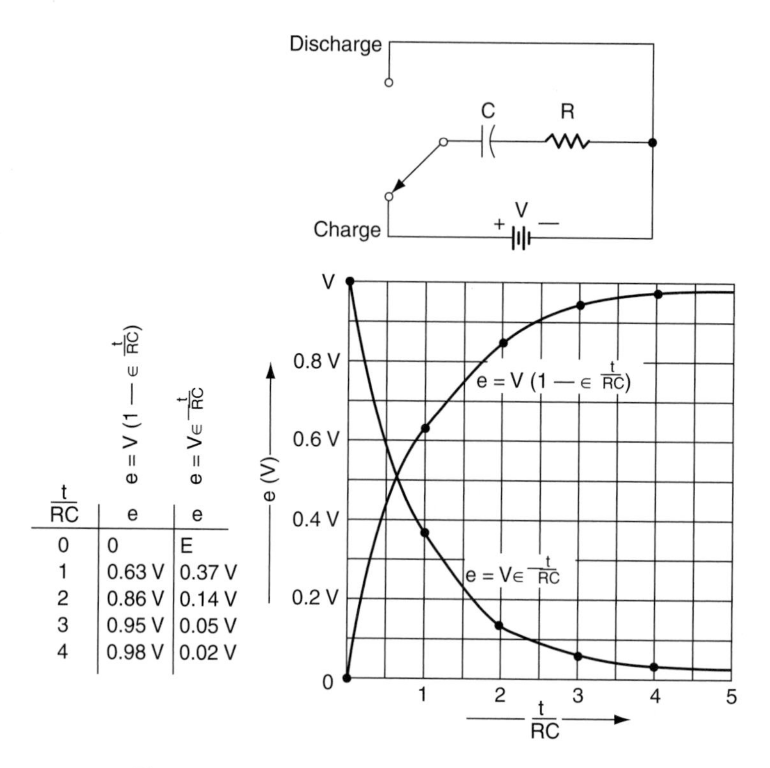

$\frac{t}{RC}$	e ($e = V(1 - \epsilon^{-\frac{t}{RC}})$)	e ($e = V\epsilon^{-\frac{t}{RC}}$)
0	0	E
1	0.63 V	0.37 V
2	0.86 V	0.14 V
3	0.95 V	0.05 V
4	0.98 V	0.02 V

Figure E–2. Capacitor charge and discharge curves.

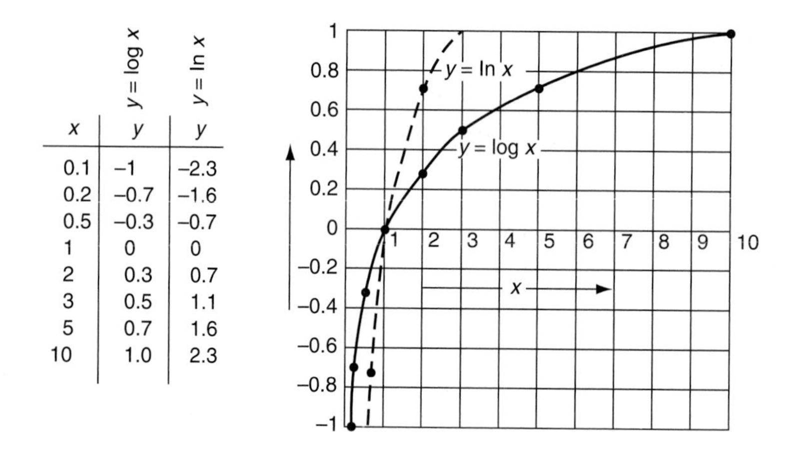

x	y ($y = \log x$)	y ($y = \ln x$)
0.1	−1	−2.3
0.2	−0.7	−1.6
0.5	−0.3	−0.7
1	0	0
2	0.3	0.7
3	0.5	1.1
5	0.7	1.6
10	1.0	2.3

Figure E–3. Logarithmic curves: equations and graph.

CALCULATOR EXAMPLES

This appendix is devoted to the effective use of electronic calculators as efficient tools for problem solving. The approach to calculator usage is that it may be used in ways that strengthen the achievement of learning. Both an understanding of the calculator and an understanding of the problem go hand in hand with getting correct answers and gaining knowledge.

Simple electronic circuit problems involving small, whole numbers are often solved in a few seconds by pencil-and-paper arithmetic. We can determine that the application of 12 V to a 6-Ω resistor results in a current of 2 A through the resistor because Ohm's law (see Chap. 3) instructs us to divide voltage by resistance to get current. As the electronic values grow larger and use more decimal places, calculations become more difficult and time consuming. Students quickly learn that an electronic calculator is a great labor-saving device, which also reduces errors and increases accuracy of answers. Multidigit numbers may be added, subtracted, multiplied, divided, squared, and more by a few simple strokes on the buttons of a calculator keypad. The fundamental advantage of using electronic calculators is to make complicated calculations easier.

The calculator procedures in this appendix illustrate how to use a typical scientific calculator to solve electronic and math problems (Fig. F–1). There are many similar calculators on the market. These procedures will work on most scientific calculators with a few changes in key labels and button sequences. One major exception is the calculator which uses RPN (reverse Polish notation) logic in the keying sequences. It will be necessary to translate the procedures given here into procedures that work on other calculators. The calculator sequences are self-explanatory. Calculator function buttons used in the book are summarized here as an aid to understanding the sequences and as a cross-reference with functions for other types of calculators. The specific sequences given here will not solve every problem that may come along. They should, however, provide enough experience for students to be able to "design" the sequences for other problems. The calculator procedures of this appendix are listed in Fig. F–2.

The remainder of this appendix shows calculator procedures for solving various types of problems (Figs. F–3 through F–20).

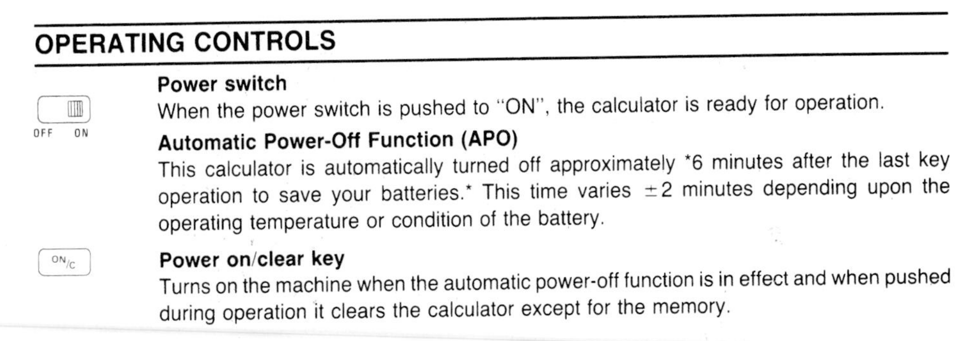
OPERATING CONTROLS

OFF ON

Power switch
When the power switch is pushed to "ON", the calculator is ready for operation.

Automatic Power-Off Function (APO)
This calculator is automatically turned off approximately *6 minutes after the last key operation to save your batteries.* This time varies ±2 minutes depending upon the operating temperature or condition of the battery.

ON/C

Power on/clear key
Turns on the machine when the automatic power-off function is in effect and when pushed during operation it clears the calculator except for the memory.

Figure F–1. Instructions for use of a scientific calculator. (Courtesy of Sharp Co.)

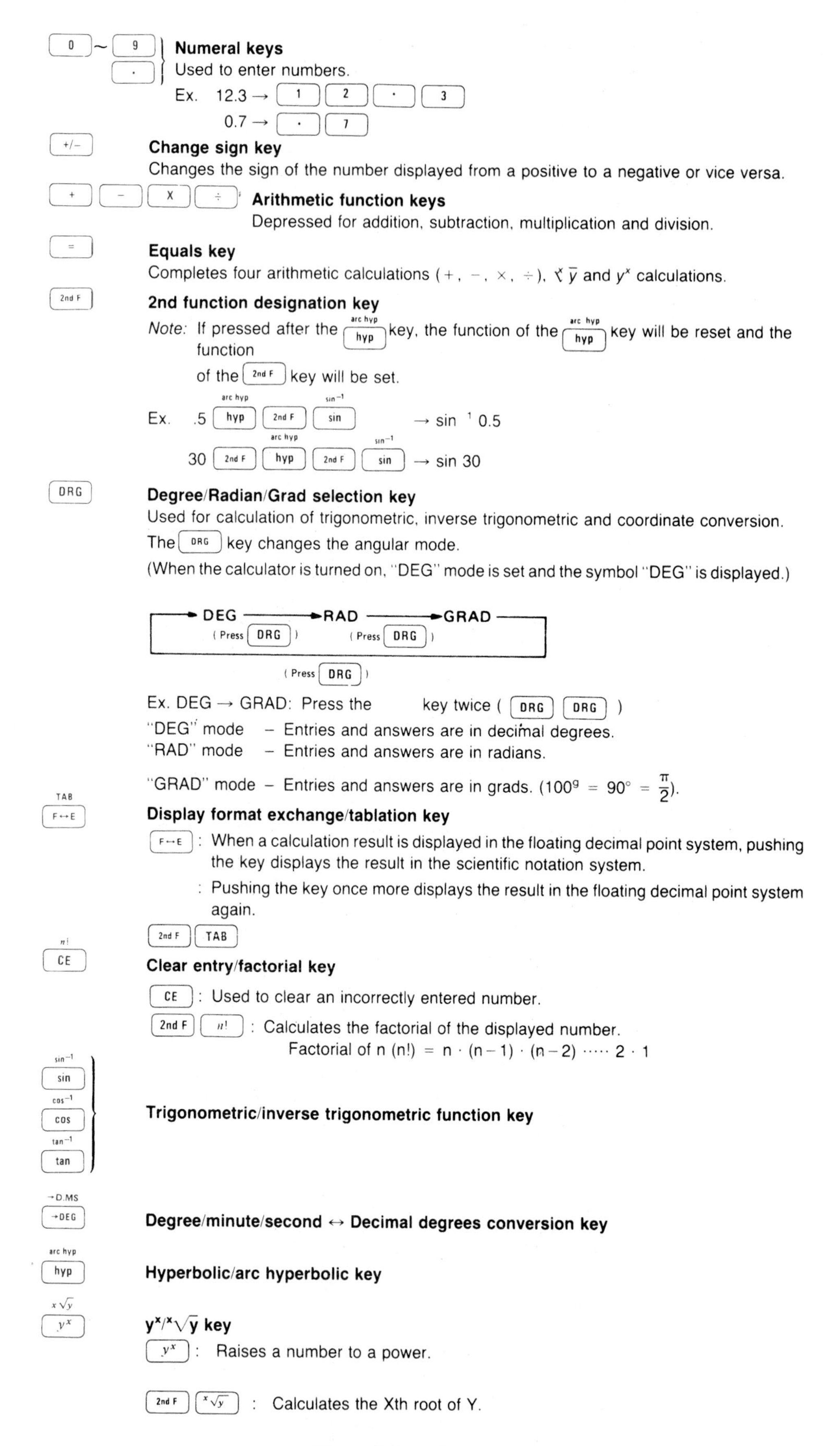

[0] ~ [9] [·] **Numeral keys**
Used to enter numbers.
Ex. 12.3 → [1] [2] [·] [3]
0.7 → [·] [7]

[+/−] **Change sign key**
Changes the sign of the number displayed from a positive to a negative or vice versa.

[+] [−] [X] [÷] **Arithmetic function keys**
Depressed for addition, subtraction, multiplication and division.

[=] **Equals key**
Completes four arithmetic calculations (+, −, ×, ÷), $\sqrt[x]{y}$ and y^x calculations.

[2nd F] **2nd function designation key**
Note: If pressed after the [hyp] key, the function of the [hyp] key will be reset and the function
of the [2nd F] key will be set.
Ex. .5 [hyp] [2nd F] [sin] → $\sin^{-1} 0.5$
30 [2nd F] [hyp] [2nd F] [sin] → sin 30

[DRG] **Degree/Radian/Grad selection key**
Used for calculation of trigonometric, inverse trigonometric and coordinate conversion.
The [DRG] key changes the angular mode.
(When the calculator is turned on, "DEG" mode is set and the symbol "DEG" is displayed.)

→ DEG (Press [DRG]) → RAD (Press [DRG]) → GRAD (Press [DRG]) →

Ex. DEG → GRAD: Press the key twice ([DRG] [DRG])
"DEG" mode – Entries and answers are in decimal degrees.
"RAD" mode – Entries and answers are in radians.
"GRAD" mode – Entries and answers are in grads. ($100^g = 90° = \frac{\pi}{2}$).

[F↔E] (TAB) **Display format exchange/tablation key**
[F↔E]: When a calculation result is displayed in the floating decimal point system, pushing the key displays the result in the scientific notation system.
: Pushing the key once more displays the result in the floating decimal point system again.
[2nd F] [TAB]

[CE] (n!) **Clear entry/factorial key**
[CE]: Used to clear an incorrectly entered number.
[2nd F] [n!]: Calculates the factorial of the displayed number.
Factorial of n (n!) = n · (n − 1) · (n − 2) ····· 2 · 1

[sin] ($\sin^{-1}$) [cos] ($\cos^{-1}$) [tan] ($\tan^{-1}$) **Trigonometric/inverse trigonometric function key**

[→DEG] (→D.MS) **Degree/minute/second ↔ Decimal degrees conversion key**

[hyp] (arc hyp) **Hyperbolic/arc hyperbolic key**

[y^x] ($\sqrt[x]{y}$) **y^x/$\sqrt[x]{y}$ key**
[y^x]: Raises a number to a power.
[2nd F] [$\sqrt[x]{y}$] : Calculates the Xth root of Y.

Figure F–1. (Continued)

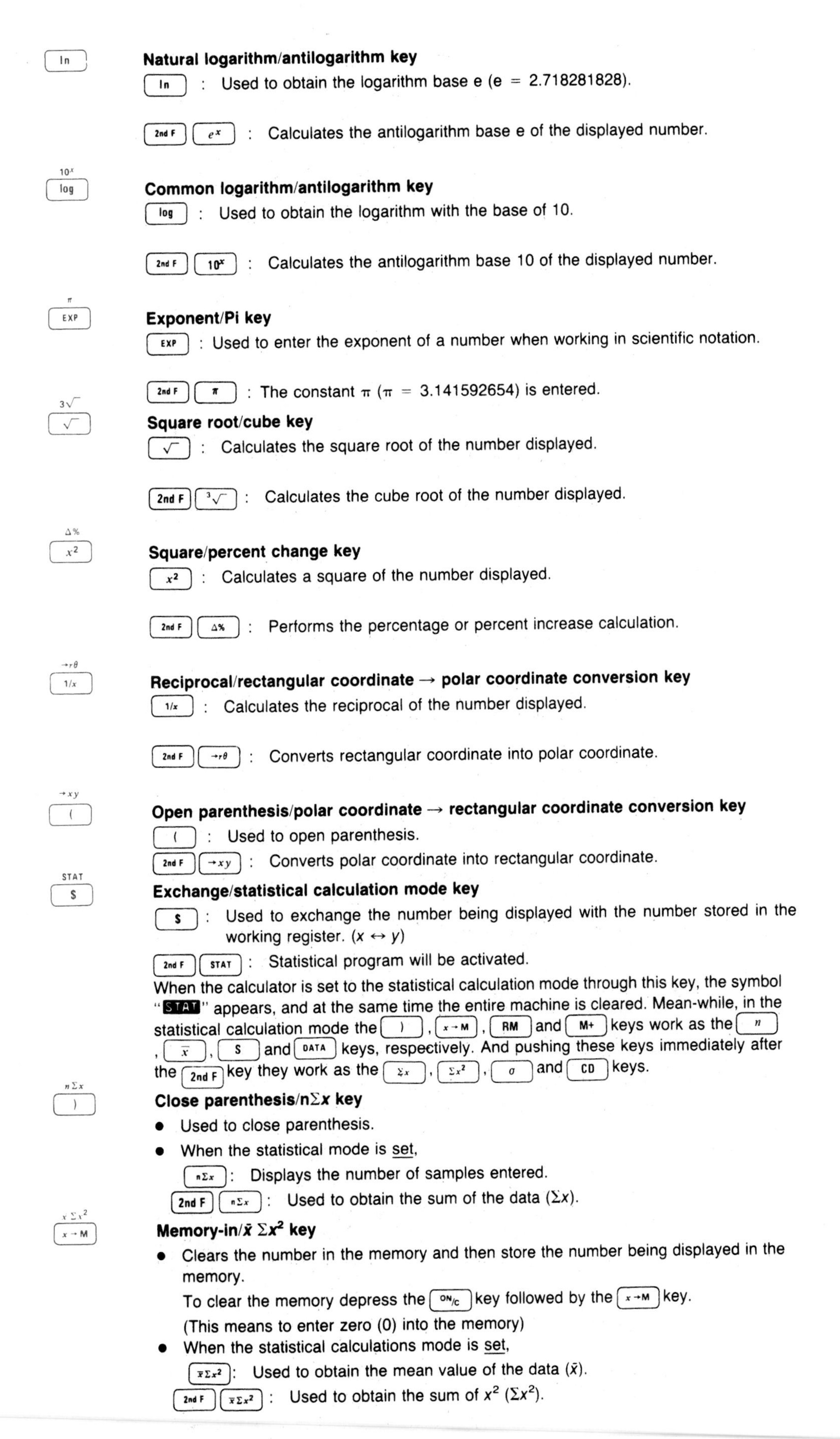

Natural logarithm/antilogarithm key

[ln] : Used to obtain the logarithm base e (e = 2.718281828).

[2nd F] [e^x] : Calculates the antilogarithm base e of the displayed number.

Common logarithm/antilogarithm key

[log] : Used to obtain the logarithm with the base of 10.

[2nd F] [10^x] : Calculates the antilogarithm base 10 of the displayed number.

Exponent/Pi key

[EXP] : Used to enter the exponent of a number when working in scientific notation.

[2nd F] [π] : The constant π (π = 3.141592654) is entered.

Square root/cube key

[$\sqrt{\ }$] : Calculates the square root of the number displayed.

[2nd F] [$\sqrt[3]{\ }$] : Calculates the cube root of the number displayed.

Square/percent change key

[x^2] : Calculates a square of the number displayed.

[2nd F] [$\Delta\%$] : Performs the percentage or percent increase calculation.

Reciprocal/rectangular coordinate → polar coordinate conversion key

[$1/x$] : Calculates the reciprocal of the number displayed.

[2nd F] [$\rightarrow r\theta$] : Converts rectangular coordinate into polar coordinate.

Open parenthesis/polar coordinate → rectangular coordinate conversion key

[( ] : Used to open parenthesis.

[2nd F] [$\rightarrow xy$] : Converts polar coordinate into rectangular coordinate.

Exchange/statistical calculation mode key

[S] : Used to exchange the number being displayed with the number stored in the working register. ($x \leftrightarrow y$)

[2nd F] [STAT] : Statistical program will be activated.

When the calculator is set to the statistical calculation mode through this key, the symbol "STAT" appears, and at the same time the entire machine is cleared. Mean-while, in the statistical calculation mode the [ )], [$x \rightarrow M$], [RM] and [M+] keys work as the [n], [$\bar{x}$], [S] and [DATA] keys, respectively. And pushing these keys immediately after the [2nd F] key they work as the [Σx], [Σx^2], [σ] and [CD] keys.

Close parenthesis/nΣx key

- Used to close parenthesis.
- When the statistical mode is set,

  [$n\Sigma x$]: Displays the number of samples entered.

  [2nd F] [$n\Sigma x$] : Used to obtain the sum of the data (Σx).

Memory-in/$\bar{x}$ Σx^2 key

- Clears the number in the memory and then store the number being displayed in the memory.

  To clear the memory depress the [ON/C] key followed by the [$x \rightarrow M$] key.

  (This means to enter zero (0) into the memory)
- When the statistical calculations mode is set,

  [$\bar{x}\Sigma x^2$]: Used to obtain the mean value of the data ($\bar{x}$).

  [2nd F] [$\bar{x}\Sigma x^2$] : Used to obtain the sum of x^2 (Σx^2).

Figure F–1. (Continued)

RM

Recall memory/standard deviation key

- Displays the contents of the memory. The contents of the memory remain unchanged after this key operation.
- When the statistical mode is set,

  Sσ : Used to obtain the standard deviation of the samples (S).

  2nd F Sσ : Used to obtain the standard deviation of the population (σ).

DATA CD
M+

Memory plus/DATA CD key

- Used to add the number being displayed or a calculated result to the contents of the memory.
  When subtracting a number from the memory, press the +/- and M+ keys in this order.
- When the statistical mode is set,

  DATA CD : Used to enter the data (numbers).

  2nd F DATA CD : Used to correct a mis-entry. (Delete function).

DISPLAY

(1) **Display format**

M DEG 1.2345678 -99

Mantissa Exponent

(2) **Symbols**

– : **Minus symbol**
Indicates that the number in the display following the "—" is a negative.

M : **Memory symbol**
Appears when a number is stored in the memory.

E : **Error symbol**
Appears when an overflow or an error is detected.

● : **Battery indicator**
The battery indicator is a grey dot located at the left side of the display. When this dot is not on, the batteries must be replaced.

2ndF : **2nd function designation symbol**
Appears when the 2nd function is designated.

HYP : **Hyperbolic function designation symbol**
Appears when hyperbolic function is designated.

DEG : **Degree mode symbol**
Appears when the degree mode is designated.

RAD : **Radian mode symbol**
Appears when the radian mode is designated.

GRAD: **Grad mode symbol**
Appears when the grad mode is designated.

STAT : **Statistical calculation mode symbol**
Appears when statistical calculation mode is set.

( ) **Parenthesis symbol**
Appears when a calculation with parenthesis is performed by pressing the (key.

(3) **Display system**

All answers will be displayed in either floating decimal point system or scientific notation system.

- The answer in the following area will be displayed in floating decimal point system.

  $0.000000001 \leq |x| \leq 9999999999$

Figure F–1. (Continued)

Problem	*Figure*
Addition	F–3
Subtraction	F–4
Multiplication	F–5
Division	F–6
Fractions	F–7
Powers and Roots	F–8
Percentages	F–9
Unit Conversion	F–10
Scientific Notation	F–11
Ohm's Law	F–12
Series Circuits	F–13
Parallel Circuits	F–14
Combination Circuits	F–15
Power	F–16
Voltage Divider	F–17
Trigonometric Functions	F–18
Series Ac Circuits	F–19
Logarithms	F–20

Figure F–2. Calculator procedures included in App. F.

Example	Operation	Readout
5 + 3 = 8	5 [+] 3 [=]	8.
2.3 + 6.8 = 9.1	2 [·] 3 [+] 6 [·] 8 [=]	9.1
10 + 14 + 5 = 29	10 [+] 14 [+] 5 [=]	29.

Figure F–3. Calculator procedures for addition.

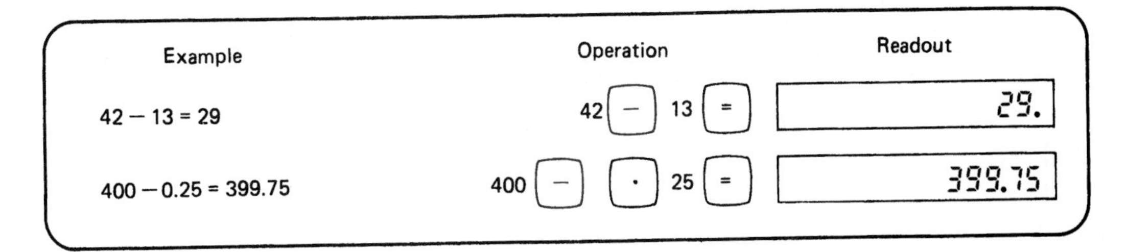

Figure F–4. Calculator procedures for subtraction.

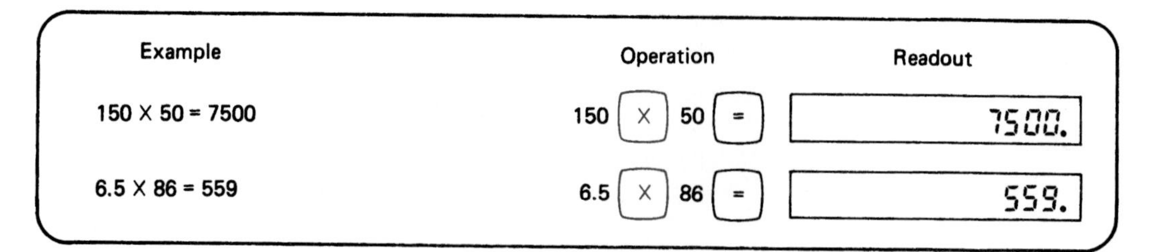

Example	Operation	Readout
150 X 50 = 7500	150 [X] 50 [=]	7500.
6.5 X 86 = 559	6.5 [X] 86 [=]	559.

Figure F–5. Calculator procedures for multiplication.

Example	Operation	Readout
$\frac{600}{3} = 200$	600 [÷] 3 [=]	200.
$\frac{487}{5} = 97.4$	487 [÷] 5 [=]	97.4

Figure F–6. Calculator procedures for division.

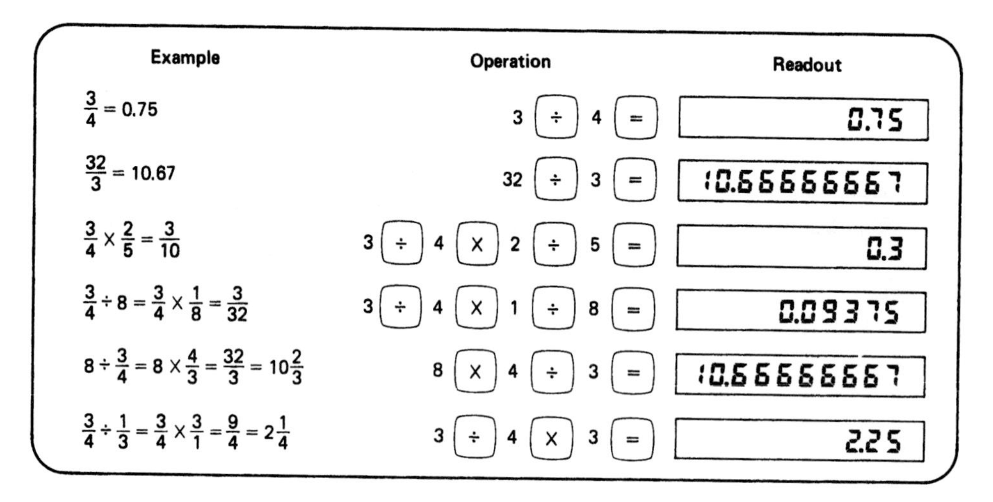

Example	Operation	Readout
$\frac{3}{4} = 0.75$	3 [÷] 4 [=]	0.75
$\frac{32}{3} = 10.67$	32 [÷] 3 [=]	10.66666667
$\frac{3}{4} \times \frac{2}{5} = \frac{3}{10}$	3 [÷] 4 [X] 2 [÷] 5 [=]	0.3
$\frac{3}{4} \div 8 = \frac{3}{4} \times \frac{1}{8} = \frac{3}{32}$	3 [÷] 4 [X] 1 [÷] 8 [=]	0.09375
$8 \div \frac{3}{4} = 8 \times \frac{4}{3} = \frac{32}{3} = 10\frac{2}{3}$	8 [X] 4 [÷] 3 [=]	10.66666667
$\frac{3}{4} \div \frac{1}{3} = \frac{3}{4} \times \frac{3}{1} = \frac{9}{4} = 2\frac{1}{4}$	3 [÷] 4 [X] 3 [=]	2.25

Figure F–7. Calculator procedures for fractions.

Example	Operation	Readout
$3^2 = 3 \times 3 = 9$	3 [x^2]	9.
$5^5 = 5 \times 5 \times 5 \times 5 \times 5 = 3125$	5 [y^x] 5 [=]	3125.
$\sqrt{64} = 8$	64 [$\sqrt{\ }$]	8.
$64^{1/2} = 64^{0.5} = 8$	64 [y^x] [·] 5 [=]	8.
$\sqrt[4]{16} = 2$	16 [2nd F] [$\sqrt[x]{y}$] 4 [=]	2.

Figure F–8. Calculator procedures for powers and roots.

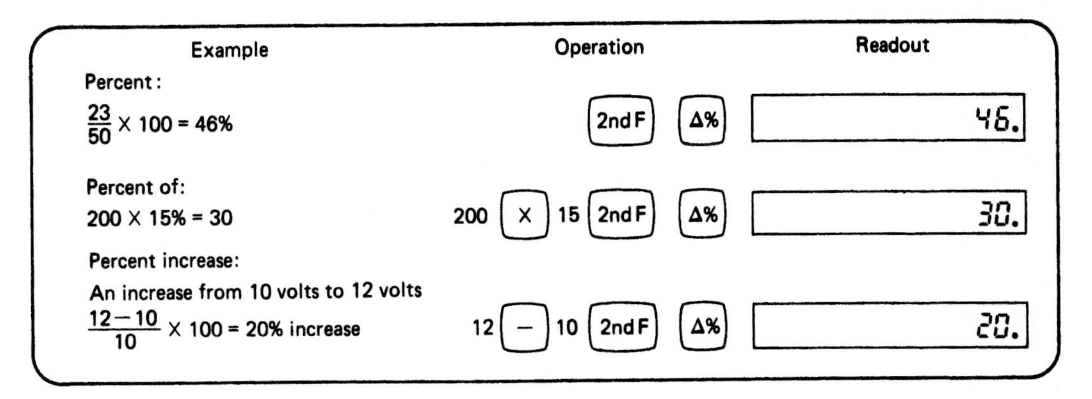

Figure F–9. Calculator procedures for percentage.

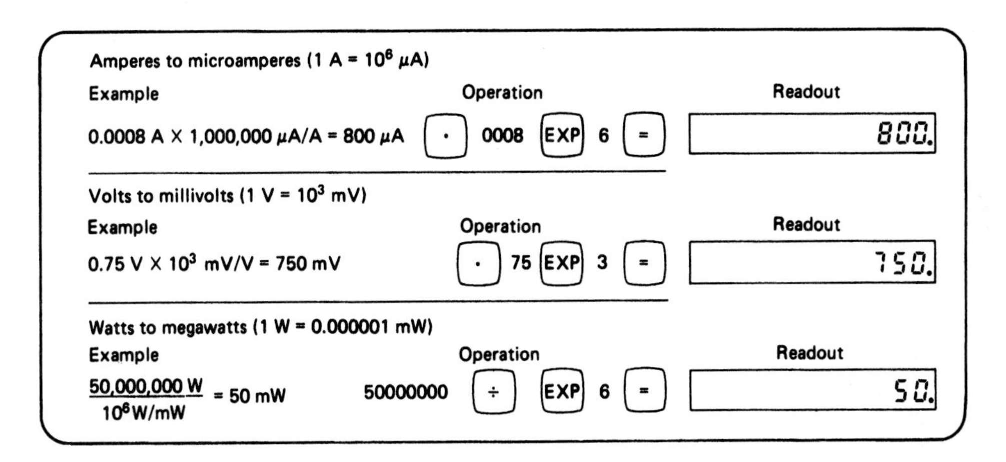

Figure F–10. Calculator procedures for conversion of electrical units.

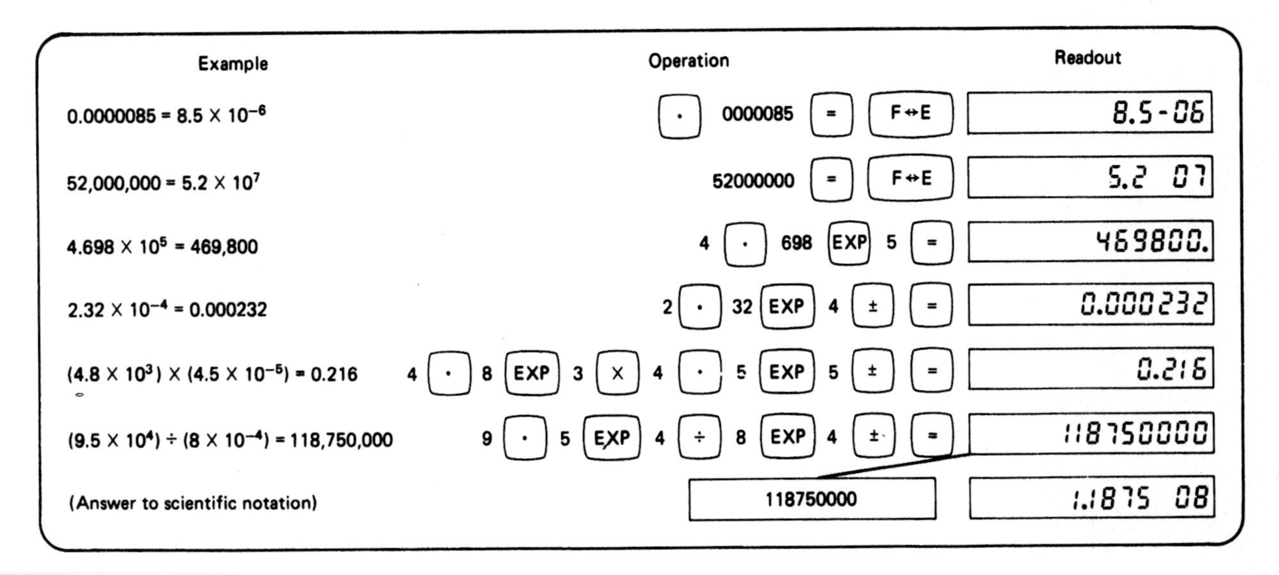

Figure F–11. Calculator procedures for scientific notation.

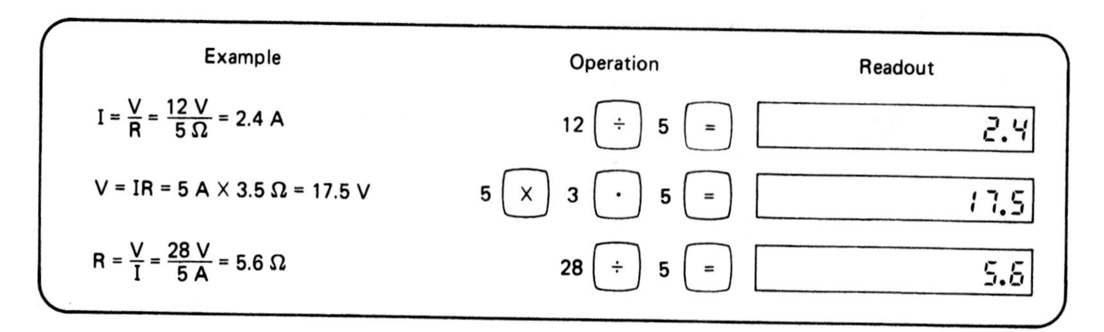

Figure F–12. Calculator procedures for Ohm's law.

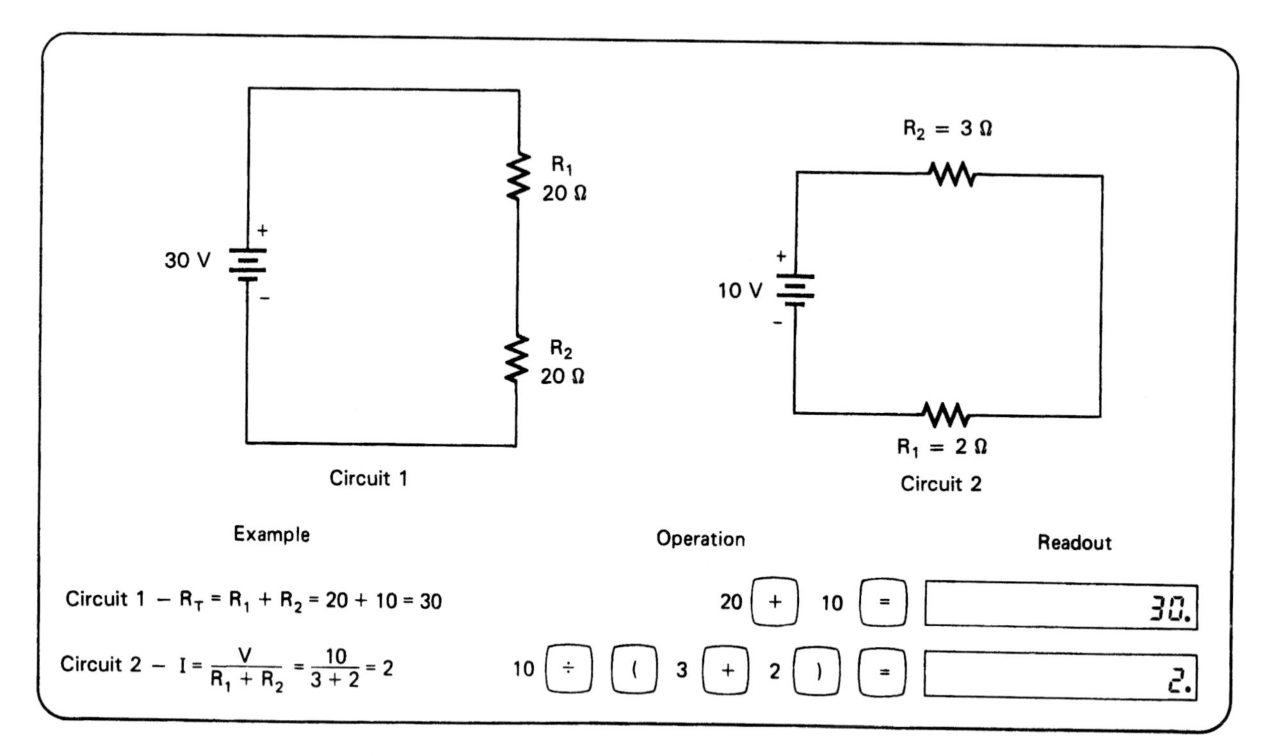

Figure F–13. Calculator procedures for series circuits.

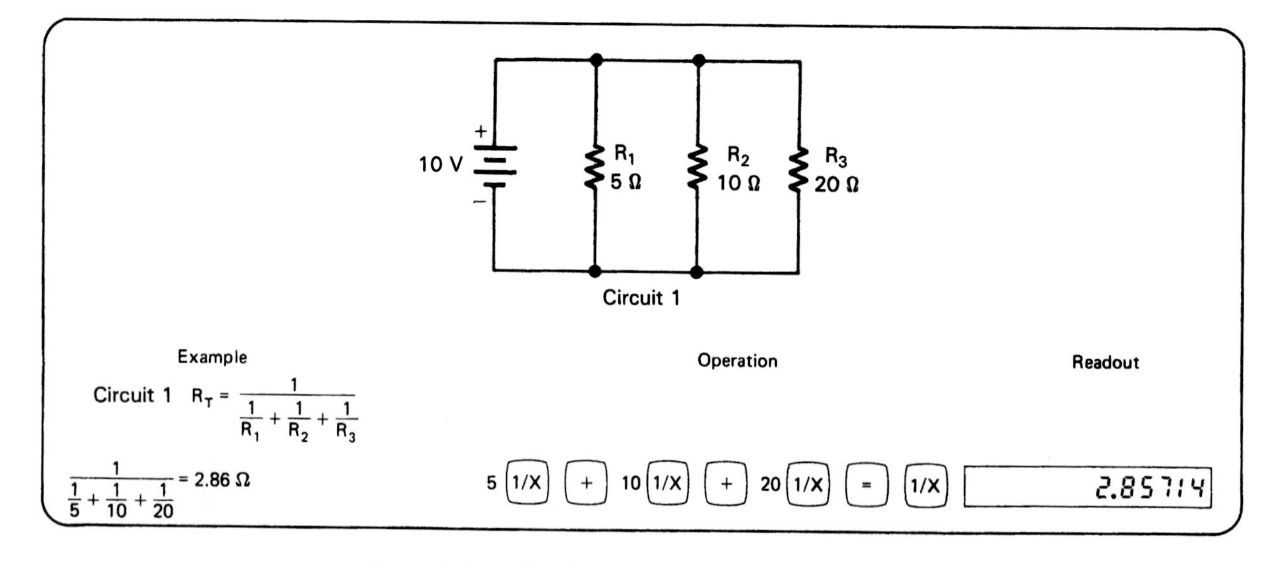

Figure F–14. Calculator procedures for parallel circuits.

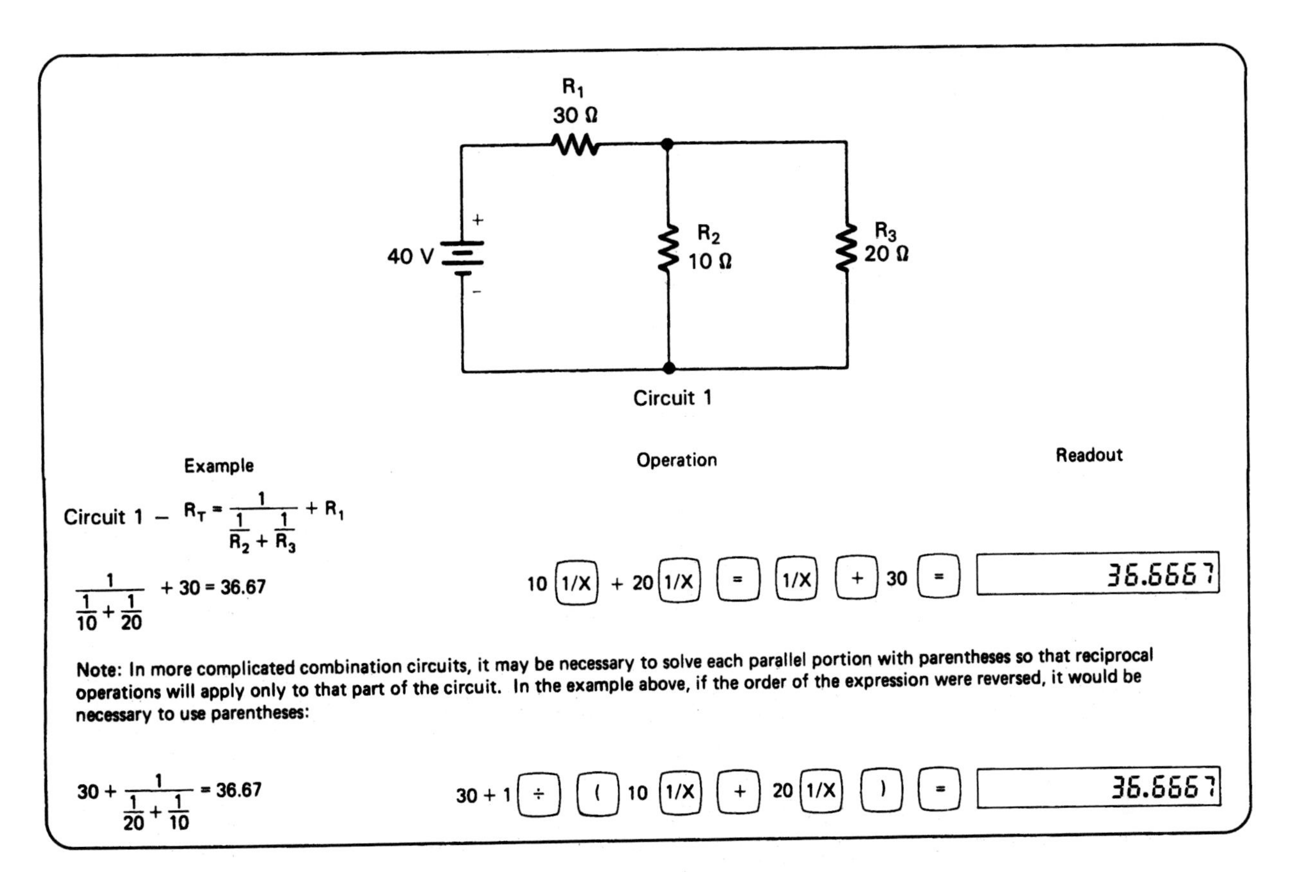

Figure F–15. Calculator procedures for combination circuits.

Example	Operation	Readout
$P = I^2R = (3A)^2 \times 10\ \Omega = 90\ W$	3 [X^2] [X] 10 [=]	90.
$P = \frac{V^2}{R} = \frac{(30\ V)^2}{10\ \Omega} = 90\ W$	30 [X^2] [÷] 10 [=]	90.

Figure F–16. Calculator procedures for electric power.

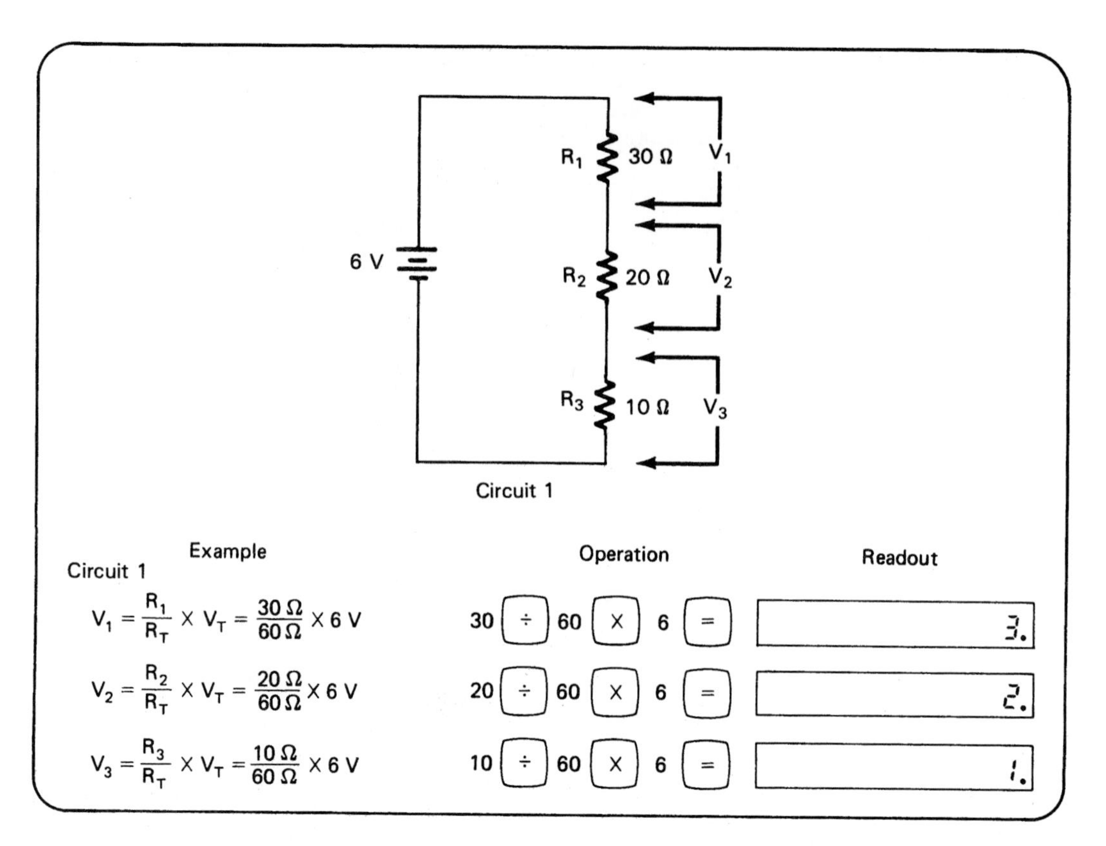

Figure F–17. Calculator procedures for voltage-divider circuits.

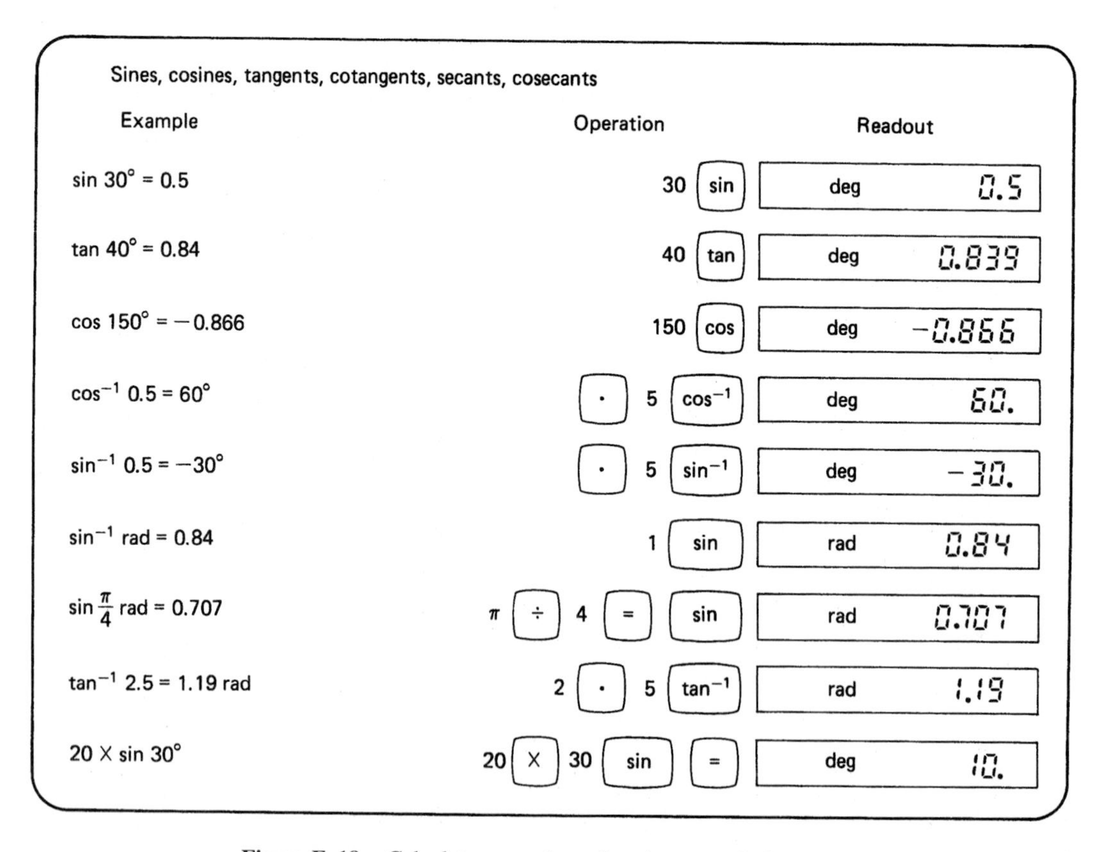

Figure F–18. Calculator procedures for trigonometric functions.

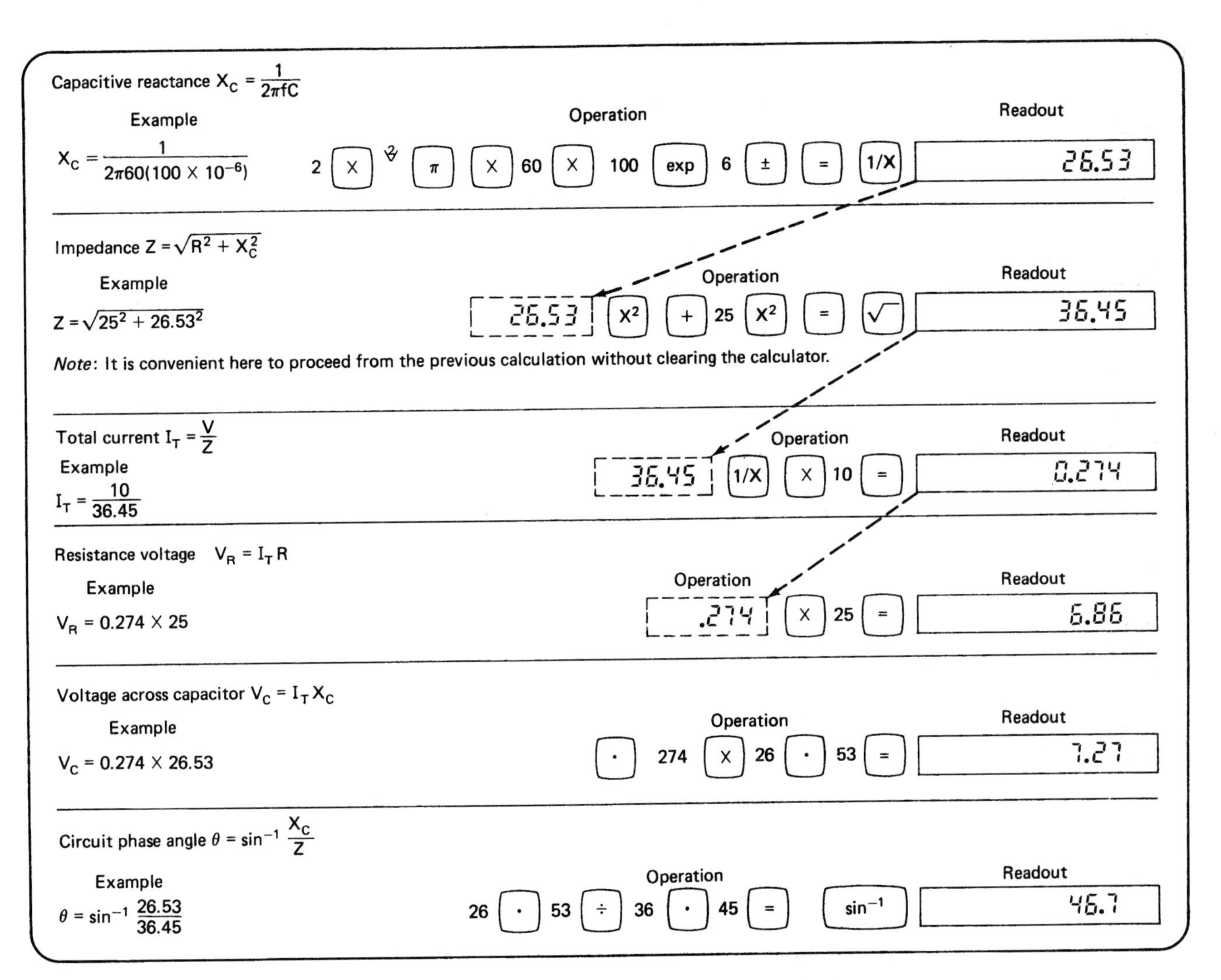

Figure F–19. Calculator procedures for series ac circuits.

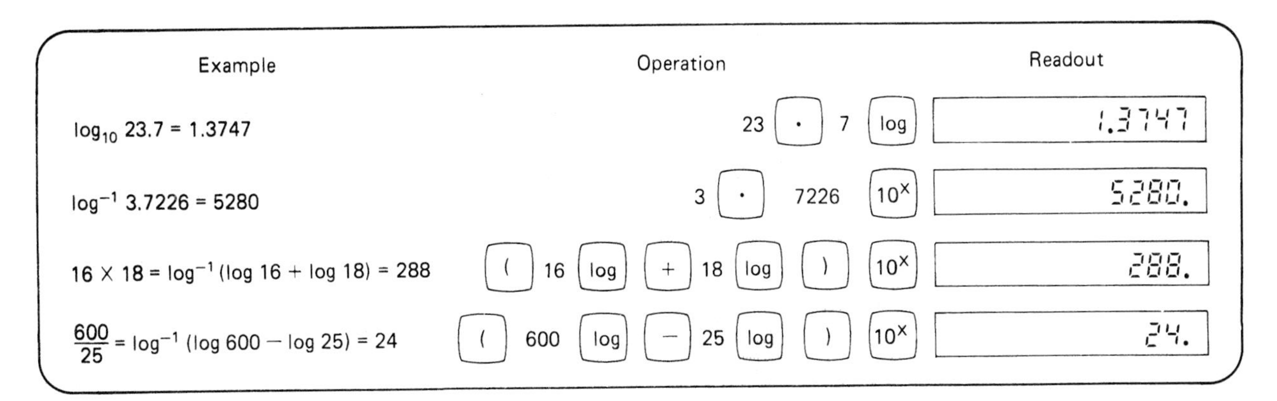

Figure F–20. Calculator procedures for logarithms.

TABLE G–4 Length

Known Quantity	*Multiply By*	*Quantity to Find*
inches (in.)	2.54	centimeters (cm)
feet (ft)	30	centimeters (cm)
yards (yd)	0.91	meters (m)
miles (mi)	1.6	kilometers (km)
millimeters (mm)	0.04	inches (in.)
centimeters (cm)	0.4	inches (in.)
meters (m)	3.3	feet (ft)
meters (m)	1.1	yards (yd)
kilometers (km)	0.6	miles (mi)
centimeters (cm)	10	millimeters (mm)
decimeters (dm)	10	centimeters (cm)
decimeters (dm)	100	millimeters (mm)
meters (m)	10	decimeters (dm)
meters (m)	1000	millimeters (mm)
dekameters (dam)	10	meters (m)
hectometers (hm)	10	dekameters (dam)
hectometers (hm)	100	meters (m)
kilometers (km)	10	hectometers (hm)
kilometers (km)	1000	meters (m)

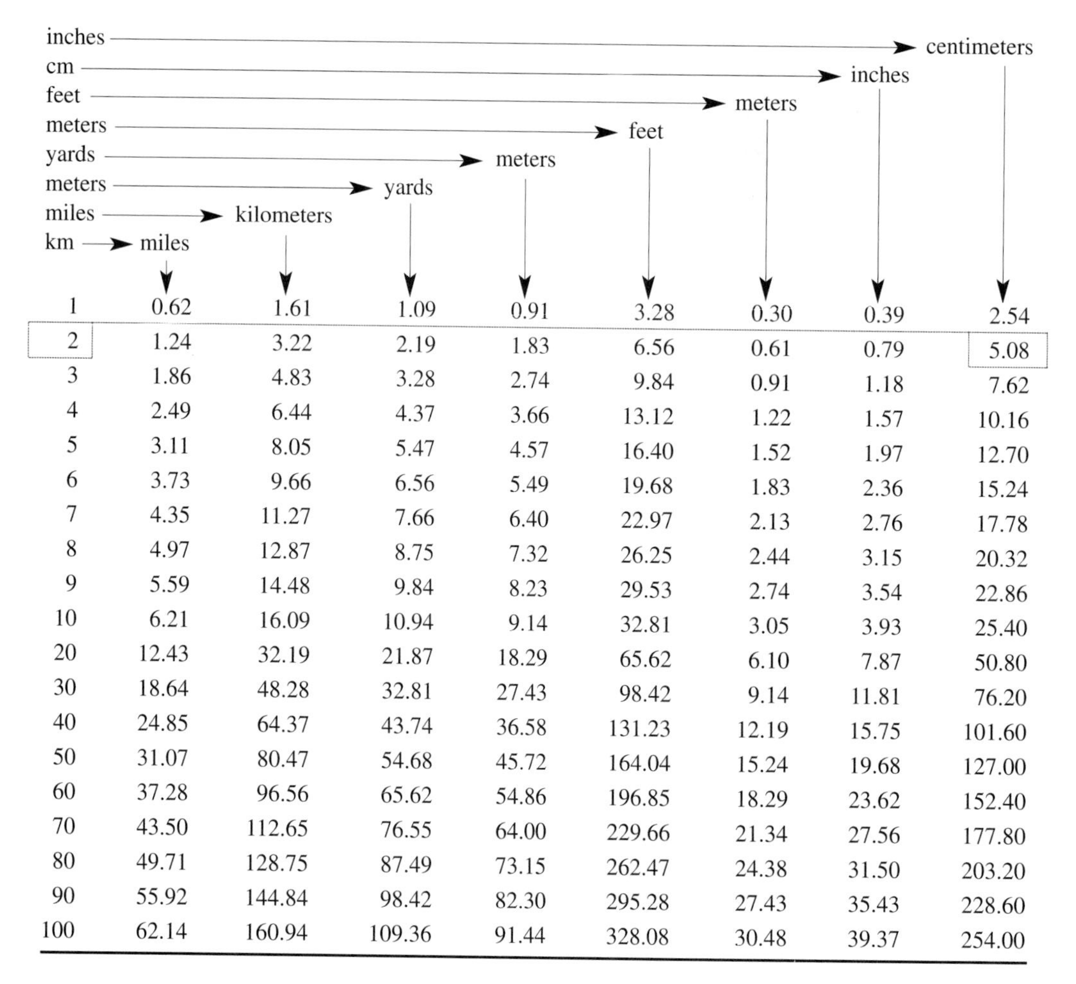

	km → miles	miles → kilometers	meters → yards	yards → meters	meters → feet	feet → meters	cm → inches	inches → centimeters
1	0.62	1.61	1.09	0.91	3.28	0.30	0.39	2.54
2	1.24	3.22	2.19	1.83	6.56	0.61	0.79	5.08
3	1.86	4.83	3.28	2.74	9.84	0.91	1.18	7.62
4	2.49	6.44	4.37	3.66	13.12	1.22	1.57	10.16
5	3.11	8.05	5.47	4.57	16.40	1.52	1.97	12.70
6	3.73	9.66	6.56	5.49	19.68	1.83	2.36	15.24
7	4.35	11.27	7.66	6.40	22.97	2.13	2.76	17.78
8	4.97	12.87	8.75	7.32	26.25	2.44	3.15	20.32
9	5.59	14.48	9.84	8.23	29.53	2.74	3.54	22.86
10	6.21	16.09	10.94	9.14	32.81	3.05	3.93	25.40
20	12.43	32.19	21.87	18.29	65.62	6.10	7.87	50.80
30	18.64	48.28	32.81	27.43	98.42	9.14	11.81	76.20
40	24.85	64.37	43.74	36.58	131.23	12.19	15.75	101.60
50	31.07	80.47	54.68	45.72	164.04	15.24	19.68	127.00
60	37.28	96.56	65.62	54.86	196.85	18.29	23.62	152.40
70	43.50	112.65	76.55	64.00	229.66	21.34	27.56	177.80
80	49.71	128.75	87.49	73.15	262.47	24.38	31.50	203.20
90	55.92	144.84	98.42	82.30	295.28	27.43	35.43	228.60
100	62.14	160.94	109.36	91.44	328.08	30.48	39.37	254.00

Example: 2 inches = 5.08 cm.

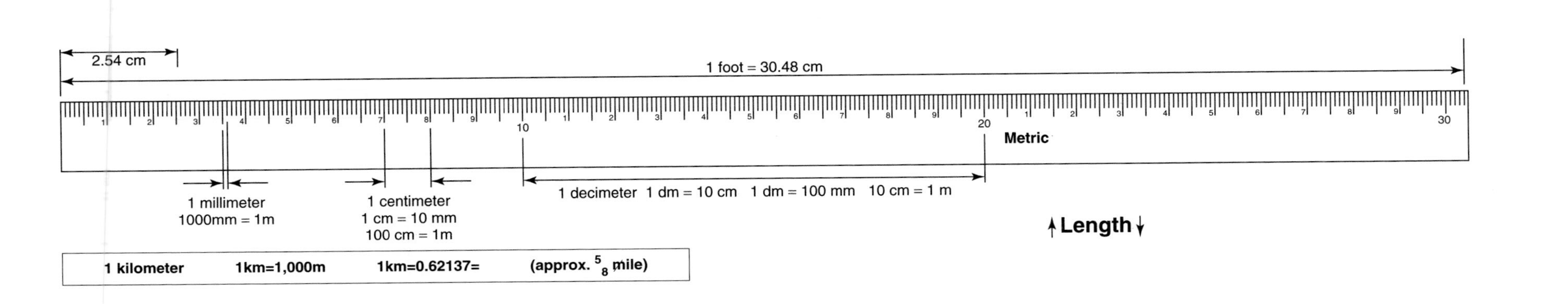
2.54 cm
1 foot = 30.48 cm
Metric
1 millimeter
1000mm = 1m
1 centimeter
1 cm = 10 mm
100 cm = 1m
1 decimeter 1 dm = 10 cm 1 dm = 100 mm 10 cm = 1 m
Length
1 kilometer 1km=1,000m 1km=0.62137= (approx. 5/8 mile)

TABLE G–9 Electrical

Known Quantity	*Multiply by*	*Quantity to Find*
Btu per minute	0.024	horsepower (hp)
Btu per minute	17.57	watts (W)
horsepower (hp)	33,000	foot-pounds per min (ft-lb/min)
horsepower (hp)	746	watts (W)
kilowatts (kW)	57	Btu per minute
kilowatts (kW)	1.34	horsepower (hp)
watts (W)	44.3	foot-pounds per min (ft-lb/min)

TABLE G–10 Time Distance Conversion

Miles per Hour	*Knots*	*Feet per Second*	*Kilometers per Hour*	*Meters per Second*
1	0.8684	1.4667	1.609	0.447
2	1.74	2.93	3.22	0.894
3	2.61	4.40	4.83	1.34
4	3.47	5.87	6.44	1.79
5	4.34	7.33	8.05	2.24
6	5.21	8.80	9.66	2.68
7	6.08	10.27	11.27	3.13
8	6.95	11.73	12.87	3.58
9	7.82	13.20	14.48	4.02
10	8.68	14.67	16.09	4.47
15	13.03	22.00	24.14	6.71
20	17.37	29.33	32.19	8.94
25	21.71	36.67	40.23	11.18
30	26.05	44.00	48.28	13.41
35	30.39	51.33	56.33	15.64
40	34.74	58.67	64.37	17.88
45	39.08	66.00	72.42	20.12
50	43.42	73.33	80.47	22.35
55	47.76	80.67	88.51	24.59
60	52.10	88.00	96.56	26.82
65	56.45	95.33	104.61	29.06
70	60.79	102.67	112.65	31.29
75	65.13	110.00	120.70	33.53
100	86.84	146.67	160.94	44.70

TABLE G–11 Sample Conversions

yards to meters	Multiply yards by 0.91
inches to centimeters	Multiply inches by 2.54
miles to kilometers	Multiply miles by 1.6
ounces to grams	Multiply ounces by 28.3
pounds to kilograms	Multiply pounds by 0.454
quarts to liters	Multiply quarts by 0.946
gallons to liters	Multiply gallons by 3.78

TABLE G–12 Common Units

Quantity	*Si Unit*	*Symbol*
Angle	radian (1 rad ≃ 57.3°)	rad
Area	square meter	m^2
Capacitance	farad	F
Conductance	siemens (mho)	S
Electrical charge	coulomb	C
Electrical current	ampere	A
Energy (work)	joule	J
Force	newton	N
Frequency	hertz	Hz
Heat	joule	J
Inductance	Henry	H
Length	meter	m
Magnetic field strength	ampere per meter	A/m
Magnetic flux	weber	Wb
Magnetic flux density	tesla (1 T = 1 Wb/m^2)	T
Magnetomotive force	ampere	A
Mass	kilogram	kg
Potential difference	volt	V
Power	watt	W
Pressure	pascal (1 Pa = 1 N/m^2)	Pa
Resistance	ohm	Ω
Resistivity	ohmmeter	Ω·m
Specific heat	joule per kilogram-kelvin	J/kg·K or J/kg·°C
Speed	meter per second	m/s
Speed of rotation	radian per second (1 rad/s = 9.55 rad/min)	rad/s
Temperature	kelvin	K
Temperature difference	kelvin or degree Celsius	K or °C
Thermal conductivity	watt per meter-kelvin	W/m·K or W/m·°C
Thermal power	watt	W
Torque	newton-meter	N·m
Volume	cubic meter	m^3
Volume	liter	L

PRINTED CIRCUIT BOARD (PCB) CONSTRUCTION

Following is a step-by-step procedure for constructing a dual-output dc power supply using a printed circuit board (PCB). If a PCB is not available, the parts may be soldered together using any type of circuit board. Materials for circuit board construction are available through many electronic parts dealers.

Power-Supply Project

1. To construct a 9-V regulated dual-output dc power supply, use the circuit of Fig. H–1.
2. A representative PCB layout is shown in Fig. H–2. The circuit board may require some modification

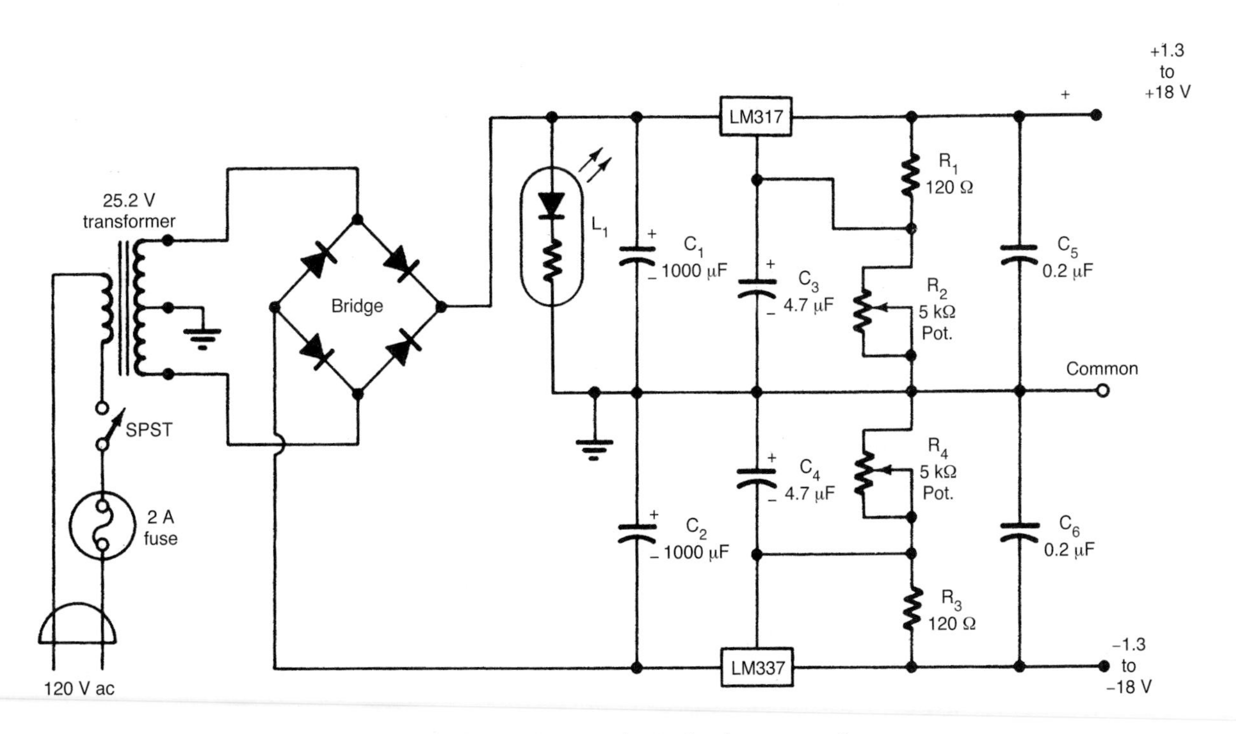

Figure H–1. Dual-output dc regulated power supply.

according to your part selection. Make a prototype layout of the PCB on a piece of card stock. Use the actual dimensions of the parts being used.

3. Lay out the PCB using tape and PC resist dots.
4. Etch the board.
5. Drill holes at the designated locations.
6. Solder components in place.
7. Wire the external circuitry.
8. Assemble the external circuitry and PCB.
9. Test the operation of the circuit with a voltmeter and/or oscilloscope. The output should be 19 V dc.
10. Assemble the entire unit in a cabinet and test it again for operation.

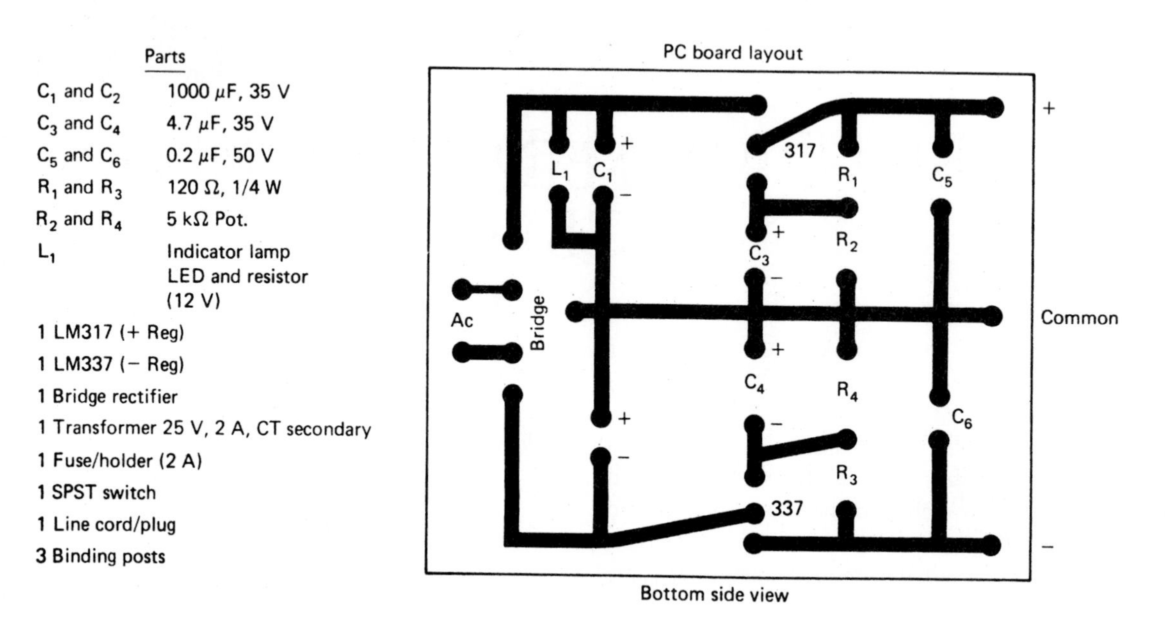

Figure H–2. PCB layout of dual-output regulated power supply.

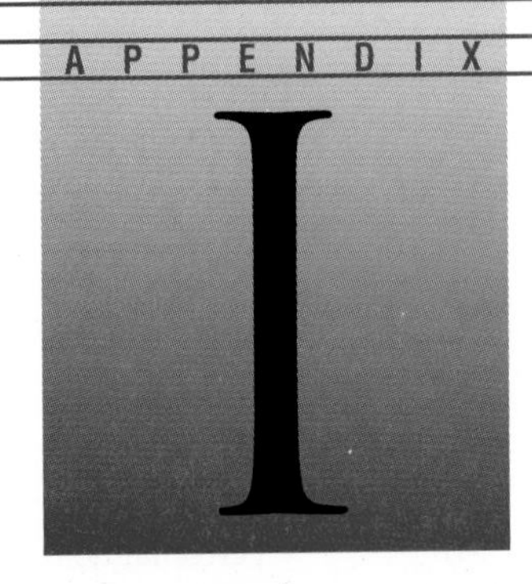

ACRONYMS AND ABBREVIATIONS

The study of electricity and electronics makes great use of acronyms and abbreviations. Many of these are difficult to remember. To help clarify some of the abbreviations and acronyms used, following is a list that should help, although it is impossible to provide a complete list of all that are used.

A/D analog-to-digital
AC alternating current
ADC analog-to-digital converter
AEA American Electronics Association
AEE Association of Energy Engineers
AF audio frequency
AFC automatic frequency control
AGC automatic gain control
ALA American Lighting Association
ALC automatic level control
AM amplitude modulation
AMLCD active-matrix liquid crystal display
ANSI American National Standards Institute
API application programming interface
AQL acceptable quality level
ASCII American (National) Standard Code for Information Interchange
ASHRAE American Society of Heating, Refrigeration and Air-conditioning Engineers
ASIC application-specific integrated circuit
ASQC American Society of Quality Control
ASME American Society of Mechanical Engineers
ASTM American Society of Testing and Materials
ATE automatic test equipment
AWG American Wire Gauge

BCD binary-coded decimal
BIOS basic input-output system
BP bandpass
BW bandwidth

CAD computer-aided design
CAE computer-aided engineering
CAT computer-aided test
CEA Canadian Electrical Association
CCD charge-coupled device
CFC current-to-frequency converter
CMRR common-mode rejection ratio
CODEC coder-decoder
COTS commercial off-the-shelf
CPS characters per second
CPS cycles per second
CPU central processing unit
CSA Canadian Standards Association
CW continuous wave

D/A digital-to-analog
DAC digital-to-analog converter
DAQ data acquisition
DAS data-acquisition system
dB decibel
dBm dB power
DC direct current
DCE data communications equipment
DCS digital cellular service
DIO digital input/output
DIP dual-in-line package

DMM	digital multimeter
DOS	disk operating system
DRAM	dynamic random access memory
DSO	digital storage oscilloscope
DSP	digital signal processor
DVM	digital voltmeter
EIA	Electronic Industries Association
ELF	extremely low frequency (<3 kHz)
EMF	electromagnetic field
EMI	electromagnetic interference
EM	electromagnetic pulse
EMR	electromagnetic radiation
EPROM	erasable programmable read-only memory
ESD	electrostatic discharge
eV	electron volt
F/V	frequency-to-voltage
FCC	Federal Communications Commission
FET	field-effect transistor
FIFO	first-in, first-out
FM	frequency modulation
FS	full scale
FSK	frequency-shift keying
FTP	file transfer protocol (Internet)
GDI	graphics device interface
GIF	graphic image file
GPIB	general-purpose interface bus
GPS	Global Positioning System
GUI	graphical user interface
HF	high frequency (3 MHz to 30 MHz)
Hipot	high-potential
HTML	hypertext markup language
http	hypertext transfer protocol
HV	high voltage
Hz	hertz
IBEW	International Brotherhood of Electrical Workers
I/O	input/output
IC	integrated circuit
IEEE	Institute of Electrical and Electronics Engineers
IES	Illuminating Engineering Society
IF	intermediate frequency
IR	infrared radiation
IRQ	interrupt request
ISA	Instrument Society of America
ISDN	integrated services digital network
ISO	International Organization for Standardization
JEDEC	Joint Electronic Device Engineering Council
LAN	local area network
laser	light amplification by stimulated emission of radiation
LCD	liquid crystal display
LED	light-emitting diode
LF	low frequency (30 kHz to 300 kHz)
LSB	least significant bit (digital) or lower sideband (communications)
LSI	large-scale integration
MFP	mini flat pack
MMI	man-machine interface
MOSFET	metal-oxide semiconductor field-effect transistor
MPU	microprocessor unit
MSB	most significant bit
MSI	medium-scale integration
MTBF	mean time between failures
MTTR	mean time to repair
MUX	multiplexer
NEMA	National Electrical Manufacturers Association
NFPA	National Fire Protection Association
NIST	National Institute of Standards and Technology
NSPE	National Society of Professional Engineers
NTC	negative temperature coefficient
OEM	original equipment manufacturer
OLE	object linking and embedding
OS	operating system
OSHA	Occupational Safety and Health Administration
P-P	peak-to-peak
PA	power amplifier
PCB	printed circuit board
PCM	pulse code modulation
PF	power factor
PID	proportional integral-differential
PIV	peak inverse volts
PLA	programmable logic array
PLD	programmable logic device
PLL	phase-locked loop
PM	phase modulation or pulse modulated
PPM	parts per million
PRF	pulse repetition frequency
PWM	pulse-width modulation
QA	quality assurance
QC	quality control
QFP	quad flat pack

RAM	random access memory
RC	resistor-capacitor
RCVR	receiver
RF	radio frequency
RFI	radio-frequency interference
RGB	red-green-blue
RH	relative humidity
RMS	root mean square
ROM	read-only memory
RTD	resistance temperature device
RTOS	real-time operating system
S/N	signal-to-noise
SI	International System of Units
SIA	Semiconductor Industry Association
SIMM	single-in-line memory module
SIP	single-in-line package
SMD	surface-mount device
SMT	surface-mount technology
SNR	signal-to-noise ratio
SPC	statistical process control
SPICE	simulation program with integrated-circuit emphasis
SPS	samples per second
SQC	statistical quality control
SRAM	static random access memory
SSB	single sideband
SWR	standing wave ratio
TC	temperature coefficient
TDM	time-division multiplex
TQM	total quality management
TSOP	thin small-outline package
TTL	transistor-transistor logic
UHF	ultrahigh frequency (300 MHz to 3 GHz)
UL	Underwriters Laboratories
UPS	uninterruptible power supply
URL	uniform resource locator
USB	upper sideband
UV	ultraviolet
VA	volt-ampere
VAC	volts alternating current
VAR	volt-ampere reactive
VB	Visual Basic (software)
VCO	voltage-controlled oscillator
VDC	volts direct current
VFC	voltage-to-frequency converter
VHF	very high frequency (30 MHz to 300 MHz)
VLF	very low frequency (below 30 kHz)
VLSI	very large-scale integration
$V_{p\text{-}p}$	peak-to-peak voltage
VSOP	very small-outline package
VSWR	voltage standing-wave ratio
VTVM	vacuum tube voltmeter
WWW	World Wide Web
XMTR	transmitter

ANSWERS TO SELECTED PROBLEMS

CHAPTER 2 ELECTRICAL COMPONENTS AND MEASUREMENTS

Resistance Measurement Problems

1. 1 Ω
3. 100 Ω
5. 10 kΩ
7. 90 Ω
9. 9 kΩ
11. 32 Ω
13. 3200 Ω
15. 320 kΩ
17. 560 Ω
19. 56 kΩ
21. 80 Ω
23. 8000 Ω
25. 800 kΩ

Voltage Measurement Problems

27. 9.6 V
29. 240 V
31. 960 V
33. 7.4 V
35. 185 V
37. 740 V
39. 4.4 V
41. 110 V
43. 440 V
45. 3.1 V
47. 75.5 V
49. 310 V
51. 2.4 V
53. 60 V
55. 240 V

Current Measurement Problems

57. 2.4 A
59. 96 mA
61. 9.6 mA
63. 240 μA
65. 2.4 A
67. 74 mA
69. 7.4 mA
71. 185 μA
73. 1.1 A
75. 44 mA
77. 4.4 mA
79. 110 μA
81. 0.755 A
83. 31 mA
85. 3.1 mA
87. 75.5 μA
89. 0.6 A
91. 24 mA
93. 2.4 mA
95. 60 μA

Electrical Unit Conversions

1. 0.65 A = 650 mA
3. 0.215 mV = 0.000215 V
5. 225 μA = 0.000225 A
7. 0.85 MΩ = 850,000 Ω
9. 68,000 V = 68 kV

Scientific Notation

1. $0.0001 = 1 \times 10^{-5}$
3. $10{,}000{,}000 = 1 \times 10^{7}$
5. $10 = 1 \times 10^{1}$
7. $10{,}000 = 1 \times 10^{4}$
9. $1.0 = 1 \times 10^{0}$
11. $0.00128 = 1.28 \times 10^{-3}$
13. $0.000632 = 6.32 \times 10^{-4}$
15. $28.2 = 2.82 \times 10^{1}$
17. $52.30 = 5.23 \times 10^{1}$
19. $0.051 = 5.1 \times 10^{-2}$

Metric Conversions

1. 100
3. 5
5. 39.37
7. 1609.4
9. 840
11. 3.18
13. 2.4
15. 0.12203

Resistor Color Code

a. 75,000 Ω; 5%
c. 5600 Ω; 20%

e. 10 Ω; 5%
g. 91,000 Ω; 5%
h. 47,000 Ω; 20%
i. 15,000 Ω; 20%
k. 56 Ω; 5%
m. 39 Ω; 5%
o. 11 Ω; 5%

CHAPTER 3 ELECTRICAL CIRCUITS

Applying Basic Electrical Theory

1. $R = \frac{12\text{ V}}{0.5\text{ A}} = 24\ \Omega$

3. $V = 10\text{ A} \times 1.25\ \Omega = 12.5\text{ V}$

5. $V = 0.6\text{ A} \times 200\ \Omega = 120\text{ V}$

7. $I = \frac{120\text{ V}}{30\ \Omega} = 4\text{ A}$

9. $I = \frac{12\text{ V}}{16\ \Omega} = 0.75\text{ A}$

Using Ohm's Law

1. $V = 2\text{ A} \times 500\ \Omega = 1000\text{ V}$

3. $V = 1.2\text{ A} \times 1000\ \Omega = 1200\text{ V}$

5. $V = 20\text{ mA} \times 120\text{ k}\Omega$
$= 0.02\text{ A} \times 120{,}000\ \Omega$
$= 2400\text{ V}$

7. $I = \frac{10\text{ V}}{1.8\text{ k}\Omega} = \frac{10\text{ V}}{1800\ \Omega} = 0.00555\text{ A} = 5.55\text{ mA}$

9. $I = \frac{12\text{ V}}{5.6\text{ k}\Omega} = \frac{12\text{ V}}{5600\ \Omega} = 0.00214\text{ A} = 2.14\text{ mA}$

11. $R = \frac{20\text{ V}}{5\text{ A}} = 4\ \Omega$

13. $R = \frac{120\text{ V}}{10\ \mu\text{A}} = \frac{120\text{ V}}{0.00001\text{ A}} = 12{,}000{,}000\ \Omega = 12\text{ M}\Omega$

15. $R = \frac{5\text{ V}}{20\ \mu\text{A}} = \frac{5\text{ V}}{0.00002\text{ A}} = 250{,}000\ \Omega = 250\text{ k}\Omega$

Working with Series Circuits

1. $I = \frac{10\text{ V}}{5\ \Omega} = 2\text{ A}$

3. $R_T = 4\ \Omega + 6\ \Omega = 10\ \Omega$

5. $I_2 = I_T = \frac{10\text{ V}}{10\ \Omega} = 1\text{ A}$

7. $V_1 = 2\text{ A} \times 4\ \Omega = 8\text{ V}$

9. $V_2 = V_T - V_1 - V_3$
$= 20\text{ V} - 8\text{ V} - 8\text{ V}$
$= 4\text{ V}$

11. $V_1 = 0.5\text{ A} \times 1\ \Omega = 0.5\text{ V}$

13. $V_3 = V_T - V_1 - V_2$
$= 5\text{ V} - 0.5\text{ V} - 2\text{ V}$
$= 2.5\text{ V}$

15. $I_3 = \frac{2.5\text{ V}}{5\ \Omega} = 0.5\text{ A}\ (= I_T)$

17. $V_1 = 1\text{ A} \times 50\ \Omega = 50\text{ V}$

19. $V_3 = 1\text{ A} \times 10\ \Omega = 10\text{ V}$

21. $P_2 = 1\text{ A} \times 40\text{ V} = 40\text{ W}$

23. $P_T = 1\text{ A} \times 100\text{ V} = 100\text{ W}$

25. $R_T = 10\ \Omega + 20\ \Omega = 30\ \Omega$

27. $I_T = \frac{18\text{ V}}{30\ \Omega} = 0.6\text{ A}$

Working with Parallel Circuits

1. $R_T = \frac{5 \times 10}{5 + 10} = \frac{50}{15} = 3.33\ \Omega$

3. $I_1 = \frac{10\text{ V}}{5\ \Omega} = 2\text{ A}$

5. $\frac{1}{R_T} = \frac{1}{2} + \frac{1}{4} + \frac{1}{6} + \frac{1}{12}$
$= 0.5 + 0.25 + 0.1666 + 0.0833$
$= 1.0$
$= 1\ \Omega$

7. $I_1 = \frac{12\text{ V}}{2\ \Omega} = 6\text{ A}$

9. $I_3 = \frac{12\text{ V}}{6\ \Omega} = 2\text{ A}$

11. $P_T = 12\text{ A} \times 12\text{ V} = 144\text{ W}$

13. $I_T = \frac{10\text{ V}}{2.857\ \Omega} = 3.5\text{ A}$

15. $I_2 = \frac{10\text{ V}}{10\ \Omega} = 1\text{ A}$

17. $P = 0.5\text{ A} \times 10\text{ V} = 5\text{ W}$

19. $P_3 = 2\text{ A} \times 10\text{ V} = 20\text{ W}$

Working with Combination Circuits

1. $R_T = 5 + 5$ in parallel with 5
$= 5 + 2.5$
$= 7.5\ \Omega$

3. $V_1 = 2.67\text{ A} \times 5\ \Omega = 13.35\text{ V}$

5. $I_2 = \frac{V_2}{R_2} = \frac{6.65\text{ V}}{5\ \Omega} = 1.33\text{ A}$

7. $R_T = 10 + 5$ in parallel with $10 + 10$ in parallel with 20
$= 10 + 3.33 + 6.67$
$= 20\ \Omega$

9. $P_T = 1\text{ A} \times 20\text{ V} = 20\text{ W}$

11. $I_2 = \frac{V_2}{R_2} = \frac{3.33\text{ V}}{5\ \Omega} = 0.67\text{ A}$; $V_2 = 1\text{ A} \times 3.33\ \Omega$
$= 3.33\text{ V}$

13. $V_5 = 1\text{ A} \times 6.67\ \Omega$
$= 6.67\text{ V}$

15. $I_T = \dfrac{50\text{ V}}{18\ \Omega} = 2.78\text{ A}$

17. $V_4 = 2.78\text{ A} \times 10\ \Omega = 27.8\text{ V}$

19. $I_3 = \dfrac{V_3}{R_3} = \dfrac{8.3\text{ V}}{12\ \Omega} = 0.69\text{ A}$

21. $I_T = \dfrac{20\text{ V}}{14.28\ \Omega} = 1.4\text{ A}$

23. $V_2 = .2\text{ A} \times 10\ \Omega$
$= 2\text{ V}$

25. $I_3 = \dfrac{6\text{ V}}{5\ \Omega} = 1.2\text{ A}$

27. $P_T = 1.4\text{ A} \times 20\text{ V}$
$= 28\text{ W}$

KVL Problems

1. $I_1 = 1\text{ A}; I_2 = 0.2\text{ A}; I_1 + I_2 = 1.2\text{ A}$

3. Loop 1 $= 75\text{ V} - I_1R_2 - V_2 - (I_1 + I_2 + I_3) R_1 = 0$
Loop 2 $= 75\text{ V} - (I_2 + I_3) R_4 - I_2R_5 - V_3 - (I_2 + I_3) R_3$
Loop 3 $= 75\text{ V} - (I_2 + I_3) R_4 - I_3 R_7 - V_4 - I_3 R_6 - (I_2 + I_3) R_3 - (I_1 + I_2 + I_3) R_1 = 0$

KCL Problems

1. $I_1 = 6\text{ A}$

3. $I_3 = 7\text{ A}$

Maximum Power Transfer Problems

1.

R_L	I_L	V_{out}	P_{out}
0 Ω	15 A	0 V	0 W
1 Ω	12 A	12 V	144.0 W
2 Ω	10 A	20 V	200.0 W
3 Ω	8.5 A	25.7 V	218.5 W
4 Ω	7.5 A	30 V	225.0 W
5 Ω	6.67 A	33.3 V	222.1 W
6 Ω	6 A	36 V	216.0 W
7 Ω	5.45 A	38.2 V	208.4 W
8 Ω	5 A	40 V	200.0 W

3.

R_L	I_L	V_{out}	P_{out}
0 Ω	6.67 A	0 V	0 W
1 Ω	5.0 A	5.0 V	25.0 W
2 Ω	4.0 A	8.0 V	32.0 W
3 Ω	3.3 A	9.9 V	32.7 W
4 Ω	2.86 A	11.43 V	32.7 W
5 Ω	2.5 A	12.5 V	31.3 W
6 Ω	2.22 A	13.3 V	29.53 W
7 Ω	2.0 A	14.0 V	28.0 W
8 Ω	1.82 A	14.5 V	26.5 W

Superposition Problems

1. $I_{R1} = 4.54\text{ A}; I_{R2} = 2.73\text{ A}; I_{R3} = 1.82\text{ A}$

Thevinin Equivalent Circuit Problems

1. $V_{TH} = 3.57\text{ V}; R_{TH} = 2.14\ \Omega$

3. $V_{TH} = 10\text{ V}; R_{TH} = 7.5\ \Omega$

5. $V_{TH} = 75\text{ V}; R_{TH} = 18.75\ \Omega$

Norton Equivalent Circuits

1. $I_N = 1.2\text{ A}; R_N = 15\ \Omega$

Bridge Circuit Simplification

1. $V_{TH} = 5\text{ V}; R_{TH} = 2.83\ \Omega$

CHAPTER 5 SOURCES OF ELECTRICAL ENERGY

Generator Problems

1. $V_1 = 320\text{ V} \times \sin 25°$
$= 320\text{ V} \times 0.4226$
$= 135.2\text{ V}$

3. $V_i = 320\text{ V} \times \sin 195°$
$= 320\text{ V} \times -0.2588$
$= -82.82\text{ V}$

5. $V_i = 320\text{ V} \times \sin 335°$
$= 320\text{ V} \times -0.4226$
$= -135.2\text{ V}$

7. $I_P = \dfrac{I_R}{1.73} = \dfrac{7.5\text{ A}}{1.73} = 4.33\text{ A}$

9. $V_P = V_L = 480\text{ V}$

11. $P_T = 3 \times V_P \times I_P$
$= 3 \times 480\text{ V} \times 4.33\text{ A}$
$= 6235.2\text{ W}$

13. $\text{f} = \dfrac{4 \times 1800}{120} = \dfrac{7200}{120} = 60\text{ Hz}$

15. $\%\text{ VR} = \dfrac{220\text{ V} - 210\text{ V}}{210\text{ V}} \times 100$
$= \dfrac{10\text{ V}}{210\text{ V}} \times 100 = 0.476 \times 100 = 4.76\%$

17. $\%\text{ efficiency} = \dfrac{12{,}000\text{ W}}{20\text{ hp}} \times 746 \times 100$
$= \dfrac{12{,}000\text{ W}}{14{,}920\text{ W}} \times 100 = 0.804 \times 100 = 80.4\%$

19. $\%\text{ efficiency} = \dfrac{3000\text{ W}}{4.25\text{ hp} \times 746} \times 100$
$= \dfrac{3000\text{ W}}{3170.5\text{ W}} \times 100 = 0.946 \times 100 = 94.6\%$

21. $P_o = I_L \times V$
$= 8.33\text{ A} \times 100\text{ V}$
$= 83.3\text{ W}$

23. $I_F = \dfrac{V_I}{R_F} = \dfrac{120\text{ V}}{40\ \Omega} = 3\text{ A}$

CHAPTER 6 ALTERNATING-CURRENT ELECTRICITY

Alternating-Current Values

	Peak Voltage		*Rms Voltage*
1.	4 V ac	=	2.828 V
3.	6 V ac	=	4.242 V
5.	1.5 V ac	=	1.06 V
7.	5 V ac	=	3.535 V
	Rms Voltage		*Peak*
9.	3 V ac	=	4.23 V
11.	8 V ac	=	11.28 V
13.	7 V ac	=	9.87 V
15.	9 V ac	=	12.69 V
17.	10 V ac	=	14.1 V
19.	15 V ac	=	21.15 V
21.	11 V ac	=	15.51 V
23.	18 V ac	=	25.38 V

Ac Measurements

1. 2.4 V
3. 1.245 V
5. 0.75 V
7. 7.6 V
9. 3.6 V
11. 48 V
13. 24.5 V
15. 15 V
17. 190 V
19. 90 V
21. 490 V
23. 247.5 V
25. 150 V
27. 960 V
29. 360 V

Alternating-Current Circuit Problems

1. $L_T = 2\text{ H} + 5\text{ H} + 6\text{ H} = 13\text{ H}$

3. $X_L = 6.28 \times 60\text{ Hz} \times 2\text{ H} = 753.6\ \Omega$

5. $C_T = 20\ \mu\text{F} + 40\ \mu\text{F} + 65\ \mu\text{F} = 125\ \mu\text{F}$

7. $t = \dfrac{20\text{ H}}{4\ \Omega} = 5\text{ s}$

9.
$$Z = \sqrt{(10^2) + (10^2)} = \sqrt{100 + 100} = \sqrt{200} = 14.14\ \Omega$$

Inductance and Inductive Reactance Problems

1. $L_T = 2\text{ H} + 3\text{ H} + 2\text{ H} = 7\text{ H}$

3. $X_L = 6.28 \times 30\text{ Hz} \times 8\text{ H} = 1507.2\ \Omega$

5. $X_L = 6.28 \times 60\text{ Hz} \times 1.0\text{ H} = 376.8\ \Omega$

7. $L_T = 2\text{ H} + 5\text{ H} + 2(0.55) = 8.1\text{ H}$

9. $L_T = 4\text{ H} + 3\text{ H} - 2(0.85) = 5.3\text{ H}$

11.
$$\frac{1}{L_T} = \frac{1}{2 + 0.55} + \frac{1}{5 + 0.55} = \frac{1}{2.55} + \frac{1}{5.55} = 0.392 + 0.180 = 0.572$$
$$L_T = \frac{1}{0.572} = 1.748\text{ H}$$

13.
$$\frac{1}{L_T} = \frac{1}{4 - 0.85} + \frac{1}{3 - 0.85} = \frac{1}{3.15} + \frac{1}{2.15} = 0.317 + 0.465 = 0.782$$
$$L_T = \frac{1}{0.782} = 1.278\text{ H}$$

Capacitance and Capacitive Reactance

1. $C = \dfrac{0.0885 \times 3 \times 400}{0.03} = 3540\text{ pF} = 0.00354\ \mu\text{F}$

3. $C = \dfrac{0.0885 \times 2 \times 68.75}{0.02} = 608.43\text{ pF}$

5.
$$\frac{1}{C_T} = \frac{1}{40} + \frac{1}{20} + \frac{1}{20} = .025 + .05 + .05 = 0.0125$$
$$C_T = \frac{1}{.125} = 8\ \mu\text{F}$$

7. $X_C = \dfrac{1}{6.28 \times 60 \times (30 \times 10^{-6})} = 88.46\ \Omega$

9.
$$C_T = C_1 + C_2 + C_3 = 50 + 20 + 30\ \mu\text{F} = 100\ \mu\text{F}$$

Series and Parallel Ac Circuit Problems

1. $X_C = \dfrac{1}{6.28 \times 60 \times (20 \times 10^{-6})} = 132.7\ \Omega$

3. $I = \dfrac{120\text{ V}}{1008\ \Omega} = 0.119\text{ A} = 119\text{ mA}$

5. $V_R = 0.119\text{ A} \times 1000\ \Omega = 119\text{ V}$

7. $\text{AP} = 0.119\text{ A} \times 120\text{ V} = 14.28\text{ VA}$

9. $\text{PF} = \dfrac{14.16\text{ W}}{14.28\text{ VA}} = 0.99 = 99\%$

10.

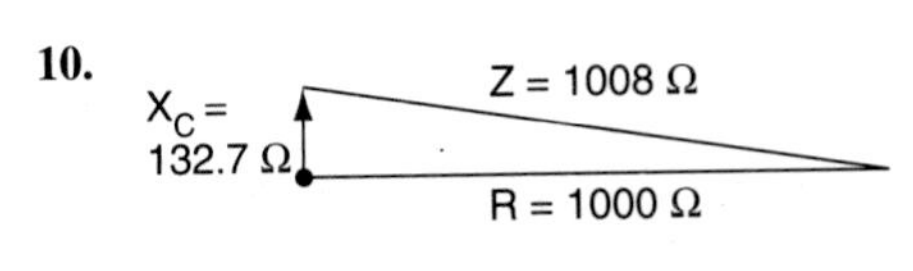

Impedance triangle

V_C =
15.79 V
V_A = 120 V
V_R = 119 V

Voltage triangle

VAR_C =
1.88
VA = 14.28
W = 14.16

Power triangle

11. AP = 240 V × 72 A = 17,280 VA

13. $\theta = \text{inv cos} \dfrac{W}{\text{VA}} = \text{inv cos } 0.69 = 46.37°$

14.

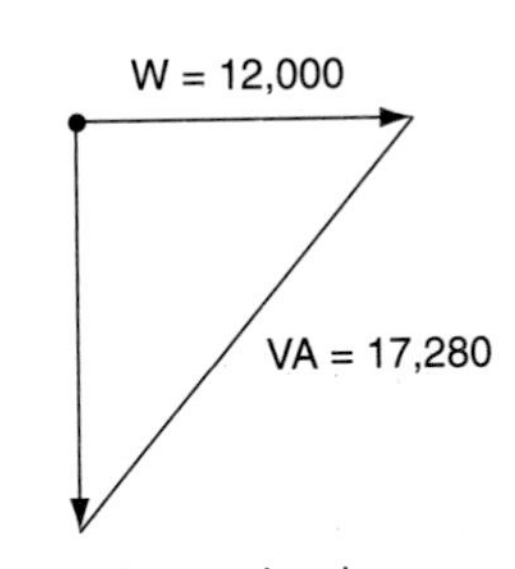

Power triangle

15. $X_C = \dfrac{1}{6.28 \times 60 \times (40 \times 10^{-6})} = 66.35\ \Omega$

17. $X_T = 66.35 - 56.52 = 9.83\ \Omega$

19. $I = \dfrac{20\text{ V}}{100.48\ \Omega} = 0.199\text{ A} = 199\text{ mA}$

21. $V_L = 0.199\text{ A} \times 56.52\ \Omega - 11.25\text{ V}$

23. $\theta = \text{inv cos} \dfrac{V_R}{V_A} = \text{inv cos} \dfrac{19.9\text{ V}}{20\text{ V}} = \text{inv cos } 0.995 = 5.73°$

24.

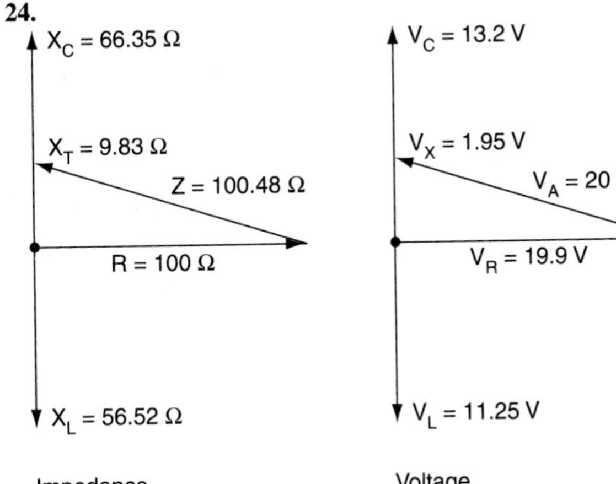

Impedance triangle

Voltage triangle

25. $I_R = \dfrac{10\text{V}}{100\ \Omega} = 0.1\text{ A} = 100\text{ mA}$

27. $I_L = \dfrac{10\text{ V}}{56.52\ \Omega} = 0.177\text{ A} = 177\text{ mA}$

29. $Z = \dfrac{10\text{ V}}{0.1036\text{ A}} = 96.52\ \Omega$

31. $B_L = \dfrac{1}{56.52\ \Omega} = 0.0177\text{ S} = 17.7\text{ mS} = 17{,}700\ \mu\text{S}$

$B_C = \dfrac{1}{66.35\ \Omega} = 0.015\text{ S} = 15\text{ mS} = 15{,}000\ \mu\text{S}$

$B_T = B_L - B_C = 17{,}700 - 15{,}000 = 2700\ \mu\text{S}$

33. $\theta = \text{inv cos} \dfrac{I_R}{I_T} = \text{inv cos} \dfrac{0.1\text{ A}}{0.1036\text{ A}} = \text{inv cos } 0.9652$

$= 15.15°$

34.

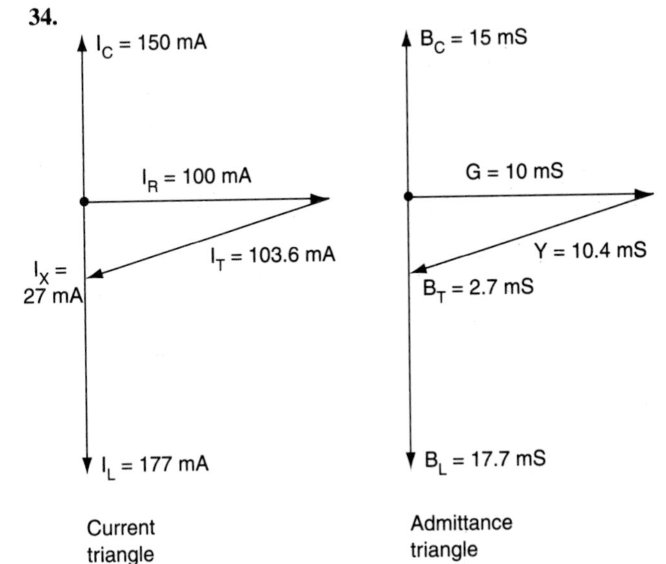

Current triangle

Admittance triangle

Transformer Problems

1. $\dfrac{V_P}{V_S} = \dfrac{N_P}{N_S}; \dfrac{120\text{ V}}{20\text{ V}} = \dfrac{1500\text{ turns}}{N_S}; N_S = 250\text{ turns}$

3. $\dfrac{V_P}{V_S} = \dfrac{N_P}{N_S}; \dfrac{4800\text{ V}}{V_S} = \dfrac{20}{1}; V_S = 240\text{ V}$

5. $\text{efficiency} = \dfrac{P_{\text{out}}}{P_{\text{in}}} = \dfrac{4800\text{ V} \times 4\text{ A}}{240\text{ V} \times 95\text{ A}} = \dfrac{19{,}200\text{ VA}}{22{,}800\text{ VA}}$

$= 0.842 = 84.2\%$

CHAPTER 7 ELECTRICAL ENERGY CONVERSION

Motor Problems

1. $I_{\text{Line}} = \dfrac{300\text{ W}}{120\text{ V}} = 2.5\text{ A}$

3. $I_A = 2.5\text{ A} - 1.41\text{ A} = 1.09\text{ A}$

5. $I_F = \dfrac{V}{R_F} = \dfrac{120}{30} = 4\text{ A}$

$I_A = I_T - I_F$
$= 40\text{ A} - 4\text{ A}$
$= 36\text{ A}$

7. $\text{hp} = \dfrac{3450\text{ rpm} \times 35\text{ ft-lb}}{5252} = \dfrac{120{,}750}{5252} = 22.99$

9. $\% \text{ slip} = \dfrac{1800 - 1725}{1800} \times 100\% = \dfrac{75}{1800} \times 100\%$

$= 0.0416 \times 100\% = 4.16\%$

or

$\% \text{ slip} = \dfrac{1800 - 1790}{1800} \times 100\% = \dfrac{10}{1800} \times 100\%$

$= 0.0055 \times 100\% = 0.55\%$

11. $\text{speed} = \dfrac{60 \text{ Hz} \times 120}{12} = \dfrac{7200}{12} = 600 \text{ rpm}$

CHAPTER 8 ELECTRICAL INSTRUMENTS

Current Meter Design

7. (a) $R_{sh} = \dfrac{1 \text{ mA} \times 200\ \Omega}{4 \text{ mA}} = 50\ \Omega$

(c) $R_{sh} = \dfrac{1 \text{ mA} \times 200\ \Omega}{49 \text{ mA}} = 4.08\ \Omega$

(e) $R_{sh} = \dfrac{1 \text{ mA} \times 200\ \Omega}{8.5 \text{ mA}} = 23.53\ \Omega$

Voltmeter Design

9. (a) $R_{mult} = \dfrac{0.9 \text{ V}}{1 \text{ mA}} = 900\ \Omega$

(c) $R_{mult} = \dfrac{14.9 \text{ V}}{1 \text{ mA}} = 14{,}900\ \Omega$

Ohmmeter Design

2. $R_T = \dfrac{1.5 \text{ V}}{1 \text{ mA}} = 1500\ \Omega$

3. $R_{lim} = 1500 - 100 = 1400\ \Omega$

INDEX

Ac. *See* Alternating current
Acceptor impurity, 312, 605
Acceptors, 321
Acronyms, 665–667
Active power, 217
Active region, of transistors, 382, 605
Actuators, 139
Admittance, 182, 605
Admittance triangle, 213, 214, 215, 216
Air-core inductor, 182, 193, 605
Air-core transformers, 223, 228
Alkaline cells, 152
Alnico, 130, 168, 605
Alpha (current gain), 410, 424, 605
Alphanumeric, definition of, 552, 605
Alternating current, 150, 605
  in amplifiers, 416
  waveforms of, 186, 187
Alternating-current circuits, 200–203
  Kirchhoff's laws for, 192
  leading and lagging currents in, 197
  Ohm's law for, 192
  power in, 212–219
  types of
    capacitive, 202
    inductive, 200–201
    parallel, 212, 213–215
    resistive, 200
    resistive-capacitive, 202–203, 208–209
    resistive-inductive, 201–202, 208
    series, 208, 209–212
  vector diagram for, 212, 213–216
Alternating-current generators. *See* Alternators
Alternating-current motors. *See* Single-phase ac motors; Three-phase ac motors
Alternating-current power control, with silicon-controlled rectifiers, 589–591. *See also* Power control systems
Alternating-current voltage, 185–189
  measurement of, 191
  single-phase, 189–190
  three-phase, 178–191
Alternation, 359, 582, 605
  in *C*-input filters, 368–370
  in power supplies, 360, 363–365, 372
Alternators, 150, 162, 167, 605
  single-phase, 162–164
  three-phase, 164–166
AM. *See* Amplitude modulation
Ammeters, 29, 38, 284, 605
  multirange, 287
  See also Volt-ohm-milliammeter
Ampere, A. M., 25
Ampere, 4, 25, 605
Ampere-hour, 150, 605
Ampere-turns, 130, 134, 605
Amplification, 410, 605
  audio-frequency, 509
  classes of, 421–423, 424
  principles of, 410–412
  radio-frequency, 507
  See also Amplifiers; Amplifying systems
Amplifiers
  basic, 412–421
    alternating-current signal in, 414, 442
    biasing in, 415
    linear and nonlinear operation of, 419
    load-line analysis in, 418
  bipolar transistors as, 412
  buffer, 504
  circuit configurations in, 410–412
  coupling of, 451–455
  field-effect transistors as, 427–434, 442
  gain of, 448–449
  integrated-circuit, 434–439, 462–464
  intermediate-frequency, 526–528, 530–548
  See also Amplification; Amplifying systems; Power amplifiers
Amplifying systems, 447
  functions of, 447–448
  gain of, 448–449
  input transducers of, 465–470
  speakers, 464–465
  See also Amplification; Amplifiers
Amplitude, 183, 605
Amplitude-modulated RF carrier, 517
Amplitude modulation (AM), 446, 518, 519, 605
  communication system, 517–526
    block diagram of, 518
    simple receivers, 520, 521–523, 546
    superheterodyne receivers, 520
    transmitters, 520–523, 545
  percentage of, 518
Analog, definition of, 552, 605
Analog Instruments, 284
Analog meter movement, 140. *See also* D'Arsonval meter movement
Analog meters, 14, 57–58, 66, 288. *See also* Specific types
AND gates, 556
Angle of lead or lag., 605
Anode-cathode current, 325
Anodes, 330, 341, 359, 361, 591
  definition of, 330, 341, 359, 605
  focusing, 539
Antennas, 514
Apparent power, 183, 606
Arithmetic-logic unit (ALU), 567
Armature, 130, 150, 246, 606
Armature reaction, 150, 174, 606
Armstrong oscillators, 480–483, 498
ASCII codes, 570
Assembly language, 552, 606
Astable multivibrators, 476, 491–492, 496, 500–501, 577, 606
  inductive-capacitive, 491–492
  integrated circuit, 495
Atomic number, 4, 15, 313, 606
Atomic theory. *See* Semiconductor theory
Atomic weight, 314
Atoms, 4, 15, 313, 606
  combinations of, 313
  donor, 321
  stable, 4, 17, 314, 315
Attenuation, 606
Audio, 455
Audio amplifiers, 410, 456, 464, 606
Audio frequency (AF), 446, 517, 606
Autoranging digital voltmeter, 201
Autotransformers, 225, 226
Avalanche breakdown, 336
Average value, 185, 359, 606
Average voltage, 183, 606

Back EMF, 606. *See also* Counterelectromotive force
Ballast, 246, 606
Bandpass filters, 183, 219–220, 606
Bandwidth, 183, 606
Bardeen, John, 7
Bar magnet, 131
Barrier potential, 330, 331–332, 606
Barrier voltage. *See* Barrier potential
Base, 552, 606
  of numbering system, 552
  of transistors, 382, 606
Base-collector biasing, 384, 385
Base current, 383–386
Batteries, 38, 150, 153, 606
  symbol for, 38, 39
  types of, 7, 43
  uses of, 153–154
  voltage of, 28
Battery connections, 154–155
Beat frequency, 504, 525, 606

Beat frequency oscillators, 504, 516–517, 544, 606
Bel, 446, 606
Bell, Alexander Graham, 7
Beta (current gain), 382, 386, 606
Beta-dependent biasing, 410, 415–417, 606
Beta-independent biasing, 422, 423
B-H curve, 143
Biasing, 416
  in amplifiers, 416
  base-collector, 384
  beta-dependent, 410, 415–417, 606
  beta-independent, 422, 423
  divider method of, 429
  emitter-base, 417
  fixed and fixed-emitter, 416, 429, 431
  forward, 330, 606
  of junction diodes, 332–334
  NPN transistor biasing 384–387
  PNP transistor biasing 386
  reverse, 303, 606
  self and self-emitter, 416, 431–432
  of transistors
    bipolar, 391–394
    field-effect, 429–432
Bidirectional triggering, 582, 606
Big picture, 7
Binary-coded decimal, 552, 556, 606
Binary-coded decimal counters, 562, 563, 578
Binary counter, 577
Binary logic functions, 552, 556–558
Binary numbering system, 554–555
Bipolar power supplies, 367–368
Bipolar transistors, 382, 383, 391, 606
  as amplifiers, 412
  biasing of, 414–415
  testing, 391–394, 404, 405
Bistable multivibrators, 493–494, 500
Bit, 552, 606
Blanking, 535
Blanking pulse, 504, 606
Block diagrams, 38, 52, 551, 606
Blocking oscillators, 476, 496, 606
Bohr, Niels, 313
Boiler, 606
Branch, 76, 606
Branch currents, 76, 86, 606
Branch resistance, 76, 86, 93 606
Branch voltage, 76, 86, 93, 606
Breakdown voltage, 317
  reverse, 337, 583, 615
Breakover voltage, 582, 606
Bridge circuit, 284, 606
Bridge-circuit simplification, 109–110, 127
Bridge rectifiers, full-wave, 365–367
Brittain, W.H., 7
Brushes, 150, 246, 606
Buffer amplifiers, 504, 520, 607
Bypass capacitor, 410, 607
Byte, 78, 552, 607
Cadmium sulfide photoresistive cell, 326
Calculator examples 639–650
Calculators, problem solving with, 77-78
Cameras, television, 535–536
  color, 539–540
Capacitance, 183, 197–199, 607
  effects of, in circuits, 197–199
Capacitive circuits, 197, 200–202
Capacitive coupling, 451, 471
Capacitive heating, 246, 607
Capacitive input filter 359, 607
Capacitive phase relationships, 198
Capacitive reactance, 183, 194–196, 607
Capacitor, 183, 193, 340, 607
  charging and discharging of, 198–199
  energy stored in, 193–196
  in parallel, 197
  in series, 197
  symbols for, 194
  testing, 200
  types of, 199–200
Capacitor divider, 476, 607
Capacitor-inductance-capacitor filter. *See* Pi filters
Capacitor-input filter, 359, 368–370
Capacitor-start motor, 264–266, 607
  capacitor-run, 264–266
Carbon brushes, 160
Carbon resistors, 47, 49–50, 52
Carbon-zinc cells, 152, 153
Carrier wave, 504, 607
Cascade, 446, 607
Cathode, 312, 342, 359, 361, 591
  definition of, 359, 607
  indirectly heated, 324, 352
Cathode-ray tube (CRT), 284, 538, 539, 607
  of oscilloscope, 191, 284
  of television receiver, 538–539
Cathode-ray tube (CRT) instruments, 296, 539. *See also* Oscilloscopes
Cells, 38. *See also* Batteries
Center frequency, 504, 607
Center tap, 183, 359, 607
Central processing unit (CPU), 552, 607
Ceramic capacitors, 191
Ceramic cartridge, 468–469, 471
Channels, 518, 607
  AM, 518
  of JFETs, 394–397. (*see also* *N*-channel JFETs; *P*-channel JFETs)
  of MOSFETs, 397–400. (*see also* *N*-channel MOSFETs; *P*-channel MOSFETs)
Characteristic, of logarithm, 446, 607
Characteristic curves
  for amplifiers, 419, 420
  collector family of, 388–390, 405
  for push-pull amplifiers, 447
  for single-ended amplifier, 455
  drain family of, 395–397, 399, 429
  for MOSFETs, 406
  for unijunction transistors, 401
Charged-coupled devices (CCDs), 531
Charging, 152
Chart recording instruments, 299, 301
Chemical sources, 152
Chemical cells, 176
Choke, 371
Choke coil, 183, 483, 607
Chroma, 504, 607
C-input filters, 368–370
Circuit breakers, 30, 274
Circuit configuration, 410, 607
Circuits, 76, 607
  capacitive effects in, 197–199
  components of, 27–29
  current flow through, 24, 93
  direct-current, power in, 95–97
  inductive effects in, 196–197
  maximum power transfer in, 97–99
  problem-solving methods for, 101–110
    bridge-circuit simplification, 109–110, 127
    Kirchhoff's voltage law, 76, 96, 123–124, 612
    Norton equivalent circuit method, 107–109
    superposition method, 101, 102, 105
    Thevinin equivalent circuit method, 106–107, 126
  troubleshooting, 83–86, 86–89
Circuit simulation software, 78
Circuit switching, 582, 607
Circular magnetic field, 131
Citizen's band (CB) radio, 506, 520
Clamp-on meters, 296, 297
CLC filter. *See* Pi filters
Closed circuits, 4, 24, 607
  current flow through, 24
Closed-loop amplification, 437, 438
Coded continuous waves, 509
Coefficient of coupling, 130, 607
Collector, in transistor, 382, 607
Collector current, 384, 386, 387
Collector family of characteristic curves, 457, 459
  for push-pull amplifiers, 458, 459
  for single-ended amplifier, 457
Color camera, 539–540
Color-coded resistors, 41–43, 46
  standard values of, 43
  symbol for, 42
Color television, 539–541
Colpitts oscillators, 484–485, 499
Combination circuits, 76, 89–90, 116–117, 607
  examples of, 89, 94
  measurements in, 90, 121–122
Combination connection of batteries, 155
Combination logic gates, 557–558
Combination starters, 273
Common. *See* Ground
Common-base amplifiers, 424–426
Common-collector amplifiers, 426–427
Common contacts, 137
Common drain amplifiers, 433–434
Common emitter amplifiers, 423–424, 425
Common-gate amplifiers, 432–433, 442
Common-source amplifiers, 429, 432, 442
Communication systems, 504, 505–506
  AM, 517–526
  continuous-wave, 508–517
  electromagnetic waves in, 506–508
  FM, 526–531
  television, 531–541
Commutation, 150, 246, 582, 607

Commutator, 150, 607
split-ring, 151
Compact discs (CDs), 78, 446, 468, 607
Comparator, 476, 607
Compass, 6
Compatibility, of monochrome televisions, 504, 607
Complementary-symmetry amplifiers, 461–462, 473
Complementary transistors, 446, 607
Complements, 557
Complete control, 582
Complex circuits. *See* Combination circuits
Complex circuit theorems, 106
Complex waveforms. *See* Nonsinusoidal waveforms
Component, 38, 607
Compounds, 4, 15, 312, 314, 607
Compound-wound dc generators, 171, 173
Compound-wound dc motors, 255–256
Computers, 565–573
data information, 569
hardware, 567–569
memory, 569–572
microcomputer systems, 572–573
operations of, 572
parts, 78, 79
problem solving with, 77–78
software, 572
Computer technician, 31
Computer terms, 80
Condenser, 183, 607
Conductance, 4, 183, 607
Conduction, of ratio detectors, 531
Conduction band, 312, 607
Conduction time, 582, 607
Conductors, 21, 38, 315, 608
current flow through, 20;np25
definition of, 2, 608
in magnetic fields (*see* Electromagnetic induction)
symbols of, 38–39
Contact bounce, 582, 608
Contact current, 139
Contactors, magnetic motor, 139
Continuity checks, 38, 56, 608
Continuous wave communication, 476, 479, 508–517, 608
receivers, 512
antennas, 512–513
audio-frequency amplification in, 517
beat frequency oscillators, 516–517
functions of, 512
heterodyne detection in, 514–516
radio-frequency amplification in, 513
signal selection in, 510
transmitters, 509–511, 544
Continuous waves, 479, 608
Control, in electrical systems, 8, 9, 12, 608. *See also* Power control systems
Control centers, 273–274
Control circuits, motor, 272–276
Control grid, 410, 608
Conventional current flow, 4, 25, 608
Copper losses, 150, 246, 608
Core, definition of, 130, 150, 246, 608
Core saturation, 130, 608
Coulomb, Charles de, 6
Coulombs, 4, 25, 608
Counter electromotive force, 359
definition of, 246, 359, 608
in direct-current motors, 253
Counters, 552, 559–561, 608
decade, 560, 561–563
frequency, 568
production line, 568
Counting systems, digital, 567
Counting timer circuit, 578
Coupling, 130, 446, 608
of amplifiers, 451–455
coefficient of, 130, 196
Covalent bonding, 312, 315, 318–319, 330, 608
in silicon crystals, 311
Crossover distortion, 446, 459–460, 608
Crystal cartridge, 468
Crystal microphones, 446, 608
Crystal oscillators, 485, 487, 499
Crystal radio receiver, 521, 523
Crystals, 320
in diac, 597
N-type, 321 (*see also N*-type materials)
piezoelectric, 158–159
P-type, 321–322, 323 (*see also P*-type materials)
in silicon-controlled rectifiers, 585
in triac, 591
Cumulative compound dc generators, 174
Current, 21, 23–25, 76, 608
alternating, 150, 181
in amplifiers, 414–415, 416
waveforms of, 185, 186
anode-cathode, 325
base, 385–387
branch, 76, 86–87
carries, 385
collector, 385–387
contact, 139
definition of, 4
direct, 150, 608, 609
measurement of, 58–59, 68, 285, 304
pulsating, 151, 166
waveforms of, 185, 186
dropout, 138, 140
eddy, 150, 251, 609
emitter, 385–386
exciting, 169
formulas for finding, 95
gate, 582
holding, 582, 587
induced, 151, 174, 611
Kirchhoff's law for, 76, 90, 123, 124, 612
lagging, 184, 197
leading, 184, 197
leakage, 332–333
maximum forward and reverse, 337
maximum plate, 353
measurement of, 304
with clamp-on meters, 296, 297
pickup, 139
total, 76
in combination circuits, 76, 89–90, 116–117, 607
in parallel circuits, 86–87
in series circuits, 83–84
See also Ohm's law
Current amplification, 412
forward biasing, 335
in *N*-type materials, 321
in *P*-type materials, 323
in transistors, 387
See also Majority current carriers; Minority current
Current flow, 23–24
amount of, 24
through circuits, 24, 25
closed, 24, 25
parallel, 76, 86–89
through conductors, 24
conventional, 6, 24
from direct-current generators, 166
direction of, 24
electron, 6, 24
in transistors, 385–387
water flow compared with, 24, 25
Current gain, 413
alpha, 410
beta, 413, 417–418
Current-limiting resistors, 335
Current ratio, of transformers, 226
Current transformers, 226
Current triangle, 213–216
Current waveforms
of capacitive circuits, 210–213
of capacitor-start motor, 265, 266
of inductive circuits, 209, 213
of resistive-capacitive circuits, 208, 210, 214
of resistive circuits, 207
of resistive-inductive circuits, 209, 213
Cutoff, 382, 389, 608
of amplifiers, 419
of transistors, 388–389
Cycle, of alternating current, 150, 162, 183, 608

Damped waves, 476, 479, 608
Damper windings, 258–259
Darlington amplifiers, 446, 452, 471, 608
Darlington transistors, 453
D'Arsonval meter movement, 140, 284, 286, 608, 613
Data information, 569
Dc. *See* Direct current
Decade counters, 552, 561–563, 608
Decay time, 183, 608
Decibels, 183, 446, 449–451, 608
Decimal numbering system, 553–554
Decoding, 552, 565, 569, 608
Deflection, 504, 608
electrostatic, 298
Deflection plates, 298
Deflection yoke, 502, 533, 608
Delay time, 582, 608
Delta connection, 150, 164, 246, 608
Demodulation, in FM receivers, 529–531

Demodulation in TV receivers, 538
Depletion region, 382, 608
Depletion-type MOSFETs, 397–399, 402
as amplifiers, 429
biasing methods for, 429–432
circuit configurations in, 432–434
operational voltages in, 428
resistance values of, 402, 403
Depletion zone, 330, 331, 608
Diac, 582, 596–598, 608
Diagrams, 39
block, 4, 38, 52, 551, 606
energy-level, 312, 315, 318, 319
schematic, 38, 52, 54, 355
vector, 203, 531
wiring, 38, 53–54
Diaphragm, 465
Dielectric, 183, 608
Dielectric barriers, 434
Dielectric constant, 183, 608
Dielectric heating, 251
Difference in potential, 76, 608
Differential amplifiers, 436
Differential compound dc generators, 174
Differentiator network, 476, 493, 609
Diffusion, 331
Digital, definition of, 552, 553, 609
Digital counting systems, 565
Digital integrated circuit, 410, 609
Digital multimeters (DMMs), 12, 60–62, 191, 299
Digital storage oscilloscopes (DSO), 300, 301
Digital systems, 553–555
binary logic functions in, 556–558
binary numbering system in, 554–555
computers, 565–573
decade counters in, 552, 561–563, 608
decimal numbering system in, 553–554
displays in, 563–565
flip-flops in, 559–561
gate circuits in, 558
operational states of, 555–556
timing and storage elements in, 558–559
Diodes, 330
characteristics of, 334–337, 354
crystal structure of, 329
junction, 330–332
biasing of, 332–334
light-emitting, 345, 355, 362, 563–564
packaging of, 341
semiconductor devices
light-emitting, 345
photodiodes, 346–347, 348, 355
photovoltaic cells, 345–346, 347
varactor, 347–349
zener, 343–345, 355
silicon, 332
specifications of, 337–339
junction capacitance, 349
switching time,340–341
temperature, 337–339
symbol of, 342
testing, 341–343, 354
vacuum-tube, 350, 352, 353
Diplexer, 504, 608
Direct coupled amplifier, 471
Direct coupling, 452–453
Direct current, 150, 608, 609
measurement of, 58–59, 68, 285, 304
pulsating, 151, 166
waveforms of, 185, 186
Direct-current circuits, power in, 104
Direct-current generators, 150, 166, 608
current flow from, 169
operating characteristics of, 174–175
parts of, 168
types of
permanent magnet, 168, 169
self-excited, 168, 170
separately excited, 168, 170
voltage output of, 166–168
Direct-current motors, 253–256
basic parts of, 252
considerations for, 272
types of
compound-wound, 255–256
permanent magnet, 254
series-wound, 254, 255
shunt-wound, 254, 255
stepping, 269–270
Direct-current operating state, 413. *See* Static state
Direct-current power control, with silicon-controlled rectifiers, 587–589, 599, 600
Direct-current voltage, 185, 186
measurement of, 58–60, 67
Direct FM transmitters, 527–529
Direct proportion, 76
Discharge, 609
Discharge transistors, 495
Displays, seven-segment, 61, 563–565
Distortion
crossover, 446, 459–460
nonlinear, 420
Divider method of biasing, 429
D-MOSFET. *See* Depletion-type MOSFET
Domain theory of magnetism, 130, 135, 137, 609
Donor atoms, 321
Donor impurity, 312, 609
Dopant, 312, 609
Doping, 312, 321, 330, 609
Double-pole single-throw (DPST) relay, 138
Drain, 382, 609
Drain family of characteristic curves, 395–397, 399, 429
Drivers, 521
Dropout current, 138, 139
Dry cells, 150, 152, 609
Dual-in-line package (DIP), 44, 410, 609
Dual power supplies, 367–368, 376
Dual resistor network, sub-miniature, 45
Dynamic characteristics of operation, 410, 427, 440, 609
Dynamic load-line analysis, 420
Dynamic microphones, 446, 466, 609
Dynamometers, 289
Eddy currents, 150, 251, 1009
Edison effect, 350
Edison, Thomas, A., 7
Effective voltage, 183, 609, 615
Efficacy, 250
Efficiency, 150, 174, 246, 330, 459, 609
of motors, 609
of single-ended amplifiers, 458
of transformers, 609
Electrical charges. *See* Charges
Electrical current. *See* Current; Current flow
Electrical energy, measurement of, 290–291, (*see also* Energy conversion; Energy source; Power supplies)
Electrical force. *See* Voltage
Electrical Hazards, 29
Electrical instruments. *See* specific types
Electrical power. *See* Power; Power control systems; Power supplies
Electrical safety, 29–30
Electrical systems
examples of, 3, 7, 8–10
parts of, 8, 9–11, 358
troubleshooting, 83
Electrical units, 47–50
large, 50
small, 49–50
Electric bell, 140
Electricity
discovery of, 5
produced by magnetism, 136–137
Electrodes, 150, 312, 609
Electrolytes, 150, 609
Electrolytic capacitors, 183, 609
Electromagnetic cartridges, 468
Electromagnetic-deflection CRT, 539
Electromagnetic fields, 476, 508, 609
Electromagnetic induction, 159–161, 177–178
Electromagnetic recording head, 141
Electromagnetic speakers, 141–142, *See* Speakers
Electromagnetic waves, 506–508
Electromagnets, 130, 132–135, 609
magnetic strength of, 134–135
Electromotive force (EMF), 4, 25, 26, 609
Electron current flow, 4, 609
Electron emission, 323–327. *See* Emission
Electronic control. *See* Power control systems
Electronics, 312
careers in, 30–31
electron emission, 323–327
semiconductor materials, 319–323
semiconductor theory, 313–323
Electronics Workbench (circuit simulation software), 78, 81
Electron pair bonding, 315, *See* Covalent bonding
Electrons, 7, 15, 19–22, 330, 609
Electron volts, 312, 609
Electroscope, 20, 21
Electrostatic charges, 19, 20
Electrostatic deflection, 298, 299

Electrostatic field, 4, 19, 183, 476, 609
Electrostatic force, 312, 609
Electrostatic generator, 19, 21
Electrovalent combinations, 315
Elements, 4, 15, 322, 609
Emission, 312, 323, 609
  photoconduction, 326–327
  photoemission, 325–326
  secondary, 313, 324–325
  thermionic, 313, 324–325
Emitter, 382, 609
Emitter-base biasing, 417
Emitter bypass capacitor, 416
Emitter current, 382, 383, 386
Emitter follower, 426
E-MOSFET. *See* Enhancement-type MOSFETs
Encoding, 552, 567, 609
Energy, 4, 12, 330, 609
  electrical, measurement of, 290–291. (*see also* Energy conversion; Energy sources; Power supplies)
  relationship to power and work, 96–97
Energy conversion, 246
  from heating systems, 250–252
  from lighting systems, 247–250
  from motor control circuits, 272–276
  from motors, 252–253, 254–260
    alternating-current, 256–266
    direct-current, 253–256
  from synchro and servo units, 247, 266–270
Energy level, 312, 609
Energy-level diagrams, 318
Energy sources, 5–7, 8, 11, 615
  alternators, 150, 167
  battery connections, 154–155
  chemical, 152–154
  direct-current generators (*see* Direct-current generators)
  electromagnetic induction, 159–161
  heat, 155–158
  light, 155
  pressure, 158–159
Engineering assistants, 31
Enhancement, 382, 609
Enhancement-type MOSFETs, 397–399
  as amplifiers, 429
  biasing methods for, 429–432
  circuit configurations in, 429
  operational voltages in, 428
  resistance values of, 401; 402
Equivalent circuits, 106. *See also* Norton equivalent circuit method; Thevinin equivalent circuit method
Equivalent resistance, 76, 609
Even harmonics, 187–188
Even-lined fields, 504, 610
Exciting current, 169
Execute cycle, 569, 570
Extrinsic material, 312, 322, 610

Farad, 183, 193, 610
Faraday, Michael, 6, 136
Faraday's law, 136–137
Feedback, 476, 610
Feedback network, 410, 610
Feedback oscillators, 477
  fundamentals of, 477–478
  inductive-capacitive circuits in, 478–486
  types of
    Armstrong, 480–483, 498
    Colpitts, 484–485, 499
    crystal, 485, 499
    Hartley, 483–484, 486, 499
    Pierce, 485
Ferrite core, 476, 610
Fetch cycle, 552, 610
Fiber optics, 541–543
Field coils, 150, 246, 610
Field-effect transistors (FETs)
  as amplifiers, 427–434
  biasing methods for, 429–432
  circuit configurations in, 432–434
  operational voltages in, 428
  See also Junction field-effect transistors; Metal-oxide semiconductor field-effect transistors
Field engineering technicians, 30
Field pole, definition of, 151, 610
Field rheostat, 256
Fields
  electromagnetic, 476, 506–507
  electrostatic, 4, 19, 183, 476
  even-lined, 504, 610
  gravitational, 19
  magnetic, 130–133
  odd-lined, 504, 610
Filament, 330, 610
Filter circuits, 219–222, 223–377
Filtering, 386–372, 377
Filters, 183, 377, 610
  bandpass, 330, 610
  in power supplies, 368–372
  vestigial sideband, 505
Firmware, 552, 610
Fixed biasing, 416, 417, 429
Fixed capacitors, 183, 194–195
Fixed-emitter bias, 417
Flashlights, as example of electrical system, 13
Flat-compounded dc generators, 171
Flat-pack resistor network, 45
Flip-flops, 476, 493, 552, 559–561, 576, 577, 610
  in digital systems, 559–561
Floppy disk, 78
Fluorescent lighting, 246, 247–249, 610
Flux, 130, 610
Flux density, 130, 134, 142, 610
FM. *See* Frequency modulation
Focusing anodes, 539
Forbidden gap, 312, 610
Force, lines of, 130, 612
Forward breakover voltage, 582
Forward biasing, 333–335, 359, 610
Forward voltage, 337, 610
Fossil fuel system, 610
Frame, 505, 610
Franklin, Benjamin, 6
Free electrons, 4, 23, 610
Free running devices, 476, 491, 610
Frequency, 151, 183, 294, 492, 610
  of alternating current, 160
  fundamental, 187–188, 516, 525
  harmonics, 187–188
  measurement of, 294
  ratio detector response to, 530–531
Frequency counters, 568
Frequency meters, 295. *See also* Oscilloscopes
Frequency modulation (FM), 526–531, 546
  communication system, 526–531, 546
Frequency response, 183, 610
Full control, 8, 12
Full-load voltage, 174
Full-wave bridge rectifiers, 365–367, 376
Full-wave control, 584
Full-wave rectification, 361, 363–367, 375
Function generators, 185
Fundamental frequencies, 516, 525
  of heterodyne detectors, 514–516
  of square waves, 187–188
Fuses, 32, 280, 610

Gain, 382, 610
  of amplifiers, 448–449
  of transistors, 386, 394
Gallium, 345
Galvanometer, 284, 610
Ganged capacitors, 505, 610
Gas lasers, 468
Gate circuits, 558, 574
Gate currents, of silicon-controlled rectifiers, 582, 585–588, 610
Gates, 552, 610
  AND, 556
  combination logic, 557–558
  in field-effect transistors, 382
  in SCRs, 582
  NOT, 557
  OR, 556–557
Gauss, 130, 610
Gauss meter, 144
Generators, 150, 610
  basics of, 161
  left-hand rule for, 161
  Van de Graaff, 19, 21
  waveform, inductive-capacitive, 494
  See also Alternators
Germanium diodes
  characteristics of, 336–337
  leakage current in, 332–333
Gilbert, 130, 610
Gilbert, William, 6
Gravitational field, 19
Ground, 428, 610
Ground-fault circuit interrupters (GFCIs), 30, 610
Ground-fault indicators (GFIs), 294
Ground waves, 507

Half-wave rectification, 361–363, 364, 375
Half-wave SCR phase shifter, 589
Hall effect, 143–144

Hard disk drive, 78
Hardware, 553, 567–569, 610
Harmonics, 187
Hartley oscillators, 483–484, 498, 517
as AM transmitter, 520, 522
as FM transmitter, 528, 529
Heat, as electrical energy source, 155–157, 176
Heating, 250–252
of cathodes, 324, 325, 351, 352
dielectric heating, 251–252
of filaments, 350, 351–352
inductive heating, 251
resistive, 251
Heat pumps, 12
Helium neon laser, 470
Henry, definition of, 183, 610
Hertz, 162, 183, 610
Hertz, Heinrich, 506
Heterodyning, 505, 515, 516, 544, 611
High-current diodes, 341
High-level language, 553, 572, 611
High-level modulation, 505, 611
High-pass filters, 219–220
History of electricity, electronics, 5–7
Holding current, 582, 592, 611
Holes, 320–322, 330, 611
Horsepower (hp), 96, 246, 611
Horseshoe magnets, 133
Hot conductor, 611
Hue, 505, 540, 611
Hydroelectric system, 611
Hydrometer, 150, 155, 611
Hysteresis, 139, 611

ICE, 611
Ideal transformer, 225
Illumination, 250
Impedance, 183, 193, 204, 410, 446, 611
Impedance ratio, 446, 611
Impedance triangle, 209, 210, 211, 212
Incandescent lighting, 246, 247, 611
Indicators, 4, 8, 11, 12, 611
Indium, 321, 323
Induced channel, 382, 611
Induced current, 150, 611
Induced poles, 131
Induced voltage, 150, 611
Inductance, 183, 192–193, 359, 611
effects of, in circuits, 183, 196–197
Inductance filtering, 370–371
Induction, 159–161
Induction motors. *See* Single-phase ac motors; Three-phase ac motors
Inductive-capacitive astable multivibrators, 491–492
Inductive-capacitive circuits, in feedback oscillators, 478–483
Inductive-capacitive waveform generators, 491–492
Inductive circuits, 205, 215, 611
Inductive heating, 246, 251, 611
Inductive phase relationships, 197
Inductive reactance, 183, 193, 196, 611
Inductor-capacitor filters, 370–371
Inductors, 183, 611
energy stored in, 192
in series and parallel, 196
symbols for, 193
types of, 194
Inkless recorders, 302
In phase, 183, 611
Input transducers, 465–470
Instantaneous voltage, 162, 184, 611
Insulation resistance, 295–296
Insulators, 4, 21–22, 315, 611
Integrated circuit, 78
Integrated-circuit amplifiers, 434–439, 462–464
Interbase resistance, 382, 400, 611
Interface, 553, 567, 611
Interlace scanning, 505, 611
Intermediate frequency (IF), 505, 523, 611
Intermediate-frequency amplifiers, 505, 523–526
Internal resistance, 582, 611
Interpoles, 174
Intrinsic material, 312, 319, 611. *See also* Crystals
Inverse, definition of, 76, 611
Inverse formula, for parallel circuits, 86–87
Inverse proportion, 76, 611
Inverting op-amp circuit, 439, 443
Ionic bonding, 312, 315, 317, 611
Ionosphere, 507, 509
Ions, 150, 312, 330, 611
Iron-core inductor, symbol for, 193
Iron-core transformers, 223
I signal, 505, 611
Isolation transformers, 29, 184, 226, 227, 611

Junction, 30, 611
Junction capacitance, 340
Junction diodes, 330–332
biasing of, 332–334
Junction field-effect transistors (JFETs), 394–397
as amplifiers, 427–434
biasing methods for, 431–432
channel construction of, 395–396. (*see also N*-channel JFETs., 394–395. *P*-channel JFETs., 395–396)
characteristic curves, 395–397, 406
circuit configurations in, 432–434
operational voltages in, 428
resistance values of, 401
Junction testing, of bipolar transistors, 391–394, 405

Keeper, for permanent magnet, 131
Keyboards, 78, 553, 567, 612
Keyed continuous waves, 509
Kilowatt-hour (kWh), 96, 284, 612
Kilowatt-hour meter, 291
Kinetic energy, 4, 13, 312, 612
Kirchhoff's current law (KCL), 76, 91, 124, 612
Kirchhoff's voltage law (KVL), 76, 90, 123, 124, 612
Lagging current, 197
Lagging phase angle, 184–612. *See also* Phase angle
Laminations, 150, 247, 612
Lamps, 38, 612
Lasers, 468
Latch, 583, 612
LC circuit, 478
LC filters, 371, 377
Lead-acid cells, 150, 153, 154, 612
Lead identification, of bipolar transistors, 392–394
Leading current, 198
Leading edge, 189
Leading phase angle, 184, 612. *See also* Phase angle
Leakage current, 332–333
Leaky capacitors, 200
Left-hand rule, 131, 135, 612
for generators, 161
for magnetic flux, 131
for polarity, 135, 612
Lenz's law, 150, 612
Light, as electrical energy source, 155, 176
Light bulbs, 12, 28, 29
Light-dependent resistor (LDR), 313, 328, 612
Light-detection circuit, 326
Light-emitting devices, 345
Light-emitting diodes (LED), 345
Lighting, 246
fluorescent, 246, 246–249, 618
incandescent, 246, 247, 618
vapor, 247, 249–250, 618
Limit switches, 272–273
Linear amplifiers, 410, 612
Linear scale, 287
Line-of-sight transmission, 505, 612
Lines of force, 130, 612
Line voltage, 165
Lithium atom, 317
Lithium cells, 150, 152, 153
Loading, 284, 612
Load-line, 410
Load-line analysis, 418, 440, 441, 612
Loads, 4, 8, 9, 12
Local area network, 31
Lodestone, 130, 612
Loadstone compass, 6
Logarithms, characteristic of, 446, 450, 612
Logarithms, common, 635–638
Logarithms, natural, 636–638
Logic, definition of, 553, 612
Logic circuits, 575, 576
Logic functions, in digital systems, 556–558
Low-current diodes, 341
Low-frequency (LF) waves, 507
Low-pass filters, 219–220
Luminance, 250
Luminous flux, 250
Luminous intensity, 250

Machine language, 553, 572, 612
Magnet, 130, 612
Magnetic circuit breakers, 140

Magnetic circuits, 130, 612
Magnetic devices, 137. *See also* specific types
Magnetic fields, 130, 132–133, 134, 612
Magnetic flux., 133, 135, 142, 612
Magnetic levitation, 144
Magnetic materials, 130, 612
Magnetic motor contactors, 139
Magnetic poles, 130, 612, 614
Magnetic recording, 141
Magnetic saturation, 130, 142, 612
Magnetic starter, 272, 275
Magnetic tape input, 465–466, 467
Magnetism, 5
  domain theory of, 130, 135–137
  electricity produced by, 136–137
  laws of, 130, 612
  nature of, 146
Magnetization curve, 143
Magnetizing force, 143
Magnetomotive force (MMF), 130, 135, 142, 612
Magnetostriction, 130, 612
Main 1 or 2, 583, 612
Majority current carriers, 313, 321, 612
  in bipolar transistors, 321
  in*PNP* transistors, 387
  reverse biasing, 332, 333–334
Mantissa, 446, 450, 613
Manual motor starter, 272
Marconi, Guglielmo, 7
Master oscillator power amplifier (MOPA) transmitters, 510, 511
Mathematics for AC, 204–208
Matter, 5, 613
  structure of, 14–15
Maximum forward current, 337
Maximum plate current, 353
Maximum power transfer, in circuits, 97–99, 125, 184, 612
Maximum reverse current, 337
Maximum reverse voltage, 337
Maxwell, 142
Measurement, units of, 651–662. *See* Electrical units
Mechanical energy, as electrical energy source, 179. *See also* Piezoelectric effect
Mechanical loads. *See* Motors
Medium-current diodes, 341
Medium-frequency (MF) waves, 507
Megger, 284, 613
Megohmmeters, 295, 296
Memory, 553, 569–572, 613
Mercuric-oxide cells, 153
Mercury vapor lamps, 249–250
Metallic bonding, 5, 24 313, 318, 613
Metal-oxide semiconductor field-effect transistors (MOSFETs), 397–400
  as amplifiers, 429
  biasing methods for, 429–432
  channel construction of, 397–399
  circuit configurations in, 429
  testing, 406
Meter movement, d'Arsonval, 284
Mho, 184, 613
Mica capacitors, 184, 613
Microcomputer systems, 573
Microfarad (MFD), 193
Microphones, 410, 465, 613
  crystal, 465–466
  dynamic, 446, 465–466
  piezoelectric principle of, 151, 465
Microcomputer system, 572–573
Microprocessor unit (MPU), 78
Midrange, 446, 465, 613
Minority current carrier, 313, 332–334, 613
  reverse biasing, 332, 333
Mixers, 514, 553, 613
Mnemonic, 553, 613
Modulating component, 505, 613
Modulation, 505, 537, 613. *See also* Amplitude modulation; Frequency modulation
Molecules, 5, 15, 613
Monitor, 78
Monochrome televisions, 505, 537, 538, 548, 613
Monostable multivibrators, 476, 492–493, 500, 613
Morse code, 509
MOSFETs. *See* Metal-oxide semiconductor field-effect transistors
Motor action, 252–253
Motor control circuits, 272–276
Motor principle, 252
Motor speed control, 602
  performance of, 270–272
  right hand rule for, 247, 253
  starters for, 247, 272–275
  universal, 260–261
  See also Direct-current motors; Single-phase ac motors; Three-phase ac motors
Motors, 13, 247, 613
Mouse, 78
Moving coil meter movement. *See* D'Arsonval movement
Multifunction meters, 38, 613. *See also* Digital multimeters; Volt-ohm-milliammeter
Multimeter, 53–54
Multiple secondary transformers, 225
Multiplier resistors, 53, 284, 287, 613
Multirange meters, 38, 191, 287, 288, 613
Multivibrators
  astable, 494, 495–496
    inductive-capacitive, 491–492
  bistable, 493–494, 500. (*see also* Flip-flops)
  monostable, 492–493
Mutual inductance, 184, 196, 613

NAND gates, 558, 574
Natural magnet, 130, 613
N-channel JFETs, 394–395, 397
  as amplifiers, 427–434
  biasing methods for, 431–432
  circuit configurations in, 432–434
  operational voltages in, 428
N-channel MOSFETs, 397–400
Negative alternation, 361, 363
Negative logic, 553, 613
Negative picture phase, 505, 613
Negative resistance, 382, 613
Negative voltage, 100
NEMA, 613
Network theorems, 106
Neutral, 613
Neutral plane, 150, 174, 613
Neutrons, 5, 15, 613
Nickel-cadmium cells, 153, 154, 156, 613
Nickel, metal-hydride cells, 153
No-load voltage, 174
Noninverting input, 410, 613
Noninverting op-amp circuit, 438, 443
Nonlinear distortion, 420
Nonlinear scale, 289
Nonsinusoidal waveforms, 187, 476, 613. *See also* Square waves
Nonvolatile memory, 553, 613
NOR gates, 574
Normally closed contacts, 137–138, 139
Normally open contacts, 137–138, 139
Norton equivalent circuit method, 107–109, 127
NOT gates, 557, 558
NPN transistors, 383
  biasing of, 383–387
  collector family of characteristic curves for, 387–390
  junction testing, 391–394
  lead identification of, 392–394
  See also Bipolar transistors
N-type materials, 321, 327
  current carriers in, 320
  forward biasing, 333–335
  in junction diodes, 320–332
  reverse biasing, 332–334
Nuclear fission system, 613
Nucleus, 5, 15, 613
Null condition, 296
Numbering systems
  binary, 552, 554–555
  decimal, 553–554
Numerical readout instruments, 299. *See also* Digital multimeters

Odd harmonics, 187
Odd-lined fields, 504, 610
Oersted, Hans Christian, 7
Off-state resistance, 583, 613
Ohm, 5, 613
Ohmmeters, 29, 38, 284, 613
  capacitor testing with, 200
  diode testing with, 341–343
  polarity of, 343
  scale of, 290
  transistor testing with, 393–396
  See also Megohmmeters; Volt-ohm-milliammeter
Ohm's law, 76, 79–83, 613
  circle, 81
    for alternating-current circuits, 192
    for magnetic circuits, 135, 137
    for parallel circuits, 86–89
    for series circuits, 83–86

Ohms-per-volt rating, 284, 613
Oil-filled capacitors, 200
One-shot multivibrators. *See* Monostable multivibrators
Onnes, Kamerlingh, 22
Open capacitors, 200
Open circuits, 5, 24, 85, 437, 614
Open-loop amplification, 410, 437, 613
Operational amplifiers, 434–439
Operational regions, 388–389, 420–421
Optic fiber, 505
Orbitals, electron, 5, 15, 313, 316, 613
OR gates, 556–557
Oscillator fundamentals, 477–478
  Oscillators, 477. *See also* Feedback oscillators; Relaxation oscillators
Oscilloscopes, 191, 284, 295, 614
  digital storage, 298, 299
  general-purpose, 298
Output transformers, 453
Overcompounded dc generator, 174
Overcurrent device, 614
Overdriven amplifiers, 421–423
Overload, 614
Overmodulation, 518

Paper capacitors, 199
Parallel capacitance, 198
Parallel circuits, 76, 86–89, 115–116, 119–120, 614
  alternating-current, 212
  examples of, 92–93
  finding power values in, 212–219
  measurements in, 87–88, 119
  troubleshooting, 88–89
Parallel connection of batteries, 155, 157
Parallel inductance, 196
Parallel resonant circuit, 184, 614
Partial control, 8, 12
Path, 5, 8, 9, 11, 614
P-channel JFETs., 395
  as amplifiers, 427
  biasing methods for, 429–431
  operational voltages in, 428
  P-channel MOSFETs, 397–9
Peak reverse voltage (PRV), of vacuum-tube diodes, 353
Peak-to-peak values, 446, 614
Peak-to-peak voltage, 184, 446, 614
Peak values, 359, 614
Peak voltage, 184, 382, 614
Pen and ink recorders, 299–302
Period, 184, 614
Permanent capacitor motor, 266
Permanent magnet, 131, 145, 614
Permanent magnet dc motors, 254
Permanent magnet generators, 170
Permeability, 131, 614
Phase, 189, 583, 614
Phase angle, 184, 614
Phase control
  SCR circuits, 590, 600
  of triac, 595–596, 597, 602
Phase inversion, 414
Phase relationships, 196–197
Phase shifter, 583, 596, 602, 614
Phonograph
  pickup cartridges, 466–468
  piezoelectric principle of, 158–159
Photoconduction, 326
Photodiodes, 346–348, 356
Photoemission, 325–327
Photons, 313, 614
Photoresistive cells, 326–327
Photovoltaic cells, 151, 345–346, 355, 614
Pickup current, 139
Picture signal, 534
Picofarads, 193
Pierce oscillators, 485, 488
Piezoelectric crystals, 158–159
Piezoelectric effect, 151, 158–159, 446, 465, 614
Pi filters, 359, 371, 378, 614
Pixels, 505, 533, 614
Plate dissipation, 353
Plate resistance, 353
Plates, 197
PNP transistors, 383
  biasing of, 391–392
  junction testing, 393–394
  lead identification of, 392–394
  in silicon-controlled rectifiers, 586
  See also Bipolar transistors
Polarity, 38, 131, 614
  left-hand rule for, 135, 612
  of ohmmeters, 343
Poles, 131
Pole shading, 266
Positive alternation, 361–363
Positive logic, 553, 614
Potential energy, 5, 13, 614
Potentiometers, 4, 38, 614
  as voltage divider, 45
Power, 12, 14, 184, 613
  in alternating-current circuits, 212–219
  definition of, 5, 76, 184
  in direct-current circuits, 95–96, 122
  formulas for finding, 95
  measurement of, 289–291
  relationship to energy and work, 95–96
  sample problem for, 95–96
  three-phase, measurement of, 291–292
Power amplification, 412, 463
Power amplifiers, 455–462, 472
  push-pull, 456–459, 472
  single-ended, 455–456, 472
Power analyzer, measuring three-phase power with, 291–292
Power control systems, 583–584, 594
  considerations for, 595
  diac, 596–598
  silicon-controlled rectifiers, 589, 598
  triac, 591–596, 598
Power curve
  of capacitive circuits, 203
  of inductive circuits, 202
  of resistive-capacitive circuits, 203
  of resistive circuits, 201
  of resistive-inductive circuits, 202
Power demand, measurement of, 293–294
Power diodes, 341
Power dissipation, 337, 410, 418, 614
Power dissipation curve, 418
Powered metal-core inductor, symbol for, 193
Powered metal-core transformers, 233
Power demand, 293–294
Power factor, 184, 614
  measurement of, 292–293
  meter, 292
  in motor performance, 270–272
Power supplies, 378
  of direct-current voltage, 190, 192
  dual, 367–368
  filters in, 368–372
  functions in, 359, 360
  rectification in, 361–368
  subsystem of, 359
  transformers in, 360–361
  voltage regulation in, 372–374
Power transfer. *See* Maximum power transfer
Power transformers, 222
Practical op-amp circuits, 438–439
Precision resistors, 38, 46
Precision-wound resistors, 39, 614
Pressure, as electrical energy source, 158–159
Primary cells, 151, 152, 614
Primary winding, 184, 614
Prime mover, 151, 161, 614
Printer, 78
Production line counter, 568
Product-over-sum method, 93
Program, 567
Programmable read-only memory (PROM), 553, 614
Protons, 5, 15, 614
Pspice (circuit simulation software), 78
P-type materials, 321–323
  current carriers in, 322, 323
  forward biasing, 333–334
  in junction diodes, 320–321
  reverse biasing, 332
Pulsating direct current, 151, 170, 359, 614
Pulse, 189
Pulse repetition frequency (PRF), 189
Pulse repetition rate (PRR), 189, 491
Pulse repetition time (PRT), 189
Pulse width, 189
Pushbutton control circuit, start-stop, 273
Pushbutton switches, 273
Push-pull amplifiers, 447, 472, 614
  class AB, 460–461
  crossover distortion in, 459–460

Q point, 411, 614
Q signal, 505, 614
Quadrant, 205, 583, 614
Quality factor, 184, 614
Quanta, definition of, 313, 325, 615
Quantum theory, 313, 325, 615

Radio, 507
Radio frequency (RF), 416, 615

Radio-frequency amplification, 416
Radio-frequency amplifiers, 454, 513
Radio-frequency choke, 476, 483, 615
Radio-frequency communication, 506
Radio-frequency oscillator coils, 481
Radio-frequency transformers, 454
Radio-frequency transmitters, 505
Radix, 553, 615
Random access memory (RAM), 78, 569–615
Rare earth magnets, 144–145
Ratio arm, 296
Ratio detectors, 530–531, 533, 548
RC charging-discharging action, 490
RC filters, 372
Reactance, 187, 196, 615
Reactive circuit, 184, 615
Reactive power, 184, 615
Reactive volt-amperes, 217
Read head, 447, 615
Reading, 553, 569, 615
Read-only memory (ROM), 553, 615
Readwrite memory, 571
Receivers
  AM
    simple, 521–523
    superheterodyne, 525
    CW, 511–514
  FM, 529
    television, 537–539
    color, 540
Reciprocal, 615. *See* Inverse
Rectification, 359, 361–368, 615
  in dual power supplies, 367–368
  full-wave, 361, 363–365
  full-wave bridge, 365–367
  half-wave, 361–363, 364, 365
Rectifiers, silicon-controlled, 584–591
Rectilinear propagation, 543
Reed relays and switches, 140–141
Regenerative feedback, 476, 478, 615
Regulation, 615
Regulated power supply, 357–373
  speed, 245, 250, 253–254
  voltage, 151, 615
Relative permeability, 143
Relaxation oscillators, 476, 487–498, 615
  astable multivibrators, 491–492
    inductive-capacitive, 479, 491–492
  bistable multivibrators, 493–494
  blocking, 496–497
  inductive-capacitive waveform generators, 487–488
  monostable multivibrators, 492–493
  resistive-capacitive circuits in, 488–489
  unijunction transistors as, 489–491
Relays, 131, 138, 615
Reluctance, 131, 135, 615
Reproduction, in amplifiers, 411
Repulsion motor, 247, 615
Residual magnetism, 131, 134, 170, 615
Resistance, 26–29, 615
  branch, 76, 86, 88, 93
  definition of, 5, 76, 184, 615
  equivalent, 76, 90
  formulas for finding, 84, 86
  high, 295–296
  insulation, 296
  interbase, 382
  internal, 582
  measurement of, 66, 288, 295, 306
  negative, 382
  off-state, 583
  plate, 353
  total, 76, 84, 86, 89
    in combination circuits, 89–90
    in parallel circuits, 86–89
    in series circuits, 83–86
  See also Ohm's law
Resistance bridge, 296
Resistance values, 46
  changes in, 85
  of transistors, 401–403
Resistive-capacitive circuits, 208
  parallel, 214
  in relaxation oscillators, 487–498
  series, 210
  time-constant, 477
Resistive circuits, 184, 202, 615
Resistive heating, 247, 615
Resistive-inductive-capacitive circuits
  parallel, 215
  series, 209–211
Resistive-inductive circuits, 201–202, 213–215
  parallel, 215
  series, 209–211
  time-constant, 477
Resistor-capacitor filters, 372
Resistors, 38, 39, 615
  color-coded, 48, 49, 65, 71
  power rating of, 47
  types of, 43–45
    carbon, 43, 47, 49–50, 52
    current-limiting, 335
    light-dependent, 313, 328, 612
    multiplier, 53, 284, 287, 613
    precision resistors, 38, 39, 614
    precision-wound resistors, 38, 614
    shunt, 284, 285, 287
    tapped, 99
    variable, 46
    wire-wound, 43, 44,
Resonance, 219–222
Resonant circuits, 184, 219–222, 615
  parallel, 219–222
  series, 184, 218, 219–222
Resonant frequency, 184, 219, 477, 479, 615
Retentivity, 131, 143, 615
Retrace, 505, 533, 615
Retrace time, 533, 535
Reverse biasing, 332, 334, 335, 359, 615
Reverse breakdown voltage, 337, 583, 615
Reverse polarity ohmmeters, 343
Reverse recovery time, 337
Rheostat, 583, 615
  field, 254, 256
  power control by, 583
Right-hand motor rule, 247, 253, 615
Ripple frequency, 365, 366
Rise time, 493
Rms. *See* Root mean square
Root mean square (rms), 184, 185, 447, 615
Root mean square voltage. *See* Effective voltage
Rotating-armature method, 151, 162, 615
Rotating-field method, 151, 162, 615
Rotors, 151, 161, 247, 615
  in single-phase ac motors, 260–261
  squirrel-cage, 247, 261, 262, 265
  in stepping motors, 269–270
  in synchro units, 269
Ruby laser, 469
Running neutral plane, 151, 174, 615
Run windings, 262, 263, 264, 265

Safety, 29–30
Safety factor, in voltage-divider circuits, 100
Sales careers, in electronics, 31
Saturation, 505, 615
  color, 540
  magnetic, 130, 142, 612
Saturation region, 477, 615
  of amplifiers, 420
  of transistors, 382
Sawtooth waveform, 184, 615
Scanning, 505, 535, 615
Schematic diagrams, 38, 52, 138, 359, 615
Scientific calculator, 77
Scientific notation, 38, 50, 51, 70, 615
Secondary cells, 151, 152, 615
Secondary emission, 313, 324–325, 615
Secondary winding, 184, 615
Seebeck, Thomas J., 155
Seebeck effect, 157
Selectivity, 184, 505, 615
Self-biasing, 416, 431
Self-emitter bias, 416–417
Self-excited dc generators
  compound-wound, 171–173
  series-wound, 170–171
  shunt-wound, 170–171
Self-starting motors, 247, 615
Semiconductors, 5, 22, 23, 315, 343
  diode devices (*see under* Diodes)
  materials in, 319–323, 331. (*see also N*-type materials; *P*-type materials)
Semiconductor laser, 469, 470
Semiconductor theory, 313, 615
  atom combinations, 313–315
  insulators, semiconductors, and conductors, 315–319
  Sensitivity, of meters, 284, 288, 615
  Separately excited dc generators, 169–170
  Series capacitance, 197
  Series circuits, 76, 83–86, 91–92, 113, 118, 615
  alternating-current
    problem solving in, 206
    types of, 205, 207–208
  examples of, 91–92
  troubleshooting, 83–86
  as voltage divider, 99, 100–101
Series connection of batteries, 155, 157
Series-fed Hartley oscillator, 484

Series inductance, 196–197
Series resonant circuits, 184, 615
Series-wound dc generators, 171
Series-wound dc motors, 254
Service technicians, 30
Servomotors, 269, 272
Servo systems, 266–270
Seven-segment displays, 345, 563, 564, 578–579
Shaded-pole motors, 247, 615
Shells, of atoms, 15, 315
Short circuits, 5, 30, 615
  in capacitors, 200
  in transformers, 227
  troubleshooting, 85
Shunt, definition of, 615
Shunt-fed Hartley oscillator, 483–485
Shunt resistors, 284, 285, 287
Shunt-wound dc generators, 169, 172
Shunt-wound dc motors, 254
Sidebands, 505, 518, 615
Siemens, 184, 615
Signals, 184, 615
  in amplifiers, 412
  in continuous-wave receivers, 513
  I, 505
  Q, 505
  sweep, 534, 535
  television, 536
  Y, 505
Silicon
  covalent bonding in, 312, 319
  extrinsic, 312, 321–322
  intrinsic, 312, 319–320
Silicon-controlled rectifiers, 584–591
  alternating-current power control with, 589–591
  characteristics of, 586–587
  construction of, 585
  direct-current power control with, 587–589
  operation of, 586
  testing, 599
Silicon diodes
  characteristics of, 334–337
  data sheet for, 338–339
  effects of temperature on, 337–339
  leakage current in, 332
Silicon transistors, input characteristic of, 460
Silver-oxide cells, 153
Sine waves, 151, 185, 615
  damped, 476, 479
Sine-wave voltage, 162–163, 185, 186. *See also* Alternating-current voltage
Single-ended power amplifiers, 455
Single-function analog meters, 284, 285. *See also* specific types
Single-in-line package (SIP) resistor network, 44
Single-phase ac generator, 151, 162, 615
Single-phase ac motors, 247, 259, 266, 615
  considerations for, 260
  induction, 260, 266, 611
    capacitor, 264, 266
    operation of, 261
    shaded-pole, 266, 267, 269
    split-phase, 260, 261
  synchronous, 266
  universal, 260, 261, 266
  Single-phase ac voltage, 191, 615
Single-phase alternators, 162–164
Single-phase wattmeters, 290, 292
Single-pole single-throw (SPST) switch, 41
Skip, 505, 615
Sky waves, 505, 615
Slip, 260
Slip rings, 151, 247, 615
Snubber, 583, 615
Sodium vapor lamps, 249–250
Software, 553, 572, 615
Solar cells, 158
Soldering, 619–622
Soldering, safety issues for, 40
Solenoids, 131, 139, 147, 615
Solid state, 313, 615
Sound waves, 465
Source, of field-effect transistor, 382. *See also* Energy sources; Power supplies
Source followers, 433
Speakers, 141, 447, 464–465, 615
  electromagnetic, 141–142
Specific gravity, 151, 615
Spectrum analyzer, 189
Speed regulation, 247, 615
Split-phase motors, 247, 262, 615
Split-ring commutator, 247. *See also* Commutator
Square waves, harmonic content of, 187
Squirrel-cage rotor, 247, 261, 615
Stability, of oscillators, 477, 615
Stable atoms, 5, 17, 313, 314, 615
Stage of amplification, 447, 615
Starters, for motors, 247, 615
Start-stop pushbutton control circuit, 272–276
Start-stop triac control, 593–594, 601
Start windings, in split-phase motors, 262–264, 265, 266
Static charges, 5, 615
Static electricity, 5, 615
Static load-line analysis, 19–420
Static state, of amplifiers, 411, 414, 440, 615
Static switch, triac as, 593, 600
Stators, 151, 161, 247, 615
  in single-phase ac motors, 259–266
  in stepping motors, 269–270
  in synchro units, 266–269
Stator windings, in split-phase motors.
Steam turbine, 262, 263, 264, 265
Step-down transformers, 184, 223, 226, 616
Stepping motors, 269–270
Step-up transformers, 184, 223, 226, 227, 617
Stereophonic, 447, 448, 469, 617
Storage battery. *See* Secondary cells
Stylus, 447, 617
Substrate, 382, 394, 617
Superconductors, 22–23
Superheterodyne circuit, 520, 521
Superheterodyne receivers, 523–526
Superposition, 101, 102, 105, 126
Susceptance, 184, 617
Sweep oscillator circuit, 298
Sweep signals, 534–535
Switches, 12, 38, 617
  limit, 272–273
  in motor control circuits, 272
  pushbutton, 273
  reed, 140–141
  single-pole single-throw, 41
  toggle, 477, 553, 617
  triac as, 591–596
Switching device, 583, 617
Switching time, of diodes, 340–341
Symbols, 38, 41, 617, 631–634
Symmetrical, definition of, 477, 617
Sync, 505, 617
Synchronization, 477, 505, 617
  of television signals, 505
Synchronous motor, 247, 617
  single-phase, 266
  three-phase, 257–259
Synchronous speed, 258
Synchro systems, 266–269
System, 7, 505–506

Tank circuits, 184, 217, 218–219, 477, 479, 617
  in feedback oscillators, 478–483
Tape, magnetic. *See* Magnetic tape input
Tapped resistors, as voltage divider, 99
Technical writers, 31
Technologists, 31
Telegraphy, 505, 509, 617
Television, 531–541
  camera tube of, 531–532, 539
  color, 539–541, 542, 549
  monochrome, 503, 530, 535
  picture signal of, 533–535
  receivers, 537–539, 541, 542, 549, 617
  transmitters, 536–537, 541–542, 617
Terminals, of triac, 583, 591, 617
Tesla, 142
Thales of Miletus, 5–6
Thermal stability, 411, 617
Thermionic emission, 313, 324, 330, 354, 539, 617
Thermocouples, 151, 157, 158, 617
Theta, 184, 617
Thevinin, M. L., 106
Thevinin equivalent circuit method, 106–107
  single-source problem, 106–107, 126
  two-source problem, 107, 127
Three-phase ac circuits, power in, 217–219
Three-phase ac generator, 151, 164–166, 617
Three-phase ac motors, 256–259
  induction, 256
  Steps for starting, 258
  synchronous, 257–259
Three-phase ac voltage, 164, 617
Three-phase alternators, 617
Three-phase power, measurement of, 291
Three-phase transformers, 227
Three-stage voltage amplifiers, 448–449
Threshold, 477, 494, 617

Time constant, 359, 477, 487, 617
Time constant circuits, 486–487, 488, 617
Time constant curves, 489
Timing elements, 558–559
Toggle, 477, 553, 617
Toggle switches, 477, 553, 617
Tolerance, 46
Tools, 624–630
Torque, 247, 252, 617
Torque action, 252–253
Total capacitance, 197
Total current, 76, 617
  in combination circuits, 89–90
  in parallel circuits, 87–88
  in series circuits, 83–84
Total inductance, 196
Total resistance, 76, 86, 617
  in combination circuits, 89–90
  in parallel circuits, 86–87
  in series circuits, 83–84, 90
Total voltage, 76, 617
Trace, 191
Trace time, 505, 533, 535, 617
Trailing edge, 189
Transducers, 447, 464, 617
  input, 465–470
Transformer action, 260
Transformer coupling, 454–455
Transformers, 184, 222–223, 243
  efficiency of, 227
  operation of, 222–223, 243
  power supply, 301–361
  testing, 227–228
  types of, 223–226
Transistors, 382
  active region of, 382
  base of, 382
  beta, 382, 386
  biasing of, 383–387
  in blocking oscillators, 497, 501
  characteristics of, 387–390
  circuit configurations in, 423–427
  collector in, 382, 384
  complementary, 446, 607
  current carriers in, 382, 385
  current flow in, 384–385
  cutoff, 382
  Darlington, 446
  discharge, 495
  gain of, 386–394
  operating regions of, 388–390, 460
  packaging of, 390–391
  resistance values of, 401, 402, 403
  saturation region of, 382, 388–389
  silicon, 412
  unijunction, 400–401
    as oscillators, 400–401, 500
    testing, 470
  unipolar, 401–403
  testing, 470 *See also* Bipolar transistors; Field-effect transistors
Transmitter-receiver systems, 506
Transmitters
  AM, 519, 520–523
  continuous-wave, 509–511
  FM, 527–529, 546
  master oscillator power amplifier, 510–511
  radio-frequency, 505
  television, 536–537
    color, 540–541
Triac, 591–596
  applications of, 592–593
  characteristics of, 592
  construction of, 591
  control, 601
  operation of, 591–592
  powercontrol, 594
  testing, 600
Triggered devices, 477, 491, 617
Triggering, definition of, 583, 617
Trigger input, 477, 494
Trigger pulse. *See* Flip-flops
Troubleshooting
  electrical systems, 83
  parallel circuits, 86–89
  series circuits, 83–86
True power, 184, 290, 617
Truth table, 553, 617
Tuned radio frequency, 513, 514
Tuner, 513
Turn-on time, 583, 617
Turns ratio, 185, 259, 617
Tutorial software, 78
Tweeters, 447, 465, 618
Two-value capacitor motors, 262–266

Ultracapacitors, 199–200
Undercompounded dc generators, 174
Undermodulation, 518, 519
Unijunction transistors (UJTs), 400–401, 407
  as oscillators, 489–491, 500
  testing, 407
Unipolar transistors, 383, 394, 401–403, 618
  testing, 407
Unity coupling, 196
Unity power factor, 216, 270
Universal motors, 247, 618
Universal time constant curves, 487

Vacuum tube, 349–354
Vacuum-tube diodes, 351, 352
  operating conditions of, 351
  representative, 352
Valence band, 313, 618
Valence electrons, 5, 15, 18, 313, 330, 618
Valence shells, 15
Valley voltage, 383, 400, 618
Van de Graaff generator, 19, 21
Vapor lighting, 247, 249–250, 618
Varactor diodes, 347–349, 350
Variable autotransformers, 225
Variable capacitors, 194, 195, 199
Variable power control, of triac, 594–595, 601
Variable resistors, 46
Variable transformers, 225, 584
Vector, 185, 618
Vector diagrams, for alternating-current circuits, 203, 531
Vertical blocking oscillators, 477, 497, 618
Very-high-frequency (VHF) waves, 507
Vestigial sideband, 505, 618
Vide, 618
Videcon, 505
V-MOSFET, 399, 400
Voice coil, 447, 497, 618
Volatile memory, 553, 618
Volt, 5, 24–27, 618
Volta, Alessandro, 6
Voltage, 25–26, 27–29, 618
  average, 183, 606
  barrier, 331–332, 606
  in battery connections, 154–155
  branch, 76, 86, 93
  breakdown, 317
    reverse, 337, 583, 615
  breakover, 582, 606
  in combination circuits, 76, 89–90
  definition of, 5, 76
  effective, 183, 609, 615
  formulas for finding, 95
  forward, 337, 610
  full-load, 174
  induced, 151, 161
  instantaneous, 162, 184, 611
  Kirchhoff's law for, 76, 90, 101–102, 104, 123
  line, 165
  maximum reverse, 337
  measurement of, 57–58, 66, 67
  negative, 100
  no-load, 174
  peak, 184, 382, 614
  peak reverse, 353
  peak-to-peak, 184, 446, 614
  in series circuits, 76, 83–86, 91–92, 113, 118
  total, 76, 617
  valley, 383, 400, 618
  working, 185, 618
  See also Alternating-current voltage; Direct-current voltage; Ohm's law
Voltage amplification, 411
Voltage-divider circuits, 99–100, 122, 447, 618
  design of, 99–100
  negative voltage derived from, 100
Voltage-divider rule. *See* Voltage-division equation
Voltage-division equation, 100
Voltage drops, 38, 76, 84, 618
  in combination circuits, 89
  in series circuits, 84–85
  in series resistive-inductive circuits, 209–211
  in voltage-divider circuits, 99–100
Voltage follower, 411, 618
Voltage gain, of amplifiers, 448
Voltage generation, 106–161, 163. *See also* Generators
Voltage output, 166
  compound-wound dc generators, 171, 173, 174
  permanent magnet dc generators, 168, 170
  separately excited dc generators, 169–170
  series-wound dc generators, 170, 171
  shunt-wound dc generators, 170, 172

Voltage regulation, 174, 372–374, 618
Voltage triangle, 203, 204, 209–212
Voltage waveforms
  of capacitive circuits, 202–203
  of capacitor-start motor, 264–266
  of inductive circuits, 200–201
  of resistive-capacitive circuits, 202–203
  of resistive circuits, 200–201
  of resistive-inductive circuits, 201–202
Voltaic cells, 151, 618
Volt-amperes, 185, 618. *See also* Apparent power
Volt-amperes reactive (VAR), 185, 618
Voltmeters, 38, 284, 618. *See also* Volt-ohm-milliammeter
Volt-ohm-milliammeter (VOM), 38, 54, 291, 618
  alternating-current voltage measurement with, 58
  controls of, 53, 54
  direct current scale of, 58
  direct-current voltage measurement with, 58–61
  ohm's scale of, 56
  resistance measurement with, 306
  scale of, 55, 66
  voltage measurement with, 57–58, 305
  See also Ammeters; Ohmmeters; Voltmeters

Walkie-talkies, 520
Watt, 5, 10, 76, 618
Watt-hour (Wh), 284, 618
Watt-hour meters, 284, 618
Wattmeters, 284, 289, 291, 618
  three-phase, 291
Watt-second, 96
Waveform generators, inductive-capacitive, 494, 495
Waveforms, 185, 618
  of alternating current, 186–187
  of astable multivibrators, 476
  of capacitive circuits, 200–208
  of capacitor-start motor, 264, 266, 607
  of *C*-input filters, 359, 368–370
  of direct current, 185, 186
  of inductive circuits, 205, 215
  of monostable multivibrators, 476, 492–493, 500, 613
  of phase control SCR circuits, 590, 600
  of resistive-capacitive circuits, 208
  of resistive circuits, 184, 202, 615
  of resistive-inductive circuits, 201–202, 213–215
  of superheterodyne receivers, 525
  viewed with DSO, 298, 300
Wavelength, 185, 505, 618
Waves
  carrier, 504, 607
  continuous, 479, 608
  damped, 476, 479, 608
  electromagnetic, 506–508
  ground, 507
  low-frequency, 507
  medium-frequency, 507
  sinusoidal, 162–163, 185, 186
  sky, 505, 615
  sound, 465
  square, 187
  very-high-frequency, 507
Weber, 142
Weber, Wilhelm, 135
Wheatstone bridge, 284, 296, 298, 307, 618
Wire-wound resistors, 48, 51, 53
Wireless microphone, 546
Wiring diagrams, 38
Woofers, 447, 465, 618
Word, 567
Work, 5, 12, 13, 618
  relationship to power and energy, 100–102
  sample problem for, 104
Working voltage, 186, 618
Writing, 553, 569, 618
Wye connection, 151, 164, 247, 618

X axis, 284, 618
X-Y recorders, 302

Y axis, 284, 618
Y signal, 505, 618

Zener breakdown, 336
Zener diodes, 343–345
Zener diode characteristics, 355
Zener diode regulator, 373, 378
Zener diode voltage regulators, 355
Zero beating, 505, 516, 618